MyLab Math Online Course for *Essentials of College Algebra*, Twelfth Edition by Lial, Hornsby, Schneider, and Daniels

(access code required)

Visualization and Conceptual Understanding

These MyLab Math resources will help you think visually and connect the concepts.

Guided Visualizations

Engaging interactive figures bring mathematical concepts to life, helping you visualize the concepts through directed explorations and purposeful manipulation. Guided Visualizations can be assigned in MyLab Math to encourage active learning, critical thinking, and conceptual understanding.

Updated Videos

Updated videos are available to support students outside of the classroom and cover all topics in the text. **Quick Review** videos cover key definitions and procedures. **Example Solution** videos offer a detailed solution process for every example in the textbook.

Video Assessment Exercises

Video assessment questions are assignable MyLab Math exercises tied to Example Solution Videos. These questions are designed to check students' understanding of the important math concepts covered in the video. Pairing the videos and Video Assessment questions provide an active learning environment where students can work at their own pace.

pearson.com/mylab/math

Connect the Concepts and Relate the Math

Student Motivation and Study Skills

Motivate and empower students with the skills needed to succeed in your College Algebra course and improve their lives.

Learning Catalytics

Generate class discussion, guide your lecture, and promote peer-to-peer learning with real-time analytics. MyLab Math now provides Learning Catalytics™—an interactive student response tool that uses students' smartphones, tablets, or laptops to engage them in more sophisticated tasks and thinking.

- Pose a variety of open-ended questions that help your students develop critical thinking skills.

- Monitor responses to find out where students are struggling.

- Use real-time data to adjust your instructional strategy and try other ways of engaging your students during class.

- Manage student interactions by automatically grouping students for discussion, teamwork, and peer-to-peer learning.

Skills for Success Modules

Skills for Success Modules help foster success in collegiate courses and prepare students for future professions. Topics such as "Time Management," "Online Learning," and "College Transition" are available within the MyLab Math course.

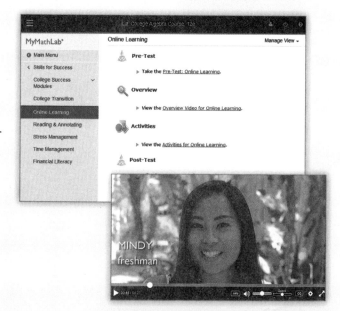

pearson.com/mylab/math

Essentials of College Algebra

TWELFTH EDITION

Essentials of College Algebra

TWELFTH EDITION

Margaret L. Lial
American River College

John Hornsby
University of New Orleans

David I. Schneider
University of Maryland

Callie J. Daniels
St. Charles Community College

 Pearson

Director, Portfolio Management: Anne Kelly

Portfolio Management Assistant: Ashley Gordon

Managing Producer: Karen Wernholm

Producer: Jonathan Wooding

Manager, Courseware QA: Mary Durnwald

Manager, Content Development: Kristina Evans

Product Marketing Manager: Stacey Sveum

Marketing Assistant: Shannon McCormack

Field Marketing Manager: Peggy Lucas

Marketing Assistant: Adrianna Valencia

Senior Author Support/Technology Specialist: Joe Vetere

Manager, Rights and Permissions: Gina Cheselka

Manufacturing Buyer: Carol Melville, LSC Communications

Text Design, Composition, Illustrations, Project Management: Cenveo® Publisher Services

Cover Design: Cenveo® Publisher Services

Cover Image: Alexander Mazurkeich/Shutterstock

TI-84 Plus C screenshots from Texas Instruments. Courtesy of Texas Instruments.

Acknowledgments of third-party content appear on page C-1, which constitutes an extension of this copyright page.

Library of Congress Cataloging-in-Publication Data

Names: Lial, Margaret L., author. | Hornsby, John, 1949- author. | Schneider,
   David I., author. | Daniels, Callie J., author.

Title: Essentials of college algebra / Margaret L. Lial, American River
   College, John Hornsby, University of New Orleans, David I. Schneider,
   University of Maryland, Callie J. Daniels, St. Charles Community College.

Description: Twelfth edition. | Boston : Pearson, 2017. | Includes index.

Identifiers: LCCN 2017039758 | ISBN 0134697022 (alk. paper)

Subjects: LCSH: Algebra—Textbooks.

Classification: LCC QA154.3 .L546 2017 | DDC 512.9—dc23 LC record available at
   https://lccn.loc.gov/2017039758

2   18

ISBN 13: 978-0-13-469702-4
ISBN 10: 0-13-469702-2

To the Kreamer ladies: Alicia, Terri, and Becky
E.J.H.

To the memory of my grandparents, Fred and Katherine Cole
C.J.D.

Contents

3 Polynomial and Rational Functions 303

Preface

WELCOME TO THE 12TH EDITION

In the twelfth edition of *Essentials of College Algebra,* we continue our ongoing commitment to providing the best possible text to help instructors teach and students succeed. In this edition, we have remained true to the pedagogical style of the past while staying focused on the needs of today's students. Support for all classroom types (traditional, hybrid, and online) may be found in this classic text and its supplements backed by the power of Pearson's MyLab Math.

In this edition, we have drawn upon the extensive teaching experience of the Lial team, with special consideration given to reviewer suggestions. General updates include enhanced readability with improved layout of examples, better use of color in displays, and language written with students in mind. Over 200 calculator screenshots have been updated and now provide color displays to enhance students' conceptual understanding. Each homework section now begins with six to ten *Concept Preview* exercises, assignable in MyLab Math, which may be used to ensure students' understanding of vocabulary and basic concepts prior to beginning the regular homework exercises.

Further enhancements include numerous current data examples and exercises that have been updated to reflect current information. Additional real-life exercises have been included to pique student interest; answers to writing exercises have been provided; better consistency has been achieved between the directions that introduce examples and those that introduce the corresponding exercises; and better guidance for rounding of answers has been provided in the exercise sets.

The Lial team believes this to be our best *Essentials of College Algebra* edition yet, and we sincerely hope that you enjoy using it as much as we have enjoyed writing it. Additional textbooks in this series are as follows.

College Algebra, Twelfth Edition
Trigonometry, Eleventh Edition
College Algebra and Trigonometry, Sixth Edition
Precalculus, Sixth Edition

HIGHLIGHTS OF NEW CONTENT

- In **Chapter R,** more detail has been added to set-builder notation, illustrations of the rules for exponents have been provided, and many exercises have been updated to better match section examples.

- Several new and updated application exercises have been inserted into the **Chapter 1** exercise sets. New objectives have been added to **Section 1.4** outlining the four methods for solving a quadratic equation, along with guidance suggesting when each method may be used efficiently.

- **Chapters 2 and 3** contain numerous new and updated application exercises, along with many updated calculator screenshots that are now provided in color. In response to reviewer suggestions, the discussion on increasing, decreasing, and constant functions in **Section 2.3** has been written to apply to open intervals of the domain.

- In **Chapter 4,** greater emphasis is given to the concept of exponential and logarithmic functions as inverses, there is a new table providing descriptions of the additional properties of exponents, and additional exercises requiring graphing logarithmic functions with translations have been included. There are also many new and updated real-life applications of exponential and logarithmic functions.

- In **Chapter 5,** special attention has been given to finding partial fraction decompositions in **Section 5.4** and to linear programing in **Section 5.6.** Examples have been rewritten to promote student understanding of these difficult topics.

- For visual learners, numbered **Figure** and **Example** references within the text are set using the same typeface as the figure number itself and bold print for the example. This makes it easier for the students to identify and connect them. We also have increased our use of a "drop down" style, when appropriate, to distinguish between simplifying expressions and solving equations, and we have added many more explanatory side comments. Guided Visualizations, with accompanying exercises and explorations, are now available and assignable in MyLab Math.

- *Essentials of College Algebra* is widely recognized for the quality of its exercises. In the twelfth edition, nearly 700 are new or modified, and hundreds present updated real-life data. Furthermore, the MyLab Math course has expanded coverage of all exercise types appearing in the exercise sets, as well as the mid-chapter Quizzes and Cumulative Reviews.

FEATURES OF THIS TEXT

SUPPORT FOR LEARNING CONCEPTS

We provide a variety of features to support students' learning of the essential topics of college algebra. Explanations that are written in understandable terms, figures and graphs that illustrate examples and concepts, graphing technology that supports and enhances algebraic manipulations, and real-life applications that enrich the topics with meaning all provide opportunities for students to deepen their understanding of mathematics. These features help students make mathematical connections and expand their own knowledge base.

- **Examples** Numbered examples that illustrate the techniques for working exercises are found in every section. We use traditional explanations, side comments, and pointers to describe the steps taken—and to warn students about common pitfalls. Some examples provide additional graphing calculator solutions, although these can be omitted if desired.

- **Now Try Exercises** Following each numbered example, the student is directed to try a corresponding odd-numbered exercise (or exercises). This feature allows for quick feedback to determine whether the student has understood the principles illustrated in the example.

- **Real-Life Applications** We have included hundreds of real-life applications, many with data updated from the previous edition. They come from fields such as business, entertainment, sports, biology, astronomy, geology, and environmental studies.

- **Function Boxes** Beginning in Chapter 2, functions provide a unifying theme throughout the text. Special function boxes offer a comprehensive, visual introduction to each type of function and also serve as an excellent resource for reference and review. Each function box includes a table of values, traditional and calculator-generated graphs, the domain, the range, and other special information about the function. These boxes are assignable in MyLab Math.

■ **Figures and Photos**   Today's students are more visually oriented than ever before, and we have updated the figures and photos in this edition to promote visual appeal. NEW Guided Visualizations with accompanying exercises and explorations are now available and assignable in MyLab Math.

■ **Use of Graphing Technology**   We have integrated the use of graphing calculators where appropriate, although *this technology is completely optional and can be omitted without loss of continuity.* We continue to stress that graphing calculators support understanding but that students must first master the underlying mathematical concepts. Exercises that require the use of a graphing calculator are marked with the icon ⊞ .

■ **Cautions and Notes**   Text that is marked **CAUTION** warns students of common errors, and NOTE comments point out explanations that should receive particular attention.

■ **Looking Ahead to Calculus**   These margin notes offer glimpses of how the topics currently being studied are used in calculus.

SUPPORT FOR PRACTICING CONCEPTS

This text offers a wide variety of exercises to help students master college algebra. The extensive exercise sets provide ample opportunity for practice, and the exercise problems increase in difficulty so that students at every level of understanding are challenged. The variety of exercise types promotes understanding of the concepts and reduces the need for rote memorization.

■ **NEW   Concept Preview**   Each exercise set now begins with a group of **CONCEPT PREVIEW** exercises designed to promote understanding of vocabulary and basic concepts of each section. These new exercises are assignable in MyLab Math and will provide support especially for hybrid, online, and flipped courses.

■ **Exercise Sets**   In addition to traditional drill exercises, this text includes writing exercises, optional graphing calculator problems ⊞, and multiple-choice, matching, true/false, and completion exercises. Those marked *Concept Check* focus on conceptual thinking. *Connecting Graphs with Equations* exercises challenge students to write equations that correspond to given graphs.

■ **Relating Concepts Exercises**   Appearing at the end of selected exercise sets, these groups of exercises are designed so that students who work them in numerical order will follow a line of reasoning that leads to an understanding of how various topics and concepts are related. All answers to these exercises appear in the student answer section, and these exercises are assignable in MyLab Math.

■ **Complete Solutions to Selected Exercises**   Complete solutions to all exercises marked ▒ are available in the eText. These are often exercises that extend the skills and concepts presented in the numbered examples.

SUPPORT FOR REVIEW AND TEST PREP

Ample opportunities for review are found within the chapters and at the ends of chapters. Quizzes that are interspersed within chapters provide a quick assessment of students' understanding of the material presented up to that point in the chapter. Chapter "Test Preps" provide comprehensive study aids to help students prepare for tests.

- **Quizzes** Students can periodically check their progress with in-chapter quizzes that appear in all chapters, beginning with Chapter 1. All answers, with corresponding section references, appear in the student answer section. These quizzes are assignable in MyLab Math.

- **Summary Exercises** These sets of in-chapter exercises give students the all-important opportunity to work *mixed* review exercises, requiring them to synthesize concepts and select appropriate solution methods.

- **End-of-Chapter Test Prep** Following the final numbered section in each chapter, the Test Prep provides a list of **Key Terms,** a list of **New Symbols** (if applicable), and a two-column **Quick Review** that includes a section-by-section summary of concepts and examples. This feature concludes with a comprehensive set of **Review Exercises** and a **Chapter Test.** The Test Prep, Review Exercises, and Chapter Test are assignable in MyLab Math.

Get the Most Out of
MyLab Math

Used by over 3 million students a year, MyLab Math is the world's leading online program for teaching and learning mathematics. MyLab Math delivers assessment, tutorials, and multimedia resources that provide engaging and personalized experiences for each student, so learning can happen in any environment. Each course is developed to accompany Pearson's best-selling content, authored by thought leaders across the math curriculum, and can be easily customized to fit any course format.

Preparedness

One of the biggest challenges in College Algebra, Trigonometry, and Precalculus is making sure students are adequately prepared with prerequisite knowledge. For a student, having the essential algebra skills upfront in a course can dramatically help increase success.

MyLab Math with Integrated Review can be used in co-requisite courses, or simply to help students who enter College Algebra without a full understanding of prerequisite skills and concepts. These courses provide videos on review topics, along with pre-made, assignable skills-check quizzes and personalized review homework assignments.

Getting Ready material provides just-in-time review, integrated throughout the course as needed to prepare students with prerequisite material to succeed. From a quick quiz, a personalized, just-in-time review assignment is generated for each student, allowing them to refresh forgotten concepts.

Resources for Success

MyLab Math Online Course for *Essentials of College Algebra,* Twelfth Edition

by Lial, Hornsby, Schneider, and Daniels *(access code required)*

MyLab Math is available to accompany Pearson's market-leading text offerings. To give students a consistent tone, voice, and teaching method each text's flavor and approach is tightly integrated throughout the accompanying MyLab Math course, making learning the material as seamless as possible.

Conceptual Understanding

It's important for students to visualize and conceptualize difficult concepts in Precalculus. Active engagement with the material helps students internalize mathematical connections.

Concept Preview Exercises

Each Homework section now begins with a group of Concept Preview Exercises, assignable in MyLab Math and in Learning Catalytics. These may be used to ensure that students understand the related vocabulary and basic concepts before beginning the regular homework problems.

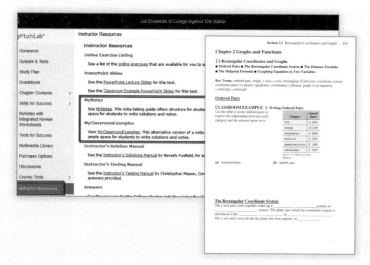

MyNotes and MyClassroomExamples

MyNotes provide a note-taking structure for students to use while they read the text or watch the MyLab Math videos. MyClassroom Examples offer structure for notes taken during lecture and are for use with the ClassroomExamples found in the Annotated Instructor's Edition.

Both sets of notes are available in MyLab Math and can be customized by the instructor.

pearson.com/mylab/math

Resources for Success

Instructor Resources

Additional resources can be downloaded from www.pearson.com or hardcopy resources can be ordered from your sales representative.

Annotated Instructor's Edition

ISBN: 0-13-467498-7 & 978-0-13-467498-8

- Answers are included on the same page beside the text exercises where possible for quick reference. Includes helpful Teaching Tips and Classroom Examples.

- Includes sample homework assignments, assignable in MyLab Math, indicated by problem numbers underlined in blue within each end-of-section exercise set.

Online Instructor's Solution Manual

By Beverly Fusfield Provides complete solutions to all text exercises.

Online Instructor's Testing Manual

By David Atwood, Rochester Community and Technical College. Testing manual includes diagnostic pretests, grouped by section, with answers provided.

TestGen®

TestGen® (www.pearsoned.com/testgen) enables instructors to build, edit, print, and administer tests using a computerized bank of questions developed to cover all the objectives of the text.

Online PowerPoint® Lecture Slides and Classroom Example PowerPoints

- The PowerPoint Lecture Slides feature presentations written and designed specifically for this text, including figures and examples from the text.

- Provide Classroom Example PowerPoints that include full worked-out solutions to all Classroom Examples.

Learning Catalytics

Generate class discussion, guide your lecture, and promote peer-to-peer learning with real-time analytics. MyLab Math now provides Learning Catalytics—an interactive student response tool that uses students' smartphones, tablets, or laptops to engage them in more sophisticated tasks and thinking.

Student Resources

Additional resources to aid student success.

Student's Solution Manual

ISBN: 0-13-467511-8 & 978-0-13-467511-4
By Beverly Fusfield Provides detailed solutions to all odd-numbered text exercises.

Video Lectures with Optional Captioning

- Quick Reviews cover key definitions and procedures from each section.

- Example Solutions walk students through the detailed solution process for every example in the textbook.

MyNotes with Integrated Review Worksheets

ISBN: 0-13-467517-7 & 978-0-13-467517-6
Available as a printed supplement and in MyLab Math. **MyNotes** offer structure for students as they watch videos or read the text.

- Includes textbook examples along with ample space for students to write solutions and notes.

- Includes key concepts along with prompts for students to read, write, and reflect on what they have just learned.

- **Customizable** so that instructors can add their own examples or remove examples that are not covered in their course.

Integrated Review Worksheets prepare students for the College Algebra material.

- Includes key terms, Guided examples with ample space for students to work and reference where in MyLab Math to get extra help.

MyClassroomExamples

- Available in MyLab Math and offer structure for students for classroom lecture.

- Includes Classroom Examples along with ample space for students to write solutions and notes.

- Includes key concepts along with fill in the blank opportunities to keep students engaged.

- **Customizable** so that instructors can add their own examples or remove Classroom Examples that are not covered in their course.

pearson.com/mylab/math

ACKNOWLEDGMENTS

We wish to thank the following individuals who provided valuable input into this edition of the text.

Barbara Aramenta – Pima Community College

Robert Bates – Honolulu Community College

Troy Brachey – Tennessee Tech University

Hugh Cornell – University of North Florida

John E. Daniels – Central Michigan University

Dan Fahringer – HACC Harrisburg

Doug Grenier – Rogers State University

Mary Hill – College of DuPage

Keith Hubbard – Stephen Austin State University

Christine Janowiak – Arapahoe Community College

Tarcia Jones – Austin Community College, Rio Grande

Rene Lumampao – Austin Community College

Rosana Maldonado – South Texas Community College, Pecan

Nilay S. Manzagol – Georgia State University

Marianna McClymonds – Phoenix College

Randy Nichols – Delta College

Preeti Parikh – SUNY Maritime

Deanna Robinson-Briedel – McLennan Community College

Sutandra Sarkar – Georgia State University

Patty Schovanec – Texas Tech University

Jimmy Vincente – El Centro College

Deanna M. Welsch – Illinois Central College

Amanda Wheeler – Amarillo College

Li Zhou – Polk State College

Our sincere thanks to those individuals at Pearson Education who have supported us throughout this revision: Anne Kelly, Christine O'Brien, Joe Vetere, and Jonathan Wooding. Terry McGinnis continues to provide behind-the-scenes guidance for both content and production. We have come to rely on her expertise during all phases of the revision process. Marilyn Dwyer of Cenveo® Publishing Services, with the assistance of Carol Merrigan, provided excellent production work. Special thanks go out to Paul Lorczak and John Morin for their excellent accuracy-checking. We thank Lucie Haskins, who provided an accurate index, and Jack Hornsby, who provided assistance in creating calculator screens, researching data updates, and proofreading. We appreciate the valuable suggestions for Chapter 5 that Mary Hill of *College of Dupage* made during our meeting with her in March 2010.

As an author team, we are committed to providing the best possible college algebra course to help instructors teach and students succeed. As we continue to work toward this goal, we welcome any comments or suggestions you might send, via e-mail, to math@pearson.com.

Margaret L. Lial

John Hornsby

David I. Schneider

Callie J. Daniels

Essentials of College Algebra

TWELFTH EDITION

R Review of Basic Concepts

Positive and negative numbers, used to represent gains and losses on a board such as this one, are examples of *real numbers* encountered in applications of mathematics.

R.1 Sets

- Basic Definitions
- Operations on Sets

Basic Definitions A **set** is a collection of objects. The objects that belong to a set are its **elements,** or **members.** In algebra, the elements of a set are usually numbers. Sets are commonly written using **set braces,** { }.

$$\{1, 2, 3, 4\} \qquad \text{The set containing the elements 1, 2, 3, and 4}$$

The order in which the elements are listed is not important. As a result, this same set can also be written as $\{4, 3, 2, 1\}$ or with any other arrangement of the four numbers.

To show that 4 is an element of the set $\{1, 2, 3, 4\}$, we use the symbol $\in$.

$$4 \in \{1, 2, 3, 4\}$$

Since 5 is *not* an element of this set, we place a slash through the symbol $\in$.

$$5 \notin \{1, 2, 3, 4\}$$

It is customary to name sets with capital letters.

$$S = \{1, 2, 3, 4\} \qquad \text{S is used to name the set.}$$

Set S was written above by listing its elements. Set S might also be described as

"the set containing the first four counting numbers."

The set F, consisting of all fractions between 0 and 1, is an example of an **infinite set**—one that has an unending list of distinct elements. A **finite set** is one that has a limited number of elements. The process of counting its elements comes to an end.

Some infinite sets can be described by listing. For example, the set of numbers N used for counting, which are the **natural numbers** or the **counting numbers,** can be written as follows.

$$N = \{1, 2, 3, 4, \ldots\} \qquad \text{Natural (counting) numbers}$$

The three dots (*ellipsis points*) show that the list of elements of the set continues according to the established pattern.

Sets are often written in **set-builder notation,** which uses a variable, such as x, to describe the elements of the set. The following set-builder notation represents the set $\{3, 4, 5, 6\}$ and is read "the set of all elements x such that x is a natural number between 2 and 7." The numbers 2 and 7 are *not* between 2 and 7.

$$\{x \mid x \text{ is a natural number between 2 and 7}\} = \{3, 4, 5, 6\} \qquad \text{Set-builder notation}$$

The set of all   such x is a natural number
elements x that   between 2 and 7

EXAMPLE 1 Using Set Notation and Terminology

Identify each set as *finite* or *infinite*. Then determine whether 10 is an element of the set.

(a) $\{7, 8, 9, \ldots, 14\}$

(b) $\left\{1, \frac{1}{4}, \frac{1}{16}, \frac{1}{64}, \ldots\right\}$

(c) $\{x \mid x \text{ is a fraction between 1 and 2}\}$

(d) $\{x \mid x \text{ is a natural number between 9 and 11}\}$

SOLUTION

(a) The set is finite, because the process of counting its elements 7, 8, 9, 10, 11, 12, 13, and 14 comes to an end. The number 10 belongs to the set.

$$10 \in \{7, 8, 9, \ldots, 14\}$$

(b) The set is infinite, because the ellipsis points indicate that the pattern continues indefinitely. In this case,

$$10 \notin \left\{1, \frac{1}{4}, \frac{1}{16}, \frac{1}{64}, \ldots\right\}.$$

(c) Between any two distinct natural numbers there are infinitely many fractions, so this set is infinite. The number 10 is not an element.

(d) There is only one natural number between 9 and 11, namely 10. So the set is finite, and 10 is an element.

✔ **Now Try Exercises 11, 13, 15, and 17.**

EXAMPLE 2 Listing the Elements of a Set

Use set notation, and list all the elements of each set.

(a) $\{x \mid x \text{ is a natural number less than } 5\}$

(b) $\{x \mid x \text{ is a natural number greater than 7 and less than 14}\}$

SOLUTION

(a) The natural numbers less than 5 form the set $\{1, 2, 3, 4\}$.

(b) This is the set $\{8, 9, 10, 11, 12, 13\}$.   ✔ **Now Try Exercise 25.**

When we are discussing a particular situation or problem, the **universal set** (whether expressed or implied) contains all the elements included in the discussion. The letter U is used to represent the universal set. The **null set,** or **empty set,** is the set containing no elements. We write the null set by either using the special symbol $\varnothing$, or else writing set braces enclosing no elements, $\{\ \}$.

CAUTION Do not combine these symbols. $\{\varnothing\}$ *is not the null set.* It is the set containing the symbol $\varnothing$.

Every element of the set $S = \{1, 2, 3, 4\}$ is a natural number. S is an example of a *subset* of the set N of natural numbers. This relationship is written using the symbol $\subseteq$.

$$S \subseteq N$$

By definition, set A is a **subset** of set B if every element of set A is also an element of set B. For example, if $A = \{2, 5, 9\}$ and $B = \{2, 3, 5, 6, 9, 10\}$, then $A \subseteq B$. However, there are some elements of B that are not in A, so B is not a subset of A. This relationship is written using the symbol $\not\subseteq$.

$$B \not\subseteq A$$

Every set is a subset of itself. Also, $\varnothing$ is a subset of every set.

If A is any set, then $A \subseteq A$ and $\varnothing \subseteq A$.

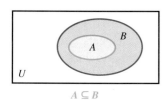

$A \subseteq B$

Figure 1

Figure 1 shows a set A that is a subset of set B. The rectangle in the drawing represents the universal set U. Such a diagram is a **Venn diagram.**

Two sets A and B are equal whenever $A \subseteq B$ and $B \subseteq A$. Equivalently, $A = B$ if the two sets contain exactly the same elements. For example,

$$\{1, 2, 3\} = \{3, 1, 2\}$$

is true because both sets contain exactly the same elements. However,

$$\{1, 2, 3\} \neq \{0, 1, 2, 3\}$$

because the set $\{0, 1, 2, 3\}$ contains the element 0, which is not an element of $\{1, 2, 3\}$.

EXAMPLE 3 Examining Subset Relationships

Let $U = \{1, 3, 5, 7, 9, 11, 13\}$, $A = \{1, 3, 5, 7, 9, 11\}$, $B = \{1, 3, 7, 9\}$, $C = \{3, 9, 11\}$, and $D = \{1, 9\}$. Determine whether each statement is *true* or *false*.

(a) $D \subseteq B$ **(b)** $B \subseteq D$ **(c)** $C \not\subseteq A$ **(d)** $U = A$

SOLUTION

(a) All elements of D, namely 1 and 9, are also elements of B, so D is a subset of B, and $D \subseteq B$ is true.

(b) There is at least one element of B (for example, 3) that is not an element of D, so B is *not* a subset of D. Thus, $B \subseteq D$ is false.

(c) C is a subset of A, because every element of C is also an element of A. Thus, $C \subseteq A$ is true, and as a result, $C \not\subseteq A$ is false.

(d) U contains the element 13, but A does not. Therefore, $U = A$ is false.

✔ **Now Try Exercises 53, 55, 63, and 65.**

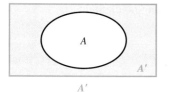

A'

Figure 2

Operations on Sets   Given a set A and a universal set U, the set of all elements of U that do **not** belong to set A is the **complement** of set A. For example, if set A is the set of all students in a class 30 years old or older, and set U is the set of all students in the class, then the complement of A would be the set of all students in the class younger than age 30.

The complement of set A is written A' (read **"A-prime"**). The Venn diagram in **Figure 2** shows a set A. Its complement, A', is in color. Using set-builder notation, the complement of set A is described as follows.

$$A' = \{x | x \in U, \ x \notin A\}$$

EXAMPLE 4 Finding Complements of Sets

Let $U = \{1, 2, 3, 4, 5, 6, 7\}$, $A = \{1, 3, 5, 7\}$, and $B = \{3, 4, 6\}$. Find each set.

(a) A' **(b)** B' **(c)** $\varnothing'$ **(d)** U'

SOLUTION

(a) Set A' contains the elements of U that are not in A. Thus, $A' = \{2, 4, 6\}$.

(b) $B' = \{1, 2, 5, 7\}$ **(c)** $\varnothing' = U$ **(d)** $U' = \varnothing$

✔ **Now Try Exercise 89.**

Given two sets A and B, the set of all elements belonging both to set A *and* to set B is the **intersection** of the two sets, written $A \cap B$. For example, if $A = \{1, 2, 4, 5, 7\}$ and $B = \{2, 4, 5, 7, 9, 11\}$, then we have the following.

$$A \cap B = \{1, 2, 4, 5, 7\} \cap \{2, 4, 5, 7, 9, 11\} = \{2, 4, 5, 7\}$$

The Venn diagram in **Figure 3** shows two sets A and B. Their intersection, $A \cap B$, is in color. Using set-builder notation, the intersection of sets A and B is described as follows.

$$A \cap B = \{x \mid x \in A \text{ and } x \in B\}$$

Two sets that have no elements in common are **disjoint sets.** If A and B are any two disjoint sets, then $A \cap B = \varnothing$. For example, there are no elements common to both $\{50, 51, 54\}$ and $\{52, 53, 55, 56\}$, so these two sets are disjoint.

$$\{50, 51, 54\} \cap \{52, 53, 55, 56\} = \varnothing$$

A ∩ B

Figure 3

EXAMPLE 5   **Finding Intersections of Two Sets**

Find each of the following. Identify any disjoint sets.

(a) $\{9, 15, 25, 36\} \cap \{15, 20, 25, 30, 35\}$

(b) $\{2, 3, 4, 5, 6\} \cap \{1, 2, 3, 4\}$

(c) $\{1, 3, 5\} \cap \{2, 4, 6\}$

SOLUTION

(a) $\{9, 15, 25, 36\} \cap \{15, 20, 25, 30, 35\} = \{15, 25\}$

The elements 15 and 25 are the only ones belonging to both sets.

(b) $\{2, 3, 4, 5, 6\} \cap \{1, 2, 3, 4\} = \{2, 3, 4\}$

(c) $\{1, 3, 5\} \cap \{2, 4, 6\} = \varnothing$ Disjoint sets

✔ **Now Try Exercises 69, 75, and 85.**

The set of all elements belonging to set A *or* to set B (or to both) is the **union** of the two sets, written $A \cup B$. For example, if $A = \{1, 3, 5\}$ and $B = \{3, 5, 7, 9\}$, then we have the following.

$$A \cup B = \{1, 3, 5\} \cup \{3, 5, 7, 9\} = \{1, 3, 5, 7, 9\}$$

The Venn diagram in **Figure 4** shows two sets A and B. Their union, $A \cup B$, is in color.

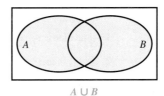

A ∪ B

Figure 4

Using set-builder notation, the union of sets A and B is described as follows.

$$A \cup B = \{x \mid x \in A \text{ or } x \in B\}$$

EXAMPLE 6 **Finding Unions of Two Sets**

Find each of the following.

(a) $\{1, 2, 5, 9, 14\} \cup \{1, 3, 4, 8\}$

(b) $\{1, 3, 5, 7\} \cup \{2, 4, 6\}$

(c) $\{1, 3, 5, 7, \ldots\} \cup \{2, 4, 6, \ldots\}$

SOLUTION

(a) Begin by listing the elements of the first set, $\{1, 2, 5, 9, 14\}$. Then include any elements from the second set that are not already listed.

$$\{1, 2, 5, 9, 14\} \cup \{1, 3, 4, 8\} = \{1, 2, 3, 4, 5, 8, 9, 14\}$$

(b) $\{1, 3, 5, 7\} \cup \{2, 4, 6\} = \{1, 2, 3, 4, 5, 6, 7\}$

(c) $\{1, 3, 5, 7, \ldots\} \cup \{2, 4, 6, \ldots\} = N$ Natural numbers

✔ **Now Try Exercises 71 and 83.**

The **set operations** are summarized below.

Set Operations

Let A and B define sets, with universal set U.

The **complement** of set A is the set A' of all elements in the universal set that do *not* belong to set A.

$$A' = \{x \,|\, x \in U, \ x \notin A\}$$

The **intersection** of sets A and B, written $A \cap B$, is made up of all the elements belonging to both set A *and* set B.

$$A \cap B = \{x \,|\, x \in A \text{ and } x \in B\}$$

The **union** of sets A and B, written $A \cup B$, is made up of all the elements belonging to set A *or* set B.

$$A \cup B = \{x \,|\, x \in A \text{ or } x \in B\}$$

R.1 Exercises

CONCEPT PREVIEW *Fill in the blank to correctly complete each sentence.*

1. The elements of the set of natural numbers are _____ .

2. Set A is a(n) _____ of set B if every element of set A is also an element of set B.

3. The set of all elements of the universal set U that do not belong to set A is the _____ of set A.

4. The _____ of sets A and B is made up of all the elements belonging to both set A *and* set B.

5. The _____ of sets A and B is made up of all the elements belonging to set A *or* set B (or both).

CONCEPT PREVIEW *Work each problem.*

6. Identify the set $\left\{1, \frac{1}{3}, \frac{1}{9}, \frac{1}{27}, \ldots\right\}$ as finite or infinite.

7. Use set notation and write the elements belonging to the set $\{x \mid x$ is a natural number less than 6$\}$.

8. Let $U = \{1, 2, 3, 4, 5\}$ and $A = \{1, 2, 3\}$. Find A'.

9. Find $\{16, 18, 21, 50\} \cap \{15, 16, 17, 18\}$.

10. Find $\{16, 18, 21, 50\} \cup \{15, 16, 17, 18\}$.

Identify each set as finite *or* infinite. *Then determine whether* 10 *is an element of the set.*
See Example 1.

11. $\{4, 5, 6, \ldots, 15\}$

12. $\{1, 2, 3, 4, 5, \ldots, 75\}$

13. $\left\{1, \frac{1}{2}, \frac{1}{4}, \frac{1}{8}, \ldots\right\}$

14. $\{4, 5, 6, \ldots\}$

15. $\{x \mid x$ is a natural number greater than 11$\}$

16. $\{x \mid x$ is a natural number greater than or equal to 10$\}$

17. $\{x \mid x$ is a fraction between 1 and 2$\}$

18. $\{x \mid x$ is an even natural number$\}$

Use set notation, and list all the elements of each set. *See Example 2.*

19. $\{12, 13, 14, \ldots, 20\}$

20. $\{8, 9, 10, \ldots, 17\}$

21. $\left\{1, \frac{1}{2}, \frac{1}{4}, \ldots, \frac{1}{32}\right\}$

22. $\{3, 9, 27, \ldots, 729\}$

23. $\{17, 22, 27, \ldots, 47\}$

24. $\{74, 68, 62, \ldots, 38\}$

25. $\{x \mid x$ is a natural number greater than 8 and less than 15$\}$

26. $\{x \mid x$ is a natural number not greater than 4$\}$

Insert $\in$ *or* $\notin$ *in each blank to make the resulting statement true.* *See Examples 1*
and 2.

27. $6 \underline{\hspace{1cm}} \{3, 4, 5, 6\}$

28. $9 \underline{\hspace{1cm}} \{2, 3, 5, 9, 8\}$

29. $5 \underline{\hspace{1cm}} \{4, 6, 8, 10\}$

30. $13 \underline{\hspace{1cm}} \{3, 5, 12, 14\}$

31. $0 \underline{\hspace{1cm}} \{0, 2, 3, 4\}$

32. $0 \underline{\hspace{1cm}} \{0, 5, 6, 7, 8, 10\}$

33. $\{3\} \underline{\hspace{1cm}} \{2, 3, 4, 5\}$

34. $\{5\} \underline{\hspace{1cm}} \{3, 4, 5, 6, 7\}$

35. $\{0\} \underline{\hspace{1cm}} \{0, 1, 2, 5\}$

36. $\{2\} \underline{\hspace{1cm}} \{2, 4, 6, 8\}$

37. $0 \underline{\hspace{1cm}} \varnothing$

38. $\varnothing \underline{\hspace{1cm}} \varnothing$

Determine whether each statement is true *or* false. *See Examples 1–3.*

39. $3 \in \{2, 5, 6, 8\}$

40. $6 \in \{2, 5, 8, 9\}$

41. $1 \in \{11, 5, 4, 3, 1\}$

42. $12 \in \{18, 17, 15, 13, 12\}$

43. $9 \notin \{8, 5, 2, 1\}$

44. $3 \notin \{7, 6, 5, 4\}$

45. $\{2, 5, 8, 9\} = \{2, 5, 9, 8\}$

46. $\{3, 0, 9, 6, 2\} = \{2, 9, 0, 3, 6\}$

47. $\{5, 8, 9\} = \{5, 8, 9, 0\}$

48. $\{3, 7, 12, 14\} = \{3, 7, 12, 14, 0\}$

49. $\{x \mid x$ is a natural number less than 3$\} = \{1, 2\}$

50. $\{x \mid x$ is a natural number greater than 10$\} = \{11, 12, 13, \ldots\}$

Let $A = \{2, 4, 6, 8, 10, 12\}$, $B = \{2, 4, 8, 10\}$, $C = \{4, 10, 12\}$, $D = \{2, 10\}$, and $U = \{2, 4, 6, 8, 10, 12, 14\}$.

Determine whether each statement is true or false. **See Example 3.**

51. $A \subseteq U$ **52.** $C \subseteq U$ **53.** $D \subseteq B$

54. $D \subseteq A$ **55.** $A \subseteq B$ **56.** $B \subseteq C$

57. $\varnothing \subseteq A$ **58.** $\varnothing \subseteq \varnothing$ **59.** $\{4, 8, 10\} \subseteq B$

60. $\{0, 2\} \subseteq D$ **61.** $B \subseteq D$ **62.** $A \subseteq C$

Insert $\subseteq$ or $\nsubseteq$ in each blank to make the resulting statement true. **See Example 3.**

63. $\{2, 4, 6\}$ _____ $\{2, 3, 4, 5, 6\}$ **64.** $\{1, 5\}$ _____ $\{0, 1, 2, 3, 5\}$

65. $\{0, 1, 2\}$ _____ $\{1, 2, 3, 4, 5\}$ **66.** $\{5, 6, 7, 8\}$ _____ $\{1, 2, 3, 4, 5, 6, 7\}$

67. $\varnothing$ _____ $\{1, 4, 6, 8\}$ **68.** $\varnothing$ _____ $\varnothing$

Determine whether each statement is true or false. **See Examples 4–6.**

69. $\{5, 7, 9, 19\} \cap \{7, 9, 11, 15\} = \{7, 9\}$

70. $\{8, 11, 15\} \cap \{8, 11, 19, 20\} = \{8, 11\}$

71. $\{1, 2, 7\} \cup \{1, 5, 9\} = \{1\}$

72. $\{6, 12, 14, 16\} \cup \{6, 14, 19\} = \{6, 14\}$

73. $\{2, 3, 5, 9\} \cap \{2, 7, 8, 10\} = \{2\}$

74. $\{6, 8, 9\} \cup \{9, 8, 6\} = \{8, 9\}$

75. $\{3, 5, 9, 10\} \cap \varnothing = \{3, 5, 9, 10\}$ **76.** $\{3, 5, 9, 10\} \cup \varnothing = \{3, 5, 9, 10\}$

77. $\{1, 2, 4\} \cup \{1, 2, 4\} = \{1, 2, 4\}$ **78.** $\{1, 2, 4\} \cap \{1, 2, 4\} = \varnothing$

79. $\varnothing \cup \varnothing = \varnothing$ **80.** $\varnothing \cap \varnothing = \varnothing$

Let $U = \{0, 1, 2, 3, 4, 5, 6, 7, 8, 9, 10, 11, 12, 13\}$, $M = \{0, 2, 4, 6, 8\}$, $N = \{1, 3, 5, 7, 9, 11, 13\}$, $Q = \{0, 2, 4, 6, 8, 10, 12\}$, and $R = \{0, 1, 2, 3, 4\}$.

Use these sets to find each of the following. Identify any disjoint sets. **See Examples 4–6.**

81. $M \cap R$ **82.** $M \cap U$ **83.** $M \cup N$

84. $M \cup R$ **85.** $M \cap N$ **86.** $U \cap N$

87. $N \cup R$ **88.** $M \cup Q$ **89.** N'

90. Q' **91.** $M' \cap Q$ **92.** $Q \cap R'$

93. $\varnothing \cap R$ **94.** $\varnothing \cap Q$ **95.** $N \cup \varnothing$

96. $R \cup \varnothing$ **97.** $(M \cap N) \cup R$ **98.** $(N \cup R) \cap M$

99. $(Q \cap M) \cup R$ **100.** $(R \cup N) \cap M'$

101. $(M' \cup Q) \cap R$ **102.** $Q \cap (M \cup N)$

103. $Q' \cap (N' \cap U)$ **104.** $(U \cap \varnothing') \cup R$

105. $\{x \mid x \in U, \ x \notin M\}$ **106.** $\{x \mid x \in U, \ x \notin R\}$

107. $\{x \mid x \in M \text{ and } x \in Q\}$ **108.** $\{x \mid x \in Q \text{ and } x \in R\}$

109. $\{x \mid x \in M \text{ or } x \in Q\}$ **110.** $\{x \mid x \in Q \text{ or } x \in R\}$

R.2 Real Numbers and Their Properties

- Sets of Numbers and the Number Line
- Exponents
- Order of Operations
- Properties of Real Numbers
- Order on the Number Line
- Absolute Value

Sets of Numbers and the Number Line   As mentioned previously, the set of **natural numbers** is written in set notation as follows.

$$\{1, 2, 3, 4, \ldots\} \quad \text{Natural numbers}$$

Including 0 with the set of natural numbers gives the set of **whole numbers.**

$$\{0, 1, 2, 3, 4, \ldots\} \quad \text{Whole numbers}$$

Including the negatives of the natural numbers with the set of whole numbers gives the set of **integers.**

$$\{\ldots, -3, -2, -1, 0, 1, 2, 3, \ldots\} \quad \text{Integers}$$

Integers can be graphed on a **number line.** See **Figure 5.** Every number corresponds to one and only one point on the number line, and each point corresponds to one and only one number. The number associated with a given point is the **coordinate** of the point. This correspondence forms a **coordinate system.**

The result of dividing two integers (with a nonzero divisor) is a *rational number*, or *fraction*. A **rational number** is an element of the set defined as follows.

$$\left\{\frac{p}{q} \,\middle|\, p \text{ and } q \text{ are integers and } q \neq 0\right\} \quad \text{Rational numbers}$$

The set of rational numbers includes the natural numbers, the whole numbers, and the integers. For example, the integer -3 is a rational number because it can be written as $\frac{-3}{1}$. Numbers that can be written as repeating or terminating decimals are also rational numbers. For example, $0.\overline{6} = 0.66666\ldots$ represents a rational number that can be expressed as the fraction $\frac{2}{3}$.

The set of all numbers that correspond to points on a number line is the **real numbers,** shown in **Figure 6.** Real numbers can be represented by decimals. Because every fraction has a decimal form—for example, $\frac{1}{4} = 0.25$—real numbers include rational numbers.

Some real numbers cannot be represented by quotients of integers. These numbers are **irrational numbers.** The set of irrational numbers includes $\sqrt{2}$ and $\sqrt{5}$. Another irrational number is π, which is *approximately* equal to 3.14159. Some rational and irrational numbers are graphed in **Figure 7.**

The sets of numbers discussed so far are summarized as follows.

Origin

Graph of the Set of Integers

Figure 5

Graph of the Set of Real Numbers

Figure 6

Graph of the Set
$$\left\{-\tfrac{2}{3}, 0, \sqrt{2}, \sqrt{5}, \pi, 4\right\}$$

$\sqrt{2}$, $\sqrt{5}$, and π are irrational. Because $\sqrt{2}$ is approximately equal to 1.41, it is located between 1 and 2, slightly closer to 1.

Figure 7

Sets of Numbers

Set	Description
Natural numbers	$\{1, 2, 3, 4, \ldots\}$
Whole numbers	$\{0, 1, 2, 3, 4, \ldots\}$
Integers	$\{\ldots, -3, -2, -1, 0, 1, 2, 3, \ldots\}$
Rational numbers	$\left\{\frac{p}{q} \,\middle|\, p \text{ and } q \text{ are integers and } q \neq 0\right\}$
Irrational numbers	$\{x \mid x \text{ is real but not rational}\}$
Real numbers	$\{x \mid x \text{ corresponds to a point on a number line}\}$

EXAMPLE 1 **Identifying Sets of Numbers**

Let $A = \left\{ -8, -6, -\frac{12}{4}, -\frac{3}{4}, 0, \frac{3}{8}, \frac{1}{2}, 1, \sqrt{2}, \sqrt{5}, 6 \right\}$. List all the elements of A that belong to each set.

(a) Natural numbers   **(b)** Whole numbers   **(c)** Integers

(d) Rational numbers   **(e)** Irrational numbers   **(f)** Real numbers

SOLUTION

(a) Natural numbers: 1 and 6   **(b)** Whole numbers: 0, 1, and 6

(c) Integers: $-8, -6, -\frac{12}{4}$ (or -3), 0, 1, and 6

(d) Rational numbers: $-8, -6, -\frac{12}{4}$ (or -3), $-\frac{3}{4}, 0, \frac{3}{8}, \frac{1}{2}$, 1, and 6

(e) Irrational numbers: $\sqrt{2}$ and $\sqrt{5}$

(f) All elements of A are real numbers.   ✔ **Now Try Exercises 11, 13, and 15.**

Figure 8 shows the relationships among the subsets of the real numbers. As shown, the natural numbers are a subset of the whole numbers, which are a subset of the integers, which are a subset of the rational numbers. The union of the rational numbers and irrational numbers is the set of real numbers.

Real Numbers

Rational numbers $\frac{4}{9}, -\frac{5}{8}, \frac{11}{7}$	**Irrational numbers** $\sqrt{2}$
Integers $-11, -6, -3, -2, -1$ **Whole numbers** 0 **Natural numbers** $1, 2, 3, 4, 5, 37, 40$	$\sqrt{15}$ $-\sqrt{8}$ π $\frac{\pi}{4}$

Figure 8

Exponents   Any collection of numbers or variables joined by the basic operations of addition, subtraction, multiplication, or division (except by 0), or the operations of raising to powers or taking roots, formed according to the rules of algebra, is an **algebraic expression.**

$$-2x^2 + 3x, \quad \frac{15y}{2y - 3}, \quad \sqrt{m^3 - 64}, \quad (3a + b)^4 \qquad \text{Algebraic expressions}$$

The expression 2^3 is an **exponential expression,** or **exponential,** where the 3 indicates that three factors of 2 appear in the corresponding product. The number 2 is the **base,** and the number 3 is the **exponent.**

Exponent: 3

$$2^3 = \underbrace{2 \cdot 2 \cdot 2}_{\text{Three factors of 2}} = 8$$

Base: 2

Exponential Notation

If n is any positive integer and a is any real number, then the nth power of a is written using exponential notation as follows.

$$a^n = \underbrace{a \cdot a \cdot a \cdot \ldots \cdot a}_{n \text{ factors of } a}$$

Read a^n as **"a to the nth power"** or simply **"a to the nth."**

EXAMPLE 2 **Evaluating Exponential Expressions**

Evaluate each exponential expression, and identify the base and the exponent.

(a) 4^3 **(b)** $(-6)^2$ **(c)** -6^2 **(d)** $4 \cdot 3^2$ **(e)** $(4 \cdot 3)^2$

SOLUTION

(a) $4^3 = \underbrace{4 \cdot 4 \cdot 4}_{3 \text{ factors of } 4} = 64$ The base is 4 and the exponent is 3.

(b) $(-6)^2 = (-6)(-6) = 36$ The base is -6 and the exponent is 2.

(c) $-6^2 = -(6 \cdot 6) = -36$ —— Notice that parts (b) and (c) are different.
The base is 6 and the exponent is 2.

(d) $4 \cdot 3^2 = 4 \cdot 3 \cdot 3 = 36$ The base is 3 and the exponent is 2.
$3^2 = 3 \cdot 3$, NOT $3 \cdot 2$

(e) $(4 \cdot 3)^2 = 12^2 = 144$ —— $(4 \cdot 3)^2 \neq 4 \cdot 3^2$
The base is $4 \cdot 3$, or 12, and the exponent is 2.

✔ **Now Try Exercises 17, 19, 21, and 23.**

Order of Operations When an expression involves more than one operation symbol, such as $5 \cdot 2 + 3$, we use the following order of operations.

Order of Operations

If grouping symbols such as parentheses, square brackets, absolute value bars, or fraction bars are present, begin as follows.

Step 1 Work separately above and below each **fraction bar.**

Step 2 Use the rules below within each set of **parentheses** or **square brackets.** Start with the innermost set and work outward.

If no grouping symbols are present, follow these steps.

Step 1 Simplify all **powers** and **roots.** *Work from left to right.*

Step 2 Do any **multiplications** or **divisions** in order. *Work from left to right.*

Step 3 Do any **negations, additions,** or **subtractions** in order. *Work from left to right.*

EXAMPLE 3 **Using Order of Operations**

Evaluate each expression.

(a) $6 \div 3 + 2^3 \cdot 5$

(b) $(8 + 6) \div 7 \cdot 3 - 6$

(c) $\dfrac{4 + 3^2}{6 - 5 \cdot 3}$

(d) $\dfrac{-(-3)^3 + (-5)}{2(-8) - 5(3)}$

SOLUTION

(a) $6 \div 3 + 2^3 \cdot 5$

$= 6 \div 3 + 8 \cdot 5$ Evaluate the exponential.

$= 2 + 8 \cdot 5$ Divide.

$= 2 + 40$ Multiply.

Multiply or divide *in order from left to right.*

$= 42$ Add.

(b) $(8 + 6) \div 7 \cdot 3 - 6$

$= 14 \div 7 \cdot 3 - 6$ Work inside the parentheses.

Be careful to divide *before* multiplying here.

$= 2 \cdot 3 - 6$ Divide.

$= 6 - 6$ Multiply.

$= 0$ Subtract.

(c) Work separately above and below the fraction bar, and then divide as a last step.

$$\dfrac{4 + 3^2}{6 - 5 \cdot 3}$$

$$= \dfrac{4 + 9}{6 - 15}$$ Evaluate the exponential and multiply.

$$= \dfrac{13}{-9}$$ Add and subtract.

$$= -\dfrac{13}{9}$$ $\dfrac{a}{-b} = -\dfrac{a}{b}$

(d) $\dfrac{-(-3)^3 + (-5)}{2(-8) - 5(3)}$

$$= \dfrac{-(-27) + (-5)}{2(-8) - 5(3)}$$ Evaluate the exponential.

$$= \dfrac{27 + (-5)}{-16 - 15}$$ Multiply.

$$= \dfrac{22}{-31}$$ Add and subtract.

$$= -\dfrac{22}{31}$$ $\dfrac{a}{-b} = -\dfrac{a}{b}$

✔ **Now Try Exercises 25, 27, and 33.**

EXAMPLE 4   **Using Order of Operations**

Evaluate each expression for $x = -2$, $y = 5$, and $z = -3$.

(a) $-4x^2 - 7y + 4z$ (b) $\dfrac{2(x - 5)^2 + 4y}{z + 4}$ (c) $\dfrac{\dfrac{x}{2} - \dfrac{y}{5}}{\dfrac{3z}{9} + \dfrac{8y}{5}}$

SOLUTION

(a) $-4x^2 - 7y + 4z$

$= -4(-2)^2 - 7(5) + 4(-3)$ Substitute: $x = -2$, $y = 5$, and $z = -3$.

Use parentheses around substituted values to avoid errors.

$= -4(4) - 7(5) + 4(-3)$ Evaluate the exponential.

$= -16 - 35 - 12$ Multiply.

$= -63$ Subtract.

(b) $\dfrac{2(x - 5)^2 + 4y}{z + 4}$

$= \dfrac{2(-2 - 5)^2 + 4(5)}{-3 + 4}$ Substitute: $x = -2$, $y = 5$, and $z = -3$.

$= \dfrac{2(-7)^2 + 4(5)}{-3 + 4}$ Work inside the parentheses.

$= \dfrac{2(49) + 4(5)}{-3 + 4}$ Evaluate the exponential.

$= \dfrac{98 + 20}{1}$ Multiply in the numerator. Add in the denominator.

$= 118$ Add; $\dfrac{a}{1} = a$.

(c) This is a *complex fraction*. Work separately above and below the main fraction bar, and then divide as a last step.

$$\dfrac{\dfrac{x}{2} - \dfrac{y}{5}}{\dfrac{3z}{9} + \dfrac{8y}{5}}$$

$= \dfrac{\dfrac{-2}{2} - \dfrac{5}{5}}{\dfrac{3(-3)}{9} + \dfrac{8(5)}{5}}$ Substitute: $x = -2$, $y = 5$, and $z = -3$.

$= \dfrac{-1 - 1}{-1 + 8}$ Simplify the fractions.

$= -\dfrac{2}{7}$ Subtract and add; $\dfrac{-a}{b} = -\dfrac{a}{b}$.

✔ **Now Try Exercises 35, 43, and 45.**

Properties of Real Numbers Recall the following basic properties.

Properties of Real Numbers

Let a, b, and c represent real numbers.

Property	Description
Closure Properties $a + b$ is a real number. ab is a real number.	The sum or product of two real numbers is a real number.
Commutative Properties $a + b = b + a$ $ab = ba$	The sum or product of two real numbers is the same regardless of their order.
Associative Properties $(a + b) + c = a + (b + c)$ $(ab)c = a(bc)$	The sum or product of three real numbers is the same no matter which two are added or multiplied first.
Identity Properties There exists a unique real number 0 such that $a + 0 = a$ and $0 + a = a.$ There exists a unique real number 1 such that $a \cdot 1 = a$ and $1 \cdot a = a.$	The sum of a real number and 0 is that real number, and the product of a real number and 1 is that real number.
Inverse Properties There exists a unique real number $-a$ such that $a + (-a) = 0$ and $-a + a = 0.$ If $a \neq 0$, there exists a unique real number $\frac{1}{a}$ such that $a \cdot \dfrac{1}{a} = 1$ and $\dfrac{1}{a} \cdot a = 1.$	The sum of any real number and its negative is 0, and the product of any nonzero real number and its reciprocal is 1.
Distributive Properties $a(b + c) = ab + ac$ $a(b - c) = ab - ac$	The product of a real number and the sum (or difference) of two real numbers equals the sum (or difference) of the products of the first number and each of the other numbers.
Multiplication Property of Zero $0 \cdot a = a \cdot 0 = 0$	The product of a real number and 0 is 0.

CAUTION With the commutative properties, the *order* changes, but with the associative properties, the *grouping* changes.

Commutative Properties	Associative Properties
$(x + 4) + 9 = (4 + x) + 9$	$(x + 4) + 9 = x + (4 + 9)$
$7 \cdot (5 \cdot 2) = (5 \cdot 2) \cdot 7$	$7 \cdot (5 \cdot 2) = (7 \cdot 5) \cdot 2$

EXAMPLE 5 **Simplifying Expressions**

Use the commutative and associative properties to simplify each expression.

(a) $6 + (9 + x)$ **(b)** $\dfrac{5}{8}(16y)$ **(c)** $-10p\left(\dfrac{6}{5}\right)$

SOLUTION

(a) $6 + (9 + x)$

$\quad = (6 + 9) + x$ Associative property

$\quad = 15 + x$ Add.

(b) $\dfrac{5}{8}(16y)$

$\quad = \left(\dfrac{5}{8} \cdot 16\right)y$ Associative property

$\quad = 10y$ Multiply.

(c) $-10p\left(\dfrac{6}{5}\right)$

$\quad = \dfrac{6}{5}(-10p)$ Commutative property

$\quad = \left[\dfrac{6}{5}(-10)\right]p$ Associative property

$\quad = -12p$ Multiply.   ✔ **Now Try Exercises 63 and 65.**

Geometric Model of the
Distributive Property

Figure 9

Figure 9 helps to explain the distributive property. The area of the entire region shown can be found in two ways, as follows.

$$4(5 + 3) = 4(8) = 32$$

or $\qquad 4(5) + 4(3) = 20 + 12 = 32$

The result is the same. This means that

$$4(5 + 3) = 4(5) + 4(3).$$

EXAMPLE 6 **Using the Distributive Property**

Rewrite each expression using the distributive property and simplify, if possible.

(a) $3(x + y)$ **(b)** $-(m - 4n)$ **(c)** $7p + 21$ **(d)** $\dfrac{1}{3}\left(\dfrac{4}{5}m - \dfrac{3}{2}n - 27\right)$

SOLUTION

(a) $3(x + y)$

$\quad = 3x + 3y$

   Distributive property

(b) $-(m - 4n)$

$\quad = -1(m - 4n)$ Be careful with the negative signs.

$\quad = -1(m) + (-1)(-4n)$

$\quad = -m + 4n$

(c) $7p + 21$

$\quad = 7p + 7 \cdot 3$

$\quad = 7 \cdot p + 7 \cdot 3$

$\quad = 7(p + 3)$

   Distributive property in reverse

(d) $\dfrac{1}{3}\left(\dfrac{4}{5}m - \dfrac{3}{2}n - 27\right)$

$\quad = \dfrac{1}{3}\left(\dfrac{4}{5}m\right) + \dfrac{1}{3}\left(-\dfrac{3}{2}n\right) + \dfrac{1}{3}(-27)$

$\quad = \dfrac{4}{15}m - \dfrac{1}{2}n - 9$

✔ **Now Try Exercises 67, 69, and 71.**

$-\sqrt{5}$ is to the left of $-\frac{11}{7}$ on the number line, so $-\sqrt{5} < -\frac{11}{7}$, and $\sqrt{20}$ is to the right of π, indicating that $\sqrt{20} > \pi$.

Figure 10

Order on the Number Line   If the real number a is to the left of the real number b on a number line, then

$$a \text{ is less than } b, \quad \text{written} \quad a < b.$$

If a is to the right of b, then

$$a \text{ is greater than } b, \quad \text{written} \quad a > b.$$

> The inequality symbol must point toward the lesser number.

See **Figure 10.** Statements involving these symbols, as well as the symbols less than or equal to, $\leq$, and greater than or equal to, $\geq$, are **inequalities.** The inequality $a < b < c$ says that b is **between** a and c because $a < b$ and $b < c$.

Distance is 5.   Distance is 5.

Figure 11

Absolute Value   The undirected distance on a number line from a number to 0 is the **absolute value** of that number. The absolute value of the number a is written $|a|$. For example, the distance on a number line from 5 to 0 is 5, as is the distance from -5 to 0. See **Figure 11.** Therefore, both of the following are true.

$$|5| = 5 \quad \text{and} \quad |-5| = 5$$

NOTE   *Because distance cannot be negative, the absolute value of a number is always positive or* **0.**

The algebraic definition of absolute value follows.

Absolute Value

Let a represent a real number.

$$|a| = \begin{cases} a & \text{if } a \geq 0 \\ -a & \text{if } a < 0 \end{cases}$$

That is, the absolute value of a positive number or **0** *equals that number, while the absolute value of a negative number equals its negative (or opposite).*

EXAMPLE 7   Evaluating Absolute Values

Evaluate each expression.

(a) $\left| -\dfrac{5}{8} \right|$ **(b)** $-|8|$ **(c)** $-|-2|$ **(d)** $|2x|$, for $x = \pi$

SOLUTION

(a) $\left| -\dfrac{5}{8} \right| = \dfrac{5}{8}$ **(b)** $-|8| = -(8) = -8$

(c) $-|-2| = -(2) = -2$ **(d)** $|2\pi| = 2\pi$

✔ **Now Try Exercises 83 and 87.**

Absolute value is useful in applications where only the *size* (or magnitude), not the *sign*, of the difference between two numbers is important.

EXAMPLE 8 **Measuring Blood Pressure Difference**

Systolic blood pressure is the maximum pressure produced by each heartbeat. Both low blood pressure and high blood pressure may be cause for medical concern. Therefore, health care professionals are interested in a patient's "pressure difference from normal," or P_d.

If 120 is considered a normal systolic pressure, then

$$P_d = |P - 120|, \quad \text{where } P \text{ is the patient's recorded systolic pressure.}$$

Find P_d for a patient with a systolic pressure, P, of 113.

SOLUTION
$$
\begin{aligned}
P_d &= |P - 120| \\
&= |113 - 120| \quad \text{Let } P = 113. \\
&= |-7| \quad\quad\quad \text{Subtract.} \\
&= 7 \quad\quad\quad\quad \text{Definition of absolute value}
\end{aligned}
$$

✔ **Now Try Exercise 89.**

Properties of Absolute Value

Let a and b represent real numbers.

Property	Description
1. $|a| \geq 0$	The absolute value of a real number is positive or 0.
2. $|-a| = |a|$	The absolute values of a real number and its opposite are equal.
3. $|a| \cdot |b| = |ab|$	The product of the absolute values of two real numbers equals the absolute value of their product.
4. $\dfrac{|a|}{|b|} = \left|\dfrac{a}{b}\right| \ (b \neq 0)$	The quotient of the absolute values of two real numbers equals the absolute value of their quotient.
5. $|a + b| \leq |a| + |b|$ (the triangle inequality)	The absolute value of the sum of two real numbers is less than or equal to the sum of their absolute values.

LOOKING AHEAD TO CALCULUS

One of the most important definitions in calculus, that of the **limit,** uses absolute value. The symbols ϵ (epsilon) and δ (delta) are often used to represent small quantities in mathematics.

Suppose that a function f is defined at every number in an open interval I containing a, except perhaps at a itself. Then the limit of $f(x)$ as x approaches a is L, written

$$\lim_{x \to a} f(x) = L,$$

if for every $\epsilon > 0$ there exists a $\delta > 0$ such that $|f(x) - L| < \epsilon$ whenever $0 < |x - a| < \delta$.

Examples of Properties 1–4:

$$|-15| = 15 \text{ and } 15 \geq 0. \qquad \text{Property 1}$$

$$|-10| = 10 \text{ and } |10| = 10, \text{ so } |-10| = |10|. \qquad \text{Property 2}$$

$$|5| \cdot |-4| = 5 \cdot 4 = 20 \text{ and } |5(-4)| = |-20| = 20,$$
$$\text{so } |5| \cdot |-4| = |5(-4)|. \qquad \text{Property 3}$$

$$\frac{|2|}{|3|} = \frac{2}{3} \text{ and } \left|\frac{2}{3}\right| = \frac{2}{3}, \text{ so } \frac{|2|}{|3|} = \left|\frac{2}{3}\right|. \qquad \text{Property 4}$$

Example of the triangle inequality:

$$|a + b| = |3 + (-7)| = |-4| = 4$$
$$\text{Let } a = 3 \text{ and } b = -7.$$
$$|a| + |b| = |3| + |-7| = 3 + 7 = 10$$

Thus, $\qquad |a + b| \leq |a| + |b|.$ \qquad Property 5

> **NOTE** As seen in **Example 9(b),** absolute value bars can also act as grouping symbols. Remember this when applying the rules for order of operations.

EXAMPLE 9 Evaluating Absolute Value Expressions

Let $x = -6$ and $y = 10$. Evaluate each expression.

(a) $|2x - 3y|$

(b) $\dfrac{2|x| - |3y|}{|xy|}$

SOLUTION

(a) $|2x - 3y|$

$\qquad = |2(-6) - 3(10)|$ Substitute: $x = -6, y = 10$.

$\qquad = |-12 - 30|$ Work inside the absolute value bars. Multiply.

$\qquad = |-42|$ Subtract.

$\qquad = 42$ Definition of absolute value

(b) $\dfrac{2|x| - |3y|}{|xy|}$

$\qquad = \dfrac{2|-6| - |3(10)|}{|-6(10)|}$ Substitute: $x = -6, y = 10$.

$\qquad = \dfrac{2 \cdot 6 - |30|}{|-60|}$ $|-6| = 6$; Multiply.

$\qquad = \dfrac{12 - 30}{60}$ Multiply; $|30| = 30, |-60| = 60$.

$\qquad = \dfrac{-18}{60}$ Subtract.

$\qquad = -\dfrac{3}{10}$ Write in lowest terms; $\dfrac{-a}{b} = -\dfrac{a}{b}$.

✔ **Now Try Exercises 93 and 95.**

Distance between Points on a Number Line

If P and Q are points on a number line with coordinates a and b, respectively, then the distance $d(P, Q)$ between them is given by the following.

$$d(P, Q) = |b - a| \quad \text{or} \quad d(P, Q) = |a - b|$$

Figure 12

That is, the distance between two points on a number line is the absolute value of the difference between their coordinates in either order. See Figure 12.

EXAMPLE 10 **Finding the Distance between Two Points**

Find the distance between -5 and 8.

SOLUTION Use the first formula in the preceding box, with $a = -5$ and $b = 8$.

$$d(P, Q) = |b - a| = |8 - (-5)| = |8 + 5| = |13| = 13$$

Using the second formula in the box, we obtain the same result.

$$d(P, Q) = |a - b| = |(-5) - 8| = |-13| = 13$$

✔ **Now Try Exercise 105.**

R.2 Exercises

CONCEPT PREVIEW *Fill in the blank(s) to correctly complete each sentence.*

1. $\{0, 1, 2, 3, \ldots\}$ describes the set of _____.

2. $\{\ldots, -3, -2, -1, 0, 1, 2, 3, \ldots\}$ describes the set of _____.

3. In the expression 6^3, 6 is the _____, and 3 is the _____.

4. If the real number a is to the left of the real number b on a number line, then a _____ b.

5. The distance on a number line from a number to 0 is the _____ of that number.

6. **CONCEPT PREVIEW** Match each number from Column I with the letter or letters of the sets of numbers from Column II to which the number belongs. There may be more than one choice, so give all choices.

I		II	
(a) 0	**(b)** 34	**A.** Natural numbers	**B.** Whole numbers
(c) $-\frac{9}{4}$	**(d)** $\sqrt{36}$	**C.** Integers	**D.** Rational numbers
(e) $\sqrt{13}$	**(f)** 2.16	**E.** Irrational numbers	**F.** Real numbers

CONCEPT PREVIEW *Evaluate each expression.*

7. 10^3 8. $2 \cdot 5 - 10 \div 2$ 9. $|-4|$

10. **CONCEPT PREVIEW** Simplify the expression $-7(x - 4y)$.

Let $A = \left\{-6, -\frac{12}{4}, -\frac{5}{8}, -\sqrt{3}, 0, \frac{1}{4}, 1, 2\pi, 3, \sqrt{12}\right\}$. *List all the elements of A that belong to each set. See Example 1.*

11. Natural numbers   12. Whole numbers

13. Integers   14. Rational numbers

15. Irrational numbers   16. Real numbers

Evaluate each expression. See Example 2.

17. -2^4 18. -3^5 19. $(-2)^4$ 20. $(-2)^6$

21. $(-3)^5$ 22. $(-2)^5$ 23. $-2 \cdot 3^4$ 24. $-4 \cdot 5^3$

*Evaluate each expression. **See Example 3.***

25. $-2 \cdot 5 + 12 \div 3$

26. $9 \cdot 3 - 16 \div 4$

27. $-4(9 - 8) + (-7)(2)^3$

28. $6(-5) - (-3)(2)^4$

29. $\left(4 - 2^3\right)\left(-2 + \sqrt{25}\right)$

30. $\left(5 - 3^2\right)\left(\sqrt{16} - 2^3\right)$

31. $\left(-\dfrac{2}{9} - \dfrac{1}{4}\right) - \left[-\dfrac{5}{18} - \left(-\dfrac{1}{2}\right)\right]$

32. $\left[-\dfrac{5}{8} - \left(-\dfrac{2}{5}\right)\right] - \left(\dfrac{3}{2} - \dfrac{11}{10}\right)$

33. $\dfrac{-8 + (-4)(-6) \div 12}{4 - (-3)}$

34. $\dfrac{15 \div 5 \cdot 4 \div 6 - 8}{-6 - (-5) - 8 \div 2}$

*Evaluate each expression for $p = -4$, $q = 8$, and $r = -10$. **See Example 4.***

35. $-p^2 - 7q + r^2$

36. $-p^2 - 2q + r$

37. $\dfrac{q + r}{q + p}$

38. $\dfrac{p + r}{p + q}$

39. $\dfrac{3q}{r} - \dfrac{5}{p}$

40. $\dfrac{3r}{q} - \dfrac{2}{r}$

41. $\dfrac{5r}{2p - 3r}$

42. $\dfrac{3q}{3p - 2r}$

43. $\dfrac{\dfrac{q}{2} - \dfrac{r}{3}}{\dfrac{3p}{4} + \dfrac{q}{8}}$

44. $\dfrac{\dfrac{q}{4} - \dfrac{r}{5}}{\dfrac{p}{2} + \dfrac{q}{2}}$

45. $\dfrac{-(p + 2)^2 - 3r}{2 - q}$

46. $\dfrac{-(q - 6)^2 - 2p}{4 - p}$

47. $\dfrac{3p + 3(4 + p)^3}{r + 8}$

48. $\dfrac{5q + 2(1 + p)^3}{r + 3}$

*Identify the property illustrated in each statement. Assume all variables represent real numbers. **See Examples 5 and 6.***

49. $6 \cdot 12 + 6 \cdot 15 = 6(12 + 15)$

50. $8(m + 4) = 8m + 32$

51. $(t - 6) \cdot \left(\dfrac{1}{t - 6}\right) = 1$, if $t - 6 \neq 0$

52. $\dfrac{2 + m}{2 - m} \cdot \dfrac{2 - m}{2 + m} = 1$, if $m \neq 2$ or -2

53. $(7.5 - y) + 0 = 7.5 - y$

54. $1 \cdot (3x - 7) = 3x - 7$

55. $5(t + 3) = (t + 3) \cdot 5$

56. $-7 + (x + 3) = (x + 3) + (-7)$

57. $(5x)\left(\dfrac{1}{x}\right) = 5\left(x \cdot \dfrac{1}{x}\right)$

58. $(38 + 99) + 1 = 38 + (99 + 1)$

59. $5 + \sqrt{3}$ is a real number.

60. 5π is a real number.

Write a short answer to each question.

61. Is there a commutative property for subtraction? That is, in general, is $a - b$ equal to $b - a$? Support your answer with an example.

62. Is there an associative property for subtraction? That is, does $(a - b) - c$ equal $a - (b - c)$ in general? Support your answer with an example.

*Simplify each expression. **See Examples 5 and 6.***

63. $\dfrac{10}{11}(22z)$

64. $\left(\dfrac{3}{4}r\right)(-12)$

65. $(m + 5) + 6$

66. $8 + (a + 7)$

67. $\dfrac{3}{8}\left(\dfrac{16}{9}y + \dfrac{32}{27}z - \dfrac{40}{9}\right)$

68. $-\dfrac{1}{4}(20m + 8y - 32z)$

Use the distributive property to rewrite sums as products and products as sums. ***See Example 6.***

69. $8p - 14p$　　　**70.** $15x - 10x$　　　**71.** $-4(z - y)$　　　**72.** $-3(m + n)$

Concept Check *Use the distributive property to calculate each value mentally.*

73. $72 \cdot 17 + 28 \cdot 17$ 　　　　　　　**74.** $32 \cdot 80 + 32 \cdot 20$

75. $123\dfrac{5}{8} \cdot 1\dfrac{1}{2} - 23\dfrac{5}{8} \cdot 1\dfrac{1}{2}$ 　　　　　**76.** $17\dfrac{2}{5} \cdot 14\dfrac{3}{4} - 17\dfrac{2}{5} \cdot 4\dfrac{3}{4}$

Concept Check *Decide whether each statement is* true *or* false. *If false, correct the statement so it is true.*

77. $|6 - 8| = |6| - |8|$ 　　　　　　　**78.** $|(-3)^3| = -|3^3|$

79. $|-5| \cdot |6| = |-5 \cdot 6|$ 　　　　　　**80.** $\dfrac{|-14|}{|2|} = \left|\dfrac{-14}{2}\right|$

81. $|a - b| = |a| - |b|$, if $b > a > 0$ 　　**82.** If a is negative, then $|a| = -a$.

Evaluate each expression. ***See Example 7.***

83. $|-10|$ 　　　　　**84.** $|-15|$ 　　　　　**85.** $-\left|\dfrac{4}{7}\right|$

86. $-\left|\dfrac{7}{2}\right|$ 　　　　　**87.** $-|-8|$ 　　　　　**88.** $-|-12|$

(Modeling) Blood Pressure Difference *Use the formula for determining blood pressure difference from normal,* $P_d = |P - 120|$, *to solve each problem.* ***See Example 8.***

89. Calculate the P_d value for a woman whose actual systolic blood pressure is 116.

90. Determine two possible values for a person's systolic blood pressure if his P_d value is 17.

Let $x = -4$ *and* $y = 2$. *Evaluate each expression.* ***See Example 9.***

91. $|3x - 2y|$ 　　　**92.** $|2x - 5y|$ 　　　**93.** $|-3x + 4y|$ 　　　**94.** $|-5y + x|$

95. $\dfrac{2|y| - 3|x|}{|xy|}$ 　　**96.** $\dfrac{4|x| + 4|y|}{|x|}$ 　　**97.** $\dfrac{|-8y + x|}{-|x|}$ 　　**98.** $\dfrac{|x| + 2|y|}{-|x|}$

Determine whether each statement is true *or* false.

99. $|25| = |-25|$ 　　　　　　　**100.** $|-8| \geq 0$

101. $|5 + (-13)| = |5| + |-13|$ 　　　　**102.** $|8 - 12| = |8| - |12|$

103. $|11| \cdot |-6| = |-66|$ 　　　　　**104.** $\left|\dfrac{10}{-2}\right| = \dfrac{|10|}{|-2|}$

Find the given distances between points P, Q, R, and S on a number line, with coordinates -4, -1, 8, *and* 12, *respectively.* ***See Example 10.***

105. $d(P, Q)$ 　　　**106.** $d(P, R)$ 　　　**107.** $d(Q, R)$ 　　　**108.** $d(Q, S)$

Concept Check *Determine what signs on values of x and y would make each statement true. Assume that x and y are not 0. (You should be able to work these mentally.)*

109. $xy > 0$

110. $x^2y > 0$

111. $\dfrac{x}{y} < 0$

112. $\dfrac{y^2}{x} < 0$

113. $\dfrac{x^3}{y} > 0$

114. $-\dfrac{x}{y} > 0$

Solve each problem.

115. *Golf Scores* Jordan Spieth won the 2015 Masters Golf Tournament with a total score that was 18 under par, and Zach Johnson won the 2007 tournament with a total score that was 1 above par. Using -18 to represent 18 below par and $+1$ to represent 1 over par, find the difference between these scores (in either order) and take the absolute value of this difference. What does this final number represent? (*Source:* www.masters.org)

116. *Total Football Yardage* As of 2015, Emmitt Smith of the Dallas Cowboys was the NFL career leader for rushing. During his 15 years in the NFL, he gained 18,355 yd rushing, 3224 yd receiving, and -15 yd returning fumbles. Find his total yardage (called *all-purpose yards*). Is this the same as the sum of the absolute values of the three categories? Explain. (*Source:* www.pro-football-reference.com)

(Modeling) Blood Alcohol Concentration The blood alcohol concentration (BAC) of a person who has been drinking is approximated by the following formula.

$$\text{BAC} = \text{number of oz} \times \% \text{ alcohol} \times 0.075 \div \text{body wt in lb} - \text{hr of drinking} \times 0.015$$

(*Source:* Lawlor, J., *Auto Math Handbook: Mathematical Calculations, Theory, and Formulas for Automotive Enthusiasts,* HP Books.)

117. Suppose a policeman stops a 190-lb man who, in 2 hr, has ingested four 12-oz beers (48 oz), each having a 3.2% alcohol content. Calculate the man's BAC to the nearest thousandth. Follow the order of operations.

118. Calculate the BAC to the nearest thousandth for a 135-lb woman who, in 3 hr, has consumed three 12-oz beers (36 oz), each having a 4.0% alcohol content.

119. Calculate the BAC to the nearest thousandth for a 200-lb man who, in 4 hr, has consumed three 20-oz beers, each having a 3.8% alcohol content. If the man's weight were greater and all other variables remained the same, how would that affect his BAC?

120. Calculate the BAC to the nearest thousandth for a 150-lb woman who, in 2 hr, has consumed two 6-oz glasses of wine, each having a 14% alcohol content. If the woman drank the same two glasses of wine over a longer period of time, how would that affect her BAC?

Archie Manning, father of NFL quarterbacks Peyton and Eli, signed this photo for author Hornsby's son, Jack.

(Modeling) Passer Rating for NFL Quarterbacks *The current system of rating passers in the National Football League is based on four performance components: completions, touchdowns, yards gained, and interceptions, as percentages of the number of passes attempted. It uses the following formula.*

$$\text{Rating} = \frac{\left(250 \cdot \dfrac{C}{A}\right) + \left(1000 \cdot \dfrac{T}{A}\right) + \left(12.5 \cdot \dfrac{Y}{A}\right) + 6.25 - \left(1250 \cdot \dfrac{I}{A}\right)}{3},$$

where A = *attempted passes,* C = *completed passes,* T = *touchdown passes,* Y = *yards gained passing, and* I = *interceptions.*

In addition to the weighting factors appearing in the formula, the four category ratios are limited to nonnegative values with the following maximums.

$$0.775 \text{ for } \frac{C}{A}, \quad 0.11875 \text{ for } \frac{T}{A}, \quad 12.5 \text{ for } \frac{Y}{A}, \quad 0.095 \text{ for } \frac{I}{A}$$

Exercises 121–132 give the 2014 regular-season statistics for the top twelve quarterbacks. Use the formula to determine the rating to the nearest tenth for each.

Quarterback, Team	A Att	C Comp	T TD	Y Yards	I Int
121. Tony Romo, DAL	435	304	34	3705	9
122. Aaron Rodgers, GB	520	341	38	4381	5
123. Ben Roethlisberger, PIT	608	408	32	4952	9
124. Peyton Manning, DEN	597	395	39	4727	15
125. Tom Brady, NE	582	373	33	4109	9
126. Drew Brees, NO	659	456	33	4952	17
127. Andrew Luck, IND	616	380	40	4761	16
128. Carson Palmer, ARI	224	141	11	1626	3
129. Ryan Fitzpatrick, HOU	312	197	17	2483	8
130. Russell Wilson, SEA	452	285	20	3475	7
131. Matt Ryan, ATL	628	415	28	4694	14
132. Philip Rivers, SD	570	379	31	4286	18

Source: www.nfl.com

Solve each problem using the passer rating formula above.

133. Peyton Manning, when he played for the Indianapolis Colts, set a full-season rating record of 121.1 in 2004 and held that record until Aaron Rodgers, of the Green Bay Packers, surpassed it in 2011. (As of 2014, Rodgers's all-time record held.) If Rodgers had 343 completions, 45 touchdowns, 6 interceptions, and 4643 yards, for 502 attempts, what was his rating in 2011?

134. Steve Young, of the San Francisco 49ers, set a full season rating record of 112.8 in 1994 and held that record until Peyton Manning surpassed it in 2004. If Manning had 336 completions, 49 touchdowns, 10 interceptions, and 4557 yards, for 497 attempts, what was his rating in 2004?

135. If Tom Brady, of the New England Patriots, during the 2010 regular season, had 324 completions, 36 touchdowns, 4 interceptions, and 3900 yards, what was his rating in 2010 for 492 attempts?

136. Refer to the passer rating formula and determine the highest rating possible (considered a "perfect" passer rating).

R.3 Polynomials

- Rules for Exponents
- Polynomials
- Addition and Subtraction
- Multiplication
- Division

Rules for Exponents Recall that the notation a^m (where m is a positive integer and a is a real number) means that a appears as a factor m times. In the same way, a^n (where n is a positive integer) means that a appears as a factor n times.

Rules for Exponents

For all positive integers m and n and all real numbers a and b, the following rules hold.

Rule	Example	Description
Product Rule $a^m \cdot a^n = a^{m+n}$	$2^2 \cdot 2^3 = (2 \cdot 2)(2 \cdot 2 \cdot 2)$ $= 2^{2+3}$ $= 2^5$	When multiplying powers of like bases, keep the base and add the exponents.
Power Rule 1 $(a^m)^n = a^{mn}$	$(4^5)^3 = 4^5 \cdot 4^5 \cdot 4^5$ $= 4^{5+5+5}$ $= 4^{5 \cdot 3}$ $= 4^{15}$	To raise a power to a power, multiply the exponents.
Power Rule 2 $(ab)^m = a^m b^m$	$(7x)^3 = (7x)(7x)(7x)$ $= (7 \cdot 7 \cdot 7)(x \cdot x \cdot x)$ $= 7^3 x^3$	To raise a product to a power, raise each factor to that power.
Power Rule 3 $\left(\dfrac{a}{b}\right)^m = \dfrac{a^m}{b^m}$ $(b \neq 0)$	$\left(\dfrac{3}{5}\right)^4 = \left(\dfrac{3}{5}\right)\left(\dfrac{3}{5}\right)\left(\dfrac{3}{5}\right)\left(\dfrac{3}{5}\right)$ $= \dfrac{3 \cdot 3 \cdot 3 \cdot 3}{5 \cdot 5 \cdot 5 \cdot 5}$ $= \dfrac{3^4}{5^4}$	To raise a quotient to a power, raise the numerator and the denominator to that power.

EXAMPLE 1 Using the Product Rule

Simplify each expression.

(a) $y^4 \cdot y^7$ **(b)** $(6z^5)(9z^3)(2z^2)$

SOLUTION

(a) $y^4 \cdot y^7 = y^{4+7} = y^{11}$ Product rule: Keep the base and add the exponents.

(b) $(6z^5)(9z^3)(2z^2)$

$= (6 \cdot 9 \cdot 2) \cdot (z^5 z^3 z^2)$ Commutative and associative properties

$= 108z^{5+3+2}$ Multiply. Apply the product rule.

$= 108z^{10}$ Add.   ✔ **Now Try Exercises 13 and 17.**

> **EXAMPLE 2**   **Using the Power Rules**

Simplify. Assume all variables represent nonzero real numbers.

(a) $(5^3)^2$ **(b)** $(3^4x^2)^3$ **(c)** $\left(\dfrac{2^5}{b^4}\right)^3$ **(d)** $\left(\dfrac{-2m^6}{t^2z}\right)^5$

SOLUTION

(a) $(5^3)^2 = 5^{3(2)} = 5^6$ Power rule 1

(b) $(3^4x^2)^3$

$\qquad = (3^4)^3(x^2)^3$ Power rule 2

$\qquad = 3^{4(3)}x^{2(3)}$ Power rule 1

$\qquad = 3^{12}x^6$

(c) $\left(\dfrac{2^5}{b^4}\right)^3$

$\qquad = \dfrac{(2^5)^3}{(b^4)^3}$ Power rule 3

$\qquad = \dfrac{2^{15}}{b^{12}}$ Power rule 1

(d) $\left(\dfrac{-2m^6}{t^2z}\right)^5$

$\qquad = \dfrac{(-2m^6)^5}{(t^2z)^5}$ Power rule 3

$\qquad = \dfrac{(-2)^5(m^6)^5}{(t^2)^5z^5}$ Power rule 2

$\qquad = \dfrac{-32m^{30}}{t^{10}z^5}$ Evaluate $(-2)^5$. Then use Power rule 1.

$\qquad = -\dfrac{32m^{30}}{t^{10}z^5}$ $\dfrac{-a}{b} = -\dfrac{a}{b}$ ✔ **Now Try Exercises 23, 25, 29, and 31.**

CAUTION   The expressions mn^2 and $(mn)^2$ are **not** equivalent. The second power rule can be used only with the second expression:

$$(mn)^2 = m^2n^2.$$

A zero exponent is defined as follows.

> **Zero Exponent**
>
> For any nonzero real number a, $a^0 = 1.$

That is, any nonzero number with a zero exponent equals 1.

To illustrate why a^0 is defined to equal 1, consider the product

$$a^n \cdot a^0, \quad \text{for} \quad a \neq 0.$$

We want the definition of a^0 to be consistent so that the product rule applies. Now apply this rule.

$$a^n \cdot a^0 = a^{n+0} = a^n$$

The product of a^n and a^0 must be a^n, and thus a^0 is acting like the identity element 1. So, for consistency, we *define* a^0 to equal 1. (**0^0 is undefined.**)

EXAMPLE 3 **Using the Definition of a^0**

Evaluate each power.

(a) 4^0 **(b)** $(-4)^0$ **(c)** -4^0

(d) $-(-4)^0$ **(e)** $(7r)^0$

SOLUTION

(a) $4^0 = 1$ Base is 4.   **(b)** $(-4)^0 = 1$ Base is -4.

(c) $-4^0 = -(4^0) = -1$ Base is 4.

(d) $-(-4)^0 = -(1) = -1$ Base is -4.

(e) $(7r)^0 = 1, r \neq 0$ Base is $7r$.   ✔ **Now Try Exercise 35.**

Polynomials   The product of a number and one or more variables raised to powers is a **term**. The number is the **numerical coefficient,** or just the **coefficient,** of the variables. The coefficient of the variable in $-3m^4$ is -3, and the coefficient in $-p^2$ is -1. **Like terms** are terms with the same variables each raised to the same powers.

$$-13x^3, \quad 4x^3, \quad -x^3 \quad \text{Like terms} \qquad 6y, \quad 6y^2, \quad 4y^3 \quad \text{Unlike terms}$$

A **polynomial** is a term or a finite sum of terms, with only positive or zero integer exponents permitted on the variables. If the terms of a polynomial contain only the variable x, then the polynomial is a **polynomial in x.**

$$5x^3 - 8x^2 + 7x - 4, \qquad 9p^5 - 3, \qquad 8r^2, \qquad 6 \quad \text{Polynomials}$$

The terms of a polynomial cannot have variables in a denominator.

$$9x^2 - 4x + \frac{6}{x} \quad \text{Not a polynomial}$$

The **degree of a term** with one variable is the exponent on the variable. For example, the degree of $2x^3$ is 3, and the degree of $17x$ (that is, $17x^1$) is 1. The greatest degree of any term in a polynomial is the **degree of the polynomial.** For example, the polynomial

$$4x^3 - 2x^2 - 3x + 7 \quad \text{has degree } 3$$

because the greatest degree of any term is 3. A nonzero constant such as -6, equivalent to $-6x^0$, has degree 0. (The polynomial 0 has no degree.)

A polynomial can have more than one variable. A term containing more than one variable has degree equal to the sum of all the exponents appearing on the variables in the term. For example,

$$-3x^4y^3z^5 \quad \text{has degree} \quad 4 + 3 + 5 = 12.$$
$$5xy^2z^7 \quad \text{has degree} \quad 1 + 2 + 7 = 10.$$

The degree of a polynomial in more than one variable is equal to the greatest degree of any term appearing in the polynomial. By this definition, the polynomial

$$2x^4y^3 - 3x^5y + x^6y^2 \quad \text{has degree } 8$$

because the x^6y^2 term has the greatest degree, 8.

A polynomial containing exactly three terms is a **trinomial.** A two-term polynomial is a **binomial.** A single-term polynomial is a **monomial.**

EXAMPLE 4 **Classifying Expressions as Polynomials**

Identify each as a *polynomial* or *not a polynomial*. For each polynomial, give the degree and identify it as a *monomial, binomial, trinomial,* or *none of these.*

(a) $9x^7 - 4x^3 + 8x^2$ **(b)** $2t^4 - \dfrac{1}{t^2}$ **(c)** $-\dfrac{4}{5}x^3y^2$

SOLUTION

(a) $9x^7 - 4x^3 + 8x^2$ is a polynomial. The first term, $9x^7$, has greatest degree, so this a polynomial of degree 7. Because it has three terms, it is a trinomial.

(b) $2t^4 - \dfrac{1}{t^2}$ is not a polynomial because it has a variable in the denominator.

(c) $-\dfrac{4}{5}x^3y^2$ is a polynomial. Add the exponents $3 + 2 = 5$ to determine that it is of degree 5. Because there is one term, it is a monomial.

✔ **Now Try Exercises 37, 39, and 45.**

Addition and Subtraction   Polynomials are added by adding coefficients of like terms. They are subtracted by subtracting coefficients of like terms.

EXAMPLE 5 **Adding and Subtracting Polynomials**

Add or subtract, as indicated.

(a) $(2y^4 - 3y^2 + y) + (4y^4 + 7y^2 + 6y)$

(b) $(-3m^3 - 8m^2 + 4) - (m^3 + 7m^2 - 3)$

(c) $(8m^4p^5 - 9m^3p^5) + (11m^4p^5 + 15m^3p^5)$

(d) $4(x^2 - 3x + 7) - 5(2x^2 - 8x - 4)$

SOLUTION

(a) $(2y^4 - 3y^2 + y) + (4y^4 + 7y^2 + 6y)$ $\quad y = 1y$

$= (2 + 4)y^4 + (-3 + 7)y^2 + (1 + 6)y$ Add coefficients of like terms.

$= 6y^4 + 4y^2 + 7y$ Work inside the parentheses.

(b) $(-3m^3 - 8m^2 + 4) - (m^3 + 7m^2 - 3)$

$= (-3 - 1)m^3 + (-8 - 7)m^2 + [4 - (-3)]$ Subtract coefficients of like terms.

$= -4m^3 - 15m^2 + 7$ Simplify.

(c) $(8m^4p^5 - 9m^3p^5) + (11m^4p^5 + 15m^3p^5)$

$= 19m^4p^5 + 6m^3p^5$

(d) $4(x^2 - 3x + 7) - 5(2x^2 - 8x - 4)$

$= 4x^2 - 4(3x) + 4(7) - 5(2x^2) - 5(-8x) - 5(-4)$ Distributive property

$= 4x^2 - 12x + 28 - 10x^2 + 40x + 20$ Multiply.

$= -6x^2 + 28x + 48$ Add like terms.

✔ **Now Try Exercises 49 and 51.**

As shown in **Examples 5(a), (b), and (d),** polynomials in one variable are often written with their terms in **descending order** (or descending degree). The term of greatest degree is first, the one of next greatest degree is next, and so on.

Multiplication One way to find the product of two polynomials, such as $3x - 4$ and $2x^2 - 3x + 5$, is to distribute each term of $3x - 4$, multiplying by each term of $2x^2 - 3x + 5$.

$(3x - 4)(2x^2 - 3x + 5)$

$= 3x(2x^2 - 3x + 5) - 4(2x^2 - 3x + 5)$ Distributive property

$= 3x(2x^2) + 3x(-3x) + 3x(5) - 4(2x^2) - 4(-3x) - 4(5)$

          Distributive property again

$= 6x^3 - 9x^2 + 15x - 8x^2 + 12x - 20$ Multiply.

$= 6x^3 - 17x^2 + 27x - 20$ Combine like terms.

Another method is to write such a product vertically, similar to the method used in arithmetic for multiplying whole numbers.

$$
\begin{array}{r}
2x^2 - 3x + 5 \\
3x - 4 \\
\hline
-8x^2 + 12x - 20 \\
6x^3 - 9x^2 + 15x \\
\hline
6x^3 - 17x^2 + 27x - 20
\end{array}
$$

Place like terms in the same column.

$\leftarrow -4(2x^2 - 3x + 5)$

$\leftarrow 3x(2x^2 - 3x + 5)$

Add in columns.

EXAMPLE 6 Multiplying Polynomials

Multiply $(3p^2 - 4p + 1)(p^3 + 2p - 8)$.

SOLUTION

This process is an ordered method of applying the distributive property.

Write like terms in columns.

$$
\begin{array}{r}
3p^2 - 4p + 1 \\
p^3 + 2p - 8 \\
\hline
-24p^2 + 32p - 8 \\
6p^3 - 8p^2 + 2p \\
3p^5 - 4p^4 + p^3 \\
\hline
3p^5 - 4p^4 + 7p^3 - 32p^2 + 34p - 8
\end{array}
$$

$\leftarrow -8(3p^2 - 4p + 1)$

$\leftarrow 2p(3p^2 - 4p + 1)$

$\leftarrow p^3(3p^2 - 4p + 1)$

Add in columns.

✔ **Now Try Exercise 63.**

The **FOIL method** is a convenient way to find the product of two binomials. The memory aid **FOIL** (for **F**irst, **O**uter, **I**nner, **L**ast) gives the pairs of terms to be multiplied when distributing each term of the first binomial, multiplying by each term of the second binomial.

EXAMPLE 7 Using the FOIL Method to Multiply Two Binomials

Find each product.

(a) $(6m + 1)(4m - 3)$ **(b)** $(2x + 7)(2x - 7)$ **(c)** $r^2(3r + 2)(3r - 2)$

SOLUTION

First Last

$$
\begin{array}{l}
\text{(a) } (6m + 1)(4m - 3) = 6m(4m) + 6m(-3) + 1(4m) + 1(-3) \\
\qquad\qquad\qquad\qquad\quad = 24m^2 - 18m + 4m - 3 \quad \text{Multiply.} \\
\qquad\qquad\qquad\qquad\quad = 24m^2 - 14m - 3 \quad \text{Combine like terms.}
\end{array}
$$

Inner

Outer

(b) $(2x + 7)(2x - 7)$

$$= 4x^2 - 14x + 14x - 49 \quad \text{FOIL method}$$
$$= 4x^2 - 49 \quad \text{Combine like terms.}$$

(c) $r^2(3r + 2)(3r - 2)$

$$= r^2(9r^2 - 6r + 6r - 4) \quad \text{FOIL method}$$
$$= r^2(9r^2 - 4) \quad \text{Combine like terms.}$$
$$= 9r^4 - 4r^2 \quad \text{Distributive property}$$

✔ **Now Try Exercises 55 and 57.**

In **Example 7(a),** the product of two binomials is a trinomial, while in **Examples 7(b) and (c)**, the product of two binomials is a binomial. The product of two binomials of the forms $x + y$ and $x - y$ is a special product form called a **difference of squares.** The squares of binomials, $(x + y)^2$ and $(x - y)^2$, are also special product forms called **perfect square trinomials.**

Special Products

Product of the Sum and Difference of Two Terms

$$(x + y)(x - y) = \underbrace{x^2 - y^2}_{\text{Difference of squares}}$$

Square of a Binomial

$$(x + y)^2 = x^2 + 2xy + y^2$$
$$(x - y)^2 = \underbrace{x^2 - 2xy + y^2}_{\text{Perfect square trinomials}}$$

EXAMPLE 8 Using the Special Products

Find each product.

(a) $(3p + 11)(3p - 11)$

(b) $(5m^3 - 3)(5m^3 + 3)$

(c) $(9k - 11r^3)(9k + 11r^3)$

(d) $(2m + 5)^2$

(e) $(3x - 7y^4)^2$

SOLUTION

(a)
$$(3p + 11)(3p - 11)$$
$$= (3p)^2 - 11^2 \quad (x + y)(x - y) = x^2 - y^2$$
$$\boxed{(3p)^2 = 3^2 p^2, \text{ not } 3p^2} = 9p^2 - 121 \quad \text{Power rule 2}$$

(b) $(5m^3 - 3)(5m^3 + 3)$
$$= (5m^3)^2 - 3^2 \quad (x - y)(x + y) = x^2 - y^2$$
$$= 25m^6 - 9 \quad \text{Power rules 2 and 1}$$

(c) $(9k - 11r^3)(9k + 11r^3)$
$$= (9k)^2 - (11r^3)^2$$
$$= 81k^2 - 121r^6 \quad \boxed{\text{Be careful applying the power rules.}}$$

(d) $(2m + 5)^2$

$= (2m)^2 + 2(2m)(5) + 5^2$ $(x + y)^2 = x^2 + 2xy + y^2$

$= 4m^2 + 20m + 25$ Power rule 2; Multiply.

(e) $(3x - 7y^4)^2$

$= (3x)^2 - 2(3x)(7y^4) + (7y^4)^2$ $(x - y)^2 = x^2 - 2xy + y^2$

$= 9x^2 - 42xy^4 + 49y^8$ Power rule 2; Multiply.

✔ **Now Try Exercises 69, 71, 73, and 75.**

CAUTION See **Examples 8(d)** and **(e)**. *The square of a binomial has three terms.* Do **not** give $x^2 + y^2$ as the result of expanding $(x + y)^2$, or $x^2 - y^2$ as the result of expanding $(x - y)^2$.

$(x + y)^2 = x^2 + 2xy + y^2$

$(x - y)^2 = x^2 - 2xy + y^2$ Remember to include the middle term.

EXAMPLE 9   **Multiplying More Complicated Binomials**

Find each product.

(a) $[(3p - 2) + 5q][(3p - 2) - 5q]$ **(b)** $(x + y)^3$ **(c)** $(2a + b)^4$

SOLUTION

(a) $[(3p - 2) + 5q][(3p - 2) - 5q]$

$= (3p - 2)^2 - (5q)^2$ Product of the sum and difference of two terms

$= 9p^2 - 12p + 4 - 25q^2$ Square both quantities.

(b) $(x + y)^3$ This does *not* equal $x^3 + y^3$.

$= (x + y)^2(x + y)$ $a^3 = a^2 \cdot a$

$= (x^2 + 2xy + y^2)(x + y)$ Square $x + y$.

$= x^3 + x^2y + 2x^2y + 2xy^2 + xy^2 + y^3$ Multiply.

$= x^3 + 3x^2y + 3xy^2 + y^3$ Combine like terms.

(c) $(2a + b)^4$

$= (2a + b)^2(2a + b)^2$ $a^4 = a^2 \cdot a^2$

$= (4a^2 + 4ab + b^2)(4a^2 + 4ab + b^2)$ Square each $2a + b$.

$= 16a^4 + 16a^3b + 4a^2b^2 + 16a^3b + 16a^2b^2$ Distributive property

    $+ 4ab^3 + 4a^2b^2 + 4ab^3 + b^4$

$= 16a^4 + 32a^3b + 24a^2b^2 + 8ab^3 + b^4$ Combine like terms.

✔ **Now Try Exercises 79, 83, and 85.**

Division   The quotient of two polynomials can be found with an algorithm (that is, a step-by-step procedure) for long division similar to that used for dividing whole numbers. ***Both polynomials must be written in descending order to use this algorithm.***

EXAMPLE 10 **Dividing Polynomials**

Divide $4m^3 - 8m^2 + 5m + 6$ by $2m - 1$.

SOLUTION

$4m^3$ divided by $2m$ is $2m^2$.

$-6m^2$ divided by $2m$ is $-3m$.

$2m$ divided by $2m$ is 1.

$$
\begin{array}{r}
2m^2 - 3m + 1 \\
2m - 1 \overline{\smash{\big)}\, 4m^3 - 8m^2 + 5m + 6}
\end{array}
$$

$\underline{4m^3 - 2m^2}$ ⟵ $2m^2(2m - 1) = 4m^3 - 2m^2$

To subtract, add the opposite.

$-6m^2 + 5m$ ⟵ Subtract. Bring down the next term.

$\underline{-6m^2 + 3m}$ ⟵ $-3m(2m - 1) = -6m^2 + 3m$

$2m + 6$ ⟵ Subtract. Bring down the next term.

$\underline{2m - 1}$ ⟵ $1(2m - 1) = 2m - 1$

7 ⟵ Subtract. The remainder is 7.

Thus, $\dfrac{4m^3 - 8m^2 + 5m + 6}{2m - 1} = 2m^2 - 3m + 1 + \dfrac{7}{2m - 1}$.

Remember to add $\dfrac{\text{remainder}}{\text{divisor}}$.

✔ **Now Try Exercise 101.**

When a polynomial has a missing term, we allow for that term by inserting a term with a 0 coefficient for it. For example,

$$3x^2 - 7 \qquad \text{is equivalent to} \quad 3x^2 + 0x - 7,$$

and $\qquad 2x^3 + x + 10 \quad \text{is equivalent to} \quad 2x^3 + 0x^2 + x + 10.$

EXAMPLE 11 **Dividing Polynomials with Missing Terms**

Divide $3x^3 - 2x^2 - 150$ by $x^2 - 4$.

SOLUTION Both polynomials have missing first-degree terms. Insert each missing term with a 0 coefficient.

$$
\begin{array}{r}
3x - 2 \\
x^2 + 0x - 4 \overline{\smash{\big)}\, 3x^3 - 2x^2 + 0x - 150}
\end{array}
$$

Missing term

$\underline{3x^3 + 0x^2 - 12x}$ —— Missing term

$-2x^2 + 12x - 150$

$\underline{-2x^2 + 0x + 8}$

$12x - 158$ ⟵ Remainder

The division process ends when the remainder is 0 or the degree of the remainder is less than that of the divisor. Because $12x - 158$ has lesser degree than the divisor $x^2 - 4$, it is the remainder. Thus, the entire quotient is written as follows.

$$\frac{3x^3 - 2x^2 - 150}{x^2 - 4} = 3x - 2 + \frac{12x - 158}{x^2 - 4}$$

✔ **Now Try Exercise 103.**

R.3 Exercises

CONCEPT PREVIEW *Fill in the blank to correctly complete each sentence.*

1. The polynomial $2x^5 - x + 4$ is a trinomial of degree _____.

2. A polynomial containing exactly one term is a(n) _____.

3. A polynomial containing exactly two terms is a(n) _____.

4. In the term $-6x^2y$, -6 is the _____.

5. A convenient way to find the product of two binomials is to use the _____ method.

CONCEPT PREVIEW *Decide whether each is* true *or* false. *If false, correct the right side of the equation.*

6. $5^0 = 1$ 7. $y^2 \cdot y^5 = y^7$ 8. $(a^2)^3 = a^5$

9. $(x + y)^2 = x^2 + y^2$ 10. $x^2 + x^2 = x^4$

Simplify each expression. **See Example 1.**

11. $(-4x^5)(4x^2)$ 12. $(3y^4)(-6y^3)$

13. $n^6 \cdot n^4 \cdot n$ 14. $a^8 \cdot a^5 \cdot a$

15. $9^3 \cdot 9^5$ 16. $4^2 \cdot 4^8$

17. $(-3m^4)(6m^2)(-4m^5)$ 18. $(-8t^3)(2t^6)(-5t^4)$

19. $(5x^2y)(-3x^3y^4)$ 20. $(-4xy^3)(7x^2y)$

21. $\left(\dfrac{1}{2}mn\right)(8m^2n^2)$ 22. $(35m^4n)\left(-\dfrac{2}{7}mn^2\right)$

Simplify each expression. Assume all variables represent nonzero real numbers. **See Examples 1–3.**

23. $(2^2)^5$ 24. $(6^4)^3$ 25. $(-6x^2)^3$

26. $(-2x^5)^5$ 27. $-(4m^3n^0)^2$ 28. $-(2x^0y^4)^3$

29. $\left(\dfrac{r^8}{s^2}\right)^3$ 30. $\left(\dfrac{p^4}{q}\right)^2$ 31. $\left(\dfrac{-4m^2}{tp^2}\right)^4$

32. $\left(\dfrac{-5n^4}{r^2}\right)^3$ 33. $-\left(\dfrac{x^3y^5}{z}\right)^0$ 34. $-\left(\dfrac{p^2q^3}{r^3}\right)^0$

Match each expression in Column I with its equivalent in Column II. **See Example 3.**

	I	II		I	II
35.	(a) 6^0	A. 0	36.	(a) $3p^0$	A. 0
	(b) -6^0	B. 1		(b) $-3p^0$	B. 1
	(c) $(-6)^0$	C. -1		(c) $(3p)^0$	C. -1
	(d) $-(-6)^0$	D. 6		(d) $(-3p)^0$	D. 3
		E. -6			E. -3

Identify each expression as a polynomial *or* not a polynomial. *For each polynomial, give the degree and identify it as a* monomial, binomial, trinomial, *or* none of these. *See Example 4.*

37. $-5x^{11}$

38. $-4y^5$

39. $6x + 3x^4$

40. $-9y + 5y^3$

41. $-7z^5 - 2z^3 + 1$

42. $-9t^4 + 8t^3 - 7$

43. $15a^2b^3 + 12a^3b^8 - 13b^5 + 12b^6$

44. $-16x^5y^7 + 12x^3y^8 - 4xy^9 + 18x^{10}$

45. $\frac{3}{8}x^5 - \frac{1}{x^2} + 9$

46. $\frac{2}{3}t^6 + \frac{3}{t^5} + 1$

47. 5

48. 9

Add or subtract, as indicated. See Example 5.

49. $(5x^2 - 4x + 7) + (-4x^2 + 3x - 5)$

50. $(3m^3 - 3m^2 + 4) + (-2m^3 - m^2 + 6)$

51. $2(12y^2 - 8y + 6) - 4(3y^2 - 4y + 2)$

52. $3(8p^2 - 5p) - 5(3p^2 - 2p + 4)$

53. $(6m^4 - 3m^2 + m) - (2m^3 + 5m^2 + 4m) + (m^2 - m)$

54. $-(8x^3 + x - 3) + (2x^3 + x^2) - (4x^2 + 3x - 1)$

Find each product. See Examples 6–8.

55. $(4r - 1)(7r + 2)$

56. $(5m - 6)(3m + 4)$

57. $x^2\left(3x - \frac{2}{3}\right)\left(5x + \frac{1}{3}\right)$

58. $m^3\left(2m - \frac{1}{4}\right)\left(3m + \frac{1}{2}\right)$

59. $4x^2(3x^3 + 2x^2 - 5x + 1)$

60. $2b^3(b^2 - 4b + 3)$

61. $(2z - 1)(-z^2 + 3z - 4)$

62. $(3w + 2)(-w^2 + 4w - 3)$

63. $(m - n + k)(m + 2n - 3k)$

64. $(r - 3s + t)(2r - s + t)$

65. $(2x + 3)(2x - 3)(4x^2 - 9)$

66. $(3y - 5)(3y + 5)(9y^2 - 25)$

67. $(x + 1)(x + 1)(x - 1)(x - 1)$

68. $(t + 4)(t + 4)(t - 4)(t - 4)$

Find each product. See Examples 8 and 9.

69. $(2m + 3)(2m - 3)$

70. $(8s - 3t)(8s + 3t)$

71. $(4x^2 - 5y)(4x^2 + 5y)$

72. $(2m^3 + n)(2m^3 - n)$

73. $(4m + 2n)^2$

74. $(a - 6b)^2$

75. $(5r - 3t^2)^2$

76. $(2z^4 - 3y)^2$

77. $[(2p - 3) + q]^2$

78. $[(4y - 1) + z]^2$

79. $[(3q + 5) - p][(3q + 5) + p]$

80. $[(9r - s) + 2][(9r - s) - 2]$

81. $[(3a + b) - 1]^2$

82. $[(2m + 7) - n]^2$

83. $(y + 2)^3$

84. $(z - 3)^3$

85. $(q - 2)^4$

86. $(r + 3)^4$

Perform the indicated operations. ***See Examples 5–9.***

87. $(p^3 - 4p^2 + p) - (3p^2 + 2p + 7)$ **88.** $(x^4 - 3x^2 + 2) - (-2x^4 + x^2 - 3)$

89. $(7m + 2n)(7m - 2n)$ **90.** $(3p + 5)^2$

91. $-3(4q^2 - 3q + 2) + 2(-q^2 + q - 4)$ **92.** $2(3r^2 + 4r + 2) - 3(-r^2 + 4r - 5)$

93. $p(4p - 6) + 2(3p - 8)$ **94.** $m(5m - 2) + 9(5 - m)$

95. $-y(y^2 - 4) + 6y^2(2y - 3)$ **96.** $-z^3(9 - z) + 4z(2 + 3z)$

Perform each division. ***See Examples 10 and 11.***

97. $\dfrac{-4x^7 - 14x^6 + 10x^4 - 14x^2}{-2x^2}$ **98.** $\dfrac{-8r^3s - 12r^2s^2 + 20rs^3}{-4rs}$

99. $\dfrac{4x^3 - 3x^2 + 1}{x - 2}$ **100.** $\dfrac{3x^3 - 2x + 5}{x - 3}$

101. $\dfrac{6m^3 + 7m^2 - 4m + 2}{3m + 2}$ **102.** $\dfrac{10x^3 + 11x^2 - 2x + 3}{5x + 3}$

103. $\dfrac{x^4 + 5x^2 + 5x + 27}{x^2 + 3}$ **104.** $\dfrac{k^4 - 4k^2 + 2k + 5}{k^2 + 1}$

(Modeling) Solve each problem.

105. *Geometric Modeling* Consider the figure, which is a square divided into two squares and two rectangles.

 (a) The length of each side of the largest square is $x + y$. Use the formula for the area of a square to write the area of the largest square as a power.

 (b) Use the formulas for the area of a square and the area of a rectangle to write the area of the largest square as a trinomial that represents the sum of the areas of the four figures that make it up.

 (c) Explain why the expressions in parts (a) and (b) must be equivalent.

 (d) What special product formula from this section does this exercise reinforce geometrically?

106. *Geometric Modeling* Use the figure to geometrically support the distributive property. Write a short paragraph explaining this process.

107. *Volume of the Great Pyramid* An amazing formula from ancient mathematics was used by the Egyptians to find the volume of the frustum of a square pyramid, as shown in the figure. Its volume is given by

$$V = \frac{1}{3}h(a^2 + ab + b^2),$$

where b is the length of the base, a is the length of the top, and h is the height. (*Source:* Freebury, H. A., *A History of Mathematics*, Macmillan Company, New York.)

 (a) When the Great Pyramid in Egypt was partially completed to a height h of 200 ft, b was 756 ft, and a was 314 ft. Calculate its volume at this stage of construction to the nearest thousand feet.

(b) Try to visualize the figure if $a = b$. What is the resulting shape? Find its volume.

(c) Let $a = b$ in the Egyptian formula and simplify. Are the results the same?

108. **Volume of the Great Pyramid** Refer to the formula and the discussion in **Exercise 107.**

(a) Use $V = \frac{1}{3}h(a^2 + ab + b^2)$ to determine a formula for the volume of a pyramid with square base of length b and height h by letting $a = 0$.

(b) The Great Pyramid in Egypt had a square base of length 756 ft and a height of 481 ft. Find the volume of the Great Pyramid to the nearest tenth million cubic feet. Compare it with the 273-ft-tall Superdome in New Orleans, which has an approximate volume of 100 million ft³. (*Source: Guinness Book of World Records.*)

(c) The Superdome covers an area of 13 acres. How many acres, to the nearest tenth, does the Great Pyramid cover? (*Hint:* 1 acre = 43,560 ft²)

(Modeling) Number of Farms in the United States *The graph shows the number of farms in the United States for selected years since 1950. We can use the formula*

$$\text{Number of farms} = 0.001259x^2 - 5.039x + 5044$$

to get a good approximation of the number of farms for these years by substituting the year for x and evaluating the polynomial. For example, if $x = 1960$, the value of the polynomial is approximately 4.1, which differs from the data in the bar graph by only 0.1.

Evaluate the polynomial for each year, and then give the difference from the value in the graph.

109. 1950

110. 1970

111. 1990

112. 2012

Number of Farms in the U.S. since 1950

Source: U.S. Department of Agriculture.

Concept Check *Perform each operation mentally.*

113. $(0.25^3)(400^3)$ 114. $(24^2)(0.5^2)$ 115. $\dfrac{4.2^5}{2.1^5}$ 116. $\dfrac{15^4}{5^4}$

Relating Concepts

For individual or collaborative investigation *(Exercises 117–120)*

The special products can be used to perform selected multiplications. On the left, we use $(x + y)(x - y) = x^2 - y^2$. On the right, $(x - y)^2 = x^2 - 2xy + y^2$.

$$
\begin{aligned}
51 \times 49 &= (50 + 1)(50 - 1) \\
&= 50^2 - 1^2 \\
&= 2500 - 1 \\
&= 2499
\end{aligned}
\qquad
\begin{aligned}
47^2 &= (50 - 3)^2 \\
&= 50^2 - 2(50)(3) + 3^2 \\
&= 2500 - 300 + 9 \\
&= 2209
\end{aligned}
$$

Use special products to evaluate each expression.

117. 99×101 118. 63×57 119. 102^2 120. 71^2

R.4 Factoring Polynomials

- Factoring Out the Greatest Common Factor
- Factoring by Grouping
- Factoring Trinomials
- Factoring Binomials
- Factoring by Substitution

The process of finding polynomials whose product equals a given polynomial is called **factoring.** Unless otherwise specified, we consider only integer coefficients when factoring polynomials. For example, because

$$4x + 12 = 4(x + 3),$$

both 4 and $x + 3$ are **factors** of $4x + 12$, and $4(x + 3)$ is a **factored form** of $4x + 12$.

A polynomial with variable terms that cannot be written as a product of two polynomials of lesser degree is a **prime polynomial.** A polynomial is **factored completely** when it is written as a product of prime polynomials.

Factoring Out the Greatest Common Factor   To factor a polynomial such as $6x^2y^3 + 9xy^4 + 18y^5$, we look for a monomial that is the **greatest common factor (GCF)** of the three terms.

$$6x^2y^3 + 9xy^4 + 18y^5$$
$$= 3y^3(2x^2) + 3y^3(3xy) + 3y^3(6y^2) \quad \text{GCF} = 3y^3$$
$$= 3y^3(2x^2 + 3xy + 6y^2) \quad \text{Distributive property}$$

EXAMPLE 1   **Factoring Out the Greatest Common Factor**

Factor out the greatest common factor from each polynomial.

(a) $9y^5 + y^2$ **(b)** $6x^2t + 8xt - 12t$

(c) $14(m + 1)^3 - 28(m + 1)^2 - 7(m + 1)$

SOLUTION

(a) $9y^5 + y^2$
$$= y^2(9y^3) + y^2(1) \quad \text{GCF} = y^2$$
$$= y^2(9y^3 + 1) \quad \boxed{\text{Remember to include the 1.}}$$

CHECK   Multiply out the factored form: $y^2(9y^3 + 1) = \overbrace{9y^5 + y^2}^{\text{Original polynomial}}$. ✓

(b) $6x^2t + 8xt - 12t$
$$= 2t(3x^2 + 4x - 6) \quad \text{GCF} = 2t$$

CHECK $2t(3x^2 + 4x - 6) = 6x^2t + 8xt - 12t$ ✓

(c) $14(m + 1)^3 - 28(m + 1)^2 - 7(m + 1)$
$$= 7(m + 1)[2(m + 1)^2 - 4(m + 1) - 1] \quad \text{GCF} = 7(m + 1)$$
$$= 7(m + 1)[2(m^2 + 2m + 1) - 4m - 4 - 1] \quad \begin{array}{l}\text{Square } m + 1; \\ \text{distributive property}\end{array}$$
$\boxed{\text{Remember the middle term.}}$
$$= 7(m + 1)(2m^2 + 4m + 2 - 4m - 4 - 1) \quad \text{Distributive property}$$
$$= 7(m + 1)(2m^2 - 3) \quad \text{Combine like terms.}$$

✔ **Now Try Exercises 13, 19, and 25.**

CAUTION In **Example 1(a),** the 1 is essential in the answer because

$$y^2(9y^3) \neq 9y^5 + y^2.$$

Factoring can always be checked by multiplying.

Factoring by Grouping When a polynomial has more than three terms, it can sometimes be factored using **factoring by grouping.** Consider this example.

$$ax + ay + 6x + 6y$$

Terms with common factor a ⎤ ⎡ Terms with common factor 6

$$= (ax + ay) + (6x + 6y) \qquad \text{Group the terms so that each group has a common factor.}$$

$$= a(x + y) + 6(x + y) \qquad \text{Factor each group.}$$

$$= (x + y)(a + 6) \qquad \text{Factor out } x + y.$$

It is not always obvious which terms should be grouped. In cases like the one above, group in pairs. Experience and repeated trials are the most reliable tools.

EXAMPLE 2 **Factoring by Grouping**

Factor each polynomial by grouping.

(a) $mp^2 + 7m + 3p^2 + 21$ **(b)** $2y^2 + az - 2z - ay^2$

(c) $4x^3 + 2x^2 - 2x - 1$

SOLUTION

(a) $mp^2 + 7m + 3p^2 + 21$

$$= (mp^2 + 7m) + (3p^2 + 21) \qquad \text{Group the terms.}$$

$$= m(p^2 + 7) + 3(p^2 + 7) \qquad \text{Factor each group.}$$

$$= (p^2 + 7)(m + 3) \qquad p^2 + 7 \text{ is a common factor.}$$

CHECK $(p^2 + 7)(m + 3)$

$$= mp^2 + 3p^2 + 7m + 21 \qquad \text{FOIL method}$$

$$= mp^2 + 7m + 3p^2 + 21 \ \checkmark \quad \text{Commutative property}$$

(b) $2y^2 + az - 2z - ay^2$

$$= 2y^2 - 2z - ay^2 + az \qquad \text{Rearrange the terms.}$$

$$= (2y^2 - 2z) + (-ay^2 + az) \quad \text{Group the terms.}$$

Be careful with signs here. $\qquad = 2(y^2 - z) - a(y^2 - z) \qquad \text{Factor out 2 and } -a \text{ so that } y^2 - z \text{ is a common factor.}$

$$= (y^2 - z)(2 - a) \qquad \text{Factor out } y^2 - z.$$

(c) $4x^3 + 2x^2 - 2x - 1$

$$= (4x^3 + 2x^2) + (-2x - 1) \quad \text{Group the terms.}$$

$$= 2x^2(2x + 1) - 1(2x + 1) \quad \text{Factor each group.}$$

$$= (2x + 1)(2x^2 - 1) \qquad \text{Factor out } 2x + 1.$$

✔ **Now Try Exercises 29 and 31.**

Factoring Trinomials   As shown in the diagram below, factoring is the opposite of multiplication.

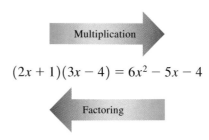

$$(2x + 1)(3x - 4) = 6x^2 - 5x - 4$$

One strategy when factoring trinomials uses the FOIL method in reverse. This strategy requires trial-and-error to find the correct arrangement of coefficients of the binomial factors.

EXAMPLE 3   Factoring Trinomials

Factor each trinomial, if possible.

(a) $4y^2 - 11y + 6$ **(b)** $6p^2 - 7p - 5$

(c) $2x^2 + 13x - 18$ **(d)** $16y^3 + 24y^2 - 16y$

SOLUTION

(a) To factor this polynomial, we must find values for integers a, b, c, and d in such a way that

$$4y^2 - 11y + 6$$
$$= (ay + b)(cy + d). \quad \text{FOIL method}$$

Using the FOIL method, we see that $ac = 4$ and $bd = 6$. The positive factors of 4 are 4 and 1 or 2 and 2. Because the middle term has a negative coefficient, we consider only negative factors of 6. The possibilities are -2 and -3 or -1 and -6.

 Now we try various arrangements of these factors until we find one that gives the correct coefficient of y.

$(2y - 1)(2y - 6)$	$(2y - 2)(2y - 3)$	$(y - 2)(4y - 3)$
$= 4y^2 - 14y + 6$	$= 4y^2 - 10y + 6$	$= 4y^2 - 11y + 6$
Incorrect	Incorrect	Correct

Therefore, $4y^2 - 11y + 6$ factors as $(y - 2)(4y - 3)$.

CHECK $(y - 2)(4y - 3)$
$$= 4y^2 - 3y - 8y + 6 \quad \text{FOIL method}$$
$$= 4y^2 - 11y + 6 \;\checkmark \quad \text{Original polynomial}$$

(b) Again, we try various possibilities to factor $6p^2 - 7p - 5$. The positive factors of 6 could be 2 and 3 or 1 and 6. As factors of -5 we have only -1 and 5 or -5 and 1.

$(2p - 5)(3p + 1)$	$(3p - 5)(2p + 1)$
$= 6p^2 - 13p - 5$ Incorrect	$= 6p^2 - 7p - 5$ Correct

Thus, $6p^2 - 7p - 5$ factors as $(3p - 5)(2p + 1)$.

(c) If we try to factor $2x^2 + 13x - 18$, we find that none of the pairs of factors gives the correct coefficient of x. Additional trials are also unsuccessful.

$(2x + 9)(x - 2)$	$(2x - 3)(x + 6)$	$(2x - 1)(x + 18)$
$= 2x^2 + 5x - 18$	$= 2x^2 + 9x - 18$	$= 2x^2 + 35x - 18$
Incorrect	Incorrect	Incorrect

This trinomial cannot be factored with integer coefficients and is prime.

(d)
$$16y^3 + 24y^2 - 16y$$
$$= 8y(2y^2 + 3y - 2) \qquad \text{Factor out the GCF, } 8y.$$
$$= 8y(2y - 1)(y + 2) \qquad \text{Factor the trinomial.}$$

Remember to include the common factor in the final form.

✔ **Now Try Exercises 35, 37, 39, and 41.**

NOTE   In **Example 3,** we chose positive factors of the positive first term (instead of two negative factors). This makes the work easier.

Each of the special patterns for multiplication can be used in reverse to obtain a pattern for factoring. Perfect square trinomials can be factored as follows.

Factoring Perfect Square Trinomials

$$x^2 + 2xy + y^2 = (x + y)^2$$
$$x^2 - 2xy + y^2 = (x - y)^2$$

EXAMPLE 4   **Factoring Perfect Square Trinomials**

Factor each trinomial.

(a) $16p^2 - 40pq + 25q^2$ **(b)** $36x^2y^2 + 84xy + 49$

SOLUTION

(a) Because $16p^2 = (4p)^2$ and $25q^2 = (5q)^2$, we use the second pattern shown in the box, with $4p$ replacing x and $5q$ replacing y.

$$16p^2 - 40pq + 25q^2$$
$$= (4p)^2 - 2(4p)(5q) + (5q)^2$$
$$= (4p - 5q)^2$$

Make sure that the middle term of the trinomial being factored, $-40pq$ here, is twice the product of the two terms in the binomial $4p - 5q$.

$$-40pq = 2(4p)(-5q)$$

Thus, $16p^2 - 40pq + 25q^2$ factors as $(4p - 5q)^2$.

CHECK $(4p - 5q)^2 = 16p^2 - 40pq + 25q^2$ ✓   Multiply.

(b) $36x^2y^2 + 84xy + 49$ factors as $(6xy + 7)^2$. ◁ 2(6xy)(7) = 84xy

CHECK   Square $6xy + 7$: $(6xy + 7)^2 = 36x^2y^2 + 84xy + 49$. ✓

✔ **Now Try Exercises 51 and 55.**

Factoring Binomials Check first to see whether the terms of a binomial have a common factor. If so, factor it out. The binomial may also fit one of the following patterns.

Factoring Binomials

Difference of Squares	$x^2 - y^2 = (x + y)(x - y)$
Difference of Cubes	$x^3 - y^3 = (x - y)(x^2 + xy + y^2)$
Sum of Cubes	$x^3 + y^3 = (x + y)(x^2 - xy + y^2)$

CAUTION *There is no factoring pattern for a sum of squares in the real number system.* In particular, for real numbers x and y,

$$x^2 + y^2 \text{ does not factor as } (x + y)^2.$$

EXAMPLE 5 **Factoring Differences of Squares**

Factor each polynomial.

(a) $4m^2 - 9$
(b) $256k^4 - 625m^4$
(c) $(a + 2b)^2 - 4c^2$
(d) $x^2 - 6x + 9 - y^4$
(e) $y^2 - x^2 + 6x - 9$

SOLUTION

(a) $4m^2 - 9$

$$= (2m)^2 - 3^2 \qquad \text{Write as a difference of squares.}$$

$$= (2m + 3)(2m - 3) \quad \text{Factor.}$$

Check by multiplying.

(b) $256k^4 - 625m^4$

$$= (16k^2)^2 - (25m^2)^2 \qquad\qquad \text{Write as a difference of squares.}$$

Don't stop here. $\quad = (16k^2 + 25m^2)(16k^2 - 25m^2) \qquad\qquad \text{Factor.}$

$$= (16k^2 + 25m^2)(4k + 5m)(4k - 5m) \quad \text{Factor } 16k^2 - 25m^2.$$

CHECK $\quad (16k^2 + 25m^2)(4k + 5m)(4k - 5m)$

$$= (16k^2 + 25m^2)(16k^2 - 25m^2) \quad \text{Multiply the last two factors.}$$

$$= 256k^4 - 625m^4 \checkmark \qquad\qquad \text{Original polynomial}$$

(c) $(a + 2b)^2 - 4c^2$

$$= (a + 2b)^2 - (2c)^2 \qquad\qquad \text{Write as a difference of squares.}$$

$$= [(a + 2b) + 2c][(a + 2b) - 2c] \quad \text{Factor.}$$

$$= (a + 2b + 2c)(a + 2b - 2c) \quad \text{Check by multiplying.}$$

(d) $x^2 - 6x + 9 - y^4$

$\qquad = (x^2 - 6x + 9) - y^4$ Group terms.

$\qquad = (x - 3)^2 - y^4$ Factor the trinomial.

$\qquad = (x - 3)^2 - (y^2)^2$ Write as a difference of squares.

$\qquad = [(x - 3) + y^2][(x - 3) - y^2]$ Factor.

$\qquad = (x - 3 + y^2)(x - 3 - y^2)$

(e) $y^2 - x^2 + 6x - 9$ $\boxed{\text{Be careful with signs. This is a perfect square trinomial.}}$

$\qquad = y^2 - (x^2 - 6x + 9)$ Factor out the negative sign, and group the last three terms.

$\qquad = y^2 - (x - 3)^2$ Write as a difference of squares.

$\qquad = [y - (x - 3)][y + (x - 3)]$ Factor.

$\qquad = (y - x + 3)(y + x - 3)$ Distributive property

✔ **Now Try Exercises 59, 61, 65, and 69.**

CAUTION When factoring as in **Example 5(e),** be careful with signs. Inserting an open parenthesis following the minus sign requires changing the signs of all of the following terms.

EXAMPLE 6   **Factoring Sums or Differences of Cubes**

Factor each polynomial.

(a) $x^3 + 27$ **(b)** $m^3 - 64n^3$ **(c)** $8q^6 + 125p^9$

SOLUTION

(a) $x^3 + 27$

$\qquad = x^3 + 3^3$ Write as a sum of cubes.

$\qquad = (x + 3)(x^2 - 3x + 3^2)$ Factor.

$\qquad = (x + 3)(x^2 - 3x + 9)$ Apply the exponent.

(b) $m^3 - 64n^3$

$\qquad = m^3 - (4n)^3$ Write as a difference of cubes.

$\qquad = (m - 4n)[m^2 + m(4n) + (4n)^2]$ Factor.

$\qquad = (m - 4n)(m^2 + 4mn + 16n^2)$ Multiply; $(4n)^2 = 4^2n^2$.

(c) $8q^6 + 125p^9$

$\qquad = (2q^2)^3 + (5p^3)^3$ Write as a sum of cubes.

$\qquad = (2q^2 + 5p^3)[(2q^2)^2 - 2q^2(5p^3) + (5p^3)^2]$ Factor.

$\qquad = (2q^2 + 5p^3)(4q^4 - 10q^2p^3 + 25p^6)$ Simplify.

✔ **Now Try Exercises 73, 75, and 77.**

Factoring by Substitution More complicated polynomials may be factored using substitution.

EXAMPLE 7 **Factoring by Substitution**

Factor each polynomial.

(a) $10(2a - 1)^2 - 19(2a - 1) - 15$ **(b)** $(2a - 1)^3 + 8$

(c) $6z^4 - 13z^2 - 5$

SOLUTION

(a) $10(2a - 1)^2 - 19(2a - 1) - 15$ *Replace $2a - 1$ with u so that*
$(2a - 1)^2$ *becomes u^2.*

$= 10u^2 - 19u - 15$

$= (5u + 3)(2u - 5)$ *Factor.*

Don't stop here.
Replace u with $2a - 1$. $= [5(2a - 1) + 3][2(2a - 1) - 5]$ *Replace u with $2a - 1$.*

$= (10a - 5 + 3)(4a - 2 - 5)$ *Distributive property*

$= (10a - 2)(4a - 7)$ *Simplify.*

$= 2(5a - 1)(4a - 7)$ *Factor out the common factor.*

(b) $(2a - 1)^3 + 8$ *Replace $2a - 1$ with u.*
Write as a sum of cubes.

$= u^3 + 2^3$

$= (u + 2)(u^2 - 2u + 4)$ *Factor.*

$= [(2a - 1) + 2][(2a - 1)^2 - 2(2a - 1) + 4]$ *Replace u with $2a - 1$.*

$= (2a + 1)(4a^2 - 4a + 1 - 4a + 2 + 4)$ *Add, and then multiply.*

$= (2a + 1)(4a^2 - 8a + 7)$ *Combine like terms.*

(c) $6z^4 - 13z^2 - 5$

$= 6u^2 - 13u - 5$ *Replace z^2 with u.*

Remember to make
the final substitution. $= (2u - 5)(3u + 1)$ *Factor the trinomial.*

$= (2z^2 - 5)(3z^2 + 1)$ *Replace u with z^2.*

✔ **Now Try Exercises 83, 87, and 91.**

R.4 Exercises

CONCEPT PREVIEW *Fill in the blank(s) to correctly complete each sentence.*

1. The process of finding polynomials whose product equals a given polynomial is called _____.

2. A polynomial is factored completely when it is written as a product of _____.

3. Factoring is the opposite of _____.

4. When a polynomial has more than three terms, it can sometimes be factored using _____.

5. There is no factoring pattern for a _____ in the real number system. In particular, $x^2 + y^2$ does not factor as $(x + y)^2$, for real numbers x and y.

CONCEPT PREVIEW *Work each problem.*

6. Match each polynomial in Column I with its factored form in Column II.

I	II
(a) $x^2 + 10xy + 25y^2$	**A.** $(x + 5y)(x - 5y)$
(b) $x^2 - 10xy + 25y^2$	**B.** $(x + 5y)^2$
(c) $x^2 - 25y^2$	**C.** $(x - 5y)^2$
(d) $25y^2 - x^2$	**D.** $(5y + x)(5y - x)$

7. Match each polynomial in Column I with its factored form in Column II.

I	II
(a) $8x^3 - 27$	**A.** $(3 - 2x)(9 + 6x + 4x^2)$
(b) $8x^3 + 27$	**B.** $(2x - 3)(4x^2 + 6x + 9)$
(c) $27 - 8x^3$	**C.** $(2x + 3)(4x^2 - 6x + 9)$

8. Which of the following is the correct factorization of $6x^2 + x - 12$?

A. $(3x + 4)(2x + 3)$ **B.** $(3x - 4)(2x - 3)$

C. $(3x + 4)(2x - 3)$ **D.** $(3x - 4)(2x + 3)$

9. Which of the following is the correct complete factorization of $x^4 - 1$?

A. $(x^2 - 1)(x^2 + 1)$ **B.** $(x^2 + 1)(x + 1)(x - 1)$

C. $(x^2 - 1)^2$ **D.** $(x - 1)^2(x + 1)^2$

10. Which of the following is the correct factorization of $x^3 + 8$?

A. $(x + 2)^3$ **B.** $(x + 2)(x^2 + 2x + 4)$

C. $(x + 2)(x^2 - 2x + 4)$ **D.** $(x + 2)(x^2 - 4x + 4)$

Factor out the greatest common factor from each polynomial. **See Examples 1 and 2.**

11. $12m + 60$ **12.** $15r - 27$ **13.** $8k^3 + 24k$

14. $9z^4 + 81z$ **15.** $xy - 5xy^2$ **16.** $5h^2j + hj$

17. $-4p^3q^4 - 2p^2q^5$ **18.** $-3z^5w^2 - 18z^3w^4$

19. $4k^2m^3 + 8k^4m^3 - 12k^2m^4$ **20.** $28r^4s^2 + 7r^3s - 35r^4s^3$

21. $2(a + b) + 4m(a + b)$ **22.** $6x(a + b) - 4y(a + b)$

23. $(5r - 6)(r + 3) - (2r - 1)(r + 3)$

24. $(4z - 5)(3z - 2) - (3z - 9)(3z - 2)$

25. $2(m - 1) - 3(m - 1)^2 + 2(m - 1)^3$ **26.** $5(a + 3)^3 - 2(a + 3) + (a + 3)^2$

27. *Concept Check* When directed to completely factor the polynomial $4x^2y^5 - 8xy^3$, a student wrote $2xy^3(2xy^2 - 4)$. When the teacher did not give him full credit, he complained because when his answer is multiplied out, the result is the original polynomial. Give the correct answer.

28. *Concept Check* Kurt factored $16a^2 - 40a - 6a + 15$ by grouping and obtained $(8a - 3)(2a - 5)$. Callie factored the same polynomial and gave an answer of $(3 - 8a)(5 - 2a)$. Which answer is correct?

Factor each polynomial by grouping. **See Example 2.**

29. $6st + 9t - 10s - 15$ **30.** $10ab - 6b + 35a - 21$

31. $2m^4 + 6 - am^4 - 3a$ **32.** $4x^6 + 36 - x^6y - 9y$

33. $p^2q^2 - 10 - 2q^2 + 5p^2$ **34.** $20z^2 - 8x + 5pz^2 - 2px$

*Factor each trinomial, if possible. **See Examples 3 and 4.***

35. $6a^2 - 11a + 4$

36. $8h^2 - 2h - 21$

37. $3m^2 + 14m + 8$

38. $9y^2 - 18y + 8$

39. $15p^2 + 24p + 8$

40. $9x^2 + 4x - 2$

41. $12a^3 + 10a^2 - 42a$

42. $36x^3 + 18x^2 - 4x$

43. $6k^2 + 5kp - 6p^2$

44. $14m^2 + 11mr - 15r^2$

45. $5a^2 - 7ab - 6b^2$

46. $12s^2 + 11st - 5t^2$

47. $12x^2 - xy - y^2$

48. $30a^2 + am - m^2$

49. $24a^4 + 10a^3b - 4a^2b^2$

50. $18x^5 + 15x^4z - 75x^3z^2$

51. $9m^2 - 12m + 4$

52. $16p^2 - 40p + 25$

53. $32a^2 + 48ab + 18b^2$

54. $20p^2 - 100pq + 125q^2$

55. $4x^2y^2 + 28xy + 49$

56. $9m^2n^2 + 12mn + 4$

57. $(a - 3b)^2 - 6(a - 3b) + 9$

58. $(2p + q)^2 - 10(2p + q) + 25$

*Factor each polynomial. **See Examples 5 and 6.***

59. $9a^2 - 16$

60. $16q^2 - 25$

61. $x^4 - 16$

62. $y^4 - 81$

63. $25s^4 - 9t^2$

64. $36z^2 - 81y^4$

65. $(a + b)^2 - 16$

66. $(p - 2q)^2 - 100$

67. $p^4 - 625$

68. $m^4 - 1296$

69. $x^2 - 8x + 16 - y^2$

70. $m^2 + 10m + 25 - n^4$

71. $y^2 - x^2 + 12x - 36$

72. $9m^2 - n^2 - 2n - 1$

73. $8 - a^3$

74. $27 - r^3$

75. $125x^3 - 27$

76. $8m^3 - 27n^3$

77. $27y^9 + 125z^6$

78. $27z^9 + 64y^{12}$

79. $(r + 6)^3 - 216$

80. $(b + 3)^3 - 27$

81. $27 - (m + 2n)^3$

82. $125 - (4a - b)^3$

*Factor each polynomial. **See Example 7.***

83. $7(3k - 1)^2 + 26(3k - 1) - 8$

84. $6(4z - 3)^2 + 7(4z - 3) - 3$

85. $9(a - 4)^2 + 30(a - 4) + 25$

86. $4(5x + 7)^2 + 12(5x + 7) + 9$

87. $(a + 1)^3 + 27$

88. $(x - 4)^3 + 64$

89. $(3x + 4)^3 - 1$

90. $(5x - 2)^3 - 8$

91. $m^4 - 3m^2 - 10$

92. $a^4 - 2a^2 - 48$

93. $12t^4 - t^2 - 35$

94. $10m^4 + 43m^2 - 9$

*Factor by any method. **See Examples 1–7.***

95. $4b^2 + 4bc + c^2 - 16$

96. $(2y - 1)^2 - 4(2y - 1) + 4$

97. $x^2 + xy - 5x - 5y$

98. $8r^2 - 3rs + 10s^2$

99. $p^4(m - 2n) + q(m - 2n)$

100. $36a^2 + 60a + 25$

101. $4z^2 + 28z + 49$

102. $6p^4 + 7p^2 - 3$

103. $1000x^3 + 343y^3$

104. $b^2 + 8b + 16 - a^2$

105. $125m^6 - 216$

106. $q^2 + 6q + 9 - p^2$

107. $64 + (3x + 2)^3$

108. $216p^3 + 125q^3$

109. $(x + y)^3 - (x - y)^3$

110. $100r^2 - 169s^2$

111. $144z^2 + 121$

112. $(3a + 5)^2 - 18(3a + 5) + 81$

113. $(x + y)^2 - (x - y)^2$

114. $4z^4 - 7z^2 - 15$

115. *Concept Check* Are there any conditions under which a sum of squares can be factored? If so, give an example.

116. *Geometric Modeling* Explain how the figures give geometric interpretation to the formula $x^2 + 2xy + y^2 = (x + y)^2$.

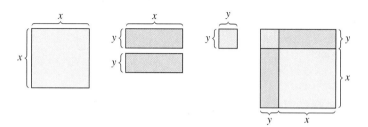

Factor each polynomial over the set of rational number coefficients.

117. $49x^2 - \dfrac{1}{25}$ **118.** $81y^2 - \dfrac{1}{49}$ **119.** $\dfrac{25}{9}x^4 - 9y^2$ **120.** $\dfrac{121}{25}y^4 - 49x^2$

Concept Check Find all values of b or c that will make the polynomial a perfect square trinomial.

121. $4z^2 + bz + 81$ **122.** $9p^2 + bp + 25$

123. $100r^2 - 60r + c$ **124.** $49x^2 + 70x + c$

Relating Concepts

For individual or collaborative investigation *(Exercises 125–130)*

*The polynomial $x^6 - 1$ can be considered either a difference of squares or a difference of cubes. Work **Exercises 125–130 in order,** to connect the results obtained when two different methods of factoring are used.*

125. Factor $x^6 - 1$ by first factoring as a difference of squares, and then factor further by using the patterns for a sum of cubes and a difference of cubes.

126. Factor $x^6 - 1$ by first factoring as a difference of cubes, and then factor further by using the pattern for a difference of squares.

127. Compare the answers in **Exercises 125 and 126.** Based on these results, what is the factorization of $x^4 + x^2 + 1$?

128. The polynomial $x^4 + x^2 + 1$ cannot be factored using the methods described in this section. However, there is a technique that enables us to factor it, as shown here. Supply the reason why each step is valid.

$$x^4 + x^2 + 1$$
$$= x^4 + 2x^2 + 1 - x^2 \qquad \rule{3cm}{0.4pt}$$
$$= (x^4 + 2x^2 + 1) - x^2 \qquad \rule{3cm}{0.4pt}$$
$$= (x^2 + 1)^2 - x^2 \qquad \rule{3cm}{0.4pt}$$
$$= (x^2 + 1 - x)(x^2 + 1 + x) \qquad \rule{3cm}{0.4pt}$$
$$= (x^2 - x + 1)(x^2 + x + 1) \qquad \rule{3cm}{0.4pt}$$

129. How does the answer in **Exercise 127** compare with the final line in **Exercise 128?**

130. Factor $x^8 + x^4 + 1$ using the technique outlined in **Exercise 128.**

R.5 Rational Expressions

- Rational Expressions
- Lowest Terms of a Rational Expression
- Multiplication and Division
- Addition and Subtraction
- Complex Fractions

Rational Expressions The quotient of two polynomials P and Q, with $Q \neq 0$, is a **rational expression.**

$$\frac{x+6}{x+2}, \quad \frac{(x+6)(x+4)}{(x+2)(x+4)}, \quad \frac{2p^2 + 7p - 4}{5p^2 + 20p} \quad \text{Rational expressions}$$

The **domain** of a rational expression is the set of real numbers for which the expression is defined. Because the denominator of a fraction cannot be 0, the domain consists of all real numbers except those that make the denominator 0. We find these numbers by setting the denominator equal to 0 and solving the resulting equation. For example, in the rational expression

$$\frac{x+6}{x+2},$$

the solution to the equation $x + 2 = 0$ is excluded from the domain. The solution is -2, so the domain is the set of all real numbers x not equal to -2.

$$\{x \mid x \neq -2\} \quad \text{Set-builder notation}$$

If the denominator of a rational expression contains a product, we determine the domain with the **zero-factor property,** which states that $ab = 0$ if and only if $a = 0$ or $b = 0$.

EXAMPLE 1 Finding the Domain

Find the domain of the rational expression.

$$\frac{(x+6)(x+4)}{(x+2)(x+4)}$$

SOLUTION

$$(x+2)(x+4) = 0 \quad \text{Set the denominator equal to zero.}$$

$$x + 2 = 0 \quad \text{or} \quad x + 4 = 0 \quad \text{Zero-factor property}$$

$$x = -2 \quad \text{or} \quad x = -4 \quad \text{Solve each equation.}$$

The domain is the set of real numbers *not equal to* -2 or -4, written

$$\{x \mid x \neq -2, -4\}.$$

✔ **Now Try Exercises 11 and 13.**

Lowest Terms of a Rational Expression A rational expression is written in **lowest terms** when the greatest common factor of its numerator and its denominator is 1. We use the following **fundamental principle of fractions** to write a rational expression in lowest terms by dividing out common factors.

Fundamental Principle of Fractions

$$\frac{ac}{bc} = \frac{a}{b} \quad (b \neq 0, c \neq 0)$$

EXAMPLE 2 **Writing Rational Expressions in Lowest Terms**

Write each rational expression in lowest terms.

(a) $\dfrac{2x^2 + 7x - 4}{5x^2 + 20x}$ (b) $\dfrac{6 - 3x}{x^2 - 4}$

SOLUTION

(a) $\dfrac{2x^2 + 7x - 4}{5x^2 + 20x}$

$= \dfrac{(2x - 1)(x + 4)}{5x(x + 4)}$ Factor.

$= \dfrac{2x - 1}{5x}$ Divide out the common factor.

To determine the domain, we find values of x that make the *original* denominator $5x^2 + 20x$ equal to 0, and exclude them.

$$5x^2 + 20x = 0 \quad \text{Set the denominator equal to 0.}$$
$$5x(x + 4) = 0 \quad \text{Factor.}$$
$$5x = 0 \quad \text{or} \quad x + 4 = 0 \quad \text{Zero-factor property}$$
$$x = 0 \quad \text{or} \quad x = -4 \quad \text{Solve each equation.}$$

The domain is $\{x \,|\, x \neq 0, -4\}$. ***From now on, we will assume such restrictions when writing rational expressions in lowest terms.***

(b) $\dfrac{6 - 3x}{x^2 - 4}$

$= \dfrac{3(2 - x)}{(x + 2)(x - 2)}$ Factor.

$= \dfrac{3(2 - x)(-1)}{(x + 2)(x - 2)(-1)}$ $2 - x$ and $x - 2$ are opposites. Multiply numerator and denominator by -1.

$= \dfrac{3(2 - x)(-1)}{(x + 2)(2 - x)}$ $(x - 2)(-1) = -x + 2 = 2 - x$

> Be careful with signs.

$= \dfrac{-3}{x + 2}$ Divide out the common factor.

Working in an alternative way would lead to the equivalent result $\dfrac{3}{-x - 2}$.

✔ **Now Try Exercises 23 and 27.**

LOOKING AHEAD TO CALCULUS

A standard problem in calculus is investigating what value an expression such as $\frac{x^2 - 1}{x - 1}$ approaches as x approaches 1. We cannot do this by simply substituting 1 for x in the expression since the result is the indeterminate form $\frac{0}{0}$. When we factor the numerator and write the expression in lowest terms, it becomes $x + 1$. Then, by substituting 1 for x, we obtain $1 + 1 = 2$, which is the **limit** of $\frac{x^2 - 1}{x - 1}$ as x approaches 1.

CAUTION The fundamental principle requires a pair of common *factors*, one in the numerator and one in the denominator. *Only after a rational expression has been factored can any common factors be divided out.* For example,

$$\frac{2x + 4}{6} = \frac{2(x + 2)}{2 \cdot 3} = \frac{x + 2}{3}. \quad \text{Factor first, and then divide.}$$

Multiplication and Division   We now multiply and divide fractions.

Multiplication and Division

For fractions $\frac{a}{b}$ and $\frac{c}{d}$ ($b \neq 0, d \neq 0$), the following hold.

$$\frac{a}{b} \cdot \frac{c}{d} = \frac{ac}{bd} \quad \text{and} \quad \frac{a}{b} \div \frac{c}{d} = \frac{a}{b} \cdot \frac{d}{c} \quad (c \neq 0)$$

That is, to find the product of two fractions, multiply their numerators to find the numerator of the product. Then multiply their denominators to find the denominator of the product.

*To divide two fractions, multiply the **dividend** (the first fraction) by the reciprocal of the **divisor** (the second fraction).*

EXAMPLE 3   Multiplying or Dividing Rational Expressions

Multiply or divide, as indicated.

(a) $\dfrac{2y^2}{9} \cdot \dfrac{27}{8y^5}$

(b) $\dfrac{3m^2 - 2m - 8}{3m^2 + 14m + 8} \cdot \dfrac{3m + 2}{3m + 4}$

(c) $\dfrac{3p^2 + 11p - 4}{24p^3 - 8p^2} \div \dfrac{9p + 36}{24p^4 - 36p^3}$

(d) $\dfrac{x^3 - y^3}{x^2 - y^2} \cdot \dfrac{2x + 2y + xz + yz}{2x^2 + 2y^2 + zx^2 + zy^2}$

SOLUTION

(a) $\dfrac{2y^2}{9} \cdot \dfrac{27}{8y^5}$

$= \dfrac{2y^2 \cdot 27}{9 \cdot 8y^5}$ Multiply fractions.

$= \dfrac{2 \cdot 9 \cdot 3 \cdot y^2}{9 \cdot 2 \cdot 4 \cdot y^2 \cdot y^3}$ Factor.

$= \dfrac{3}{4y^3}$ Lowest terms

Although we usually factor first and then multiply the fractions (see parts (b)–(d)), we did the opposite here. Either order is acceptable.

(b) $\dfrac{3m^2 - 2m - 8}{3m^2 + 14m + 8} \cdot \dfrac{3m + 2}{3m + 4}$

$= \dfrac{(m - 2)(3m + 4)}{(m + 4)(3m + 2)} \cdot \dfrac{3m + 2}{3m + 4}$ Factor.

$= \dfrac{(m - 2)(3m + 4)(3m + 2)}{(m + 4)(3m + 2)(3m + 4)}$ Multiply fractions.

$= \dfrac{m - 2}{m + 4}$ Lowest terms

R.5 Rational Expressions **49**

(c) $\dfrac{3p^2 + 11p - 4}{24p^3 - 8p^2} \div \dfrac{9p + 36}{24p^4 - 36p^3}$

$\qquad = \dfrac{(p + 4)(3p - 1)}{8p^2(3p - 1)} \div \dfrac{9(p + 4)}{12p^3(2p - 3)}$ Factor.

$\qquad = \dfrac{(p + 4)(3p - 1)}{8p^2(3p - 1)} \cdot \dfrac{12p^3(2p - 3)}{9(p + 4)}$ Multiply by the reciprocal of the divisor.

$\qquad = \dfrac{12p^3(2p - 3)}{9 \cdot 8p^2}$ Divide out common factors. Multiply fractions.

$\qquad = \dfrac{3 \cdot 4 \cdot p^2 \cdot p(2p - 3)}{3 \cdot 3 \cdot 4 \cdot 2 \cdot p^2}$ Factor.

$\qquad = \dfrac{p(2p - 3)}{6}$ Lowest terms

(d) $\dfrac{x^3 - y^3}{x^2 - y^2} \cdot \dfrac{2x + 2y + xz + yz}{2x^2 + 2y^2 + zx^2 + zy^2}$

$\qquad = \dfrac{(x - y)(x^2 + xy + y^2)}{(x + y)(x - y)} \cdot \dfrac{2(x + y) + z(x + y)}{2(x^2 + y^2) + z(x^2 + y^2)}$ Factor. Group terms and factor.

$\qquad = \dfrac{(x - y)(x^2 + xy + y^2)}{(x + y)(x - y)} \cdot \dfrac{(x + y)(2 + z)}{(x^2 + y^2)(2 + z)}$ Factor by grouping.

$\qquad = \dfrac{x^2 + xy + y^2}{x^2 + y^2}$ Divide out common factors. Multiply fractions.

✔ **Now Try Exercises 33, 43, and 47.**

Addition and Subtraction   We add and subtract rational expressions in the same way that we add and subtract fractions.

Addition and Subtraction

For fractions $\dfrac{a}{b}$ and $\dfrac{c}{d}$ ($b \neq 0$, $d \neq 0$), the following hold.

$$\frac{a}{b} + \frac{c}{d} = \frac{ad + bc}{bd} \quad \text{and} \quad \frac{a}{b} - \frac{c}{d} = \frac{ad - bc}{bd}$$

That is, to add (or subtract) two fractions in practice, find their least common denominator (LCD) and change each fraction to one with the LCD as denominator. The sum (or difference) of their numerators is the numerator of their sum (or difference), and the LCD is the denominator of their sum (or difference).

Finding the Least Common Denominator (LCD)

Step 1   Write each denominator as a product of prime factors.

Step 2   Form a product of all the different prime factors. Each factor should have as exponent the *greatest* exponent that appears on that factor.

EXAMPLE 4 **Adding or Subtracting Rational Expressions**

Add or subtract, as indicated.

(a) $\dfrac{5}{9x^2} + \dfrac{1}{6x}$

(b) $\dfrac{y}{y-2} + \dfrac{8}{2-y}$

(c) $\dfrac{3}{(x-1)(x+2)} - \dfrac{1}{(x+3)(x-4)}$

SOLUTION

(a) $\dfrac{5}{9x^2} + \dfrac{1}{6x}$

Step 1 Write each denominator as a product of prime factors.

$$9x^2 = 3^2 \cdot x^2$$

$$6x = 2^1 \cdot 3^1 \cdot x^1$$

Step 2 For the LCD, form the product of all the prime factors, with each factor having the greatest exponent that appears on it.

Greatest exponent on 3 is 2.─┐   ┌─Greatest exponent on x is 2.

$$\text{LCD} = 2^1 \cdot 3^2 \cdot x^2$$

$$= 18x^2$$

Write the given expressions with this denominator, and then add.

$$\dfrac{5}{9x^2} + \dfrac{1}{6x}$$

$$= \dfrac{5 \cdot 2}{9x^2 \cdot 2} + \dfrac{1 \cdot 3x}{6x \cdot 3x} \qquad \text{LCD} = 18x^2$$

$$= \dfrac{10}{18x^2} + \dfrac{3x}{18x^2} \qquad \text{Multiply.}$$

$$= \dfrac{10 + 3x}{18x^2} \qquad \text{Add the numerators.}$$

Always check to see that the answer is in lowest terms.

(b) $\dfrac{y}{y-2} + \dfrac{8}{2-y}$ We arbitrarily choose $y-2$ as the LCD.

$$= \dfrac{y}{y-2} + \dfrac{8(-1)}{(2-y)(-1)} \qquad \begin{array}{l}\text{Multiply the second expression by } -1 \text{ in} \\ \text{both the numerator and the denominator.}\end{array}$$

$$= \dfrac{y}{y-2} + \dfrac{-8}{y-2} \qquad \text{Simplify.}$$

$$= \dfrac{y-8}{y-2} \qquad \text{Add the numerators.}$$

We could use $2-y$ as the common denominator instead of $y-2$.

$$\frac{y(-1)}{(y-2)(-1)} + \frac{8}{2-y}$$

Multiply the first expression by -1 in both the numerator and the denominator.

$$= \frac{-y}{2-y} + \frac{8}{2-y}$$

Simplify.

$$= \frac{8-y}{2-y}$$

This equivalent expression results.

(c) $\dfrac{3}{(x-1)(x+2)} - \dfrac{1}{(x+3)(x-4)}$

The LCD is $(x-1)(x+2)(x+3)(x-4)$.

$$= \frac{3(x+3)(x-4)}{(x-1)(x+2)(x+3)(x-4)} - \frac{1(x-1)(x+2)}{(x+3)(x-4)(x-1)(x+2)}$$

$$= \frac{3(x^2 - x - 12) - (x^2 + x - 2)}{(x-1)(x+2)(x+3)(x-4)}$$

Multiply in the numerators, and then subtract them.

Be careful with signs.

$$= \frac{3x^2 - 3x - 36 - x^2 - x + 2}{(x-1)(x+2)(x+3)(x-4)}$$

Distributive property

$$= \frac{2x^2 - 4x - 34}{(x-1)(x+2)(x+3)(x-4)}$$

Combine like terms in the numerator.

✔ **Now Try Exercises 57, 63, and 69.**

CAUTION *When subtracting fractions where the second fraction has more than one term in the numerator, as in Example 4(c), be sure to distribute the negative sign to each term.* Use parentheses as in the second step to avoid an error.

Complex Fractions The quotient of two rational expressions is a **complex fraction**. There are two methods for simplifying a complex fraction.

EXAMPLE 5 Simplifying Complex Fractions

Simplify each complex fraction. In part (b), use two methods.

(a) $\dfrac{6 - \dfrac{5}{k}}{1 + \dfrac{5}{k}}$

(b) $\dfrac{\dfrac{a}{a+1} + \dfrac{1}{a}}{\dfrac{1}{a} + \dfrac{1}{a+1}}$

SOLUTION

(a) Method 1 for simplifying uses the identity property for multiplication. We multiply both numerator and denominator by the LCD of all the fractions, k.

Distribute k to *all* terms within the parentheses.

$$\frac{6 - \dfrac{5}{k}}{1 + \dfrac{5}{k}} = \frac{k\left(6 - \dfrac{5}{k}\right)}{k\left(1 + \dfrac{5}{k}\right)} = \frac{6k - k\left(\dfrac{5}{k}\right)}{k + k\left(\dfrac{5}{k}\right)} = \frac{6k - 5}{k + 5}$$

(b) $\dfrac{\dfrac{a}{a+1}+\dfrac{1}{a}}{\dfrac{1}{a}+\dfrac{1}{a+1}} = \dfrac{\left(\dfrac{a}{a+1}+\dfrac{1}{a}\right)a(a+1)}{\left(\dfrac{1}{a}+\dfrac{1}{a+1}\right)a(a+1)}$

For Method 1, multiply both numerator and denominator by the LCD of all the fractions, $a(a+1)$.

(Method 1)

$= \dfrac{\dfrac{a}{a+1}(a)(a+1)+\dfrac{1}{a}(a)(a+1)}{\dfrac{1}{a}(a)(a+1)+\dfrac{1}{a+1}(a)(a+1)}$

Distributive property

$= \dfrac{a^2+(a+1)}{(a+1)+a}$

Multiply.

$= \dfrac{a^2+a+1}{2a+1}$

Combine like terms.

$\dfrac{\dfrac{a}{a+1}+\dfrac{1}{a}}{\dfrac{1}{a}+\dfrac{1}{a+1}} = \dfrac{\dfrac{a^2+1(a+1)}{a(a+1)}}{\dfrac{1(a+1)+1(a)}{a(a+1)}}$

For Method 2, find the LCD, and add terms in the numerator and denominator of the complex fraction.

(Method 2)

$= \dfrac{\dfrac{a^2+a+1}{a(a+1)}}{\dfrac{2a+1}{a(a+1)}}$

Combine terms in the numerator and denominator.

$= \dfrac{a^2+a+1}{a(a+1)} \cdot \dfrac{a(a+1)}{2a+1}$

Multiply by the reciprocal of the divisor.

The result is the same as in Method 1.

$= \dfrac{a^2+a+1}{2a+1}$

Multiply fractions, and write in lowest terms.

✔ **Now Try Exercises 71 and 83.**

R.5 Exercises

CONCEPT PREVIEW *Fill in the blank(s) to correctly complete each sentence.*

1. The quotient of two polynomials in which the denominator is not equal to zero is a _____.

2. The domain of a rational expression consists of all real numbers except those that make the _____ equal to 0.

3. In the rational expression $\dfrac{x+1}{x-5}$, the domain cannot include the number _____.

4. A rational expression is in lowest terms when the greatest common factor of its numerator and its denominator is _____.

CONCEPT PREVIEW *Perform the indicated operation, and write each answer in lowest terms.*

5. $\dfrac{2x}{5} \cdot \dfrac{10}{x^2}$

6. $\dfrac{y^3}{8} \div \dfrac{y}{4}$

7. $\dfrac{3}{x} + \dfrac{7}{x}$

8. $\dfrac{4}{x-y} - \dfrac{9}{x-y}$ **9.** $\dfrac{2x}{5} + \dfrac{x}{4}$ **10.** $\dfrac{7}{x^2} - \dfrac{8}{y}$

Find the domain of each rational expression. **See Example 1.**

11. $\dfrac{x+3}{x-6}$ **12.** $\dfrac{2x-4}{x+7}$ **13.** $\dfrac{3x+7}{(4x+2)(x-1)}$

14. $\dfrac{9x+12}{(2x+3)(x-5)}$ **15.** $\dfrac{12}{x^2+5x+6}$ **16.** $\dfrac{3}{x^2-5x-6}$

17. $\dfrac{x^2-1}{x+1}$ **18.** $\dfrac{x^2-25}{x-5}$ **19.** $\dfrac{x^3-1}{x-1}$

20. *Concept Check* Use specific values for x and y to show that in general, $\frac{1}{x} + \frac{1}{y}$ is not equivalent to $\frac{1}{x+y}$.

Write each rational expression in lowest terms. **See Example 2.**

21. $\dfrac{8x^2+16x}{4x^2}$ **22.** $\dfrac{36y^2+72y}{9y^2}$ **23.** $\dfrac{3(3-t)}{(t+5)(t-3)}$

24. $\dfrac{-8(4-y)}{(y+2)(y-4)}$ **25.** $\dfrac{8k+16}{9k+18}$ **26.** $\dfrac{20r+10}{30r+15}$

27. $\dfrac{m^2-4m+4}{m^2+m-6}$ **28.** $\dfrac{r^2-r-6}{r^2+r-12}$ **29.** $\dfrac{8m^2+6m-9}{16m^2-9}$

30. $\dfrac{6y^2+11y+4}{3y^2+7y+4}$ **31.** $\dfrac{x^3+64}{x+4}$ **32.** $\dfrac{y^3-27}{y-3}$

Multiply or divide, as indicated. **See Example 3.**

33. $\dfrac{15p^3}{9p^2} \div \dfrac{6p}{10p^2}$ **34.** $\dfrac{8r^3}{6r} \div \dfrac{5r^2}{9r^3}$ **35.** $\dfrac{2k+8}{6} \div \dfrac{3k+12}{2}$

36. $\dfrac{5m+25}{10} \div \dfrac{6m+30}{12}$ **37.** $\dfrac{x^2+x}{5} \cdot \dfrac{25}{xy+y}$ **38.** $\dfrac{y^3+y^2}{7} \cdot \dfrac{49}{y^4+y^3}$

39. $\dfrac{4a+12}{2a-10} \div \dfrac{a^2-9}{a^2-a-20}$ **40.** $\dfrac{6r-18}{9r^2+6r-24} \div \dfrac{4r-12}{12r-16}$

41. $\dfrac{p^2-p-12}{p^2-2p-15} \cdot \dfrac{p^2-9p+20}{p^2-8p+16}$ **42.** $\dfrac{x^2+2x-15}{x^2+11x+30} \cdot \dfrac{x^2+2x-24}{x^2-8x+15}$

43. $\dfrac{m^2+3m+2}{m^2+5m+4} \div \dfrac{m^2+5m+6}{m^2+10m+24}$ **44.** $\dfrac{y^2+y-2}{y^2+3y-4} \div \dfrac{y^2+3y+2}{y^2+4y+3}$

45. $\dfrac{x^3+y^3}{x^3-y^3} \cdot \dfrac{x^2-y^2}{x^2+2xy+y^2}$ **46.** $\dfrac{x^2-y^2}{(x-y)^2} \cdot \dfrac{x^2-xy+y^2}{x^2-2xy+y^2} \div \dfrac{x^3+y^3}{(x-y)^4}$

47. $\dfrac{xz-xw+2yz-2yw}{z^2-w^2} \cdot \dfrac{4z+4w+xz+wx}{16-x^2}$

48. $\dfrac{ac+ad+bc+bd}{a^2-b^2} \cdot \dfrac{a^3-b^3}{2a^2+2ab+2b^2}$

49. *Concept Check* Which of the following rational expressions is equivalent to -1? In choices A, B, and D, $x \ne -4$, and in choice C, $x \ne 4$. (*Hint:* There may be more than one answer.)

A. $\dfrac{x-4}{x+4}$ **B.** $\dfrac{-x-4}{x+4}$ **C.** $\dfrac{x-4}{4-x}$ **D.** $\dfrac{x-4}{-x-4}$

50. Explain how to find the least common denominator of several fractions.

*Add or subtract, as indicated. **See Example 4.***

51. $\dfrac{3}{2k} + \dfrac{5}{3k}$

52. $\dfrac{8}{5p} + \dfrac{3}{4p}$

53. $\dfrac{1}{6m} + \dfrac{2}{5m} + \dfrac{4}{m}$

54. $\dfrac{8}{3p} + \dfrac{5}{4p} + \dfrac{9}{2p}$

55. $\dfrac{1}{a} - \dfrac{b}{a^2}$

56. $\dfrac{3}{z} + \dfrac{x}{z^2}$

57. $\dfrac{5}{12x^2y} - \dfrac{11}{6xy}$

58. $\dfrac{7}{18a^3b^2} - \dfrac{2}{9ab}$

59. $\dfrac{17y + 3}{9y + 7} - \dfrac{-10y - 18}{9y + 7}$

60. $\dfrac{7x + 8}{3x + 2} - \dfrac{x + 4}{3x + 2}$

61. $\dfrac{1}{x + z} + \dfrac{1}{x - z}$

62. $\dfrac{m + 1}{m - 1} + \dfrac{m - 1}{m + 1}$

63. $\dfrac{3}{a - 2} - \dfrac{1}{2 - a}$

64. $\dfrac{4}{p - q} - \dfrac{2}{q - p}$

65. $\dfrac{x + y}{2x - y} - \dfrac{2x}{y - 2x}$

66. $\dfrac{m - 4}{3m - 4} - \dfrac{5m}{4 - 3m}$

67. $\dfrac{4}{x + 1} + \dfrac{1}{x^2 - x + 1} - \dfrac{12}{x^3 + 1}$

68. $\dfrac{5}{x + 2} + \dfrac{2}{x^2 - 2x + 4} - \dfrac{60}{x^3 + 8}$

69. $\dfrac{3x}{x^2 + x - 12} - \dfrac{x}{x^2 - 16}$

70. $\dfrac{p}{2p^2 - 9p - 5} - \dfrac{2p}{6p^2 - p - 2}$

*Simplify each complex fraction. **See Example 5.***

71. $\dfrac{1 + \dfrac{1}{x}}{1 - \dfrac{1}{x}}$

72. $\dfrac{2 - \dfrac{2}{y}}{2 + \dfrac{2}{y}}$

73. $\dfrac{\dfrac{1}{x + 1} - \dfrac{1}{x}}{\dfrac{1}{x}}$

74. $\dfrac{\dfrac{1}{y + 3} - \dfrac{1}{y}}{\dfrac{1}{y}}$

75. $\dfrac{1 + \dfrac{1}{1 - b}}{1 - \dfrac{1}{1 + b}}$

76. $\dfrac{2 + \dfrac{2}{1 + x}}{2 - \dfrac{2}{1 - x}}$

77. $\dfrac{\dfrac{1}{a^3 + b^3}}{\dfrac{1}{a^2 + 2ab + b^2}}$

78. $\dfrac{\dfrac{1}{x^3 - y^3}}{\dfrac{1}{x^2 - y^2}}$

79. $\dfrac{m - \dfrac{1}{m^2 - 4}}{\dfrac{1}{m + 2}}$

80. $\dfrac{y + \dfrac{1}{y^2 - 9}}{\dfrac{1}{y + 3}}$

81. $\dfrac{\dfrac{3}{p^2 - 16} + p}{\dfrac{1}{p - 4}}$

82. $\dfrac{\dfrac{6}{x^2 - 25} + x}{\dfrac{1}{x - 5}}$

83. $\dfrac{\dfrac{y + 3}{y} - \dfrac{4}{y - 1}}{\dfrac{y}{y - 1} + \dfrac{1}{y}}$

84. $\dfrac{\dfrac{x + 4}{x} - \dfrac{3}{x - 2}}{\dfrac{x}{x - 2} + \dfrac{1}{x}}$

85. $\dfrac{\dfrac{1}{x + h} - \dfrac{1}{x}}{h}$

86. $\dfrac{\dfrac{-2}{x + h} - \dfrac{-2}{x}}{h}$

87. $\dfrac{\dfrac{1}{(x + h)^2 + 9} - \dfrac{1}{x^2 + 9}}{h}$

88. $\dfrac{\dfrac{2}{(x + h)^2 + 16} - \dfrac{2}{x^2 + 16}}{h}$

(Modeling) Distance from the Origin of the Nile River *The Nile River in Africa is about 4000 mi long. The Nile begins as an outlet of Lake Victoria at an altitude of 7000 ft above sea level and empties into the Mediterranean Sea at sea level (0 ft). The distance from its origin in thousands of miles is related to its height above sea level in thousands of feet (x) by the following formula.*

$$\text{Distance} = \frac{7 - x}{0.639x + 1.75}$$

For example, when the river is at an altitude of 600 ft, x = 0.6 (thousand), and the distance from the origin is

$$\text{Distance} = \frac{7 - 0.6}{0.639(0.6) + 1.75} \approx 3, \quad \text{which represents } 3000 \text{ mi.}$$

(Source: World Almanac and Book of Facts.)

89. What is the distance from the origin of the Nile when the river has an altitude of 7000 ft?

90. What is the distance, to the nearest mile, from the origin of the Nile when the river has an altitude of 1200 ft?

(Modeling) Cost-Benefit Model for a Pollutant *In situations involving environmental pollution, a **cost-benefit model** expresses cost in terms of the percentage of pollutant removed from the environment. Suppose a cost-benefit model is expressed as*

$$y = \frac{6.7x}{100 - x},$$

where y is the cost in thousands of dollars of removing x percent of a certain pollutant. Find the value of y for each given value of x.

91. $x = 75$ (75%)   **92.** $x = 95$ (95%)

R.6 Rational Exponents

- Negative Exponents and the Quotient Rule
- Rational Exponents
- Complex Fractions Revisited

Negative Exponents and the Quotient Rule   Suppose that n is a positive integer, and we wish to define a^{-n} to be consistent with the application of the product rule. Consider the product $a^n \cdot a^{-n}$, and apply the rule.

$$a^n \cdot a^{-n}$$

$$= a^{n+(-n)} \quad \text{Product rule: Add exponents.}$$

$$= a^0 \quad \quad n \text{ and } -n \text{ are additive inverses.}$$

$$= 1 \quad \quad \text{Definition of } a^0$$

The expression a^{-n} acts as the *reciprocal* of a^n, which is written $\frac{1}{a^n}$. Thus, these two expressions must be equivalent.

Negative Exponent

Let a be a nonzero real number and n be any integer.

$$a^{-n} = \frac{1}{a^n}$$

EXAMPLE 1 **Using the Definition of a Negative Exponent**

Write each expression without negative exponents, and evaluate if possible. Assume all variables represent nonzero real numbers.

(a) 4^{-2} (b) -4^{-2} (c) $\left(\dfrac{2}{5}\right)^{-3}$ (d) $(xy)^{-3}$ (e) xy^{-3}

SOLUTION

(a) $4^{-2} = \dfrac{1}{4^2} = \dfrac{1}{16}$ (b) $-4^{-2} = -\dfrac{1}{4^2} = -\dfrac{1}{16}$

(c) $\left(\dfrac{2}{5}\right)^{-3} = \dfrac{1}{\left(\frac{2}{5}\right)^3} = \dfrac{1}{\frac{8}{125}} = 1 \div \dfrac{8}{125} = 1 \cdot \dfrac{125}{8} = \dfrac{125}{8}$

Multiply by the reciprocal of the divisor.

(d) $(xy)^{-3} = \dfrac{1}{(xy)^3} = \dfrac{1}{x^3 y^3}$ (e) $xy^{-3} = x \cdot \dfrac{1}{y^3} = \dfrac{x}{y^3}$

↑   ↑
Base is xy.   Base is y.

✔ **Now Try Exercises 11, 13, 15, 17, and 19.**

CAUTION *A negative exponent indicates a reciprocal, not a sign change of the expression.*

Example 1(c) showed the following.

$$\left(\frac{2}{5}\right)^{-3} = \frac{125}{8} = \left(\frac{5}{2}\right)^3$$

We can generalize this result. If $a \neq 0$ and $b \neq 0$, then for any integer n, the following is true.

$$\left(\frac{a}{b}\right)^{-n} = \left(\frac{b}{a}\right)^n$$

The following example suggests the **quotient rule** for exponents.

$$\frac{5^6}{5^2} = \frac{5 \cdot 5 \cdot 5 \cdot 5 \cdot 5 \cdot 5}{5 \cdot 5} = 5^4 \leftarrow$$ This exponent is the result of dividing common factors, or, essentially, subtracting the original exponents.

Quotient Rule

Let m and n be integers and a be a nonzero real number.

$$\frac{a^m}{a^n} = a^{m-n}$$

That is, when dividing powers of like bases, keep the same base and subtract the exponent of the denominator from the exponent of the numerator.

CAUTION When applying the quotient rule, be sure to subtract the exponents in the correct order. Be careful especially when the exponent in the denominator is negative, and avoid sign errors.

EXAMPLE 2 **Using the Quotient Rule**

Simplify each expression. Assume all variables represent nonzero real numbers.

(a) $\dfrac{12^5}{12^2}$

(b) $\dfrac{a^5}{a^{-8}}$

(c) $\dfrac{16m^{-9}}{12m^{11}}$

(d) $\dfrac{25r^7z^5}{10r^9z}$

SOLUTION

> Use parentheses to avoid errors.

(a) $\dfrac{12^5}{12^2} = 12^{5-2} = 12^3$

(b) $\dfrac{a^5}{a^{-8}} = a^{5-(-8)} = a^{13}$

(c) $\dfrac{16m^{-9}}{12m^{11}}$

$= \dfrac{16}{12} \cdot m^{-9-11}$

$= \dfrac{4}{3}m^{-20}$

$= \dfrac{4}{3} \cdot \dfrac{1}{m^{20}}$

$= \dfrac{4}{3m^{20}}$

(d) $\dfrac{25r^7z^5}{10r^9z}$

$= \dfrac{25}{10} \cdot \dfrac{r^7}{r^9} \cdot \dfrac{z^5}{z^1}$

$= \dfrac{5}{2}r^{7-9}z^{5-1}$

$= \dfrac{5}{2}r^{-2}z^4$

$= \dfrac{5z^4}{2r^2}$

✔ **Now Try Exercises 23, 29, 31, and 33.**

The previous rules for exponents were stated for positive integer exponents and for zero as an exponent. Those rules continue to apply in expressions involving negative exponents, as seen in the next example.

EXAMPLE 3 **Using the Rules for Exponents**

Simplify each expression. Write answers without negative exponents. Assume all variables represent nonzero real numbers.

(a) $3x^{-2}(4^{-1}x^{-5})^2$

(b) $\dfrac{12p^3q^{-1}}{8p^{-2}q}$

(c) $\dfrac{(3x^2)^{-1}(3x^5)^{-2}}{(3^{-1}x^{-2})^2}$

SOLUTION

(a) $3x^{-2}(4^{-1}x^{-5})^2$

$= 3x^{-2}(4^{-2}x^{-10})$ Power rules

$= 3 \cdot 4^{-2} \cdot x^{-2+(-10)}$ Rearrange factors; product rule

$= 3 \cdot 4^{-2} \cdot x^{-12}$ Simplify the exponent on x.

$= \dfrac{3}{16x^{12}}$ Write with positive exponents, and multiply.

(b) $\dfrac{12p^3q^{-1}}{8p^{-2}q}$

$\qquad = \dfrac{12}{8} \cdot \dfrac{p^3}{p^{-2}} \cdot \dfrac{q^{-1}}{q^1}$ Write as separate factors.

$\qquad = \dfrac{3}{2} \cdot p^{3-(-2)}q^{-1-1}$ Write $\frac{12}{8}$ in lowest terms; quotient rule

$\qquad = \dfrac{3}{2}p^5q^{-2}$ Simplify the exponents.

$\qquad = \dfrac{3p^5}{2q^2}$ Write with positive exponents, and multiply.

(c) $\dfrac{(3x^2)^{-1}(3x^5)^{-2}}{(3^{-1}x^{-2})^2}$

$\qquad = \dfrac{3^{-1}x^{-2}3^{-2}x^{-10}}{3^{-2}x^{-4}}$ Power rules

$\qquad = \dfrac{3^{-1+(-2)}x^{-2+(-10)}}{3^{-2}x^{-4}}$ Product rule

$\qquad = \dfrac{3^{-3}x^{-12}}{3^{-2}x^{-4}}$ Simplify the exponents.

$\qquad = 3^{-3-(-2)}x^{-12-(-4)}$ Quotient rule

Be careful with signs. $\qquad = 3^{-1}x^{-8}$ Simplify the exponents.

$\qquad = \dfrac{1}{3x^8}$ Write with positive exponents, and multiply.

✔ **Now Try Exercises 37, 43, and 45.**

CAUTION   Notice the use of the power rule $(ab)^n = a^nb^n$ in **Example 3(c).** *Remember to apply the exponent to the numerical coefficient 3.*

$$(3x^2)^{-1} = 3^{-1}(x^2)^{-1} = 3^{-1}x^{-2}$$

Rational Exponents   The definition of a^n can be extended to rational values of n by defining $a^{1/n}$ to be the nth root of a. By one of the power rules of exponents (extended to a rational exponent), we have the following.

$$(a^{1/n})^n = a^{(1/n)n} = a^1 = a$$

This suggests that $a^{1/n}$ is a number whose nth power is a.

The Expression $a^{1/n}$

$a^{1/n}$, n Even   If n is an *even* positive integer, and if $a > 0$, then $a^{1/n}$ is the positive real number whose nth power is a. That is, $(a^{1/n})^n = a$. (In this case, $a^{1/n}$ is the principal nth root of a.)

$a^{1/n}$, n Odd   If n is an *odd* positive integer, and a is *any nonzero real number*, then $a^{1/n}$ is the positive or negative real number whose nth power is a. That is, $(a^{1/n})^n = a$.

For all positive integers n, $0^{1/n} = 0$.

EXAMPLE 4 **Using the Definition of $a^{1/n}$**

Evaluate each expression.

(a) $36^{1/2}$ **(b)** $-100^{1/2}$ **(c)** $-(225)^{1/2}$ **(d)** $625^{1/4}$

(e) $(-1296)^{1/4}$ **(f)** $-1296^{1/4}$ **(g)** $(-27)^{1/3}$ **(h)** $-32^{1/5}$

SOLUTION

(a) $36^{1/2} = 6$ because $6^2 = 36$.   **(b)** $-100^{1/2} = -10$

(c) $-(225)^{1/2} = -15$ **(d)** $625^{1/4} = 5$

(e) $(-1296)^{1/4}$ is not a real number.   **(f)** $-1296^{1/4} = -6$

(g) $(-27)^{1/3} = -3$ **(h)** $-32^{1/5} = -2$

✔ **Now Try Exercises 47, 49, and 53.**

The notation $a^{m/n}$ must be defined in such a way that all the previous rules for exponents still hold. For the power rule to hold, $(a^{1/n})^m$ must equal $a^{m/n}$. Therefore, $a^{m/n}$ is defined as follows.

The Expression $a^{m/n}$

Let m be any integer, n be any positive integer, and a be any real number for which $a^{1/n}$ is a real number.

$$a^{m/n} = (a^{1/n})^m$$

EXAMPLE 5 **Using the Definition of $a^{m/n}$**

Evaluate each expression.

(a) $125^{2/3}$ **(b)** $32^{7/5}$ **(c)** $-81^{3/2}$ **(d)** $(-27)^{2/3}$ **(e)** $16^{-3/4}$ **(f)** $(-4)^{5/2}$

SOLUTION

(a) $125^{2/3} = (125^{1/3})^2$ **(b)** $32^{7/5} = (32^{1/5})^7$

$\phantom{125^{2/3}} = 5^2$ $= 2^7$

$\phantom{125^{2/3}} = 25$ $= 128$

(c) $-81^{3/2} = -(81^{1/2})^3$ **(d)** $(-27)^{2/3} = [(-27)^{1/3}]^2$

$\phantom{-81^{3/2}} = -9^3$ $= (-3)^2$

$\phantom{-81^{3/2}} = -729$ $= 9$

(e) $16^{-3/4} = \dfrac{1}{16^{3/4}}$ **(f)** $(-4)^{5/2}$ is not a real number. This is because $(-4)^{1/2}$ is not a real number.

$\phantom{16^{-3/4}} = \dfrac{1}{(16^{1/4})^3}$

$\phantom{16^{-3/4}} = \dfrac{1}{2^3}$

$\phantom{16^{-3/4}} = \dfrac{1}{8}$ ✔ **Now Try Exercises 55, 59, and 61.**

NOTE For all real numbers a, integers m, and positive integers n for which $a^{1/n}$ is a real number, $a^{m/n}$ can be interpreted as follows.

$$a^{m/n} = \left(a^{1/n}\right)^m \quad \text{or} \quad a^{m/n} = \left(a^m\right)^{1/n}$$

So $a^{m/n}$ can be evaluated either as $\left(a^{1/n}\right)^m$ or as $\left(a^m\right)^{1/n}$.

$$27^{4/3} = \left(27^{1/3}\right)^4 = 3^4 = 81$$

or
$$27^{4/3} = \left(27^4\right)^{1/3} = 531{,}441^{1/3} = 81$$

The result is the same.

The earlier results for integer exponents also apply to rational exponents.

Definitions and Rules for Exponents

Let r and s be rational numbers. The following results are valid for all positive numbers a and b.

Product rule $\quad a^r \cdot a^s = a^{r+s}$ $\qquad$ **Power rules** $\quad (a^r)^s = a^{rs}$

Quotient rule $\quad \dfrac{a^r}{a^s} = a^{r-s}$ $\qquad\qquad\qquad\quad (ab)^r = a^r b^r$

Negative exponent $\quad a^{-r} = \dfrac{1}{a^r}$ $\qquad\qquad\qquad \left(\dfrac{a}{b}\right)^r = \dfrac{a^r}{b^r}$

EXAMPLE 6 Using the Rules for Exponents

Simplify each expression. Assume all variables represent positive real numbers.

(a) $\dfrac{27^{1/3} \cdot 27^{5/3}}{27^3}$ $\qquad$ **(b)** $81^{5/4} \cdot 4^{-3/2}$ $\qquad$ **(c)** $6y^{2/3} \cdot 2y^{1/2}$

(d) $\left(\dfrac{3m^{5/6}}{y^{3/4}}\right)^2 \left(\dfrac{8y^3}{m^6}\right)^{2/3}$ $\qquad$ **(e)** $m^{2/3}\left(m^{7/3} + 2m^{1/3}\right)$

SOLUTION

(a) $\dfrac{27^{1/3} \cdot 27^{5/3}}{27^3}$

$= \dfrac{27^{1/3+5/3}}{27^3}$ $\quad$ Product rule

$= \dfrac{27^2}{27^3}$ $\quad$ Simplify.

$= 27^{2-3}$ $\quad$ Quotient rule

$= 27^{-1}$ $\quad$ Simplify the exponent.

$= \dfrac{1}{27}$ $\quad$ Negative exponent

(b) $81^{5/4} \cdot 4^{-3/2}$

$= \left(81^{1/4}\right)^5 \left(4^{1/2}\right)^{-3}$

$= 3^5 \cdot 2^{-3}$

$= \dfrac{3^5}{2^3}$

$= \dfrac{243}{8}$

(c) $6y^{2/3} \cdot 2y^{1/2}$

$= 6 \cdot 2y^{2/3+1/2}$ $\quad$ Product rule

$= 12y^{7/6}$ $\quad$ Multiply. Simplify the exponent.

(d) $\left(\dfrac{3m^{5/6}}{y^{3/4}}\right)^2\left(\dfrac{8y^3}{m^6}\right)^{2/3}$

$\qquad = \dfrac{9m^{5/3}}{y^{3/2}} \cdot \dfrac{4y^2}{m^4} \qquad$ Power rules

$\qquad = 36m^{5/3-4}y^{2-3/2} \qquad$ Quotient rule

$\qquad = 36m^{-7/3}y^{1/2} \qquad$ Simplify the exponents.

$\qquad = \dfrac{36y^{1/2}}{m^{7/3}} \qquad$ Simplify.

(e) $m^{2/3}(m^{7/3} + 2m^{1/3})$ Do *not* multiply the exponents.

$\qquad = m^{2/3} \cdot m^{7/3} + m^{2/3} \cdot 2m^{1/3} \qquad$ Distributive property

$\qquad = m^{2/3+7/3} + 2m^{2/3+1/3} \qquad$ Product rule

$\qquad = m^3 + 2m \qquad$ Simplify the exponents.

✔ **Now Try Exercises 65, 67, 75, and 81.**

EXAMPLE 7   **Factoring Expressions with Negative or Rational Exponents**

Factor out the least power of the variable or variable expression. Assume all variables represent positive real numbers.

(a) $12x^{-2} - 8x^{-3}$ **(b)** $4m^{1/2} + 3m^{3/2}$

(c) $(y - 2)^{-1/3} + (y - 2)^{2/3}$

SOLUTION

(a) The least exponent of the variable x in $12x^{-2} - 8x^{-3}$ is -3. Because 4 is a common numerical factor, factor out $4x^{-3}$.

$\qquad 12x^{-2} - 8x^{-3}$

$\qquad = 4x^{-3}(3x^{-2-(-3)} - 2x^{-3-(-3)}) \qquad$ Factor.

$\qquad = 4x^{-3}(3x - 2) \qquad$ Simplify the exponents.

$\qquad$ CHECK $4x^{-3}(3x - 2) = 12x^{-2} - 8x^{-3}$ ✓ Multiply.

(b) $4m^{1/2} + 3m^{3/2}$

$\qquad = m^{1/2}(4 + 3m) \qquad$ The least exponent is $1/2$. Factor out $m^{1/2}$.

(c) $(y - 2)^{-1/3} + (y - 2)^{2/3}$

$\qquad = (y - 2)^{-1/3}[1 + (y - 2)] \qquad$ Factor out $(y - 2)^{-1/3}$.

$\qquad = (y - 2)^{-1/3}(y - 1) \qquad$ Simplify.

✔ **Now Try Exercises 89, 95, and 99.**

LOOKING AHEAD TO CALCULUS

The technique of **Example 7(c)** is used often in calculus.

Complex Fractions Revisited   Negative exponents are sometimes used to write complex fractions. Recall that complex fractions are simplified either by first multiplying the numerator and denominator by the LCD of all the denominators, or by performing any indicated operations in the numerator and the denominator and then using the definition of division for fractions.

EXAMPLE 8 Simplifying a Fraction with Negative Exponents

Simplify $\dfrac{(x + y)^{-1}}{x^{-1} + y^{-1}}$. Write the result with only positive exponents.

SOLUTION　$\dfrac{(x + y)^{-1}}{x^{-1} + y^{-1}}$

$$= \dfrac{\dfrac{1}{x + y}}{\dfrac{1}{x} + \dfrac{1}{y}} \qquad \text{Definition of negative exponent}$$

$$= \dfrac{\dfrac{1}{x + y}}{\dfrac{y + x}{xy}} \qquad \text{Add fractions in the denominator.}$$

$$= \dfrac{1}{x + y} \cdot \dfrac{xy}{x + y} \qquad \begin{array}{l}\text{Multiply by the reciprocal of the}\\ \text{denominator of the complex fraction.}\end{array}$$

$$= \dfrac{xy}{(x + y)^2} \qquad \text{Multiply fractions.}$$

✔ **Now Try Exercise 105.**

CAUTION Remember that if $r \neq 1$, then $(x + y)^r \neq x^r + y^r$. In particular, this means that $(x + y)^{-1} \neq x^{-1} + y^{-1}$.

R.6 Exercises

CONCEPT PREVIEW *Decide whether each statement is true or false. If false, correct the right side of the equation.*

1. $5^{-2} = \dfrac{1}{5^2}$

2. $\left(\dfrac{2}{3}\right)^{-2} = \left(\dfrac{3}{2}\right)^2$

3. $\dfrac{a^6}{a^4} = a^{-2}$

4. $(3x^2)^{-1} = 3x^{-2}$

5. $(x + y)^{-1} = x^{-1} + y^{-1}$

6. $m^{2/3} \cdot m^{1/3} = m^{2/9}$

CONCEPT PREVIEW *Match each expression in Column I with its equivalent expression in Column II. Choices may be used once, more than once, or not at all.*

	I	II		I	II
7.	**(a)** 4^{-2}	**A.** 16	**8.**	**(a)** 5^{-3}	**A.** 125
	(b) -4^{-2}	**B.** $\dfrac{1}{16}$		**(b)** -5^{-3}	**B.** -125
	(c) $(-4)^{-2}$	**C.** -16		**(c)** $(-5)^{-3}$	**C.** $\dfrac{1}{125}$
	(d) $-(-4)^{-2}$	**D.** $-\dfrac{1}{16}$		**(d)** $-(-5)^{-3}$	**D.** $-\dfrac{1}{125}$

I **II**

9. (a) $\left(\dfrac{4}{9}\right)^{3/2}$ 10. (a) $\left(\dfrac{8}{27}\right)^{2/3}$ **A.** $\dfrac{9}{4}$ **B.** $-\dfrac{9}{4}$

(b) $\left(\dfrac{4}{9}\right)^{-3/2}$ (b) $\left(\dfrac{8}{27}\right)^{-2/3}$ **C.** $-\dfrac{4}{9}$ **D.** $\dfrac{4}{9}$

(c) $-\left(\dfrac{9}{4}\right)^{3/2}$ (c) $-\left(\dfrac{27}{8}\right)^{2/3}$ **E.** $\dfrac{8}{27}$ **F.** $-\dfrac{27}{8}$

(d) $-\left(\dfrac{4}{9}\right)^{-3/2}$ (d) $-\left(\dfrac{27}{8}\right)^{-2/3}$ **G.** $\dfrac{27}{8}$ **H.** $-\dfrac{8}{27}$

Write each expression without negative exponents, and evaluate if possible. Assume all variables represent nonzero real numbers. ***See Example 1.***

11. $(-4)^{-3}$ **12.** $(-5)^{-2}$ **13.** -5^{-4}

14. -7^{-2} **15.** $\left(\dfrac{1}{3}\right)^{-2}$ **16.** $\left(\dfrac{4}{3}\right)^{-3}$

17. $(4x)^{-2}$ **18.** $(5t)^{-3}$ **19.** $4x^{-2}$

20. $5t^{-3}$ **21.** $-a^{-3}$ **22.** $-b^{-4}$

Simplify each expression. Write answers without negative exponents. Assume all variables represent nonzero real numbers. ***See Examples 2 and 3.***

23. $\dfrac{4^8}{4^6}$ **24.** $\dfrac{5^9}{5^7}$ **25.** $\dfrac{x^{12}}{x^8}$ **26.** $\dfrac{y^{14}}{y^{10}}$

27. $\dfrac{r^7}{r^{10}}$ **28.** $\dfrac{y^8}{y^{12}}$ **29.** $\dfrac{6^4}{6^{-2}}$ **30.** $\dfrac{7^5}{7^{-3}}$

31. $\dfrac{4r^{-3}}{6r^{-6}}$ **32.** $\dfrac{15s^{-4}}{5s^{-8}}$ **33.** $\dfrac{16m^{-5}n^4}{12m^2n^{-3}}$ **34.** $\dfrac{15a^{-5}b^{-1}}{25a^{-2}b^4}$

35. $-4r^{-2}(r^4)^2$ **36.** $-2m^{-1}(m^3)^2$ **37.** $(5a^{-1})^4(a^2)^{-3}$ **38.** $(3p^{-4})^2(p^3)^{-1}$

39. $\dfrac{(p^{-2})^0}{5p^{-4}}$ **40.** $\dfrac{(m^4)^0}{9m^{-3}}$ **41.** $\dfrac{(3pq)q^2}{6p^2q^4}$ **42.** $\dfrac{(-8xy)y^3}{4x^5y^4}$

43. $\dfrac{4a^5(a^{-1})^3}{(a^{-2})^{-2}}$ **44.** $\dfrac{12k^{-2}(k^{-3})^{-4}}{6k^5}$ **45.** $\dfrac{(5x)^{-2}(5x^3)^{-3}}{(5^{-2}x^{-3})^3}$ **46.** $\dfrac{(8y^2)^{-4}(8y^5)^{-2}}{(8^{-3}y^{-4})^2}$

Evaluate each expression. ***See Example 4.***

47. $169^{1/2}$ **48.** $121^{1/2}$ **49.** $16^{1/4}$ **50.** $625^{1/4}$

51. $\left(-\dfrac{64}{27}\right)^{1/3}$ **52.** $\left(-\dfrac{8}{27}\right)^{1/3}$ **53.** $(-4)^{1/2}$ **54.** $(-64)^{1/4}$

Simplify each expression. Write answers without negative exponents. Assume all variables represent positive real numbers. ***See Examples 5 and 6.***

55. $8^{2/3}$ **56.** $27^{4/3}$ **57.** $100^{3/2}$

58. $64^{3/2}$ **59.** $-81^{3/4}$ **60.** $(-32)^{-4/5}$

61. $\left(\dfrac{27}{64}\right)^{-4/3}$ **62.** $\left(\dfrac{121}{100}\right)^{-3/2}$ **63.** $3^{1/2} \cdot 3^{3/2}$

64. $6^{4/3} \cdot 6^{2/3}$ **65.** $\dfrac{64^{5/3}}{64^{4/3}}$ **66.** $\dfrac{125^{7/3}}{125^{5/3}}$

67. $y^{7/3} \cdot y^{-4/3}$

68. $r^{-8/9} \cdot r^{17/9}$

69. $\dfrac{k^{1/3}}{k^{2/3} \cdot k^{-1}}$

70. $\dfrac{z^{3/4}}{z^{5/4} \cdot z^{-2}}$

71. $\dfrac{(x^{1/4}y^{2/5})^{20}}{x^2}$

72. $\dfrac{(r^{1/5}s^{2/3})^{15}}{r^2}$

73. $\dfrac{(x^{2/3})^2}{(x^2)^{7/3}}$

74. $\dfrac{(p^3)^{1/4}}{(p^{5/4})^2}$

75. $\left(\dfrac{16m^3}{n}\right)^{1/4}\left(\dfrac{9n^{-1}}{m^2}\right)^{1/2}$

76. $\left(\dfrac{25^4a^3}{b^2}\right)^{1/8}\left(\dfrac{4^2b^{-5}}{a^2}\right)^{1/4}$

77. $\dfrac{p^{1/5}p^{7/10}p^{1/2}}{(p^3)^{-1/5}}$

78. $\dfrac{z^{1/3}z^{-2/3}z^{1/6}}{(z^{-1/6})^3}$

(Modeling) Solve each problem.

79. *Holding Time of Athletes* A group of ten athletes were tested for isometric endurance by measuring the length of time they could resist a load pulling on their legs while seated. The approximate amount of time (called the **holding time**) that they could resist the load was given by the formula

$$t = 31{,}293w^{-1.5},$$

where w is the weight of the load in pounds and the holding time t is measured in seconds. (*Source:* Townend, M. Stewart, *Mathematics in Sport*, Chichester, Ellis Horwood Limited.)

(a) Determine the holding time, to the nearest second, for a load of 25 lb.

(b) When the weight of the load is doubled, by what factor is the holding time changed?

80. *Duration of a Storm* Suppose that meteorologists approximate the duration of a particular storm by using the formula

$$T = 0.07D^{3/2},$$

where T is the time (in hours) that a storm of diameter D (in miles) lasts.

(a) The National Weather Service reports that a storm 4 mi in diameter is headed toward New Haven. How many minutes is the storm expected to last?

(b) A thunderstorm is predicted for a farming community. The crops need at least 1.5 hr of rain. Local radar shows that the storm is 7 mi in diameter. Will it rain long enough to meet the farmers' need?

*Find each product. Assume all variables represent positive real numbers. **See Example 6(e).***

81. $y^{5/8}(y^{3/8} - 10y^{11/8})$

82. $p^{11/5}(3p^{4/5} + 9p^{19/5})$

83. $-4k(k^{7/3} - 6k^{1/3})$

84. $-5y(3y^{9/10} + 4y^{3/10})$

85. $(x + x^{1/2})(x - x^{1/2})$

86. $(2z^{1/2} + z)(z^{1/2} - z)$

87. $(r^{1/2} - r^{-1/2})^2$

88. $(p^{1/2} - p^{-1/2})(p^{1/2} + p^{-1/2})$

*Factor out the least power of the variable or variable expression. Assume all variables represent positive real numbers. **See Example 7.***

89. $4k^{-1} + k^{-2}$

90. $y^{-5} - 3y^{-3}$

91. $4t^{-2} + 8t^{-4}$

92. $5r^{-6} - 10r^{-8}$

93. $9z^{-1/2} + 2z^{1/2}$

94. $3m^{2/3} - 4m^{-1/3}$

95. $p^{-3/4} - 2p^{-7/4}$ **96.** $6r^{-2/3} - 5r^{-5/3}$

97. $-4a^{-2/5} + 16a^{-7/5}$ **98.** $-3p^{-3/4} - 30p^{-7/4}$

99. $(p + 4)^{-3/2} + (p + 4)^{-1/2} + (p + 4)^{1/2}$

100. $(3r + 1)^{-2/3} + (3r + 1)^{1/3} + (3r + 1)^{4/3}$

101. $2(3x + 1)^{-3/2} + 4(3x + 1)^{-1/2} + 6(3x + 1)^{1/2}$

102. $7(5t + 3)^{-5/3} + 14(5t + 3)^{-2/3} - 21(5t + 3)^{1/3}$

103. $4x(2x + 3)^{-5/9} + 6x^2(2x + 3)^{4/9} - 8x^3(2x + 3)^{13/9}$

104. $6y^3(4y - 1)^{-3/7} - 8y^2(4y - 1)^{4/7} + 16y(4y - 1)^{11/7}$

Perform all indicated operations, and write each answer with positive integer exponents.
See Example 8.

105. $\dfrac{a^{-1} + b^{-1}}{(ab)^{-1}}$ **106.** $\dfrac{p^{-1} - q^{-1}}{(pq)^{-1}}$

107. $\dfrac{r^{-1} + q^{-1}}{r^{-1} - q^{-1}} \cdot \dfrac{r - q}{r + q}$ **108.** $\dfrac{x^{-2} + y^{-2}}{x^{-2} - y^{-2}} \cdot \dfrac{x + y}{x - y}$

109. $\dfrac{x - 9y^{-1}}{(x - 3y^{-1})(x + 3y^{-1})}$ **110.** $\dfrac{a - 16b^{-1}}{(a + 4b^{-1})(a - 4b^{-1})}$

Simplify each rational expression. Assume all variable expressions represent positive real numbers. (Hint: Use factoring and divide out any common factors as a first step.)

111. $\dfrac{(x^2 + 1)^4(2x) - x^2(4)(x^2 + 1)^3(2x)}{(x^2 + 1)^8}$ **112.** $\dfrac{(y^2 + 2)^5(3y) - y^3(6)(y^2 + 2)^4(3y)}{(y^2 + 2)^7}$

113. $\dfrac{4(x^2 - 1)^3 + 8x(x^2 - 1)^4}{16(x^2 - 1)^3}$ **114.** $\dfrac{10(4x^2 - 9)^2 - 25x(4x^2 - 9)^3}{15(4x^2 - 9)^6}$

115. $\dfrac{2(2x - 3)^{1/3} - (x - 1)(2x - 3)^{-2/3}}{(2x - 3)^{2/3}}$ **116.** $\dfrac{7(3t + 1)^{1/4} - (t - 1)(3t + 1)^{-3/4}}{(3t + 1)^{3/4}}$

Concept Check *Answer each question.*

117. If the lengths of the sides of a cube are tripled, by what factor will the volume change?

118. If the radius of a circle is doubled, by what factor will the area change?

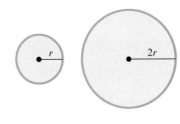

Concept Check *Calculate each value mentally.*

119. $0.2^{2/3} \cdot 40^{2/3}$ **120.** $0.1^{3/2} \cdot 90^{3/2}$ **121.** $\dfrac{2^{2/3}}{2000^{2/3}}$ **122.** $\dfrac{20^{3/2}}{5^{3/2}}$

R.7 Radical Expressions

- Radical Notation
- Simplified Radicals
- Operations with Radicals
- Rationalizing Denominators

Radical Notation Previously, we used rational exponents to express roots. An alternative notation for roots is **radical notation.**

Radical Notation for $a^{1/n}$

Let a be a real number, n be a positive integer, and $a^{1/n}$ be a real number.

$$\sqrt[n]{a} = a^{1/n}$$

Radical Notation for $a^{m/n}$

Let a be a real number, m be an integer, n be a positive integer, and $\sqrt[n]{a}$ be a real number.

$$a^{m/n} = \left(\sqrt[n]{a}\right)^m = \sqrt[n]{a^m}$$

We use the familiar notation $\sqrt{a}$ instead of $\sqrt[2]{a}$ for the square root. For even values of n (square roots, fourth roots, and so on), when a is positive, there are two nth roots, one positive and one negative.

$\sqrt[n]{a}$ represents the positive root, the **principal nth root.**

$-\sqrt[n]{a}$ represents the **negative root.**

EXAMPLE 1 Evaluating Roots

Write each root using exponents and evaluate.

(a) $\sqrt[4]{16}$ **(b)** $-\sqrt[4]{16}$ **(c)** $\sqrt[5]{-32}$

(d) $\sqrt[3]{1000}$ **(e)** $\sqrt[6]{\dfrac{64}{729}}$ **(f)** $\sqrt[4]{-16}$

SOLUTION

(a) $\sqrt[4]{16} = 16^{1/4} = 2$ **(b)** $-\sqrt[4]{16} = -16^{1/4} = -2$

(c) $\sqrt[5]{-32} = (-32)^{1/5} = -2$ **(d)** $\sqrt[3]{1000} = 1000^{1/3} = 10$

(e) $\sqrt[6]{\dfrac{64}{729}} = \left(\dfrac{64}{729}\right)^{1/6} = \dfrac{2}{3}$ **(f)** $\sqrt[4]{-16}$ is not a real number.

✔ **Now Try Exercises 11, 13, 17, and 21.**

EXAMPLE 2 **Converting from Rational Exponents to Radicals**

Write in radical form and simplify. Assume all variable expressions represent positive real numbers.

(a) $8^{2/3}$ **(b)** $(-32)^{4/5}$ **(c)** $-16^{3/4}$ **(d)** $x^{5/6}$

(e) $3x^{2/3}$ **(f)** $2p^{1/2}$ **(g)** $(3a+b)^{1/4}$

SOLUTION

(a) $8^{2/3} = \left(\sqrt[3]{8}\right)^2 = 2^2 = 4$ **(b)** $(-32)^{4/5} = \left(\sqrt[5]{-32}\right)^4 = (-2)^4 = 16$

(c) $-16^{3/4} = -\left(\sqrt[4]{16}\right)^3 = -(2)^3 = -8$

(d) $x^{5/6} = \sqrt[6]{x^5}$ **(e)** $3x^{2/3} = 3\sqrt[3]{x^2}$

(f) $2p^{1/2} = 2\sqrt{p}$ **(g)** $(3a+b)^{1/4} = \sqrt[4]{3a+b}$

✔ **Now Try Exercises 23 and 25.**

LOOKING AHEAD TO CALCULUS

In calculus, the "power rule" for derivatives requires converting radicals to rational exponents.

CAUTION It is not possible to "distribute" exponents over a sum, so in **Example 2(g),** $(3a+b)^{1/4}$ *cannot be written as* $(3a)^{1/4} + b^{1/4}$.

$$\sqrt[n]{x^n + y^n} \ \textit{ is not equivalent to } \ x + y.$$

(For example, let $n = 2$, $x = 3$, and $y = 4$ to see this.)

EXAMPLE 3 **Converting from Radicals to Rational Exponents**

Write in exponential form. Assume all variable expressions represent positive real numbers.

(a) $\sqrt[4]{x^5}$ **(b)** $\sqrt{3y}$ **(c)** $10\left(\sqrt[5]{z}\right)^2$

(d) $5\sqrt[3]{(2x^4)^7}$ **(e)** $\sqrt{p^2 + q}$

SOLUTION

(a) $\sqrt[4]{x^5} = x^{5/4}$ **(b)** $\sqrt{3y} = (3y)^{1/2}$ **(c)** $10\left(\sqrt[5]{z}\right)^2 = 10z^{2/5}$

(d) $5\sqrt[3]{(2x^4)^7} = 5(2x^4)^{7/3} = 5 \cdot 2^{7/3}x^{28/3}$

(e) $\sqrt{p^2 + q} = (p^2 + q)^{1/2}$ ✔ **Now Try Exercises 27 and 29.**

We *cannot* simply write $\sqrt{x^2} = x$ for all real numbers x. For example, what if x represents a negative number such as -5?

$$\sqrt{x^2} = \sqrt{(-5)^2} = \sqrt{25} = 5 \neq x$$

To take care of the fact that a negative value of x can produce a positive result, we use absolute value. For any real number a, the following holds.

$$\sqrt{a^2} = |a|$$

Examples: $\sqrt{(-9)^2} = |-9| = 9$ and $\sqrt{13^2} = |13| = 13$

We can generalize this result to any *even* nth root.

Evaluating $\sqrt[n]{a^n}$

If n is an *even* positive integer, then $\sqrt[n]{a^n} = |a|$.

If n is an *odd* positive integer, then $\sqrt[n]{a^n} = a$.

EXAMPLE 4 **Using Absolute Value to Simplify Roots**

Simplify.

(a) $\sqrt{p^4}$ **(b)** $\sqrt[4]{p^4}$ **(c)** $\sqrt{16m^8 r^6}$

(d) $\sqrt[6]{(-2)^6}$ **(e)** $\sqrt[5]{m^5}$ **(f)** $\sqrt{(2k+3)^2}$

(g) $\sqrt{x^2 - 4x + 4}$

SOLUTION

(a) $\sqrt{p^4} = \sqrt{(p^2)^2} = |p^2| = p^2$ **(b)** $\sqrt[4]{p^4} = |p|$

(c) $\sqrt{16m^8 r^6} = |4m^4 r^3| = 4m^4 |r^3|$ **(d)** $\sqrt[6]{(-2)^6} = |-2| = 2$

(e) $\sqrt[5]{m^5} = m$ **(f)** $\sqrt{(2k+3)^2} = |2k+3|$

(g) $\sqrt{x^2 - 4x + 4} = \sqrt{(x-2)^2} = |x-2|$

✔ **Now Try Exercises 35, 37, and 39.**

NOTE When working with variable radicands, we will *usually* assume that all variables in radicands represent only nonnegative real numbers.

The following rules for working with radicals are simply the power rules for exponents written in radical notation.

Rules for Radicals

Suppose that a and b represent real numbers, and m and n represent positive integers for which the indicated roots are real numbers.

Rule	Description
Product rule $\sqrt[n]{a} \cdot \sqrt[n]{b} = \sqrt[n]{ab}$	The product of two roots is the root of the product.
Quotient rule $\sqrt[n]{\dfrac{a}{b}} = \dfrac{\sqrt[n]{a}}{\sqrt[n]{b}} \quad (b \neq 0)$	The root of a quotient is the quotient of the roots.
Power rule $\sqrt[m]{\sqrt[n]{a}} = \sqrt[mn]{a}$	The index of the root of a root is the product of their indexes.

EXAMPLE 5 **Simplifying Radical Expressions**

Simplify. Assume all variable expressions represent positive real numbers.

(a) $\sqrt{6} \cdot \sqrt{54}$

(b) $\sqrt[3]{m} \cdot \sqrt[3]{m^2}$

(c) $\sqrt{\dfrac{7}{64}}$

(d) $\sqrt[4]{\dfrac{a}{b^4}}$

(e) $\sqrt[7]{\sqrt[3]{2}}$

(f) $\sqrt[4]{\sqrt{3}}$

SOLUTION

(a) $\sqrt{6} \cdot \sqrt{54}$

$= \sqrt{6 \cdot 54}$ Product rule

$= \sqrt{324}$ Multiply.

$= 18$

(b) $\sqrt[3]{m} \cdot \sqrt[3]{m^2}$

$= \sqrt[3]{m^3}$

$= m$

(c) $\sqrt{\dfrac{7}{64}} = \dfrac{\sqrt{7}}{\sqrt{64}} = \dfrac{\sqrt{7}}{8}$ Quotient rule

(d) $\sqrt[4]{\dfrac{a}{b^4}} = \dfrac{\sqrt[4]{a}}{\sqrt[4]{b^4}} = \dfrac{\sqrt[4]{a}}{b}$

(e) $\sqrt[7]{\sqrt[3]{2}} = \sqrt[21]{2}$ Power rule

(f) $\sqrt[4]{\sqrt{3}} = \sqrt[4 \cdot 2]{3} = \sqrt[8]{3}$

✔ **Now Try Exercises 45, 49, 53, and 73.**

NOTE   Converting to rational exponents shows why these rules work.

$$\sqrt[7]{\sqrt[3]{2}} = (2^{1/3})^{1/7} = 2^{1/3(1/7)} = 2^{1/21} = \sqrt[21]{2} \quad \text{Example 5(e)}$$

$$\sqrt[4]{\sqrt{3}} = (3^{1/2})^{1/4} = 3^{1/2(1/4)} = 3^{1/8} = \sqrt[8]{3} \quad \text{Example 5(f)}$$

Simplified Radicals   In working with numbers, we generally prefer to write a number in its simplest form. For example, $\dfrac{10}{2}$ is written as 5, and $-\dfrac{9}{6}$ is written as $-\dfrac{3}{2}$. Similarly, expressions with radicals can be written in their simplest forms.

Simplified Radicals

An expression with radicals is simplified when all of the following conditions are satisfied.

1. The radicand has no factor raised to a power greater than or equal to the index.

2. The radicand has no fractions.

3. No denominator contains a radical.

4. Exponents in the radicand and the index of the radical have greatest common factor 1.

5. All indicated operations have been performed (if possible).

EXAMPLE 6 Simplifying Radicals

Simplify each radical.

(a) $\sqrt{175}$ **(b)** $-3\sqrt[5]{32}$ **(c)** $\sqrt[3]{81x^5y^7z^6}$

SOLUTION

(a) $\sqrt{175}$

$= \sqrt{25 \cdot 7}$ Factor.

$= \sqrt{25} \cdot \sqrt{7}$ Product rule

$= 5\sqrt{7}$ Square root

(b) $-3\sqrt[5]{32}$

$= -3\sqrt[5]{2^5}$ Exponential form

$= -3 \cdot 2$ $\sqrt[n]{a^n} = a$ if n is odd.

$= -6$ Multiply.

(c) $\sqrt[3]{81x^5y^7z^6}$

$= \sqrt[3]{27 \cdot 3 \cdot x^3 \cdot x^2 \cdot y^6 \cdot y \cdot z^6}$ Factor.

$= \sqrt[3]{27x^3y^6z^6(3x^2y)}$ Group all perfect cubes.

$= 3xy^2z^2\sqrt[3]{3x^2y}$ Remove all perfect cubes from the radical.

✔ **Now Try Exercises 41 and 59.**

Operations with Radicals   Radicals with the same radicand and the same index, such as $3\sqrt[4]{11pq}$ and $-7\sqrt[4]{11pq}$, are **like radicals.** On the other hand, examples of **unlike radicals** are as follows.

$2\sqrt{5}$ and $2\sqrt{3}$ Radicands are different.

$2\sqrt{3}$ and $2\sqrt[3]{3}$ Indexes are different.

We add or subtract like radicals using the distributive property. *Only like radicals can be combined.*

EXAMPLE 7 Adding and Subtracting Radicals

Add or subtract, as indicated. Assume all variables represent positive real numbers.

(a) $3\sqrt[4]{11pq} - 7\sqrt[4]{11pq}$ **(b)** $\sqrt{98x^3y} + 3x\sqrt{32xy}$

(c) $\sqrt[3]{64m^4n^5} - \sqrt[3]{-27m^{10}n^{14}}$

SOLUTION

(a) $3\sqrt[4]{11pq} - 7\sqrt[4]{11pq}$

$= (3 - 7)\sqrt[4]{11pq}$ Distributive property

$= -4\sqrt[4]{11pq}$ Subtract.

(b) $\sqrt{98x^3y} + 3x\sqrt{32xy}$

$= \sqrt{49 \cdot 2 \cdot x^2 \cdot x \cdot y} + 3x\sqrt{16 \cdot 2 \cdot x \cdot y}$ Factor.

$= 7x\sqrt{2xy} + 3x(4)\sqrt{2xy}$ Remove all perfect squares from the radicals.

$= 7x\sqrt{2xy} + 12x\sqrt{2xy}$ Multiply.

$= (7x + 12x)\sqrt{2xy}$ Distributive property

$= 19x\sqrt{2xy}$ Add.

(c) $\sqrt[3]{64m^4n^5} - \sqrt[3]{-27m^{10}n^{14}}$

$\quad = \sqrt[3]{64m^3n^3(mn^2)} - \sqrt[3]{-27m^9n^{12}(mn^2)}$ Factor.

$\quad = 4mn\sqrt[3]{mn^2} - (-3m^3n^4)\sqrt[3]{mn^2}$ Remove all perfect cubes from the radicals.

$\quad = 4mn\sqrt[3]{mn^2} + 3m^3n^4\sqrt[3]{mn^2}$ $-(-a) = a$

$\quad = (4mn + 3m^3n^4)\sqrt[3]{mn^2}$ ── This *cannot* be simplified further.

✔ **Now Try Exercises 77 and 81.**

CAUTION The terms $4mn\sqrt[3]{mn^2}$ and $3m^3n^4\sqrt[3]{mn^2}$ in **Example 7(c)** are rewritten in the final line using the distributive property.

If the index of the radical and an exponent in the radicand have a common factor, we can simplify the radical by first writing it in exponential form. We simplify the rational exponent, and then write the result as a radical again, as shown in **Example 8**.

EXAMPLE 8 **Simplifying Radicals**

Simplify each radical. Assume all variables represent positive real numbers.

(a) $\sqrt[6]{3^2}$ **(b)** $\sqrt[6]{x^{12}y^3}$ **(c)** $\sqrt[9]{\sqrt{6^3}}$

SOLUTION

(a) $\sqrt[6]{3^2}$

$\quad = 3^{2/6}$

$\quad = 3^{1/3}$

$\quad = \sqrt[3]{3}$

(b) $\sqrt[6]{x^{12}y^3}$

$\quad = (x^{12}y^3)^{1/6}$

$\quad = x^{12/6}y^{3/6}$

$\quad = x^2y^{1/2}$

$\quad = x^2\sqrt{y}$

(c) $\sqrt[9]{\sqrt{6^3}}$

$\quad = \sqrt[9]{6^{3/2}}$

$\quad = (6^{3/2})^{1/9}$

$\quad = 6^{1/6}$

$\quad = \sqrt[6]{6}$

✔ **Now Try Exercises 71 and 75.**

In **Example 8(a)**, we simplified $\sqrt[6]{3^2}$ as $\sqrt[3]{3}$. However, to simplify $\left(\sqrt[6]{x}\right)^2$, the variable x must represent a nonnegative number. For example, consider the statement

$$(-8)^{2/6} = [(-8)^{1/6}]^2.$$

This result is not a real number because $(-8)^{1/6}$ is not a real number. On the other hand,

$$(-8)^{1/3} = -2.$$

Here, even though $\frac{2}{6} = \frac{1}{3}$,

$$\left(\sqrt[6]{x}\right)^2 \neq \sqrt[3]{x}.$$

If a is nonnegative, then it is always true that $a^{m/n} = a^{(mp)/(np)}$. Simplifying rational exponents on negative bases should be considered case by case.

EXAMPLE 9 **Multiplying Radical Expressions**

Find each product.

(a) $\left(\sqrt{7} - \sqrt{10}\right)\left(\sqrt{7} + \sqrt{10}\right)$ **(b)** $\left(\sqrt{2} + 3\right)\left(\sqrt{8} - 5\right)$

SOLUTION

(a) $\left(\sqrt{7} - \sqrt{10}\right)\left(\sqrt{7} + \sqrt{10}\right)$

$= \left(\sqrt{7}\right)^2 - \left(\sqrt{10}\right)^2$ Product of the sum and difference of two terms

$= 7 - 10$ $\left(\sqrt{a}\right)^2 = a$

$= -3$ Subtract.

(b) $\left(\sqrt{2} + 3\right)\left(\sqrt{8} - 5\right)$

$= \sqrt{2}\left(\sqrt{8}\right) - \sqrt{2}(5) + 3\sqrt{8} - 3(5)$ FOIL method

$= \sqrt{16} - 5\sqrt{2} + 3\left(2\sqrt{2}\right) - 15$ Multiply; $\sqrt{8} = 2\sqrt{2}$.

$= 4 - 5\sqrt{2} + 6\sqrt{2} - 15$ Simplify.

$= -11 + \sqrt{2}$ Combine like terms.

✔ **Now Try Exercises 85 and 91.**

Rationalizing Denominators   Condition 3 for a simplified radical requires that no denominator contain a radical. We achieve this by **rationalizing the denominator**—that is, multiplying by a form of 1.

EXAMPLE 10 **Rationalizing Denominators**

Rationalize each denominator.

(a) $\dfrac{4}{\sqrt{3}}$ **(b)** $\sqrt[4]{\dfrac{3}{5}}$

SOLUTION

(a) $\dfrac{4}{\sqrt{3}} = \dfrac{4}{\sqrt{3}} \cdot \dfrac{\sqrt{3}}{\sqrt{3}} = \dfrac{4\sqrt{3}}{3}$ Multiply by $\frac{\sqrt{3}}{\sqrt{3}}$ (which equals 1). In the denominator, $\sqrt{a} \cdot \sqrt{a} = a$.

(b) $\sqrt[4]{\dfrac{3}{5}} = \dfrac{\sqrt[4]{3}}{\sqrt[4]{5}}$ Quotient rule

The denominator will be a rational number if it equals $\sqrt[4]{5^4}$. That is, four factors of 5 are needed under the radical. We multiply by $\dfrac{\sqrt[4]{5^3}}{\sqrt[4]{5^3}}$.

$\dfrac{\sqrt[4]{3}}{\sqrt[4]{5}}$ Because $\sqrt[4]{5}$ has just one factor of 5, three additional factors are needed.

$= \dfrac{\sqrt[4]{3} \cdot \sqrt[4]{5^3}}{\sqrt[4]{5} \cdot \sqrt[4]{5^3}}$ Multiply by $\frac{\sqrt[4]{5^3}}{\sqrt[4]{5^3}}$.

$= \dfrac{\sqrt[4]{3 \cdot 5^3}}{\sqrt[4]{5^4}}$ Product rule

$= \dfrac{\sqrt[4]{375}}{5}$ Simplify.

✔ **Now Try Exercises 63 and 67.**

EXAMPLE 11 Simplifying Radical Expressions with Fractions

Simplify each expression. Assume all variables represent positive real numbers.

(a) $\dfrac{\sqrt[4]{xy^3}}{\sqrt[4]{x^3y^2}}$

(b) $\sqrt[3]{\dfrac{5}{x^6}} - \sqrt[3]{\dfrac{4}{x^9}}$

SOLUTION

(a) $\dfrac{\sqrt[4]{xy^3}}{\sqrt[4]{x^3y^2}}$

$= \sqrt[4]{\dfrac{xy^3}{x^3y^2}}$ Quotient rule

$= \sqrt[4]{\dfrac{y}{x^2}}$ Simplify the radicand.

$= \dfrac{\sqrt[4]{y}}{\sqrt[4]{x^2}}$ Quotient rule

$= \dfrac{\sqrt[4]{y}}{\sqrt[4]{x^2}} \cdot \dfrac{\sqrt[4]{x^2}}{\sqrt[4]{x^2}}$ Rationalize the denominator.

$= \dfrac{\sqrt[4]{x^2y}}{x}$ $\sqrt[4]{x^2} \cdot \sqrt[4]{x^2} = \sqrt[4]{x^4} = x$

(b) $\sqrt[3]{\dfrac{5}{x^6}} - \sqrt[3]{\dfrac{4}{x^9}}$

$= \dfrac{\sqrt[3]{5}}{\sqrt[3]{x^6}} - \dfrac{\sqrt[3]{4}}{\sqrt[3]{x^9}}$ Quotient rule

$= \dfrac{\sqrt[3]{5}}{x^2} - \dfrac{\sqrt[3]{4}}{x^3}$ Simplify the denominators.

$= \dfrac{x\sqrt[3]{5}}{x^3} - \dfrac{\sqrt[3]{4}}{x^3}$ Write with a common denominator.

$= \dfrac{x\sqrt[3]{5} - \sqrt[3]{4}}{x^3}$ Subtract the numerators.

✔ **Now Try Exercises 93 and 95.**

In **Example 9(a),** we saw that the product

$$\left(\sqrt{7} - \sqrt{10}\right)\left(\sqrt{7} + \sqrt{10}\right) \quad \text{equals} \quad -3, \quad \text{a rational number.}$$

This suggests a way to rationalize a denominator that is a binomial in which one or both terms is a square root radical. The expressions $a - b$ and $a + b$ are **conjugates.**

EXAMPLE 12 Rationalizing a Binomial Denominator

Rationalize the denominator of $\dfrac{1}{1 - \sqrt{2}}$.

SOLUTION $\dfrac{1}{1 - \sqrt{2}}$ $\boxed{\dfrac{1 + \sqrt{2}}{1 + \sqrt{2}} = 1}$

$= \dfrac{1\left(1 + \sqrt{2}\right)}{\left(1 - \sqrt{2}\right)\left(1 + \sqrt{2}\right)}$ Multiply numerator and denominator by the conjugate of the denominator, $1 + \sqrt{2}$.

$= \dfrac{1 + \sqrt{2}}{1 - 2}$ $(x - y)(x + y) = x^2 - y^2$

$= \dfrac{1 + \sqrt{2}}{-1}$ Subtract.

$= -1 - \sqrt{2}$ Divide by -1.

✔ **Now Try Exercise 101.**

R.7 Exercises

CONCEPT PREVIEW *Work each problem.*

1. Write $\sqrt[3]{64}$ using exponents and evaluate.

2. Write $27^{2/3}$ in radical form and simplify.

CONCEPT PREVIEW *Match the rational exponent expression in Column I with the equivalent radical expression in Column II. Assume that x is not 0.*

I **II**

3. (a) $(-3x)^{1/3}$ **4. (a)** $-3x^{1/3}$ **A.** $\dfrac{3}{\sqrt[3]{x}}$ **B.** $-3\sqrt[3]{x}$

(b) $(-3x)^{-1/3}$ **(b)** $-3x^{-1/3}$ **C.** $\dfrac{1}{\sqrt[3]{3x}}$ **D.** $\dfrac{-3}{\sqrt[3]{x}}$

(c) $(3x)^{1/3}$ **(c)** $3x^{-1/3}$ **E.** $3\sqrt[3]{x}$ **F.** $\sqrt[3]{-3x}$

(d) $(3x)^{-1/3}$ **(d)** $3x^{1/3}$ **G.** $\sqrt[3]{3x}$ **H.** $\dfrac{1}{\sqrt[3]{-3x}}$

CONCEPT PREVIEW *Perform the operation and/or simplify each of the following. Assume all variables represent positive real numbers.*

5. $\sqrt[5]{t^5}$ **6.** $\sqrt{6}\cdot\sqrt{24}$ **7.** $\sqrt{50}$

8. $\sqrt{\dfrac{7}{36}}$ **9.** $3\sqrt{xy}-8\sqrt{xy}$ **10.** $\left(2+\sqrt{3}\right)\left(2-\sqrt{3}\right)$

Write each root using exponents and evaluate. **See Example 1.**

11. $\sqrt[3]{125}$ **12.** $\sqrt[3]{216}$ **13.** $\sqrt[4]{81}$ **14.** $\sqrt[4]{256}$

15. $\sqrt[3]{-125}$ **16.** $\sqrt[3]{-343}$ **17.** $\sqrt[4]{-81}$ **18.** $\sqrt[4]{-256}$

19. $\sqrt[5]{32}$ **20.** $\sqrt[7]{128}$ **21.** $-\sqrt[5]{-32}$ **22.** $-\sqrt[3]{-343}$

If the expression is in exponential form, write it in radical form. If it is in radical form, write it in exponential form. Assume all variables represent positive real numbers. **See Examples 2 and 3.**

23. $m^{2/3}$ **24.** $p^{5/4}$ **25.** $(2m+p)^{2/3}$ **26.** $(5r+3t)^{4/7}$

27. $\sqrt[5]{k^2}$ **28.** $\sqrt[4]{z^5}$ **29.** $-3\sqrt{5p^3}$ **30.** $-m\sqrt{2y^5}$

Concept Check *Answer each question.*

31. For which of the following cases is $\sqrt{ab}=\sqrt{a}\cdot\sqrt{b}$ a true statement?

   A. *a* and *b* both positive **B.** *a* and *b* both negative

32. For which positive integers *n* greater than or equal to 2 is $\sqrt[n]{a^n}=a$ always a true statement?

33. For what values of *x* is $\sqrt{9ax^2}=3x\sqrt{a}$ a true statement? Assume $a\ge 0$.

34. Which of the following expressions is *not* simplified? Give the simplified form.

   A. $\sqrt[3]{2y}$ **B.** $\dfrac{\sqrt{5}}{2}$ **C.** $\sqrt[4]{m^3}$ **D.** $\sqrt{\dfrac{3}{4}}$

*Simplify. **See Example 4.***

35. $\sqrt[4]{x^4}$

36. $\sqrt[6]{x^6}$

37. $\sqrt{25k^4m^2}$

38. $\sqrt[4]{81p^{12}q^4}$

39. $\sqrt{(4x-y)^2}$

40. $\sqrt[4]{(5+2m)^4}$

*Simplify each expression. Assume all variables represent positive real numbers. **See Examples 1, 4–6, and 8–11.***

41. $\sqrt[3]{81}$

42. $\sqrt[3]{250}$

43. $-\sqrt[4]{32}$

44. $-\sqrt[4]{243}$

45. $\sqrt{14}\cdot\sqrt{3pqr}$

46. $\sqrt{7}\cdot\sqrt{5xt}$

47. $\sqrt[3]{7x}\cdot\sqrt[3]{2y}$

48. $\sqrt[3]{9x}\cdot\sqrt[3]{4y}$

49. $-\sqrt{\dfrac{9}{25}}$

50. $-\sqrt{\dfrac{16}{49}}$

51. $-\sqrt[3]{\dfrac{5}{8}}$

52. $-\sqrt[4]{\dfrac{3}{16}}$

53. $\sqrt[4]{\dfrac{m}{n^4}}$

54. $\sqrt[6]{\dfrac{r}{s^6}}$

55. $3\sqrt[5]{-3125}$

56. $5\sqrt[3]{-343}$

57. $\sqrt[3]{16(-2)^4(2)^8}$

58. $\sqrt[3]{25(-3)^4(5)^3}$

59. $\sqrt{8x^5z^8}$

60. $\sqrt{24m^6n^5}$

61. $\sqrt[4]{x^4+y^4}$

62. $\sqrt[3]{27+a^3}$

63. $\sqrt{\dfrac{2}{3x}}$

64. $\sqrt{\dfrac{5}{3p}}$

65. $\sqrt{\dfrac{x^5y^3}{z^2}}$

66. $\sqrt{\dfrac{g^3h^5}{r^3}}$

67. $\sqrt[3]{\dfrac{8}{x^4}}$

68. $\sqrt[3]{\dfrac{9}{16p^4}}$

69. $\sqrt[4]{\dfrac{g^3h^5}{9r^6}}$

70. $\sqrt[4]{\dfrac{32x^5}{y^5}}$

71. $\sqrt[8]{3^4}$

72. $\sqrt[9]{5^3}$

73. $\sqrt[3]{\sqrt{4}}$

74. $\sqrt[4]{\sqrt{25}}$

75. $\sqrt[4]{\sqrt[3]{2}}$

76. $\sqrt[5]{\sqrt[3]{9}}$

*Perform the indicated operations. Assume all variables represent positive real numbers. **See Examples 7, 9, and 11.***

77. $8\sqrt{2x}-\sqrt{8x}+\sqrt{72x}$

78. $4\sqrt{18k}-\sqrt{72k}+\sqrt{50k}$

79. $2\sqrt[3]{3}+4\sqrt[3]{24}-\sqrt[3]{81}$

80. $\sqrt[3]{32}-5\sqrt[3]{4}+2\sqrt[3]{108}$

81. $\sqrt[4]{81x^6y^3}-\sqrt[4]{16x^{10}y^3}$

82. $\sqrt[4]{256x^5y^6}+\sqrt[4]{625x^9y^2}$

83. $5\sqrt{6}+2\sqrt{10}$

84. $3\sqrt{11}-5\sqrt{13}$

85. $\left(\sqrt{2}+3\right)\left(\sqrt{2}-3\right)$

86. $\left(\sqrt{5}+\sqrt{2}\right)\left(\sqrt{5}-\sqrt{2}\right)$

87. $\left(\sqrt[3]{11}-1\right)\left(\sqrt[3]{11^2}+\sqrt[3]{11}+1\right)$

88. $\left(\sqrt[3]{7}+3\right)\left(\sqrt[3]{7^2}-3\sqrt[3]{7}+9\right)$

89. $\left(\sqrt{3}+\sqrt{8}\right)^2$

90. $\left(\sqrt{5}+\sqrt{10}\right)^2$

91. $\left(3\sqrt{2}+\sqrt{3}\right)\left(2\sqrt{3}-\sqrt{2}\right)$

92. $\left(4\sqrt{5}+\sqrt{2}\right)\left(3\sqrt{2}-\sqrt{5}\right)$

93. $\dfrac{\sqrt[3]{mn}\cdot\sqrt[3]{m^2}}{\sqrt[3]{n^2}}$

94. $\dfrac{\sqrt[3]{8m^2n^3}\cdot\sqrt[3]{2m^2}}{\sqrt[3]{32m^4n^3}}$

95. $\sqrt[3]{\dfrac{2}{x^6}}-\sqrt[3]{\dfrac{5}{x^9}}$

96. $\sqrt[4]{\dfrac{7}{t^{12}}}+\sqrt[4]{\dfrac{9}{t^4}}$

97. $\dfrac{1}{\sqrt{2}}+\dfrac{3}{\sqrt{8}}+\dfrac{1}{\sqrt{32}}$

98. $\dfrac{2}{\sqrt{12}}-\dfrac{1}{\sqrt{27}}-\dfrac{5}{\sqrt{48}}$

99. $\dfrac{-4}{\sqrt[3]{3}}+\dfrac{1}{\sqrt[3]{24}}-\dfrac{2}{\sqrt[3]{81}}$

100. $\dfrac{5}{\sqrt[3]{2}}-\dfrac{2}{\sqrt[3]{16}}+\dfrac{1}{\sqrt[3]{54}}$

Rationalize each denominator. Assume all variables represent nonnegative numbers and that no denominators are 0. **See Example 12.**

101. $\dfrac{\sqrt{3}}{\sqrt{5} + \sqrt{3}}$

102. $\dfrac{\sqrt{7}}{\sqrt{3} - \sqrt{7}}$

103. $\dfrac{\sqrt{7} - 1}{2\sqrt{7} + 4\sqrt{2}}$

104. $\dfrac{1 + \sqrt{3}}{3\sqrt{5} + 2\sqrt{3}}$

105. $\dfrac{p - 4}{\sqrt{p} + 2}$

106. $\dfrac{9 - r}{3 - \sqrt{r}}$

107. $\dfrac{3m}{2 + \sqrt{m + n}}$

108. $\dfrac{a}{\sqrt{a + b} - 1}$

109. $\dfrac{5\sqrt{x}}{2\sqrt{x} + \sqrt{y}}$

110. *Concept Check* By what number should the numerator and denominator of

$$\frac{1}{\sqrt[3]{3} - \sqrt[3]{5}}$$

be multiplied in order to rationalize the denominator? Write this fraction with a rationalized denominator.

(Modeling) Solve each problem.

111. *Rowing Speed* Olympic rowing events have one-, two-, four-, or eight-person crews, with each person pulling a single oar. Increasing the size of the crew increases the speed of the boat. An analysis of Olympic rowing events concluded that the approximate speed, *s*, of the boat (in feet per second) was given by the formula

$$s = 15.18\sqrt[9]{n},$$

where *n* is the number of oarsmen. Estimate the speed of a boat with a four-person crew. (*Source:* Townend, M. Stewart, *Mathematics in Sport*, Chichester, Ellis Horwood Limited.)

112. *Rowing Speed* **See Exercise 111.** Estimate the speed of a boat with an eight-person crew.

(Modeling) Windchill *The National Weather Service has used the formula*

$$\text{Windchill temperature} = 35.74 + 0.6215T - 35.75V^{0.16} + 0.4275TV^{0.16},$$

where T is the temperature in °F and V is the wind speed in miles per hour, to calculate windchill. (Source: National Oceanic and Atmospheric Administration, National Weather Service.) Use the formula to calculate the windchill to the nearest degree given the following conditions.

113. 10°F, 30 mph wind

114. 30°F, 15 mph wind

Concept Check *Simplify each expression mentally.*

115. $\sqrt[4]{8} \cdot \sqrt[4]{2}$

116. $\dfrac{\sqrt[3]{54}}{\sqrt[3]{2}}$

117. $\dfrac{\sqrt[5]{320}}{\sqrt[5]{10}}$

118. $\sqrt{0.1} \cdot \sqrt{40}$

119. $\dfrac{\sqrt[3]{15}}{\sqrt[3]{5}} \cdot \sqrt[3]{9}$

120. $\sqrt[6]{2} \cdot \sqrt[6]{4} \cdot \sqrt[6]{8}$

*The screen in **Figure A** seems to indicate that π and $\sqrt[4]{\frac{2143}{22}}$ are exactly equal, since the eight decimal values given by the calculator agree. However, as shown in **Figure B** using one more decimal place in the display, they differ in the ninth decimal place. **The radical expression is a very good approximation for π, but it is still only an approximation.***

Figure A   **Figure B**

*Use your calculator to answer each question. Refer to the display for π in **Figure B**.*

121. The Chinese of the fifth century used $\frac{355}{113}$ as an approximation for π. How many decimal places of accuracy does this fraction give?

122. A value for π that the Greeks used circa 150 CE is equivalent to $\frac{377}{120}$. In which decimal place does this value first differ from π?

123. The Hindu mathematician Bhaskara used $\frac{3927}{1250}$ as an approximation for π circa 1150 CE. In which decimal place does this value first differ from π?

Chapter R Test Prep

Key Terms

R.1 set
elements (members)
infinite set
finite set
Venn diagram
disjoint sets
R.2 number line
coordinate
coordinate system
algebraic expression
exponential expression
(exponential)

base
exponent
absolute value
R.3 term
numerical coefficient
(coefficient)
like terms
polynomial
polynomial in x
degree of a term
degree of a
polynomial

trinomial
binomial
monomial
descending order
FOIL method
R.4 factoring
factored form
prime polynomial
factored completely
greatest common
factor (GCF)

R.5 rational
expression
domain of a
rational
expression
lowest terms
complex fraction
R.7 radicand
principal nth root
like radicals
unlike radicals
conjugates

New Symbols

{ }   set braces
$\in$ is an element of
$\notin$ is not an element of
$\{x \mid x \text{ has property } p\}$ set-builder notation
U universal set
$\varnothing$, or { }   null (empty) set
$\subseteq$ is a subset of
$\nsubseteq$ is not a subset of
A' complement of a set A

$\cap$ set intersection
$\cup$ set union
a^n n factors of a
$<$ is less than
$>$ is greater than
$\leq$ is less than or equal to
$\geq$ is greater than or equal to
$|a|$ absolute value of a
$\sqrt[n]{\ }$ radical symbol with index n

Quick Review

Concepts	Examples

R.1 Sets

Set Operations

For all sets A and B, with universal set U:

The **complement** of set A is the set A' of all elements in U that do *not* belong to set A.

$$A' = \{x \mid x \in U, \ x \notin A\}$$

The **intersection** of sets A and B, written $A \cap B$, is made up of all the elements belonging to both set A *and* set B.

$$A \cap B = \{x \mid x \in A \text{ and } x \in B\}$$

The **union** of sets A and B, written $A \cup B$, is made up of all the elements belonging to set A *or* set B.

$$A \cup B = \{x \mid x \in A \text{ or } x \in B\}$$

Examples:

Let $U = \{1, 2, 3, 4, 5, 6\}$, $A = \{1, 2, 3, 4\}$, and $B = \{3, 4, 6\}$.

$$A' = \{5, 6\}$$

$$A \cap B = \{3, 4\}$$

$$A \cup B = \{1, 2, 3, 4, 6\}$$

R.2 Real Numbers and Their Properties

Sets of Numbers

Natural numbers

$$\{1, 2, 3, 4, \ldots\}$$

$5, 17, 142$

Whole numbers

$$\{0, 1, 2, 3, 4, \ldots\}$$

$0, 27, 96$

Integers

$$\{\ldots, -3, -2, -1, 0, 1, 2, 3, \ldots\}$$

$-24, 0, 19$

Rational numbers

$$\left\{ \frac{p}{q} \,\middle|\, p \text{ and } q \text{ are integers and } q \neq 0 \right\}$$

$-\frac{3}{4}, -0.28, 0, 7, \frac{9}{16}, 0.66\overline{6}$

Irrational numbers

$$\{x \mid x \text{ is real but not rational}\}$$

$-\sqrt{15}, 0.101101110\ldots, \sqrt{2}, \pi$

Real numbers

$$\{x \mid x \text{ corresponds to a point on a number line}\}$$

$-46, 0.7, \pi, \sqrt{19}, \frac{8}{5}$

Properties of Real Numbers

For all real numbers a, b, and c, the following hold.

Closure Properties

$$a + b \text{ is a real number.}$$
$$ab \text{ is a real number.}$$

$1 + \sqrt{2}$ is a real number.
$3\sqrt{7}$ is a real number.

Commutative Properties

$$a + b = b + a$$
$$ab = ba$$

$5 + 18 = 18 + 5$
$-4 \cdot 8 = 8 \cdot (-4)$

Associative Properties

$$(a + b) + c = a + (b + c)$$
$$(ab)c = a(bc)$$

$[6 + (-3)] + 5 = 6 + (-3 + 5)$
$(7 \cdot 6)20 = 7(6 \cdot 20)$

Concepts	Examples

Identity Properties

There exists a unique real number 0 such that

$$a + 0 = a \quad \text{and} \quad 0 + a = a.$$

$$145 + 0 = 145 \quad \text{and} \quad 0 + 145 = 145$$

There exists a unique real number 1 such that

$$a \cdot 1 = a \quad \text{and} \quad 1 \cdot a = a.$$

$$-60 \cdot 1 = -60 \quad \text{and} \quad 1 \cdot (-60) = -60$$

Inverse Properties

There exists a unique real number $-a$ such that

$$a + (-a) = 0 \quad \text{and} \quad -a + a = 0.$$

$$17 + (-17) = 0 \quad \text{and} \quad -17 + 17 = 0$$

If $a \neq 0$, there exists a unique real number $\frac{1}{a}$ such that

$$a \cdot \frac{1}{a} = 1 \quad \text{and} \quad \frac{1}{a} \cdot a = 1.$$

$$22 \cdot \frac{1}{22} = 1 \quad \text{and} \quad \frac{1}{22} \cdot 22 = 1$$

Distributive Properties

$$a(b + c) = ab + ac$$
$$a(b - c) = ab - ac$$

$$3(5 + 8) = 3 \cdot 5 + 3 \cdot 8$$
$$6(4 - 2) = 6 \cdot 4 - 6 \cdot 2$$

Multiplication Property of Zero

$$0 \cdot a = a \cdot 0 = 0$$

$$0 \cdot 4 = 4 \cdot 0 = 0$$

Order

$a > b$ if a is to the right of b on a number line.

$a < b$ if a is to the left of b on a number line.

$$7 > -5$$
$$0 < 15$$

Absolute Value

$$|a| = \begin{cases} a & \text{if } a \geq 0 \\ -a & \text{if } a < 0 \end{cases}$$

$$|3| = 3 \quad \text{and} \quad |-3| = 3$$

R.3 Polynomials

Operations

To add or subtract polynomials, add or subtract the coefficients of like terms.

$$(2x^2 + 3x + 1) - (x^2 - x + 2)$$
$$= (2 - 1)x^2 + (3 + 1)x + (1 - 2)$$
$$= x^2 + 4x - 1$$

To multiply polynomials, distribute each term of the first polynomial, multiplying by each term of the second polynomial.

$$(x - 5)(x^2 + 5x + 25)$$
$$= x^3 + 5x^2 + 25x - 5x^2 - 25x - 125$$
$$= x^3 - 125$$

To divide polynomials when the divisor has two or more terms, use a process of long division similar to that for dividing whole numbers.

$$
\begin{array}{r}
4x^2 - 10x + 21 \\
x + 2 \overline{)4x^3 - 2x^2 + x - 1} \\
\underline{4x^3 + 8x^2} \\
-10x^2 + x \\
\underline{-10x^2 - 20x} \\
21x - 1 \\
\underline{21x + 42} \\
-43 \leftarrow \text{Remainder}
\end{array}
$$

$$\frac{4x^3 - 2x^2 + x - 1}{x + 2} = 4x^2 - 10x + 21 + \frac{-43}{x + 2}$$

Concepts	Examples

Special Products

Product of the Sum and Difference of Two Terms

$$(x + y)(x - y) = x^2 - y^2$$

Square of a Binomial

$$(x + y)^2 = x^2 + 2xy + y^2$$

$$(x - y)^2 = x^2 - 2xy + y^2$$

$$(7 - x)(7 + x) = 7^2 - x^2$$
$$= 49 - x^2$$

$$(3a + b)^2 = (3a)^2 + 2(3a)(b) + b^2$$
$$= 9a^2 + 6ab + b^2$$

$$(2m - 5)^2 = (2m)^2 - 2(2m)(5) + 5^2$$
$$= 4m^2 - 20m + 25$$

R.4 Factoring Polynomials

Factoring Patterns

Perfect Square Trinomial

$$x^2 + 2xy + y^2 = (x + y)^2$$
$$x^2 - 2xy + y^2 = (x - y)^2$$

Difference of Squares

$$x^2 - y^2 = (x + y)(x - y)$$

Difference of Cubes

$$x^3 - y^3 = (x - y)(x^2 + xy + y^2)$$

Sum of Cubes

$$x^3 + y^3 = (x + y)(x^2 - xy + y^2)$$

$$p^2 + 4pq + 4q^2 = (p + 2q)^2$$
$$9m^2 - 12mn + 4n^2 = (3m - 2n)^2$$

$$4t^2 - 9 = (2t + 3)(2t - 3)$$

$$r^3 - 8 = (r - 2)(r^2 + 2r + 4)$$

$$27x^3 + 64 = (3x + 4)(9x^2 - 12x + 16)$$

R.5 Rational Expressions

Operations

Let $\frac{a}{b}$ and $\frac{c}{d}$ ($b \neq 0, d \neq 0$) represent fractions.

$$\frac{a}{b} \pm \frac{c}{d} = \frac{ad \pm bc}{bd}$$

$$\frac{a}{b} \cdot \frac{c}{d} = \frac{ac}{bd} \quad \text{and} \quad \frac{a}{b} \div \frac{c}{d} = \frac{ad}{bc} \quad (c \neq 0)$$

$$\frac{2}{x} + \frac{5}{y} = \frac{2y + 5x}{xy} \qquad \frac{x}{6} - \frac{2y}{5} = \frac{5x - 12y}{30}$$

$$\frac{3}{q} \cdot \frac{3}{2p} = \frac{9}{2pq} \qquad \frac{z}{4} \div \frac{z}{2t} = \frac{z}{4} \cdot \frac{2t}{z} = \frac{2zt}{4z} = \frac{t}{2}$$

R.6 Rational Exponents

Rules for Exponents

Let r and s be rational numbers. The following results are valid for all positive numbers a and b.

Product rule $\qquad a^r \cdot a^s = a^{r+s}$

Quotient rule $\qquad \dfrac{a^r}{a^s} = a^{r-s}$

Negative exponent $\qquad a^{-r} = \dfrac{1}{a^r}$

Power rules $\qquad (a^r)^s = a^{rs} \qquad \left(\dfrac{a}{b}\right)^r = \dfrac{a^r}{b^r}$

$$(ab)^r = a^r b^r$$

$$6^2 \cdot 6^3 = 6^5$$

$$\frac{p^5}{p^2} = p^3$$

$$4^{-3} = \frac{1}{4^3}$$

$$(m^2)^3 = m^6 \qquad \left(\frac{x}{3}\right)^2 = \frac{x^2}{3^2}$$

$$(3x)^4 = 3^4 x^4$$

Concepts	Examples

R.7 Radical Expressions

Radical Notation

Let a be a real number, n be a positive integer, and $a^{1/n}$ be a real number.

$$\sqrt[n]{a} = a^{1/n}$$

$$\sqrt[4]{16} = 16^{1/4} = 2$$

Let a be a real number, m be an integer, n be a positive integer, and $\sqrt[n]{a}$ be a real number.

$$a^{m/n} = \left(\sqrt[n]{a}\right)^m = \sqrt[n]{a^m}$$

$$8^{2/3} = \left(\sqrt[3]{8}\right)^2 = \sqrt[3]{8^2} = 4$$

Operations

Operations with radical expressions are performed like operations with polynomials.

$$\sqrt{8x} + \sqrt{32x}$$
$$= 2\sqrt{2x} + 4\sqrt{2x}$$
$$= 6\sqrt{2x}$$

$$\left(\sqrt{5} - \sqrt{3}\right)\left(\sqrt{5} + \sqrt{3}\right)$$
$$= 5 - 3$$
$$= 2$$

$$\left(\sqrt{2} + \sqrt{7}\right)\left(\sqrt{3} - \sqrt{6}\right)$$
$$= \sqrt{6} - 2\sqrt{3} + \sqrt{21} - \sqrt{42}$$

FOIL method; $\sqrt{12} = 2\sqrt{3}$

Rationalizing the Denominator

Rationalize the denominator by multiplying numerator and denominator by a form of 1.

$$\frac{\sqrt{7y}}{\sqrt{5}} = \frac{\sqrt{7y}}{\sqrt{5}} \cdot \frac{\sqrt{5}}{\sqrt{5}} = \frac{\sqrt{35y}}{5}$$

Chapter R Review Exercises

1. Use set notation to list all the elements of the set $\{6, 8, 10, \ldots, 20\}$.

2. Is the set $\{x \mid x$ is a decimal between 0 and 1$\}$ finite or infinite?

3. *Concept Check* *True* or *false:* The set of negative integers and the set of whole numbers are disjoint sets.

4. *Concept Check* *True* or *false:* 9 is an element of the set $\{999\}$.

Determine whether each statement is true *or* false.

5. $1 \in \{6, 2, 5, 1\}$

6. $7 \notin \{1, 3, 5, 7\}$

7. $\{8, 11, 4\} = \{8, 11, 4, 0\}$

8. $\{0\} = \varnothing$

Let $A = \{1, 3, 4, 5, 7, 8\}$, $B = \{2, 4, 6, 8\}$, $C = \{1, 3, 5, 7\}$, $D = \{1, 2, 3\}$, $E = \{3, 7\}$, *and* $U = \{1, 2, 3, 4, 5, 6, 7, 8, 9, 10\}$.

Determine whether each statement is true *or* false.

9. $\varnothing \subseteq A$

10. $E \subseteq C$

11. $D \nsubseteq B$

12. $E \nsubseteq A$

Refer to the sets given for Exercises 9–12. Specify each set.

13. A'

14. $B \cap A$

15. $B \cap E$

16. $C \cup E$

17. $D \cap \varnothing$

18. $B \cup \varnothing$

19. $(C \cap D) \cup B$

20. $(D' \cap U) \cup E$

21. $\varnothing'$

22. *Concept Check* *True* or *false:* For all sets A and B, $(A \cap B) \subseteq (A \cup B)$.

Let $K = \left\{ -12, -6, -0.9, -\sqrt{7}, -\sqrt{4}, 0, \frac{1}{8}, \frac{\pi}{4}, 6, \sqrt{11} \right\}$. List all the elements of K that belong to each set.

23. Integers

24. Rational numbers

Choose all words from the following list that apply to each number.

> natural number   whole number   integer
> rational number   irrational number   real number

25. $\dfrac{4\pi}{5}$

26. $\dfrac{\pi}{0}$

27. 0

28. $-\sqrt{36}$

Write each algebraic identity (true statement) as a complete English sentence without using the names of the variables. For instance, $z(x + y) = zx + zy$ can be stated as "The multiple of a sum is the sum of the multiples."

29. $\dfrac{1}{xy} = \dfrac{1}{x} \cdot \dfrac{1}{y}$

30. $a(b - c) = ab - ac$

31. $(ab)^n = a^n b^n$

32. $a^2 - b^2 = (a + b)(a - b)$

33. $\left(\dfrac{a}{b} \right)^n = \dfrac{a^n}{b^n}$

34. $|st| = |s| \cdot |t|$

Identify the property illustrated in each statement.

35. $8(5 + 9) = (5 + 9)8$

36. $4 \cdot 6 + 4 \cdot 12 = 4(6 + 12)$

37. $3 \cdot (4 \cdot 2) = (3 \cdot 4) \cdot 2$

38. $-8 + 8 = 0$

39. $(9 + p) + 0 = 9 + p$

40. $\dfrac{1}{\sqrt{2}} \cdot \dfrac{\sqrt{2}}{\sqrt{2}} = \dfrac{\sqrt{2}}{2}$

41. *(Modeling) Online College Courses* The number of students (in millions) taking at least one online college course between the years 2002 and 2012 can be approximated by the formula

$$\text{Number of students} = 0.0112x^2 + 0.4663x + 1.513,$$

where $x = 0$ corresponds to 2002, $x = 1$ corresponds to 2003, and so on. According to this model, how many students took at least one online college course in 2012? (*Source:* Babson Survey Research Group.)

42. *Counting Marshmallows* Recently, there were media reports about students providing a correction to the following question posed on boxes of Swiss Miss Chocolate: *On average, how many mini-marshmallows are in one serving?*

$$3 + 2 \times 4 \div 2 - 3 \times 7 - 4 + 47 = \underline{\qquad}$$

The company provided 92 as the answer. What is the *correct* calculation provided by the students? (*Source:* Swiss Miss Chocolate box.)

Simplify each expression.

43. $(-4 - 1)(-3 - 5) - 2^3$

44. $(6 - 9)(-2 - 7) \div (-4)$

45. $\left(-\dfrac{5}{9} - \dfrac{2}{3}\right) - \dfrac{5}{6}$

46. $\left(-\dfrac{2^3}{5} - \dfrac{3}{4}\right) - \left(-\dfrac{1}{2}\right)$

47. $\dfrac{6(-4) - 3^2(-2)^3}{-5[-2 - (-6)]}$

48. $\dfrac{(-7)(-3) - (-2^3)(-5)}{(-2^2 - 2)(-1 - 6)}$

Evaluate each expression for $a = -1$, $b = -2$, and $c = 4$.

49. $-c(2a - 5b)$

50. $(a - 2) \div 5 \cdot b + c$

51. $\dfrac{9a + 2b}{a + b + c}$

52. $\dfrac{3|b| - 4|c|}{|ac|}$

Perform the indicated operations.

53. $(3q^3 - 9q^2 + 6) + (4q^3 - 8q + 3)$

54. $2(3y^6 - 9y^2 + 2y) - (5y^6 - 4y)$

55. $(8y - 7)(2y^2 + 7y - 3)$

56. $(2r + 11s)(4r - 9s)$

57. $(3k - 5m)^2$

58. $(4a - 3b)^2$

Perform each division.

59. $\dfrac{30m^3 - 9m^2 + 22m + 5}{5m + 1}$

60. $\dfrac{72r^2 + 59r + 12}{8r + 3}$

61. $\dfrac{3b^3 - 8b^2 + 12b - 30}{b^2 + 4}$

62. $\dfrac{5m^3 - 7m^2 + 14}{m^2 - 2}$

Factor as completely as possible.

63. $3(z - 4)^2 + 9(z - 4)^3$

64. $7z^2 - 9z^3 + z$

65. $z^2 - 6zk - 16k^2$

66. $r^2 + rp - 42p^2$

67. $48a^8 - 12a^7b - 90a^6b^2$

68. $6m^2 - 13m - 5$

69. $49m^8 - 9n^2$

70. $169y^4 - 1$

71. $6(3r - 1)^2 + (3r - 1) - 35$

72. $8y^3 - 1000z^6$

73. $xy + 2x - y - 2$

74. $15mp + 9mq - 10np - 6nq$

Factor each expression. (These expressions arise in calculus from a technique called the product rule that is used to determine the shape of a curve.)

75. $(3x - 4)^2 + (x - 5)(2)(3x - 4)(3)$

76. $(5 - 2x)(3)(7x - 8)^2(7) + (7x - 8)^3(-2)$

Perform the indicated operations.

77. $\dfrac{k^2 + k}{8k^3} \cdot \dfrac{4}{k^2 - 1}$

78. $\dfrac{3r^3 - 9r^2}{r^2 - 9} \div \dfrac{8r^3}{r + 3}$

79. $\dfrac{x^2 + x - 2}{x^2 + 5x + 6} \div \dfrac{x^2 + 3x - 4}{x^2 + 4x + 3}$

80. $\dfrac{27m^3 - n^3}{3m - n} \div \dfrac{9m^2 + 3mn + n^2}{9m^2 - n^2}$

81. $\dfrac{p^2 - 36q^2}{p^2 - 12pq + 36q^2} \cdot \dfrac{p^2 - 5pq - 6q^2}{p^2 + 2pq + q^2}$

82. $\dfrac{1}{4y} + \dfrac{8}{5y}$

83. $\dfrac{m}{4-m} + \dfrac{3m}{m-4}$

84. $\dfrac{3}{x^2 - 4x + 3} - \dfrac{2}{x^2 - 1}$

85. $\dfrac{p^{-1} + q^{-1}}{1 - (pq)^{-1}}$

86. $\dfrac{3 + \dfrac{2m}{m^2 - 4}}{\dfrac{5}{m-2}}$

Simplify each expression. Write answers without negative exponents. Assume all variables represent positive real numbers.

87. $\left(-\dfrac{5}{4}\right)^{-2}$

88. $3^{-1} - 4^{-1}$

89. $(5z^3)(-2z^5)$

90. $(8p^2q^3)(-2p^5q^{-4})$

91. $(-6p^5w^4m^{12})^0$

92. $(-6x^2y^{-3}z^2)^{-2}$

93. $\dfrac{-8y^7p^{-2}}{y^{-4}p^{-3}}$

94. $\dfrac{a^{-6}(a^{-8})}{a^{-2}(a^{11})}$

95. $\dfrac{(p+q)^4(p+q)^{-3}}{(p+q)^6}$

96. $\dfrac{[p^2(m+n)^3]^{-2}}{p^{-2}(m+n)^{-5}}$

97. $(7r^{1/2})(2r^{3/4})(-r^{1/6})$

98. $(a^{3/4}b^{2/3})(a^{5/8}b^{-5/6})$

99. $\dfrac{y^{5/3} \cdot y^{-2}}{y^{-5/6}}$

100. $\left(\dfrac{25m^3n^5}{m^{-2}n^6}\right)^{-1/2}$

101. $\dfrac{(p^{15}q^{12})^{-4/3}}{(p^{24}q^{16})^{-3/4}}$

102. Simplify the product $-m^{3/4}(8m^{1/2} + 4m^{3/2})$. Assume the variable represents a positive real number.

Simplify each expression. Assume all variables represent positive real numbers.

103. $\sqrt{200}$

104. $\sqrt[3]{16}$

105. $\sqrt[4]{1250}$

106. $-\sqrt{\dfrac{16}{3}}$

107. $-\sqrt[3]{\dfrac{2}{5p^2}}$

108. $\sqrt{\dfrac{27y^8}{m^3}}$

109. $\sqrt[4]{\sqrt[3]{m}}$

110. $\dfrac{\sqrt[4]{8p^2q^5} \cdot \sqrt[4]{2p^3q}}{\sqrt[4]{p^5q^2}}$

111. $\left(\sqrt[3]{2} + 4\right)\left(\sqrt[3]{2^2} - 4\sqrt[3]{2} + 16\right)$

112. $\dfrac{3}{\sqrt{5}} - \dfrac{2}{\sqrt{45}} + \dfrac{6}{\sqrt{80}}$

113. $\sqrt{18m^3} - 3m\sqrt{32m} + 5\sqrt{m^3}$

114. $\dfrac{2}{7 - \sqrt{3}}$

115. $\dfrac{6}{3 - \sqrt{2}}$

116. $\dfrac{k}{\sqrt{k} - 3}$

Concept Check *Correct each **INCORRECT** statement by changing the right side of the equation.*

117. $x(x^2 + 5) = x^3 + 5$

118. $-3^2 = 9$

119. $(m^2)^3 = m^5$

120. $(3x)(3y) = 3xy$

121. $\dfrac{\left(\dfrac{a}{b}\right)}{2} = \dfrac{2a}{b}$

122. $\dfrac{m}{r} \cdot \dfrac{n}{r} = \dfrac{mn}{r}$

123. $\dfrac{1}{(-2)^3} = 2^{-3}$

124. $(-5)^2 = -5^2$

125. $\left(\dfrac{8}{7} + \dfrac{a}{b}\right)^{-1} = \dfrac{7}{8} + \dfrac{b}{a}$

Chapter R Test

Let $U = \{1, 2, 3, 4, 5, 6, 7, 8\}$, $A = \{1, 2, 3, 4, 5, 6\}$, $B = \{1, 3, 5\}$, $C = \{1, 6\}$, and $D = \{4\}$. Decide whether each statement is true or false.

1. $B' = \{2, 4, 6, 8\}$

2. $C \subseteq A$

3. $(B \cap C) \cup D = \{1, 3, 4, 5, 6\}$

4. $(A' \cup C) \cap B' = \{6, 7, 8\}$

5. Let $A = \left\{-13, -\frac{12}{4}, 0, \frac{3}{5}, \frac{\pi}{4}, 5.9, \sqrt{49}\right\}$. List all the elements of A that belong to each set.

    (a) Integers   **(b)** Rational numbers   **(c)** Real numbers

6. Evaluate the expression $\left|\dfrac{x^2 + 2yz}{3(x + z)}\right|$ for $x = -2$, $y = -4$, and $z = 5$.

7. Identify each property illustrated. Let a, b, and c represent any real numbers.

    (a) $a + (b + c) = (a + b) + c$ **(b)** $a + (c + b) = a + (b + c)$

    (c) $a(b + c) = ab + ac$ **(d)** $a + [b + (-b)] = a + 0$

8. *(Modeling) Passer Rating for NFL Quarterbacks* Approximate the quarterback rating (to the nearest tenth) of Drew Brees of the New Orleans Saints during the 2013 regular season. He attempted 650 passes, completed 446, had 5162 total yards, threw for 39 touchdowns, and had 12 interceptions. (*Source:* www.nfl.com)

$$\text{Rating} = \frac{\left(250 \cdot \dfrac{C}{A}\right) + \left(1000 \cdot \dfrac{T}{A}\right) + \left(12.5 \cdot \dfrac{Y}{A}\right) + 6.25 - \left(1250 \cdot \dfrac{I}{A}\right)}{3},$$

where $A = $ attempted passes, $C = $ completed passes, $T = $ touchdown passes, $Y = $ yards gained passing, and $I = $ interceptions.

In addition to the weighting factors that appear in the formula, the four category ratios are limited to nonnegative values with the following maximums.

$$0.775 \text{ for } \frac{C}{A}, \quad 0.11875 \text{ for } \frac{T}{A}, \quad 12.5 \text{ for } \frac{Y}{A}, \quad 0.095 \text{ for } \frac{I}{A}$$

Perform the indicated operations.

9. $(x^2 - 3x + 2) - (x - 4x^2) + 3x(2x + 1)$

10. $(6r - 5)^2$ **11.** $(t + 2)(3t^2 - t + 4)$ **12.** $\dfrac{2x^3 - 11x^2 + 28}{x - 5}$

(Modeling) Adjusted Poverty Threshold The adjusted poverty threshold for a single person between the years 1999 and 2013 can be approximated by the formula

$$y = 2.719x^2 + 196.1x + 8718,$$

where $x = 0$ corresponds to 1999, $x = 1$ corresponds to 2000, and so on, and the adjusted poverty threshold amount, y, is in dollars. According to this model, what was the adjusted poverty threshold, to the nearest dollar, in each given year? (*Source: U.S. Census Bureau.*)

13. 2005   **14.** 2012

Factor completely.

15. $6x^2 - 17x + 7$ **16.** $x^4 - 16$

17. $24m^3 - 14m^2 - 24m$ **18.** $x^3y^2 - 9x^3 - 8y^2 + 72$

19. $(a - b)^2 + 2(a - b)$ **20.** $1 - 27x^6$

Perform the indicated operations.

21. $\dfrac{5x^2 - 9x - 2}{30x^3 + 6x^2} \div \dfrac{x^4 - 3x^2 - 4}{2x^8 + 6x^7 + 4x^6}$

22. $\dfrac{x}{x^2 + 3x + 2} + \dfrac{2x}{2x^2 - x - 3}$

23. $\dfrac{a + b}{2a - 3} - \dfrac{a - b}{3 - 2a}$

24. $\dfrac{y - 2}{y - \dfrac{4}{y}}$

Simplify or evaluate as appropriate. Assume all variables represent positive real numbers.

25. $\sqrt{18x^5 y^8}$

26. $\sqrt{32x} + \sqrt{2x} - \sqrt{18x}$

27. $\left(\sqrt{x} - \sqrt{y}\right)\left(\sqrt{x} + \sqrt{y}\right)$

28. $\dfrac{14}{\sqrt{11} - \sqrt{7}}$

29. $\left(\dfrac{x^{-2}y^{-1/3}}{x^{-5/3}y^{-2/3}}\right)^3$

30. $\left(-\dfrac{64}{27}\right)^{-2/3}$

31. *Concept Check* *True* or *false*: For all real numbers x, $\sqrt{x^2} = x$.

32. (*Modeling*) *Period of a Pendulum* The period t, in seconds, of the swing of a pendulum is given by the formula

$$t = 2\pi\sqrt{\dfrac{L}{32}},$$

where L is the length of the pendulum in feet. Find the period of a pendulum 3.5 ft long. Use a calculator, and round the answer to the nearest tenth.

1 Equations and Inequalities

Balance, as seen in this natural setting, is a critical component of life and provides the key to solving mathematical *equations*.

1.1 Linear Equations

- Basic Terminology of Equations
- Linear Equations
- Identities, Conditional Equations, and Contradictions
- Solving for a Specified Variable (Literal Equations)

Basic Terminology of Equations An **equation** is a statement that two expressions are equal.

$$x + 2 = 9, \quad 11x = 5x + 6x, \quad x^2 - 2x - 1 = 0 \quad \text{Equations}$$

To *solve* an equation means to find all numbers that make the equation a true statement. These numbers are the **solutions,** or **roots,** of the equation. A number that is a solution of an equation is said to *satisfy* the equation, and the solutions of an equation make up its **solution set.** Equations with the same solution set are **equivalent equations.** For example,

$$x = 4, \quad x + 1 = 5, \quad \text{and} \quad 6x + 3 = 27 \quad \text{are equivalent equations}$$

because they have the same solution set, $\{4\}$. However, the equations

$$x^2 = 9 \quad \text{and} \quad x = 3 \quad \text{are } not \text{ equivalent}$$

because the first has solution set $\{-3, 3\}$ while the solution set of the second is $\{3\}$.

One way to solve an equation is to rewrite it as a series of simpler equivalent equations using the **addition and multiplication properties of equality.**

Addition and Multiplication Properties of Equality

Let a, b, and c represent real numbers.

$$\textbf{If } a = b, \textbf{ then } a + c = b + c.$$

That is, the same number may be added to each side of an equation without changing the solution set.

$$\textbf{If } a = b \textbf{ and } c \neq 0, \textbf{ then } ac = bc.$$

That is, each side of an equation may be multiplied by the same nonzero number without changing the solution set. (Multiplying each side by zero leads to $0 = 0$.)

These properties can be extended: The same number may be subtracted from each side of an equation, and each side may be divided by the same non-zero number, without changing the solution set.

Linear Equations We use the properties of equality to solve *linear equations.*

Linear Equation in One Variable

A **linear equation in one variable** is an equation that can be written in the form

$$ax + b = 0,$$

where a and b are real numbers and $a \neq 0$.

A linear equation is a **first-degree equation** because the greatest degree of the variable is 1.

$$3x + \sqrt{2} = 0, \quad \frac{3}{4}x = 12, \quad 0.5(x + 3) = 2x - 6 \quad \text{Linear equations}$$

$$\sqrt{x} + 2 = 5, \quad \frac{1}{x} = -8, \quad x^2 + 3x + 0.2 = 0 \quad \text{Nonlinear equations}$$

EXAMPLE 1 Solving a Linear Equation

Solve $3(2x - 4) = 7 - (x + 5)$.

SOLUTION

$$3(2x - 4) = 7 - (x + 5)$$

Be careful with signs.

$$6x - 12 = 7 - x - 5 \qquad \text{Distributive property}$$

$$6x - 12 = 2 - x \qquad \text{Combine like terms.}$$

$$6x - 12 + x = 2 - x + x \qquad \text{Add } x \text{ to each side.}$$

$$7x - 12 = 2 \qquad \text{Combine like terms.}$$

$$7x - 12 + 12 = 2 + 12 \qquad \text{Add 12 to each side.}$$

$$7x = 14 \qquad \text{Combine like terms.}$$

$$\frac{7x}{7} = \frac{14}{7} \qquad \text{Divide each side by 7.}$$

$$x = 2$$

CHECK

A check of the solution is recommended.

$$3(2x - 4) = 7 - (x + 5) \qquad \text{Original equation}$$

$$3(2 \cdot 2 - 4) \stackrel{?}{=} 7 - (2 + 5) \qquad \text{Let } x = 2.$$

$$3(4 - 4) \stackrel{?}{=} 7 - (7) \qquad \text{Work inside the parentheses.}$$

$$0 = 0 \checkmark \qquad \text{True}$$

Replacing x with 2 results in a true statement, so 2 is a solution of the given equation. The solution set is $\{2\}$. ✔ **Now Try Exercise 13.**

EXAMPLE 2 Solving a Linear Equation with Fractions

Solve $\dfrac{2x + 4}{3} + \dfrac{1}{2}x = \dfrac{1}{4}x - \dfrac{7}{3}$.

SOLUTION

$$\frac{2x + 4}{3} + \frac{1}{2}x = \frac{1}{4}x - \frac{7}{3}$$

Distribute to all terms within the parentheses.

$$12\left(\frac{2x + 4}{3} + \frac{1}{2}x\right) = 12\left(\frac{1}{4}x - \frac{7}{3}\right) \qquad \begin{array}{l}\text{Multiply by 12, the}\\\text{LCD of the fractions.}\end{array}$$

$$12\left(\frac{2x + 4}{3}\right) + 12\left(\frac{1}{2}x\right) = 12\left(\frac{1}{4}x\right) - 12\left(\frac{7}{3}\right) \qquad \text{Distributive property}$$

$$4(2x + 4) + 6x = 3x - 28 \qquad \text{Multiply.}$$

$$8x + 16 + 6x = 3x - 28 \qquad \text{Distributive property}$$

$$14x + 16 = 3x - 28 \qquad \text{Combine like terms.}$$

$$11x = -44 \qquad \text{Subtract } 3x. \text{ Subtract 16.}$$

$$x = -4 \qquad \text{Divide each side by 11.}$$

CHECK $\qquad$ $\dfrac{2x+4}{3} + \dfrac{1}{2}x = \dfrac{1}{4}x - \dfrac{7}{3}$ $\qquad$ Original equation

$$\dfrac{2(-4)+4}{3} + \dfrac{1}{2}(-4) \overset{?}{=} \dfrac{1}{4}(-4) - \dfrac{7}{3} \qquad \text{Let } x = -4.$$

$$\dfrac{-4}{3} + (-2) \overset{?}{=} -1 - \dfrac{7}{3} \qquad \text{Simplify on each side.}$$

$$-\dfrac{10}{3} = -\dfrac{10}{3} \;\checkmark \qquad \text{True}$$

The solution set is $\{-4\}$. ✔ **Now Try Exercise 21.**

Identities, Conditional Equations, and Contradictions An equation satisfied by every number that is a meaningful replacement for the variable is an **identity.**

$$3(x+1) = 3x + 3 \qquad \text{Identity}$$

An equation that is satisfied by some numbers but not others is a **conditional equation.**

$$2x = 4 \qquad \text{Conditional equation}$$

The equations in **Examples 1 and 2** are conditional equations. An equation that has no solution is a **contradiction.**

$$x = x + 1 \qquad \text{Contradiction}$$

EXAMPLE 3 **Identifying Types of Equations**

Determine whether each equation is an *identity*, a *conditional equation*, or a *contradiction.* Give the solution set.

(a) $-2(x+4) + 3x = x - 8$ $\qquad$ **(b)** $5x - 4 = 11$ $\qquad$ **(c)** $3(3x-1) = 9x + 7$

SOLUTION

(a) $-2(x+4) + 3x = x - 8$

$\qquad -2x - 8 + 3x = x - 8$ $\qquad$ Distributive property

$\qquad\qquad x - 8 = x - 8$ $\qquad$ Combine like terms.

$\qquad\qquad\qquad 0 = 0$ $\qquad$ Subtract x. Add 8.

When a *true* statement such as $0 = 0$ results, the equation is an identity, and the solution set is **{all real numbers}.**

(b) $5x - 4 = 11$

$\qquad\qquad 5x = 15$ $\qquad$ Add 4 to each side.

$\qquad\qquad x = 3$ $\qquad$ Divide each side by 5.

This is a conditional equation, and its solution set is $\{3\}$.

(c) $3(3x-1) = 9x + 7$

$\qquad 9x - 3 = 9x + 7$ $\qquad$ Distributive property

$\qquad\qquad -3 = 7$ $\qquad$ Subtract $9x$.

When a *false* statement such as $-3 = 7$ results, the equation is a contradiction, and the solution set is the **empty set,** or **null set,** symbolized $\varnothing$.

✔ **Now Try Exercises 31, 33, and 35.**

Identifying Types of Linear Equations

1. If solving a linear equation leads to a true statement such as $0 = 0$, the equation is an **identity**. Its solution set is **{all real numbers}**. **(See Example 3(a).)**

2. If solving a linear equation leads to a single solution such as $x = 3$, the equation is **conditional.** Its solution set consists of a single element. **(See Example 3(b).)**

3. If solving a linear equation leads to a false statement such as $-3 = 7$, the equation is a **contradiction**. Its solution set is $\varnothing$. **(See Example 3(c).)**

Solving for a Specified Variable (Literal Equations) A formula is an example of a **literal equation** (an equation involving letters).

EXAMPLE 4 Solving for a Specified Variable

Solve each formula or equation for the specified variable.

(a) $I = Prt$, for t

(b) $A - P = Prt$, for P

(c) $3(2x - 5a) + 4b = 4x - 2$, for x

SOLUTION

(a) This is the formula for **simple interest** I on a principal amount of P dollars at an annual interest rate r for t years. To solve for t, we treat t as if it were the only variable, and the other variables as if they were constants.

$$I = Prt \quad \text{Goal: Isolate } t \text{ on one side.}$$

$$\frac{I}{Pr} = \frac{Prt}{Pr} \quad \text{Divide each side by } Pr.$$

$$\frac{I}{Pr} = t, \quad \text{or} \quad t = \frac{I}{Pr}$$

(b) The formula $A = P(1 + rt)$, which can also be written $A - P = Prt$, gives the **future value,** or **maturity value,** A of P dollars invested for t years at annual simple interest rate r.

$$A - P = Prt \quad \text{Goal: Isolate } P, \text{ the specified variable.}$$

$$A = P + Prt \quad \text{Transform so that all terms involving } P \text{ are on one side.}$$

Pay close attention to this step. $\longrightarrow A = P(1 + rt)$ Factor out P.

$$\frac{A}{1 + rt} = P, \quad \text{or} \quad P = \frac{A}{1 + rt} \quad \text{Divide by } 1 + rt.$$

(c) $3(2x - 5a) + 4b = 4x - 2$ Solve for x.

$6x - 15a + 4b = 4x - 2$ Distributive property

$6x - 4x = 15a - 4b - 2$ Isolate the x-terms on one side.

$2x = 15a - 4b - 2$ Combine like terms.

$x = \dfrac{15a - 4b - 2}{2}$ Divide each side by 2.

✔ **Now Try Exercises 39, 47, and 49.**

EXAMPLE 5 Applying the Simple Interest Formula

A woman borrowed $5240 for new furniture. She will pay it off in 11 months at an annual simple interest rate of 4.5%. How much interest will she pay?

SOLUTION Use the simple interest formula $I = Prt$.

$$I = 5240(0.045)\left(\frac{11}{12}\right) = \$216.15 \qquad \begin{array}{l} P = 5240, r = 0.045, \\ \text{and } t = \frac{11}{12} \text{ (year)} \end{array}$$

She will pay $216.15 interest on her purchase. ✔ **Now Try Exercise 59.**

1.1 Exercises

CONCEPT PREVIEW *Fill in the blank to correctly complete each sentence.*

1. A(n) _____ is a statement that two expressions are equal.

2. To _____ an equation means to find all numbers that make the equation a true statement.

3. A linear equation is a(n) _____ because the greatest degree of the variable is 1.

4. A(n) _____ is an equation satisfied by every number that is a meaningful replacement for the variable.

5. A(n) _____ is an equation that has no solution.

CONCEPT PREVIEW *Decide whether each statement is* true *or* false.

6. The solution set of $2x + 5 = x - 3$ is $\{-8\}$.

7. The equation $5(x - 8) = 5x - 40$ is an example of an identity.

8. The equation $5x = 4x$ is an example of a contradiction.

9. Solving the literal equation $A = \frac{1}{2} bh$ for the variable h gives $h = \frac{A}{2b}$.

10. **CONCEPT PREVIEW** Which one is *not* a linear equation?

    A. $5x + 7(x - 1) = -3x$ **B.** $9x^2 - 4x + 3 = 0$

    C. $7x + 8x = 13x$ **D.** $0.04x - 0.08x = 0.40$

*Solve each equation. **See Examples 1 and 2.***

11. $5x + 4 = 3x - 4$ **12.** $9x + 11 = 7x + 1$

13. $6(3x - 1) = 8 - (10x - 14)$ **14.** $4(-2x + 1) = 6 - (2x - 4)$

15. $\frac{5}{6}x - 2x + \frac{4}{3} = \frac{5}{3}$ **16.** $\frac{7}{4} + \frac{1}{5}x - \frac{3}{2} = \frac{4}{5}x$

17. $3x + 5 - 5(x + 1) = 6x + 7$ **18.** $5(x + 3) + 4x - 3 = -(2x - 4) + 2$

19. $2[x - (4 + 2x) + 3] = 2x + 2$ **20.** $4[2x - (3 - x) + 5] = -6x - 28$

21. $\frac{1}{14}(3x - 2) = \frac{x + 10}{10}$ **22.** $\frac{1}{15}(2x + 5) = \frac{x + 2}{9}$

23. $0.2x - 0.5 = 0.1x + 7$ **24.** $0.01x + 3.1 = 2.03x - 2.96$

25. $-4(2x - 6) + 8x = 5x + 24 + x$ **26.** $-8(3x + 4) + 6x = 4(x - 8) + 4x$

27. $0.5x + \dfrac{4}{3}x = x + 10$ **28.** $\dfrac{2}{3}x + 0.25x = x + 2$

29. $0.08x + 0.06(x + 12) = 7.72$ **30.** $0.04(x - 12) + 0.06x = 1.52$

Determine whether each equation is an identity, *a* conditional equation, *or a* contradiction. *Give the solution set.* **See Example 3.**

31. $4(2x + 7) = 2x + 22 + 3(2x + 2)$ **32.** $\dfrac{1}{2}(6x + 20) = x + 4 + 2(x + 3)$

33. $2(x - 8) = 3x - 16$ **34.** $-8(x + 5) = -8x - 5(x + 8)$

35. $4(x + 7) = 2(x + 12) + 2(x + 1)$ **36.** $-6(2x + 1) - 3(x - 4) = -15x + 1$

37. $0.3(x + 2) - 0.5(x + 2) = -0.2x - 0.4$

38. $-0.6(x - 5) + 0.8(x - 6) = 0.2x - 1.8$

Solve each formula for the specified variable. Assume that the denominator is not 0 if variables appear in the denominator. **See Examples 4(a) and (b).**

39. $V = lwh$,   for l (volume of a rectangular box)

40. $I = Prt$,   for P (simple interest)

41. $P = a + b + c$,   for c (perimeter of a triangle)

42. $P = 2l + 2w$,   for w (perimeter of a rectangle)

43. $\mathcal{A} = \dfrac{1}{2}h(B + b)$,   for B (area of a trapezoid)

44. $\mathcal{A} = \dfrac{1}{2}h(B + b)$,   for h (area of a trapezoid)

45. $S = 2\pi rh + 2\pi r^2$,   for h (surface area of a right circular cylinder)

46. $s = \dfrac{1}{2}gt^2$,   for g (distance traveled by a falling object)

47. $S = 2lw + 2wh + 2hl$,   for h (surface area of a rectangular box)

48. $z = \dfrac{x - \mu}{\sigma}$,   for x (standardized value)

Solve each equation for x. **See Example 4(c).**

49. $2(x - a) + b = 3x + a$ **50.** $5x - (2a + c) = 4(x + c)$

51. $ax + b = 3(x - a)$ **52.** $4a - ax = 3b + bx$

53. $\dfrac{x}{a - 1} = ax + 3$ **54.** $\dfrac{x - 1}{2a} = 2x - a$

55. $a^2x + 3x = 2a^2$ **56.** $ax + b^2 = bx - a^2$

57. $3x = (2x - 1)(m + 4)$ **58.** $-x = (5x + 3)(3k + 1)$

Simple Interest   *Work each problem.* **See Example 5.**

59. Elmer borrowed $3150 from his brother Julio to pay for books and tuition. He agreed to repay Julio in 6 months with simple annual interest at 4%.

 (a) How much will the interest amount to?

 (b) What amount must Elmer pay Julio at the end of the 6 months?

60. Levada borrows $30,900 from her bank to open a florist shop. She agrees to repay the money in 18 months with simple annual interest of 5.5%.

 (a) How much must she pay the bank in 18 months?

 (b) How much of the amount in part (a) is interest?

Celsius and Fahrenheit Temperatures In the metric system of weights and measures, temperature is measured in degrees Celsius (°C) instead of degrees Fahrenheit (°F). To convert between the two systems, we use the equations

$$C = \frac{5}{9}(F - 32) \quad \text{and} \quad F = \frac{9}{5}C + 32.$$

In each exercise, convert to the other system. Round answers to the nearest tenth of a degree if necessary.

61. 20°C   **62.** 200°C   **63.** 50°F

64. 77°F   **65.** 100°F   **66.** 350°F

Work each problem. Round to the nearest tenth of a degree, if necessary.

67. *Temperature of Venus* Venus is the hottest planet, with a surface temperature of 867°F. What is this temperature in Celsius? (*Source: World Almanac and Book of Facts.*)

68. *Temperature at Soviet Antarctica Station* A record low temperature of −89.4°C was recorded at the Soviet Antarctica Station of Vostok on July 21, 1983. Find the corresponding Fahrenheit temperature. (*Source: World Almanac and Book of Facts.*)

69. *Temperature in South Carolina* A record high temperature of 113°F was recorded for the state of South Carolina on June 29, 2012. What is the corresponding Celsius temperature? (*Source:* U.S. National Oceanic and Atmospheric Administration.)

70. *Temperature in Haiti* The average annual temperature in Port-au-Prince, Haiti, is approximately 28.1°C. What is the corresponding Fahrenheit temperature? (*Source:* www.haiti.climatemps.com)

1.2 Applications and Modeling with Linear Equations

- Solving Applied Problems
- Geometry Problems
- Motion Problems
- Mixture Problems
- Modeling with Linear Equations

Solving Applied Problems   One of the main reasons for learning mathematics is to be able use it to solve application problems. While there is no one method that enables us to solve all types of applied problems, the following six steps provide a useful guide.

Solving an Applied Problem

Step 1 **Read** the problem carefully until you understand what is given and what is to be found.

Step 2 **Assign a variable** to represent the unknown value, using diagrams or tables as needed. Write down what the variable represents. If necessary, express any other unknown values in terms of the variable.

Step 3 **Write an equation** using the variable expression(s).

Step 4 **Solve** the equation.

Step 5 **State the answer** to the problem. Does it seem reasonable?

Step 6 **Check** the answer in the words of the original problem.

Geometry Problems

> ### EXAMPLE 1 Finding the Dimensions of a Square
>
> If the length of each side of a square is increased by 3 cm, the perimeter of the new square is 40 cm more than twice the length of each side of the original square. Find the dimensions of the original square.

SOLUTION

Step 1 **Read** the problem. We must find the length of each side of the original square.

Step 2 **Assign a variable.** Since the length of a side of the original square is to be found, let the variable represent this length.

Let x = the length of a side of the original square in centimeters.

The length of a side of the new square is 3 cm more than the length of a side of the old square.

Then $x + 3$ = the length of a side of the new square.

See **Figure 1.** Now write a variable expression for the perimeter of the new square. The perimeter of a square is 4 times the length of a side.

Thus, $4(x + 3)$ = the perimeter of the new square.

Original square   Side is increased by 3.

x and $x + 3$ are in centimeters.

Figure 1

Step 3 **Write an equation.** Translate the English sentence that follows into its equivalent algebraic equation.

The new perimeter	is	40	more than	twice the length of each side of the original square.
$4(x + 3)$	$=$	40	$+$	$2x$

Step 4 **Solve** the equation.

$$4x + 12 = 40 + 2x \qquad \text{Distributive property}$$

$$2x = 28 \qquad \text{Subtract } 2x \text{ and } 12.$$

$$x = 14 \qquad \text{Divide by 2.}$$

Step 5 **State the answer.** Each side of the original square measures 14 cm.

Step 6 **Check.** Go back to the words of the original problem to see that all necessary conditions are satisfied. The length of a side of the new square would be $14 + 3 = 17$ cm. The perimeter of the new square would be $4(17) = 68$ cm. Twice the length of a side of the original square would be $2(14) = 28$ cm. Because $40 + 28 = 68$, the answer checks.

✔ **Now Try Exercise 15.**

Motion Problems

LOOKING AHEAD TO CALCULUS
In calculus the concept of the **definite integral** is used to find the distance traveled by an object traveling at a *non-constant* velocity.

PROBLEM SOLVING HINT   In a motion problem, the three components *distance*, *rate*, and *time* are denoted by the letters d, r, and t, respectively. (The *rate* is also called the *speed* or *velocity*. Here, rate is understood to be constant.) These variables are related by the following equations.

$$d = rt, \quad \text{and its related forms} \quad r = \frac{d}{t} \quad \text{and} \quad t = \frac{d}{r}$$

EXAMPLE 2 **Solving a Motion Problem**

Maria and Eduardo are traveling to a business conference. The trip takes 2 hr for Maria and 2.5 hr for Eduardo because he lives 40 mi farther away. Eduardo travels 5 mph faster than Maria. Find their average rates.

SOLUTION

Step 1 **Read** the problem. We must find Maria's and Eduardo's average rates.

Step 2 **Assign a variable.** Because average rates are to be found, we let the variable represent one of these rates.

$$\text{Let } x = \text{Maria's rate.}$$

Because Eduardo travels 5 mph faster than Maria, we can express his average rate using the same variable.

$$\text{Then } x + 5 = \text{Eduardo's rate.}$$

Make a table. The expressions in the last column were found by multiplying the corresponding rates and times.

	r	t	d	
Maria	x	2	$2x$	Summarize the given information in a table.
Eduardo	$x + 5$	2.5	$2.5(x + 5)$	Use $d = rt$.

Step 3 **Write an equation.** Eduardo's distance traveled exceeds Maria's distance by 40 mi. Translate this into an equation.

Eduardo's distance	is	40 more than Maria's.
$2.5(x + 5)$	$=$	$2x + 40$

Step 4 **Solve.**

$$2.5x + 12.5 = 2x + 40 \quad \text{Distributive property}$$
$$0.5x = 27.5 \quad \text{Subtract } 2x \text{ and } 12.5.$$
$$x = 55 \quad \text{Divide by } 0.5.$$

Step 5 **State the answer.** Maria's rate of travel is 55 mph, and Eduardo's rate is

$$55 + 5 = 60 \text{ mph.}$$

Step 6 **Check.** The conditions of the problem are satisfied, as shown below.

Distance traveled by Maria: $2(55) = 110 \text{ mi}$ $150 - 110 = 40$
Distance traveled by Eduardo: $2.5(60) = 150 \text{ mi}$ as required.

✔ **Now Try Exercise 19.**

George Polya (1887–1985)

Polya, a native of Budapest, Hungary, wrote more than 250 papers and a number of books. He proposed a general outline for solving applied problems in his classic book *How to Solve It*.

Mixture Problems   Problems involving mixtures of two types of the same substance, salt solution, candy, and so on, often involve percentages.

PROBLEM-SOLVING HINT   In mixture problems involving solutions,

$$\begin{array}{c}\text{rate (percent)}\\\text{of concentration}\end{array} \cdot \text{quantity} = \begin{array}{c}\text{amount of pure}\\\text{substance present.}\end{array}$$

The concentration of the final mixture must be between the concentrations of the two solutions making up the mixture.

EXAMPLE 3 **Solving a Mixture Problem**

A chemist needs a 20% solution of alcohol. She has a 15% solution on hand, as well as a 30% solution. How many liters of the 15% solution should she add to 3 L of the 30% solution to obtain the 20% solution?

SOLUTION

Step 1 **Read** the problem. We must find the required number of liters of 15% alcohol solution.

Step 2 **Assign a variable.**

Let x = the number of liters of 15% solution to be added.

Figure 2 and the table show what is happening in the problem. The numbers in the last column were found by multiplying the strengths and the numbers of liters.

Figure 2

Strength	Liters of Solution	Liters of Pure Alcohol	
15%	x	$0.15x$	
30%	3	$0.30(3)$	Sum must equal
20%	$x + 3$	$0.20(x + 3)$	

Step 3 **Write an equation.** The number of liters of pure alcohol in the 15% solution plus the number of liters in the 30% solution must equal the number of liters in the final 20% solution.

$$\underset{\text{in 15\% solution}}{\underbrace{\text{Liters of pure alcohol}}} + \underset{\text{in 30\% solution}}{\underbrace{\text{Liters of pure alcohol}}} = \underset{\text{in 20\% solution}}{\underbrace{\text{Liters of pure alcohol}}}$$

$$0.15x \quad + \quad 0.30(3) \quad = \quad 0.20(x + 3)$$

Step 4 **Solve.**

$$0.15x + 0.90 = 0.20x + 0.60 \quad \text{Distributive property}$$

$$0.30 = 0.05x \quad \text{Subtract 0.60 and 0.15}x.$$

$$6 = x \quad \text{Divide by 0.05.}$$

Step 5 **State the answer.** Thus, 6 L of 15% solution should be mixed with 3 L of 30% solution, giving $6 + 3 = 9$ L of 20% solution.

Step 6 **Check.** The answer checks because the amount of alcohol in the two solutions is equal to the amount of alcohol in the mixture.

$$0.15(6) + 0.9 = 0.9 + 0.9 = 1.8 \quad \text{Solutions}$$

$$0.20(6 + 3) = 0.20(9) = 1.8 \quad \text{Mixture}$$

✔ **Now Try Exercise 29.**

PROBLEM-SOLVING HINT In mixed investment problems, multiply the principal amount P by the interest rate r, expressed as a decimal, and the time t, in years, to find the amount of interest earned I.

$$I = Prt \quad \text{Simple interest formula}$$

EXAMPLE 4 Solving an Investment Problem

An artist has sold a painting for $410,000. He invests a portion of the money for 6 months at 2.65% and the rest for a year at 2.91%. His broker tells him the two investments will earn a total of $8761. How much should be invested at each rate to obtain that amount of interest?

SOLUTION

Step 1 **Read** the problem. We must find the amount to be invested at each rate.

Step 2 **Assign a variable.**

Let x = the dollar amount to be invested for 6 months at 2.65%.

$410,000 - x$ = the dollar amount to be invested for 1 yr at 2.91%.

P Invested Amount	r Interest Rate (%)	t Time (in years)	I Interest Earned
x	2.65	0.5	$x(0.0265)(0.5)$
$410,000 - x$	2.91	1	$(410,000 - x)(0.0291)(1)$

Summarize the information in a table using the formula $I = Prt$.

Step 3 **Write an equation.** The sum of the two interest amounts must equal the total interest earned.

Interest from 2.65% investment $+$ Interest from 2.91% investment $=$ Total interest

$$0.5x(0.0265) \quad + \quad 0.0291(410,000 - x) \quad = \quad 8761$$

Step 4 **Solve.**

$$0.01325x + 11,931 - 0.0291x = 8761 \qquad \text{Distributive property}$$

$$11,931 - 0.01585x = 8761 \qquad \text{Combine like terms.}$$

$$-0.01585x = -3170 \qquad \text{Subtract 11,931.}$$

$$x = 200,000 \qquad \text{Divide by } -0.01585.$$

Step 5 **State the answer.** The artist should invest $200,000 at 2.65% for 6 months and

$$\$410,000 - \$200,000 = \$210,000$$

at 2.91% for 1 yr to earn $8761 in interest.

Step 6 **Check.** The 6-month investment earns

$$\$200,000(0.0265)(0.5) = \$2650,$$

and the 1-yr investment earns

$$\$210,000(0.0291)(1) = \$6111.$$

The total amount of interest earned is

$$\$2650 + \$6111 = \$8761, \quad \text{as required.}$$

✔ **Now Try Exercise 35.**

Modeling with Linear Equations A **mathematical model** is an equation (or inequality) that describes the relationship between two quantities. A **linear model** is a linear equation. The next example shows how a linear model is applied.

EXAMPLE 5 **Modeling Prevention of Indoor Pollutants**

If a vented range hood removes contaminants such as carbon monoxide and nitrogen dioxide from the air at a rate of F liters of air per second, then the percent P of contaminants that are also removed from the surrounding air can be modeled by the linear equation

$$P = 1.06F + 7.18, \quad \text{where} \quad 10 \le F \le 75.$$

What flow F (to the nearest hundredth) must a range hood have to remove 50% of the contaminants from the air? (*Source: Proceedings of the Third International Conference on Indoor Air Quality and Climate.*)

SOLUTION Replace P with 50 in the linear model, and solve for F.

$$P = 1.06F + 7.18 \qquad \text{Given model}$$
$$50 = 1.06F + 7.18 \qquad \text{Let } P = 50.$$
$$42.82 = 1.06F \qquad \text{Subtract 7.18.}$$
$$F \approx 40.40 \qquad \text{Divide by 1.06.}$$

Therefore, to remove 50% of the contaminants, the flow rate must be 40.40 L of air per second.

✔ **Now Try Exercise 41.**

EXAMPLE 6 **Modeling Health Care Costs**

The projected per capita health care expenditures in the United States, where y is in dollars, and x is years after 2000, are given by the following linear equation.

$$y = 331x + 5091 \qquad \text{Linear model}$$

(*Source:* Centers for Medicare and Medicaid Services.)

(a) What were the per capita health care expenditures in the year 2010?

(b) If this model continues to describe health care expenditures, when will the per capita expenditures reach $11,000?

SOLUTION In part (a) we are given information to determine a value for x and asked to find the corresponding value of y, whereas in part (b) we are given a value for y and asked to find the corresponding value of x.

(a) The year 2010 is 10 yr after the year 2000. Let $x = 10$ and find the value of y.

$$y = 331x + 5091 \qquad \text{Given model}$$
$$y = 331(10) + 5091 \qquad \text{Let } x = 10.$$
$$y = 8401 \qquad \text{Multiply and then add.}$$

In 2010, the estimated per capita health care expenditures were $8401.

(b) Let $y = 11,000$ in the given model, and find the value of x.

$$11,000 = 331x + 5091 \qquad \text{Let } y = 11,000.$$
$$5909 = 331x \qquad \text{Subtract 5091.}$$
$$x \approx 17.9 \qquad \text{Divide by 331.}$$

17 corresponds to 2000 + 17 = 2017.

The x-value of 17.9 indicates that per capita health care expenditures are projected to reach $11,000 during the 17th year after 2000—that is, 2017.

✔ **Now Try Exercise 45.**

1.2 Exercises

CONCEPT PREVIEW *Solve each problem.*

1. *Time Traveled* How long will it take a car to travel 400 mi at an average rate of 50 mph?

2. *Distance Traveled* If a train travels at 80 mph for 15 min, what is the distance traveled?

3. *Investing* If a person invests $500 at 2% simple interest for 4 yr, how much interest is earned?

4. *Value of Coins* If a jar of coins contains 40 half-dollars and 200 quarters, what is the monetary value of the coins?

5. *Acid Mixture* If 120 L of an acid solution is 75% acid, how much pure acid is there in the mixture?

6. *Sale Price* Suppose that a computer that originally sold for x dollars has been discounted 60%. Which one of the following expressions does not represent its sale price?

   A. $x - 0.60x$ **B.** $0.40x$ **C.** $\dfrac{4}{10}x$ **D.** $x - 0.60$

7. *Acid Mixture* Suppose two acid solutions are mixed. One is 26% acid and the other is 34% acid. Which one of the following concentrations cannot possibly be the concentration of the mixture?

   A. 24%   **B.** 30%   **C.** 31%   **D.** 33%

8. *Unknown Numbers* Consider the following problem.

   The difference between seven times a number and 9 is equal to five times the sum of the number and 2. Find the number.

   If x represents the number, which equation is correct for solving this problem?

   A. $7x - 9 = 5(x + 2)$ **B.** $9 - 7x = 5(x + 2)$

   C. $7x - 9 = 5x + 2$ **D.** $9 - 7x = 5x + 2$

9. *Unknown Numbers* Consider the following problem.

   One number is 3 less than 6 times a second number. Their sum is 46. Find the numbers.

   If x represents the second number, which equation is correct for solving this problem?

   A. $46 - (x + 3) = 6x$ **B.** $(3 - 6x) + x = 46$

   C. $46 - (3 - 6x) = x$ **D.** $(6x - 3) + x = 46$

10. *Dimensions of a Rectangle* Which one or more of the following cannot be a correct equation to solve a geometry problem, if x represents the length of a rectangle? (*Hint:* Solve each equation and consider the solution.)

    A. $2x + 2(x - 1) = 14$ **B.** $-2x + 7(5 - x) = 52$

    C. $5(x + 2) + 5x = 10$ **D.** $2x + 2(x - 3) = 22$

Note: **Geometry formulas can be found on the back inside cover of this book.**

Solve each problem. **See Example 1.**

11. *Perimeter of a Rectangle* The perimeter of a rectangle is 294 cm. The width is 57 cm. Find the length.

12. *Perimeter of a Storage Shed* Michael must build a rectangular storage shed. He wants the length to be 6 ft greater than the width, and the perimeter will be 44 ft. Find the length and the width of the shed.

13. *Dimensions of a Puzzle Piece* A puzzle piece in the shape of a triangle has perimeter 30 cm. Two sides of the triangle are each twice as long as the shortest side. Find the length of the shortest side. (Side lengths in the figure are in centimeters.)

14. *Dimensions of a Label* The length of a rectangular label is 2.5 cm less than twice the width. The perimeter is 40.6 cm. Find the width. (Side lengths in the figure are in centimeters.)

15. *Perimeter of a Plot of Land* The perimeter of a triangular plot of land is 2400 ft. The longest side is 200 ft less than twice the shortest. The middle side is 200 ft less than the longest side. Find the lengths of the three sides of the triangular plot.

16. *World Largest Ice Cream Cake* The world's largest ice cream cake, a rectangular cake made by a Dairy Queen in Toronto, Ontario, Canada, on May 10, 2011, had length 0.39 m greater than its width. Its perimeter was 17.02 m. What were the length and width of this 10-ton cake? (*Source:* www.guinnessworldrecords.com)

17. *Storage Bin Dimensions* A storage bin is in the shape of a rectangular box. Find the height of the box if its length is 18 ft, its width is 8 ft, and its surface area is 496 ft². (In the figure, h = height. Assume that the given surface area includes that of the top lid of the box.)

18. *Cylinder Dimensions* A right circular cylinder has radius 6 in. and volume 144π in.³. What is its height? (In the figure, h = height.)

h is in inches.

Solve each problem. See Example 2.

19. *Distance to an Appointment* Margaret drove to a business appointment at 50 mph. Her average speed on the return trip was 40 mph. The return trip took $\frac{1}{4}$ hr longer because of heavy traffic. How far did she travel to the appointment?

	r	t	d
Morning	50	x	
Afternoon	40	$x + \frac{1}{4}$	

20. *Distance between Cities* Elwyn averaged 50 mph traveling from Denver to Minneapolis. Returning by a different route that covered the same number of miles, he averaged 55 mph. What is the distance between the two cities to the nearest ten miles if his total traveling time was 32 hr?

	r	t	d
Going	50	x	
Returning	55	$32 - x$	

21. *Distance to Work* David gets to work in 20 min when he drives his car. Riding his bike (by the same route) takes him 45 min. His average driving speed is 4.5 mph greater than his average speed on his bike. How far does he travel to work?

22. *Speed of a Plane* Two planes leave Los Angeles at the same time. One heads south to San Diego, while the other heads north to San Francisco. The San Diego plane flies 50 mph slower than the San Francisco plane. In $\frac{1}{2}$ hr, the planes are 275 mi apart. What are their speeds?

23. *Running Times* Mary and Janet are running in the Apple Hill Fun Run. Mary runs at 7 mph, Janet at 5 mph. If they start at the same time, how long will it be before they are 1.5 mi apart?

24. *Running Times* If the run in **Exercise 23** has a staggered start, and Janet starts first, with Mary starting 10 min later, how long will it be before Mary catches up with Janet?

25. *Track Event Speeds* At the 2008 Summer Olympics in Beijing, China, Usain Bolt (Jamaica) set a new Olympic and world record in the 100-m dash with a time of 9.69 sec. If this pace could be maintained for an entire 26-mi marathon, what would his time be? How would this time compare to the fastest time for a marathon, which is 2 hr, 3 min, 23 sec, set in 2013? (*Hint:* 1 m $\approx$ 3.281 ft.) (*Source: Sports Illustrated Almanac.*)

26. *Track Event Speeds* On August 16, 2009, at the World Track and Field Championship in Berlin, Usain Bolt set a new world record in the 100-m dash with a time of 9.58 sec. Refer to **Exercise 25** and answer the questions using Bolt's 2009 time. (*Source: Sports Illustrated Almanac.*)

27. *Boat Speed* Callie took 20 min to drive her boat upstream to water-ski at her favorite spot. Coming back later in the day, at the same boat speed, took her 15 min. If the current in that part of the river is 5 km per hr, what was her boat speed?

28. *Wind Speed* Joe traveled against the wind in a small plane for 3 hr. The return trip with the wind took 2.8 hr. Find the speed of the wind to the nearest tenth if the speed of the plane in still air is 180 mph.

Solve each problem. **See Example 3.**

29. *Acid Mixture* How many gallons of a 5% acid solution must be mixed with 5 gal of a 10% solution to obtain a 7% solution?

Strength	Gallons of Solution	Gallons of Pure Acid
5%	x	
10%	5	
7%	$x + 5$	

30. *Acid Mixture* A student needs 10% hydrochloric acid for a chemistry experiment. How much 5% acid should she mix with 60 mL of 20% acid to get a 10% solution?

Strength	mL of Solution	mL of Pure Acid
5%	x	
20%	60	
10%	$x + 60$	

31. *Alcohol Mixture* Beau wishes to strengthen a mixture from 10% alcohol to 30% alcohol. How much pure alcohol should be added to 7 L of the 10% mixture?

32. *Alcohol Mixture* How many gallons of pure alcohol should be mixed with 20 gal of a 15% alcohol solution to obtain a mixture that is 25% alcohol?

33. *Saline Solution* How much water should be added to 8 mL of 6% saline solution to reduce the concentration to 4%?

34. *Acid Mixture* How much pure acid should be added to 18 L of 30% acid to increase the concentration to 50% acid?

Solve each problem. See Example 4.

35. *Real Estate Financing* Cody wishes to sell a piece of property for $240,000. He wants the money to be paid off in two ways: a short-term note at 2% interest and a long-term note at 2.5%. Find the amount of each note if the total annual interest paid is $5500.

P Note Amount	r Interest Rate (%)	t Time (in years)	I Interest Paid
x	2	1	$x(0.02)(1)$
$240,000 - x$	2.5	1	$(240,000 - x)(0.025)(1)$

36. *Buying and Selling Land* Roger bought two plots of land for a total of $120,000. When he sold the first plot, he made a profit of 15%. When he sold the second, he lost 10%. His total profit was $5500. How much did he pay for each piece of land?

Land Price	Rate of Return (%)	Profit (or Loss)
x	15%	
$120,000 - x$	10%	

37. *Retirement Planning* In planning her retirement, Janet deposits some money at 2.5% interest, with twice as much deposited at 3%. Find the amount deposited at each rate if the total annual interest income is $850.

38. *Investing a Building Fund* A church building fund has invested some money in two ways: part of the money at 3% interest and four times as much at 2.75%. Find the amount invested at each rate if the total annual income from interest is $2800.

39. *Lottery Winnings* Linda won $200,000 in a state lottery. She first paid income tax of 30% on the winnings. She invested some of the rest at 1.5% and some at 4%, earning $4350 interest per year. How much did she invest at each rate?

40. *Cookbook Royalties* Becky earned $48,000 from royalties on her cookbook. She paid a 28% income tax on these royalties. The balance was invested in two ways, some of it at 3.25% interest and some at 1.75%. The investments produced $904.80 interest per year. Find the amount invested at each rate.

(Modeling) Solve each problem. See Examples 5 and 6.

41. *Warehouse Club Membership* If the annual fee for a warehouse club membership is $100 and the reward rate is 2% on club purchases for the year, then the linear equation

$$y = 100 - 0.02x$$

models the actual annual cost of the membership y, in dollars. Here x represents the annual amount of club purchases, also in dollars.

(a) Determine the actual annual cost of the membership if club purchases for the year are $2400.

(b) What amount of club purchases would reduce the actual annual cost of the membership to $50?

(c) How much would a member have to spend in annual club purchases to reduce the annual membership cost to $0?

42. *Warehouse Club Membership* Suppose that the annual fee for a warehouse club membership is $50 and that the reward rate on club purchases for the year is 1.6%. Then the actual annual cost of a membership y, in dollars, for an amount of annual club purchases x, in dollars, can be modeled by the following linear equation.

$$y = 50 - 0.016x$$

(a) Determine the actual annual cost of the membership if club purchases for the year are $1500.

(b) What amount of club purchases would reduce the actual annual cost of the membership to $0?

(c) If club purchases for the year exceed $3125, how is the actual annual membership cost affected?

43. *Indoor Air Pollution* Formaldehyde is an indoor air pollutant formerly found in plywood, foam insulation, and carpeting. When concentrations in the air reach 33 micrograms per cubic foot ($\mu g/ft^3$), eye irritation can occur. One square foot of new plywood could emit 140 μg per hr. (*Source:* A. Hines, *Indoor Air Quality & Control.*)

(a) A room has 100 ft^2 of new plywood flooring. Find a linear equation F that computes the amount of formaldehyde, in micrograms, emitted in x hours.

(b) The room contains 800 ft^3 of air and has no ventilation. Determine how long it would take for concentrations to reach 33 $\mu g/ft^3$. (Round to the nearest tenth.)

44. *Classroom Ventilation* According to the American Society of Heating, Refrigerating and Air-Conditioning Engineers, Inc. (ASHRAE), a nonsmoking classroom should have a ventilation rate of 15 ft^3 per min for each person in the room.

(a) Write an equation that models the total ventilation V (in cubic feet per hour) necessary for a classroom with x students.

(b) A common unit of ventilation is air change per hour (ach). One ach is equivalent to exchanging all the air in a room every hour. If x students are in a classroom having volume 15,000 ft^3, determine how many air exchanges per hour (A) are necessary to keep the room properly ventilated.

(c) Find the necessary number of ach (A) if the classroom has 40 students in it.

(d) In areas like bars and lounges that allow smoking, the ventilation rate should be increased to 50 ft^3 per min per person. Compared to classrooms, ventilation should be increased by what factor in heavy smoking areas?

45. *College Enrollments* The graph shows the projections in total enrollment at degree-granting institutions from fall 2014 to fall 2021.

Enrollments at Degree-Granting Institutions

Source: U.S. Department of Education, National Center for Education Statistics.

The following linear model provides the approximate enrollment, in millions, between the years 2014 and 2021, where $x = 0$ corresponds to 2014, $x = 1$ to 2015, and so on, and y is in millions of students.

$$y = 0.3143x + 21.95$$

(a) Use the model to determine projected enrollment for fall 2018.

(b) Use the model to determine the year in which enrollment is projected to reach 24 million.

(c) How do your answers to parts (a) and (b) compare to the corresponding values shown in the graph?

(d) The actual enrollment in fall 2000 was 15.3 million. The model here is based on data from 2014 to 2021. If we were to use the model for 2000, what would the projected enrollment be?

(e) Compare the actual value and the value based on the model in part (d). Discuss the pitfalls of using the model to predict enrollment for years preceding 2014.

46. *Baby Boom* U.S. population during the years between 1946 and 1964, commonly known as the Baby Boom, can be modeled by the following linear equation.

$$y = 2.8370x + 140.83$$

Here y represents the population in millions as of July 1 of a given year, and x represents number of years after 1946. Thus, $x = 0$ corresponds to 1946, $x = 1$ corresponds to 1947, and so on. (*Source:* U.S. Census Bureau.)

(a) According to the model, what was the U.S. population on July 1, 1952?

(b) In what year did the U.S. population reach 150 million?

1.3 Complex Numbers

- **Basic Concepts of Complex Numbers**
- **Operations on Complex Numbers**

Basic Concepts of Complex Numbers   The set of real numbers does not include all the numbers needed in algebra. For example, there is no real number solution of the equation

$$x^2 = -1$$

because no real number, when squared, gives -1. To extend the real number system to include solutions of equations of this type, the number i is defined.

Imaginary Unit i

$$i = \sqrt{-1}, \quad \text{and therefore,} \quad i^2 = -1.$$

(Note that $-i$ is also a square root of -1.)

Complex numbers are formed by adding real numbers and multiples of i.

Complex Number

If a and b are real numbers, then any number of the form

$$a + bi$$

is a **complex number.** In the complex number $a + bi$, a is the **real part** and b is the **imaginary part.***

Square roots of negative numbers were not incorporated into an integrated number system until the 16th century. They were then used as solutions of equations and later (in the 18th century) in surveying. Today, such numbers are used extensively in science and engineering.

Two complex numbers $a + bi$ and $c + di$ are equal provided that their real parts are equal and their imaginary parts are equal—that is, they are equal if and only if $a = c$ and $b = d$.

*In some texts, the term bi is defined to be the imaginary part.

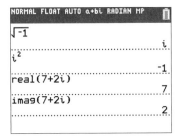

The calculator is in complex number mode. This screen supports the definition of i. It also shows how the calculator returns the real and imaginary parts of the complex number $7 + 2i$.

Figure 3

Some graphing calculators, such as the TI-84 Plus, are capable of working with complex numbers, as seen in **Figure 3.** ■

The following important concepts apply to a complex number $a + bi$.

1. If $b = 0$, then $a + bi = a$, which is a real number. (This means that the set of real numbers is a subset of the set of complex numbers. See **Figure 4.**)

2. If $b \neq 0$, then $a + bi$ is a **nonreal complex number.**
   Examples: $7 + 2i$, $-1 - i$

3. If $a = 0$ and $b \neq 0$, then the nonreal complex number is a **pure imaginary number.**
   Examples: $3i$, $-16i$

The form $a + bi$ (or $a + ib$) is **standard form.** (The form $a + ib$ is used to write expressions such as $i\sqrt{5}$, because $\sqrt{5}i$ could be mistaken for $\sqrt{5i}$.)

The relationships among the subsets of the complex numbers are shown in **Figure 4.**

Figure 4

For a positive real number a, the expression $\sqrt{-a}$ is defined as follows.

Meaning of $\sqrt{-a}$

For $a > 0$, $\qquad\qquad\qquad \sqrt{-a} = i\sqrt{a}.$

EXAMPLE 1 Writing $\sqrt{-a}$ as $i\sqrt{a}$

Write each number as the product of a real number and i.

(a) $\sqrt{-16}$ **(b)** $\sqrt{-70}$ **(c)** $\sqrt{-48}$

SOLUTION

(a) $\sqrt{-16} = i\sqrt{16} = 4i$ **(b)** $\sqrt{-70} = i\sqrt{70}$

(c) $\sqrt{-48} = i\sqrt{48} = i\sqrt{16 \cdot 3} = 4i\sqrt{3}$ Product rule for radicals

✔ **Now Try Exercises 21, 23, and 25.**

Operations on Complex Numbers Products or quotients with negative radicands are simplified by first rewriting $\sqrt{-a}$ as $i\sqrt{a}$ for a positive number a.

CAUTION *When working with negative radicands, use the definition* $\sqrt{-a} = i\sqrt{a}$ *before using any of the other rules for radicals.* In particular, the rule $\sqrt{c} \cdot \sqrt{d} = \sqrt{cd}$ is valid only when c and d are *not* both negative. For example, consider the following.

$$\sqrt{-4} \cdot \sqrt{-9} = 2i \cdot 3i = 6i^2 = -6 \quad \text{Correct}$$

$$\sqrt{-4} \cdot \sqrt{-9} = \sqrt{(-4)(-9)} = \sqrt{36} = 6 \quad \text{Incorrect}$$

EXAMPLE 2 Finding Products and Quotients Involving $\sqrt{-a}$

Find each product or quotient. Simplify the answers.

(a) $\sqrt{-7} \cdot \sqrt{-7}$ **(b)** $\sqrt{-6} \cdot \sqrt{-10}$ **(c)** $\dfrac{\sqrt{-20}}{\sqrt{-2}}$ **(d)** $\dfrac{\sqrt{-48}}{\sqrt{24}}$

SOLUTION

(a) $\sqrt{-7} \cdot \sqrt{-7}$

$= i\sqrt{7} \cdot i\sqrt{7}$ ← First write all square roots in terms of i.

$= i^2 \cdot \left(\sqrt{7}\right)^2$

$= -1 \cdot 7$ $i^2 = -1; \left(\sqrt{a}\right)^2 = a$

$= -7$ Multiply.

(b) $\sqrt{-6} \cdot \sqrt{-10}$

$= i\sqrt{6} \cdot i\sqrt{10}$

$= i^2 \cdot \sqrt{60}$

$= -1\sqrt{4 \cdot 15}$

$= -1 \cdot 2\sqrt{15}$

$= -2\sqrt{15}$

(c) $\dfrac{\sqrt{-20}}{\sqrt{-2}} = \dfrac{i\sqrt{20}}{i\sqrt{2}} = \sqrt{\dfrac{20}{2}} = \sqrt{10}$ Quotient rule for radicals

(d) $\dfrac{\sqrt{-48}}{\sqrt{24}} = \dfrac{i\sqrt{48}}{\sqrt{24}} = i\sqrt{\dfrac{48}{24}} = i\sqrt{2}$ Quotient rule for radicals

✔ **Now Try Exercises 29, 31, 33, and 35.**

EXAMPLE 3 Simplifying a Quotient Involving $\sqrt{-a}$

Write $\dfrac{-8 + \sqrt{-128}}{4}$ in standard form $a + bi$.

SOLUTION $\dfrac{-8 + \sqrt{-128}}{4}$

$= \dfrac{-8 + \sqrt{-64 \cdot 2}}{4}$ Product rule for radicals

$= \dfrac{-8 + 8i\sqrt{2}}{4}$ $\sqrt{-64} = 8i$

Be sure to factor before simplifying. $= \dfrac{4\left(-2 + 2i\sqrt{2}\right)}{4}$ Factor.

$= -2 + 2i\sqrt{2}$ Lowest terms; standard form   ✔ **Now Try Exercise 41.**

With the definitions $i^2 = -1$ and $\sqrt{-a} = i\sqrt{a}$ for $a > 0$, all properties of real numbers are extended to complex numbers. As a result, complex numbers are added, subtracted, multiplied, and divided using real number properties and the following definitions.

Addition and Subtraction of Complex Numbers

For complex numbers $a + bi$ and $c + di$,

$$(a + bi) + (c + di) = (a + c) + (b + d)i$$

and $\qquad (a + bi) - (c + di) = (a - c) + (b - d)i.$

That is, to add or subtract complex numbers, add or subtract the real parts and add or subtract the imaginary parts.

EXAMPLE 4 **Adding and Subtracting Complex Numbers**

Find each sum or difference. Write answers in standard form.

(a) $(3 - 4i) + (-2 + 6i)$ **(b)** $(-4 + 3i) - (6 - 7i)$

SOLUTION

(a) $(3 - 4i) + (-2 + 6i)$

$\qquad$ Add real $\qquad$ Add imaginary
$\qquad\quad$ parts. $\qquad\qquad$ parts.

$\qquad = \overbrace{[3 + (-2)]} + \overbrace{[-4 + 6]}i \qquad$ Commutative, associative, distributive properties

$\qquad = 1 + 2i \qquad$ Standard form

(b) $(-4 + 3i) - (6 - 7i)$

$\qquad = (-4 - 6) + [3 - (-7)]i \qquad$ Subtract real parts. Subtract imaginary parts.

$\qquad = -10 + 10i \qquad$ Standard form

✔ **Now Try Exercises 47 and 49.**

The product of two complex numbers is found by multiplying as though the numbers were binomials and using the fact that $i^2 = -1$, as follows.

$\qquad (a + bi)(c + di)$

$\qquad = ac + adi + bic + bidi \qquad$ FOIL method

$\qquad = ac + adi + bci + bdi^2 \qquad$ Commutative property; Multiply.

$\qquad = ac + (ad + bc)i + bd(-1) \qquad$ Distributive property; $i^2 = -1$

$\qquad = (ac - bd) + (ad + bc)i \qquad$ Group like terms.

Multiplication of Complex Numbers

For complex numbers $a + bi$ and $c + di$,

$$(a + bi)(c + di) = (ac - bd) + (ad + bc)i.$$

To find a given product in routine calculations, it is often easier just to multiply as with binomials and use the fact that $i^2 = -1$.

EXAMPLE 5 Multiplying Complex Numbers

Find each product. Write answers in standard form.

(a) $(2 - 3i)(3 + 4i)$ **(b)** $(4 + 3i)^2$ **(c)** $(6 + 5i)(6 - 5i)$

SOLUTION

(a) $(2 - 3i)(3 + 4i)$

$\quad = 2(3) + 2(4i) - 3i(3) - 3i(4i)$ FOIL method

$\quad = 6 + 8i - 9i - 12i^2$ Multiply.

$\quad = 6 - i - 12(-1)$ Combine like terms; $i^2 = -1$

$\quad = 18 - i$ Standard form

(b) $(4 + 3i)^2$

$\quad = 4^2 + 2(4)(3i) + (3i)^2$ Square of a binomial

Remember to add twice the product of the two terms.

$\quad = 16 + 24i + 9i^2$ Multiply.

$\quad = 16 + 24i + 9(-1)$ $i^2 = -1$

$\quad = 7 + 24i$ Standard form

NORMAL FLOAT AUTO REAL RADIAN MP

(2-3i)(3+4i)

$\qquad$ 18-i

(4+3i)²

$\qquad$ 7+24i

(6+5i)(6-5i)

$\qquad$ 61

This screen shows how the TI-84 Plus displays the results found in **Example 5.**

(c) $(6 + 5i)(6 - 5i)$

$\quad = 6^2 - (5i)^2$ Product of the sum and difference of two terms

$\quad = 36 - 25(-1)$ Square 6; $(5i)^2 = 5^2 i^2 = 25(-1)$.

$\quad = 36 + 25$ Multiply.

$\quad = 61,$ or $61 + 0i$ Standard form

✔ **Now Try Exercises 55, 59, and 63.**

Example 5(c) showed that $(6 + 5i)(6 - 5i) = 61$. The numbers $6 + 5i$ and $6 - 5i$ differ only in the sign of their imaginary parts and are **complex conjugates.** *The product of a complex number and its conjugate is always a real number.* This product is the sum of the squares of the real and imaginary parts.

Property of Complex Conjugates

For real numbers a and b,

$$(a + bi)(a - bi) = a^2 + b^2.$$

To find the quotient of two complex numbers in standard form, we multiply both the numerator and the denominator by the complex conjugate of the denominator.

EXAMPLE 6 **Dividing Complex Numbers**

Find each quotient. Write answers in standard form.

(a) $\dfrac{3 + 2i}{5 - i}$ (b) $\dfrac{3}{i}$

SOLUTION

(a) $\dfrac{3 + 2i}{5 - i}$

$= \dfrac{(3 + 2i)(5 + i)}{(5 - i)(5 + i)}$ Multiply by the complex conjugate of the denominator in both the numerator and the denominator.

$= \dfrac{15 + 3i + 10i + 2i^2}{25 - i^2}$ Multiply.

$= \dfrac{13 + 13i}{26}$ Combine like terms; $i^2 = -1$

$= \dfrac{13}{26} + \dfrac{13i}{26}$ $\dfrac{a + bi}{c} = \dfrac{a}{c} + \dfrac{bi}{c}$

$= \dfrac{1}{2} + \dfrac{1}{2}i$ Write in lowest terms and standard form.

CHECK $\left(\dfrac{1}{2} + \dfrac{1}{2}i\right)(5 - i) = 3 + 2i$ ✓ Quotient × Divisor = Dividend

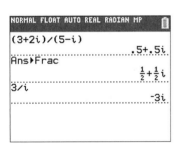

(b) $\dfrac{3}{i}$

$= \dfrac{3(-i)}{i(-i)}$ $-i$ is the conjugate of i.

$= \dfrac{-3i}{-i^2}$ Multiply.

$= \dfrac{-3i}{1}$ $-i^2 = -(-1) = 1$

$= -3i$, or $0 - 3i$ Standard form ✔ **Now Try Exercises 73 and 79.**

This screen supports the results in **Example 6.**

Powers of i can be simplified using the facts

$$i^2 = -1 \quad \text{and} \quad i^4 = (i^2)^2 = (-1)^2 = 1.$$

Consider the following powers of i.

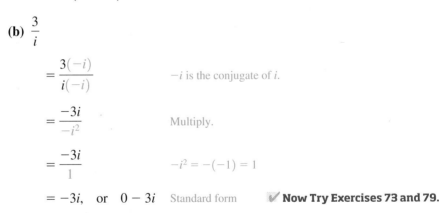

$i^1 = i$ $\qquad$ $i^5 = i^4 \cdot i = 1 \cdot i = i$

$i^2 = -1$ $\qquad$ $i^6 = i^4 \cdot i^2 = 1(-1) = -1$

$i^3 = i^2 \cdot i = (-1) \cdot i = -i$ $\qquad$ $i^7 = i^4 \cdot i^3 = 1 \cdot (-i) = -i$

$i^4 = i^2 \cdot i^2 = (-1)(-1) = 1$ $\qquad$ $i^8 = i^4 \cdot i^4 = 1 \cdot 1 = 1$ and so on.

Powers of i can be found on the TI-84 Plus calculator.

Powers of i cycle through the same four outcomes $(i, -1, -i,$ and $1)$ because i^4 has the same multiplicative property as 1. Also, any power of i with an exponent that is a multiple of 4 has value 1. As with real numbers, $i^0 = 1$.

EXAMPLE 7 **Simplifying Powers of *i***

Simplify each power of *i*.

(a) i^{15} **(b)** i^{-3}

SOLUTION

(a) Because $i^4 = 1$, write the given power as a product involving i^4.

$$i^{15} = i^{12} \cdot i^3 = (i^4)^3 \cdot i^3 = 1^3(-i) = -i$$

(b) Multiply i^{-3} by 1 in the form of i^4 to create the least positive exponent for *i*.

$$i^{-3} = i^{-3} \cdot 1 = i^{-3} \cdot i^4 = i \qquad i^4 = 1$$

✔ **Now Try Exercises 89 and 97.**

1.3 Exercises

CONCEPT PREVIEW *Fill in the blank to correctly complete each sentence.*

1. By definition, $i = $ _____, and therefore, $i^2 = $ _____.

2. If *a* and *b* are real numbers, then any number of the form $a + bi$ is a(n) _____.

3. The numbers $6 + 5i$ and $6 - 5i$, which differ only in the sign of their imaginary parts, are _____.

4. The product of a complex number and its conjugate is always a(n) _____.

5. To find the quotient of two complex numbers in standard form, multiply both the numerator and the denominator by the complex conjugate of the _____.

CONCEPT PREVIEW *Decide whether each statement is* true *or* false. *If false, correct the right side of the equation.*

6. $\sqrt{-25} = 5i$ **7.** $\sqrt{-4} \cdot \sqrt{-9} = -6$ **8.** $i^{12} = 1$

9. $(-2 + 7i) - (10 - 6i) = -12 + i$ **10.** $(5 + 3i)^2 = 16$

Concept Check *Identify each number as* real, complex, pure imaginary, *or* nonreal complex. *(More than one of these descriptions will apply.)*

11. -4 **12.** 0 **13.** $13i$ **14.** $-7i$ **15.** $5 + i$

16. $-6 - 2i$ **17.** π **18.** $\sqrt{24}$ **19.** $\sqrt{-25}$ **20.** $\sqrt{-36}$

Write each number as the product of a real number and i. ***See Example 1.***

21. $\sqrt{-25}$ **22.** $\sqrt{-36}$ **23.** $\sqrt{-10}$ **24.** $\sqrt{-15}$

25. $\sqrt{-288}$ **26.** $\sqrt{-500}$ **27.** $-\sqrt{-18}$ **28.** $-\sqrt{-80}$

Find each product or quotient. Simplify the answers. ***See Example 2.***

29. $\sqrt{-13} \cdot \sqrt{-13}$ **30.** $\sqrt{-17} \cdot \sqrt{-17}$ **31.** $\sqrt{-3} \cdot \sqrt{-8}$

32. $\sqrt{-5} \cdot \sqrt{-15}$ **33.** $\dfrac{\sqrt{-30}}{\sqrt{-10}}$ **34.** $\dfrac{\sqrt{-70}}{\sqrt{-7}}$

35. $\dfrac{\sqrt{-24}}{\sqrt{8}}$

36. $\dfrac{\sqrt{-54}}{\sqrt{27}}$

37. $\dfrac{\sqrt{-10}}{\sqrt{-40}}$

38. $\dfrac{\sqrt{-8}}{\sqrt{-72}}$

39. $\dfrac{\sqrt{-6} \cdot \sqrt{-2}}{\sqrt{3}}$

40. $\dfrac{\sqrt{-12} \cdot \sqrt{-6}}{\sqrt{8}}$

Write each number in standard form a + bi. ***See Example 3.***

41. $\dfrac{-6 - \sqrt{-24}}{2}$

42. $\dfrac{-9 - \sqrt{-18}}{3}$

43. $\dfrac{10 + \sqrt{-200}}{5}$

44. $\dfrac{20 + \sqrt{-8}}{2}$

45. $\dfrac{-3 + \sqrt{-18}}{24}$

46. $\dfrac{-5 + \sqrt{-50}}{10}$

Find each sum or difference. Write answers in standard form. ***See Example 4.***

47. $(3 + 2i) + (9 - 3i)$

48. $(4 - i) + (8 + 5i)$

49. $(-2 + 4i) - (-4 + 4i)$

50. $(-3 + 2i) - (-4 + 2i)$

51. $(2 - 5i) - (3 + 4i) - (-2 + i)$

52. $(-4 - i) - (2 + 3i) + (-4 + 5i)$

53. $-i\sqrt{2} - 2 - \left(6 - 4i\sqrt{2}\right) - \left(5 - i\sqrt{2}\right)$

54. $3\sqrt{7} - \left(4\sqrt{7} - i\right) - 4i + \left(-2\sqrt{7} + 5i\right)$

Find each product. Write answers in standard form. ***See Example 5.***

55. $(2 + i)(3 - 2i)$

56. $(-2 + 3i)(4 - 2i)$

57. $(2 + 4i)(-1 + 3i)$

58. $(1 + 3i)(2 - 5i)$

59. $(3 - 2i)^2$

60. $(2 + i)^2$

61. $(3 + i)(3 - i)$

62. $(5 + i)(5 - i)$

63. $(-2 - 3i)(-2 + 3i)$

64. $(6 - 4i)(6 + 4i)$

65. $\left(\sqrt{6} + i\right)\left(\sqrt{6} - i\right)$

66. $\left(\sqrt{2} - 4i\right)\left(\sqrt{2} + 4i\right)$

67. $i(3 - 4i)(3 + 4i)$

68. $i(2 + 7i)(2 - 7i)$

69. $3i(2 - i)^2$

70. $-5i(4 - 3i)^2$

71. $(2 + i)(2 - i)(4 + 3i)$

72. $(3 - i)(3 + i)(2 - 6i)$

Find each quotient. Write answers in standard form. ***See Example 6.***

73. $\dfrac{6 + 2i}{1 + 2i}$

74. $\dfrac{14 + 5i}{3 + 2i}$

75. $\dfrac{2 - i}{2 + i}$

76. $\dfrac{4 - 3i}{4 + 3i}$

77. $\dfrac{1 - 3i}{1 + i}$

78. $\dfrac{-3 + 4i}{2 - i}$

79. $\dfrac{-5}{i}$

80. $\dfrac{-6}{i}$

81. $\dfrac{8}{-i}$

82. $\dfrac{12}{-i}$

83. $\dfrac{2}{3i}$

84. $\dfrac{5}{9i}$

*(Modeling) **Alternating Current*** *Complex numbers are used to describe current I, voltage E, and impedance Z (the opposition to current). These three quantities are related by the equation*

$$E = IZ, \quad \text{which is known as } \textbf{Ohm's Law.}$$

Thus, if any two of these quantities are known, the third can be found. In each exercise, solve the equation E = IZ for the remaining value.

85. $I = 5 + 7i, \ Z = 6 + 4i$

86. $I = 20 + 12i, \ Z = 10 - 5i$

87. $I = 10 + 4i, \ E = 88 + 128i$

88. $E = 57 + 67i, \ Z = 9 + 5i$

Simplify each power of i. See Example 7.

89. i^{25} **90.** i^{29} **91.** i^{22} **92.** i^{26}

93. i^{23} **94.** i^{27} **95.** i^{32} **96.** i^{40}

97. i^{-13} **98.** i^{-14} **99.** $\dfrac{1}{i^{-11}}$ **100.** $\dfrac{1}{i^{-12}}$

Work each problem.

101. Show that $\dfrac{\sqrt{2}}{2} + \dfrac{\sqrt{2}}{2}i$ is a square root of i.

102. Show that $-\dfrac{\sqrt{2}}{2} - \dfrac{\sqrt{2}}{2}i$ is a square root of i.

103. Show that $\dfrac{\sqrt{3}}{2} + \dfrac{1}{2}i$ is a cube root of i.

104. Show that $-\dfrac{\sqrt{3}}{2} + \dfrac{1}{2}i$ is a cube root of i.

105. Show that $-2 + i$ is a solution of the equation $x^2 + 4x + 5 = 0$.

106. Show that $-2 - i$ is a solution of the equation $x^2 + 4x + 5 = 0$.

107. Show that $-3 + 4i$ is a solution of the equation $x^2 + 6x + 25 = 0$.

108. Show that $-3 - 4i$ is a solution of the equation $x^2 + 6x + 25 = 0$.

1.4 Quadratic Equations

- The Zero-Factor Property
- The Square Root Property
- Completing the Square
- The Quadratic Formula
- Solving for a Specified Variable
- The Discriminant

A *quadratic equation* is defined as follows.

Quadratic Equation in One Variable

An equation that can be written in the form

$$ax^2 + bx + c = 0,$$

where a, b, and c are real numbers with $a \neq 0$, is a **quadratic equation.** The given form is called **standard form.**

A quadratic equation is a **second-degree equation**—that is, an equation with a squared variable term and no terms of greater degree.

$$x^2 = 25, \quad 4x^2 + 4x - 5 = 0, \quad 3x^2 = 4x - 8 \qquad \text{Quadratic equations}$$

The Zero-Factor Property   When the expression $ax^2 + bx + c$ in a quadratic equation is easily factorable over the real numbers, it is efficient to factor and then apply the following **zero-factor property.**

Zero-Factor Property

If a and b are complex numbers with $ab = 0$, then $a = 0$ or $b = 0$ or both equal zero.

EXAMPLE 1 Using the Zero-Factor Property

Solve $6x^2 + 7x = 3$.

SOLUTION

$$6x^2 + 7x = 3 \quad \boxed{\text{Don't factor out } x \text{ here.}}$$

$$6x^2 + 7x - 3 = 0 \qquad \text{Standard form}$$

$$(3x - 1)(2x + 3) = 0 \qquad \text{Factor.}$$

$$3x - 1 = 0 \quad \text{or} \quad 2x + 3 = 0 \qquad \text{Zero-factor property}$$

$$3x = 1 \quad \text{or} \qquad 2x = -3 \qquad \text{Solve each equation.}$$

$$x = \frac{1}{3} \quad \text{or} \qquad x = -\frac{3}{2}$$

CHECK
$$6x^2 + 7x = 3 \qquad \text{Original equation}$$

$$6\left(\frac{1}{3}\right)^2 + 7\left(\frac{1}{3}\right) \overset{?}{=} 3 \quad \text{Let } x = \tfrac{1}{3}. \qquad \bigg| \qquad 6\left(-\frac{3}{2}\right)^2 + 7\left(-\frac{3}{2}\right) \overset{?}{=} 3 \quad \text{Let } x = -\tfrac{3}{2}.$$

$$\frac{6}{9} + \frac{7}{3} \overset{?}{=} 3 \qquad\qquad\qquad\qquad \frac{54}{4} - \frac{21}{2} \overset{?}{=} 3$$

$$3 = 3 \ \checkmark \ \text{True} \qquad\qquad\qquad\qquad 3 = 3 \ \checkmark \ \text{True}$$

Both values check because true statements result. The solution set is $\left\{\frac{1}{3}, -\frac{3}{2}\right\}$.

✔ **Now Try Exercise 15.**

The Square Root Property When a quadratic equation can be written in the form $x^2 = k$, where k is a constant, the equation can be solved as follows.

$$x^2 = k$$

$$x^2 - k = 0 \qquad \text{Subtract } k.$$

$$\left(x - \sqrt{k}\right)\left(x + \sqrt{k}\right) = 0 \qquad \text{Factor.}$$

$$x - \sqrt{k} = 0 \quad \text{or} \quad x + \sqrt{k} = 0 \qquad \text{Zero-factor property}$$

$$x = \sqrt{k} \quad \text{or} \qquad x = -\sqrt{k} \qquad \text{Solve each equation.}$$

This proves the **square root property.**

Square Root Property

If $x^2 = k$, then $x = \sqrt{k}$ or $x = -\sqrt{k}$.

That is, the solution set of $x^2 = k$ is

$$\left\{\sqrt{k}, -\sqrt{k}\right\}, \quad \textit{which may be abbreviated} \quad \left\{\pm\sqrt{k}\right\}.$$

Both solutions $\sqrt{k}$ and $-\sqrt{k}$ of $x^2 = k$ are real if $k > 0$. Both are pure imaginary if $k < 0$. If $k < 0$, then we write the solution set as

$$\left\{\pm i\sqrt{|k|}\right\}.$$

If $k = 0$, then there is only one distinct solution, 0, sometimes called a **double solution.**

EXAMPLE 2 Using the Square Root Property

Solve each quadratic equation.

(a) $x^2 = 17$ 　　　　**(b)** $x^2 = -25$ 　　　　**(c)** $(x-4)^2 = 12$

SOLUTION

(a) $x^2 = 17$

$\qquad x = \pm\sqrt{17}$ 　Square root property

The solution set is $\left\{\pm\sqrt{17}\right\}$.

(b) $x^2 = -25$

$\qquad x = \pm\sqrt{-25}$ 　Square root property

$\qquad x = \pm 5i$ 　　　　$\sqrt{-1} = i$

The solution set is $\{\pm 5i\}$.

(c) $\qquad (x-4)^2 = 12$

$\qquad x - 4 = \pm\sqrt{12}$ 　　Generalized square root property

$\qquad x = 4 \pm \sqrt{12}$ 　Add 4.

$\qquad x = 4 \pm 2\sqrt{3}$ 　$\sqrt{12} = \sqrt{4 \cdot 3} = 2\sqrt{3}$

CHECK $\qquad\qquad (x-4)^2 = 12$ 　Original equation

$\left(4 + 2\sqrt{3} - 4\right)^2 \stackrel{?}{=} 12$ 　Let $x = 4 + 2\sqrt{3}$.

$\left(2\sqrt{3}\right)^2 \stackrel{?}{=} 12$

$2^2 \cdot \left(\sqrt{3}\right)^2 \stackrel{?}{=} 12$

$12 = 12$ ✓ True

$\left(4 - 2\sqrt{3} - 4\right)^2 \stackrel{?}{=} 12$ 　Let $x = 4 - 2\sqrt{3}$.

$\left(-2\sqrt{3}\right)^2 \stackrel{?}{=} 12$

$(-2)^2 \cdot \left(\sqrt{3}\right)^2 \stackrel{?}{=} 12$

$12 = 12$ ✓ True

The solution set is $\left\{4 \pm 2\sqrt{3}\right\}$. 　　　　✔ **Now Try Exercises 27, 29, and 31.**

Completing the Square 　 Any quadratic equation can be solved by the method of **completing the square,** summarized in the box below. While this method may seem tedious, it has several useful applications, including analyzing the graph of a parabola and developing a general formula for solving quadratic equations.

Solving a Quadratic Equation Using Completing the Square

To solve $ax^2 + bx + c = 0$, where $a \neq 0$, using completing the square, follow these steps.

Step 1 If $a \neq 1$, divide each side of the equation by a.

Step 2 Rewrite the equation so that the constant term is alone on one side of the equality symbol.

Step 3 Square half the coefficient of x, and add this square to each side of the equation.

Step 4 Factor the resulting trinomial as a perfect square and combine like terms on the other side.

Step 5 Use the square root property to complete the solution.

EXAMPLE 3 **Using Completing the Square ($a = 1$)**

Solve $x^2 - 4x - 14 = 0$.

SOLUTION $\quad x^2 - 4x - 14 = 0$

Step 1 This step is not necessary because $a = 1$.

Step 2 $\qquad\qquad x^2 - 4x = 14 \qquad\qquad$ Add 14 to each side.

Step 3 $\qquad x^2 - 4x + 4 = 14 + 4 \qquad \left[\frac{1}{2}(-4)\right]^2 = 4$; Add 4 to each side.

Step 4 $\qquad\qquad (x - 2)^2 = 18 \qquad\qquad$ Factor. Combine like terms.

Step 5 $\qquad\qquad x - 2 = \pm\sqrt{18} \qquad\qquad$ Square root property

$\boxed{\text{Take } both \text{ roots.}}$

$\qquad\qquad\qquad x = 2 \pm \sqrt{18} \qquad\qquad$ Add 2 to each side.

$\qquad\qquad\qquad x = 2 \pm 3\sqrt{2} \qquad\qquad$ Simplify the radical.

The solution set is $\left\{2 \pm 3\sqrt{2}\right\}$. ✔ **Now Try Exercise 41.**

EXAMPLE 4 **Using Completing the Square ($a \neq 1$)**

Solve $9x^2 - 12x + 9 = 0$.

SOLUTION $\quad 9x^2 - 12x + 9 = 0$

Step 1 $\qquad x^2 - \dfrac{4}{3}x + 1 = 0 \qquad\qquad$ Divide by 9 so that $a = 1$.

Step 2 $\qquad x^2 - \dfrac{4}{3}x = -1 \qquad\qquad$ Subtract 1 from each side.

Step 3 $\qquad x^2 - \dfrac{4}{3}x + \dfrac{4}{9} = -1 + \dfrac{4}{9} \qquad \left[\frac{1}{2}\left(-\frac{4}{3}\right)\right]^2 = \frac{4}{9}$; Add $\frac{4}{9}$ to each side.

Step 4 $\qquad \left(x - \dfrac{2}{3}\right)^2 = -\dfrac{5}{9} \qquad\qquad$ Factor. Combine like terms.

Step 5 $\qquad x - \dfrac{2}{3} = \pm\sqrt{-\dfrac{5}{9}} \qquad\qquad$ Square root property

$\qquad x - \dfrac{2}{3} = \pm\dfrac{\sqrt{5}}{3}i \qquad \sqrt{-\frac{5}{9}} = \frac{\sqrt{-5}}{\sqrt{9}} = \frac{i\sqrt{5}}{3}$, or $\frac{\sqrt{5}}{3}i$

$\qquad\qquad x = \dfrac{2}{3} \pm \dfrac{\sqrt{5}}{3}i \qquad$ Add $\frac{2}{3}$ to each side.

The solution set is $\left\{\dfrac{2}{3} \pm \dfrac{\sqrt{5}}{3}i\right\}$. ✔ **Now Try Exercise 47.**

The Quadratic Formula If we start with the equation $ax^2 + bx + c = 0$, for $a > 0$, and complete the square to solve for x in terms of the constants a, b, and c, the result is a general formula for solving any quadratic equation.

$$ax^2 + bx + c = 0$$

$$x^2 + \frac{b}{a}x + \frac{c}{a} = 0 \qquad \text{Divide each side by } a. \text{ (Step 1)}$$

$$x^2 + \frac{b}{a}x = -\frac{c}{a} \qquad \text{Subtract } \tfrac{c}{a} \text{ from each side. (Step 2)}$$

Square half the coefficient of x: $\left[\frac{1}{2}\left(\frac{b}{a}\right)\right]^2 = \left(\frac{b}{2a}\right)^2 = \frac{b^2}{4a^2}$.

$$x^2 + \frac{b}{a}x + \frac{b^2}{4a^2} = -\frac{c}{a} + \frac{b^2}{4a^2}$$ Add $\frac{b^2}{4a^2}$ to each side. (Step 3)

$$\left(x + \frac{b}{2a}\right)^2 = \frac{b^2}{4a^2} + \frac{-c}{a}$$ Factor. Use the commutative property. (Step 4)

$$\left(x + \frac{b}{2a}\right)^2 = \frac{b^2}{4a^2} + \frac{-4ac}{4a^2}$$ Write fractions with a common denominator.

$$\left(x + \frac{b}{2a}\right)^2 = \frac{b^2 - 4ac}{4a^2}$$ Add fractions.

$$x + \frac{b}{2a} = \pm\sqrt{\frac{b^2 - 4ac}{4a^2}}$$ Square root property (Step 5)

$$x + \frac{b}{2a} = \frac{\pm\sqrt{b^2 - 4ac}}{2a}$$ Since $a > 0$, $\sqrt{4a^2} = 2a$.

$$x = \frac{-b}{2a} \pm \frac{\sqrt{b^2 - 4ac}}{2a}$$ Subtract $\frac{b}{2a}$ from each side.

Quadratic Formula
This result is also $\longrightarrow$ $x = \dfrac{-b \pm \sqrt{b^2 - 4ac}}{2a}$ Combine terms on the right.
true for $a < 0$.

Quadratic Formula

The solutions of the quadratic equation $ax^2 + bx + c = 0$, where $a \neq 0$, are given by the quadratic formula.

$$x = \frac{-b \pm \sqrt{b^2 - 4ac}}{2a}$$

EXAMPLE 5   Using the Quadratic Formula (Real Solutions)

Solve $x^2 - 4x = -2$.

SOLUTION $x^2 - 4x + 2 = 0$ Write in standard form. Here $a = 1$, $b = -4$, and $c = 2$.

$$x = \frac{-b \pm \sqrt{b^2 - 4ac}}{2a}$$ Quadratic formula

$$x = \frac{-(-4) \pm \sqrt{(-4)^2 - 4(1)(2)}}{2(1)}$$ Substitute $a = 1$, $b = -4$, and $c = 2$.

The fraction bar extends *under* $-b$.

$$x = \frac{4 \pm \sqrt{16 - 8}}{2}$$ Simplify.

$$x = \frac{4 \pm 2\sqrt{2}}{2}$$ $\sqrt{16 - 8} = \sqrt{8} = \sqrt{4 \cdot 2} = 2\sqrt{2}$

$$x = \frac{2(2 \pm \sqrt{2})}{2}$$ Factor out 2 in the numerator.

Factor first, then divide.

$$x = 2 \pm \sqrt{2}$$ Lowest terms

The solution set is $\left\{2 \pm \sqrt{2}\right\}$.

✔ **Now Try Exercise 53.**

CAUTION *Remember to extend the fraction bar in the quadratic formula under the* $-b$ *term in the numerator.*

Throughout this text, unless otherwise specified, we use the set of complex numbers as the domain when solving equations of degree 2 or greater.

EXAMPLE 6 **Using the Quadratic Formula (Nonreal Complex Solutions)**

Solve $2x^2 = x - 4$.

SOLUTION

$$2x^2 - x + 4 = 0 \qquad \text{Write in standard form.}$$

$$x = \frac{-(-1) \pm \sqrt{(-1)^2 - 4(2)(4)}}{2(2)} \qquad \begin{array}{l}\text{Quadratic formula with} \\ a = 2, b = -1, c = 4\end{array}$$

Use parentheses and substitute carefully to avoid errors.

$$x = \frac{1 \pm \sqrt{1 - 32}}{4}$$

$$x = \frac{1 \pm \sqrt{-31}}{4} \qquad \text{Simplify.}$$

$$x = \frac{1 \pm i\sqrt{31}}{4} \qquad \sqrt{-1} = i$$

The solution set is $\left\{\frac{1}{4} \pm \frac{\sqrt{31}}{4}i\right\}$. ✔ **Now Try Exercise 57.**

The equation $x^3 + 8 = 0$ is a **cubic equation** because the greatest degree of the terms is 3. While a quadratic equation (degree 2) can have as many as two solutions, a cubic equation (degree 3) can have as many as three solutions. The maximum possible number of solutions corresponds to the degree of the equation.

EXAMPLE 7 **Solving a Cubic Equation**

Solve $x^3 + 8 = 0$ using factoring and the quadratic formula.

SOLUTION $\qquad x^3 + 8 = 0$

$$(x + 2)(x^2 - 2x + 4) = 0 \qquad \begin{array}{l}\text{Factor as a sum} \\ \text{of cubes.}\end{array}$$

$$x + 2 = 0 \quad \text{or} \quad x^2 - 2x + 4 = 0 \qquad \text{Zero-factor property}$$

$$x = -2 \quad \text{or} \quad x = \frac{-(-2) \pm \sqrt{(-2)^2 - 4(1)(4)}}{2(1)} \qquad \begin{array}{l}\text{Quadratic formula with} \\ a = 1, b = -2, c = 4\end{array}$$

$$x = \frac{2 \pm \sqrt{-12}}{2} \qquad \text{Simplify.}$$

$$x = \frac{2 \pm 2i\sqrt{3}}{2} \qquad \text{Simplify the radical.}$$

$$x = \frac{2\left(1 \pm i\sqrt{3}\right)}{2} \qquad \begin{array}{l}\text{Factor out 2 in the} \\ \text{numerator.}\end{array}$$

$$x = 1 \pm i\sqrt{3} \qquad \begin{array}{l}\text{Divide out the} \\ \text{common factor.}\end{array}$$

The solution set is $\left\{-2, 1 \pm i\sqrt{3}\right\}$. ✔ **Now Try Exercise 67.**

Solving for a Specified Variable   To solve a quadratic equation for a specified variable, we usually apply the square root property or the quadratic formula.

EXAMPLE 8   **Solving for a Quadratic Variable in a Formula**

Solve each equation for the specified variable. Use $\pm$ when taking square roots.

(a) $\mathcal{A} = \dfrac{\pi d^2}{4}$,   for d

(b) $rt^2 - st = k\ (r \neq 0)$,   for t

SOLUTION

(a)

$$\mathcal{A} = \dfrac{\pi d^2}{4} \quad \text{Goal: Isolate } d \text{, the specified variable.}$$

$$4\mathcal{A} = \pi d^2 \qquad \text{Multiply each side by 4.}$$

$$\dfrac{4\mathcal{A}}{\pi} = d^2 \qquad \text{Divide each side by } \pi.$$

$$d = \pm\sqrt{\dfrac{4\mathcal{A}}{\pi}} \qquad \begin{array}{l}\text{Interchange sides;}\\ \text{square root property}\end{array}$$

See the Note following this example.

$$d = \dfrac{\pm\sqrt{4\mathcal{A}}}{\sqrt{\pi}} \cdot \dfrac{\sqrt{\pi}}{\sqrt{\pi}} \qquad \text{Multiply by } \tfrac{\sqrt{\pi}}{\sqrt{\pi}}.$$

$$d = \dfrac{\pm\sqrt{4\mathcal{A}\pi}}{\pi} \qquad \begin{array}{l}\text{Multiply numerators.}\\ \text{Multiply denominators.}\end{array}$$

$$d = \dfrac{\pm 2\sqrt{\mathcal{A}\pi}}{\pi} \qquad \text{Simplify the radical.}$$

(b) Because $rt^2 - st = k$ has terms with t^2 and t, use the quadratic formula.

$$rt^2 - st - k = 0 \qquad \text{Write in standard form.}$$

$$t = \dfrac{-b \pm \sqrt{b^2 - 4ac}}{2a} \qquad \text{Quadratic formula}$$

$$t = \dfrac{-(-s) \pm \sqrt{(-s)^2 - 4(r)(-k)}}{2(r)} \qquad \text{Here, } a = r,\ b = -s, \text{ and } c = -k.$$

$$t = \dfrac{s \pm \sqrt{s^2 + 4rk}}{2r} \qquad \text{Simplify.}$$

✔ **Now Try Exercises 71 and 77.**

NOTE   In **Example 8,** we took both positive and negative square roots. However, if the variable represents time or length in an application, we consider only the *positive* square root.

The Discriminant   The quantity under the radical in the quadratic formula, $b^2 - 4ac$, is the **discriminant.**

$$x = \dfrac{-b \pm \sqrt{b^2 - 4ac}}{2a} \leftarrow \text{Discriminant}$$

When the numbers a, b, and c are *integers* (but not necessarily otherwise), the value of the discriminant $b^2 - 4ac$ can be used to determine whether the solutions of a quadratic equation are rational, irrational, or nonreal complex numbers. The number and type of solutions based on the value of the discriminant are shown in the following table.

Solutions of Quadratic Equations

Discriminant	Number of Solutions	Type of Solutions	
Positive, perfect square	Two	Rational	
Positive, but not a perfect square	Two	Irrational	← As seen in Example 5
Zero	One (a double solution)	Rational	
Negative	Two	Nonreal complex	← As seen in Example 6

CAUTION *The restriction on a, b, and c is important.* For example,

$$x^2 - \sqrt{5}x - 1 = 0 \quad \text{has discriminant} \quad b^2 - 4ac = 5 + 4 = 9,$$

which would indicate two rational solutions *if the coefficients were integers.* By the quadratic formula, the two solutions $\dfrac{\sqrt{5} \pm 3}{2}$ are *irrational* numbers.

EXAMPLE 9 **Using the Discriminant**

Evaluate the discriminant for each equation. Then use it to determine the number of distinct solutions, and tell whether they are *rational, irrational*, or *nonreal complex* numbers.

(a) $5x^2 + 2x - 4 = 0$ **(b)** $x^2 - 10x = -25$ **(c)** $2x^2 - x + 1 = 0$

SOLUTION

(a) For $5x^2 + 2x - 4 = 0$, use $a = 5$, $b = 2$, and $c = -4$.

$$b^2 - 4ac = 2^2 - 4(5)(-4) = 84 \leftarrow \text{Discriminant}$$

The discriminant 84 is positive and not a perfect square, so there are two distinct irrational solutions.

(b) First, write the equation in standard form as

$$x^2 - 10x + 25 = 0.$$

Thus, $a = 1$, $b = -10$, and $c = 25$.

$$b^2 - 4ac = (-10)^2 - 4(1)(25) = 0 \leftarrow \text{Discriminant}$$

There is one distinct rational solution, a double solution.

(c) For $2x^2 - x + 1 = 0$, use $a = 2$, $b = -1$, and $c = 1$.

$$b^2 - 4ac = (-1)^2 - 4(2)(1) = -7 \leftarrow \text{Discriminant}$$

There are two distinct nonreal complex solutions. (They are complex conjugates.)

✔ Now Try Exercises 83, 85, and 89.

1.4 Exercises

CONCEPT PREVIEW *Match the equation in Column I with its solution(s) in Column II.*

I

1. $x^2 = 25$ **2.** $x^2 = -25$

3. $x^2 + 5 = 0$ **4.** $x^2 - 5 = 0$

5. $x^2 = -20$ **6.** $x^2 = 20$

7. $x - 5 = 0$ **8.** $x + 5 = 0$

II

A. $\pm 5i$ **B.** $\pm 2\sqrt{5}$

C. $\pm i\sqrt{5}$ **D.** 5

E. $\pm \sqrt{5}$ **F.** -5

G. ± 5 **H.** $\pm 2i\sqrt{5}$

CONCEPT PREVIEW *Use Choices A–D to answer each question.*

A. $3x^2 - 17x - 6 = 0$ **B.** $(2x + 5)^2 = 7$

C. $x^2 + x = 12$ **D.** $(3x - 1)(x - 7) = 0$

9. Which equation is set up for direct use of the zero-factor property? Solve it.

10. Which equation is set up for direct use of the square root property? Solve it.

11. Only one of the equations does not require Step 1 of the method for completing the square described in this section. Which one is it? Solve it.

12. Only one of the equations is set up so that the values of a, b, and c can be determined immediately. Which one is it? Solve it.

Solve each equation using the zero-factor property. ***See Example 1.***

13. $x^2 - 5x + 6 = 0$ **14.** $x^2 + 2x - 8 = 0$ **15.** $5x^2 - 3x - 2 = 0$

16. $2x^2 - x - 15 = 0$ **17.** $-4x^2 + x = -3$ **18.** $-6x^2 + 7x = -10$

19. $x^2 - 100 = 0$ **20.** $x^2 - 64 = 0$ **21.** $4x^2 - 4x + 1 = 0$

22. $9x^2 - 12x + 4 = 0$ **23.** $25x^2 + 30x + 9 = 0$ **24.** $36x^2 + 60x + 25 = 0$

Solve each equation using the square root property. ***See Example 2.***

25. $x^2 = 16$ **26.** $x^2 = 121$ **27.** $27 - x^2 = 0$

28. $48 - x^2 = 0$ **29.** $x^2 = -81$ **30.** $x^2 = -400$

31. $(3x - 1)^2 = 12$ **32.** $(4x + 1)^2 = 20$ **33.** $(x + 5)^2 = -3$

34. $(x - 4)^2 = -5$ **35.** $(5x - 3)^2 = -3$ **36.** $(-2x + 5)^2 = -8$

Solve each equation using completing the square. ***See Examples 3 and 4.***

37. $x^2 - 4x + 3 = 0$ **38.** $x^2 - 7x + 12 = 0$ **39.** $2x^2 - x - 28 = 0$

40. $4x^2 - 3x - 10 = 0$ **41.** $x^2 - 2x - 2 = 0$ **42.** $x^2 - 10x + 18 = 0$

43. $2x^2 + x = 10$ **44.** $3x^2 + 2x = 5$ **45.** $-2x^2 + 4x + 3 = 0$

46. $-3x^2 + 6x + 5 = 0$ **47.** $-4x^2 + 8x = 7$ **48.** $-3x^2 + 9x = 7$

Concept Check Answer each question.

49. Francisco claimed that the equation

$$x^2 - 8x = 0$$

cannot be solved by the quadratic formula since there is no value for c. Is he correct?

50. Francesca, Francisco's twin sister, claimed that the equation

$$x^2 - 19 = 0$$

cannot be solved by the quadratic formula since there is no value for b. Is she correct?

*Solve each equation using the quadratic formula. **See Examples 5 and 6.***

51. $x^2 - x - 1 = 0$ **52.** $x^2 - 3x - 2 = 0$ **53.** $x^2 - 6x = -7$

54. $x^2 - 4x = -1$ **55.** $x^2 = 2x - 5$ **56.** $x^2 = 2x - 10$

57. $-4x^2 = -12x + 11$ **58.** $-6x^2 = 3x + 2$ **59.** $\frac{1}{2}x^2 + \frac{1}{4}x - 3 = 0$

60. $\frac{2}{3}x^2 + \frac{1}{4}x = 3$ **61.** $0.2x^2 + 0.4x - 0.3 = 0$

62. $0.1x^2 - 0.1x = 0.3$ **63.** $(4x - 1)(x + 2) = 4x$

64. $(3x + 2)(x - 1) = 3x$ **65.** $(x - 9)(x - 1) = -16$

66. *Concept Check* Why do the following two equations have the same solution set? (Do not solve.)

$$-2x^2 + 3x - 6 = 0 \quad \text{and} \quad 2x^2 - 3x + 6 = 0$$

*Solve each cubic equation using factoring and the quadratic formula. **See Example 7.***

67. $x^3 - 8 = 0$ **68.** $x^3 - 27 = 0$

69. $x^3 + 27 = 0$ **70.** $x^3 + 64 = 0$

*Solve each equation for the specified variable. (Assume no denominators are 0.) **See Example 8.***

71. $s = \frac{1}{2}gt^2$,   for t **72.** $\mathcal{A} = \pi r^2$,   for r

73. $F = \frac{kMv^2}{r}$,   for v **74.** $E = \frac{e^2 k}{2r}$,   for e

75. $r = r_0 + \frac{1}{2}at^2$,   for t **76.** $s = s_0 + gt^2 + k$,   for t

77. $h = -16t^2 + v_0 t + s_0$,   for t **78.** $S = 2\pi rh + 2\pi r^2$,   for r

*For each equation, **(a)** solve for x in terms of y, and **(b)** solve for y in terms of x. **See Example 8.***

79. $4x^2 - 2xy + 3y^2 = 2$ **80.** $3y^2 + 4xy - 9x^2 = -1$

81. $2x^2 + 4xy - 3y^2 = 2$ **82.** $5x^2 - 6xy + 2y^2 = 1$

Evaluate the discriminant for each equation. Then use it to determine the number of distinct solutions, and tell whether they are rational, irrational, *or* nonreal complex *numbers. (Do not solve the equation.) **See Example 9.***

83. $x^2 - 8x + 16 = 0$ **84.** $x^2 + 4x + 4 = 0$ **85.** $3x^2 + 5x + 2 = 0$

86. $8x^2 = -14x - 3$ **87.** $4x^2 = -6x + 3$ **88.** $2x^2 + 4x + 1 = 0$

89. $9x^2 + 11x + 4 = 0$ **90.** $3x^2 = 4x - 5$ **91.** $8x^2 - 72 = 0$

Concept Check Answer each question.

92. Show that the discriminant for the equation

$$\sqrt{2}x^2 + 5x - 3\sqrt{2} = 0$$

is 49. If this equation is completely solved, it can be shown that the solution set is $\left\{-3\sqrt{2}, \frac{\sqrt{2}}{2}\right\}$. We have a discriminant that is positive and a perfect square, yet the two solutions are irrational. Does this contradict the discussion in this section?

93. Is it possible for the solution set of a quadratic equation with integer coefficients to consist of a single irrational number?

94. Is it possible for the solution set of a quadratic equation with real coefficients to consist of one real number and one nonreal complex number?

Find the values of a, b, and c for which the quadratic equation

$$ax^2 + bx + c = 0$$

has the given numbers as solutions. (Hint: Use the zero-factor property in reverse.)

95. $4, 5$ **96.** $-3, 2$ **97.** $1 + \sqrt{2}, 1 - \sqrt{2}$ **98.** $i, -i$

Chapter 1 | Quiz (Sections 1.1–1.4)

1. Solve the linear equation $3(x - 5) + 2 = 1 - (4 + 2x)$.

2. Determine whether each equation is an *identity*, a *conditional equation*, or a *contradiction*. Give the solution set.

  (a) $4x - 5 = -2(3 - 2x) + 3$

  (b) $5x - 9 = 5(-2 + x) + 1$

  (c) $5x - 4 = 3(6 - x)$

3. Solve the equation $ay + 2x = y + 5x$ for y. (Assume $a \neq 1$.)

4. *Earning Interest* Johnny deposits some money at 2.5% annual interest and twice as much at 3.0%. Find the amount deposited at each rate if his total annual interest income is $850.

5. *(Modeling) Minimum Hourly Wage* One model for the minimum hourly wage in the United States for the period 1979–2014 is

$$y = 0.128x - 250.43,$$

where x represents the year and y represents the wage, in dollars. (*Source:* Bureau of Labor Statistics.) The actual 2008 minimum wage was $6.55. What does this model predict as the wage? What is the difference between the actual wage and the predicted wage?

6. Write $\dfrac{-4 + \sqrt{-24}}{8}$ in standard form $a + bi$.

7. Write the quotient $\dfrac{7 - 2i}{2 + 4i}$ in standard form $a + bi$.

Solve each equation.

8. $3x^2 - x = -1$ **9.** $x^2 - 29 = 0$ **10.** $A = \dfrac{1}{2}r^2\theta$, for r

1.5 Applications and Modeling with Quadratic Equations

- Geometry Problems
- The Pythagorean Theorem
- Height of a Projected Object
- Modeling with Quadratic Equations

Geometry Problems To solve these applications, we continue to use a six-step problem-solving strategy.

EXAMPLE 1 Solving a Problem Involving Volume

A piece of machinery produces rectangular sheets of metal such that the length is three times the width. Equal-sized squares measuring 5 in. on a side can be cut from the corners so that the resulting piece of metal can be shaped into an open box by folding up the flaps. If specifications call for the volume of the box to be 1435 in.^{3}, find the dimensions of the original piece of metal.

SOLUTION

Step 1 **Read** the problem. We must find the dimensions of the original piece of metal.

Step 2 **Assign a variable.** We know that the length is three times the width.

Let x = the width (in inches) and thus, $3x$ = the length.

The box is formed by cutting $5 + 5 = 10$ in. from both the length and the width. See **Figure 5**. The width of the bottom of the box is $x - 10$, the length of the bottom of the box is $3x - 10$, and the height is 5 in. (the length of the side of each cut-out square). See **Figure 6**.

Figure 5

Figure 6

Step 3 **Write an equation.** The formula for volume of a box is $V = lwh$.

$$\underset{\downarrow}{\text{Volume}} = \underset{\downarrow}{\text{length}} \times \underset{\downarrow}{\text{width}} \times \underset{\downarrow}{\text{height}}$$

$$1435 = (3x - 10)(x - 10)(5)$$

(Note that the dimensions of the box must be positive numbers, so $3x - 10$ and $x - 10$ must be greater than 0, which implies $x > \frac{10}{3}$ and $x > 10$. These are both satisfied when $x > 10$.)

Step 4 **Solve** the equation from Step 3.

$$1435 = 15x^2 - 200x + 500 \qquad \text{Multiply.}$$

$$0 = 15x^2 - 200x - 935 \qquad \text{Subtract 1435 from each side.}$$

$$0 = 3x^2 - 40x - 187 \qquad \text{Divide each side by 5.}$$

$$0 = (3x + 11)(x - 17) \qquad \text{Factor.}$$

$$3x + 11 = 0 \qquad \text{or} \qquad x - 17 = 0 \qquad \text{Zero-factor property}$$

The width cannot be negative. $\quad x = -\dfrac{11}{3} \qquad \text{or} \qquad x = 17 \qquad \text{Solve each equation.}$

Step 5 **State the answer.** Only 17 satisfies the restriction $x > 10$. Thus, the dimensions of the original piece should be 17 in. by $3(17) = 51$ in.

Step 6 **Check.** The length and width of the bottom of the box are

$$51 - 2(5) = 41 \text{ in.} \quad \text{Length}$$

and

$$17 - 2(5) = 7 \text{ in.} \quad \text{Width}$$

The height is 5 in. (the amount cut on each corner), so the volume is

$$V = lwh = 41 \times 7 \times 5 = 1435 \text{ in.}^3, \quad \text{as required.}$$

✔ **Now Try Exercise 27.**

PROBLEM-SOLVING HINT   As seen in **Example 1,** discard any solution that does not satisfy the physical constraints of a problem.

The Pythagorean Theorem   **Example 2** requires the use of the **Pythagorean theorem** for right triangles. Recall that the **legs** of a right triangle form the right angle, and the **hypotenuse** is the side opposite the right angle.

Pythagorean Theorem

In a right triangle, the sum of the squares of the lengths of the legs is equal to the square of the length of the hypotenuse.

$$a^2 + b^2 = c^2$$

EXAMPLE 2   **Applying the Pythagorean Theorem**

A piece of property has the shape of a right triangle. The longer leg is 20 m longer than twice the length of the shorter leg. The hypotenuse is 10 m longer than the length of the longer leg. Find the lengths of the sides of the triangular lot.

SOLUTION

Step 1 **Read** the problem. We must find the lengths of the three sides.

Step 2 **Assign a variable.**

Let $x =$ the length of the shorter leg (in meters).

Then $2x + 20 =$ the length of the longer leg, and

$(2x + 20) + 10,$ or $2x + 30 =$ the length of the hypotenuse.

$2x + 30$

x

$2x + 20$

x is in meters.

Figure 7

See **Figure 7.**

Step 3 **Write an equation.**

$$
\underset{\downarrow}{a^2} \quad + \quad \underset{\downarrow}{b^2} \quad = \quad \underset{\downarrow}{c^2}
$$

The hypotenuse is c.

$$x^2 \quad + \quad (2x + 20)^2 \quad = \quad (2x + 30)^2$$

Substitute into the Pythagorean theorem.

Step 4 **Solve** the equation.

$$x^2 + (4x^2 + 80x + 400) = 4x^2 + 120x + 900 \qquad \text{Square the binomials. Remember the middle terms.}$$

$$x^2 - 40x - 500 = 0 \qquad \text{Standard form}$$

$$(x - 50)(x + 10) = 0 \qquad \text{Factor.}$$

$$x - 50 = 0 \quad \text{or} \quad x + 10 = 0 \qquad \text{Zero-factor property}$$

$$x = 50 \quad \text{or} \qquad x = -10 \qquad \text{Solve each equation.}$$

Step 5 **State the answer.** Because x represents a length, -10 is not reasonable. The lengths of the sides of the triangular lot are

$$50 \text{ m}, \quad 2(50) + 20 = 120 \text{ m}, \quad \text{and} \quad 2(50) + 30 = 130 \text{ m}.$$

Step 6 **Check.** The lengths 50, 120, and 130 satisfy the words of the problem and also satisfy the Pythagorean theorem.

✔ **Now Try Exercise 35.**

Galileo Galilei (1564–1642)

According to legend, Galileo dropped objects of different weights from the Leaning Tower of Pisa to disprove the Aristotelian view that heavier objects fall faster than lighter objects. He developed the formula $d = 16t^2$ for freely falling objects, where d is the distance in feet that an object falls (neglecting air resistance) in t seconds, regardless of weight.

Height of a Projected Object If air resistance is neglected, the height s (in feet) of an object projected directly upward from an initial height of s_0 feet, with initial velocity v_0 feet per second, is given by the following equation.

$$s = -16t^2 + v_0t + s_0$$

Here t represents the number of seconds after the object is projected. The coefficient of t^2, -16, is a constant based on the gravitational force of Earth. This constant varies on other surfaces, such as the moon and other planets.

EXAMPLE 3 Solving a Problem Involving Projectile Height

If a projectile is launched vertically upward from the ground with an initial velocity of 100 ft per sec, neglecting air resistance, its height s (in feet) above the ground t seconds after projection is given by

$$s = -16t^2 + 100t.$$

(a) After how many seconds will it be 50 ft above the ground?

(b) How long will it take for the projectile to return to the ground?

SOLUTION

(a) We must find value(s) of t so that height s is 50 ft.

$$s = -16t^2 + 100t$$

$$50 = -16t^2 + 100t \qquad \text{Let } s = 50.$$

$$0 = -16t^2 + 100t - 50 \qquad \text{Standard form}$$

$$0 = 8t^2 - 50t + 25 \qquad \text{Divide by } -2.$$

$$t = \frac{-b \pm \sqrt{b^2 - 4ac}}{2a} \qquad \text{Quadratic formula}$$

> Substitute carefully.

$$t = \frac{-(-50) \pm \sqrt{(-50)^2 - 4(8)(25)}}{2(8)} \qquad \text{Substitute } a = 8, \\ b = -50, \text{ and } c = 25.$$

$$t = \frac{50 \pm \sqrt{1700}}{16} \qquad \text{Simplify.}$$

$$t \approx 0.55 \quad \text{or} \quad t \approx 5.70 \qquad \text{Use a calculator.}$$

Both solutions are acceptable. The projectile reaches 50 ft twice—once on its way up (after 0.55 sec) and once on its way down (after 5.70 sec).

(b) When the projectile returns to the ground, the height s will be 0 ft.

$$s = -16t^2 + 100t$$

$$0 = -16t^2 + 100t \qquad \text{Let } s = 0.$$

$$0 = -4t(4t - 25) \qquad \text{Factor.}$$

$$-4t = 0 \quad \text{or} \quad 4t - 25 = 0 \qquad \text{Zero-factor property}$$

$$t = 0 \quad \text{or} \quad t = 6.25 \qquad \text{Solve each equation.}$$

The first solution, 0, represents the time at which the projectile was on the ground prior to being launched, so it does not answer the question. The projectile will return to the ground 6.25 sec after it is launched.

✔ **Now Try Exercise 47.**

Modeling with Quadratic Equations

EXAMPLE 4 **Analyzing Trolley Ridership**

The I-Ride Trolley service carries passengers along the International Drive resort area of Orlando, Florida. The bar graph in **Figure 8** shows I-Ride Trolley ridership data in millions. The quadratic equation

$$y = -0.00525x^2 + 0.0913x + 1.64$$

models ridership from 2000 to 2013, where y represents ridership in millions, and $x = 0$ represents 2000, $x = 1$ represents 2001, and so on.

Ridership

Source: I-Ride Trolley, International Drive Master Transit, www.itrolley.com

Figure 8

(a) Use the model to determine ridership in 2011. Compare the result to the actual ridership figure of 2.1 million.

(b) According to the model, in what year did ridership reach 1.8 million?

SOLUTION

(a) Because $x = 0$ represents the year 2000, use $x = 11$ to represent 2011.

$$y = -0.00525x^2 + 0.0913x + 1.64 \qquad \text{Given model}$$

$$y = -0.00525(11)^2 + 0.0913(11) + 1.64 \qquad \text{Let } x = 11.$$

$$y \approx 2.0 \text{ million} \qquad \text{Use a calculator.}$$

The prediction is about 0.1 million (that is, 100,000) less than the actual figure of 2.1 million.

(b)

$$y = -0.00525x^2 + 0.0913x + 1.64 \qquad \text{Given model}$$

$$1.8 = -0.00525x^2 + 0.0913x + 1.64 \qquad \text{Let } y = 1.8.$$

Solve this equation for x.

$$0 = -0.00525x^2 + 0.0913x - 0.16 \qquad \text{Standard form}$$

$$x = \frac{-0.0913 \pm \sqrt{(0.0913)^2 - 4(-0.00525)(-0.16)}}{2(-0.00525)} \qquad \text{Quadratic formula}$$

$$x \approx 2.0 \quad \text{or} \quad x \approx 15.4 \qquad \text{Use a calculator.}$$

The year 2002 corresponds to $x = 2.0$. Thus, according to the model, ridership reached 1.8 million in the year 2002. This outcome closely matches the bar graph and seems reasonable.

The year 2015 corresponds to $x = 15.4$. Round down to the year 2015 because 15.4 yr from 2000 occurs during 2015. There is no value on the bar graph to compare this to, because the last data value is for the year 2013. Always view results that are *beyond* the data in a model with skepticism, and realistically consider whether the model will continue as given. The model *predicts* that ridership will be 1.8 million again in the year 2015.

✔ **Now Try Exercise 49.**

1.5 **Exercises**

CONCEPT PREVIEW *Answer each question.*

1. *Area of a Parking Lot* For the rectangular parking area of the shopping center shown, with x in yards, which one of the following equations says that the area is 40,000 yd^2?

2x + 200

 A. $x(2x + 200) = 40{,}000$ **B.** $2x + 2(2x + 200) = 40{,}000$

 C. $x + (2x + 200) = 40{,}000$ **D.** $x^2 + (2x + 200)^2 = 40{,}000^2$

2. *Diagonal of a Rectangle* If a rectangle is r feet long and s feet wide, which expression represents the length of its diagonal in terms of r and s?

 A. $\sqrt{rs}$ **B.** $r + s$ **C.** $\sqrt{r^2 + s^2}$ **D.** $r^2 + s^2$

3. *Sides of a Right Triangle* To solve for the lengths of the right triangle sides, which equation is correct?

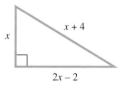

 A. $x^2 = (2x - 2)^2 + (x + 4)^2$

 B. $x^2 + (x + 4)^2 = (2x - 2)^2$

 C. $x^2 = (2x - 2)^2 - (x + 4)^2$

 D. $x^2 + (2x - 2)^2 = (x + 4)^2$

4. *Area of a Picture* The mat and frame around the picture shown measure x inches across. Which equation says that the area of the picture itself is 600 in.^{2}?

34 in.

21 in.

x in.

x in.

 A. $2(34 - 2x) + 2(21 - 2x) = 600$

 B. $(34 - 2x)(21 - 2x) = 600$

 C. $(34 - x)(21 - x) = 600$

 D. $x(34)(21) = 600$

5. *Volume of a Box* A rectangular piece of metal is 5 in. longer than it is wide. Squares with sides 2 in. long are cut from the four corners, and the flaps are folded upward to form an open box. Which equation indicates that the volume of the box is 64 in.^{3}?

 A. $(x + 1)(x - 4)(2) = 64$ **B.** $x(x + 5)(2) = 64$

 C. $(x + 1)(x - 4) = 64$ **D.** $x(x + 5) = 64$

6. *Height of a Projectile* If a projectile is launched vertically upward from the ground with an initial velocity of 60 ft per sec, neglecting air resistance, its height s (in feet) above the ground t seconds after projection is given by

$$s = -16t^2 + 60t.$$

Which equation should be used to determine the time at which the height of the projectile reaches 40 ft?

A. $s = -16(40)^2 + 60$ **B.** $s = -16(40)^2 + 60(40)$

C. $40 = -16t^2 + 60t$ **D.** $40 = -16t^2$

7. *Height of a Projectile* If a projectile is launched vertically upward from the ground with an initial velocity of 45 ft per sec, neglecting air resistance, its height s (in feet) above the ground t seconds after projection is given by

$$s = -16t^2 + 45t.$$

Which equation should be used to determine the height of the projectile after 2 sec?

A. $s = 2(-16t^2 + 45t)$ **B.** $s = -16(2)^2 + 45(2)$

C. $2 = -16t^2 + 45t$ **D.** $2 = -16t^2$

8. *New Car Sales* Suppose that the quadratic equation

$$S = 0.0538x^2 - 0.807x + 8.84$$

models sales of new cars, where S represents sales in millions, and $x = 0$ represents 2000, $x = 1$ represents 2001, and so on. Which equation should be used to determine sales in 2010?

A. $10 = 0.0538x^2 - 0.807x + 8.84$

B. $2010 = 0.0538x^2 - 0.807x + 8.84$

C. $S = 0.0538(10)^2 - 0.807(10) + 8.84$

D. $S = 0.0538(2010)^2 - 0.807(2010) + 8.84$

To prepare for the applications that come later, work the following basic problems that lead to quadratic equations.

Unknown Numbers In Exercises 9–18, use the following facts.

> *If x represents an integer, then $x + 1$ represents the next consecutive integer.*
>
> *If x represents an even integer, then $x + 2$ represents the next consecutive even integer.*
>
> *If x represents an odd integer, then $x + 2$ represents the next consecutive odd integer.*

9. Find two consecutive integers whose product is 56.

10. Find two consecutive integers whose product is 110.

11. Find two consecutive even integers whose product is 168.

12. Find two consecutive even integers whose product is 224.

13. Find two consecutive odd integers whose product is 63.

14. Find two consecutive odd integers whose product is 143.

15. The sum of the squares of two consecutive odd integers is 202. Find the integers.

16. The sum of the squares of two consecutive even integers is 52. Find the integers.

17. The difference of the squares of two positive consecutive even integers is 84. Find the integers.

18. The difference of the squares of two positive consecutive odd integers is 32. Find the integers.

Solve each problem. **See Examples 1 and 2.**

19. *Dimensions of a Right Triangle* The lengths of the sides of a right triangle are consecutive even integers. Find these lengths. (*Hint:* Use the Pythagorean theorem.)

20. *Dimensions of a Right Triangle* The lengths of the sides of a right triangle are consecutive positive integers. Find these lengths. (*Hint:* Use the Pythagorean theorem.)

21. *Dimensions of a Square* The length of each side of a square is 3 in. more than the length of each side of a smaller square. The sum of the areas of the squares is 149 in.^{2}. Find the lengths of the sides of the two squares.

22. *Dimensions of a Square* The length of each side of a square is 5 in. more than the length of each side of a smaller square. The difference of the areas of the squares is 95 in.^{2}. Find the lengths of the sides of the two squares.

Solve each problem. **See Example 1.**

23. *Dimensions of a Parking Lot* A parking lot has a rectangular area of 40,000 yd^2. The length is 200 yd more than twice the width. Find the dimensions of the lot.

24. *Dimensions of a Garden* An ecology center wants to set up an experimental garden using 300 m of fencing to enclose a rectangular area of 5000 m^2. Find the dimensions of the garden.

$150 - x$

x

x is in meters.

25. *Dimensions of a Rug* Zachary wants to buy a rug for a room that is 12 ft wide and 15 ft long. He wants to leave a uniform strip of floor around the rug. He can afford to buy 108 ft^2 of carpeting. What dimensions should the rug have?

108 ft^2

12 ft

15 ft

26. *Width of a Flower Border* A landscape architect has included a rectangular flower bed measuring 9 ft by 5 ft in her plans for a new building. She wants to use two colors of flowers in the bed: one in the center and the other for a border of the same width on all four sides. If she has enough plants to cover 24 ft^2 for the border, how wide can the border be?

27. *Volume of a Box* A rectangular piece of metal is 10 in. longer than it is wide. Squares with sides 2 in. long are cut from the four corners, and the flaps are folded upward to form an open box. If the volume of the box is 832 in.^{3}, what were the original dimensions of the piece of metal?

28. *Volume of a Box* In **Exercise 27,** suppose that the piece of metal has length twice the width, and 4-in. squares are cut from the corners. If the volume of the box is 1536 in.^{3}, what were the original dimensions of the piece of metal?

29. *Manufacturing to Specifications* A manufacturing firm wants to package its product in a cylindrical container 3 ft high with surface area 8π ft^2. What should the radius of the circular top and bottom of the container be? (*Hint:* The surface area consists of the circular top and bottom and a rectangle that represents the side cut open vertically and unrolled.)

30. *Manufacturing to Specifications* In **Exercise 29,** what radius would produce a container with a volume of π times the radius? (*Hint:* The volume is the area of the circular base times the height.)

31. *Dimensions of a Square* What is the length of the side of a square if its area and perimeter are numerically equal?

32. *Dimensions of a Rectangle* A rectangle has an area that is numerically twice its perimeter. If the length is twice the width, what are its dimensions?

33. *Radius of a Can* A can of Blue Runner Red Kidney Beans has surface area 371 cm^2. Its height is 12 cm. What is the radius of the circular top? Round to the nearest hundredth.

34. *Dimensions of a Cereal Box* The volume of a 15-oz cereal box is 180.4 in.^{3}. The length of the box is 3.2 in. less than the height, and its width is 2.3 in. Find the height and length of the box to the nearest tenth.

Solve each problem. See Example 2.

35. *Height of a Dock* A boat is being pulled into a dock with a rope attached to the boat at water level. When the boat is 12 ft from the dock, the length of the rope from the boat to the dock is 3 ft longer than twice the height of the dock above the water. Find the height of the dock.

36. *Height of a Kite* Grady is flying a kite on 50 ft of string. Its vertical distance from his hand is 10 ft more than its horizontal distance from his hand. Assuming that the string is being held 5 ft above ground level, find its horizontal distance from Grady and its vertical distance from the ground.

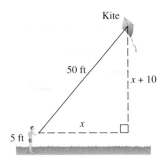

37. *Radius Covered by a Circular Lawn Sprinkler* A square lawn has area 800 ft². A sprinkler placed at the center of the lawn sprays water in a circular pattern as shown in the figure. What is the radius of the circle?

38. *Dimensions of a Solar Panel Frame* Molly has a solar panel with a width of 26 in. To get the proper inclination for her climate, she needs a right triangular support frame that has one leg twice as long as the other. To the nearest tenth of an inch, what dimensions should the frame have?

39. *Length of a Ladder* A building is 2 ft from a 9-ft fence that surrounds the property. A worker wants to wash a window in the building 13 ft from the ground. He plans to place a ladder over the fence so it rests against the building. (See the figure.) He decides he should place the ladder 8 ft from the fence for stability. To the nearest tenth of a foot, how long a ladder will he need?

40. *Range of Receivers* Tanner and Sheldon have received communications receivers for Christmas. If they leave from the same point at the same time, Tanner walking north at 2.5 mph and Sheldon walking east at 3 mph, how long will they be able to talk to each other if the range of the communications receivers is 4 mi? Round the answer to the nearest minute.

41. *Length of a Walkway* A nature conservancy group decides to construct a raised wooden walkway through a wetland area. To enclose the most interesting part of the wetlands, the walkway will have the shape of a right triangle with one leg 700 yd longer than the other and the hypotenuse 100 yd longer than the longer leg. Find the total length of the walkway.

42. *Broken Bamboo* Problems involving the Pythagorean theorem have appeared in mathematics for thousands of years. This one is taken from the ancient Chinese work *Arithmetic in Nine Sections:*

> *There is a bamboo 10 ft high, the upper end of which, being broken, reaches the ground 3 ft from the stem. Find the height of the break.*

(Modeling) Solve each problem. See Example 3.

Height of a Projectile A projectile is launched from ground level with an initial velocity of v_0 feet per second. Neglecting air resistance, its height in feet t seconds after launch is given by

$$s = -16t^2 + v_0 t.$$

*In Exercises 43–46, find the time(s) that the projectile will **(a)** reach a height of 80 ft and **(b)** return to the ground for the given value of v_0. Round answers to the nearest hundredth if necessary.*

43. $v_0 = 96$ **44.** $v_0 = 128$ **45.** $v_0 = 32$ **46.** $v_0 = 16$

47. *Height of a Projected Ball* An astronaut on the moon throws a baseball upward. The astronaut is 6 ft, 6 in. tall, and the initial velocity of the ball is 30 ft per sec. The height *s* of the ball in feet is given by the equation

$$s = -2.7t^2 + 30t + 6.5,$$

where *t* is the number of seconds after the ball was thrown.

(a) After how many seconds is the ball 12 ft above the moon's surface? Round to the nearest hundredth.

(b) How many seconds will it take for the ball to hit the moon's surface? Round to the nearest hundredth.

48. *Concept Check* The ball in **Exercise 47** will never reach a height of 100 ft. How can this be determined algebraically?

(Modeling) Solve each problem. *See Example 4.*

49. *NFL Salary Cap* In 1994, the National Football League introduced a salary cap that limits the amount of money spent on players' salaries. The quadratic model

$$y = 0.2313x^2 + 2.600x + 35.17$$

approximates this cap in millions of dollars for the years 1994–2009, where $x = 0$ represents 1994, $x = 1$ represents 1995, and so on. (*Source:* www.businessinsider.com)

(a) Approximate the NFL salary cap in 2007 to the nearest tenth of a million dollars.

(b) According to the model, in what year did the salary cap reach 90 million dollars?

50. *NFL Rookie Wage Scale* Salaries, in millions of dollars, for rookies selected in the first round of the NFL 2014 draft can be approximated by the quadratic model

$$y = 0.0258x^2 - 1.30x + 23.3,$$

where *x* represents draft pick order. Players selected earlier in the round have higher salaries than those selected later in the round. (*Source:* www.forbes.com)

(a) Use the model to estimate the salary of the player selected first overall to the nearest tenth of a million dollars.

(b) What is the estimated salary of the player selected 10th overall? Round to the nearest tenth of a million dollars.

51. *Carbon Monoxide Exposure* Carbon monoxide (CO) combines with the hemoglobin of the blood to form carboxyhemoglobin (COHb), which reduces transport of oxygen to tissues. Smokers routinely have a 4% to 6% COHb level in their blood. The quadratic model

$$T = 0.00787x^2 - 1.528x + 75.89$$

approximates the exposure time in hours necessary to reach this 4% to 6% level, where $50 \leq x \leq 100$ is the amount of carbon monoxide present in the air in parts per million (ppm). (*Source: Indoor Air Quality Environmental Information Handbook: Combustion Sources.*)

(a) A kerosene heater or a room full of smokers is capable of producing 50 ppm of carbon monoxide. How long would it take for a nonsmoking person to start feeling the above symptoms? Round to the nearest tenth.

(b) Find the carbon monoxide concentration necessary for a person to reach the 4% to 6% COHb level in 3 hr. Round to the nearest tenth.

52. *Carbon Monoxide Exposure* Refer to **Exercise 51.** High concentrations of carbon monoxide (CO) can cause coma and death. The time required for a person to reach a COHb level capable of causing a coma can be approximated by the quadratic model

$$T = 0.0002x^2 - 0.316x + 127.9,$$

where *T* is the exposure time in hours necessary to reach this level and $500 \le x \le 800$ is the amount of carbon monoxide present in the air in parts per million (ppm). (*Source: Indoor Air Quality Environmental Information Handbook: Combustion Sources.*)

(a) What is the exposure time when $x = 600$ ppm?

(b) Find the concentration of CO necessary to produce a coma in 4 hr. Round to the nearest tenth part per million.

53. *Methane Gas Emissions* The table gives methane gas emissions from all sources in the United States, in millions of metric tons. The quadratic model

$$y = 0.0429x^2 - 9.73x + 606$$

approximates the emissions for these years. In the model, *x* represents the number of years since 2008, so $x = 0$ represents 2008, $x = 1$ represents 2009, and so on.

Year	Millions of Metric Tons of Methane
2008	606.0
2009	596.6
2010	585.5
2011	578.4
2012	567.3

Source: U.S. Environmental Protection Agency.

(a) According to the model, what would emissions be in 2014? Round to the nearest tenth of a million metric tons.

(b) Find the nearest year beyond 2008 for which this model predicts that emissions will reach 500 million metric tons.

54. *Cost of Public Colleges* The average cost, in dollars, for tuition and fees for in-state students at four-year public colleges over the period 2000–2014 can be modeled by the equation

$$y = 4.065x^2 + 370.1x + 3450$$

where $x = 0$ corresponds to 2000, $x = 1$ corresponds to 2001, and so on. Based on this model, for what year after 2000 was the average cost $8605? (*Source:* The College Board, *Annual Survey of Colleges.*)

55. *Internet Publishing* Estimated revenue from Internet publishing and web search portals in the United States during the years 2007 through 2012 can be modeled by the equation

$$y = 710.55x^2 + 1333.7x + 32{,}399$$

where $x = 0$ corresponds to the year 2007, $x = 1$ corresponds to 2008, and so on, and *y* is in millions of dollars. Approximate the revenue from these services in 2010 to the nearest million. (*Source:* U.S. Census Bureau.)

56. *Cable's Top Internet Speeds* The top cable Internet speeds during the years 2007 through 2013 can be modeled by the equation

$$y = 23.09x^2 - 62.12x + 32.78,$$

where $x = 0$ corresponds to 2007, $x = 1$ corresponds to 2008, and so on, and *y* is in megabits per second (MBS). Based on this model, what was cable TV's top Internet speed in 2012? (*Source:* National Cable & Telecommunications Association.)

Relating Concepts

For individual or collaborative investigation *(Exercises 57–60)*

*If p units of an item are sold for x dollars per unit, the revenue is R = px. Use this idea to analyze the following problem, **working Exercises 57–60 in order.***

Number of Apartments Rented *The manager of an 80-unit apartment complex knows from experience that at a rent of* $300, *all the units will be full. On the average, one additional unit will remain vacant for each* $20 *increase in rent over* $300. *Furthermore, the manager must keep at least 30 units rented due to other financial considerations. Currently, the revenue from the complex is* $35,000. *How many apartments are rented?*

57. Suppose that x represents the number of $20 increases over $300. Represent the number of apartment units that will be rented in terms of x.

58. Represent the rent per unit in terms of x.

59. Use the answers in **Exercises 57 and 58** to write an equation that defines the revenue generated when there are x increases of $20 over $300.

60. According to the problem, the revenue currently generated is $35,000. Substitute this value for revenue into the equation from **Exercise 59.** Solve for x to answer the question in the problem.

*Solve each problem. (**See Exercises 57–60.**)*

61. ***Number of Airline Passengers*** The cost of a charter flight to Miami is $225 each for 75 passengers, with a refund of $5 per passenger for each passenger in excess of 75. How many passengers must take the flight to produce a revenue of $16,000?

62. ***Number of Bus Passengers*** A charter bus company charges a fare of $40 per person, plus $2 per person for each unsold seat on the bus. If the bus holds 100 passengers and x represents the number of unsold seats, how many passengers must ride the bus to produce revenue of $5950? (*Note:* Because of the company's commitment to efficient fuel use, the charter will not run unless filled to at least half-capacity.)

63. ***Harvesting a Cherry Orchard*** The manager of a cherry orchard wants to schedule the annual harvest. If the cherries are picked now, the average yield per tree will be 100 lb, and the cherries can be sold for 40 cents per pound. Past experience shows that the yield per tree will increase about 5 lb per week, while the price will decrease about 2 cents per pound per week. How many weeks should the manager wait to get an average revenue of $38.40 per tree?

64. ***Recycling Aluminum Cans*** A local group of scouts has been collecting old aluminum cans for recycling. The group has already collected 12,000 lb of cans, for which they could currently receive $4 per hundred pounds. The group can continue to collect cans at the rate of 400 lb per day. However, a glut in the old-can market has caused the recycling company to announce that it will lower its price, starting immediately, by $0.10 per hundred pounds per day. The scouts can make only one trip to the recycling center. How many days should they wait in order to receive $490 for their cans?

1.6 Other Types of Equations and Applications

- Rational Equations
- Work Rate Problems
- Equations with Radicals
- Equations with Rational Exponents
- Equations Quadratic in Form

Rational Equations   A **rational equation** is an equation that has a rational expression for one or more terms. To solve a rational equation, multiply each side by the least common denominator (LCD) of the terms of the equation to eliminate fractions, and then solve the resulting equation.

A value of the variable that appears to be a solution after each side of a rational equation is multiplied by a variable expression (the LCD) is called a **proposed solution.** Because a rational expression is not defined when its denominator is 0, **proposed solutions for which any denominator equals 0 are excluded from the solution set.**

Be sure to check all proposed solutions in the original equation.

EXAMPLE 1   **Solving Rational Equations That Lead to Linear Equations**

Solve each equation.

(a) $\dfrac{3x-1}{3} - \dfrac{2x}{x-1} = x$

(b) $\dfrac{x}{x-2} = \dfrac{2}{x-2} + 2$

SOLUTION

(a) The least common denominator is $3(x-1)$, which is equal to 0 if $x=1$. Therefore, 1 cannot possibly be a solution of this equation.

$$\frac{3x-1}{3} - \frac{2x}{x-1} = x$$

$$3(x-1)\left(\frac{3x-1}{3}\right) - 3(x-1)\left(\frac{2x}{x-1}\right) = 3(x-1)x \qquad \text{Multiply by the LCD. } 3(x-1), \text{ where } x \neq 1.$$

$$(x-1)(3x-1) - 3(2x) = 3x(x-1) \qquad \text{Divide out common factors.}$$

$$3x^2 - 4x + 1 - 6x = 3x^2 - 3x \qquad \text{Multiply.}$$

$$1 - 10x = -3x \qquad \begin{array}{l}\text{Subtract } 3x^2. \\ \text{Combine like terms.}\end{array}$$

$$1 = 7x \qquad \text{Solve the linear equation.}$$

$$x = \frac{1}{7} \qquad \text{Proposed solution}$$

The proposed solution $\frac{1}{7}$ meets the requirement that $x \neq 1$ and does not cause any denominator to equal 0. Substitute to check for correct algebra.

CHECK $\dfrac{3x-1}{3} - \dfrac{2x}{x-1} = x$ Original equation

$$\frac{3\left(\frac{1}{7}\right)-1}{3} - \frac{2\left(\frac{1}{7}\right)}{\frac{1}{7}-1} \stackrel{?}{=} \frac{1}{7} \qquad \text{Let } x = \tfrac{1}{7}.$$

$$-\frac{4}{21} - \left(-\frac{1}{3}\right) \stackrel{?}{=} \frac{1}{7} \qquad \text{Simplify the complex fractions.}$$

$$\frac{1}{7} = \frac{1}{7} \quad \checkmark \quad \text{True}$$

The solution set is $\left\{\frac{1}{7}\right\}$.

(b)

$$\frac{x}{x-2} = \frac{2}{x-2} + 2$$

$$(x-2)\left(\frac{x}{x-2}\right) = (x-2)\left(\frac{2}{x-2}\right) + (x-2)2 \qquad \text{Multiply by the LCD, } x-2, \text{ where } x \neq 2.$$

$$x = 2 + 2(x-2) \qquad \text{Divide out common factors.}$$

$$x = 2 + 2x - 4 \qquad \text{Distributive property}$$

$$-x = -2 \qquad \text{Solve the linear equation.}$$

$$x = 2 \qquad \text{Proposed solution}$$

The proposed solution is 2. However, the variable is restricted to real numbers except 2. If $x = 2$, then not only does it cause a zero denominator, but also multiplying by $x - 2$ in the first step is multiplying both sides by 0, which is not valid. Thus, the solution set is $\emptyset$.

✔ **Now Try Exercises 17 and 19.**

EXAMPLE 2 **Solving Rational Equations That Lead to Quadratic Equations**

Solve each equation.

(a) $\dfrac{3x+2}{x-2} + \dfrac{1}{x} = \dfrac{-2}{x^2-2x}$
(b) $\dfrac{-4x}{x-1} + \dfrac{4}{x+1} = \dfrac{-8}{x^2-1}$

SOLUTION

(a)

$$\frac{3x+2}{x-2} + \frac{1}{x} = \frac{-2}{x^2-2x}$$

$$\frac{3x+2}{x-2} + \frac{1}{x} = \frac{-2}{x(x-2)} \qquad \text{Factor the last denominator.}$$

$$x(x-2)\left(\frac{3x+2}{x-2}\right) + x(x-2)\left(\frac{1}{x}\right) = x(x-2)\left(\frac{-2}{x(x-2)}\right) \qquad \text{Multiply by } x(x-2), x \neq 0, 2.$$

$$x(3x+2) + (x-2) = -2 \qquad \text{Divide out common factors.}$$

$$3x^2 + 2x + x - 2 = -2 \qquad \text{Distributive property}$$

$$3x^2 + 3x = 0 \qquad \text{Standard form}$$

$$3x(x+1) = 0 \qquad \text{Factor.}$$

Set *each* factor equal to 0. → $3x = 0 \quad \text{or} \quad x+1 = 0 \qquad \text{Zero-factor property}$

$$x = 0 \quad \text{or} \quad x = -1 \qquad \text{Proposed solutions}$$

Because of the restriction $x \neq 0$, the only valid proposed solution is -1. Check -1 in the original equation. The solution set is $\{-1\}$.

(b)

$$\frac{-4x}{x-1} + \frac{4}{x+1} = \frac{-8}{x^2-1}$$

$$\frac{-4x}{x-1} + \frac{4}{x+1} = \frac{-8}{(x+1)(x-1)} \qquad \text{Factor.}$$

The restrictions on x are $x \neq \pm 1$. Multiply by the LCD, $(x+1)(x-1)$.

$$(x + 1)(x - 1)\left(\frac{-4x}{x - 1}\right) + (x + 1)(x - 1)\left(\frac{4}{x + 1}\right) = (x + 1)(x - 1)\left(\frac{-8}{(x + 1)(x - 1)}\right)$$

$$-4x(x + 1) + 4(x - 1) = -8 \qquad \text{Divide out common factors.}$$

$$-4x^2 - 4x + 4x - 4 = -8 \qquad \text{Distributive property}$$

$$-4x^2 + 4 = 0 \qquad \text{Standard form}$$

$$x^2 - 1 = 0 \qquad \text{Divide by } -4.$$

$$(x + 1)(x - 1) = 0 \qquad \text{Factor.}$$

$$x + 1 = 0 \quad \text{or} \quad x - 1 = 0 \qquad \text{Zero-factor property}$$

$$x = -1 \quad \text{or} \qquad x = 1 \qquad \text{Proposed solutions}$$

Neither proposed solution is valid, so the solution set is $\varnothing$.

✔ **Now Try Exercises 25 and 27.**

Work Rate Problems   If a job can be completed in 3 hr, then the rate of work is $\frac{1}{3}$ of the job per hr. After 1 hr the job would be $\frac{1}{3}$ complete, and after 2 hr the job would be $\frac{2}{3}$ complete. In 3 hr the job would be $\frac{3}{3}$ complete, meaning that 1 complete job had been accomplished.

PROBLEM-SOLVING HINT   If a job can be completed in t units of time, then the rate of work, r, is $\frac{1}{t}$ of the job per unit time.

$$r = \frac{1}{t}$$

The amount of work completed, A, is found by multiplying the rate of work, r, and the amount of time worked, t. This formula is similar to the distance formula $d = rt$.

Amount of work completed = rate of work × amount of time worked

or $\qquad\qquad\qquad\qquad A = rt$

EXAMPLE 3   **Solving a Work Rate Problem**

One printer can do a job twice as fast as another. Working together, both printers can do the job in 2 hr. How long would it take each printer, working alone, to do the job?

SOLUTION

Step 1 **Read** the problem. We must find the time it would take each printer, working alone, to do the job.

Step 2 **Assign a variable.** Let x represent the number of hours it would take the faster printer, working alone, to do the job. The time for the slower printer to do the job alone is then $2x$ hours.

Therefore, $\qquad\qquad \dfrac{1}{x} =$ the rate of the faster printer (job per hour)

and $\qquad\qquad\qquad \dfrac{1}{2x} =$ the rate of the slower printer (job per hour).

The time for the printers to do the job together is 2 hr. Multiplying each rate by the time will give the fractional part of the job completed by each.

	Rate	Time	Part of the Job Completed
Faster Printer	$\frac{1}{x}$	2	$2\left(\frac{1}{x}\right) = \frac{2}{x}$
Slower Printer	$\frac{1}{2x}$	2	$2\left(\frac{1}{2x}\right) = \frac{1}{x}$

$$\Big\}\ A = rt$$

Step 3 **Write an equation.** The sum of the two parts of the job completed is 1 because one whole job is done.

$$\underbrace{\text{Part of the job done by the faster printer}}\ +\ \underbrace{\text{Part of the job done by the slower printer}}\ =\ \underbrace{\text{One whole job}}.$$

$$\frac{2}{x}\ +\ \frac{1}{x}\ =\ 1$$

Step 4 **Solve.**

$$x\left(\frac{2}{x} + \frac{1}{x}\right) = x(1) \qquad \text{Multiply each side by } x, \text{ where } x \neq 0.$$

$$x\left(\frac{2}{x}\right) + x\left(\frac{1}{x}\right) = x(1) \qquad \text{Distributive property}$$

$$2 + 1 = x \qquad \text{Multiply.}$$

$$3 = x \qquad \text{Add.}$$

Step 5 **State the answer.** The faster printer would take 3 hr to do the job alone. The slower printer would take $2(3) = 6$ hr. Give *both* answers here.

Step 6 **Check.** The answer is reasonable because the time working together (2 hr, as stated in the problem) is less than the time it would take the faster printer working alone (3 hr, as found in Step 4).

✔ **Now Try Exercise 39.**

NOTE   **Example 3** can also be solved by using the fact that the sum of the rates of the individual printers is equal to their rate working together. Because the printers can complete the job together in 2 hr, their combined rate is $\frac{1}{2}$ of the job per hr.

$$\frac{1}{x} + \frac{1}{2x} = \frac{1}{2}$$

$$2x\left(\frac{1}{x} + \frac{1}{2x}\right) = 2x\left(\frac{1}{2}\right) \qquad \text{Multiply each side by } 2x.$$

$$2 + 1 = x \qquad \text{Distributive property}$$

$$3 = x \qquad \text{Same solution found earlier}$$

Equations with Radicals   To solve an equation such as

$$x - \sqrt{15 - 2x} = 0,$$

in which the variable appears in a radicand, we use the following **power property** to eliminate the radical.

Power Property

If P and Q are algebraic expressions, then every solution of the equation $P = Q$ is also a solution of the equation $P^n = Q^n$, for any positive integer n.

When the power property is used to solve equations, the new equation may have *more* solutions than the original equation. For example, the equation

$$x = -2 \quad \text{has solution set} \quad \{-2\}.$$

If we square each side of the equation $x = -2$, we obtain the new equation

$$x^2 = 4, \quad \text{which has solution set} \quad \{-2, 2\}.$$

Because the solution sets are not equal, the equations are not equivalent. **When we use the power property to solve an equation, it is essential to check all proposed solutions in the original equation.**

CAUTION *Be very careful when using the power property.* It does *not* say that the equations $P = Q$ and $P^n = Q^n$ are equivalent. It says only that each solution of the original equation $P = Q$ is also a solution of the new equation $P^n = Q^n$.

Solving an Equation Involving Radicals

Step 1 Isolate the radical on one side of the equation.

Step 2 Raise each side of the equation to a power that is the same as the index of the radical so that the radical is eliminated.

If the equation still contains a radical, repeat Steps 1 and 2.

Step 3 Solve the resulting equation.

Step 4 Check each proposed solution in the *original* equation.

EXAMPLE 4 **Solving an Equation Containing a Radical (Square Root)**

Solve $x - \sqrt{15 - 2x} = 0$.

SOLUTION

$$x - \sqrt{15 - 2x} = 0$$

Step 1 $\qquad\qquad x = \sqrt{15 - 2x}$ Isolate the radical.

Step 2 $\qquad\qquad x^2 = \left(\sqrt{15 - 2x}\right)^2$ Square each side.

$\qquad\qquad\qquad x^2 = 15 - 2x$ $\left(\sqrt{a}\right)^2 = a$, for $a \geq 0$.

Step 3 $\qquad x^2 + 2x - 15 = 0$ Write in standard form.

$\qquad\quad (x + 5)(x - 3) = 0$ Factor.

$\qquad x + 5 = 0 \quad \text{or} \quad x - 3 = 0$ Zero-factor property

$\qquad\quad x = -5 \quad \text{or} \qquad x = 3$ Proposed solutions

Step 4

CHECK $\qquad x - \sqrt{15 - 2x} = 0$ $\qquad$ Original equation

$$-5 - \sqrt{15 - 2(-5)} \overset{?}{=} 0 \quad \text{Let } x = -5. \quad \Big| \quad 3 - \sqrt{15 - 2(3)} \overset{?}{=} 0 \qquad \text{Let } x = 3.$$

$$-5 - \sqrt{25} \overset{?}{=} 0 \qquad\qquad\qquad 3 - \sqrt{9} \overset{?}{=} 0$$

$$-5 - 5 \overset{?}{=} 0 \qquad\qquad\qquad\qquad 3 - 3 \overset{?}{=} 0$$

$$-10 = 0 \quad \text{False} \qquad\qquad\qquad 0 = 0 \ \checkmark \ \text{True}$$

As the check shows, only 3 is a solution, so the solution set is $\{3\}$.

✔ **Now Try Exercise 45.**

EXAMPLE 5 **Solving an Equation Containing Two Radicals**

Solve $\sqrt{2x + 3} - \sqrt{x + 1} = 1$.

SOLUTION

$$\sqrt{2x + 3} - \sqrt{x + 1} = 1 \quad \boxed{\text{Isolate one of the radicals on one side of the equation.}}$$

Step 1 $\qquad \sqrt{2x + 3} = 1 + \sqrt{x + 1}$ $\qquad\qquad$ Isolate $\sqrt{2x + 3}$.

Step 2 $\qquad \left(\sqrt{2x + 3}\right)^2 = \left(1 + \sqrt{x + 1}\right)^2$ $\qquad\qquad$ Square each side.

$$2x + 3 = 1 + 2\sqrt{x + 1} + (x + 1) \quad \begin{array}{l}\textbf{Be careful:}\\ (a + b)^2 = a^2 + 2ab + b^2\end{array}$$

$\boxed{\text{Don't forget this term when squaring.}}$

Step 1 $\qquad x + 1 = 2\sqrt{x + 1}$ $\qquad\qquad$ Isolate the remaining radical.

Step 2 $\qquad (x + 1)^2 = \left(2\sqrt{x + 1}\right)^2$ $\qquad\qquad$ Square again.

$$x^2 + 2x + 1 = 4(x + 1) \quad \boxed{(ab)^2 = a^2 b^2} \quad \text{Apply the exponents.}$$

$$x^2 + 2x + 1 = 4x + 4 \qquad\qquad\qquad \text{Distributive property}$$

Step 3 $\qquad x^2 - 2x - 3 = 0$ $\qquad\qquad\qquad$ Write in standard form.

$$(x - 3)(x + 1) = 0 \qquad\qquad\qquad \text{Factor.}$$

$$x - 3 = 0 \quad \text{or} \quad x + 1 = 0 \qquad\qquad \text{Zero-factor property}$$

$$x = 3 \quad \text{or} \qquad x = -1 \qquad\qquad \text{Proposed solutions}$$

Step 4

CHECK $\qquad \sqrt{2x + 3} - \sqrt{x + 1} = 1$ $\qquad$ Original equation

$$\sqrt{2(3) + 3} - \sqrt{3 + 1} \overset{?}{=} 1 \quad \text{Let } x = 3. \quad \Big| \quad \sqrt{2(-1) + 3} - \sqrt{-1 + 1} \overset{?}{=} 1 \quad \text{Let } x = -1.$$

$$\sqrt{9} - \sqrt{4} \overset{?}{=} 1 \qquad\qquad\qquad\qquad \sqrt{1} - \sqrt{0} \overset{?}{=} 1$$

$$3 - 2 \overset{?}{=} 1 \qquad\qquad\qquad\qquad\qquad 1 - 0 \overset{?}{=} 1$$

$$1 = 1 \ \checkmark \ \text{True} \qquad\qquad\qquad\qquad 1 = 1 \ \checkmark \ \text{True}$$

Both 3 and -1 are solutions of the original equation, so $\{-1, 3\}$ is the solution set.

✔ **Now Try Exercise 57.**

CAUTION Remember to isolate a radical in Step 1. It would be incorrect to square each term individually as the first step in **Example 5.**

> **EXAMPLE 6** **Solving an Equation Containing a Radical (Cube Root)**

Solve $\sqrt[3]{4x^2 - 4x + 1} - \sqrt[3]{x} = 0$.

SOLUTION

$$\sqrt[3]{4x^2 - 4x + 1} - \sqrt[3]{x} = 0$$

Step 1 $\sqrt[3]{4x^2 - 4x + 1} = \sqrt[3]{x}$ Isolate a radical.

Step 2 $\left(\sqrt[3]{4x^2 - 4x + 1}\right)^3 = \left(\sqrt[3]{x}\right)^3$ Cube each side.

$$4x^2 - 4x + 1 = x \qquad \text{Apply the exponents.}$$

Step 3 $4x^2 - 5x + 1 = 0$ Write in standard form.

$$(4x - 1)(x - 1) = 0 \qquad \text{Factor.}$$

$$4x - 1 = 0 \quad \text{or} \quad x - 1 = 0 \qquad \text{Zero-factor property}$$

$$x = \frac{1}{4} \quad \text{or} \qquad x = 1 \qquad \text{Proposed solutions}$$

Step 4

CHECK $\sqrt[3]{4x^2 - 4x + 1} - \sqrt[3]{x} = 0$ Original equation

$\sqrt[3]{4\left(\frac{1}{4}\right)^2 - 4\left(\frac{1}{4}\right) + 1} - \sqrt[3]{\frac{1}{4}} \stackrel{?}{=} 0$ Let $x = \frac{1}{4}$. $\sqrt[3]{4(1)^2 - 4(1) + 1} - \sqrt[3]{1} \stackrel{?}{=} 0$ Let $x = 1$..

$\sqrt[3]{\frac{1}{4}} - \sqrt[3]{\frac{1}{4}} \stackrel{?}{=} 0$ $\sqrt[3]{1} - \sqrt[3]{1} \stackrel{?}{=} 0$

$0 = 0 \checkmark$ True $0 = 0 \checkmark$ True

Both are valid solutions, and the solution set is $\left\{\frac{1}{4}, 1\right\}$.

✔ **Now Try Exercise 69.**

Equations with Rational Exponents   An equation with a rational exponent contains a variable, or variable expression, raised to an exponent that is a rational number. For example, the radical equation

$$\left(\sqrt[5]{x}\right)^3 = 27 \quad \text{can be written with a rational exponent as} \quad x^{3/5} = 27$$

and solved by raising each side to the reciprocal of the exponent, with care taken regarding signs as seen in **Example 7(b).**

> **EXAMPLE 7** **Solving Equations with Rational Exponents**

Solve each equation.

(a) $x^{3/5} = 27$ **(b)** $(x - 4)^{2/3} = 16$

SOLUTION

(a) $x^{3/5} = 27$

$\left(x^{3/5}\right)^{5/3} = 27^{5/3}$ Raise each side to the power $\frac{5}{3}$, the reciprocal of the exponent of x.

$x = 243$ $27^{5/3} = \left(\sqrt[3]{27}\right)^5 = 3^5 = 243$

CHECK   Let $x = 243$ in the original equation.

$$x^{3/5} = 243^{3/5} = \left(\sqrt[5]{243}\right)^3 = 3^3 = 27 \checkmark \text{ True}$$

The solution set is $\{243\}$.

(b) $(x - 4)^{2/3} = 16$ Raise each side to the power $\frac{3}{2}$. Insert $\pm$ because this involves an even root, as indicated by the 2 in the denominator.

$$\left[(x - 4)^{2/3}\right]^{3/2} = \pm 16^{3/2}$$

$$x - 4 = \pm 64 \qquad \pm 16^{3/2} = \pm\left(\sqrt{16}\right)^3 = \pm 4^3 = \pm 64$$

$$x = 4 \pm 64 \qquad \text{Add 4 to each side.}$$

$$x = -60 \quad \text{or} \quad x = 68 \qquad \text{Proposed solutions}$$

CHECK $(x - 4)^{2/3} = 16$ Original equation

$(-60 - 4)^{2/3} \stackrel{?}{=} 16$ Let $x = -60$. | $(68 - 4)^{2/3} \stackrel{?}{=} 16$ Let $x = 68$.

$(-64)^{2/3} \stackrel{?}{=} 16$ | $64^{2/3} \stackrel{?}{=} 16$

$\left(\sqrt[3]{-64}\right)^2 \stackrel{?}{=} 16$ | $\left(\sqrt[3]{64}\right)^2 \stackrel{?}{=} 16$

$16 = 16 \checkmark \text{ True}$ | $16 = 16 \checkmark \text{ True}$

Both proposed solutions check, so the solution set is $\{-60, 68\}$.

✔ **Now Try Exercises 75 and 79.**

Equations Quadratic in Form   Many equations that are not quadratic equations can be solved using similar methods. The equation

$$(x + 1)^{2/3} - (x + 1)^{1/3} - 2 = 0$$

is not a quadratic equation in x. However, with the substitutions

$$u = (x + 1)^{1/3} \quad \text{and} \quad u^2 = \left[(x + 1)^{1/3}\right]^2 = (x + 1)^{2/3},$$

the equation becomes

$$u^2 - u - 2 = 0,$$

which is a quadratic equation in u. This quadratic equation can be solved to find u, and then $u = (x + 1)^{1/3}$ can be used to find the values of x, the solutions to the original equation.

Equation Quadratic in Form

An equation is **quadratic in form** if it can be written as

$$au^2 + bu + c = 0,$$

where $a \neq 0$ and u is some algebraic expression.

EXAMPLE 8 **Solving Equations Quadratic in Form**

Solve each equation.

(a) $(x + 1)^{2/3} - (x + 1)^{1/3} - 2 = 0$ **(b)** $6x^{-2} + x^{-1} = 2$

SOLUTION

(a) $(x + 1)^{2/3} - (x + 1)^{1/3} - 2 = 0$ $(x + 1)^{2/3} = [(x + 1)^{1/3}]^2$, so

$u^2 - u - 2 = 0$ let $u = (x + 1)^{1/3}$.

$(u - 2)(u + 1) = 0$ Factor.

$u - 2 = 0$ or $u + 1 = 0$ Zero-factor property

$u = 2$ or $u = -1$ Solve each equation.

$\boxed{\text{Don't forget this step.}}$ $(x + 1)^{1/3} = 2$ or $(x + 1)^{1/3} = -1$ Replace u with $(x + 1)^{1/3}$.

$[(x + 1)^{1/3}]^3 = 2^3$ or $[(x + 1)^{1/3}]^3 = (-1)^3$ Cube each side.

$x + 1 = 8$ or $x + 1 = -1$ Apply the exponents.

$x = 7$ or $x = -2$ Proposed solutions

CHECK $(x + 1)^{2/3} - (x + 1)^{1/3} - 2 = 0$ Original equation

$(7 + 1)^{2/3} - (7 + 1)^{1/3} - 2 \stackrel{?}{=} 0$ Let $x = 7$. │ $(-2 + 1)^{2/3} - (-2 + 1)^{1/3} - 2 \stackrel{?}{=} 0$ Let $x = -2$.

$8^{2/3} - 8^{1/3} - 2 \stackrel{?}{=} 0$ │ $(-1)^{2/3} - (-1)^{1/3} - 2 \stackrel{?}{=} 0$

$4 - 2 - 2 \stackrel{?}{=} 0$ │ $1 + 1 - 2 \stackrel{?}{=} 0$

$0 = 0$ ✓ True   │ $0 = 0$ ✓ True

Both proposed solutions check, so the solution set is $\{-2, 7\}$.

(b) $6x^{-2} + x^{-1} = 2$

$6x^{-2} + x^{-1} - 2 = 0$ Subtract 2 from each side.

$6u^2 + u - 2 = 0$ Let $u = x^{-1}$. Then $u^2 = x^{-2}$.

$(3u + 2)(2u - 1) = 0$ Factor.

$3u + 2 = 0$ or $2u - 1 = 0$ Zero-factor property

$\boxed{\text{Don't stop here. Substitute for } u.}$ $u = -\dfrac{2}{3}$ or $u = \dfrac{1}{2}$ Solve each equation.

$x^{-1} = -\dfrac{2}{3}$ or $x^{-1} = \dfrac{1}{2}$ Replace u with x^{-1}.

$x = -\dfrac{3}{2}$ or $x = 2$ x^{-1} is the reciprocal of x.

Both proposed solutions check, so the solution set is $\left\{ -\frac{3}{2}, 2 \right\}$.

✔ **Now Try Exercises 93 and 99.**

CAUTION *When using a substitution variable in solving an equation that is quadratic in form, do not forget the step that gives the solution in terms of the original variable.*

EXAMPLE 9 Solving an Equation Quadratic in Form

Solve $12x^4 - 11x^2 + 2 = 0$.

SOLUTION

$$12x^4 - 11x^2 + 2 = 0$$

$$12(x^2)^2 - 11x^2 + 2 = 0 \qquad x^4 = (x^2)^2$$

$$12u^2 - 11u + 2 = 0 \qquad \text{Let } u = x^2. \text{ Then } u^2 = x^4.$$

$$(3u - 2)(4u - 1) = 0 \qquad \text{Solve the quadratic equation.}$$

$3u - 2 = 0$	or $\quad 4u - 1 = 0$	Zero-factor property
$u = \dfrac{2}{3}$	or $\qquad u = \dfrac{1}{4}$	Solve each equation.
$x^2 = \dfrac{2}{3}$	or $\qquad x^2 = \dfrac{1}{4}$	Replace u with x^2.
$x = \pm\sqrt{\dfrac{2}{3}}$	or $\qquad x = \pm\sqrt{\dfrac{1}{4}}$	Square root property
$x = \dfrac{\pm\sqrt{2}}{\sqrt{3}} \cdot \dfrac{\sqrt{3}}{\sqrt{3}}$	or $\qquad x = \pm\dfrac{1}{2}$	Simplify radicals.
$x = \pm\dfrac{\sqrt{6}}{3}$		

Check that the solution set is $\left\{ \pm\dfrac{\sqrt{6}}{3}, \pm\dfrac{1}{2} \right\}$. ✔ **Now Try Exercise 87.**

NOTE To solve the equation from **Example 9,**

$$12x^4 - 11x^2 + 2 = 0,$$

we could factor $12x^4 - 11x^2 + 2$ directly as $(3x^2 - 2)(4x^2 - 1)$, set each factor equal to zero, and then solve the resulting two quadratic equations. *Which method to use is a matter of personal preference.*

1.6 Exercises

CONCEPT PREVIEW *Fill in the blank to correctly complete each sentence.*

1. A(n) _____ is an equation that has a rational expression for one or more terms.

2. Proposed solutions for which any denominator equals _____ are excluded from the solution set of a rational equation.

3. If a job can be completed in 4 hr, then the rate of work is _____ of the job per hour.

4. When the power property is used to solve an equation, it is essential to check all proposed solutions in the _____ .

5. An equation such as $x^{3/2} = 8$ is an equation with a(n) _____ , because it contains a variable raised to an exponent that is a rational number.

CONCEPT PREVIEW *Match each equation in Column I with the correct first step for solving it in Column II.*

I

II

6. $\dfrac{2x+3}{x} + \dfrac{5}{x+5} = 7$

A. Cube each side of the equation.

7. $\sqrt{x+5} = 7$

B. Multiply each side of the equation by $x(x+5)$.

8. $(x+5)^{5/2} = 32$

C. Raise each side of the equation to the power $\frac{2}{5}$.

9. $(x+5)^{2/3} - (x+5)^{1/3} - 6 = 0$

D. Square each side of the equation.

10. $\sqrt[3]{x(x+5)} = \sqrt[3]{-6}$

E. Let $u = (x+5)^{1/3}$ and $u^2 = (x+5)^{2/3}$.

Decide what values of the variable cannot possibly be solutions for each equation. Do not solve. **See Examples 1 and 2.**

11. $\dfrac{5}{2x+3} - \dfrac{1}{x-6} = 0$

12. $\dfrac{2}{x+1} + \dfrac{3}{5x-2} = 0$

13. $\dfrac{3}{x-2} + \dfrac{1}{x+1} = \dfrac{3}{x^2-x-2}$

14. $\dfrac{2}{x+3} - \dfrac{5}{x-1} = \dfrac{-5}{x^2+2x-3}$

15. $\dfrac{1}{4x} - \dfrac{2}{x} = 3$

16. $\dfrac{5}{2x} + \dfrac{2}{x} = 6$

Solve each equation. **See Example 1.**

17. $\dfrac{2x+5}{2} - \dfrac{3x}{x-2} = x$

18. $\dfrac{4x+3}{4} - \dfrac{2x}{x+1} = x$

19. $\dfrac{x}{x-3} = \dfrac{3}{x-3} + 3$

20. $\dfrac{x}{x-4} = \dfrac{4}{x-4} + 4$

21. $\dfrac{-2}{x-3} + \dfrac{3}{x+3} = \dfrac{-12}{x^2-9}$

22. $\dfrac{3}{x-2} + \dfrac{1}{x+2} = \dfrac{12}{x^2-4}$

23. $\dfrac{4}{x^2+x-6} - \dfrac{1}{x^2-4} = \dfrac{2}{x^2+5x+6}$

24. $\dfrac{3}{x^2+x-2} - \dfrac{1}{x^2-1} = \dfrac{7}{2x^2+6x+4}$

Solve each equation. **See Example 2.**

25. $\dfrac{2x+1}{x-2} + \dfrac{3}{x} = \dfrac{-6}{x^2-2x}$

26. $\dfrac{4x+3}{x+1} + \dfrac{2}{x} = \dfrac{1}{x^2+x}$

27. $\dfrac{x}{x-1} - \dfrac{1}{x+1} = \dfrac{2}{x^2-1}$

28. $\dfrac{-x}{x+1} - \dfrac{1}{x-1} = \dfrac{-2}{x^2-1}$

29. $\dfrac{5}{x^2} - \dfrac{43}{x} = 18$

30. $\dfrac{7}{x^2} + \dfrac{19}{x} = 6$

31. $2 = \dfrac{3}{2x-1} + \dfrac{-1}{(2x-1)^2}$

32. $6 = \dfrac{7}{2x-3} + \dfrac{3}{(2x-3)^2}$

33. $\dfrac{2x-5}{x} = \dfrac{x-2}{3}$

34. $\dfrac{x+4}{2x} = \dfrac{x-1}{3}$

35. $\dfrac{2x}{x-2} = 5 + \dfrac{4x^2}{x-2}$

36. $\dfrac{3x^2}{x-1} + 2 = \dfrac{x}{x-1}$

Solve each problem. **See Example 3.**

37. *Painting a House* (This problem appears in the 1994 movie *Little Big League*.) If Joe can paint a house in 3 hr, and Sam can paint the same house in 5 hr, how long does it take them to do it together?

38. *Painting a House* Repeat **Exercise 37,** but assume that Joe takes 6 hr working alone, and Sam takes 8 hr working alone.

39. *Pollution in a River* Two chemical plants are polluting a river. If plant A produces a predetermined maximum amount of pollutant twice as fast as plant B, and together they produce the maximum pollutant in 26 hr, how long will it take plant B alone?

	Rate	Time	Part of Job Completed	
Pollution from A	$\frac{1}{x}$	26	$26\left(\frac{1}{x}\right)$	*x* represents the number of hours it takes plant A, working alone, to produce the maximum pollutant.
Pollution from B		26		

40. *Filling a Settling Pond* A sewage treatment plant has two inlet pipes to its settling pond. One pipe can fill the pond 3 times as fast as the other pipe, and together they can fill the pond in 12 hr. How long will it take the faster pipe to fill the pond alone?

	Rate	Time	Part of Job Completed	
Faster Inlet Pipe	$\frac{1}{x}$	12		*x* represents the number of hours it takes the faster inlet pipe, working alone, to fill the settling pond.
Slower Inlet Pipe	$\frac{1}{3x}$	12		

41. *Filling a Pool* An inlet pipe can fill Blake's pool in 5 hr, and an outlet pipe can empty it in 8 hr. In his haste to surf the Internet, Blake left both pipes open. How long did it take to fill the pool?

42. *Filling a Pool* Suppose Blake discovered his error (see **Exercise 41**) after an hour-long surf. If he then closed the outlet pipe, how much more time would be needed to fill the pool?

43. *Filling a Sink* With both taps open, Robert can fill his kitchen sink in 5 min. When full, the sink drains in 10 min. How long will it take to fill the sink if Robert forgets to put in the stopper?

44. *Filling a Sink* If Robert (see **Exercise 43**) remembers to put in the stopper after 1 min, how much longer will it take to fill the sink?

Solve each equation. **See Examples 4–6.**

45. $x - \sqrt{2x + 3} = 0$ **46.** $x - \sqrt{3x + 18} = 0$

47. $\sqrt{3x + 7} = 3x + 5$ **48.** $\sqrt{4x + 13} = 2x - 1$

49. $\sqrt{4x + 5} - 6 = 2x - 11$ **50.** $\sqrt{6x + 7} - 9 = x - 7$

51. $\sqrt{4x} - x + 3 = 0$

52. $\sqrt{2x} - x + 4 = 0$

53. $\sqrt{x} - \sqrt{x-5} = 1$

54. $\sqrt{x} - \sqrt{x-12} = 2$

55. $\sqrt{x+7} + 3 = \sqrt{x-4}$

56. $\sqrt{x+5} + 2 = \sqrt{x-1}$

57. $\sqrt{2x+5} - \sqrt{x+2} = 1$

58. $\sqrt{4x+1} - \sqrt{x-1} = 2$

59. $\sqrt{3x} = \sqrt{5x+1} - 1$

60. $\sqrt{2x} = \sqrt{3x+12} - 2$

61. $\sqrt{x+2} = 1 - \sqrt{3x+7}$

62. $\sqrt{2x-5} = 2 + \sqrt{x-2}$

63. $\sqrt{2\sqrt{7x+2}} = \sqrt{3x+2}$

64. $\sqrt{3\sqrt{2x+3}} = \sqrt{5x-6}$

65. $3 - \sqrt{x} = \sqrt{2\sqrt{x}-3}$

66. $\sqrt{x} + 2 = \sqrt{4+7\sqrt{x}}$

67. $\sqrt[3]{4x+3} = \sqrt[3]{2x-1}$

68. $\sqrt[3]{2x} = \sqrt[3]{5x+2}$

69. $\sqrt[3]{5x^2 - 6x + 2} - \sqrt[3]{x} = 0$

70. $\sqrt[3]{3x^2 - 9x + 8} = \sqrt[3]{x}$

71. $\sqrt[4]{x-15} = 2$

72. $\sqrt[4]{3x+1} = 1$

73. $\sqrt[4]{x^2 + 2x} = \sqrt[4]{3}$

74. $\sqrt[4]{x^2 + 6x} = 2$

Solve each equation. ***See Example 7.***

75. $x^{3/2} = 125$

76. $x^{5/4} = 32$

77. $(x^2 + 24x)^{1/4} = 3$

78. $(3x^2 + 52x)^{1/4} = 4$

79. $(x-3)^{2/5} = 4$

80. $(x+200)^{2/3} = 36$

81. $(2x+5)^{1/3} - (6x-1)^{1/3} = 0$

82. $(3x+7)^{1/3} - (4x+2)^{1/3} = 0$

83. $(2x-1)^{2/3} = x^{1/3}$

84. $(x-3)^{2/5} = (4x)^{1/5}$

85. $x^{2/3} = 2x^{1/3}$

86. $3x^{3/4} = x^{1/2}$

Solve each equation. ***See Examples 8 and 9.***

87. $2x^4 - 7x^2 + 5 = 0$

88. $4x^4 - 8x^2 + 3 = 0$

89. $x^4 + 2x^2 - 15 = 0$

90. $3x^4 + 10x^2 - 25 = 0$

91. $(x-1)^{2/3} + (x-1)^{1/3} - 12 = 0$

92. $(2x-1)^{2/3} + 2(2x-1)^{1/3} - 3 = 0$

93. $(x+1)^{2/5} - 3(x+1)^{1/5} + 2 = 0$

94. $(x+5)^{2/3} + (x+5)^{1/3} - 20 = 0$

95. $4(x+1)^4 - 13(x+1)^2 = -9$

96. $25(x-5)^4 - 116(x-5)^2 = -64$

97. $6(x+2)^4 - 11(x+2)^2 = -4$

98. $8(x-4)^4 - 10(x-4)^2 = -3$

99. $10x^{-2} + 33x^{-1} - 7 = 0$

100. $7x^{-2} - 10x^{-1} - 8 = 0$

101. $x^{-2/3} + x^{-1/3} - 6 = 0$

102. $2x^{-2/5} - x^{-1/5} - 1 = 0$

103. $16x^{-4} - 65x^{-2} + 4 = 0$

104. $625x^{-4} - 125x^{-2} + 4 = 0$

Solve each equation for the specified variable. (Assume all denominators are nonzero.)

105. $d = k\sqrt{h}$, for h

106. $x^{2/3} + y^{2/3} = a^{2/3}$, for y

107. $m^{3/4} + n^{3/4} = 1$, for m

108. $\dfrac{1}{R} = \dfrac{1}{r_1} + \dfrac{1}{r_2}$, for R

109. $\dfrac{E}{e} = \dfrac{R+r}{r}$, for e

110. $a^2 + b^2 = c^2$, for b

Relating Concepts

For individual or collaborative investigation *(Exercises 111–114)*

In this section we introduced methods of solving equations quadratic in form by substitution and solving equations involving radicals by raising each side of the equation to a power. Suppose we wish to solve

$$x - \sqrt{x} - 12 = 0.$$

We can solve this equation using either of the two methods. **Work Exercises 111–114 in order,** *to see how both methods apply.*

111. Let $u = \sqrt{x}$ and solve the equation by substitution.

112. Solve the equation by isolating $\sqrt{x}$ on one side and then squaring.

113. Which one of the methods used in **Exercises 111 and 112** do you prefer? Why?

114. Solve $3x - 2\sqrt{x} - 8 = 0$ using one of the two methods described.

Summary Exercises on Solving Equations

This section of miscellaneous equations provides practice in solving all the types introduced in this chapter so far. Solve each equation.

1. $4x - 3 = 2x + 3$

2. $5 - (6x + 3) = 2(2 - 2x)$

3. $x(x + 6) = 9$

4. $x^2 = 8x - 12$

5. $\sqrt{x + 2} + 5 = \sqrt{x + 15}$

6. $\dfrac{5}{x + 3} - \dfrac{6}{x - 2} = \dfrac{3}{x^2 + x - 6}$

7. $\dfrac{3x + 4}{3} - \dfrac{2x}{x - 3} = x$

8. $\dfrac{x}{2} + \dfrac{4}{3}x = x + 5$

9. $5 - \dfrac{2}{x} + \dfrac{1}{x^2} = 0$

10. $(2x + 1)^2 = 9$

11. $x^{-2/5} - 2x^{-1/5} - 15 = 0$

12. $\sqrt{x + 2} + 1 = \sqrt{2x + 6}$

13. $x^4 - 3x^2 - 4 = 0$

14. $1.2x + 0.3 = 0.7x - 0.9$

15. $\sqrt[3]{2x + 1} = \sqrt[3]{9}$

16. $3x^2 - 2x = -1$

17. $3[2x - (6 - 2x) + 1] = 5x$

18. $\sqrt{x + 1} = \sqrt{11 - \sqrt{x}}$

19. $(14 - 2x)^{2/3} = 4$

20. $-x^{-2} + 2x^{-1} = 1$

21. $\dfrac{3}{x - 3} = \dfrac{3}{x - 3}$

22. $a^2 + b^2 = c^2$, for a

1.7 Inequalities

- **Linear Inequalities**
- **Three-Part Inequalities**
- **Quadratic Inequalities**
- **Rational Inequalities**

An **inequality** says that one expression is greater than, greater than or equal to, less than, or less than or equal to another. As with equations, a value of the variable for which the inequality is true is a solution of the inequality, and the set of all solutions is the solution set of the inequality. Two inequalities with the same solution set are equivalent.

Inequalities are solved with the properties of inequality, which are similar to the properties of equality.

Properties of Inequality

Let a, b, and c represent real numbers.

1. If $a < b$, then $a + c < b + c$.

2. If $a < b$ and if $c > 0$, then $ac < bc$.

3. If $a < b$ and if $c < 0$, then $ac > bc$.

Replacing $<$ with $>$, $\leq$, or $\geq$ results in similar properties. (Restrictions on c remain the same.)

NOTE Multiplication may be replaced by division in Properties 2 and 3. *Always remember to reverse the direction of the inequality symbol when multiplying or dividing by a negative number.*

Linear Inequalities The definition of a *linear inequality* is similar to the definition of a linear equation.

Linear Inequality in One Variable

A **linear inequality in one variable** is an inequality that can be written in the form

$$ax + b > 0,*$$

where a and b are real numbers and $a \neq 0$.

*The symbol $>$ can be replaced with $<$, $\leq$, or $\geq$.

EXAMPLE 1 Solving a Linear Inequality

Solve $-3x + 5 > -7$.

SOLUTION

$$-3x + 5 > -7$$

$$-3x + 5 - 5 > -7 - 5 \quad \text{Subtract 5.}$$

$$-3x > -12 \quad \text{Combine like terms.}$$

$$\frac{-3x}{-3} < \frac{-12}{-3} \quad \begin{array}{l}\text{Divide by } -3. \text{ Reverse the direction of the} \\ \text{inequality symbol when multiplying or} \\ \text{dividing by a negative number.}\end{array}$$

> Don't forget to reverse the inequality symbol here.

$$x < 4$$

Figure 9

Thus, the original inequality $-3x + 5 > -7$ is satisfied by any real number less than 4. The solution set can be written $\{x \mid x < 4\}$.

A graph of the solution set is shown in **Figure 9,** where the parenthesis is used to show that 4 itself does not belong to the solution set. As shown below, testing values from the solution set in the original inequality will produces true statements. Testing values outside the solution set produces false statements.

CHECK $\qquad -3x + 5 > -7 \quad$ Original inequality

$$-3(0) + 5 \overset{?}{>} -7 \quad \text{Let } x = 0. \qquad \qquad -3(5) + 5 \overset{?}{>} -7 \quad \text{Let } x = 5.$$

$$5 > -7 \ \checkmark \ \text{True} \qquad \qquad \qquad -10 > -7 \quad \text{False}$$

The solution set of the inequality,

$$\{x \mid x < 4\}, \quad \text{Set-builder notation}$$

is an example of an **interval.** We use **interval notation** to write intervals. With this notation, we write the above interval as

$$(-\infty, 4). \quad \text{Interval notation}$$

The symbol $-\infty$ does not represent an actual number. Rather, it is used to show that the interval includes all real numbers less than 4. The interval $(-\infty, 4)$ is an example of an **open interval** because the endpoint, 4, is not part of the interval. An interval that includes both its endpoints is a **closed interval.** A square bracket indicates that a number *is* part of an interval, and a parenthesis indicates that a number *is not* part of an interval.

> ✔ **Now Try Exercise 13.**

In the table that follows, we assume that $a < b$.

Summary of Types of Intervals

Type of Interval	Set	Interval Notation	Graph
Open interval	$\{x \mid x > a\}$	(a, ∞)	
	$\{x \mid a < x < b\}$	(a, b)	
	$\{x \mid x < b\}$	$(-\infty, b)$	
Other intervals	$\{x \mid x \geq a\}$	$[a, \infty)$	
	$\{x \mid a < x \leq b\}$	$(a, b]$	
	$\{x \mid a \leq x < b\}$	$[a, b)$	
	$\{x \mid x \leq b\}$	$(-\infty, b]$	
Closed interval	$\{x \mid a \leq x \leq b\}$	$[a, b]$	
Disjoint interval	$\{x \mid x < a \text{ or } x > b\}$	$(-\infty, a) \cup (b, \infty)$	
All real numbers	$\{x \mid x \text{ is a real number}\}$	$(-\infty, \infty)$	

EXAMPLE 2 Solving a Linear Inequality

Solve $4 - 3x \leq 7 + 2x$. Give the solution set in interval notation.

SOLUTION

$$4 - 3x \leq 7 + 2x$$

$$4 - 3x - 4 \leq 7 + 2x - 4 \qquad \text{Subtract 4.}$$

$$-3x \leq 3 + 2x \qquad \text{Combine like terms.}$$

$$-3x - 2x \leq 3 + 2x - 2x \qquad \text{Subtract } 2x.$$

$$-5x \leq 3 \qquad \text{Combine like terms.}$$

$$\frac{-5x}{-5} \geq \frac{3}{-5} \qquad \begin{array}{l}\text{Divide by } -5. \text{ Reverse the direction of}\\ \text{the inequality symbol.}\end{array}$$

$$x \geq -\frac{3}{5} \qquad \frac{a}{-b} = -\frac{a}{b}$$

Figure 10

In interval notation, the solution set is $\left[-\frac{3}{5}, \infty\right)$. See **Figure 10** for the graph.

✔ **Now Try Exercise 15.**

A product will break even, or begin to produce a profit, only if the revenue from selling the product at least equals the cost of producing it. If R represents revenue and C is cost, then the **break-even point** is the point where $R = C$.

EXAMPLE 3 Finding the Break-Even Point

If the revenue and cost of a certain product are given by

$$R = 4x \quad \text{and} \quad C = 2x + 1000,$$

where x is the number of units produced and sold, at what production level does R *at least equal* C?

SOLUTION Set $R \geq C$ and solve for x.

At least equal to translates as ≥.

$$R \geq C$$

$$4x \geq 2x + 1000 \qquad \text{Substitute.}$$

$$2x \geq 1000 \qquad \text{Subtract } 2x.$$

$$x \geq 500 \qquad \text{Divide by 2.}$$

The break-even point is at $x = 500$. This product will at least break even if the number of units produced and sold is in the interval $[500, \infty)$.

✔ **Now Try Exercise 25.**

Three-Part Inequalities The inequality $-2 < 5 + 3x < 20$ says that

$$5 + 3x \text{ is } between -2 \text{ and } 20.$$

This inequality is solved using an extension of the properties of inequality given earlier, working with all three expressions at the same time.

EXAMPLE 4 Solving a Three-Part Inequality

Solve $-2 < 5 + 3x < 20$. Give the solution set in interval notation.

SOLUTION

$$-2 < 5 + 3x < 20$$

$$-2 - 5 < 5 + 3x - 5 < 20 - 5 \quad \text{Subtract 5 from each part.}$$

$$-7 < 3x < 15 \quad \text{Combine like terms in each part.}$$

$$\frac{-7}{3} < \frac{3x}{3} < \frac{15}{3} \quad \text{Divide each part by 3.}$$

$$-\frac{7}{3} < x < 5$$

Figure 11

The solution set, graphed in **Figure 11,** is the interval $\left(-\frac{7}{3}, 5\right)$.

☑ **Now Try Exercise 29.**

Quadratic Inequalities   We can distinguish a *quadratic inequality* from a linear inequality by noticing that it is of degree 2.

Quadratic Inequality

A **quadratic inequality** is an inequality that can be written in the form

$$ax^2 + bx + c < 0,^*$$

where a, b, and c are real numbers and $a \neq 0$.

*The symbol $<$ can be replaced with $>$, $\leq$, or $\geq$.

One method of solving a quadratic inequality involves finding the solutions of the corresponding quadratic equation and then testing values in the intervals on a number line determined by those solutions.

Solving a Quadratic Inequality

Step 1 Solve the corresponding quadratic equation.

Step 2 Identify the intervals determined by the solutions of the equation.

Step 3 Use a test value from each interval to determine which intervals form the solution set.

EXAMPLE 5 Solving a Quadratic Inequality

Solve $x^2 - x - 12 < 0$.

SOLUTION

Step 1 Find the values of x that satisfy $x^2 - x - 12 = 0$.

$$x^2 - x - 12 = 0 \quad \text{Corresponding quadratic equation}$$

$$(x + 3)(x - 4) = 0 \quad \text{Factor.}$$

$$x + 3 = 0 \quad \text{or} \quad x - 4 = 0 \quad \text{Zero-factor property}$$

$$x = -3 \quad \text{or} \quad x = 4 \quad \text{Solve each equation.}$$

Step 2 The two numbers -3 and 4 cause the expression $x^2 - x - 12$ to ***equal*** zero and can be used to divide the number line into three intervals, as shown in **Figure 12**. The expression $x^2 - x - 12$ will take on a value that is either ***less than*** zero or ***greater than*** zero on each of these intervals. We are looking for x-values that make the expression ***less than*** zero, so we use open circles at -3 and 4 to indicate that they are not included in the solution set.

Use open circles because the inequality symbol does not include *equality*. -3 and 4 do not satisfy the *inequality*.

Figure 12

Step 3 Choose a test value in each interval to see whether it satisfies the original inequality, $x^2 - x - 12 < 0$. If the test value makes the statement true, then the entire interval belongs to the solution set.

Interval	Test Value	Is $x^2 - x - 12 < 0$ True or False?
A: $(-\infty, -3)$	-4	$(-4)^2 - (-4) - 12 \overset{?}{<} 0$ $8 < 0$ False
B: $(-3, 4)$	0	$0^2 - 0 - 12 \overset{?}{<} 0$ $-12 < 0$ True
C: $(4, \infty)$	5	$5^2 - 5 - 12 \overset{?}{<} 0$ $8 < 0$ False

Figure 13

Because the values in Interval B make the inequality true, the solution set is the interval $(-3, 4)$. See **Figure 13**.   ✔ **Now Try Exercise 41.**

EXAMPLE 6   **Solving a Quadratic Inequality**

Solve $2x^2 + 5x - 12 \geq 0$.

SOLUTION

Step 1 Find the values of x that satisfy $2x^2 + 5x - 12 = 0$.

$$2x^2 + 5x - 12 = 0 \qquad \text{Corresponding quadratic equation}$$

$$(2x - 3)(x + 4) = 0 \qquad \text{Factor.}$$

$$2x - 3 = 0 \quad \text{or} \quad x + 4 = 0 \qquad \text{Zero-factor property}$$

$$x = \frac{3}{2} \quad \text{or} \qquad x = -4 \quad \text{Solve each equation.}$$

Step 2 The values $\frac{3}{2}$ and -4 cause the expression $2x^2 + 5x - 12$ to equal 0 and can be used to form the intervals $(-\infty, -4)$, $\left(-4, \frac{3}{2}\right)$, and $\left(\frac{3}{2}, \infty\right)$ on the number line, as seen in **Figure 14**.

Use closed circles because the inequality symbol includes *equality*. -4 and $\frac{3}{2}$ satisfy the *inequality*.

Figure 14

Step 3 Choose a test value in each interval.

Interval	Test Value	Is $2x^2 + 5x - 12 \geq 0$ True or False?
A: $(-\infty, -4)$	-5	$2(-5)^2 + 5(-5) - 12 \overset{?}{\geq} 0$ $13 \geq 0$ True
B: $\left(-4, \frac{3}{2}\right)$	0	$2(0)^2 + 5(0) - 12 \overset{?}{\geq} 0$ $-12 \geq 0$ False
C: $\left(\frac{3}{2}, \infty\right)$	2	$2(2)^2 + 5(2) - 12 \overset{?}{\geq} 0$ $6 \geq 0$ True

The values in Intervals A and C make the inequality true, so the solution set is a disjoint interval: the *union* of the two intervals, written

$$\left(-\infty, -4\right] \cup \left[\frac{3}{2}, \infty\right).$$

-4   0 $\frac{3}{2}$

Figure 15

The graph of the solution set is shown in **Figure 15.**

✔ **Now Try Exercise 39.**

NOTE   Inequalities that use the symbols $<$ and $>$ are **strict inequalities**, while $\leq$ and $\geq$ are used in **nonstrict inequalities**. The solutions of the equation in **Example 5** were not included in the solution set because the inequality was a *strict* inequality. In **Example 6,** the solutions of the equation *were* included in the solution set because of the nonstrict inequality.

EXAMPLE 7   **Finding Projectile Height**

If a projectile is launched from ground level with an initial velocity of 96 ft per sec, its height s in feet t seconds after launching is given by the following equation.

$$s = -16t^2 + 96t$$

When will the projectile be greater than 80 ft above ground level?

SOLUTION

$$-16t^2 + 96t > 80 \qquad \text{Set } s \text{ greater than 80.}$$
$$-16t^2 + 96t - 80 > 0 \qquad \text{Subtract 80.}$$
$$t^2 - 6t + 5 < 0 \qquad \text{Divide by } -16.$$

> Reverse the direction of the inequality symbol.

Now solve the corresponding *equation*.

$$t^2 - 6t + 5 = 0$$
$$(t - 1)(t - 5) = 0 \qquad \text{Factor.}$$
$$t - 1 = 0 \quad \text{or} \quad t - 5 = 0 \qquad \text{Zero-factor property}$$
$$t = 1 \quad \text{or} \qquad t = 5 \qquad \text{Solve each equation.}$$

Use these values to determine the intervals

$$(-\infty, 1), \quad (1, 5), \quad \text{and} \quad (5, \infty).$$

We are solving a strict inequality, so solutions of the equation $t^2 - 6t + 5 = 0$ are **not** included. Choose a test value from each interval to see whether it satisfies the inequality $t^2 - 6t + 5 < 0$. See **Figure 16** on the next page.

Figure 16

The values in Interval B, $(1, 5)$, make the inequality true. The projectile is greater than 80 ft above ground level between 1 and 5 sec after it is launched.

✔ **Now Try Exercise 81.**

Rational Inequalities Inequalities involving one or more rational expressions are **rational inequalities**.

$$\frac{5}{x + 4} \geq 1 \quad \text{and} \quad \frac{2x - 1}{3x + 4} < 5 \qquad \text{Rational inequalities}$$

Solving a Rational Inequality

Step 1 Rewrite the inequality, if necessary, so that 0 is on one side and there is a single fraction on the other side.

Step 2 Determine the values that will cause either the numerator or the denominator of the rational expression to equal 0. These values determine the intervals on the number line to consider.

Step 3 Use a test value from each interval to determine which intervals form the solution set.

A value causing a denominator to equal zero will never be included in the solution set. If the inequality is strict, any value causing the numerator to equal zero will be excluded. If the inequality is nonstrict, any such value will be included.

CAUTION Solving a rational inequality such as $\frac{5}{x + 4} \geq 1$ by multiplying each side by $x + 4$ requires considering *two cases*, because the sign of $x + 4$ depends on the value of x. If $x + 4$ is negative, then the inequality symbol must be reversed. The procedure described in the preceding box eliminates the need for considering separate cases.

EXAMPLE 8   Solving a Rational Inequality

Solve $\dfrac{5}{x + 4} \geq 1$.

SOLUTION

Step 1
$$\frac{5}{x + 4} - 1 \geq 0 \qquad \text{Subtract 1 so that 0 is on one side.}$$

> Note the careful use of parentheses.

$$\frac{5}{x + 4} - \frac{x + 4}{x + 4} \geq 0 \qquad \text{Use } x + 4 \text{ as the common denominator.}$$

$$\frac{5 - (x + 4)}{x + 4} \geq 0 \qquad \text{Write as a single fraction.}$$

$$\frac{1 - x}{x + 4} \geq 0 \qquad \text{Combine like terms in the numerator, being careful with signs.}$$

Step 2 The quotient possibly changes sign only where x-values make the numerator or denominator 0. This occurs at

$$1 - x = 0 \quad \text{or} \quad x + 4 = 0$$
$$x = 1 \quad \text{or} \quad x = -4.$$

These values form the intervals $(-\infty, -4)$, $(-4, 1)$, and $(1, \infty)$ on the number line, as seen in **Figure 17.**

Use a solid circle on 1 because the symbol is $\geq$. The value -4 cannot be in the solution set because it causes the denominator to equal 0. Use an open circle on -4.

Figure 17

Step 3 Choose test values.

Interval	Test Value	Is $\dfrac{5}{x + 4} \geq 1$ True or False?
A: $(-\infty, -4)$	-5	$\dfrac{5}{-5 + 4} \overset{?}{\geq} 1$ $-5 \geq 1$ False
B: $(-4, 1)$	0	$\dfrac{5}{0 + 4} \overset{?}{\geq} 1$ $\dfrac{5}{4} \geq 1$ True
C: $(1, \infty)$	2	$\dfrac{5}{2 + 4} \overset{?}{\geq} 1$ $\dfrac{5}{6} \geq 1$ False

The values in Interval B, $(-4, 1)$, satisfy the original inequality. The value 1 makes the nonstrict inequality true, so it must be included in the solution set. Because -4 makes the denominator 0, it must be excluded. The solution set is the interval $(-4, 1]$. ✔ **Now Try Exercise 59.**

CAUTION *Be careful with the endpoints of the intervals when solving rational inequalities.*

EXAMPLE 9 **Solving a Rational Inequality**

Solve $\dfrac{2x - 1}{3x + 4} < 5$.

SOLUTION $\qquad \dfrac{2x - 1}{3x + 4} - 5 < 0$ Subtract 5.

$$\dfrac{2x - 1}{3x + 4} - \dfrac{5(3x + 4)}{3x + 4} < 0 \quad \text{The common denominator is } 3x + 4.$$

$$\dfrac{2x - 1 - 5(3x + 4)}{3x + 4} < 0 \quad \text{Write as a single fraction.}$$

Be careful with signs. $\dfrac{2x - 1 - 15x - 20}{3x + 4} < 0 \quad \text{Distributive property}$

$$\dfrac{-13x - 21}{3x + 4} < 0 \quad \text{Combine like terms in the numerator.}$$

Set the numerator and denominator of $\frac{-13x - 21}{3x + 4}$ equal to 0 and solve the resulting equations to find the values of x where sign changes may occur.

$$-13x - 21 = 0 \quad \text{or} \quad 3x + 4 = 0$$

$$x = -\frac{21}{13} \quad \text{or} \quad x = -\frac{4}{3}$$

Use these values to form intervals on the number line. Use an open circle at $-\frac{21}{13}$ because of the strict inequality, and use an open circle at $-\frac{4}{3}$ because it causes the denominator to equal 0. See **Figure 18.**

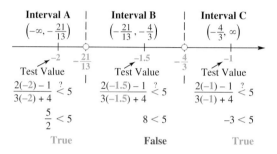

Figure 18

Choosing a test value from each interval shows that the values in Intervals A and C satisfy the original inequality, $\frac{2x - 1}{3x + 4} < 5$. So the solution set is the union of these intervals.

$$\left(-\infty, -\frac{21}{13}\right) \cup \left(-\frac{4}{3}, \infty\right)$$

✔ **Now Try Exercise 71.**

1.7 Exercises

CONCEPT PREVIEW *Match the inequality in each exercise in Column I with its equivalent interval notation in Column II.*

I	II
1. $x < -6$	**A.** $(-2, 6]$
2. $x \leq 6$	**B.** $[-2, 6)$
3. $-2 < x \leq 6$	**C.** $(-\infty, -6]$
4. $x^2 \geq 0$	**D.** $[6, \infty)$
5. $x \geq -6$	**E.** $(-\infty, -3) \cup (3, \infty)$
6. $6 \leq x$	**F.** $(-\infty, -6)$

G. $(0, 8)$

H. $(-\infty, \infty)$

I. $[-6, \infty)$

J. $(-\infty, 6]$

11. Explain how to determine whether to use a parenthesis or a square bracket when writing the solution set of a linear inequality in interval notation.

12. *Concept Check* The three-part inequality $a < x < b$ means "a is less than x and x is less than b." Which inequality is *not* satisfied by some real number x?

A. $-3 < x < 10$ **B.** $0 < x < 6$

C. $-3 < x < -1$ **D.** $-8 < x < -10$

Solve each inequality. Give the solution set in interval notation. ***See Examples 1 and 2.***

13. $-2x + 8 \leq 16$ **14.** $-3x - 8 \leq 7$

15. $-2x - 2 \leq 1 + x$ **16.** $-4x + 3 \geq -2 + x$

17. $3(x + 5) + 1 \geq 5 + 3x$ **18.** $6x - (2x + 3) \geq 4x - 5$

19. $8x - 3x + 2 < 2(x + 7)$ **20.** $2 - 4x + 5(x - 1) < -6(x - 2)$

21. $\dfrac{4x + 7}{-3} \leq 2x + 5$ **22.** $\dfrac{2x - 5}{-8} \leq 1 - x$

23. $\dfrac{1}{3}x + \dfrac{2}{5}x - \dfrac{1}{2}(x + 3) \leq \dfrac{1}{10}$ **24.** $-\dfrac{2}{3}x - \dfrac{1}{6}x + \dfrac{2}{3}(x + 1) \leq \dfrac{4}{3}$

Break-Even Interval *Find all intervals where each product will at least break even.* ***See Example 3.***

25. The cost to produce x units of picture frames is $C = 50x + 5000$, while the revenue is $R = 60x$.

26. The cost to produce x units of baseball caps is $C = 100x + 6000$, while the revenue is $R = 500x$.

27. The cost to produce x units of coffee cups is $C = 105x + 900$, while the revenue is $R = 85x$.

28. The cost to produce x units of briefcases is $C = 70x + 500$, while the revenue is $R = 60x$.

Solve each inequality. Give the solution set in interval notation. ***See Example 4.***

29. $-5 < 5 + 2x < 11$ **30.** $-7 < 2 + 3x < 5$

31. $10 \leq 2x + 4 \leq 16$ **32.** $-6 \leq 6x + 3 \leq 21$

33. $-11 > -3x + 1 > -17$ **34.** $2 > -6x + 3 > -3$

35. $-4 \leq \dfrac{x + 1}{2} \leq 5$ **36.** $-5 \leq \dfrac{x - 3}{3} \leq 1$

37. $-3 \leq \dfrac{3x - 4}{-5} < 4$ **38.** $1 \leq \dfrac{4x - 5}{-2} < 9$

Solve each quadratic inequality. Give the solution set in interval notation. ***See Examples 5 and 6.***

39. $x^2 - x - 6 > 0$ **40.** $x^2 - 7x + 10 > 0$

41. $2x^2 - 9x \leq 18$ **42.** $3x^2 + x \leq 4$

43. $-x^2 - 4x - 6 \leq -3$ **44.** $-x^2 - 6x - 16 > -8$

45. $x(x - 1) \leq 6$ **46.** $x(x + 1) < 12$

47. $x^2 \leq 9$ **48.** $x^2 > 16$

49. $x^2 + 5x + 7 < 0$

50. $4x^2 + 3x + 1 \leq 0$

51. $x^2 - 2x \leq 1$

52. $x^2 + 4x > -1$

53. *Concept Check* Which inequality has solution set $(-\infty, \infty)$?

    A. $(x - 3)^2 \geq 0$ **B.** $(5x - 6)^2 \leq 0$

    C. $(6x + 4)^2 > 0$ **D.** $(8x + 7)^2 < 0$

54. *Concept Check* Which inequality in **Exercise 53** has solution set $\varnothing$?

Solve each rational inequality. Give the solution set in interval notation. ***See Examples 8 and 9.***

55. $\dfrac{x - 3}{x + 5} \leq 0$

56. $\dfrac{x + 1}{x - 4} > 0$

57. $\dfrac{1 - x}{x + 2} < -1$

58. $\dfrac{6 - x}{x + 2} > 1$

59. $\dfrac{3}{x - 6} \leq 2$

60. $\dfrac{3}{x - 2} < 1$

61. $\dfrac{-4}{1 - x} < 5$

62. $\dfrac{-6}{3x - 5} \leq 2$

63. $\dfrac{10}{3 + 2x} \leq 5$

64. $\dfrac{1}{x + 2} \geq 3$

65. $\dfrac{7}{x + 2} \geq \dfrac{1}{x + 2}$

66. $\dfrac{5}{x + 1} > \dfrac{12}{x + 1}$

67. $\dfrac{3}{2x - 1} > \dfrac{-4}{x}$

68. $\dfrac{-5}{3x + 2} \geq \dfrac{5}{x}$

69. $\dfrac{4}{2 - x} \geq \dfrac{3}{1 - x}$

70. $\dfrac{4}{x + 1} < \dfrac{2}{x + 3}$

71. $\dfrac{x + 3}{x - 5} \leq 1$

72. $\dfrac{x + 2}{3 + 2x} \leq 5$

Solve each rational inequality. Give the solution set in interval notation.

73. $\dfrac{2x - 3}{x^2 + 1} \geq 0$

74. $\dfrac{9x - 8}{4x^2 + 25} < 0$

75. $\dfrac{(5 - 3x)^2}{(2x - 5)^3} > 0$

76. $\dfrac{(5x - 3)^3}{(25 - 8x)^2} \leq 0$

77. $\dfrac{(2x - 3)(3x + 8)}{(x - 6)^3} \geq 0$

78. $\dfrac{(9x - 11)(2x + 7)}{(3x - 8)^3} > 0$

(Modeling) *Solve each problem.*

79. *Box Office Receipts* U.S. movie box office receipts, in billions of dollars, are shown in 5-year increments from 1993 to 2013. (*Source:* www.boxofficemojo.com)

Year	Receipts
1993	5.154
1998	6.949
2003	9.240
2008	9.631
2013	10.924

These receipts R are reasonably approximated by the linear model

$$R = 0.2844x + 5.535,$$

where $x = 0$ corresponds to 1993, $x = 5$ corresponds to 1998, and so on. Using the model, calculate the year in which the receipts first exceed each amount.

(a) $7.6 billion   (b) $10 billion

80. *Recovery of Solid Waste* The percent W of municipal solid waste recovered is shown in the bar graph. The linear model

$$W = 0.33x + 33.1,$$

where $x = 1$ represents 2008, $x = 2$ represents 2009, and so on, fits the data reasonably well.

(a) Based on this model, when did the percent of waste recovered first exceed 34%?

(b) In what years was it between 33.9% and 34.5%?

Municipal Solid Waste Recovered

Source: U.S. Environmental Protection Agency.

Solve each problem. See Example 7.

81. *Height of a Projectile* A projectile is fired straight up from ground level. After t seconds, its height above the ground is s feet, where

$$s = -16t^2 + 220t.$$

For what time period is the projectile at least 624 ft above the ground?

82. *Height of a Projectile* See **Exercise 81.** For what time period is the projectile at least 744 ft above the ground?

83. *Height of a Baseball* A baseball is hit so that its height, s, in feet after t seconds is

$$s = -16t^2 + 44t + 4.$$

For what time period is the ball at least 32 ft above the ground?

84. *Height of a Baseball* See **Exercise 83.** For what time period is the ball greater than 28 ft above the ground?

85. *Velocity of an Object* Suppose the velocity, v, of an object is given by

$$v = 2t^2 - 5t - 12,$$

where t is time in seconds. (Here t can be positive or negative.) Find the intervals where the velocity is negative.

86. *Velocity of an Object* The velocity of an object, v, after t seconds is given by

$$v = 3t^2 - 18t + 24.$$

Find the interval where the velocity is negative.

Relating Concepts

For individual or collaborative investigation (Exercises 87–90)

Inequalities that involve more than two factors, such as

$$(3x - 4)(x + 2)(x + 6) \leq 0,$$

can be solved using an extension of the method shown in **Examples 5 and 6.** **Work Exercises 87–90 in order,** *to see how the method is extended.*

87. Use the zero-factor property to solve $(3x - 4)(x + 2)(x + 6) = 0$.

88. Plot the three solutions in **Exercise 87** on a number line, using closed circles because of the nonstrict inequality, $\leq$.

89. The number line from **Exercise 88** should show four intervals formed by the three points. For each interval, choose a test value from the interval and decide whether it satisfies the original inequality.

90. On a single number line, do the following.

(a) Graph the intervals that satisfy the inequality, including endpoints. This is the graph of the solution set of the inequality.

(b) Write the solution set in interval notation.

Use the technique described in **Exercises 87–90** *to solve each inequality. Write each solution set in interval notation.*

91. $(2x - 3)(x + 2)(x - 3) \geq 0$ **92.** $(x + 5)(3x - 4)(x + 2) \geq 0$

93. $4x - x^3 \geq 0$ **94.** $16x - x^3 \geq 0$

95. $(x + 1)^2(x - 3) < 0$ **96.** $(x - 5)^2(x + 1) < 0$

97. $x^3 + 4x^2 - 9x \geq 36$ **98.** $x^3 + 3x^2 - 16x \leq 48$

99. $x^2(x + 4)^2 \geq 0$ **100.** $-x^2(2x - 3)^2 \leq 0$

1.8   Absolute Value Equations and Inequalities

■ **Basic Concepts**

■ **Absolute Value Equations**

■ **Absolute Value Inequalities**

■ **Special Cases**

■ **Absolute Value Models for Distance and Tolerance**

Basic Concepts   Recall that the **absolute value** of a number a, written $|a|$, gives the undirected distance from a to 0 on a number line. By this definition, the equation $|x| = 3$ can be solved by finding all real numbers at a distance of 3 units from 0. As shown in **Figure 19,** two numbers satisfy this equation, -3 and 3, so the solution set is $\{-3, 3\}$.

Figure 19

Similarly, $|x| < 3$ is satisfied by all real numbers whose undirected distances from 0 are less than 3. As shown in **Figure 19,** this is the interval

$$-3 < x < 3, \quad \text{or} \quad (-3, 3).$$

Finally, $|x| > 3$ is satisfied by all real numbers whose undirected distances from 0 are greater than 3. These numbers are less than -3 or greater than 3, so the solution set is

$$(-\infty, -3) \cup (3, \infty).$$

Notice in **Figure 19** that the union of the solution sets of $|x| = 3$, $|x| < 3$, and $|x| > 3$ is the set of real numbers.

These observations support the cases for solving absolute value equations and inequalities summarized in the table that follows. If the equation or inequality fits the form of Case 1, 2, or 3, change it to its equivalent form and solve. The solution set and its graph will look similar to those shown.

Solving Absolute Value Equations and Inequalities

Absolute Value Equation or Inequality*	Equivalent Form	Graph of the Solution Set	Solution Set
Case 1: $\|x\| = k$	$x = k$ or $x = -k$	$-k \quad k$	$\{-k, k\}$
Case 2: $\|x\| < k$	$-k < x < k$	$-k \quad k$	$(-k, k)$
Case 3: $\|x\| > k$	$x < -k$ or $x > k$	$-k \quad k$	$(-\infty, -k) \cup (k, \infty)$

*For each equation or inequality in Cases 1–3, assume that $k > 0$.

In Cases 2 and 3, the strict inequality may be replaced by its nonstrict form. Additionally, if an absolute value equation takes the form $|a| = |b|$, then a and b must be equal in value or opposite in value.

Thus, the equivalent form of $|a| = |b|$ is $a = b$ or $a = -b$.

Absolute Value Equations   Because absolute value represents undirected distance from 0 on a number line, solving an absolute value equation requires solving two possibilities, as shown in the examples that follow.

EXAMPLE 1   **Solving Absolute Value Equations (Case 1 and the Special Case $|a| = |b|$)**

Solve each equation.

(a) $|5 - 3x| = 12$ **(b)** $|4x - 3| = |x + 6|$

SOLUTION

(a) For the given expression $5 - 3x$ to have absolute value 12, it must represent either 12 or -12. This equation fits the form of Case 1.

$$|5 - 3x| = 12 \qquad \text{Don't forget this second possibility.}$$

$$5 - 3x = 12 \quad \text{or} \quad 5 - 3x = -12 \qquad \text{Case 1}$$

$$-3x = 7 \quad \text{or} \quad -3x = -17 \qquad \text{Subtract 5.}$$

$$x = -\frac{7}{3} \quad \text{or} \quad x = \frac{17}{3} \qquad \text{Divide by } -3.$$

Check the solutions $-\frac{7}{3}$ and $\frac{17}{3}$ by substituting them in the original absolute value equation. The solution set is $\left\{-\frac{7}{3}, \frac{17}{3}\right\}$.

LOOKING AHEAD TO CALCULUS

The precise definition of a **limit** in calculus requires writing absolute value inequalities.

A standard problem in calculus is to find the "interval of convergence" of a **power series** by solving the following inequality.

$$|x - a| < r$$

This inequality says that x can be any number within r units of a on the number line, so its solution set is indeed an interval—namely the interval $(a - r, a + r)$.

(b) If the absolute values of two expressions are equal, then those expressions are either equal in value or opposite in value.

$$|4x - 3| = |x + 6|$$

$4x - 3 = x + 6$ or $4x - 3 = -(x + 6)$ Consider both possibilities.

$3x = 9$ or $4x - 3 = -x - 6$ Solve each linear equation.

$x = 3$ or $5x = -3$

$$x = -\frac{3}{5}$$

CHECK $|4x - 3| = |x + 6|$ Original equation

$\left|4\left(-\frac{3}{5}\right) - 3\right| \overset{?}{=} \left|-\frac{3}{5} + 6\right|$ Let $x = -\frac{3}{5}$. $\quad |4(3) - 3| \overset{?}{=} |3 + 6|$ Let $x = 3$.

$\left|-\frac{12}{5} - 3\right| \overset{?}{=} \left|-\frac{3}{5} + 6\right|$ $\qquad\qquad |12 - 3| \overset{?}{=} |3 + 6|$

$\left|-\frac{27}{5}\right| = \left|\frac{27}{5}\right|$ ✓   True $\qquad\qquad |9| = |9|$ ✓   True

Both solutions check. The solution set is $\left\{-\frac{3}{5}, 3\right\}$.

✔ **Now Try Exercises 9 and 19.**

Absolute Value Inequalities

EXAMPLE 2   **Solving Absolute Value Inequalities (Cases 2 and 3)**

Solve each inequality.

(a) $|2x + 1| < 7$ **(b)** $|2x + 1| > 7$

SOLUTION

(a) This inequality fits Case 2. If the absolute value of an expression is less than 7, then the value of the expression is *between* -7 and 7.

$$|2x + 1| < 7$$

$-7 < 2x + 1 < 7$ Case 2

$-8 < \quad 2x \quad < 6$ Subtract 1 from each part.

$-4 < \quad x \quad < 3$ Divide each part by 2.

The final inequality gives the solution set $(-4, 3)$ in interval notation.

(b) This inequality fits Case 3. If the absolute value of an expression is greater than 7, then the value of the expression is either less than -7 *or* greater than 7.

$$|2x + 1| > 7$$

$2x + 1 < -7$ or $2x + 1 > 7$ Case 3

$2x < -8$ or $\qquad 2x > 6$ Subtract 1 from each side.

$x < -4$ or $\qquad x > 3$ Divide each side by 2.

The solution set written in interval notation is $(-\infty, -4) \cup (3, \infty)$.

✔ **Now Try Exercises 27 and 29.**

Cases 1, 2, and 3 require that the absolute value expression be isolated on one side of the equation or inequality.

EXAMPLE 3 Solving an Absolute Value Inequality (Case 3)

Solve $|2 - 7x| - 1 > 4$.

SOLUTION
$$|2 - 7x| - 1 > 4$$

$$|2 - 7x| > 5 \qquad \text{Add 1 to each side.}$$

$$2 - 7x < -5 \quad \text{or} \quad 2 - 7x > 5 \qquad \text{Case 3}$$

$$-7x < -7 \quad \text{or} \quad -7x > 3 \qquad \text{Subtract 2 from each side.}$$

$$x > 1 \quad \text{or} \quad x < -\frac{3}{7} \qquad \text{Divide by } -7. \text{ Reverse the direction of each inequality.}$$

The solution set written in interval notation is $\left(-\infty, -\frac{3}{7}\right) \cup (1, \infty)$.

✔ **Now Try Exercise 51.**

Special Cases The three cases given in this section require the constant k to be positive. **When $k \leq 0$, use the fact that the absolute value of any expression must be nonnegative, and consider the conditions necessary for the statement to be true.**

EXAMPLE 4 Solving Special Cases

Solve each equation or inequality.

(a) $|2 - 5x| \geq -4$ **(b)** $|4x - 7| < -3$ **(c)** $|5x + 15| = 0$

SOLUTION

(a) Since the absolute value of a number is always nonnegative, the inequality

$$|2 - 5x| \geq -4 \text{ is always true.}$$

The solution set includes all real numbers, written $(-\infty, \infty)$.

(b) There is no number whose absolute value is less than -3 (or less than *any* negative number).

$$\text{The solution set of } |4x - 7| < -3 \text{ is } \varnothing.$$

(c) The absolute value of a number will be 0 only if that number is 0. Therefore, $|5x + 15| = 0$ is equivalent to

$$5x + 15 = 0, \quad \text{which has solution set} \quad \{-3\}.$$

CHECK   Substitute -3 into the original equation.

$$|5x + 15| = 0 \qquad \text{Original equation}$$

$$|5(-3) + 15| \stackrel{?}{=} 0 \qquad \text{Let } x = -3.$$

$$0 = 0 \checkmark \text{ True}$$

✔ **Now Try Exercises 55, 57, and 59.**

Absolute Value Models for Distance and Tolerance   If a and b represent two real numbers, then the absolute value of their difference,

$$\text{either} \quad |a - b| \quad \text{or} \quad |b - a|,$$

represents the undirected distance between them.

EXAMPLE 5 Using Absolute Value Inequalities with Distances

Write each statement using an absolute value inequality.

(a) k is no less than 5 units from 8.   **(b)** n is within 0.001 unit of 6.

SOLUTION

(a) Since the distance from k to 8, written $|k - 8|$ or $|8 - k|$, is no less than 5, the distance is greater than or equal to 5. This can be written as

$$|k - 8| \geq 5, \quad \text{or, equivalently,} \quad |8 - k| \geq 5. \quad \text{Either form is acceptable.}$$

(b) This statement indicates that the distance between n and 6 is less than 0.001.

$$|n - 6| < 0.001, \quad \text{or, equivalently,} \quad |6 - n| < 0.001$$

✔ **Now Try Exercises 69 and 71.**

EXAMPLE 6 Using Absolute Value to Model Tolerance

In quality control situations, such as filling bottles on an assembly line, we often wish to keep the difference between two quantities within some predetermined amount, called the **tolerance.**

Suppose $y = 2x + 1$ and we want y to be within 0.01 unit of 4. For what values of x will this be true?

SOLUTION

$$|y - 4| < 0.01 \quad \text{Write an absolute value inequality.}$$
$$|2x + 1 - 4| < 0.01 \quad \text{Substitute } 2x + 1 \text{ for } y.$$
$$|2x - 3| < 0.01 \quad \text{Combine like terms.}$$
$$-0.01 < 2x - 3 < 0.01 \quad \text{Case 2}$$
$$2.99 < 2x < 3.01 \quad \text{Add 3 to each part.}$$
$$1.495 < x < 1.505 \quad \text{Divide each part by 2.}$$

Reversing these steps shows that keeping x in the interval $(1.495, 1.505)$ ensures that the difference between y and 4 is within 0.01 unit.

✔ **Now Try Exercise 75.**

1.8 Exercises

CONCEPT PREVIEW *Match each equation or inequality in Column I with the graph of its solution set in Column II.*

I   **II**

1. $|x| = 7$ **A.**

2. $|x| = -7$ **B.**

3. $|x| > -7$ **C.**

4. $|x| > 7$ **D.**

5. $|x| < 7$ **E.**

6. $|x| \geq 7$ **F.**

7. $|x| \leq 7$ **G.**

8. $|x| \neq 7$ **H.**

Solve each equation. **See Example 1.**

9. $|3x - 1| = 2$ **10.** $|4x + 2| = 5$ **11.** $|5 - 3x| = 3$

12. $|7 - 3x| = 3$ **13.** $\left|\dfrac{x - 4}{2}\right| = 5$ **14.** $\left|\dfrac{x + 2}{2}\right| = 7$

15. $\left|\dfrac{5}{x - 3}\right| = 10$ **16.** $\left|\dfrac{3}{2x - 1}\right| = 4$ **17.** $\left|\dfrac{6x + 1}{x - 1}\right| = 3$

18. $\left|\dfrac{2x + 3}{3x - 4}\right| = 1$ **19.** $|2x - 3| = |5x + 4|$ **20.** $|x + 1| = |1 - 3x|$

21. $|4 - 3x| = |2 - 3x|$ **22.** $|3 - 2x| = |5 - 2x|$ **23.** $|5x - 2| = |2 - 5x|$

24. The equation $|5x - 6| = 3x$ cannot have a negative solution. Why?

25. The equation $|7x + 3| = -5x$ cannot have a positive solution. Why?

26. *Concept Check* Determine the solution set of each equation by inspection.

    (a) $-|x| = |x|$ **(b)** $|-x| = |x|$ **(c)** $|x^2| = |x|$ **(d)** $-|x| = 9$

Solve each inequality. Give the solution set in interval notation. **See Example 2.**

27. $|2x + 5| < 3$ **28.** $|3x - 4| < 2$ **29.** $|2x + 5| \geq 3$

30. $|3x - 4| \geq 2$ **31.** $\left|\dfrac{1}{2} - x\right| < 2$ **32.** $\left|\dfrac{3}{5} + x\right| < 1$

33. $4|x - 3| > 12$ **34.** $5|x + 1| > 10$ **35.** $|5 - 3x| > 7$

36. $|7 - 3x| > 4$ **37.** $|5 - 3x| \leq 7$ **38.** $|7 - 3x| \leq 4$

39. $\left|\dfrac{2}{3}x + \dfrac{1}{2}\right| \leq \dfrac{1}{6}$ **40.** $\left|\dfrac{5}{3} - \dfrac{1}{2}x\right| > \dfrac{2}{9}$ **41.** $|0.01x + 1| < 0.01$

42. Explain why the equation $|x| = \sqrt{x^2}$ has infinitely many solutions.

Solve each equation or inequality. ***See Examples 3 and 4.***

43. $|4x + 3| - 2 = -1$ **44.** $|8 - 3x| - 3 = -2$ **45.** $|6 - 2x| + 1 = 3$

46. $|4 - 4x| + 2 = 4$ **47.** $|3x + 1| - 1 < 2$ **48.** $|5x + 2| - 2 < 3$

49. $\left|5x + \dfrac{1}{2}\right| - 2 < 5$ **50.** $\left|2x + \dfrac{1}{3}\right| + 1 < 4$ **51.** $|10 - 4x| + 1 \geq 5$

52. $|12 - 6x| + 3 \geq 9$ **53.** $|3x - 7| + 1 < -2$ **54.** $|-5x + 7| - 4 < -6$

Solve each equation or inequality. ***See Example 4.***

55. $|10 - 4x| \geq -4$ **56.** $|12 - 9x| \geq -12$ **57.** $|6 - 3x| < -11$

58. $|18 - 3x| < -13$ **59.** $|8x + 5| = 0$ **60.** $|7 + 2x| = 0$

61. $|4.3x + 9.8| < 0$ **62.** $|1.5x - 14| < 0$ **63.** $|2x + 1| \leq 0$

64. $|3x + 2| \leq 0$ **65.** $|3x + 2| > 0$ **66.** $|4x + 3| > 0$

67. *Concept Check* Write an equation involving absolute value that says the distance between p and q is 2 units.

68. *Concept Check* Write an equation involving absolute value that says the distance between r and s is 6 units.

Write each statement using an absolute value equation or inequality. ***See Example 5.***

69. m is no more than 2 units from 7. **70.** z is no less than 5 units from 4.

71. p is within 0.0001 unit of 9. **72.** k is within 0.0002 unit of 10.

73. r is no less than 1 unit from 29. **74.** q is no more than 8 units from 22.

(Modeling) *Solve each problem.* ***See Example 6.***

75. *Tolerance* Suppose that $y = 5x + 1$ and we want y to be within 0.002 unit of 6. For what values of x will this be true?

76. *Tolerance* Repeat **Exercise 75**, but let $y = 10x + 2$.

77. *Weights of Babies* Dr. Tydings has found that, over the years, 95% of the babies he has delivered weighed x pounds, where

$$|x - 8.2| \leq 1.5.$$

What range of weights corresponds to this inequality?

78. *Temperatures on Mars* The temperatures on the surface of Mars in degrees Celsius approximately satisfy the inequality $|C + 84| \leq 56$. What range of temperatures corresponds to this inequality?

79. *Conversion of Methanol to Gasoline* The industrial process that is used to convert methanol to gasoline is carried out at a temperature range of 680°F to 780°F. Using F as the variable, write an absolute value inequality that corresponds to this range.

80. *Wind Power Extraction Tests* When a model kite was flown in crosswinds in tests to determine its limits of power extraction, it attained speeds of 98 to 148 ft per sec in winds of 16 to 26 ft per sec. Using x as the variable in each case, write absolute value inequalities that correspond to these ranges.

(Modeling) Carbon Dioxide Emissions When humans breathe, carbon dioxide is emitted. In one study, the emission rates of carbon dioxide by college students were measured during both lectures and exams. The average individual rate R_L (in grams per hour) during a lecture class satisfied the inequality

$$|R_L - 26.75| \le 1.42,$$

whereas during an exam the rate R_E satisfied the inequality

$$|R_E - 38.75| \le 2.17.$$

(Source: Wang, T. C., ASHRAE Trans., 81 (Part 1), 32.)

Use this information to solve each problem.

81. Find the range of values for R_L and R_E.

82. The class had 225 students. If T_L and T_E represent the total amounts of carbon dioxide in grams emitted during a 1-hour lecture and a 1-hour exam, respectively, write inequalities that model the ranges for T_L and T_E.

Relating Concepts

For individual or collaborative investigation *(Exercises 83–86)*

To see how to solve an equation that involves the absolute value of a quadratic polynomial, such as $|x^2 - x| = 6$, work Exercises 83–86 in order.

83. For $x^2 - x$ to have an absolute value equal to 6, what are the two possible values that it may be? (*Hint:* One is positive and the other is negative.)

84. Write an equation stating that $x^2 - x$ is equal to the positive value found in **Exercise 83,** and solve it using the zero-factor property.

85. Write an equation stating that $x^2 - x$ is equal to the negative value found in **Exercise 83,** and solve it using the quadratic formula. (*Hint:* The solutions are not real numbers.)

86. Give the complete solution set of $|x^2 - x| = 6$, using the results from **Exercises 84 and 85.**

*Use the method described in **Exercises 83–86,** if applicable, and properties of absolute value to solve each equation or inequality. (Hint: Exercises 93 and 94 can be solved by inspection.)*

87. $|3x^2 + x| = 14$

88. $|2x^2 - 3x| = 5$

89. $|4x^2 - 23x - 6| = 0$

90. $|6x^3 + 23x^2 + 7x| = 0$

91. $|x^2 + 1| - |2x| = 0$

92. $\left| \dfrac{x^2 + 2}{x} \right| - \dfrac{11}{3} = 0$

93. $|x^4 + 2x^2 + 1| < 0$

94. $|x^2 + 10| < 0$

95. $\left| \dfrac{x - 4}{3x + 1} \right| \ge 0$

96. $\left| \dfrac{9 - x}{7 + 8x} \right| \ge 0$

Chapter 1 Test Prep

Key Terms

1.1 equation
solution (root)
solution set
equivalent
   equations
linear equation in
   one variable
first-degree
   equation
identity
conditional
   equation
contradiction
literal equation

simple interest
future value
   (maturity value)
1.2 mathematical
   model
linear model
1.3 imaginary unit
complex number
real part
imaginary part
pure imaginary
   number
nonreal complex
   number

standard form
complex conjugate
1.4 quadratic equation
standard form
second-degree equation
double solution
cubic equation
discriminant
1.5 leg
hypotenuse
1.6 rational equation
proposed solution
equation quadratic
   in form

1.7 inequality
linear inequality in
   one variable
interval
interval notation
open interval
closed interval
break-even point
quadratic inequality
strict inequality
nonstrict inequality
rational inequality
1.8 tolerance

New Symbols

$\varnothing$	empty or null set	
i	imaginary unit	
∞	infinity	
$-\infty$	negative infinity	

$$\left.\begin{array}{l}(a, b) \\ (-\infty, a] \\ [a, b)\end{array}\right\} \text{interval notation}$$

$|a|$ absolute value of a

Quick Review

Concepts	Examples

1.1 Linear Equations

Addition and Multiplication Properties of Equality
Let a, b, and c represent real numbers.

    If $a = b$, then $a + c = b + c$.

    If $a = b$ and $c \neq 0$, then $ac = bc$.

Solve.

$$5(x + 3) = 3x + 7$$
$$5x + 15 = 3x + 7 \qquad \text{Distributive property}$$
$$2x = -8 \qquad \text{Subtract } 3x. \text{ Subtract } 15.$$
$$x = -4 \qquad \text{Divide by 2.}$$

The solution set is $\{-4\}$.

1.2 Applications and Modeling with Linear Equations

Solving an Applied Problem
Step 1   Read the problem.

How many liters of 30% alcohol solution and 80% alcohol solution must be mixed to obtain 50 L of 50% alcohol solution?

Concepts	Examples

Step 2 Assign a variable.

Let x = the number of liters of 30% solution.
$50 - x$ = the number of liters of 80% solution.
Summarize the information of the problem in a table.

Strength	Liters of Solution	Liters of Pure Alcohol
30%	x	$0.30x$
80%	$50 - x$	$0.80(50 - x)$
50%	50	$0.50(50)$

Step 3 Write an equation.

The equation is $0.30x + 0.80(50 - x) = 0.50(50)$.

Step 4 Solve the equation.

Solve the equation to obtain $x = 30$.

Step 5 State the answer.

Therefore, 30 L of the 30% solution and $50 - 30 = 20$ L of the 80% solution must be mixed.

Step 6 Check.

CHECK

$$0.30(30) + 0.80(50 - 30) \stackrel{?}{=} 0.50(50)$$
$$25 = 25 \ \checkmark \ \text{True}$$

1.3 Complex Numbers

Definition of i

$$i = \sqrt{-1}, \quad \text{and therefore,} \quad i^2 = -1$$

Definition of Complex Number (a and b real)

$$\underset{\substack{\uparrow \\ \text{Real} \\ \text{part}}}{a} + \underset{\substack{\uparrow \\ \text{Imaginary} \\ \text{part}}}{b i}$$

In the complex number $-6 + 2i$, the real part is -6 and the imaginary part is 2.

Definition of $\sqrt{-a}$

For $a > 0$, $\qquad \sqrt{-a} = i\sqrt{a}.$

Simplify.

$$\sqrt{-4} = 2i$$
$$\sqrt{-12} = i\sqrt{12} = i\sqrt{4 \cdot 3} = 2i\sqrt{3}$$

Adding and Subtracting Complex Numbers
Add or subtract the real parts, and add or subtract the imaginary parts.

$$(2 + 3i) + (3 + i) - (2 - i)$$
$$= (2 + 3 - 2) + (3 + 1 + 1)i$$
$$= 3 + 5i$$

Multiplying and Dividing Complex Numbers
Multiply complex numbers as with binomials, and use the fact that $i^2 = -1$.

$$(6 + i)(3 - 2i)$$
$$= 18 - 12i + 3i - 2i^2 \qquad \text{FOIL method}$$
$$= (18 + 2) + (-12 + 3)i \quad i^2 = -1$$
$$= 20 - 9i$$

Concepts	Examples
Divide complex numbers by multiplying the numerator and denominator by the complex conjugate of the denominator.	$\dfrac{3+i}{1+i}$

$$= \dfrac{(3+i)(1-i)}{(1+i)(1-i)} \qquad \text{Multiply by } \tfrac{1-i}{1-i}.$$

$$= \dfrac{3-3i+i-i^2}{1-i^2} \qquad \text{Multiply.}$$

$$= \dfrac{4-2i}{2} \qquad \text{Combine like terms; } i^2 = -1$$

$$= \dfrac{2(2-i)}{2} \qquad \text{Factor in the numerator.}$$

$$= 2 - i \qquad \text{Divide out the common factor.}$$

1.4 Quadratic Equations

Zero-Factor Property
If a and b are complex numbers with $ab = 0$, then $a = 0$ or $b = 0$ or both equal zero.

Solve.

$$6x^2 + x - 1 = 0$$

$$(3x - 1)(2x + 1) = 0 \qquad \text{Factor.}$$

$$3x - 1 = 0 \quad \text{or} \quad 2x + 1 = 0 \qquad \begin{array}{l}\text{Zero-factor}\\ \text{property}\end{array}$$

$$x = \dfrac{1}{3} \quad \text{or} \qquad x = -\dfrac{1}{2}$$

The solution set is $\left\{-\tfrac{1}{2}, \tfrac{1}{3}\right\}$.

Square Root Property
The solution set of $x^2 = k$ is

$$\left\{\sqrt{k}, -\sqrt{k}\right\}, \quad \text{abbreviated} \quad \left\{\pm\sqrt{k}\right\}.$$

$$x^2 = 12$$

$$x = \pm\sqrt{12} = \pm 2\sqrt{3}$$

The solution set is $\left\{\pm 2\sqrt{3}\right\}$.

Quadratic Formula
The solutions of the quadratic equation $ax^2 + bx + c = 0$, where $a \neq 0$, are given by the quadratic formula.

$$x = \dfrac{-b \pm \sqrt{b^2 - 4ac}}{2a}$$

$$x^2 + 2x + 3 = 0$$

$$x = \dfrac{-2 \pm \sqrt{2^2 - 4(1)(3)}}{2(1)} \qquad a = 1, b = 2, c = 3$$

$$x = \dfrac{-2 \pm \sqrt{-8}}{2} \qquad \text{Simplify.}$$

$$x = \dfrac{-2 \pm 2i\sqrt{2}}{2} \qquad \text{Simplify the radical.}$$

$$x = \dfrac{2\left(-1 \pm i\sqrt{2}\right)}{2} \qquad \begin{array}{l}\text{Factor out 2 in the}\\ \text{numerator.}\end{array}$$

$$x = -1 \pm i\sqrt{2} \qquad \begin{array}{l}\text{Divide out the}\\ \text{common factor.}\end{array}$$

The solution set is $\left\{-1 \pm i\sqrt{2}\right\}$.

Concepts	Examples

1.5 Applications and Modeling with Quadratic Equations

Pythagorean Theorem

In a right triangle, the sum of the squares of the lengths of legs a and b is equal to the square of the length of hypotenuse c.

$$a^2 + b^2 = c^2$$

In a right triangle, the shorter leg is 7 in. less than the longer leg, and the hypotenuse is 2 in. greater than the longer leg. What are the lengths of the sides?

Let x = the length of the longer leg.

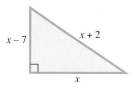

$$(x - 7)^2 + x^2 = (x + 2)^2 \qquad \text{Substitute into the Pythagorean theorem.}$$

$$x^2 - 14x + 49 + x^2 = x^2 + 4x + 4 \qquad \text{Square the binomials.}$$

$$x^2 - 18x + 45 = 0 \qquad \text{Standard form}$$

$$(x - 15)(x - 3) = 0 \qquad \text{Factor.}$$

$$x - 15 = 0 \quad \text{or} \quad x - 3 = 0 \qquad \text{Zero-factor property}$$

$$x = 15 \quad \text{or} \qquad x = 3 \qquad \text{Solve each equation.}$$

The value 3 must be rejected because the height would be negative. The lengths of the sides are 15 in., 8 in., and 17 in. Check to see that the conditions of the problem are satisfied.

Height of a Projected Object

The height s (in feet) of an object projected directly upward from an initial height of s_0 feet, with initial velocity v_0 feet per second, is

$$s = -16t^2 + v_0 t + s_0,$$

where t is the number of seconds after the object is projected.

The height of an object projected upward from ground level with an initial velocity of 64 ft per sec is given by

$$s = -16t^2 + 64t.$$

Find the time(s) that the projectile will reach a height of 56 ft.

$$56 = -16t^2 + 64t \qquad \text{Let } s = 56.$$

$$0 = -16t^2 + 64t - 56 \qquad \text{Subtract 56.}$$

$$0 = 2t^2 - 8t + 7 \qquad \text{Divide by } -8.$$

$$t = \frac{-(-8) \pm \sqrt{(-8)^2 - 4(2)(7)}}{2(2)} \qquad \text{Quadratic formula}$$

$$t = \frac{8 \pm \sqrt{8}}{4} \qquad \text{Simplify.}$$

$$t \approx 1.29 \quad \text{or} \quad t \approx 2.71 \qquad \text{Use a calculator.}$$

The object reaches a height of 56 ft twice—once on its way up (after 1.29 sec) and once on its way down (after 2.71 sec).

Concepts	Examples

1.6 Other Types of Equations and Applications

Power Property

If P and Q are algebraic expressions, then every solution of the equation $P = Q$ is also a solution of the equation $P^n = Q^n$, for any positive integer n.

Quadratic in Form

An equation in the form $au^2 + bu + c = 0$, where $a \neq 0$ and u is an algebraic expression, can be solved by using a substitution variable.

If the power property is applied, or if both sides of an equation are multiplied by a variable expression, check all proposed solutions.

Solve.

$$(x + 1)^{2/3} + (x + 1)^{1/3} - 6 = 0$$
$$u^2 + u - 6 = 0$$
$$\text{Let } u = (x + 1)^{1/3}.$$
$$(u + 3)(u - 2) = 0$$

$$u + 3 = 0 \quad \text{or} \quad u - 2 = 0$$
$$u = -3 \quad \text{or} \quad u = 2$$
$$(x + 1)^{1/3} = -3 \quad \text{or} \quad (x + 1)^{1/3} = 2$$
$$x + 1 = -27 \quad \text{or} \quad x + 1 = 8 \quad \text{Cube.}$$
$$x = -28 \quad \text{or} \quad x = 7 \quad \text{Subtract 1.}$$

Both solutions check. The solution set is $\{-28, 7\}$.

1.7 Inequalities

Properties of Inequality

Let a, b, and c represent real numbers.

1. If $a < b$, then $a + c < b + c$.

2. If $a < b$ and if $c > 0$, then $ac < bc$.

3. If $a < b$ and if $c < 0$, then $ac > bc$.

Solving a Quadratic Inequality

Step 1 Solve the corresponding quadratic equation.

Step 2 Identify the intervals determined by the solutions of the equation.

Step 3 Use a test value from each interval to determine which intervals form the solution set.

Solving a Rational Inequality

Step 1 Rewrite the inequality so that 0 is on one side and a single fraction is on the other.

Step 2 Find the values that make either the numerator or the denominator 0.

Step 3 Use a test value from each interval to determine which intervals form the solution set.

Solve.

$$-3(x + 4) + 2x < 6$$
$$-3x - 12 + 2x < 6$$
$$-x < 18$$
$$x > -18 \quad \begin{array}{l} \text{Multiply by } -1. \\ \text{Change } < \text{ to } >. \end{array}$$

The solution set is $(-18, \infty)$.

$$x^2 + 6x \leq 7$$
$$x^2 + 6x - 7 = 0 \qquad \text{Corresponding equation}$$
$$(x + 7)(x - 1) = 0 \qquad \text{Factor.}$$
$$x + 7 = 0 \quad \text{or} \quad x - 1 = 0 \quad \text{Zero-factor property}$$
$$x = -7 \quad \text{or} \quad x = 1 \quad \text{Solve each equation.}$$

The intervals formed are $(-\infty, -7)$, $(-7, 1)$, and $(1, \infty)$. Test values show that values in the intervals $(-\infty, -7)$ and $(1, \infty)$ do not satisfy the original inequality, while those in $(-7, 1)$ do. Because the symbol $\leq$ includes equality, the endpoints are included.

The solution set is $[-7, 1]$.

$$\frac{x}{x + 3} \geq \frac{5}{x + 3}$$
$$\frac{x}{x + 3} - \frac{5}{x + 3} \geq 0$$
$$\frac{x - 5}{x + 3} \geq 0$$

The values -3 and 5 make either the numerator or the denominator 0. The intervals formed are

$$(-\infty, -3), (-3, 5), \text{ and } (5, \infty).$$

The value -3 must be excluded and 5 must be included. Test values show that values in the intervals $(-\infty, -3)$ and $(5, \infty)$ yield true statements.

The solution set is $(-\infty, -3) \cup [5, \infty)$.

Concepts	Examples

1.8 Absolute Value Equations and Inequalities

Solving Absolute Value Equations and Inequalities
For each equation or inequality in Cases 1–3, assume that $k > 0$.

Case 1: To solve $|x| = k$, use the equivalent form

$$x = k \quad \text{or} \quad x = -k.$$

Case 2: To solve $|x| < k$, use the equivalent form

$$-k < x < k.$$

Case 3: To solve $|x| > k$, use the equivalent form

$$x < -k \quad \text{or} \quad x > k.$$

Solve.

$$|5x - 2| = 3$$
$$5x - 2 = 3 \quad \text{or} \quad 5x - 2 = -3$$
$$5x = 5 \quad \text{or} \quad 5x = -1$$
$$x = 1 \quad \text{or} \quad x = -\frac{1}{5}$$

The solution set is $\left\{ -\frac{1}{5}, 1 \right\}$.

$$|5x - 2| < 3$$
$$-3 < 5x - 2 < 3$$
$$-1 < 5x < 5$$
$$-\frac{1}{5} < x < 1$$

The solution set is $\left(-\frac{1}{5}, 1 \right)$.

$$|5x - 2| \geq 3$$
$$5x - 2 \leq -3 \quad \text{or} \quad 5x - 2 \geq 3$$
$$5x \leq -1 \quad \text{or} \quad 5x \geq 5$$
$$x \leq -\frac{1}{5} \quad \text{or} \quad x \geq 1$$

The solution set is $\left(-\infty, -\frac{1}{5} \right] \cup [1, \infty)$.

Chapter 1 Review Exercises

Solve each equation.

1. $2x + 8 = 3x + 2$

2. $\frac{1}{6}x - \frac{1}{12}(x - 1) = \frac{1}{2}$

3. $5x - 2(x + 4) = 3(2x + 1)$

4. $9x - 11(k + p) = x(a - 1)$, for x

5. $A = \dfrac{24f}{B(p + 1)}$, for f (approximate annual interest rate)

Solve each problem.

6. *Concept Check* Which of the following cannot be a correct equation to solve a geometry problem, if x represents the measure of a side of a rectangle? (*Hint:* Solve the equations and consider the solutions.)

A. $2x + 2(x + 2) = 20$

B. $2x + 2(5 + x) = -2$

C. $8(x + 2) + 4x = 16$

D. $2x + 2(x - 3) = 10$

7. *Concept Check* If x represents the number of pennies in a jar in an applied problem, which of the following equations cannot be a correct equation for finding x? (*Hint:* Solve the equations and consider the solutions.)

A. $5x + 3 = 11$ **B.** $12x + 6 = -4$

C. $100x = 50(x + 3)$ **D.** $6(x + 4) = x + 24$

8. *Airline Carry-On Baggage Size* Carry-on rules for domestic economy-class travel differ from one airline to another, as shown in the table.

Airline	Size (linear inches)
Alaska	51
American	45
Delta	45
Southwest	50
United	45
USAirways	45

Source: Individual airline websites.

To determine the number of linear inches for a carry-on, add the length, width, and height of the bag.

(a) One Samsonite rolling bag measures 9 in. by 12 in. by 21 in. Are there any airlines that would not allow it as a carry-on?

(b) A Lark wheeled bag measures 10 in. by 14 in. by 22 in. On which airlines does it qualify as a carry-on?

Solve each problem.

9. *Dimensions of a Square* If the length of each side of a square is decreased by 4 in., the perimeter of the new square is 10 in. more than half the perimeter of the original square. What are the dimensions of the original square?

10. *Distance from a Library* Becky can ride her bike to the university library in 20 min. The trip home, which is all uphill, takes her 30 min. If her rate is 8 mph faster on her trip there than her trip home, how far does she live from the library?

11. *Alcohol Mixture* Alan wishes to strengthen a mixture that is 10% alcohol to one that is 30% alcohol. How much pure alcohol should he add to 12 L of the 10% mixture?

12. *Loan Interest Rates* A realtor borrowed $90,000 to develop some property. He was able to borrow part of the money at 5.5% interest and the rest at 6%. The annual interest on the two loans amounts to $5125. How much was borrowed at each rate?

13. *Speed of a Plane* Mary Lynn left by plane to visit her mother in Louisiana, 420 km away. Fifteen minutes later, her mother left to meet her at the airport. She drove the 20 km to the airport at 40 km per hr, arriving just as the plane taxied in. What was the speed of the plane?

14. *Toxic Waste* Two chemical plants are releasing toxic waste into a holding tank. Plant I releases waste twice as fast as Plant II. Together they fill the tank in 3 hr. How long would it take the slower plant to fill the tank working alone?

15. *(Modeling) Lead Intake* As directed by the "Safe Drinking Water Act" of December 1974, the EPA proposed a maximum lead level in public drinking water of 0.05 mg per liter. This standard assumed an individual consumption of two liters of water per day.

(a) If EPA guidelines are followed, write an equation that models the maximum amount of lead A ingested in x years. Assume that there are 365.25 days in a year.

(b) If the average life expectancy is 72 yr, find the EPA maximum lead intake from water over a lifetime.

16. *(Modeling) Online Retail Sales* Projected e-commerce sales (in billions of dollars) for the years 2010–2018 can be modeled by the equation

$$y = 40.892x + 150.53,$$

where $x = 0$ corresponds to 2010, $x = 1$ corresponds to 2011, and so on. Based on this model, what would expected retail e-commerce sales be in 2018? (*Source:* Statistics Portal.)

17. *(Modeling) Minimum Wage* U.S. minimum hourly wage, in dollars, for selected years from 1956–2009 is shown in the table. The linear model

$$y = 0.1132x + 0.4609$$

approximates the minimum wage during this time period, where x is the number of years after 1956 and y is the minimum wage in dollars.

Year	Minimum Wage	Year	Minimum Wage
1956	1.00	1996	4.75
1963	1.25	1997	5.15
1975	2.10	2007	5.85
1981	3.35	2008	6.55
1990	3.80	2009	7.25

Source: Bureau of Labor Statistics.

(a) Use the model to approximate the minimum wage in 1990. How does it compare to the data in the table?

(b) Use the model to approximate the year in which the minimum wage was $5.85. How does the answer compare to the data in the table?

18. *(Modeling) New York State Population* The U.S. population, in millions, for selected years is given in the table. The bar graph shows the percentages of the U.S. population that lived in New York State during those years.

Year	U.S. Population (in millions)
1980	226.5
1990	248.7
2000	281.4
2010	308.7
2014	318.9

Source: U.S. Census Bureau.

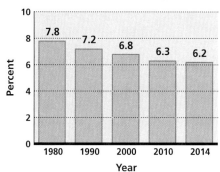

New York State Population as Percent of U.S. Population

Source: U.S. Census Bureau.

(a) Find the number of Americans, to the nearest tenth of a million, living in New York State for each year given in the table.

(b) The percentages given in the bar graph decrease each year, while the populations given in the table increase each year. From the answers to part (a), is the number of Americans living in New York State increasing or decreasing?

Perform each operation. Write answers in standard form.

19. $(6 - i) + (7 - 2i)$

20. $(-11 + 2i) - (8 - 7i)$

21. $15i - (3 + 2i) - 11$

22. $-6 + 4i - (8i - 2)$

23. $(5 - i)(3 + 4i)$

24. $(-8 + 2i)(-1 + i)$

25. $(5 - 11i)(5 + 11i)$

26. $(4 - 3i)^2$

27. $-5i(3 - i)^2$

28. $4i(2 + 5i)(2 - i)$

29. $\dfrac{-12 - i}{-2 - 5i}$

30. $\dfrac{-7 + i}{-1 - i}$

Simplify each power of i.

31. i^{11}

32. i^{60}

33. i^{1001}

34. i^{110}

35. i^{-27}

36. $\dfrac{1}{i^{17}}$

Solve each equation.

37. $(x + 7)^2 = 5$

38. $(2 - 3x)^2 = 8$

39. $2x^2 + x - 15 = 0$

40. $12x^2 = 8x - 1$

41. $-2x^2 + 11x = -21$

42. $-x(3x + 2) = 5$

43. $(2x + 1)(x - 4) = x$

44. $\sqrt{2}x^2 - 4x + \sqrt{2} = 0$

45. $x^2 - \sqrt{5}\,x - 1 = 0$

46. $(x + 4)(x + 2) = 2x$

47. *Concept Check* Which equation has two real, distinct solutions? Do not actually solve.

    A. $(3x - 4)^2 = -9$

    B. $(4 - 7x)^2 = 0$

    C. $(5x - 9)(5x - 9) = 0$

    D. $(7x + 4)^2 = 11$

48. *Concept Check* See **Exercise 47.**

    (a) Which equations have only one distinct real solution?

    (b) Which equation has two nonreal complex solutions?

Evaluate the discriminant for each equation. Then use it to determine the number and type of solutions.

49. $-6x^2 + 2x = -3$

50. $8x^2 = -2x - 6$

51. $-8x^2 + 10x = 7$

52. $16x^2 + 3 = -26x$

53. $x(9x + 6) = -1$

54. $25x^2 + 110x + 121 = 0$

Solve each problem.

55. *(Modeling) Height of a Projectile* A projectile is fired straight up from ground level. After t seconds its height s, in feet above the ground, is given by

$$s = -16t^2 + 220t.$$

At what times is the projectile exactly 750 ft above the ground?

56. *Dimensions of a Picture Frame* Zach went into a frame-it-yourself shop. He wanted a frame 3 in. longer than it was wide. The frame he chose extended 1.5 in. beyond the picture on each side. Find the outside dimensions of the frame if the area of the unframed picture is 70 in.².

57. *Kitchen Flooring* Paula plans to replace the vinyl floor covering in her 10-ft by 12-ft kitchen. She wants to have a border of even width of a special material. She can afford only 21 ft² of this material. How wide a border can she have?

58. *(Modeling) Airplane Landing Speed* To determine the appropriate landing speed of a small airplane, the formula

$$D = 0.1s^2 - 3s + 22$$

is used, where s is the initial landing speed in feet per second and D is the length of the runway in feet. If the landing speed is too fast, the pilot may run out of runway. If the speed is too slow, the plane may stall. If the runway is 800 ft long, what is the appropriate landing speed? Round to the nearest tenth.

59. *(Modeling) U.S. Government Spending on Medical Care* The amount spent in billions of dollars by the U.S. government on medical care during the period 1990–2013 can be approximated by the equation

$$y = 1.016x^2 + 12.49x + 197.8$$

where $x = 0$ corresponds to 1990, $x = 1$ corresponds to 1991, and so on. According to this model, about how much was spent by the U.S. government on medical care in 2009? Round to the nearest tenth of a billion. (*Source: U.S. Office of Management and Budget.*)

60. *Dimensions of a Right Triangle* The lengths of the sides of a right triangle are such that the shortest side is 7 in. shorter than the middle side, while the longest side (the hypotenuse) is 1 in. longer than the middle side. Find the lengths of the sides.

Solve each equation.

61. $4x^4 + 3x^2 - 1 = 0$

62. $x^2 - 2x^4 = 0$

63. $\dfrac{2}{x} - \dfrac{4}{3x} = 8 + \dfrac{3}{x}$

64. $2 - \dfrac{5}{x} = \dfrac{3}{x^2}$

65. $\dfrac{10}{4x - 4} = \dfrac{1}{1 - x}$

66. $\dfrac{13}{x^2 + 10} = \dfrac{2}{x}$

67. $(x - 4)^{2/5} = 9$

68. $(x^2 - 6x)^{1/4} = 2$

69. $(x - 2)^{2/3} = x^{1/3}$

70. $\sqrt{2x + 3} = x + 2$

71. $\sqrt{x + 2} - x = 2$

72. $\sqrt{x} - \sqrt{x + 3} = -1$

73. $\sqrt{4x - 2} = \sqrt{3x + 1}$

74. $\sqrt{5x - 15} - \sqrt{x + 1} = 2$

75. $\sqrt{x + 3} - \sqrt{3x + 10} = 1$

76. $\sqrt[5]{2x} = \sqrt[5]{3x + 2}$

77. $\sqrt[3]{6x + 2} - \sqrt[3]{4x} = 0$

78. $\sqrt{x^2 + 3x} - 2 = 0$

79. $\dfrac{x}{x + 2} + \dfrac{1}{x} + 3 = \dfrac{2}{x^2 + 2x}$

80. $\dfrac{2}{x + 2} + \dfrac{1}{x + 4} = \dfrac{4}{x^2 + 6x + 8}$

81. $(2x + 3)^{2/3} + (2x + 3)^{1/3} - 6 = 0$

82. $(x + 3)^{-2/3} - 2(x + 3)^{-1/3} = 3$

Solve each inequality. Give the solution set using interval notation.

83. $-9x + 3 < 4x + 10$

84. $11x \geq 2(x - 4)$

85. $-5x - 4 \geq 3(2x - 5)$

86. $7x - 2(x - 3) \leq 5(2 - x)$

87. $5 \leq 2x - 3 \leq 7$

88. $-8 > 3x - 5 > -12$

89. $x^2 + 3x - 4 \leq 0$

90. $x^2 + 4x - 21 > 0$

91. $6x^2 - 11x < 10$

92. $x^2 - 3x \geq 5$

93. $x^3 - 16x \leq 0$

94. $2x^3 - 3x^2 - 5x < 0$

95. $\dfrac{3x + 6}{x - 5} > 0$

96. $\dfrac{x + 7}{2x + 1} \leq 1$

97. $\dfrac{3x - 2}{x} - 4 > 0$

98. $\dfrac{5x + 2}{x} + 1 < 0$

99. $\dfrac{3}{x - 1} \leq \dfrac{5}{x + 3}$

100. $\dfrac{3}{x + 2} > \dfrac{2}{x - 4}$

(Modeling) Solve each problem.

101. *Ozone Concentration* Guideline levels for indoor ozone are less than 50 parts per billion (ppb). In a scientific study, a Purafil air filter was used to reduce an initial ozone concentration of 140 ppb. The filter removed 43% of the ozone. (*Source:* Parmar and Grosjean, *Removal of Air Pollutants from Museum Display Cases*, Getty Conservation Institute, Marina del Rey, CA.)

 (a) What is the ozone concentration after the Purafil air filter is used?

 (b) What is the maximum initial concentration of ozone that this filter will reduce to an acceptable level? Round the answer to the nearest tenth part per billion.

102. *Break-Even Interval* A company produces earbuds. The revenue from the sale of x units of these earbuds is

$$R = 8x.$$

The cost to produce x units of earbuds is

$$C = 3x + 1500.$$

In what interval will the company at least break even?

103. *Height of a Projectile* A projectile is launched upward from the ground. Its height s in feet above the ground after t seconds is given by the following equation.

$$s = -16t^2 + 320t$$

 (a) After how many seconds in the air will it hit the ground?

 (b) During what time interval is the projectile more than 576 ft above the ground?

104. *Social Security* The total amount paid by the U.S. government to individuals for Social Security retirement and disability insurance benefits during the period 2004–2013 can be approximated by the linear model

$$y = 35.7x + 486,$$

where $x = 0$ corresponds to 2004, $x = 1$ corresponds to 2005, and so on. The variable y is in billions of dollars. Based on this model, during what year did the amount paid by the government first exceed \$800 billion? Round the answer to the nearest year. Compare the answer to the bar graph.

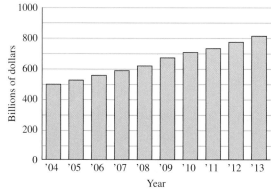

Social Security Retirement and Disability Insurance Benefits

Source: U.S. Office of Management and Budget.

105. Without actually solving the inequality, explain why 3 cannot be in the solution set of $\frac{14x + 9}{x - 3} < 0$.

106. Without actually solving the inequality, explain why -4 must be in the solution set of $\frac{x + 4}{2x + 1} \geq 0$.

Solve each equation or inequality.

107. $|x + 4| = 7$

108. $|2 - x| = 3$

109. $\left| \dfrac{7}{2 - 3x} \right| - 9 = 0$

110. $\left| \dfrac{8x - 1}{3x + 2} \right| - 7 = 0$

111. $|5x - 1| = |2x + 3|$

112. $|x + 10| = |x - 11|$

113. $|2x + 9| \le 3$

114. $|8 - 5x| \ge 2$

115. $|7x - 3| > 4$

116. $\left| \dfrac{1}{2}x + \dfrac{2}{3} \right| < 3$

117. $|3x + 7| - 5 < 5$

118. $|7x + 8| - 6 > -3$

119. $|4x - 12| \ge -3$

120. $|7 - 2x| \le -9$

121. $|x^2 + 4x| \le 0$

122. $|x^2 + 4x| > 0$

Write each statement using an absolute value equation or inequality.

123. k is 12 units from 6.

124. p is at least 3 units from 1.

125. t is no less than 0.01 unit from 5.

Chapter 1 | Test

Solve each equation.

1. $3(x - 4) - 5(x + 2) = 2 - (x + 24)$

2. $\dfrac{2}{3}x + \dfrac{1}{2}(x - 4) = x - 4$

3. $6x^2 - 11x - 7 = 0$

4. $(3x + 1)^2 = 8$

5. $3x^2 + 2x = -2$

6. $\dfrac{12}{x^2 - 9} = \dfrac{2}{x - 3} - \dfrac{3}{x + 3}$

7. $\dfrac{4x}{x - 2} + \dfrac{3}{x} = \dfrac{-6}{x^2 - 2x}$

8. $\sqrt{3x + 4} + 5 = 2x + 1$

9. $\sqrt{-2x + 3} + \sqrt{x + 3} = 3$

10. $\sqrt[3]{3x - 8} = \sqrt[3]{9x + 4}$

11. $x^4 - 17x^2 + 16 = 0$

12. $(x + 3)^{2/3} + (x + 3)^{1/3} - 6 = 0$

13. $|4x + 3| = 7$

14. $|2x + 1| = |5 - x|$

15. *Surface Area of a Rectangular Solid* The formula for the surface area of a rectangular solid is

$$S = 2HW + 2LW + 2LH,$$

where S, H, W, and L represent surface area, height, width, and length, respectively. Solve this formula for W.

16. Perform each operation. Write answers in standard form.

(a) $(9 - 3i) - (4 + 5i)$

(b) $(4 + 3i)(-5 + 3i)$

(c) $(8 + 3i)^2$

(d) $\dfrac{3 + 19i}{1 + 3i}$

17. Simplify each power of i.

(a) i^{42}

(b) i^{-31}

(c) $\dfrac{1}{i^{19}}$

Solve each problem.

18. *(Modeling) Water Consumption for Snowmaking* Ski resorts require large amounts of water in order to make snow. Snowmass Ski Area in Colorado plans to pump between 1120 and 1900 gal of water per minute at least 12 hr per day from Snowmass Creek between mid-October and late December. (*Source:* York Snow Incorporated.)

(a) Determine an equation that will calculate the *minimum* amount of water A (in gallons) pumped after x days during mid-October to late December.

(b) Find the minimum amount of water pumped in 30 days.

(c) Suppose the water being pumped from Snowmass Creek was used to fill swimming pools. The average backyard swimming pool holds 20,000 gal of water. Determine an equation that will give the minimum number of pools P that could be filled after x days. How many pools could be filled each day (to the nearest whole number)?

(d) To the nearest day, in how many days could a minimum of 1000 pools be filled?

19. *Dimensions of a Rectangle* The perimeter of a rectangle is 620 m. The length is 20 m less than twice the width. What are the length and width?

20. *Nut Mixture* To make a special mix, the owner of a fruit and nut stand wants to combine cashews that sell for $7.00 per lb with walnuts that sell for $5.50 per lb to obtain 35 lb of a mixture that sells for $6.50 per lb. How many pounds of each type of nut should be used in the mixture?

21. *Speed of an Excursion Boat* An excursion boat travels upriver to a landing and then returns to its starting point. The trip upriver takes 1.2 hr, and the trip back takes 0.9 hr. If the average speed on the return trip is 5 mph faster than on the trip upriver, what is the boat's speed upriver?

22. *(Modeling) Cigarette Use* The percentage of college freshmen who smoke declined substantially from the year 2004 to the year 2014 and can be modeled by the linear equation

$$y = -0.461x + 6.32,$$

where x represents the number of years since 2004. Thus, $x = 0$ represents 2004, $x = 1$ represents 2005, and so on, (*Source:* Higher Education Research Institute, UCLA.)

(a) Use the model to determine the percentage of college freshmen who smoked in the year 2014. Round the answer to the nearest tenth of a percent.

(b) According to the model, in what year did 4.9% of college freshmen smoke?

23. *(Modeling) Height of a Projectile* A projectile is launched straight up from ground level with an initial velocity of 96 ft per sec. Its height in feet, s, after t seconds is given by the equation

$$s = -16t^2 + 96t.$$

(a) At what time(s) will it reach a height of 80 ft?

(b) After how many seconds will it return to the ground?

Solve each inequality. Give the answer using interval notation.

24. $-2(x - 1) - 12 < 2(x + 1)$ **25.** $-3 \le \frac{1}{2}x + 2 \le 3$

26. $2x^2 - x \ge 3$ **27.** $\frac{x + 1}{x - 3} < 5$

28. $|2x - 5| < 9$ **29.** $|2x + 1| - 11 \ge 0$ **30.** $|3x + 7| \le 0$

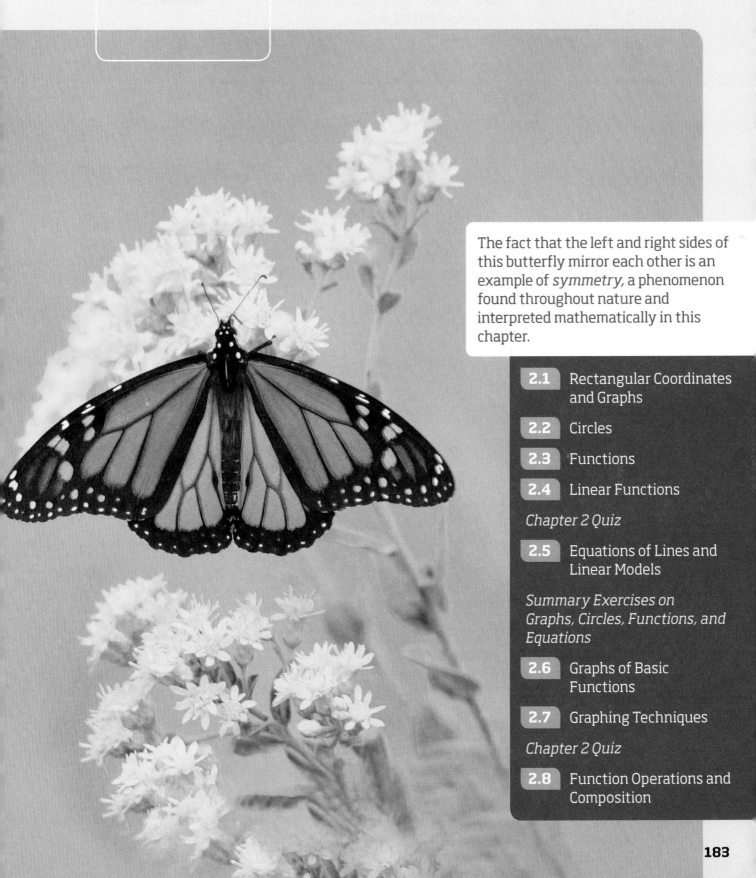

2 Graphs and Functions

The fact that the left and right sides of this butterfly mirror each other is an example of *symmetry*, a phenomenon found throughout nature and interpreted mathematically in this chapter.

2.1 Rectangular Coordinates and Graphs

- Ordered Pairs
- The Rectangular Coordinate System
- The Distance Formula
- The Midpoint Formula
- Equations in Two Variables

Ordered Pairs   The idea of pairing one quantity with another is often encountered in everyday life.

- A numerical score in a mathematics course is paired with a corresponding letter grade.

- The number of gallons of gasoline pumped into a tank is paired with the amount of money needed to purchase it.

- Expense categories are paired with dollars spent by the average American household in 2013. (See the table in the margin.)

Pairs of related quantities, such as a 96 determining a grade of A, 3 gallons of gasoline costing $10.50, and 2013 spending on food of $8506, can be expressed as *ordered pairs:* (96, A), (3, $10.50), (food, $8506). An **ordered pair** consists of two components, written inside parentheses.

Category	Amount Spent
food	$ 8506
housing	$21,374
transportation	$12,153
health care	$ 4917
apparel and services	$ 2076
entertainment	$ 3240

Source: U.S. Bureau of Labor Statistics.

EXAMPLE 1   Writing Ordered Pairs

Use the table to write ordered pairs to express the relationship between each category and the amount spent on it.

(a) housing   **(b)** entertainment

SOLUTION

(a) Use the data in the second row: (housing, $21,374).

(b) Use the data in the last row: (entertainment, $3240).

✔ **Now Try Exercise 13.**

In mathematics, we are most often interested in ordered pairs whose components are numbers. The ordered pairs (a, b) and (c, d) are equal provided that $a = c$ *and* $b = d$.

NOTE   Notation such as (2, 4) is used to show an interval on a number line, and the same notation is used to indicate an ordered pair of numbers. The intended use is usually clear from the context of the discussion.

The Rectangular Coordinate System   Each real number corresponds to a point on a number line. This idea is extended to ordered pairs of real numbers by using two perpendicular number lines, one horizontal and one vertical, that intersect at their zero-points. The point of intersection is the **origin.** The horizontal line is the **x-axis,** and the vertical line is the **y-axis.**

The x-axis and y-axis together make up a **rectangular coordinate system,** or **Cartesian coordinate system** (named for one of its coinventors, René Descartes. The other coinventor was Pierre de Fermat). The plane into which the coordinate system is introduced is the **coordinate plane,** or **xy-plane.** See **Figure 1.** The x-axis and y-axis divide the plane into four regions, or **quadrants,** labeled as shown. The points on the x-axis or the y-axis belong to no quadrant.

Each point P in the xy-plane corresponds to a unique ordered pair (a, b) of real numbers. The point P corresponding to the ordered pair (a, b) often is written $P(a, b)$ as in **Figure 1** and referred to as "the point (a, b)." The numbers a and b are the **coordinates** of point P.

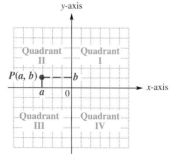

Rectangular (Cartesian) Coordinate System

Figure 1

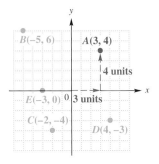

Figure 2

To locate on the *xy*-plane the point corresponding to the ordered pair $(3, 4)$, for example, start at the origin, move 3 units in the positive *x*-direction, and then move 4 units in the positive *y*-direction. See **Figure 2**. Point *A* corresponds to the ordered pair $(3, 4)$.

The Distance Formula Recall that the distance on a number line between points *P* and *Q* with coordinates x_1 and x_2 is

$$d(P, Q) = |x_1 - x_2| = |x_2 - x_1|. \quad \text{Definition of distance}$$

By using the coordinates of their ordered pairs, we can extend this idea to find the distance between any two points in a plane.

Figure 3 shows the points $P(-4, 3)$ and $R(8, -2)$. If we complete a right triangle that has its 90° angle at $Q(8, 3)$ as in the figure, the legs have lengths

$$d(P, Q) = |8 - (-4)| = 12$$

and $$d(Q, R) = |3 - (-2)| = 5.$$

By the Pythagorean theorem, the hypotenuse has length

$$\sqrt{12^2 + 5^2} = \sqrt{144 + 25} = \sqrt{169} = 13.$$

Thus, the distance between $(-4, 3)$ and $(8, -2)$ is 13.

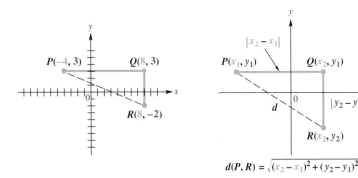

Figure 3   **Figure 4**

To obtain a general formula, let $P(x_1, y_1)$ and $R(x_2, y_2)$ be any two distinct points in a plane, as shown in **Figure 4**. Complete a triangle by locating point *Q* with coordinates (x_2, y_1). The Pythagorean theorem gives the distance between *P* and *R*.

$$d(P, R) = \sqrt{(x_2 - x_1)^2 + (y_2 - y_1)^2}$$

Absolute value bars are not necessary in this formula because, for all real numbers *a* and *b*,

$$|a - b|^2 = (a - b)^2.$$

The **distance formula** can be summarized as follows.

Distance Formula

Suppose that $P(x_1, y_1)$ and $R(x_2, y_2)$ are two points in a coordinate plane. The distance between *P* and *R*, written $d(P, R)$, is given by the following formula.

$$d(P, R) = \sqrt{(x_2 - x_1)^2 + (y_2 - y_1)^2}$$

René Descartes (1596–1650)

The initial flash of *analytic geometry* may have come to Descartes as he was watching a fly crawling about on the ceiling near a corner of his room. It struck him that the path of the fly on the ceiling could be described if only one knew the relation connecting the fly's distances from two adjacent walls.

Source: An Introduction to the History of Mathematics by Howard Eves.

LOOKING AHEAD TO CALCULUS

In analytic geometry and calculus, the distance formula is extended to two points in space. Points in space can be represented by **ordered triples.** The distance between the two points

$$(x_1, y_1, z_1) \quad \text{and} \quad (x_2, y_2, z_2)$$

is given by the following expression.

$$\sqrt{(x_2 - x_1)^2 + (y_2 - y_1)^2 + (z_2 - z_1)^2}$$

The distance formula can be stated in words.

The distance between two points in a coordinate plane is the square root of the sum of the square of the difference between their x-coordinates and the square of the difference between their y-coordinates.

Although our derivation of the distance formula assumed that P and R are not on a horizontal or vertical line, the result is true for any two points.

EXAMPLE 2 Using the Distance Formula

Find the distance between $P(-8, 4)$ and $Q(3, -2)$.

SOLUTION Use the distance formula.

$$d(P, Q) = \sqrt{(x_2 - x_1)^2 + (y_2 - y_1)^2} \quad \text{Distance formula}$$
$$= \sqrt{[3 - (-8)]^2 + (-2 - 4)^2} \quad x_1 = -8, y_1 = 4, x_2 = 3, y_2 = -2$$
$$= \sqrt{11^2 + (-6)^2} \quad \text{Be careful when subtracting a negative number.}$$
$$= \sqrt{121 + 36}$$
$$= \sqrt{157} \quad \checkmark \text{ Now Try Exercise 15(a).}$$

A statement of the form "If p, then q" is a **conditional statement.** The related statement "If q, then p" is its **converse.** The *converse* of the Pythagorean theorem is also a true statement.

If the sides a, b, and c of a triangle satisfy $a^2 + b^2 = c^2$, then the triangle is a right triangle with legs having lengths a and b and hypotenuse having length c.

EXAMPLE 3 Applying the Distance Formula

Determine whether the points $M(-2, 5)$, $N(12, 3)$, and $Q(10, -11)$ are the vertices of a right triangle.

SOLUTION A triangle with the three given points as vertices, shown in **Figure 5,** is a right triangle if the square of the length of the longest side equals the sum of the squares of the lengths of the other two sides. Use the distance formula to find the length of each side of the triangle.

$$d(M, N) = \sqrt{[12 - (-2)]^2 + (3 - 5)^2} = \sqrt{196 + 4} = \sqrt{200}$$
$$d(M, Q) = \sqrt{[10 - (-2)]^2 + (-11 - 5)^2} = \sqrt{144 + 256} = \sqrt{400} = 20$$
$$d(N, Q) = \sqrt{(10 - 12)^2 + (-11 - 3)^2} = \sqrt{4 + 196} = \sqrt{200}$$

The longest side, of length 20 units, is chosen as the hypotenuse. Because

$$\left(\sqrt{200}\right)^2 + \left(\sqrt{200}\right)^2 = 400 = 20^2$$

Figure 5

is true, the triangle is a right triangle with hypotenuse joining M and Q.

$\checkmark$ **Now Try Exercise 23.**

$d(P, Q) + d(Q, R) = d(P, R)$

Figure 6

Using a similar procedure, we can tell whether three points are **collinear** (that is, lying on a straight line). See **Figure 6**.

> *Three points are collinear if the sum of the distances between two pairs of the points is equal to the distance between the remaining pair of points.*

EXAMPLE 4 Applying the Distance Formula

Determine whether the points $P(-1, 5)$, $Q(2, -4)$, and $R(4, -10)$ are collinear.

SOLUTION Use the distance formula.

$$d(P, Q) = \sqrt{(-1-2)^2 + [5-(-4)]^2} = \sqrt{9+81} = \sqrt{90} = 3\sqrt{10}$$

$$\sqrt{90} = \sqrt{9 \cdot 10} = 3\sqrt{10}$$

$$d(Q, R) = \sqrt{(2-4)^2 + [-4-(-10)]^2} = \sqrt{4+36} = \sqrt{40} = 2\sqrt{10}$$

$$d(P, R) = \sqrt{(-1-4)^2 + [5-(-10)]^2} = \sqrt{25+225} = \sqrt{250} = 5\sqrt{10}$$

Because $3\sqrt{10} + 2\sqrt{10} = 5\sqrt{10}$ is true, the three points are collinear.

✔ **Now Try Exercise 29.**

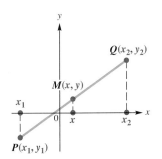

Figure 7

The Midpoint Formula The midpoint of a line segment is equidistant from the endpoints of the segment. The **midpoint formula** is used to find the coordinates of the midpoint of a line segment. To develop the midpoint formula, let $P(x_1, y_1)$ and $Q(x_2, y_2)$ be any two distinct points in a plane. (Although **Figure 7** shows $x_1 < x_2$, no particular order is required.) Let $M(x, y)$ be the midpoint of the segment joining P and Q. Draw vertical lines from each of the three points to the x-axis, as shown in **Figure 7**.

The ordered pair $M(x, y)$ is the midpoint of the line segment joining P and Q, so the distance between x and x_1 equals the distance between x and x_2.

$$x_2 - x = x - x_1$$

$$x_2 + x_1 = 2x \qquad \text{Add } x \text{ and } x_1 \text{ to each side.}$$

$$x = \frac{x_1 + x_2}{2} \qquad \text{Divide by 2 and rewrite.}$$

Similarly, the y-coordinate is $\dfrac{y_1 + y_2}{2}$, yielding the following formula.

Midpoint Formula

The coordinates of the midpoint M of the line segment with endpoints $P(x_1, y_1)$ and $Q(x_2, y_2)$ are given by the following.

$$M = \left(\frac{x_1 + x_2}{2}, \frac{y_1 + y_2}{2} \right)$$

That is, the x-coordinate of the midpoint of a line segment is the **average** *of the x-coordinates of the segment's endpoints, and the y-coordinate is the* **average** *of the y-coordinates of the segment's endpoints.*

EXAMPLE 5 Using the Midpoint Formula

Use the midpoint formula to do each of the following.

(a) Find the coordinates of the midpoint M of the line segment with endpoints $(8, -4)$ and $(-6, 1)$.

(b) Find the coordinates of the other endpoint Q of a line segment with one endpoint $P(-6, 12)$ and midpoint $M(8, -2)$.

SOLUTION

(a) The coordinates of M are found using the midpoint formula.

$$M = \left(\frac{8 + (-6)}{2}, \frac{-4 + 1}{2} \right) = \left(1, -\frac{3}{2} \right) \quad \text{Substitute in } M = \left(\frac{x_1 + x_2}{2}, \frac{y_1 + y_2}{2} \right).$$

The coordinates of midpoint M are $\left(1, -\frac{3}{2} \right)$.

(b) Let (x, y) represent the coordinates of Q. Use both parts of the midpoint formula.

x-value of P	x-value of M		y-value of P	y-value of M
↓	↓		↓	↓

Substitute carefully.

$$\frac{x + (-6)}{2} = 8 \qquad\qquad \frac{y + 12}{2} = -2$$

$$x - 6 = 16 \qquad\qquad y + 12 = -4$$

$$x = 22 \qquad\qquad y = -16$$

The coordinates of endpoint Q are $(22, -16)$.

✔ **Now Try Exercises 15(b) and 35.**

EXAMPLE 6 Applying the Midpoint Formula

Figure 8 depicts how a graph might indicate the increase in the revenue generated by fast-food restaurants in the United States from $69.8 billion in 1990 to $195.1 billion in 2014. Use the midpoint formula and the two given points to estimate the revenue from fast-food restaurants in 2002, and compare it to the actual figure of $138.3 billion.

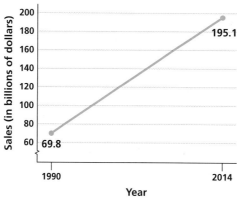

Revenue of U.S. Fast-Food Restaurants

Source: National Restaurant Association.

Figure 8

SOLUTION The year 2002 lies halfway between 1990 and 2014, so we must find the coordinates of the midpoint of the line segment that has endpoints

$$(1990, 69.8) \quad \text{and} \quad (2014, 195.1).$$

(Here, the second component is in billions of dollars.)

$$M = \left(\frac{1990 + 2014}{2}, \frac{69.8 + 195.1}{2} \right) = (2002, 132.5) \quad \text{Use the midpoint formula.}$$

Our estimate is $132.5 billion, which is less than the actual figure of $138.3 billion. Models are used to predict outcomes. They rarely give exact values.

✔ **Now Try Exercise 41.**

Equations in Two Variables Ordered pairs are used to express the solutions of equations in two variables. When an ordered pair represents the solution of an equation with the variables x and y, the x-value is written first. For example, we say that

$$(1, 2) \quad \text{is a solution of} \quad 2x - y = 0.$$

Substituting 1 for x and 2 for y in the equation gives a true statement.

$$2x - y = 0$$
$$2(1) - 2 \overset{?}{=} 0 \qquad \text{Let } x = 1 \text{ and } y = 2.$$
$$0 = 0 \checkmark \quad \text{True}$$

EXAMPLE 7 **Finding Ordered-Pair Solutions of Equations**

For each equation, find at least three ordered pairs that are solutions.

(a) $y = 4x - 1$ **(b)** $x = \sqrt{y - 1}$ **(c)** $y = x^2 - 4$

SOLUTION

(a) Choose any real number for x or y, and substitute in the equation to obtain the corresponding value of the other variable. For example, let $x = -2$ and then let $y = 3$.

$$
\begin{array}{ll}
y = 4x - 1 & \\
y = 4(-2) - 1 \quad \text{Let } x = -2. &
\end{array}
\qquad
\begin{array}{ll}
y = 4x - 1 & \\
3 = 4x - 1 \quad \text{Let } y = 3. &
\end{array}
$$

$$
\begin{array}{ll}
y = -8 - 1 \quad \text{Multiply.} & \qquad 4 = 4x \quad \text{Add 1.} \\
y = -9 \quad \text{Subtract.} & \qquad 1 = x \quad \text{Divide by 4.}
\end{array}
$$

This gives the ordered pairs $(-2, -9)$ and $(1, 3)$. Verify that the ordered pair $(0, -1)$ is also a solution.

(b)

$$x = \sqrt{y - 1} \qquad \text{Given equation}$$
$$1 = \sqrt{y - 1} \qquad \text{Let } x = 1.$$
$$1 = y - 1 \qquad \text{Square each side.}$$
$$2 = y \qquad \text{Add 1.}$$

One ordered pair is $(1, 2)$. Verify that the ordered pairs $(0, 1)$ and $(2, 5)$ are also solutions of the equation.

(c) A table provides an organized method for determining ordered pairs. Here, we let x equal $-2, -1, 0, 1,$ and 2 in $y = x^2 - 4$ and determine the corresponding y-values.

x	y	
-2	0	$(-2)^2 - 4 = 4 - 4 = 0$
-1	-3	$(-1)^2 - 4 = 1 - 4 = -3$
0	-4	$0^2 - 4 = -4$
1	-3	$1^2 - 4 = -3$
2	0	$2^2 - 4 = 0$

Five ordered pairs are $(-2, 0), (-1, -3), (0, -4), (1, -3),$ and $(2, 0)$.

✔ **Now Try Exercises 47(a), 51(a), and 53(a).**

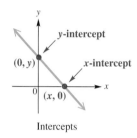

Intercepts

The **graph** of an equation is found by plotting ordered pairs that are solutions of the equation. The **intercepts** of the graph are good points to plot first. An ***x*-intercept** is a point where the graph intersects the x-axis. A ***y*-intercept** is a point where the graph intersects the y-axis. In other words, the x-intercept is represented by an ordered pair with y-coordinate 0, and the y-intercept is an ordered pair with x-coordinate 0.

A general algebraic approach for graphing an equation using intercepts and point-plotting follows.

Graphing an Equation by Point Plotting

Step 1 Find the intercepts.

Step 2 Find as many additional ordered pairs as needed.

Step 3 Plot the ordered pairs from Steps 1 and 2.

Step 4 Join the points from Step 3 with a smooth line or curve.

EXAMPLE 8 **Graphing Equations**

Graph each of the equations here, from **Example 7.**

(a) $y = 4x - 1$ **(b)** $x = \sqrt{y - 1}$ **(c)** $y = x^2 - 4$

SOLUTION

(a) ***Step 1*** Let $y = 0$ to find the x-intercept, and let $x = 0$ to find the y-intercept.

$$y = 4x - 1 \qquad\qquad y = 4x - 1$$
$$0 = 4x - 1 \quad \text{Let } y = 0. \qquad y = 4(0) - 1 \quad \text{Let } x = 0.$$
$$1 = 4x \qquad\qquad y = 0 - 1$$
$$\frac{1}{4} = x \qquad\qquad y = -1$$

The intercepts are $\left(\frac{1}{4}, 0\right)$ and $(0, -1)$.* Note that the y-intercept is one of the ordered pairs we found in **Example 7(a).**

---

*Intercepts are sometimes defined as numbers, such as x-intercept $\frac{1}{4}$ and y-intercept -1. In this text, we define them as ordered pairs, such as $\left(\frac{1}{4}, 0\right)$ and $(0, -1)$.

Step 2 We use the other ordered pairs found in **Example 7(a):**

$$(-2, -9) \quad \text{and} \quad (1, 3).$$

Step 3 Plot the four ordered pairs from Steps 1 and 2 as shown in **Figure 9.**

Step 4 Join the points plotted in Step 3 with a straight line. This line, also shown in **Figure 9,** is the graph of the equation $y = 4x - 1$.

Figure 9   **Figure 10**

(b) For $x = \sqrt{y - 1}$, the y-intercept $(0, 1)$ was found in **Example 7(b).** Solve

$$x = \sqrt{0 - 1} \quad \text{Let } y = 0.$$

to find the x-intercept. When $y = 0$, the quantity under the radical symbol is negative, so there is no x-intercept. In fact, $y - 1$ must be greater than or equal to 0, so y must be greater than or equal to 1.

We start by plotting the ordered pairs from **Example 7(b)** and then join the points with a smooth curve as in **Figure 10.** To confirm the direction the curve will take as x increases, we find another solution, $(3, 10)$. (Point plotting for graphs other than lines is often inefficient. We will examine other graphing methods later.)

(c) In **Example 7(c),** we made a table of five ordered pairs that satisfy the equation $y = x^2 - 4$.

$$(-2, 0), \quad (-1, -3), \quad (0, -4), \quad (1, -3), \quad (2, 0)$$

        ↑   ↑   ↑
    x-intercept   *y*-intercept   *x*-intercept

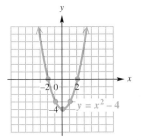

Figure 11

Plotting the points and joining them with a smooth curve gives the graph in **Figure 11.** This curve is called a **parabola.**

✔ **Now Try Exercises 47(b), 51(b), and 53(b).**

To graph an equation on a calculator, such as

$$y = 4x - 1, \quad \text{Equation from \textbf{Example 8(a)}}$$

we must first solve it for y (if necessary). Here the equation is already in the correct form, $y = 4x - 1$, so we enter $4x - 1$ for y_1.*

The intercepts can help determine an appropriate window, since we want them to appear in the graph. A good choice is often the **standard viewing window** for the TI-84 Plus, which has x minimum $= -10$, x maximum $= 10$, y minimum $= -10$, y maximum $= 10$, with x scale $= 1$ and y scale $= 1$. (The x and y scales determine the spacing of the tick marks.) Because the intercepts here are very close to the origin, we have chosen the x and y minimum and maximum to be -3 and 3 instead. See **Figure 12.** ∎

Figure 12

*In this text, we use lowercase letters for variables when referencing graphing calculators. (Some models use uppercase letters.)

2.1 Exercises

CONCEPT PREVIEW *Fill in the blank to correctly complete each sentence.*

1. The point $(-1, 3)$ lies in quadrant _____ in the rectangular coordinate system.

2. The point $(4,$ __$)$ lies on the graph of the equation $y = 3x - 6$.

3. Any point that lies on the x-axis has y-coordinate equal to _____.

4. The y-intercept of the graph of $y = -2x + 6$ is _____.

5. The x-intercept of the graph of $2x + 5y = 10$ is _____.

6. The distance from the origin to the point $(-3, 4)$ is _____.

CONCEPT PREVIEW *Determine whether each statement is* true *or* false. *If false, explain why.*

7. The graph of $y = x^2 + 2$ has no x-intercepts.

8. The graph of $y = x^2 - 2$ has two x-intercepts.

9. The midpoint of the segment joining $(0, 0)$ and $(4, 4)$ is 2.

10. The distance between the points $(0, 0)$ and $(4, 4)$ is 4.

Give three ordered pairs from each table. **See Example 1.**

11.

x	y
2	-5
-1	7
3	-9
5	-17
6	-21

12.

x	y
3	3
-5	-21
8	18
4	6
0	-6

13. *Percent of High School Students Who Smoke*

Year	Percent
1999	35
2001	29
2003	22
2005	23
2007	20
2009	20

Source: Centers for Disease Control and Prevention.

14. *Number of U.S. Viewers of the Super Bowl*

Year	Viewers (millions)
2002	86.8
2004	89.8
2006	90.7
2008	97.4
2010	106.5
2012	111.4
2014	111.5

Source: www.tvbythenumbers.com

For the points P and Q, find **(a)** *the distance* $d(P, Q)$ *and* **(b)** *the coordinates of the midpoint M of line segment PQ.* **See Examples 2 and 5(a).**

15. $P(-5, -6)$, $Q(7, -1)$

16. $P(-4, 3)$, $Q(2, -5)$

17. $P(8, 2)$, $Q(3, 5)$

18. $P(-8, 4)$, $Q(3, -5)$

19. $P(-6, -5)$, $Q(6, 10)$

20. $P(6, -2)$, $Q(4, 6)$

21. $P(3\sqrt{2}, 4\sqrt{5})$, $Q(\sqrt{2}, -\sqrt{5})$

22. $P(-\sqrt{7}, 8\sqrt{3})$, $Q(5\sqrt{7}, -\sqrt{3})$

Determine whether the three points are the vertices of a right triangle. ***See Example 3.***

23. $(-6, -4), (0, -2), (-10, 8)$ **24.** $(-2, -8), (0, -4), (-4, -7)$

25. $(-4, 1), (1, 4), (-6, -1)$ **26.** $(-2, -5), (1, 7), (3, 15)$

27. $(-4, 3), (2, 5), (-1, -6)$ **28.** $(-7, 4), (6, -2), (0, -15)$

Determine whether the three points are collinear. ***See Example 4.***

29. $(0, -7), (-3, 5), (2, -15)$ **30.** $(-1, 4), (-2, -1), (1, 14)$

31. $(0, 9), (-3, -7), (2, 19)$ **32.** $(-1, -3), (-5, 12), (1, -11)$

33. $(-7, 4), (6, -2), (-1, 1)$ **34.** $(-4, 3), (2, 5), (-1, 4)$

Find the coordinates of the other endpoint of each line segment, given its midpoint and one endpoint. ***See Example 5(b).***

35. midpoint $(5, 8)$, endpoint $(13, 10)$ **36.** midpoint $(-7, 6)$, endpoint $(-9, 9)$

37. midpoint $(12, 6)$, endpoint $(19, 16)$ **38.** midpoint $(-9, 8)$, endpoint $(-16, 9)$

39. midpoint (a, b), endpoint (p, q) **40.** midpoint $(6a, 6b)$, endpoint $(3a, 5b)$

Solve each problem. ***See Example 6.***

41. *Bachelor's Degree Attainment* The graph shows a straight line that approximates the percentage of Americans 25 years and older who had earned bachelor's degrees or higher for the years 1990–2012. Use the midpoint formula and the two given points to estimate the percent in 2001. Compare the answer with the actual percent of 26.2.

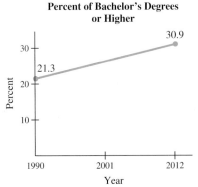

Source: U.S. Census Bureau.

42. *Newspaper Advertising Revenue* The graph shows a straight line that approximates national advertising revenue, in millions of dollars, for newspapers in the United States for the years 2006–2012. Use the midpoint formula and the two given points to estimate revenue in 2009. Compare the answer with the actual figure of 4424 million dollars.

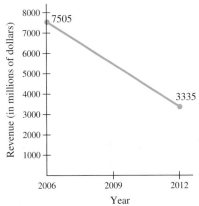

Source: Newspaper Association of America.

43. *Poverty Level Income Cutoffs* The table lists how poverty level income cutoffs (in dollars) for a family of four have changed over time. Use the midpoint formula to approximate the poverty level cutoff in 2012 to the nearest dollar.

Year	Income (in dollars)
1990	13,359
2000	17,604
2010	22,315
2011	23,021
2013	23,834

Source: U.S. Census Bureau.

44. *Public College Enrollment* Enrollments in public colleges for recent years are shown in the table. Assuming a linear relationship, estimate the enrollments for **(a)** 2003 and **(b)** 2009. Give answers to the nearest tenth of thousands if applicable.

Year	Enrollment (in thousands)
2000	11,753
2006	13,180
2012	14,880

Source: National Center for Education Statistics.

45. Show that if M is the midpoint of the line segment with endpoints $P(x_1, y_1)$ and $Q(x_2, y_2)$, then

$$d(P, M) + d(M, Q) = d(P, Q) \quad \text{and} \quad d(P, M) = d(M, Q).$$

46. Write the distance formula $d = \sqrt{(x_2 - x_1)^2 + (y_2 - y_1)^2}$ using a rational exponent.

*For each equation, **(a)** give a table with at least three ordered pairs that are solutions, and **(b)** graph the equation. See Examples 7 and 8.*

47. $y = \dfrac{1}{2}x - 2$

48. $y = -\dfrac{1}{2}x + 2$

49. $2x + 3y = 5$

50. $3x - 2y = 6$

51. $y = x^2$

52. $y = x^2 + 2$

53. $y = \sqrt{x - 3}$

54. $y = \sqrt{x} - 3$

55. $y = |x - 2|$

56. $y = -|x + 4|$

57. $y = x^3$

58. $y = -x^3$

Concept Check Answer the following.

59. If a vertical line is drawn through the point $(4, 3)$, at what point will it intersect the x-axis?

60. If a horizontal line is drawn through the point $(4, 3)$, at what point will it intersect the y-axis?

61. If the point (a, b) is in the second quadrant, then in what quadrant is $(a, -b)$? $(-a, b)$? $(-a, -b)$? (b, a)?

62. Show that the points $(-2, 2)$, $(13, 10)$, $(21, -5)$, and $(6, -13)$ are the vertices of a rhombus (all sides equal in length).

63. Are the points $A(1, 1)$, $B(5, 2)$, $C(3, 4)$, and $D(-1, 3)$ the vertices of a parallelogram (opposite sides equal in length)? of a rhombus (all sides equal in length)?

64. Find the coordinates of the points that divide the line segment joining $(4, 5)$ and $(10, 14)$ into three equal parts.

2.2 Circles

- Center-Radius Form
- General Form
- An Application

Center-Radius Form   By definition, a **circle** is the set of all points in a plane that lie a given distance from a given point. The given distance is the **radius** of the circle, and the given point is the **center.**

We can find the equation of a circle from its definition using the distance formula. Suppose that the point (h, k) is the center and the circle has radius r, where $r > 0$. Let (x, y) represent any point on the circle. See **Figure 13.**

LOOKING AHEAD TO CALCULUS

The circle $x^2 + y^2 = 1$ is called the **unit circle.** It is important in interpreting the *trigonometric* or *circular* functions that appear in the study of calculus.

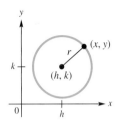

Figure 13

$$\sqrt{(x_2 - x_1)^2 + (y_2 - y_1)^2} = d \quad \text{Distance formula}$$

$$\sqrt{(x - h)^2 + (y - k)^2} = r \quad (x_1, y_1) = (h, k), (x_2, y_2) = (x, y), \text{ and } d = r$$

$$(x - h)^2 + (y - k)^2 = r^2 \quad \text{Square each side.}$$

Center-Radius Form of the Equation of a Circle

A circle with center (h, k) and radius r has equation

$$(x - h)^2 + (y - k)^2 = r^2,$$

which is the **center-radius form** of the equation of the circle. As a special case, a circle with center $(0, 0)$ and radius r has the following equation.

$$x^2 + y^2 = r^2$$

EXAMPLE 1   Finding the Center-Radius Form

Find the center-radius form of the equation of each circle described.

(a) center $(-3, 4)$, radius 6

(b) center $(0, 0)$, radius 3

SOLUTION

(a)
$$(x - h)^2 + (y - k)^2 = r^2 \quad \text{Center-radius form}$$

$$[x - (-3)]^2 + (y - 4)^2 = 6^2 \quad \text{Substitute. Let } (h, k) = (-3, 4) \text{ and } r = 6.$$

Be careful with signs here.
$$(x + 3)^2 + (y - 4)^2 = 36 \quad \text{Simplify.}$$

(b) The center is the origin and $r = 3$.

$$x^2 + y^2 = r^2 \quad \text{Special case of the center-radius form}$$

$$x^2 + y^2 = 3^2 \quad \text{Let } r = 3.$$

$$x^2 + y^2 = 9 \quad \text{Apply the exponent.}$$

✔ **Now Try Exercises 11(a) and 17(a).**

EXAMPLE 2 **Graphing Circles**

Graph each circle discussed in **Example 1**.

(a) $(x + 3)^2 + (y - 4)^2 = 36$ **(b)** $x^2 + y^2 = 9$

SOLUTION

(a) Writing the given equation in center-radius form

$$[x - (-3)]^2 + (y - 4)^2 = 6^2$$

gives $(-3, 4)$ as the center and 6 as the radius. See **Figure 14**.

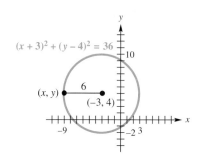

Figure 14   Figure 15

(b) The graph with center $(0, 0)$ and radius 3 is shown in **Figure 15**.

✔ **Now Try Exercises 11(b) and 17(b).**

The circles graphed in **Figures 14 and 15** of **Example 2** can be generated on a graphing calculator by first solving for y and then entering two expressions for y_1 and y_2. See **Figures 16 and 17**. In both cases, the plot of y_1 yields the top half of the circle, and that of y_2 yields the bottom half. It is necessary to use a **square viewing window** to avoid distortion when graphing circles.

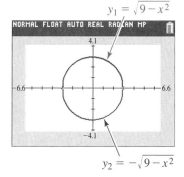

The graph of this circle has equation $(x + 3)^2 + (y - 4)^2 = 36$.

Figure 16

The graph of this circle has equation $x^2 + y^2 = 9$.

Figure 17

General Form Consider the center-radius form of the equation of a circle, and rewrite it so that the binomials are expanded and the right side equals 0.

Don't forget the middle term when squaring each binomial.

$$(x - h)^2 + (y - k)^2 = r^2 \quad \text{Center-radius form}$$

$$x^2 - 2xh + h^2 + y^2 - 2yk + k^2 - r^2 = 0 \quad \text{Square each binomial, and subtract } r^2.$$

$$x^2 + y^2 + \underbrace{(-2h)}_{D}x + \underbrace{(-2k)}_{E}y + \underbrace{(h^2 + k^2 - r^2)}_{F} = 0 \quad \text{Properties of real numbers}$$

If $r > 0$, then the graph of this equation is a circle with center (h, k) and radius r.

This form is the **general form of the equation of a circle.**

General Form of the Equation of a Circle

For some real numbers D, E, and F, the equation

$$x^2 + y^2 + Dx + Ey + F = 0$$

can have a graph that is a circle or a point, or is nonexistent.

Starting with an equation in this general form, we can complete the square to get an equation of the form

$$(x - h)^2 + (y - k)^2 = c, \quad \text{for some number } c.$$

There are three possibilities for the graph based on the value of c.

1. If $c > 0$, then $r^2 = c$, and the graph of the equation is a circle with radius $\sqrt{c}$.

2. If $c = 0$, then the graph of the equation is the single point (h, k).

3. If $c < 0$, then no points satisfy the equation, and the graph is nonexistent.

The next example illustrates the procedure for finding the center and radius.

EXAMPLE 3 Finding the Center and Radius by Completing the Square

Show that $x^2 - 6x + y^2 + 10y + 18 = 0$ has a circle as its graph. Find the center and radius.

SOLUTION   We complete the square twice, once for x and once for y. Begin by subtracting 18 from each side.

$$x^2 - 6x + y^2 + 10y + 18 = 0$$

$$(x^2 - 6x \quad) + (y^2 + 10y \quad) = -18$$

Think: $\quad \left[\dfrac{1}{2}(-6)\right]^2 = (-3)^2 = 9 \quad \text{and} \quad \left[\dfrac{1}{2}(10)\right]^2 = 5^2 = 25$

Add 9 and 25 on the left to complete the two squares, and to compensate, add 9 and 25 on the right.

$$(x^2 - 6x + 9) + (y^2 + 10y + 25) = -18 + 9 + 25 \qquad \text{Complete the square.}$$

Add 9 and 25 on *both* sides.

$$(x - 3)^2 + (y + 5)^2 = 16 \qquad \begin{array}{l}\text{Factor.}\\ \text{Add on the right.}\end{array}$$

$$(x - 3)^2 + [y - (-5)]^2 = 4^2 \qquad \text{Center-radius form}$$

Because $4^2 = 16$ and $16 > 0$, the equation represents a circle with center $(3, -5)$ and radius 4.

✔ **Now Try Exercise 27.**

EXAMPLE 4 **Finding the Center and Radius by Completing the Square**

Show that $2x^2 + 2y^2 - 6x + 10y = 1$ has a circle as its graph. Find the center and radius.

SOLUTION To complete the square, the coefficient of the x^2-term and that of the y^2-term must be 1. In this case they are both 2, so begin by dividing each side by 2.

$$2x^2 + 2y^2 - 6x + 10y = 1$$

$$x^2 + y^2 - 3x + 5y = \frac{1}{2} \qquad \text{Divide by 2.}$$

$$(x^2 - 3x \quad) + (y^2 + 5y \quad) = \frac{1}{2} \qquad \begin{array}{l}\text{Rearrange and regroup terms} \\ \text{in anticipation of completing} \\ \text{the square.}\end{array}$$

$$\left(x^2 - 3x + \frac{9}{4}\right) + \left(y^2 + 5y + \frac{25}{4}\right) = \frac{1}{2} + \frac{9}{4} + \frac{25}{4} \qquad \begin{array}{l}\text{Complete the square for } both \\ x \text{ and } y; \left[\frac{1}{2}(-3)\right]^2 = \frac{9}{4} \text{ and} \\ \left[\frac{1}{2}(5)\right]^2 = \frac{25}{4}.\end{array}$$

$$\left(x - \frac{3}{2}\right)^2 + \left(y + \frac{5}{2}\right)^2 = 9 \qquad \text{Factor and add.}$$

$$\left(x - \frac{3}{2}\right)^2 + \left[y - \left(-\frac{5}{2}\right)\right]^2 = 3^2 \qquad \text{Center-radius form}$$

The equation has a circle with center $\left(\frac{3}{2}, -\frac{5}{2}\right)$ and radius 3 as its graph.

✔ **Now Try Exercise 31.**

EXAMPLE 5 **Determining Whether a Graph Is a Point or Nonexistent**

The graph of the equation $x^2 + 10x + y^2 - 4y + 33 = 0$ either is a point or is nonexistent. Which is it?

SOLUTION

$$x^2 + 10x + y^2 - 4y + 33 = 0$$

$$x^2 + 10x + y^2 - 4y = -33 \qquad \text{Subtract 33.}$$

Think: $\left[\frac{1}{2}(10)\right]^2 = 25$ and $\left[\frac{1}{2}(-4)\right]^2 = 4$ $\quad$ $\begin{array}{l}\text{Prepare to complete the} \\ \text{square for both } x \text{ and } y.\end{array}$

$$(x^2 + 10x + 25) + (y^2 - 4y + 4) = -33 + 25 + 4 \qquad \text{Complete the square.}$$

$$(x + 5)^2 + (y - 2)^2 = -4 \qquad \text{Factor and add.}$$

Because $-4 < 0$, there are *no* ordered pairs (x, y), with x and y both real numbers, satisfying the equation. The graph of the given equation is nonexistent—it contains no points. (If the constant on the right side were 0, the graph would consist of the single point $(-5, 2)$.)

✔ **Now Try Exercise 33.**

An Application Seismologists can locate the epicenter of an earthquake by determining the intersection of three circles. The radii of these circles represent the distances from the epicenter to each of three receiving stations. The centers of the circles represent the receiving stations.

EXAMPLE 6 **Locating the Epicenter of an Earthquake**

Suppose receiving stations *A*, *B*, and *C* are located on a coordinate plane at the points

$$(1, 4), \quad (-3, -1), \quad \text{and} \quad (5, 2).$$

Let the distances from the earthquake epicenter to these stations be 2 units, 5 units, and 4 units, respectively. Where on the coordinate plane is the epicenter located?

SOLUTION Graph the three circles as shown in **Figure 18.** From the graph it appears that the epicenter is located at $(1, 2)$. To check this algebraically, determine the equation for each circle and substitute $x = 1$ and $y = 2$.

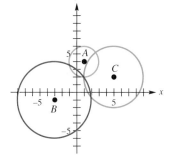

Figure 18

Station *A*:

$$(x - 1)^2 + (y - 4)^2 = 2^2 \qquad \text{Equation of a circle with center } (1, 4) \text{ and radius 2}$$

$$(1 - 1)^2 + (2 - 4)^2 \stackrel{?}{=} 4 \qquad \text{Let } x = 1 \text{ and } y = 2.$$

$$0 + 4 \stackrel{?}{=} 4$$

$$4 = 4 \checkmark$$

Station *B*:

$$(x + 3)^2 + (y + 1)^2 = 5^2 \qquad \text{Equation of a circle with center } (-3, -1) \text{ and radius 5}$$

$$(1 + 3)^2 + (2 + 1)^2 \stackrel{?}{=} 25 \qquad \text{Let } x = 1 \text{ and } y = 2.$$

$$16 + 9 \stackrel{?}{=} 25$$

$$25 = 25 \checkmark$$

Station *C*:

$$(x - 5)^2 + (y - 2)^2 = 4^2 \qquad \text{Equation of a circle with center } (5, 2) \text{ and radius 4}$$

$$(1 - 5)^2 + (2 - 2)^2 \stackrel{?}{=} 16 \qquad \text{Let } x = 1 \text{ and } y = 2.$$

$$16 + 0 \stackrel{?}{=} 16$$

$$16 = 16 \checkmark$$

The point $(1, 2)$ lies on all three graphs. Thus, we can conclude that the epicenter of the earthquake is at $(1, 2)$.

✔ **Now Try Exercise 39.**

2.2 Exercises

CONCEPT PREVIEW *Fill in the blank(s) to correctly complete each sentence.*

1. The circle with equation $x^2 + y^2 = 49$ has center with coordinates _____ and radius equal to _____.

2. The circle with center $(3, 6)$ and radius 4 has equation _____.

3. The graph of $(x - 4)^2 + (y + 7)^2 = 9$ has center with coordinates _____.

4. The graph of $x^2 + (y - 5)^2 = 9$ has center with coordinates _____.

CONCEPT PREVIEW *Match each equation in Column I with its graph in Column II.*

I

II

5. $(x - 3)^2 + (y - 2)^2 = 25$

A.

B.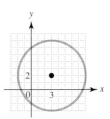

6. $(x - 3)^2 + (y + 2)^2 = 25$

7. $(x + 3)^2 + (y - 2)^2 = 25$

C.

D.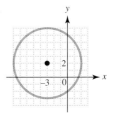

8. $(x + 3)^2 + (y + 2)^2 = 25$

CONCEPT PREVIEW *Answer each question.*

9. How many points lie on the graph of $x^2 + y^2 = 0$?

10. How many points lie on the graph of $x^2 + y^2 = -100$?

*In the following exercises, (**a**) find the center-radius form of the equation of each circle described, and (**b**) graph it. See **Examples 1 and 2**.*

11. center $(0, 0)$, radius 6

12. center $(0, 0)$, radius 9

13. center $(2, 0)$, radius 6

14. center $(3, 0)$, radius 3

15. center $(0, 4)$, radius 4

16. center $(0, -3)$, radius 7

17. center $(-2, 5)$, radius 4

18. center $(4, 3)$, radius 5

19. center $(5, -4)$, radius 7

20. center $(-3, -2)$, radius 6

21. center $\left(\sqrt{2}, \sqrt{2}\right)$, radius $\sqrt{2}$

22. center $\left(-\sqrt{3}, -\sqrt{3}\right)$, radius $\sqrt{3}$

*Connecting Graphs with Equations Use each graph to determine an equation of the circle in (**a**) center-radius form and (**b**) general form.*

23.

24.

25.

26.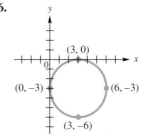

*Decide whether or not each equation has a circle as its graph. If it does, give the center and radius. If it does not, describe the graph. **See Examples 3–5.***

27. $x^2 + y^2 + 6x + 8y + 9 = 0$

28. $x^2 + y^2 + 8x - 6y + 16 = 0$

29. $x^2 + y^2 - 4x + 12y = -4$

30. $x^2 + y^2 - 12x + 10y = -25$

31. $4x^2 + 4y^2 + 4x - 16y - 19 = 0$

32. $9x^2 + 9y^2 + 12x - 18y - 23 = 0$

33. $x^2 + y^2 + 2x - 6y + 14 = 0$

34. $x^2 + y^2 + 4x - 8y + 32 = 0$

35. $x^2 + y^2 - 6x - 6y + 18 = 0$

36. $x^2 + y^2 + 4x + 4y + 8 = 0$

37. $9x^2 + 9y^2 - 6x + 6y - 23 = 0$

38. $4x^2 + 4y^2 + 4x - 4y - 7 = 0$

*Epicenter of an Earthquake Solve each problem. To visualize the situation, use graph paper and a compass to carefully graph each circle. **See Example 6.***

39. Suppose that receiving stations X, Y, and Z are located on a coordinate plane at the points

$$(7, 4), \quad (-9, -4), \quad \text{and} \quad (-3, 9),$$

respectively. The epicenter of an earthquake is determined to be 5 units from X, 13 units from Y, and 10 units from Z. Where on the coordinate plane is the epicenter located?

40. Suppose that receiving stations P, Q, and R are located on a coordinate plane at the points

$$(3, 1), \quad (5, -4), \quad \text{and} \quad (-1, 4),$$

respectively. The epicenter of an earthquake is determined to be $\sqrt{5}$ units from P, 6 units from Q, and $2\sqrt{10}$ units from R. Where on the coordinate plane is the epicenter located?

41. The locations of three receiving stations and the distances to the epicenter of an earthquake are contained in the following three equations:

$$(x - 2)^2 + (y - 1)^2 = 25, \quad (x + 2)^2 + (y - 2)^2 = 16,$$
$$\text{and} \quad (x - 1)^2 + (y + 2)^2 = 9.$$

Determine the location of the epicenter.

42. The locations of three receiving stations and the distances to the epicenter of an earthquake are contained in the following three equations:

$$(x - 2)^2 + (y - 4)^2 = 25, \quad (x - 1)^2 + (y + 3)^2 = 25,$$
$$\text{and} \quad (x + 3)^2 + (y + 6)^2 = 100.$$

Determine the location of the epicenter.

Concept Check Work each of the following.

43. Find the center-radius form of the equation of a circle with center $(3, 2)$ and tangent to the x-axis. (*Hint:* A line **tangent** to a circle touches it at exactly one point.)

44. Find the equation of a circle with center at $(-4, 3)$, passing through the point $(5, 8)$. Write it in center-radius form.

45. Find all points (x, y) with $x = y$ that are 4 units from $(1, 3)$.

46. Find all points satisfying $x + y = 0$ that are 8 units from $(-2, 3)$.

47. Find the coordinates of all points whose distance from $(1, 0)$ is $\sqrt{10}$ and whose distance from $(5, 4)$ is $\sqrt{10}$.

48. Find the equation of the circle of least radius that contains the points $(1, 4)$ and $(-3, 2)$ within or on its boundary.

49. Find all values of y such that the distance between $(3, y)$ and $(-2, 9)$ is 12.

50. Suppose that a circle is tangent to both axes, is in the third quadrant, and has radius $\sqrt{2}$. Find the center-radius form of its equation.

51. Find the shortest distance from the origin to the graph of the circle with equation

$$x^2 - 16x + y^2 - 14y + 88 = 0.$$

52. Phlash Phelps, the morning radio personality on SiriusXM Satellite Radio's *Sixties on Six* Decades channel, is an expert on U.S. geography. He loves traveling around the country to strange, out-of-the-way locations. The photo shows Phlash seated in front of a sign in a small Arizona settlement called *Nothing*. (Nothing is so small that it's not named on current maps.) The sign indicates that Nothing is 50 mi from Wickenburg, AZ, 75 mi from Kingman, AZ, 105 mi from Phoenix, AZ, and 180 mi from Las Vegas, NV. Explain how the concepts of **Example 6** can be used to locate Nothing, AZ, on a map of Arizona and southern Nevada.

Relating Concepts

For individual or collaborative investigation *(Exercises 53-58)*

The distance formula, midpoint formula, and center-radius form of the equation of a circle are closely related in the following problem.

    A circle has a diameter with endpoints $(-1, 3)$ and $(5, -9)$. Find the center-radius form of the equation of this circle.

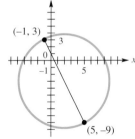

Work Exercises 53–58 in order, *to see the relationships among these concepts.*

53. To find the center-radius form, we must find both the radius and the coordinates of the center. Find the coordinates of the center using the midpoint formula. (The center of the circle must be the midpoint of the diameter.)

54. There are several ways to find the radius of the circle. One way is to find the distance between the center and the point $(-1, 3)$. Use the result from **Exercise 53** and the distance formula to find the radius.

55. Another way to find the radius is to repeat **Exercise 54,** but use the point $(5, -9)$ rather than $(-1, 3)$. Do this to obtain the same answer found in **Exercise 54.**

56. There is yet another way to find the radius. Because the radius is half the diameter, it can be found by finding half the length of the diameter. Using the endpoints of the diameter given in the problem, find the radius in this manner. The same answer found in **Exercise 54** should be obtained.

57. Using the center found in **Exercise 53** and the radius found in **Exercises 54–56,** give the center-radius form of the equation of the circle.

58. Use the method described in **Exercises 53–57** to find the center-radius form of the equation of the circle with diameter having endpoints $(3, -5)$ and $(-7, 3)$.

*Find the center-radius form of the circle described or graphed. (**See Exercises 53–58.**)*

59. a circle having a diameter with endpoints $(-1, 2)$ and $(11, 7)$

60. a circle having a diameter with endpoints $(5, 4)$ and $(-3, -2)$

61.

62.

2.3 Functions

- **Relations and Functions**
- **Domain and Range**
- **Determining Whether Relations Are Functions**
- **Function Notation**
- **Increasing, Decreasing, and Constant Functions**

Relations and Functions   As we saw previously, one quantity can sometimes be described in terms of another.

- The letter grade a student receives in a mathematics course depends on a numerical score.

- The amount paid (in dollars) for gas at a gas station depends on the number of gallons pumped.

- The dollars spent by the average American household depends on the expense category.

We used ordered pairs to represent these corresponding quantities. For example, $(3, \$10.50)$ indicates that we pay \$10.50 for 3 gallons of gas. Since the amount we pay *depends* on the number of gallons pumped, the amount (in dollars) is called the *dependent variable*, and the number of gallons pumped is called the *independent variable*.

Generalizing, if the value of the second component y depends on the value of the first component x, then y is the **dependent variable** and x is the **independent variable.**

Independent variable ⌐ ⌐ Dependent variable
$$(x, y)$$

A set of ordered pairs such as $\{(3, 10.50), (8, 28.00), (10, 35.00)\}$ is a *relation*. A *function* is a special kind of relation.

Relation and Function

A **relation** is a set of ordered pairs. A **function** is a relation in which, for each distinct value of the first component of the ordered pairs, there is *exactly one* value of the second component.

NOTE   The relation from the beginning of this section representing the number of gallons of gasoline and the corresponding cost is a function because each x-value is paired with exactly one y-value.

EXAMPLE 1 **Deciding Whether Relations Define Functions**

Decide whether each relation defines a function.

$$F = \{(1, 2), (-2, 4), (3, 4)\}$$

$$G = \{(1, 1), (1, 2), (1, 3), (2, 3)\}$$

$$H = \{(-4, 1), (-2, 1), (-2, 0)\}$$

SOLUTION Relation F is a function because for each different x-value there is exactly one y-value. We can show this correspondence as follows.

$$\{1, -2, 3\} \quad \text{\small x-values of F}$$
$$\downarrow \quad \downarrow \quad \downarrow$$
$$\{2, \quad 4, \quad 4\} \quad \text{\small y-values of F}$$

As the correspondence below shows, relation G is not a function because one first component corresponds to *more than one* second component.

$$\{1, 2\} \quad \text{\small x-values of G}$$

$$\{1, 2, 3\} \quad \text{\small y-values of G}$$

In relation H the last two ordered pairs have the same x-value paired with two different y-values (-2 is paired with both 1 and 0), so H is a relation but not a function. ***In a function, no two ordered pairs can have the same first component and different second components.***

Different y-values
$$H = \{(-4, 1), (-2, 1), (-2, 0)\} \quad \text{Not a function}$$
Same x-value

✔ **Now Try Exercises 11 and 13.**

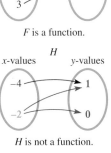

x-values   *F*   *y*-values

F is a function.

x-values   *H*   *y*-values

H is not a function.

Figure 19

Relations and functions can also be expressed as a correspondence or *mapping* from one set to another, as shown in **Figure 19** for function F and relation H from **Example 1.** The arrow from 1 to 2 indicates that the ordered pair $(1, 2)$ belongs to F—each first component is paired with exactly one second component. In the mapping for relation H, which is not a function, the first component -2 is paired with two different second components, 1 and 0.

Because relations and functions are sets of ordered pairs, we can represent them using tables and graphs. A table and graph for function F are shown in **Figure 20.**

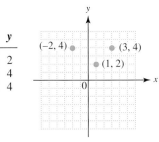

x	y
1	2
-2	4
3	4

Graph of F

Figure 20

Finally, we can describe a relation or function using a rule that tells how to determine the dependent variable for a specific value of the independent variable. The rule may be given in words: for instance, "the dependent variable is twice the independent variable." Usually the rule is an equation, such as the one below.

Dependent variable $\longrightarrow y = 2x \longleftarrow$ Independent variable

On this particular day, an *input* of pumping 7.870 gallons of gasoline led to an *output* of $29.58 from the purchaser's wallet. This is an example of a function whose domain consists of numbers of gallons pumped, and whose range consists of amounts from the purchaser's wallet. Dividing the dollar amount by the number of gallons pumped gives the exact price of gasoline that day. Use a calculator to check this. Was this pump fair? (Later we will see that this price is an example of the slope *m* of a linear function of the form $y = mx$.)

In a function, there is exactly one value of the dependent variable, the second component, for each value of the independent variable, the first component.

NOTE Another way to think of a function relationship is to think of the independent variable as an **input** and of the dependent variable as an **output.** This is illustrated by the **input-output (function) machine** for the function $y = 2x$.

Function machine

Domain and Range We now consider two important concepts concerning relations.

Domain and Range

For every relation consisting of a set of ordered pairs (x, y), there are two important sets of elements.

- The set of all values of the independent variable (x) is the **domain.**
- The set of all values of the dependent variable (y) is the **range.**

EXAMPLE 2 **Finding Domains and Ranges of Relations**

Give the domain and range of each relation. Tell whether the relation defines a function.

(a) $\{(3, -1), (4, 2), (4, 5), (6, 8)\}$

(b)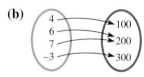

(c)

x	y
−5	2
0	2
5	2

SOLUTION

(a) The domain is the set of *x*-values, $\{3, 4, 6\}$. The range is the set of *y*-values, $\{-1, 2, 5, 8\}$. This relation is not a function because the same *x*-value, 4, is paired with two different *y*-values, 2 and 5.

(b) The domain is $\{4, 6, 7, -3\}$ and the range is $\{100, 200, 300\}$. This mapping defines a function. Each *x*-value corresponds to exactly one *y*-value.

(c) This relation is a set of ordered pairs, so the domain is the set of *x*-values $\{-5, 0, 5\}$ and the range is the set of *y*-values $\{2\}$. The table defines a function because each different *x*-value corresponds to exactly one *y*-value (even though it is the same *y*-value).

✔ **Now Try Exercises 19, 21, and 23.**

EXAMPLE 3 **Finding Domains and Ranges from Graphs**

Give the domain and range of each relation.

(a)

(b)

(c)

(d)

SOLUTION

(a) The domain is the set of x-values, $\{-1, 0, 1, 4\}$. The range is the set of y-values, $\{-3, -1, 1, 2\}$.

(b) The x-values of the points on the graph include all numbers between -4 and 4, inclusive. The y-values include all numbers between -6 and 6, inclusive.

> The domain is $[-4, 4]$.   *Use interval notation.*
> The range is $[-6, 6]$.

(c) The arrowheads indicate that the line extends indefinitely left and right, as well as up and down. Therefore, both the domain and the range include all real numbers, which is written $(-\infty, \infty)$.

(d) The arrowheads indicate that the graph extends indefinitely left and right, as well as upward. The domain is $(-\infty, \infty)$. Because there is a least y-value, -3, the range includes all numbers greater than or equal to -3, written $[-3, \infty)$.

✔ **Now Try Exercises 27 and 29.**

Relations are often defined by equations, such as $y = 2x + 3$ and $y^2 = x$, so we must sometimes determine the domain of a relation from its equation. In this book, we assume the following agreement on the domain of a relation.

Agreement on Domain

Unless specified otherwise, the domain of a relation is assumed to be all real numbers that produce real numbers when substituted for the independent variable.

To illustrate this agreement, because any real number can be used as a replacement for x in $y = 2x + 3$, the domain of this function is the set of all real numbers.

This is the graph of a function.
Each *x*-value corresponds
to only one *y*-value.

(a)

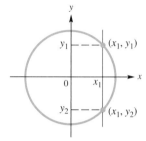

This is not the graph of a function.
The same *x*-value corresponds to
two different *y*-values.

(b)

Figure 21

As another example, the function $y = \frac{1}{x}$ has the set of all real numbers *except* 0 as domain because *y* is undefined if $x = 0$.

> *In general, the domain of a function defined by an algebraic expression is the set of all real numbers, except those numbers that lead to division by **0** or to an even root of a negative number.*

(There are also exceptions for logarithmic and trigonometric functions. They are covered in further treatment of precalculus mathematics.)

Determining Whether Relations Are Functions Because each value of *x* leads to only one value of *y* in a function, any vertical line must intersect the graph in at most one point. This is the **vertical line test** for a function.

Vertical Line Test

If every vertical line intersects the graph of a relation in no more than one point, then the relation is a function.

The graph in **Figure 21(a)** represents a function because each vertical line intersects the graph in no more than one point. The graph in **Figure 21(b)** is not the graph of a function because there exists a vertical line that intersects the graph in more than one point.

EXAMPLE 4 **Using the Vertical Line Test**

Use the vertical line test to determine whether each relation graphed in **Example 3** is a function.

SOLUTION We repeat each graph from **Example 3,** this time with vertical lines drawn through the graphs.

(a)

(b)

(c)

(d)

- The graphs of the relations in parts (a), (c), and (d) pass the vertical line test because every vertical line intersects each graph no more than once. Thus, these graphs represent functions.

- The graph of the relation in part (b) fails the vertical line test because the same *x*-value corresponds to two different *y*-values. Therefore, it is not the graph of a function.

✔ **Now Try Exercises 27 and 29.**

The vertical line test is a simple method for identifying a function defined by a graph. Deciding whether a relation defined by an equation or an inequality is a function, as well as determining the domain and range, is more difficult. The next example gives some hints that may help.

EXAMPLE 5 Identifying Functions, Domains, and Ranges

Decide whether each relation defines y as a function of x, and give the domain and range.

(a) $y = x + 4$ **(b)** $y = \sqrt{2x - 1}$ **(c)** $y^2 = x$

(d) $y \leq x - 1$ **(e)** $y = \dfrac{5}{x - 1}$

SOLUTION

(a) In the defining equation (or rule), $y = x + 4$, y is always found by adding 4 to x. Thus, each value of x corresponds to just one value of y, and the relation defines a function. The variable x can represent any real number, so the domain is

$$\{x \mid x \text{ is a real number}\}, \quad \text{or} \quad (-\infty, \infty).$$

Because y is always 4 more than x, y also may be any real number, and so the range is $(-\infty, \infty)$.

(b) For any choice of x in the domain of $y = \sqrt{2x - 1}$, there is exactly one corresponding value for y (the radical is a nonnegative number), so this equation defines a function. The equation involves a square root, so the quantity under the radical sign cannot be negative.

$$2x - 1 \geq 0 \qquad \text{Solve the inequality.}$$
$$2x \geq 1 \qquad \text{Add 1.}$$
$$x \geq \frac{1}{2} \qquad \text{Divide by 2.}$$

Figure 22

The domain of the function is $\left[\frac{1}{2}, \infty\right)$. Because the radical must represent a nonnegative number, as x takes values greater than or equal to $\frac{1}{2}$, the range is $\{y \mid y \geq 0\}$, or $[0, \infty)$. See **Figure 22.**

(c) The ordered pairs $(16, 4)$ and $(16, -4)$ both satisfy the equation $y^2 = x$. There exists at least one value of x—for example, 16—that corresponds to two values of y, 4 and -4, so this equation does not define a function.

    Because x is equal to the square of y, the values of x must always be nonnegative. The domain of the relation is $[0, \infty)$. Any real number can be squared, so the range of the relation is $(-\infty, \infty)$. See **Figure 23.**

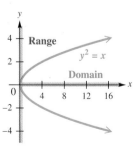

Figure 23

(d) By definition, y is a function of x if every value of x leads to exactly one value of y. Substituting a particular value of x, say 1, into $y \leq x - 1$ corresponds to many values of y. The ordered pairs

$$(1, 0), \quad (1, -1), \quad (1, -2), \quad (1, -3), \quad \text{and} \quad \text{so on}$$

all satisfy the inequality, so y is not a function of x here. Any number can be used for x or for y, so the domain and the range of this relation are both the set of real numbers, $(-\infty, \infty)$.

(e) Given any value of x in the domain of

$$y = \frac{5}{x - 1},$$

we find y by subtracting 1 from x, and then dividing the result into 5. This process produces exactly one value of y for each value in the domain, so this equation defines a function.

The domain of $y = \frac{5}{x - 1}$ includes all real numbers except those that make the denominator 0. We find these numbers by setting the denominator equal to 0 and solving for x.

$$x - 1 = 0$$

$$x = 1 \quad \text{Add 1.}$$

Thus, the domain includes all real numbers except 1, written as the interval $(-\infty, 1) \cup (1, \infty)$. Values of y can be positive or negative, but never 0, because a fraction cannot equal 0 unless its numerator is 0. Therefore, the range is the interval $(-\infty, 0) \cup (0, \infty)$, as shown in **Figure 24.**

✔ **Now Try Exercises 37, 39, and 45.**

Figure 24

Variations of the Definition of Function

1. A **function** is a relation in which, for each distinct value of the first component of the ordered pairs, there is exactly one value of the second component.

2. A **function** is a set of ordered pairs in which no first component is repeated.

3. A **function** is a rule or correspondence that assigns exactly one range value to each distinct domain value.

Function Notation When a function f is defined with a rule or an equation using x and y for the independent and dependent variables, we say, "y is *a function of x*" to emphasize that y *depends on x.* We use the notation

$$y = f(x),$$

called **function notation,** to express this and read $f(x)$ as **"f of x,"** or **"f at x."** The letter f is the name given to this function.

For example, if $y = 3x - 5$, we can name the function f and write

$$f(x) = 3x - 5.$$

Note that $f(x)$ is just another name for the dependent variable y. For example, if $y = f(x) = 3x - 5$ and $x = 2$, then we find y, or $f(2)$, by replacing x with 2.

$$f(2) = 3 \cdot 2 - 5 \quad \text{Let } x = 2.$$

$$f(2) = 1 \quad \text{Multiply, and then subtract.}$$

The statement "In the function f, if $x = 2$, then $y = 1$" represents the ordered pair $(2, 1)$ and is abbreviated with function notation as follows.

$$f(2) = 1$$

The symbol $f(2)$ is read "f of 2" or "f at 2."

Function notation can be illustrated as follows.

Name of the function

Defining expression

$$y = f(x) = 3x - 5$$

Value of the function   Name of the independent variable

CAUTION *The symbol f(x) does not indicate "f times x,"* but repre-sents the *y*-value associated with the indicated *x*-value. As just shown, $f(2)$ is the *y*-value that corresponds to the *x*-value 2.

EXAMPLE 6 Using Function Notation

Let $f(x) = -x^2 + 5x - 3$ and $g(x) = 2x + 3$. Find each of the following.

(a) $f(2)$ **(b)** $f(q)$ **(c)** $g(a + 1)$

SOLUTION

(a) $f(x) = -x^2 + 5x - 3$

$f(2) = -2^2 + 5 \cdot 2 - 3$ Replace *x* with 2.

$f(2) = -4 + 10 - 3$ Apply the exponent and multiply.

$f(2) = 3$ Add and subtract.

Thus, $f(2) = 3$, and the ordered pair $(2, 3)$ belongs to f.

(b) $f(x) = -x^2 + 5x - 3$

$f(q) = -q^2 + 5q - 3$ Replace *x* with *q*.

(c) $g(x) = 2x + 3$

$g(a + 1) = 2(a + 1) + 3$ Replace *x* with *a* + 1.

$g(a + 1) = 2a + 2 + 3$ Distributive property

$g(a + 1) = 2a + 5$ Add.

The replacement of one variable with another variable or expression, as in parts (b) and (c), is important in later courses.

✔ **Now Try Exercises 51, 59, and 65.**

Functions can be evaluated in a variety of ways, as shown in **Example 7.**

EXAMPLE 7 Using Function Notation

For each function, find $f(3)$.

(a) $f(x) = 3x - 7$ **(b)** $f = \{(-3, 5), (0, 3), (3, 1), (6, -1)\}$

(c)

(d)

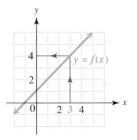

Figure 25

Figure 26

SOLUTION

(a) $f(x) = 3x - 7$

$f(3) = 3(3) - 7$ Replace x with 3.

$f(3) = 2$ Simplify.

$f(3) = 2$ indicates that the ordered pair $(3, 2)$ belongs to f.

(b) For $f = \{(-3, 5), (0, 3), (3, 1), (6, -1)\}$, we want $f(3)$, the y-value of the ordered pair where $x = 3$. As indicated by the ordered pair $(3, 1)$, when $x = 3$, $y = 1$, so $f(3) = 1$.

(c) In the mapping, repeated in **Figure 25,** the domain element 3 is paired with 5 in the range, so $f(3) = 5$.

(d) To evaluate $f(3)$ using the graph, find 3 on the x-axis. See **Figure 26.** Then move up until the graph of f is reached. Moving horizontally to the y-axis gives 4 for the corresponding y-value. Thus, $f(3) = 4$.

✔ **Now Try Exercises 67, 69, and 71.**

If a function f is defined by an equation with x and y (and not with function notation), use the following steps to find $f(x)$.

Finding an Expression for $f(x)$

Consider an equation involving x and y. Assume that y can be expressed as a function f of x. To find an expression for $f(x)$, use the following steps.

Step 1 Solve the equation for y.

Step 2 Replace y with $f(x)$.

EXAMPLE 8 **Writing Equations Using Function Notation**

Assume that y is a function f of x. Rewrite each equation using function notation. Then find $f(-2)$ and $f(p)$.

(a) $y = x^2 + 1$ **(b)** $x - 4y = 5$

SOLUTION

(a) ***Step 1*** $y = x^2 + 1$ This equation is already solved for y.

Step 2 $f(x) = x^2 + 1$ Let $y = f(x)$.

Now find $f(-2)$ and $f(p)$.

$f(-2) = (-2)^2 + 1$ Let $x = -2$. $\qquad$ $f(p) = p^2 + 1$ Let $x = p$.

$f(-2) = 4 + 1$

$f(-2) = 5$

(b) ***Step 1*** $x - 4y = 5$ Given equation.

$-4y = -x + 5$ Add $-x$.

$y = \dfrac{x - 5}{4}$ Multiply by -1. Divide by 4.

Step 2 $f(x) = \dfrac{1}{4}x - \dfrac{5}{4}$ Let $y = f(x)$; $\dfrac{a - b}{c} = \dfrac{a}{c} - \dfrac{b}{c}$.

Now find $f(-2)$ and $f(p)$.

$$f(x) = \frac{1}{4}x - \frac{5}{4}$$

$f(-2) = \frac{1}{4}(-2) - \frac{5}{4}$ Let $x = -2$. $f(p) = \frac{1}{4}p - \frac{5}{4}$ Let $x = p$.

$f(-2) = -\frac{7}{4}$

✔ **Now Try Exercises 77 and 81.**

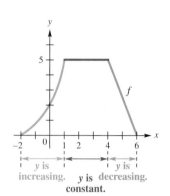

Figure 27

Increasing, Decreasing, and Constant Functions   Informally speaking, a function *increases* over an open interval of its domain if its graph rises from left to right on the interval. It *decreases* over an open interval of its domain if its graph falls from left to right on the interval. It is *constant* over an open interval of its domain if its graph is horizontal on the interval.

For example, consider **Figure 27.**

- The function increases over the open interval $(-2, 1)$ because the y-values continue to get larger for x-values in that interval.

- The function is constant over the open interval $(1, 4)$ because the y-values are always 5 for all x-values there.

- The function decreases over the open interval $(4, 6)$ because in that interval the y-values continuously get smaller.

The intervals refer to the x-values where the y-values either increase, decrease, or are constant.

The formal definitions of these concepts follow.

Increasing, Decreasing, and Constant Functions

Suppose that a function f is defined over an *open* interval I and x_1 and x_2 are in I.

(a) f **increases** over I if, whenever $x_1 < x_2$, $f(x_1) < f(x_2)$.

(b) f **decreases** over I if, whenever $x_1 < x_2$, $f(x_1) > f(x_2)$.

(c) f is **constant** over I if, for every x_1 and x_2, $f(x_1) = f(x_2)$.

Figure 28 illustrates these ideas.

Whenever $x_1 < x_2$, and $f(x_1) < f(x_2)$, f is increasing.

(a)

Whenever $x_1 < x_2$, and $f(x_1) > f(x_2)$, f is decreasing.

(b)

For every x_1 and x_2, if $f(x_1) = f(x_2)$, then f is constant.

(c)

Figure 28

NOTE To decide whether a function is increasing, decreasing, or constant over an interval, ask yourself, *"What does y do as x goes from left to right?"* Our definition of *increasing, decreasing,* and *constant* function behavior applies to open intervals of the domain, not to individual points.

EXAMPLE 9 **Determining Increasing, Decreasing, and Constant Intervals**

Figure 29 shows the graph of a function. Determine the largest open intervals of the domain over which the function is increasing, decreasing, or constant.

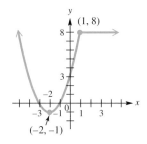

Figure 29

SOLUTION We observe the domain and ask, *"What is happening to the y-values as the x-values are getting larger?"* Moving from left to right on the graph, we see the following:

- On the open interval $(-\infty, -2)$, the y-values are *decreasing*.

- On the open interval $(-2, 1)$, the y-values are *increasing*.

- On the open interval $(1, \infty)$, the y-values are *constant* (and equal to 8).

Therefore, the function is decreasing on $(-\infty, -2)$, increasing on $(-2, 1)$, and constant on $(1, \infty)$.

✔ **Now Try Exercise 91.**

EXAMPLE 10 **Interpreting a Graph**

Figure 30 shows the relationship between the number of gallons, $g(t)$, of water in a small swimming pool and time in hours, t. By looking at this graph of the function, we can answer questions about the water level in the pool at various times. For example, at time 0 the pool is empty. The water level then increases, stays constant for a while, decreases, and then becomes constant again.

Figure 30

Use the graph to respond to the following.

(a) What is the maximum number of gallons of water in the pool? When is the maximum water level first reached?

(b) For how long is the water level increasing? decreasing? constant?

(c) How many gallons of water are in the pool after 90 hr?

(d) Describe a series of events that could account for the water level changes shown in the graph.

**Swimming Pool
Water Level**

Figure 30 (repeated)

SOLUTION

(a) The maximum range value is 3000, as indicated by the horizontal line segment for the hours 25 to 50. This maximum number of gallons, 3000, is first reached at $t = 25$ hr.

(b) The water level is increasing for $25 - 0 = 25$ hr. The water level is decreasing for $75 - 50 = 25$ hr. It is constant for

$$(50 - 25) + (100 - 75)$$
$$= 25 + 25$$
$$= 50 \text{ hr.}$$

(c) When $t = 90$, $y = g(90) = 2000$. There are 2000 gal after 90 hr.

(d) Looking at the graph in **Figure 30,** we might write the following description.

The pool is empty at the beginning and then is filled to a level of 3000 gal during the first 25 hr. For the next 25 hr, the water level remains the same. At 50 hr, the pool starts to be drained, and this draining lasts for 25 hr, until only 2000 gal remain. For the next 25 hr, the water level is unchanged.

✔ **Now Try Exercise 93.**

2.3 Exercises

CONCEPT PREVIEW *Fill in the blank(s) to correctly complete each sentence.*

1. The domain of the relation $\{(3, 5), (4, 9), (10, 13)\}$ is _____.

2. The range of the relation in **Exercise 1** is _____.

3. The equation $y = 4x - 6$ defines a function with independent variable _____ and dependent variable _____.

4. The function in **Exercise 3** includes the ordered pair $(6, \text{__})$.

5. For the function $f(x) = -4x + 2$, $f(-2) = $ _____.

6. For the function $g(x) = \sqrt{x}$, $g(9) = $ _____.

7. The function in **Exercise 6** has domain _____.

8. The function in **Exercise 6** has range _____.

9. The largest open interval over which the function graphed here increases is _____.

10. The largest open interval over which the function graphed here decreases is _____.

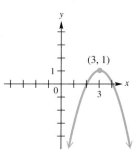

*Decide whether each relation defines a function. **See Example 1.***

11. $\{(5, 1), (3, 2), (4, 9), (7, 8)\}$

12. $\{(8, 0), (5, 7), (9, 3), (3, 8)\}$

13. $\{(2, 4), (0, 2), (2, 6)\}$

14. $\{(9, -2), (-3, 5), (9, 1)\}$

15. $\{(-3, 1), (4, 1), (-2, 7)\}$

16. $\{(-12, 5), (-10, 3), (8, 3)\}$

17.

x	y
3	−4
7	−4
10	−4

18.

x	y
−4	$\sqrt{2}$
0	$\sqrt{2}$
4	$\sqrt{2}$

Decide whether each relation defines a function, and give the domain and range. See Examples 1–4.

19. $\{(1, 1), (1, -1), (0, 0), (2, 4), (2, -4)\}$

20. $\{(2, 5), (3, 7), (3, 9), (5, 11)\}$

21.

22.

23.

x	y
0	0
−1	1
−2	2

24.

x	y
0	0
1	−1
2	−2

25. *Number of Visits to U.S. National Parks*

Year (x)	Number of Visits (y) (millions)
2010	64.9
2011	63.0
2012	65.1
2013	63.5

Source: National Park Service.

26. *Attendance at NCAA Women's College Basketball Games*

Season* (x)	Attendance (y)
2011	11,159,999
2012	11,210,832
2013	11,339,285
2014	11,181,735

Source: NCAA.

*Each season overlaps the given year with the previous year.

27.

28.

29.

30.

31.

32.

Decide whether each relation defines y as a function of x. Give the domain and range.
See Example 5.

33. $y = x^2$ **34.** $y = x^3$ **35.** $x = y^6$

36. $x = y^4$ **37.** $y = 2x - 5$ **38.** $y = -6x + 4$

39. $x + y < 3$ **40.** $x - y < 4$ **41.** $y = \sqrt{x}$

42. $y = -\sqrt{x}$ **43.** $xy = 2$ **44.** $xy = -6$

45. $y = \sqrt{4x + 1}$ **46.** $y = \sqrt{7 - 2x}$

47. $y = \dfrac{2}{x - 3}$ **48.** $y = \dfrac{-7}{x - 5}$

49. *Concept Check* Choose the correct answer: For function f, the notation $f(3)$ means
  A. the variable f times 3, or $3f$.
  B. the value of the dependent variable when the independent variable is 3.
  C. the value of the independent variable when the dependent variable is 3.
  D. f equals 3.

50. *Concept Check* Give an example of a function from everyday life. (*Hint:* Fill in the blanks: _____ depends on _____, so _____ is a function of _____.)

*Let $f(x) = -3x + 4$ and $g(x) = -x^2 + 4x + 1$. Find each of the following. Simplify if necessary. **See Example 6.***

51. $f(0)$ **52.** $f(-3)$ **53.** $g(-2)$

54. $g(10)$ **55.** $f\left(\dfrac{1}{3}\right)$ **56.** $f\left(-\dfrac{7}{3}\right)$

57. $g\left(\dfrac{1}{2}\right)$ **58.** $g\left(-\dfrac{1}{4}\right)$ **59.** $f(p)$

60. $g(k)$ **61.** $f(-x)$ **62.** $g(-x)$

63. $f(x + 2)$ **64.** $f(a + 4)$ **65.** $f(2m - 3)$ **66.** $f(3t - 2)$

*For each function, find **(a)** $f(2)$ and **(b)** $f(-1)$. **See Example 7.***

67. $f = \{(-1, 3), (4, 7), (0, 6), (2, 2)\}$ **68.** $f = \{(2, 5), (3, 9), (-1, 11), (5, 3)\}$

69.

70.

71.

72.

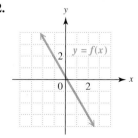

Use the graph of y = f(x) to find each function value: (a) f(−2), (b) f(0), (c) f(1), and (d) f(4). See Example 7(d).

73.

74.

75.

76.

An equation that defines y as a function of x is given. (a) Rewrite each equation using function notation f(x). (b) Find f(3). See Example 8.

77. $x + 3y = 12$

78. $x - 4y = 8$

79. $y + 2x^2 = 3 - x$

80. $y - 3x^2 = 2 + x$

81. $4x - 3y = 8$

82. $-2x + 5y = 9$

Concept Check *Answer each question.*

83. If $(3, 4)$ is on the graph of $y = f(x)$, which one of the following must be true: $f(3) = 4$ or $f(4) = 3$?

84. The figure shows a portion of the graph of

$$f(x) = x^2 + 3x + 1$$

and a rectangle with its base on the *x*-axis and a vertex on the graph. What is the area of the rectangle? (*Hint:* $f(0.2)$ is the height.)

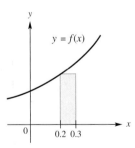

85. The graph of $y_1 = f(x)$ is shown with a display at the bottom. What is $f(3)$?

86. The graph of $y_1 = f(x)$ is shown with a display at the bottom. What is $f(-2)$?

Determine the largest open intervals of the domain over which each function is **(a)** *increasing,* **(b)** *decreasing, and* **(c)** *constant.* ***See Example 9.***

87.

88.

89.

90.

91.

92.

Solve each problem. ***See Example 10.***

93. *Electricity Usage* The graph shows the daily megawatts of electricity used on a record-breaking summer day in Sacramento, California.

  (a) Is this the graph of a function?

  (b) What is the domain?

  (c) Estimate the number of megawatts used at 8 A.M.

  (d) At what time was the most electricity used? the least electricity?

  (e) Call this function f. What is $f(12)$? Interpret your answer.

  (f) During what time intervals is usage increasing? decreasing?

Electricity Use

Source: Sacramento Municipal Utility District.

94. *Height of a Ball* A ball is thrown straight up into the air. The function $y = h(t)$ in the graph gives the height of the ball (in feet) at t seconds. (*Note:* The graph does *not* show the path of the ball. The ball is rising straight up and then falling straight down.)

  (a) What is the height of the ball at 2 sec?

  (b) When will the height be 192 ft?

  (c) During what time intervals is the ball going up? down?

  (d) How high does the ball go? When does the ball reach its maximum height?

  (e) After how many seconds does the ball hit the ground?

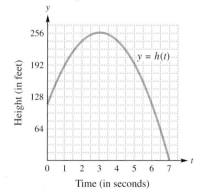

Height of a Thrown Ball

95. *Temperature* The graph shows temperatures on a given day in Bratenahl, Ohio.

    (a) At what times during the day was the temperature over 55°?

    (b) When was the temperature at or below 40°?

    (c) Greenville, South Carolina, is 500 mi south of Bratenahl, Ohio, and its temperature is 7° higher all day long. At what time was the temperature in Greenville the same as the temperature at noon in Bratenahl?

    (d) Use the graph to give a word description of the 24-hr period in Bratenahl.

Temperature in Bratenahl, Ohio

96. *Drug Levels in the Bloodstream* When a drug is taken orally, the amount of the drug in the bloodstream after t hours is given by the function $y = f(t)$, as shown in the graph.

    (a) How many units of the drug are in the bloodstream at 8 hr?

    (b) During what time interval is the level of the drug in the bloodstream increasing? decreasing?

    (c) When does the level of the drug in the bloodstream reach its maximum value, and how many units are in the bloodstream at that time?

    (d) When the drug reaches its maximum level in the bloodstream, how many additional hours are required for the level to drop to 16 units?

    (e) Use the graph to give a word description of the 12-hr period.

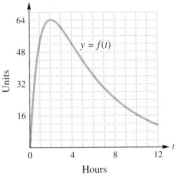

Drug Levels in the Bloodstream

2.4 Linear Functions

- Basic Concepts of Linear Functions
- Standard Form $Ax + By = C$
- Slope
- Average Rate of Change
- Linear Models

Basic Concepts of Linear Functions We begin our study of specific functions by looking at *linear* functions.

Linear Function

A function f is a **linear function** if, for real numbers a and b,

$$f(x) = ax + b.$$

If $a \neq 0$, then the domain and the range of f are both $(-\infty, \infty)$.

Lines can be graphed by finding ordered pairs and plotting them. Although only two points are necessary to graph a linear function, we usually plot a third point as a check. The intercepts are often good points to choose for graphing lines.

EXAMPLE 1 **Graphing a Linear Function Using Intercepts**

Graph $f(x) = -2x + 6$. Give the domain and range.

SOLUTION The x-intercept is found by letting $f(x) = 0$ and solving for x.

$$f(x) = -2x + 6$$

$$0 = -2x + 6 \quad \text{Let } f(x) = 0.$$

$$x = 3 \quad \text{Add } 2x \text{ and divide by 2.}$$

We plot the x-intercept $(3, 0)$. The y-intercept is found by evaluating $f(0)$.

$$f(0) = -2(0) + 6 \quad \text{Let } x = 0.$$

$$f(0) = 6 \quad \text{Simplify.}$$

Therefore, another point on the graph is the y-intercept, $(0, 6)$. We plot this point and join the two points with a straight-line graph. We use the point $(2, 2)$ as a check. See **Figure 31.** The domain and the range are both $(-\infty, \infty)$.

The corresponding calculator graph with $f(x) = y_1$ is shown in **Figure 32.**

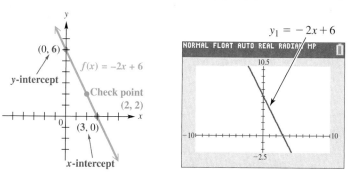

Figure 31   **Figure 32**

✔ **Now Try Exercise 13.**

If $a = 0$ in the definition of linear function, then the equation becomes $f(x) = b$. In this case, the domain is $(-\infty, \infty)$ and the range is $\{b\}$. A function of the form $\boldsymbol{f(x) = b}$ is a **constant function,** and its graph is a horizontal line.

EXAMPLE 2 **Graphing a Horizontal Line**

Graph $f(x) = -3$. Give the domain and range.

SOLUTION Because $f(x)$, or y, always equals -3, the value of y can never be 0 and the graph has no x-intercept. If a straight line has no x-intercept then it must be parallel to the x-axis, as shown in **Figure 33.** The domain of this linear function is $(-\infty, \infty)$ and the range is $\{-3\}$. **Figure 34** shows the calculator graph.

Figure 33   **Figure 34**

✔ **Now Try Exercise 17.**

Vertical line

(−3, 0)

x = −3

Figure 35

EXAMPLE 3 **Graphing a Vertical Line**

Graph $x = -3$. Give the domain and range of this relation.

SOLUTION Because x always equals -3, the value of x can never be 0, and the graph has no y-intercept. Using reasoning similar to that of **Example 2,** we find that this graph is parallel to the y-axis, as shown in **Figure 35.** The domain of this relation, which is *not* a function, is $\{-3\}$, while the range is $(-\infty, \infty)$.

✔ **Now Try Exercise 25.**

Standard Form $Ax + By = C$ Equations of lines are often written in the form $Ax + By = C$, known as **standard form.**

NOTE The definition of "standard form" is, ironically, not standard from one text to another. Any linear equation can be written in infinitely many different, but equivalent, forms. For example, the equation $2x + 3y = 8$ can be written equivalently as

$$2x + 3y - 8 = 0, \quad 3y = 8 - 2x, \quad x + \frac{3}{2}y = 4, \quad 4x + 6y = 16,$$

and so on. In this text we will agree that if the coefficients and constant in a linear equation are rational numbers, then we will consider the standard form to be $Ax + By = C$, where $A \geq 0$, A, B, and C are integers, and the greatest common factor of A, B, and C is 1. If $A = 0$, then we choose $B > 0$. (If two or more integers have a greatest common factor of 1, they are said to be **relatively prime**.)

EXAMPLE 4 **Graphing $Ax + By = C$ ($C = 0$)**

Graph $4x - 5y = 0$. Give the domain and range.

SOLUTION Find the intercepts.

$4x - 5y = 0$	$4x - 5y = 0$
$4(0) - 5y = 0$ Let $x = 0$.	$4x - 5(0) = 0$ Let $y = 0$.
$y = 0$ The y-intercept is $(0, 0)$.	$x = 0$ The x-intercept is $(0, 0)$.

The graph of this function has just one intercept—the origin $(0, 0)$. We need to find an additional point to graph the function by choosing a different value for x (or y).

$$4(5) - 5y = 0 \quad \text{We choose } x = 5.$$
$$20 - 5y = 0 \quad \text{Multiply.}$$
$$4 = y \quad \text{Add 5y. Divide by 5.}$$

This leads to the ordered pair $(5, 4)$. Complete the graph using the two points $(0, 0)$ and $(5, 4)$, with a third point as a check. The domain and range are both $(-\infty, \infty)$. See **Figure 36.**

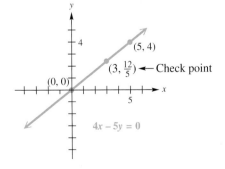

Figure 36

✔ **Now Try Exercise 23.**

Figure 37

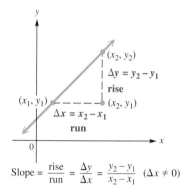

$$\text{Slope} = \frac{\text{rise}}{\text{run}} = \frac{\Delta y}{\Delta x} = \frac{y_2 - y_1}{x_2 - x_1} \quad (\Delta x \neq 0)$$

Figure 38

To use a graphing calculator to graph a linear function, as in **Figure 37,** we must first solve the defining equation for y.

$$4x - 5y = 0 \qquad \text{Equation from \textbf{Example 4}}$$

$$y = \frac{4}{5}x \qquad \text{Subtract } 4x. \text{ Divide by } -5.$$

Slope   Slope is a numerical measure of the steepness and orientation of a straight line. (Geometrically, this may be interpreted as the ratio of **rise** to **run.**) The slope of a highway (sometimes called the *grade*) is often given as a percent. For example, a 10% $\left(\text{or } \frac{10}{100} = \frac{1}{10}\right)$ slope means the highway rises 1 unit for every 10 horizontal units.

To find the slope of a line, start with two distinct points (x_1, y_1) and (x_2, y_2) on the line, as shown in **Figure 38,** where $x_1 \neq x_2$. As we move along the line from (x_1, y_1) to (x_2, y_2), the horizontal difference

$$\Delta x = x_2 - x_1$$

is the **change in x,** denoted by Δx (read **"delta x"**), where Δ is the Greek capital letter **delta.** The vertical difference, the **change in y,** can be written

$$\Delta y = y_2 - y_1.$$

The *slope* of a nonvertical line is defined as the quotient (ratio) of the change in y and the change in x, as follows.

Slope

The **slope** m of the line through the points (x_1, y_1) and (x_2, y_2) is given by the following.

$$m = \frac{\text{rise}}{\text{run}} = \frac{\Delta y}{\Delta x} = \frac{y_2 - y_1}{x_2 - x_1}, \quad \text{where } \Delta x \neq 0$$

That is, the slope of a line is the change in y divided by the corresponding change in x, where the change in x is not 0.

CAUTION *When using the slope formula, it makes no difference which point is (x_1, y_1) or (x_2, y_2). However, be consistent.* Start with the x- and y-values of *one* point (either one), and subtract the corresponding values of the *other* point.

$$Use \quad \frac{y_2 - y_1}{x_2 - x_1} \quad or \quad \frac{y_1 - y_2}{x_1 - x_2}, \quad not \quad \frac{y_2 - y_1}{x_1 - x_2} \quad or \quad \frac{y_1 - y_2}{x_2 - x_1}.$$

Be sure to write the difference of the y-values in the numerator and the difference of the x-values in the denominator.

The slope of a line can be found only if the line is nonvertical. This guarantees that $x_2 \neq x_1$ so that the denominator $x_2 - x_1 \neq 0$.

Undefined Slope

The slope of a vertical line is undefined.

EXAMPLE 5 Finding Slopes with the Slope Formula

Find the slope of the line through the given points.

(a) $(-4, 8), (2, -3)$ **(b)** $(2, 7), (2, -4)$ **(c)** $(5, -3), (-2, -3)$

SOLUTION

(a) Let $x_1 = -4$, $y_1 = 8$, and $x_2 = 2$, $y_2 = -3$.

$$m = \frac{\text{rise}}{\text{run}} = \frac{\Delta y}{\Delta x} \qquad \text{Definition of slope}$$

$$= \frac{-3 - 8}{2 - (-4)} \qquad \boxed{\text{Substitute carefully.}}$$

$$= \frac{-11}{6}, \quad \text{or} \quad -\frac{11}{6} \qquad \text{Subtract; } \frac{-a}{b} = -\frac{a}{b}.$$

We can also subtract in the opposite order, letting $x_1 = 2, y_1 = -3$ and $x_2 = -4, y_2 = 8$. The same slope results.

$$m = \frac{8 - (-3)}{-4 - 2} = \frac{11}{-6}, \quad \text{or} \quad -\frac{11}{6}$$

(b) If we attempt to use the slope formula with the points $(2, 7)$ and $(2, -4)$, we obtain a zero denominator.

$$m = \frac{-4 - 7}{2 - 2} = \frac{-11}{0} \qquad \textbf{Undefined}$$

The formula is not valid here because $\Delta x = x_2 - x_1 = 2 - 2 = 0$. A sketch would show that the line through $(2, 7)$ and $(2, -4)$ is vertical. As mentioned above, the slope of a vertical line is undefined.

(c) For $(5, -3)$ and $(-2, -3)$, the slope equals 0.

$$m = \frac{-3 - (-3)}{-2 - 5} = \frac{0}{-7} = 0$$

A sketch would show that the line through $(5, -3)$ and $(-2, -3)$ is horizontal.

✔ **Now Try Exercises 41, 47, and 49.**

The results in **Example 5(c)** suggest the following generalization.

LOOKING AHEAD TO CALCULUS

The **derivative** of a function provides a formula for determining the slope of a line tangent to a curve. If the slope is positive on a given interval, then the function is increasing there. If it is negative, then the function is decreasing. If it is 0, then the function is constant.

Slope Equal to Zero

The slope of a horizontal line is 0.

Theorems for similar triangles can be used to show that the slope of a line is independent of the choice of points on the line. *That is, slope is the same no matter which pair of distinct points on the line are used to find it.*

If the equation of a line is in the form

$$y = ax + b,$$

we can show that the slope of the line is a. To do this, we use function notation and the definition of slope.

$$m = \frac{f(x_2) - f(x_1)}{x_2 - x_1}$$ 　　　Slope formula

$$m = \frac{[a(x + 1) + b] - (ax + b)}{(x + 1) - x}$$ 　　　Let $f(x) = ax + b$, $x_1 = x$, and $x_2 = x + 1$.

$$m = \frac{ax + a + b - ax - b}{x + 1 - x}$$ 　　　Distributive property

$$m = \frac{a}{1}$$ 　　　Combine like terms.

$$m = a$$ 　　　The slope is a.

This discussion enables us to find the slope of the graph of any linear equation by solving for y and identifying the coefficient of x, which is the slope.

EXAMPLE 6 　Finding Slope from an Equation

Find the slope of the line $4x + 3y = 12$.

SOLUTION 　Solve the equation for y.

$$4x + 3y = 12$$

$$3y = -4x + 12 \quad \text{Subtract } 4x.$$

> Be careful to divide *each* term by 3.

$$y = -\frac{4}{3}x + 4 \quad \text{Divide by 3.}$$

The slope is $-\frac{4}{3}$, which is the coefficient of x when the equation is solved for y.

✔ **Now Try Exercise 55(a).**

Because the slope of a line is the ratio of vertical change (rise) to horizontal change (run), if we know the slope of a line and the coordinates of a point on the line, we can draw the graph of the line.

EXAMPLE 7 　Graphing a Line Using a Point and the Slope

Graph the line passing through the point $(-1, 5)$ and having slope $-\frac{5}{3}$.

SOLUTION 　First locate the point $(-1, 5)$ as shown in **Figure 39**. The slope of this line is $\frac{-5}{3}$, so a change of -5 units vertically (that is, 5 units *down*) corresponds to a change of 3 units horizontally (that is, 3 units to the *right*). This gives a second point, $(2, 0)$, which can then be used to complete the graph.

Because $\frac{-5}{3} = \frac{5}{-3}$, another point could be obtained by starting at $(-1, 5)$ and moving 5 units *up* and 3 units to the *left*. We would reach a different second point, $(-4, 10)$, but the graph would be the same line. Confirm this in **Figure 39**.

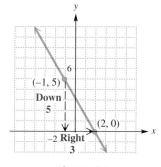

Figure 39

✔ **Now Try Exercise 59.**

Figure 40 shows lines with various slopes.

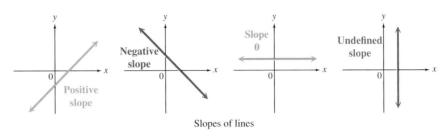

Slopes of lines

Figure 40

Notice the following important concepts.

- A line with a positive slope rises from left to right. The corresponding linear function is increasing on its entire domain.

- A line with a negative slope falls from left to right. The corresponding linear function is decreasing on its entire domain.

- A line with slope 0 neither rises nor falls. The corresponding linear function is constant on its entire domain.

- The slope of a vertical line is undefined.

Average Rate of Change   We know that the slope of a line is the ratio of the vertical change in *y* to the horizontal change in *x*. ***Thus, slope gives the average rate of change in y per unit of change in x, where the value of y depends on the value of x.*** If *f* is a linear function defined on the interval $[a, b]$, then we have the following.

$$\text{Average rate of change on } [a, b] = \frac{f(b) - f(a)}{b - a}$$

This is simply another way to write the slope formula, using function notation.

EXAMPLE 8   **Interpreting Slope as Average Rate of Change**

In 2009, Google spent $2800 million on research and development. In 2013, Google spent $8000 million on research and development. Assume a linear relationship, and find the average rate of change in the amount of money spent on R&D per year. Graph as a line segment, and interpret the result. (*Source:* MIT Technology Review.)

SOLUTION   To use the slope formula, we need two ordered pairs. Here, If *x* = 2009, then *y* = 2800, and if *x* = 2013, then *y* = 8000. This gives the two ordered pairs (2009, 2800) and (2013, 8000). (Here *y* is in millions of dollars.)

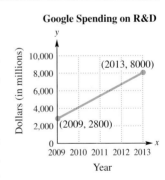

Google Spending on R&D

Figure 41

$$\text{Average rate of change} = \frac{8000 - 2800}{2013 - 2009} = \frac{5200}{4} = 1300$$

The graph in **Figure 41** confirms that the line through the ordered pairs rises from left to right and therefore has positive slope. Thus, the annual amount of money spent by Google on R&D *increased* by an average of about of $1300 million each year from 2009 to 2013.

✔ **Now Try Exercise 75.**

Linear Models   In **Example 8**, we used the graph of a line to approximate real data, a process known as **mathematical modeling.** Points on the straight-line graph model, or approximate, the actual points that correspond to the data.

A **linear cost function** has the form

$$C(x) = mx + b, \quad \text{Linear cost function}$$

where x represents the number of items produced, m represents the **cost** per item, and b represents the **fixed cost.** The fixed cost is constant for a particular product and does not change as more items are made. The value of mx, which increases as more items are produced, covers labor, materials, packaging, shipping, and so on.

The **revenue function** for selling a product depends on the price per item p and the number of items sold x. It is given by the following function.

$$R(x) = px \quad \text{Revenue function}$$

Profit is found by subtracting cost from revenue and is described by the **profit function.**

$$P(x) = R(x) - C(x) \quad \text{Profit function}$$

In applications we are often interested in values of x that will assure that profit is a positive number. In such cases we solve $R(x) - C(x) > 0$.

EXAMPLE 9   **Writing Linear Cost, Revenue, and Profit Functions**

Assume that the cost to produce an item is a linear function and all items produced are sold. The fixed cost is $1500, the variable cost per item is $100, and the item sells for $125. Write linear functions to model each of the following.

(a) cost   **(b)** revenue   **(c)** profit

(d) How many items must be sold for the company to make a profit?

SOLUTION

(a) The cost function is linear, so it will have the following form.

$$C(x) = mx + b \quad \text{Cost function}$$
$$C(x) = 100x + 1500 \quad \text{Let } m = 100 \text{ and } b = 1500.$$

(b) The revenue function is defined by the product of 125 and x.

$$R(x) = px \quad \text{Revenue function}$$
$$R(x) = 125x \quad \text{Let } p = 125.$$

(c) The profit function is found by subtracting the cost function from the revenue function.

$$P(x) = R(x) - C(x)$$
$$P(x) = 125x - (100x + 1500) \quad \text{Use parentheses here.}$$
$$P(x) = 125x - 100x - 1500 \quad \text{Distributive property}$$
$$P(x) = 25x - 1500 \quad \text{Combine like terms.}$$

ALGEBRAIC SOLUTION

(d) To make a profit, $P(x)$ must be positive.

$$P(x) = 25x - 1500 \quad \text{Profit function from part (c)}$$

Set $P(x) > 0$ and solve.

$$P(x) > 0$$
$$25x - 1500 > 0 \quad P(x) = 25x - 1500$$
$$25x > 1500 \quad \text{Add 1500 to each side.}$$
$$x > 60 \quad \text{Divide by 25.}$$

The number of items must be a whole number, so at least 61 items must be sold for the company to make a profit.

GRAPHING CALCULATOR SOLUTION

(d) Define y_1 as $25x - 1500$ and graph the line. Use the capability of a calculator to locate the x-intercept. See **Figure 42.** As the graph shows, y-values for x less than 60 are negative, and y-values for x greater than 60 are positive, so at least 61 items must be sold for the company to make a profit.

Figure 42

✔ **Now Try Exercise 77.**

CAUTION In problems involving $R(x) - C(x)$ like **Example 9(c),** pay attention to the use of parentheses around the expression for $C(x)$.

2.4 Exercises

CONCEPT PREVIEW *Match the description in Column I with the correct response in Column II. Some choices may not be used.*

I	II
1. a linear function whose graph has y-intercept $(0, 6)$	**A.** $f(x) = 5x$
2. a vertical line	**B.** $f(x) = 3x + 6$
3. a constant function	**C.** $f(x) = -8$
4. a linear function whose graph has x-intercept $(-2, 0)$ and y-intercept $(0, 4)$	**D.** $f(x) = x^2$
	E. $x + y = -6$
5. a linear function whose graph passes through the origin	**F.** $f(x) = 3x + 4$
	G. $2x - y = -4$
6. a function that is not linear	**H.** $x = 9$

CONCEPT PREVIEW *For each given slope, identify the line in A–D that could have this slope.*

7. -3 **8.** 0 **9.** 3 **10.** undefined

A.   **B.**   **C.**   **D.**

Graph each linear function. Give the domain and range. Identify any constant functions.
See Examples 1 and 2.

11. $f(x) = x - 4$ **12.** $f(x) = -x + 4$ **13.** $f(x) = \dfrac{1}{2}x - 6$

14. $f(x) = \dfrac{2}{3}x + 2$ **15.** $f(x) = 3x$ **16.** $f(x) = -2x$

17. $f(x) = -4$ **18.** $f(x) = 3$ **19.** $f(x) = 0$

20. *Concept Check* Write the equation of the linear function f with graph having slope 9 and passing through the origin. Give the domain and range.

Graph each line. Give the domain and range. **See Examples 3 and 4.**

21. $-4x + 3y = 12$ **22.** $2x + 5y = 10$ **23.** $3y - 4x = 0$

24. $3x + 2y = 0$ **25.** $x = 3$ **26.** $x = -4$

27. $2x + 4 = 0$ **28.** $-3x + 6 = 0$

29. $-x + 5 = 0$ **30.** $3 + x = 0$

Match each equation with the sketch that most closely resembles its graph. **See Examples 2 and 3.**

31. $y = 5$ **32.** $y = -5$ **33.** $x = 5$ **34.** $x = -5$

A.   **B.**   **C.**   **D.**

Use a graphing calculator to graph each equation in the standard viewing window. **See Examples 1, 2, and 4.**

35. $y = 3x + 4$ **36.** $y = -2x + 3$

37. $3x + 4y = 6$ **38.** $-2x + 5y = 10$

39. *Concept Check* If a walkway rises 2.5 ft for every 10 ft on the horizontal, which of the following express its slope (or grade)? (There are several correct choices.)

    A. 0.25   **B.** 4   **C.** $\dfrac{2.5}{10}$ **D.** 25%

    E. $\dfrac{1}{4}$ **F.** $\dfrac{10}{2.5}$ **G.** 400%   **H.** 2.5%

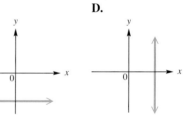

40. *Concept Check* If the pitch of a roof is $\frac{1}{4}$, how many feet in the horizontal direction correspond to a rise of 4 ft?

Find the slope of the line satisfying the given conditions. **See Example 5.**

41. through $(2, -1)$ and $(-3, -3)$ **42.** through $(-3, 4)$ and $(2, -8)$

43. through $(5, 8)$ and $(3, 12)$ **44.** through $(5, -3)$ and $(1, -7)$

45. through $(5, 9)$ and $(-2, 9)$ **46.** through $(-2, 4)$ and $(6, 4)$

47. horizontal, through $(5, 1)$ **48.** horizontal, through $(3, 5)$

49. vertical, through $(4, -7)$ **50.** vertical, through $(-8, 5)$

For each line, **(a)** *find the slope and* **(b)** *sketch the graph.* **See Examples 6 and 7.**

51. $y = 3x + 5$ **52.** $y = 2x - 4$ **53.** $2y = -3x$

54. $-4y = 5x$ **55.** $5x - 2y = 10$ **56.** $4x + 3y = 12$

Graph the line passing through the given point and having the indicated slope. Plot two points on the line. **See Example 7.**

57. through $(-1, 3)$, $m = \frac{3}{2}$ **58.** through $(-2, 8)$, $m = \frac{2}{5}$

59. through $(3, -4)$, $m = -\frac{1}{3}$ **60.** through $(-2, -3)$, $m = -\frac{3}{4}$

61. through $\left(-\frac{1}{2}, 4\right)$, $m = 0$ **62.** through $\left(\frac{3}{2}, 2\right)$, $m = 0$

63. through $\left(-\frac{5}{2}, 3\right)$, undefined slope   **64.** through $\left(\frac{9}{4}, 2\right)$, undefined slope

Concept Check *Find and interpret the average rate of change illustrated in each graph.*

65.

66.

67.

68.

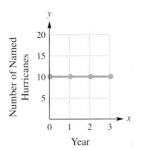

Solve each problem. **See Example 8.**

69. ***Dropouts*** In 1980, the number of high school dropouts in the United States was 5085 thousand. By 2012, this number had decreased to 2562 thousand. Find and interpret the average rate of change per year in the number of high school dropouts. Round the answer to the nearest tenth. (*Source: 2013 Digest of Education Statistics.*)

70. ***Plasma Flat-Panel TV Sales*** The total amount spent on plasma flat-panel TVs in the United States changed from $5302 million in 2006 to $1709 million in 2013. Find and interpret the average rate of change in sales, in millions of dollars per year. Round the answer to the nearest hundredth. (*Source: Consumer Electronics Association.*)

71. *(Modeling) Olympic Times for 5000-Meter Run* The graph shows the winning times (in minutes) at the Olympic Games for the men's 5000-m run, together with a linear approximation of the data.

Olympic Times for 5000-Meter Run (in minutes)

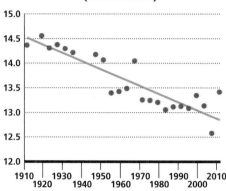

Source: World Almanac and Book of Facts.

(a) An equation for a linear model, based on data from 1912–2012 (where *x* represents the year), is

$$y = -0.0167x + 46.45.$$

Determine the slope. (**See Example 6.**) What does the slope of this line represent? Why is the slope negative?

(b) What reason might explain why there are no data points for the years 1916, 1940, and 1944?

(c) The winning time for the 2000 Olympic Games was 13.35 min. What does the model predict to the nearest hundredth? How far is the prediction from the actual value?

72. *(Modeling) U.S. Radio Stations* The graph shows the number of U.S. radio stations on the air, along with the graph of a linear function that models the data.

U.S. Radio Stations

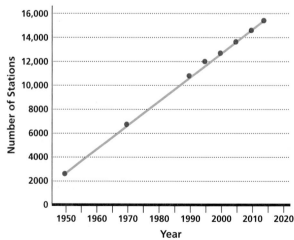

Source: Federal Communications Commission.

(a) An equation for a linear model, based on data from 1950–2014 (where $x = 0$ represents 1950, $x = 10$ represents 1960, and so on) is

$$y = 200.02x + 2727.7.$$

Determine the slope. (**See Example 6.**) What does the slope of this line represent? Why is the slope positive?

(b) Use the model in part (a) to predict the number of stations in 2018.

73. *Cellular Telephone Subscribers* The table gives the number of cellular telephone subscribers in the U.S. (in thousands) from 2008 through 2013. Find the average annual rate of change during this time period. Round to the nearest unit.

Year	Subscribers (in thousands)
2008	270,334
2009	285,646
2010	296,286
2011	315,964
2012	326,475
2013	335,652

Source: CTIA-The Wireless Association.

74. *Earned Run Average* In 2006, in an effort to end the so-called "steroid era," Major League Baseball introduced a strict drug-testing policy in order to discourage players from using performance-enhancing drugs. The table shows how overall earned run average, or ERA, changed from 2006 through 2014. Find the average annual rate of change, to the nearest thousandth, during this period.

Year	ERA
2006	4.53
2007	4.47
2008	4.32
2009	4.32
2010	4.08
2011	3.94
2012	4.01
2013	3.87
2014	3.74

Source: www.baseball-reference.com

75. *Mobile Homes* The graph provides a good approximation of the number of mobile homes (in thousands) placed in use in the United States from 2003 through 2013.

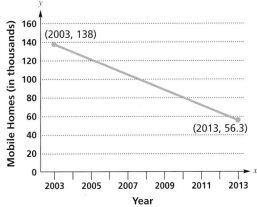

Mobile Homes Placed in Use

Source: U.S. Census Bureau.

(a) Use the given ordered pairs to find the average rate of change in the number of mobile homes per year during this period.

(b) Interpret what a negative slope means in this situation.

76. ***Teen Birth Rates*** In 1991, there were 61.8 births per thousand for adolescent females aged 15–19. By 2013, this number had decreased to 26.6 births per thousand. Find and interpret the average annual rate of change in teen births per year for this period. Round the answer to the nearest tenth. (*Source:* U.S. Department of Health and Human Services.)

(Modeling) Cost, Revenue, and Profit Analysis A firm will break even (no profit and no loss) as long as revenue just equals cost. The value of x (the number of items produced and sold) where C(x) = R(x) is the **break-even point.** *Assume that each of the following can be expressed as a linear function. Find*

(a) *the cost function,*   (b) *the revenue function, and*   (c) *the profit function.*

(d) *Find the break-even point and decide whether the product should be produced, given the restrictions on sales.*

See Example 9.

	Fixed Cost	Variable Cost	Price of Item	
77.	$ 500	$ 10	$ 35	No more than 18 units can be sold.
78.	$2700	$150	$280	No more than 25 units can be sold.
79.	$1650	$400	$305	All units produced can be sold.
80.	$ 180	$ 11	$ 20	No more than 30 units can be sold.

(Modeling) Break-Even Point The manager of a small company that produces roof tile has determined that the total cost in dollars, C(x), of producing x units of tile is given by

$$C(x) = 200x + 1000,$$

while the revenue in dollars, R(x), from the sale of x units of tile is given by

$$R(x) = 240x.$$

81. Find the break-even point and the cost and revenue at the break-even point.

82. Suppose the variable cost is actually $220 per unit, instead of $200. How does this affect the break-even point? Is the manager better off or not?

Relating Concepts

For individual or collaborative investigation *(Exercises 83-92)*

The table shows several points on the graph of a linear function. **Work Exercises 83–92 in order,** *to see connections between the slope formula, distance formula, midpoint formula, and linear functions.*

x	y
0	−6
1	−3
2	0
3	3
4	6
5	9
6	12

83. Use the first two points in the table to find the slope of the line.

84. Use the second and third points in the table to find the slope of the line.

85. Make a conjecture by filling in the blank: If we use any two points on a line to find its slope, we find that the slope is _____ in all cases.

86. Find the distance between the first two points in the table. (*Hint:* Use the distance formula.)

87. Find the distance between the second and fourth points in the table.

88. Find the distance between the first and fourth points in the table.

89. Add the results in **Exercises 86 and 87,** and compare the sum to the answer found in **Exercise 88.** What do you notice?

90. Fill in each blank, basing the answers on observations in **Exercises 86–89:**

If points A, B, and C lie on a line in that order, then the distance between A and B added to the distance between _____ and _____ is equal to the distance between _____ and _____.

91. Find the midpoint of the segment joining $(0, -6)$ and $(6, 12)$. Compare the answer to the middle entry in the table. What do you notice?

92. If the table were set up to show an x-value of 4.5, what would be the corresponding y-value?

Chapter 2   Quiz (Sections 2.1–2.4)

1. For $A(-4, 2)$ and $B(-8, -3)$, find $d(A, B)$, the distance between A and B.

2. *Two-Year College Enrollment* Enrollments in two-year colleges for selected years are shown in the table. Use the midpoint formula to estimate the enrollments for 2006 and 2010.

Year	Enrollment (in millions)
2004	6.55
2008	6.97
2012	7.50

Source: National Center for Education Statistics.

3. Graph $y = -x^2 + 4$ by plotting points.

4. Graph $x^2 + y^2 = 16$.

5. Determine the radius and the coordinates of the center of the circle with equation

$$x^2 + y^2 - 4x + 8y + 3 = 0.$$

For Exercises 6–8, refer to the graph of $f(x) = |x + 3|$.

6. Find $f(-1)$.

7. Give the domain and the range of f.

8. Give the largest open interval over which the function f is
  (a) decreasing,   **(b)** increasing,   **(c)** constant.

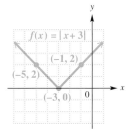

9. Find the slope of the line through the given points.
  (a) $(1, 5)$ and $(5, 11)$ **(b)** $(-7, 4)$ and $(-1, 4)$ **(c)** $(6, 12)$ and $(6, -4)$

10. *Motor Vehicle Sales* The graph shows a straight line segment that approximates new motor vehicle sales in the United States from 2009 to 2013. Determine the average rate of change from 2009 to 2013, and interpret the results.

Source: U.S. Bureau of Economic Analysis.

2.5 Equations of Lines and Linear Models

- Point-Slope Form
- Slope-Intercept Form
- Vertical and Horizontal Lines
- Parallel and Perpendicular Lines
- Modeling Data
- Graphical Solution of Linear Equations in One Variable

Point-Slope Form The graph of a linear function is a straight line. We now develop various forms for the equation of a line.

Figure 43 shows the line passing through the fixed point (x_1, y_1) having slope m. (Assuming that the line has a slope guarantees that it is not vertical.) Let (x, y) be any other point on the line. Because the line is not vertical, $x - x_1 \neq 0$. Now use the definition of slope.

Figure 43

$$m = \frac{y - y_1}{x - x_1} \qquad \text{Slope formula}$$

$$m(x - x_1) = y - y_1 \qquad \text{Multiply each side by } x - x_1.$$

$$y - y_1 = m(x - x_1) \qquad \text{Interchange sides.}$$

This result is the *point-slope form* of the equation of a line.

LOOKING AHEAD TO CALCULUS
A standard problem in calculus is to find the equation of the line tangent to a curve at a given point. The derivative (see *Looking Ahead to Calculus* earlier in this chapter) is used to find the slope of the desired line, and then the slope and the given point are used in the point-slope form to solve the problem.

Point-Slope Form

The **point-slope form** of the equation of the line with slope m passing through the point (x_1, y_1) is given as follows.

$$y - y_1 = m(x - x_1)$$

EXAMPLE 1 Using the Point-Slope Form (Given a Point and the Slope)

Write an equation of the line through the point $(-4, 1)$ having slope -3.

SOLUTION Here $x_1 = -4$, $y_1 = 1$, and $m = -3$.

$$y - y_1 = m(x - x_1) \qquad \text{Point-slope form}$$

$$y - 1 = -3[x - (-4)] \qquad x_1 = -4, y_1 = 1, m = -3$$

$$y - 1 = -3(x + 4) \qquad \text{Be careful with signs.}$$

$$y - 1 = -3x - 12 \qquad \text{Distributive property}$$

$$y = -3x - 11 \qquad \text{Add 1.} \qquad \checkmark \text{ Now Try Exercise 29.}$$

EXAMPLE 2 Using the Point-Slope Form (Given Two Points)

Write an equation of the line through the points $(-3, 2)$ and $(2, -4)$. Write the result in standard form $Ax + By = C$.

SOLUTION Find the slope first.

$$m = \frac{-4 - 2}{2 - (-3)} = -\frac{6}{5} \qquad \text{Definition of slope}$$

The slope m is $-\frac{6}{5}$. Either the point $(-3, 2)$ or the point $(2, -4)$ can be used for (x_1, y_1). We choose $(-3, 2)$.

$$y - y_1 = m(x - x_1) \qquad \text{Point-slope form}$$

$$y - 2 = -\frac{6}{5}[x - (-3)] \quad x_1 = -3, y_1 = 2, m = -\frac{6}{5}$$

$$5(y - 2) = -6(x + 3) \qquad \text{Multiply by 5.}$$

$$5y - 10 = -6x - 18 \qquad \text{Distributive property}$$

$$6x + 5y = -8 \qquad \text{Standard form}$$

Verify that we obtain the same equation if we use $(2, -4)$ instead of $(-3, 2)$ in the point-slope form.

✔ **Now Try Exercise 19.**

NOTE The lines in **Examples 1 and 2** both have negative slopes. Keep in mind that a slope of the form $-\frac{A}{B}$ may be interpreted as either $\frac{-A}{B}$ or $\frac{A}{-B}$.

Slope-Intercept Form   As a special case of the point-slope form of the equation of a line, suppose that a line has y-intercept $(0, b)$. If the line has slope m, then using the point-slope form with $x_1 = 0$ and $y_1 = b$ gives the following.

$$y - y_1 = m(x - x_1) \qquad \text{Point-slope form}$$

$$y - b = m(x - 0) \qquad x_1 = 0, y_1 = b$$

$$y - b = mx \qquad \text{Distributive property}$$

$$y = mx + b \qquad \text{Solve for } y.$$

Slope ⟶ ⟵ The y-intercept is $(0, b)$.

Because this result shows the slope of the line and indicates the y-intercept, it is known as the *slope-intercept form* of the equation of the line.

Slope-Intercept Form

The **slope-intercept form** of the equation of the line with slope m and y-intercept $(0, b)$ is given as follows.

$$y = mx + b$$

EXAMPLE 3   **Finding Slope and y-Intercept from an Equation of a Line**

Find the slope and y-intercept of the line with equation $4x + 5y = -10$.

SOLUTION   Write the equation in slope-intercept form.

$$4x + 5y = -10$$

$$5y = -4x - 10 \qquad \text{Subtract } 4x.$$

$$y = -\frac{4}{5}x - 2 \qquad \text{Divide by 5.}$$

↑ ↑
m b

The slope is $-\frac{4}{5}$, and the y-intercept is $(0, -2)$.   ✔ **Now Try Exercise 37.**

NOTE   Generalizing from **Example 3,** we see that the slope m of the graph of the equation

$$Ax + By = C$$

is $-\frac{A}{B}$, and the y-intercept is $\left(0, \frac{C}{B}\right)$.

EXAMPLE 4 **Using the Slope-Intercept Form (Given Two Points)**

Write an equation of the line through the points $(1, 1)$ and $(2, 4)$. Then graph the line using the slope-intercept form.

SOLUTION In **Example 2,** we used the *point-slope form* in a similar problem. Here we show an alternative method using the *slope-intercept form*. First, find the slope.

$$m = \frac{4 - 1}{2 - 1} = \frac{3}{1} = 3 \quad \text{Definition of slope}$$

Now substitute 3 for m in $y = mx + b$ and choose one of the given points, say $(1, 1)$, to find the value of b.

$$y = mx + b \quad \text{Slope-intercept form}$$
$$1 = 3(1) + b \quad m = 3, x = 1, y = 1$$

The y-intercept is $(0, b)$. $\rightarrow b = -2$ \quad Solve for b.

The slope-intercept form is

$$y = 3x - 2.$$

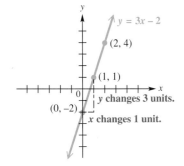

Figure 44

The graph is shown in **Figure 44.** We can plot $(0, -2)$ and then use the definition of slope to arrive at $(1, 1)$. Verify that $(2, 4)$ also lies on the line.

✔ **Now Try Exercise 19.**

EXAMPLE 5 **Finding an Equation from a Graph**

Use the graph of the linear function f shown in **Figure 45** to complete the following.

(a) Identify the slope, y-intercept, and x-intercept.

(b) Write an equation that defines f.

SOLUTION

Figure 45

(a) The line falls 1 unit each time the x-value increases 3 units. Therefore, the slope is $\frac{-1}{3} = -\frac{1}{3}$. The graph intersects the y-axis at the y-intercept $(0, -1)$ and the x-axis at the x-intercept $(-3, 0)$.

(b) The slope is $m = -\frac{1}{3}$, and the y-intercept is $(0, -1)$.

$$y = f(x) = mx + b \quad \text{Slope-intercept form}$$
$$f(x) = -\frac{1}{3}x - 1 \quad m = -\frac{1}{3}, b = -1$$

✔ **Now Try Exercise 45.**

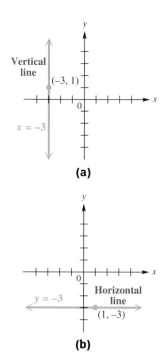

(a)

(b)

Figure 46

Vertical and Horizontal Lines   The vertical line through the point (a, b) passes through all points of the form (a, y), for any value of y. Consequently, the equation of a vertical line through (a, b) is $x = a$. For example, the vertical line through $(-3, 1)$ has equation $x = -3$. See **Figure 46(a).** Because each point on the y-axis has x-coordinate 0, *the equation of the y-axis is $x = 0$.*

The horizontal line through the point (a, b) passes through all points of the form (x, b), for any value of x. Therefore, the equation of a horizontal line through (a, b) is $y = b$. For example, the horizontal line through $(1, -3)$ has equation $y = -3$. See **Figure 46(b).** Because each point on the x-axis has y-coordinate 0, *the equation of the x-axis is $y = 0$.*

Equations of Vertical and Horizontal Lines

An equation of the **vertical line** through the point (a, b) is $x = a.$

An equation of the **horizontal line** through the point (a, b) is $y = b.$

Parallel and Perpendicular Lines   Two parallel lines are equally "steep," so they should have the same slope. Also, two distinct lines with the same "steepness" are parallel. The following result summarizes this discussion. (The statement "p if and only if q" means "if p then q *and* if q then p.")

Parallel Lines

Two distinct nonvertical lines are parallel if and only if they have the same slope.

When two lines have slopes with a product of -1, the lines are perpendicular.

Perpendicular Lines

Two lines, neither of which is vertical, are perpendicular if and only if their slopes have a product of -1. Thus, the slopes of perpendicular lines, neither of which is vertical, are *negative reciprocals*.

Example: If the slope of a line is $-\frac{3}{4}$, then the slope of any line perpendicular to it is $\frac{4}{3}$ because

$$-\frac{3}{4}\left(\frac{4}{3}\right) = -1.$$

(Numbers like $-\frac{3}{4}$ and $\frac{4}{3}$ are **negative reciprocals** of each other.) A proof of this result is outlined in **Exercises 79–85.**

NOTE   Because a vertical line has *undefined* slope, it does not follow the *mathematical* rules for parallel and perpendicular lines. We intuitively know that all vertical lines are parallel and that a vertical line and a horizontal line are perpendicular.

EXAMPLE 6 Finding Equations of Parallel and Perpendicular Lines

Write an equation in both slope-intercept and standard form of the line that passes through the point $(3, 5)$ and satisfies the given condition.

(a) parallel to the line $2x + 5y = 4$

(b) perpendicular to the line $2x + 5y = 4$

SOLUTION

(a) We know that the point $(3, 5)$ is on the line, so we need only find the slope to use the point-slope form. We find the slope by writing the equation of the given line in slope-intercept form. (That is, we solve for y.)

$$2x + 5y = 4$$

$$5y = -2x + 4 \qquad \text{Subtract } 2x.$$

$$y = -\frac{2}{5}x + \frac{4}{5} \qquad \text{Divide by 5.}$$

The slope is $-\frac{2}{5}$. Because the lines are parallel, $-\frac{2}{5}$ is also the slope of the line whose equation is to be found. Now substitute this slope and the given point $(3, 5)$ in the point-slope form.

$$y - y_1 = m(x - x_1) \qquad \text{Point-slope form}$$

$$y - 5 = -\frac{2}{5}(x - 3) \qquad m = -\tfrac{2}{5}, x_1 = 3, y_1 = 5$$

$$y - 5 = -\frac{2}{5}x + \frac{6}{5} \qquad \text{Distributive property}$$

$$\text{Slope-intercept form} \longrightarrow y = -\frac{2}{5}x + \frac{31}{5} \qquad \text{Add } 5 = \tfrac{25}{5}.$$

$$5y = -2x + 31 \qquad \text{Multiply by 5.}$$

$$\text{Standard form} \longrightarrow 2x + 5y = 31 \qquad \text{Add } 2x.$$

(b) There is no need to find the slope again—in part (a) we found that the slope of the line $2x + 5y = 4$ is $-\frac{2}{5}$. The slope of any line perpendicular to it is $\frac{5}{2}$.

$$y - y_1 = m(x - x_1) \qquad \text{Point-slope form}$$

$$y - 5 = \frac{5}{2}(x - 3) \qquad m = \tfrac{5}{2}, x_1 = 3, y_1 = 5$$

$$y - 5 = \frac{5}{2}x - \frac{15}{2} \qquad \text{Distributive property}$$

$$\text{Slope-intercept form} \longrightarrow y = \frac{5}{2}x - \frac{5}{2} \qquad \text{Add } 5 = \tfrac{10}{2}.$$

$$2y = 5x - 5 \qquad \text{Multiply by 2.}$$

$$-5x + 2y = -5 \qquad \text{Subtract } 5x.$$

$$\text{Standard form} \longrightarrow 5x - 2y = 5 \qquad \text{Multiply by } -1 \text{ so that } A > 0.$$

✔ **Now Try Exercises 51 and 53.**

We can use a graphing calculator to support the results of **Example 6.** In **Figure 47(a),** we graph the equations of the parallel lines

$$y_1 = -\frac{2}{5}x + \frac{4}{5} \quad \text{and} \quad y_2 = -\frac{2}{5}x + \frac{31}{5}. \quad \text{See Example 6(a)}.$$

The lines appear to be parallel, giving visual support for our result. We must use caution, however, when viewing such graphs, as the limited resolution of a graphing calculator screen may cause two lines to *appear* to be parallel even when they are not. For example, **Figure 47(b)** shows the graphs of the equations

$$y_1 = 2x + 6 \quad \text{and} \quad y_2 = 2.01x - 3$$

in the standard viewing window, and they appear to be parallel. This is not the case, however, because their slopes, 2 and 2.01, are different.

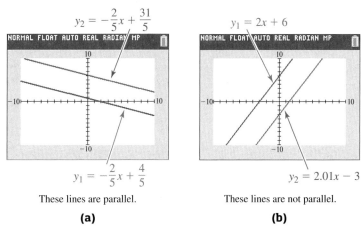

These lines are parallel.	These lines are not parallel.
(a)	**(b)**

Figure 47

Now we graph the equations of the perpendicular lines

$$y_1 = -\frac{2}{5}x + \frac{4}{5} \quad \text{and} \quad y_2 = \frac{5}{2}x - \frac{5}{2}. \quad \text{See Example 6(b)}.$$

If we use the standard viewing window, the lines do not appear to be perpendicular. See **Figure 48(a).** To obtain the correct perspective, we must use a square viewing window, as in **Figure 48(b).**

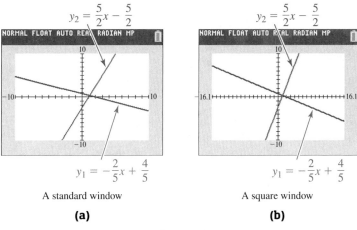

A standard window	A square window
(a)	**(b)**

Figure 48

A summary of the various forms of linear equations follows.

Summary of Forms of Linear Equations

Equation	Description	When to Use
$y = mx + b$	**Slope-Intercept Form** Slope is m. y-intercept is $(0, b)$.	The slope and y-intercept can be easily identified and used to quickly graph the equation. This form can also be used to find the equation of a line given a point and the slope.
$y - y_1 = m(x - x_1)$	**Point-Slope Form** Slope is m. Line passes through (x_1, y_1).	This form is ideal for finding the equation of a line if the slope and a point on the line or two points on the line are known.
$Ax + By = C$	**Standard Form** (If the coefficients and constant are rational, then A, B, and C are expressed as relatively prime integers, with $A \geq 0$.) Slope is $-\frac{A}{B}$ $(B \neq 0)$. x-intercept is $\left(\frac{C}{A}, 0\right)$ $(A \neq 0)$. y-intercept is $\left(0, \frac{C}{B}\right)$ $(B \neq 0)$.	The x- and y-intercepts can be found quickly and used to graph the equation. The slope must be calculated.
$y = b$	**Horizontal Line** Slope is 0. y-intercept is $(0, b)$.	If the graph intersects only the y-axis, then y is the only variable in the equation.
$x = a$	**Vertical Line** Slope is undefined. x-intercept is $(a, 0)$.	If the graph intersects only the x-axis, then x is the only variable in the equation.

Modeling Data   We can write equations of lines that mathematically describe, or model, real data if the data change at a fairly constant rate. In this case, the data fit a linear pattern, and the rate of change is the slope of the line.

EXAMPLE 7 **Finding an Equation of a Line That Models Data**

Average annual tuition and fees for in-state students at public four-year colleges are shown in the table for selected years and graphed as ordered pairs of points in **Figure 49**, where $x = 0$ represents 2009, $x = 1$ represents 2010, and so on, and y represents the cost in dollars. This graph of ordered pairs of data is a **scatter diagram.**

Year	Cost (in dollars)
2009	6312
2010	6695
2011	7136
2012	7703
2013	8070

Source: National Center for Education Statistics.

Figure 49

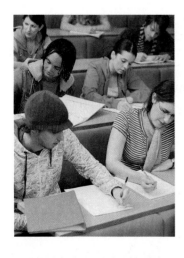

(a) Find an equation that models the data.

(b) Use the equation from part (a) to estimate the cost of tuition and fees at public four-year colleges in 2015.

SOLUTION

(a) The points in **Figure 49** lie approximately on a straight line, so we can write a linear equation that models the relationship between year x and cost y. We choose two data points, $(0, 6312)$ and $(4, 8070)$, to find the slope of the line.

$$m = \frac{8070 - 6312}{4 - 0} = \frac{1758}{4} = 439.5$$

The slope 439.5 indicates that the cost of tuition and fees increased by about $440 per year from 2009 to 2013. We use this slope, the y-intercept $(0, 6312)$, and the slope-intercept form to write an equation of the line.

$$y = mx + b \qquad \text{Slope-intercept form}$$

$$y = 439.5x + 6312 \qquad \text{Substitute for } m \text{ and } b.$$

(b) The value $x = 6$ corresponds to the year 2015, so we substitute 6 for x.

$$y = 439.5x + 6312 \qquad \text{Model from part (a)}$$

$$y = 439.5(6) + 6312 \qquad \text{Let } x = 6.$$

$$y = 8949 \qquad \text{Multiply, and then add.}$$

The model estimates that average tuition and fees for in-state students at public four-year colleges in 2015 were about $8949.

✔ **Now Try Exercise 63(a) and (b).**

NOTE   In **Example 7,** if we had chosen different data points, we would have obtained a slightly different equation.

Guidelines for Modeling

Step 1 Make a scatter diagram of the data.

Step 2 Find an equation that models the data. For a line, this involves selecting two data points and finding the equation of the line through them.

Linear regression is a technique from statistics that provides the line of "best fit." **Figure 50** shows how a TI-84 Plus calculator accepts the data points, calculates the equation of this line of best fit (in this case, $y_1 = 452.4x + 6278.4$), and plots the data points and line on the same screen.

(a)

(b)

(c)

Figure 50

Graphical Solution of Linear Equations in One Variable Suppose that y_1 and y_2 are linear expressions in x. We can solve the equation $y_1 = y_2$ graphically as follows (assuming it has a unique solution).

1. Rewrite the equation as $y_1 - y_2 = 0$.

2. Graph the linear function $y_3 = y_1 - y_2$.

3. Find the x-intercept of the graph of the function y_3. This x-value is the solution of $y_1 = y_2$.

Some calculators use the term *zero* to identify the x-value of an x-intercept. In general, if $f(a) = 0$, then a is a **zero** of f.

Figure 51

EXAMPLE 8 **Solving an Equation with a Graphing Calculator**

Use a graphing calculator to solve $-2x - 4(2 - x) = 3x + 4$.

SOLUTION We write an equivalent equation with 0 on one side.

$$-2x - 4(2 - x) - 3x - 4 = 0 \quad \text{Subtract } 3x \text{ and } 4.$$

Then we graph $y = -2x - 4(2 - x) - 3x - 4$ to find the x-intercept. The standard viewing window cannot be used because the x-intercept does not lie in the interval $[-10, 10]$. As seen in **Figure 51,** the solution of the equation is -12, and the solution set is $\{-12\}$.

✔ **Now Try Exercise 69.**

2.5 **Exercises**

CONCEPT PREVIEW *Fill in the blank(s) to correctly complete each sentence.*

1. The graph of the line $y - 3 = 4(x - 8)$ has slope _____ and passes through the point $(8, $ ___$)$.

2. The graph of the line $y = -2x + 7$ has slope _____ and y-intercept _____.

3. The vertical line through the point $(-4, 8)$ has equation _____ $= -4$.

4. The horizontal line through the point $(-4, 8)$ has equation _____ $= 8$.

5. Any line parallel to the graph of $6x + 7y = 9$ must have slope _____.

6. Any line perpendicular to the graph of $6x + 7y = 9$ must have slope _____.

CONCEPT PREVIEW *Match each equation with its graph in A–D.*

7. $y = \dfrac{1}{4}x + 2$

8. $4x + 3y = 12$

9. $y - (-1) = \dfrac{3}{2}(x - 1)$

10. $y = 4$

A.

B.

C.

D.

Write an equation for each line described. Give answers in standard form for Exercises 11–20 and in slope-intercept form (if possible) for Exercises 21–32. ***See Examples 1–4.***

11. through $(1, 3)$, $m = -2$

12. through $(2, 4)$, $m = -1$

13. through $(-5, 4)$, $m = -\frac{3}{2}$

14. through $(-4, 3)$, $m = \frac{3}{4}$

15. through $(-8, 4)$, undefined slope

16. through $(5, 1)$, undefined slope

17. through $(5, -8)$, $m = 0$

18. through $(-3, 12)$, $m = 0$

19. through $(-1, 3)$ and $(3, 4)$

20. through $(2, 3)$ and $(-1, 2)$

21. x-intercept $(3, 0)$, y-intercept $(0, -2)$

22. x-intercept $(-4, 0)$, y-intercept $(0, 3)$

23. vertical, through $(-6, 4)$

24. vertical, through $(2, 7)$

25. horizontal, through $(-7, 4)$

26. horizontal, through $(-8, -2)$

27. $m = 5$, $b = 15$

28. $m = -2$, $b = 12$

29. through $(-2, 5)$ having slope -4

30. through $(4, -7)$ having slope -2

31. slope 0, y-intercept $\left(0, \frac{3}{2}\right)$

32. slope 0, y-intercept $\left(0, -\frac{5}{4}\right)$

33. *Concept Check* Fill in each blank with the appropriate response:

The line $x + 2 = 0$ has x-intercept _____ . It _____ have a
            (does/does not)
y-intercept. The slope of this line is _____ .
            (0/undefined)
The line $4y = 2$ has y-intercept _____ . It _____ have an
            (does/does not)
x-intercept. The slope of this line is _____ .
            (0/undefined)

34. *Concept Check* Match each equation with the line that would most closely resemble its graph. (*Hint:* Consider the signs of m and b in the slope-intercept form.)

(a) $y = 3x + 2$ **(b)** $y = -3x + 2$ **(c)** $y = 3x - 2$ **(d)** $y = -3x - 2$

A.   **B.**   **C.**   **D.**

Find the slope and y-intercept of each line, and graph it. ***See Example 3.***

35. $y = 3x - 1$

36. $y = -2x + 7$

37. $4x - y = 7$

38. $2x + 3y = 16$

39. $4y = -3x$

40. $2y = x$

41. $x + 2y = -4$

42. $x + 3y = -9$

43. $y - \frac{3}{2}x - 1 = 0$

44. *Concept Check* The table represents a linear function f.

(a) Find the slope of the graph of $y = f(x)$.

(b) Find the y-intercept of the line.

(c) Write an equation for this line in slope-intercept form.

x	y
-2	-11
-1	-8
0	-5
1	-2
2	1
3	4

Connecting Graphs with Equations The graph of a linear function f is shown. *(a)* Identify the slope, y-intercept, and x-intercept. *(b)* Write an equation that defines f. *See Example 5.*

45.

46.

47.

48.

49.

50.

Write an equation **(a)** *in standard form and* **(b)** *in slope-intercept form for each line described.* *See Example 6.*

51. through $(-1, 4)$, parallel to $x + 3y = 5$

52. through $(3, -2)$, parallel to $2x - y = 5$

53. through $(1, 6)$, perpendicular to $3x + 5y = 1$

54. through $(-2, 0)$, perpendicular to $8x - 3y = 7$

55. through $(4, 1)$, parallel to $y = -5$

56. through $(-2, -2)$, parallel to $y = 3$

57. through $(-5, 6)$, perpendicular to $x = -2$

58. through $(4, -4)$, perpendicular to $x = 4$

Work each problem.

59. Find k so that the line through $(4, -1)$ and $(k, 2)$ is

    (a) parallel to $3y + 2x = 6$ **(b)** perpendicular to $2y - 5x = 1$.

60. Find r so that the line through $(2, 6)$ and $(-4, r)$ is

    (a) parallel to $2x - 3y = 4$ **(b)** perpendicular to $x + 2y = 1$.

(Modeling) *Solve each problem.* *See Example 7.*

61. *Annual Tuition and Fees* Refer to the table that accompanies **Figure 49** in **Example 7.**

    (a) Use the data points $(0, 6312)$ and $(3, 7703)$ to find a linear equation that models the data.

    (b) Use the equation from part (a) to estimate average tuition and fees for in-state students at public four-year colleges in 2013. How does the result compare to the actual figure given in the table, $8070?

62. *Annual Tuition and Fees* Refer to the table that accompanies **Figure 49** in **Example 7.**

   (a) Use the data points for the years 2009 and 2011 to find a linear equation that models the data.

   (b) Use the equation from part (a) to estimate average tuition and fees for in-state students at public four-year colleges in 2013. How does the result compare to the actual figure given in the table, $8070?

63. *Cost of Private College Education* The table lists average annual cost (in dollars) of tuition and fees at private four-year colleges for selected years.

Year	Cost (in dollars)
2009	22,036
2010	21,908
2011	22,771
2012	23,460
2013	24,525

Source: National Center for Education Statistics.

   (a) Determine a linear function $f(x) = ax + b$ that models the data, where $x = 0$ represents 2009, $x = 1$ represents 2010, and so on. Use the points $(0, 22{,}036)$ and $(4, 24{,}525)$ to graph f and a scatter diagram of the data on the same coordinate axes. (Use a graphing calculator if desired.) What does the slope of the graph indicate?

   (b) Use the function from part (a) to approximate average tuition and fees in 2012. Compare the approximation to the actual figure given in the table, $23,460.

   (c) Use the linear regression feature of a graphing calculator to find the equation of the line of best fit.

64. *Distances and Velocities of Galaxies* The table lists the distances (in megaparsecs; 1 megaparsec $= 3.085 \times 10^{24}$ cm, and 1 megaparsec $= 3.26$ million light-years) and velocities (in kilometers per second) of four galaxies moving rapidly away from Earth.

Galaxy	Distance	Velocity
Virgo	15	1600
Ursa Minor	200	15,000
Corona Borealis	290	24,000
Bootes	520	40,000

Source: Acker, A., and C. Jaschek, *Astronomical Methods and Calculations,* John Wiley and Sons. Karttunen, H. (editor), *Fundamental Astronomy,* Springer-Verlag.

   (a) Plot the data using distances for the *x*-values and velocities for the *y*-values. What type of relationship seems to hold between the data?

   (b) Find a linear equation in the form $y = mx$ that models these data using the points $(520, 40{,}000)$ and $(0, 0)$. Graph the equation with the data on the same coordinate axes.

   (c) The galaxy Hydra has a velocity of 60,000 km per sec. How far away, to the nearest megaparsec, is it according to the model in part (b)?

   (d) The value of m is the **Hubble constant.** The Hubble constant can be used to estimate the age of the universe A (in years) using the formula

$$A = \frac{9.5 \times 10^{11}}{m}.$$

   Approximate A using the value of m. Round to the nearest hundredth of a billion years.

   (e) Astronomers currently place the value of the Hubble constant between 50 and 100. What is the range for the age of the universe A?

65. *Celsius and Fahrenheit Temperatures* When the Celsius temperature is 0°, the corresponding Fahrenheit temperature is 32°. When the Celsius temperature is 100°, the corresponding Fahrenheit temperature is 212°. Let C represent the Celsius temperature and F the Fahrenheit temperature.

(a) Express F as an exact linear function of C.

(b) Solve the equation in part (a) for C, thus expressing C as a function of F.

(c) For what temperature is $F = C$ a true statement?

66. *Water Pressure on a Diver* The pressure P of water on a diver's body is a linear function of the diver's depth, x. At the water's surface, the pressure is 1 atmosphere. At a depth of 100 ft, the pressure is about 3.92 atmospheres.

(a) Find a linear function that relates P to x.

(b) Compute the pressure, to the nearest hundredth, at a depth of 10 fathoms (60 ft).

67. *Consumption Expenditures* In Keynesian macroeconomic theory, total consumption expenditure on goods and services, C, is assumed to be a linear function of national personal income, I. The table gives the values of C and I for 2009 and 2013 in the United States (in billions of dollars).

Year	2009	2013
Total consumption (C)	$10,089	$11,484
National income (I)	$12,026	$14,167

Source: U.S. Bureau of Economic Analysis.

(a) Find a formula for C as a function of I.

(b) The slope of the linear function found in part (a) is the **marginal propensity to consume.** What is the marginal propensity to consume for the United States from 2009–2013?

68. *Concept Check* A graph of $y = f(x)$ is shown in the standard viewing window. Which is the only value of x that could possibly be the solution of the equation $f(x) = 0$?

A. -15 **B.** 0 **C.** 5 **D.** 15

 Use a graphing calculator to solve each linear equation. ***See Example 8.***

69. $2x + 7 - x = 4x - 2$ **70.** $7x - 2x + 4 - 5 = 3x + 1$

71. $3(2x + 1) - 2(x - 2) = 5$

72. $4x - 3(4 - 2x) = 2(x - 3) + 6x + 2$

73. (a) Solve $-2(x - 5) = -x - 2$ using traditional paper-and-pencil methods.

 (b) Explain why the standard viewing window of a graphing calculator cannot graphically support the solution found in part (a). What minimum and maximum x-values would make it possible for the solution to be seen?

74. Use a graphing calculator to try to solve

$$-3(2x + 6) = -4x + 8 - 2x.$$

Explain what happens. What is the solution set?

If three distinct points A, B, and C in a plane are such that the slopes of nonvertical line segments AB, AC, and BC are equal, then A, B, and C are collinear. Otherwise, they are not. Use this fact to determine whether the three points given are collinear.

75. $(-1, 4), (-2, -1), (1, 14)$ **76.** $(0, -7), (-3, 5), (2, -15)$

77. $(-1, -3), (-5, 12), (1, -11)$ **78.** $(0, 9), (-3, -7), (2, 19)$

Relating Concepts

For individual or collaborative investigation *(Exercises 79–85)*

In this section we state that two lines, neither of which is vertical, are perpendicular if and only if their slopes have a product of -1. In Exercises 79–85, we outline a partial proof of this for the case where the two lines intersect at the origin. **Work these exercises in order,** *and refer to the figure as needed.*

By the converse of the Pythagorean theorem, if

$$[d(O, P)]^2 + [d(O, Q)]^2 = [d(P, Q)]^2,$$

then triangle POQ is a right triangle with right angle at O.

79. Find an expression for the distance $d(O, P)$.

80. Find an expression for the distance $d(O, Q)$.

81. Find an expression for the distance $d(P, Q)$.

82. Use the results from **Exercises 79–81,** and substitute into the equation from the Pythagorean theorem. Simplify to show that this leads to the equation

$$-2m_1m_2x_1x_2 - 2x_1x_2 = 0.$$

83. Factor $-2x_1x_2$ from the final form of the equation in **Exercise 82.**

84. Use the property that if $ab = 0$ then $a = 0$ or $b = 0$ to solve the equation in **Exercise 83,** showing that $m_1m_2 = -1$.

85. State a conclusion based on the results of **Exercises 79–84.**

Summary Exercises on Graphs, Circles, Functions, and Equations

These summary exercises provide practice with some of the concepts covered so far in this chapter.

    For the points P and Q, find **(a)** *the distance $d(P, Q)$,* **(b)** *the coordinates of the midpoint of the segment PQ, and* **(c)** *an equation for the line through the two points. Write the equation in slope-intercept form if possible.*

1. $P(3, 5), Q(2, -3)$ **2.** $P(-1, 0), Q(4, -2)$

3. $P(-2, 2), Q(3, 2)$ **4.** $P\left(2\sqrt{2}, \sqrt{2}\right), Q\left(\sqrt{2}, 3\sqrt{2}\right)$

5. $P(5, -1), Q(5, 1)$ **6.** $P(1, 1), Q(-3, -3)$

7. $P\left(2\sqrt{3}, 3\sqrt{5}\right), Q\left(6\sqrt{3}, 3\sqrt{5}\right)$ **8.** $P(0, -4), Q(3, 1)$

Write an equation for each of the following, and sketch the graph.

9. the line through $(-2, 1)$ and $(4, -1)$

10. the horizontal line through $(2, 3)$

11. the circle with center $(2, -1)$ and radius 3

12. the circle with center $(0, 2)$ and tangent to the *x*-axis

13. the line through $(3, -5)$ with slope $-\frac{5}{6}$

14. the vertical line through $(-4, 3)$

15. the line through $(-3, 2)$ and parallel to the line $2x + 3y = 6$

16. the line through the origin and perpendicular to the line $3x - 4y = 2$

Decide whether or not each equation has a circle as its graph. If it does, give the center and the radius.

17. $x^2 + y^2 - 4x + 2y = 4$ **18.** $x^2 + y^2 + 6x + 10y + 36 = 0$

19. $x^2 + y^2 - 12x + 20 = 0$ **20.** $x^2 + y^2 + 2x + 16y = -61$

21. $x^2 + y^2 - 2x + 10 = 0$ **22.** $x^2 + y^2 - 8y - 9 = 0$

Solve each problem.

23. Find the coordinates of the points of intersection of the line $y = 2$ and the circle with center at $(4, 5)$ and radius 4.

24. Find the shortest distance from the origin to the graph of the circle with equation

$$x^2 + y^2 - 10x - 24y + 144 = 0.$$

*For each relation, **(a)** find the domain and range, and **(b)** if the relation defines y as a function f of x, rewrite the relation using function notation and find $f(-2)$.*

25. $x - 4y = -6$ **26.** $y^2 - x = 5$

27. $(x + 2)^2 + y^2 = 25$ **28.** $x^2 - 2y = 3$

2.6   Graphs of Basic Functions

- Continuity
- The Identity, Squaring, and Cubing Functions
- The Square Root and Cube Root Functions
- The Absolute Value Function
- Piecewise-Defined Functions
- The Relation $x = y^2$

Continuity   The graph of a linear function—a straight line—may be drawn by hand over any interval of its domain without picking the pencil up from the paper. In mathematics we say that a function with this property is *continuous* over any interval. The formal definition of continuity requires concepts from calculus, but we can give an informal definition at the college algebra level.

Continuity (Informal Definition)

A function is **continuous** over an interval of its domain if its hand-drawn graph over that interval can be sketched without lifting the pencil from the paper.

The function is
discontinuous at $x = 2$.

Figure 52

If a function is not continuous at a *point,* then it has a *discontinuity* there. **Figure 52** shows the graph of a function with a discontinuity at the point where $x = 2$.

EXAMPLE 1 **Determining Intervals of Continuity**

Describe the intervals of continuity for each function in **Figure 53.**

(a)   (b)

Figure 53

SOLUTION   The function in **Figure 53(a)** is continuous over its entire domain, $(-\infty, \infty)$. The function in **Figure 53(b)** has a point of discontinuity at $x = 3$. Thus, it is continuous over the intervals

$$(-\infty, 3) \quad \text{and} \quad (3, \infty).$$

✔ **Now Try Exercises 11 and 15.**

Graphs of the basic functions studied in college algebra can be sketched by careful point plotting or generated by a graphing calculator. As you become more familiar with these graphs, you should be able to provide quick rough sketches of them.

The Identity, Squaring, and Cubing Functions   The **identity function** $f(x) = x$ pairs every real number with itself. See **Figure 54.**

Identity Function $f(x) = x$

Domain: $(-\infty, \infty)$ Range: $(-\infty, \infty)$

x	y
-2	-2
-1	-1
0	0
1	1
2	2

Figure 54

• $f(x) = x$ is increasing on its entire domain, $(-\infty, \infty)$.

• It is continuous on its entire domain, $(-\infty, \infty)$.

The **squaring function** $f(x) = x^2$ pairs each real number with its square. Its graph is a **parabola.** The point $(0, 0)$ at which the graph changes from decreasing to increasing is the **vertex** of the parabola. See **Figure 55.** (For a parabola that opens downward, the vertex is the point at which the graph changes from increasing to decreasing.)

Figure 55

- $f(x) = x^2$ decreases on the open interval $(-\infty, 0)$ and increases on the open interval $(0, \infty)$.

- It is continuous on its entire domain, $(-\infty, \infty)$.

The function $f(x) = x^3$ is the **cubing function.** It pairs each real number with the cube of the number. See **Figure 56.** The point $(0, 0)$ at which the graph changes from "opening downward" to "opening upward" is an **inflection point.**

Figure 56

- $f(x) = x^3$ increases on its entire domain, $(-\infty, \infty)$.

- It is continuous on its entire domain, $(-\infty, \infty)$.

The Square Root and Cube Root Functions   The function $f(x) = \sqrt{x}$ is the **square root function.** It pairs each real number with its principal square root. See **Figure 57.** For the function value to be a real number, the domain must be restricted to $[0, \infty)$.

Square Root Function $f(x) = \sqrt{x}$

Domain: $[0, \infty)$ Range: $[0, \infty)$

Figure 57

- $f(x) = \sqrt{x}$ increases on the open interval, $(0, \infty)$.
- It is continuous on its entire domain, $[0, \infty)$.

The **cube root function** $f(x) = \sqrt[3]{x}$ pairs each real number with its cube root. See **Figure 58.** The cube root function differs from the square root function in that *any* real number has a real number cube root. Thus, the domain is $(-\infty, \infty)$.

Cube Root Function $f(x) = \sqrt[3]{x}$

Domain: $(-\infty, \infty)$ Range: $(-\infty, \infty)$

Figure 58

- $f(x) = \sqrt[3]{x}$ increases on its entire domain, $(-\infty, \infty)$.
- It is continuous on its entire domain, $(-\infty, \infty)$.

The Absolute Value Function   The **absolute value function,** $f(x) = |x|$, which pairs every real number with its absolute value, is graphed in **Figure 59** on the next page and is defined as follows.

$$f(x) = |x| = \begin{cases} x & \text{if } x \geq 0 \\ -x & \text{if } x < 0 \end{cases}$$ Absolute value function

That is, we use $|x| = x$ if x is positive or 0, and we use $|x| = -x$ if x is negative.

Absolute Value Function $f(x) = |x|$

Domain: $(-\infty, \infty)$ Range: $[0, \infty)$

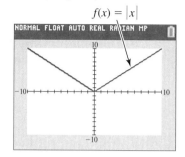

$f(x) = |x|$

Figure 59

- $f(x) = |x|$ decreases on the open interval $(-\infty, 0)$ and increases on the open interval $(0, \infty)$.

- It is continuous on its entire domain, $(-\infty, \infty)$.

Piecewise-Defined Functions   The absolute value function is a **piecewise-defined function.** It is defined by different rules over different intervals of its domain.

EXAMPLE 2   Graphing Piecewise-Defined Functions

Graph each function.

(a) $f(x) = \begin{cases} -2x + 5 & \text{if } x \le 2 \\ x + 1 & \text{if } x > 2 \end{cases}$ **(b)** $f(x) = \begin{cases} 2x + 3 & \text{if } x \le 0 \\ -x^2 + 3 & \text{if } x > 0 \end{cases}$

ALGEBRAIC SOLUTION

(a) We graph each interval of the domain separately. If $x \le 2$, the graph of $f(x) = -2x + 5$ has an endpoint at $x = 2$. We find the corresponding y-value by substituting 2 for x in $-2x + 5$ to obtain $y = 1$. To find another point on this part of the graph, we choose $x = 0$, so $y = 5$. We draw the graph through $(2, 1)$ and $(0, 5)$ as a partial line with endpoint $(2, 1)$.

We graph the function for $x > 2$ similarly, using $f(x) = x + 1$. This partial line has an open endpoint at $(2, 3)$. We use $y = x + 1$ to find another point with x-value greater than 2 to complete the graph. See **Figure 60.**

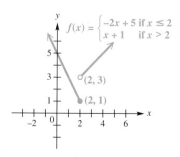

Figure 60

GRAPHING CALCULATOR SOLUTION

(a) We use the TEST feature of the TI-84 Plus to graph the piecewise-defined function. (Press [2ND] [MATH] to display a list containing inequality symbols.) The result of a true statement is 1, and the result of a false statement is 0. We choose x with the appropriate inequality based on how the function is defined. Next we multiply each defining expression by the test condition result. We then add these products to obtain the complete function.

The expression for the function in part (a) is shown at the top of the screen in **Figure 61.**

Figure 61

(b) First graph $f(x) = 2x + 3$ for $x \leq 0$. Then for $x > 0$, graph $f(x) = -x^2 + 3$. The two graphs meet at the point $(0, 3)$. See **Figure 62**.

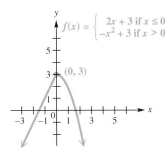

Figure 62

(b) Use the procedure described in part (a). The expression for the function is shown at the top of the screen in **Figure 63**.

Figure 63

☑️ **Now Try Exercises 23 and 29.**

Another piecewise-defined function is the *greatest integer function.*

LOOKING AHEAD TO CALCULUS

The **greatest integer function** is used in calculus as a classic example of how the limit of a function may not exist at a particular value in its domain. For a limit to exist, the left- and right-hand limits must be equal. We can see from the graph of the greatest integer function that for an integer value such as 3, as x approaches 3 from the left, function values are all 2, while as x approaches 3 from the right, function values are all 3. Because the left- and right-hand limits are different, the limit as x approaches 3 does not exist.

$f(x) = [\![x]\!]$

The **greatest integer function** $f(x) = [\![x]\!]$ pairs every real number x with the greatest integer less than or equal to x.

Examples: $[\![8.4]\!] = 8$, $[\![-5]\!] = -5$, $[\![\pi]\!] = 3$, $[\![-6.9]\!] = -7$.

The graph is shown in **Figure 64**. In general, if $f(x) = [\![x]\!]$, then

$$\begin{aligned}
\text{for } -2 \leq x < -1, \quad & f(x) = -2, \\
\text{for } -1 \leq x < 0, \quad & f(x) = -1, \\
\text{for } 0 \leq x < 1, \quad & f(x) = 0, \\
\text{for } 1 \leq x < 2, \quad & f(x) = 1, \\
\text{for } 2 \leq x < 3, \quad & f(x) = 2, \quad \text{and so on.}
\end{aligned}$$

Greatest Integer Function $\quad f(x) = [\![x]\!]$

Domain: $(-\infty, \infty)$

Range: $\{y \mid y \text{ is an integer}\} = \{\ldots, -3, -2, -1, 0, 1, 2, 3, \ldots\}$

x	y
-2	-2
-1.5	-2
-0.99	-1
0	0
0.001	0
3	3
3.99	3

The dots indicate that the graph continues indefinitely in the same pattern.

Figure 64

- $f(x) = [\![x]\!]$ is constant on the open intervals $\ldots, (-2, -1), (-1, 0), (0, 1), (1, 2), (2, 3), \ldots$.

- It is discontinuous at all integer values in its domain, $(-\infty, \infty)$.

EXAMPLE 3 Graphing a Greatest Integer Function

Graph $f(x) = \left[\!\left[\frac{1}{2}x + 1\right]\!\right]$.

Figure 65

SOLUTION If x is in the interval $[0, 2)$, then $y = 1$. For x in $[2, 4)$, $y = 2$, and so on. Some sample ordered pairs are given in the table.

x	0	$\frac{1}{2}$	1	$\frac{3}{2}$	2	3	4	-1	-2	-3
y	1	1	1	1	2	2	3	0	0	-1

These ordered pairs suggest the graph shown in **Figure 65.** The domain is $(-\infty, \infty)$. The range is $\{\ldots, -2, -1, 0, 1, 2, \ldots\}$.

✔ **Now Try Exercise 45.**

The greatest integer function is an example of a **step function,** a function with a graph that looks like a series of steps.

EXAMPLE 4 Applying the Greatest Integer Function Concept

An express mail company charges $25 for a package weighing up to 2 lb. For each additional pound or fraction of a pound, there is an additional charge of $3. Let $y = D(x)$ represent the cost to send a package weighing x pounds. Graph $y = D(x)$ for x in the interval $(0, 6]$.

SOLUTION For x in the interval $(0, 2]$, we obtain $y = 25$. For x in $(2, 3]$, $y = 25 + 3 = 28$. For x in $(3, 4]$, $y = 28 + 3 = 31$, and so on. The graph, which is that of a step function, is shown in **Figure 66.** In this case, the first step has a different length.

Figure 66

✔ **Now Try Exercise 47.**

The Relation $x = y^2$ *Recall that a function is a relation where every domain value is paired with one and only one range value.* Consider the relation defined by the equation $x = y^2$, which is not a function. Notice from the table of selected ordered pairs on the next page that this relation has two different y-values for each positive value of x.

If we plot the points from the table and join them with a smooth curve, we find that the graph of $x = y^2$ is a parabola opening to the right with vertex $(0, 0)$. See **Figure 67(a).** The domain is $[0, \infty)$ and the range is $(-\infty, \infty)$.

⊞ To use a calculator in function mode to graph the relation $x = y^2$, we graph the two functions $y_1 = \sqrt{x}$ (to generate the top half of the parabola) and $y_2 = -\sqrt{x}$ (to generate the bottom half). See **Figure 67(b).** ∎

Selected Ordered Pairs
for $x = y^2$

x	y
0	0
1	±1
4	±2
9	±3

There are two different y-values for the same x-value.

(a)

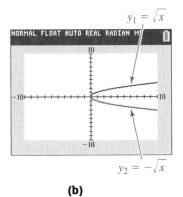

(b)

Figure 67

Exercises

CONCEPT PREVIEW *To answer each question, refer to the following basic graphs.*

A.

B.

C.

D.

E.

F.

G.

H.

I.

1. Which one is the graph of $f(x) = x^2$? What is its domain?

2. Which one is the graph of $f(x) = |x|$? On what open interval is it increasing?

3. Which one is the graph of $f(x) = x^3$? What is its range?

4. Which one is not the graph of a function? What is its equation?

5. Which one is the identity function? What is its equation?

6. Which one is the graph of $f(x) = [\![x]\!]$? What is the function value when $x = 1.5$?

7. Which one is the graph of $f(x) = \sqrt[3]{x}$? Is there any open interval over which the function is decreasing?

8. Which one is the graph of $f(x) = \sqrt{x}$? What is its domain?

9. Which one is discontinuous at many points? What is its range?

10. Which graphs of functions decrease over part of the domain and increase over the rest of the domain? On what open intervals do they increase? decrease?

*Determine the intervals of the domain over which each function is continuous. **See Example 1.***

11.

12.

13.

14.

15.

16.

*For each piecewise-defined function, find **(a)** $f(-5)$, **(b)** $f(-1)$, **(c)** $f(0)$, and **(d)** $f(3)$. **See Example 2.***

17. $f(x) = \begin{cases} 2x & \text{if } x \le -1 \\ x - 1 & \text{if } x > -1 \end{cases}$

18. $f(x) = \begin{cases} x - 2 & \text{if } x < 3 \\ 5 - x & \text{if } x \ge 3 \end{cases}$

19. $f(x) = \begin{cases} 2 + x & \text{if } x < -4 \\ -x & \text{if } -4 \le x \le 2 \\ 3x & \text{if } x > 2 \end{cases}$

20. $f(x) = \begin{cases} -2x & \text{if } x < -3 \\ 3x - 1 & \text{if } -3 \le x \le 2 \\ -4x & \text{if } x > 2 \end{cases}$

*Graph each piecewise-defined function. **See Example 2.***

21. $f(x) = \begin{cases} x - 1 & \text{if } x \le 3 \\ 2 & \text{if } x > 3 \end{cases}$

22. $f(x) = \begin{cases} 6 - x & \text{if } x \le 3 \\ 3 & \text{if } x > 3 \end{cases}$

23. $f(x) = \begin{cases} 4 - x & \text{if } x < 2 \\ 1 + 2x & \text{if } x \ge 2 \end{cases}$

24. $f(x) = \begin{cases} 2x + 1 & \text{if } x \ge 0 \\ x & \text{if } x < 0 \end{cases}$

25. $f(x) = \begin{cases} -3 & \text{if } x \le 1 \\ -1 & \text{if } x > 1 \end{cases}$

26. $f(x) = \begin{cases} -2 & \text{if } x \le 1 \\ 2 & \text{if } x > 1 \end{cases}$

27. $f(x) = \begin{cases} 2 + x & \text{if } x < -4 \\ -x & \text{if } -4 \le x \le 5 \\ 3x & \text{if } x > 5 \end{cases}$

28. $f(x) = \begin{cases} -2x & \text{if } x < -3 \\ 3x - 1 & \text{if } -3 \le x \le 2 \\ -4x & \text{if } x > 2 \end{cases}$

29. $f(x) = \begin{cases} -\dfrac{1}{2}x^2 + 2 & \text{if } x \le 2 \\ \dfrac{1}{2}x & \text{if } x > 2 \end{cases}$

30. $f(x) = \begin{cases} x^3 + 5 & \text{if } x \le 0 \\ -x^2 & \text{if } x > 0 \end{cases}$

31. $f(x) = \begin{cases} 2x & \text{if } -5 \le x < -1 \\ -2 & \text{if } -1 \le x < 0 \\ x^2 - 2 & \text{if } 0 \le x \le 2 \end{cases}$

32. $f(x) = \begin{cases} 0.5x^2 & \text{if } -4 \le x \le -2 \\ x & \text{if } -2 < x < 2 \\ x^2 - 4 & \text{if } 2 \le x \le 4 \end{cases}$

33. $f(x) = \begin{cases} x^3 + 3 & \text{if } -2 \le x \le 0 \\ x + 3 & \text{if } 0 < x < 1 \\ 4 + x - x^2 & \text{if } 1 \le x \le 3 \end{cases}$

34. $f(x) = \begin{cases} -2x & \text{if } -3 \le x < -1 \\ x^2 + 1 & \text{if } -1 \le x \le 2 \\ \dfrac{1}{2}x^3 + 1 & \text{if } 2 < x \le 3 \end{cases}$

Connecting Graphs with Equations *Give a rule for each piecewise-defined function. Also give the domain and range.*

35.

36.

37.

38.

39.

40.

41.

42.

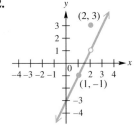

Graph each function. Give the domain and range. ***See Example 3.***

43. $f(x) = [\![-x]\!]$

44. $f(x) = -[\![x]\!]$

45. $f(x) = [\![2x]\!]$

46. $g(x) = [\![2x - 1]\!]$

(Modeling) Solve each problem. **See Example 4.**

47. *Postage Charges* Assume that postage rates are $0.49 for the first ounce, plus $0.21 for each additional ounce, and that each letter carries one $0.49 stamp and as many $0.21 stamps as necessary. Graph the function f that models the number of stamps on a letter weighing x ounces over the interval $(0, 5]$.

48. *Parking Charges* The cost of parking a car at an hourly parking lot is $3 for the first half-hour and $2 for each additional half-hour or fraction of a half-hour. Graph the function f that models the cost of parking a car for x hours over the interval $(0, 2]$.

49. *Water in a Tank* Sketch a graph that depicts the amount of water in a 100-gal tank. The tank is initially empty and then filled at a rate of 5 gal per minute. Immediately after it is full, a pump is used to empty the tank at 2 gal per minute.

50. *Distance from Home* Sketch a graph showing the distance a person is from home after x hours if he or she drives on a straight road at 40 mph to a park 20 mi away, remains at the park for 2 hr, and then returns home at a speed of 20 mph.

51. *New Truck Market Share* The new vehicle market share (in percent) in the United States for trucks is shown in the graph. Let $x = 0$ represent 2000, $x = 8$ represent 2008, and so on.

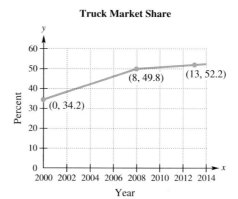

Truck Market Share

Source: U.S. Bureau of Economic Analysis.

(a) Use the points on the graph to write equations for the graphs in the intervals $[0, 8]$ and $(8, 13]$.

(b) Define this graph as a piecewise-defined function f.

52. *Flow Rates* A water tank has an inlet pipe with a flow rate of 5 gal per minute and an outlet pipe with a flow rate of 3 gal per minute. A pipe can be either closed or completely open. The graph shows the number of gallons of water in the tank after x minutes. Use the concept of slope to interpret each piece of this graph.

Water in a Tank

53. *Swimming Pool Levels* The graph of $y = f(x)$ represents the amount of water in thousands of gallons remaining in a swimming pool after x days.

Water in a Swimming Pool

(a) Estimate the initial and final amounts of water contained in the pool.

(b) When did the amount of water in the pool remain constant?

(c) Approximate $f(2)$ and $f(4)$.

(d) At what rate was water being drained from the pool when $1 \leq x \leq 3$?

54. *Gasoline Usage* The graph shows the gallons of gasoline y in the gas tank of a car after x hours.

(a) Estimate how much gasoline was in the gas tank when $x = 3$.

(b) When did the car burn gasoline at the greatest rate?

Gasoline Use

55. *Lumber Costs* Lumber that is used to frame walls of houses is frequently sold in multiples of 2 ft. If the length of a board is not exactly a multiple of 2 ft, there is often no charge for the additional length. For example, if a board measures at least 8 ft, but less than 10 ft, then the consumer is charged for only 8 ft.

(a) Suppose that the cost of lumber is $0.80 every 2 ft. Find a formula for a function f that computes the cost of a board x feet long for $6 \leq x \leq 18$.

(b) Determine the costs of boards with lengths of 8.5 ft and 15.2 ft.

56. *Snow Depth* The snow depth in a particular location varies throughout the winter. In a typical winter, the snow depth in inches might be approximated by the following function.

$$f(x) = \begin{cases} 6.5x & \text{if } 0 \leq x \leq 4 \\ -5.5x + 48 & \text{if } 4 < x \leq 6 \\ -30x + 195 & \text{if } 6 < x \leq 6.5 \end{cases}$$

Here, x represents the time in months with $x = 0$ representing the beginning of October, $x = 1$ representing the beginning of November, and so on.

(a) Graph $y = f(x)$.

(b) In what month is the snow deepest? What is the deepest snow depth?

(c) In what months does the snow begin and end?

2.7 Graphing Techniques

- Stretching and Shrinking
- Reflecting
- Symmetry
- Even and Odd Functions
- Translations

Graphing techniques presented in this section show how to graph functions that are defined by altering the equation of a basic function.

Stretching and Shrinking We begin by considering how the graphs of

$$y = af(x) \quad \text{and} \quad y = f(ax)$$

compare to the graph of $y = f(x)$, where $a > 0$.

EXAMPLE 1 Stretching or Shrinking Graphs

Graph each function.

(a) $g(x) = 2|x|$ (b) $h(x) = \dfrac{1}{2}|x|$ (c) $k(x) = |2x|$

SOLUTION

(a) Comparing the tables of values for $f(x) = |x|$ and $g(x) = 2|x|$ in **Figure 68,** we see that for corresponding x-values, the y-values of g are each twice those of f. The graph of $f(x) = |x|$ is *vertically stretched*. The graph of $g(x)$, shown in blue in **Figure 68,** is narrower than that of $f(x)$, shown in red for comparison.

x	$f(x) = \lvert x \rvert$	$g(x) = 2\lvert x \rvert$
-2	2	4
-1	1	2
0	0	0
1	1	2
2	2	4

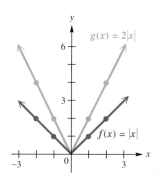

Figure 68

(b) The graph of $h(x) = \dfrac{1}{2}|x|$ is also the same general shape as that of $f(x)$, but here the coefficient $\dfrac{1}{2}$ is between 0 and 1 and causes a *vertical shrink*. The graph of $h(x)$ is wider than the graph of $f(x)$, as we see by comparing the tables of values. See **Figure 69.**

x	$f(x) = \lvert x \rvert$	$h(x) = \frac{1}{2}\lvert x \rvert$
-2	2	1
-1	1	$\frac{1}{2}$
0	0	0
1	1	$\frac{1}{2}$
2	2	1

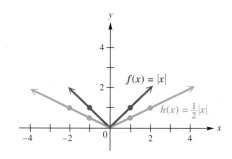

Figure 69

(c) Use the property of absolute value that states $|ab| = |a| \cdot |b|$ to rewrite $|2x|$.

$$k(x) = |2x| = |2| \cdot |x| = 2|x| \quad \text{Property 3}$$

Therefore, the graph of $k(x) = |2x|$ is the same as the graph of $g(x) = 2|x|$ in part (a). This is a *horizontal shrink* of the graph of $f(x) = |x|$. See **Figure 68** on the previous page.

✔ **Now Try Exercises 17 and 19.**

Vertical Stretching or Shrinking of the Graph of a Function

Suppose that $a > 0$. If a point (x, y) lies on the graph of $y = f(x)$, then the point (x, ay) lies on the graph of $y = af(x)$.

(a) If $a > 1$, then the graph of $y = af(x)$ is a **vertical stretching** of the graph of $y = f(x)$.

(b) If $0 < a < 1$, then the graph of $y = af(x)$ is a **vertical shrinking** of the graph of $y = f(x)$.

Figure 70 shows graphical interpretations of vertical stretching and shrinking. *In both cases, the x-intercepts of the graph remain the same but the y-intercepts are affected.*

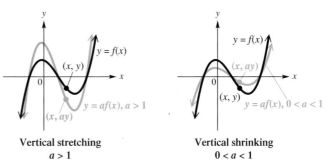

Vertical stretching
$a > 1$ Vertical shrinking
$0 < a < 1$

Figure 70

Graphs of functions can also be stretched and shrunk horizontally.

Horizontal Stretching or Shrinking of the Graph of a Function

Suppose that $a > 0$. If a point (x, y) lies on the graph of $y = f(x)$, then the point $\left(\frac{x}{a}, y\right)$ lies on the graph of $y = f(ax)$.

(a) If $0 < a < 1$, then the graph of $y = f(ax)$ is a **horizontal stretching** of the graph of $y = f(x)$.

(b) If $a > 1$, then the graph of $y = f(ax)$ is a **horizontal shrinking** of the graph of $y = f(x)$.

Horizontal stretching
$0 < a < 1$

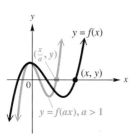

Horizontal shrinking
$a > 1$

Figure 71

See **Figure 71** for graphical interpretations of horizontal stretching and shrinking. *In both cases, the y-intercept remains the same but the x-intercepts are affected.*

Reflecting   Forming the mirror image of a graph across a line is called *reflecting the graph across the line.*

EXAMPLE 2   **Reflecting Graphs across Axes**

Graph each function.

(a) $g(x) = -\sqrt{x}$ **(b)** $h(x) = \sqrt{-x}$

SOLUTION

(a) The tables of values for $g(x) = -\sqrt{x}$ and $f(x) = \sqrt{x}$ are shown with their graphs in **Figure 72.** As the tables suggest, every y-value of the graph of $g(x) = -\sqrt{x}$ is the negative of the corresponding y-value of $f(x) = \sqrt{x}$. This has the effect of reflecting the graph across the x-axis.

x	$f(x) = \sqrt{x}$	$g(x) = -\sqrt{x}$
0	0	0
1	1	−1
4	2	−2

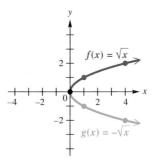

Figure 72

(b) The domain of $h(x) = \sqrt{-x}$ is $(-\infty, 0]$, while the domain of $f(x) = \sqrt{x}$ is $[0, \infty)$. Choosing x-values for $h(x)$ that are negatives of those used for $f(x)$, we see that corresponding y-values are the same. The graph of h is a reflection of the graph of f across the y-axis. See **Figure 73.**

x	$f(x) = \sqrt{x}$	$h(x) = \sqrt{-x}$
−4	undefined	2
−1	undefined	1
0	0	0
1	1	undefined
4	2	undefined

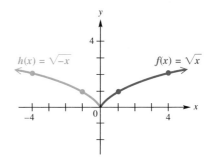

Figure 73

✔ **Now Try Exercises 27 and 33.**

The graphs in **Example 2** suggest the following generalizations.

Reflecting across an Axis

The graph of $y = -f(x)$ is the same as the graph of $y = f(x)$ reflected across the x-axis. (If a point (x, y) lies on the graph of $y = f(x)$, then $(x, -y)$ lies on this reflection.)

The graph of $y = f(-x)$ is the same as the graph of $y = f(x)$ reflected across the y-axis. (If a point (x, y) lies on the graph of $y = f(x)$, then $(-x, y)$ lies on this reflection.)

Symmetry   The graph of f shown in **Figure 74(a)** is cut in half by the y-axis, with each half the mirror image of the other half. Such a graph is *symmetric with respect to the y-axis*. **In general, for a graph to be symmetric with respect to the y-axis, the point $(-x, y)$ must be on the graph whenever the point (x, y) is on the graph.**

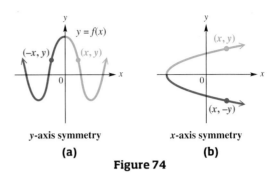

y-axis symmetry
(a)

x-axis symmetry
(b)

Figure 74

Similarly, if the graph in **Figure 74(b)** were folded in half along the x-axis, the portion at the top would exactly match the portion at the bottom. Such a graph is *symmetric with respect to the x-axis*. **In general, for a graph to be symmetric with respect to the x-axis, the point $(x, -y)$ must be on the graph whenever the point (x, y) is on the graph.**

Symmetry with Respect to an Axis

The graph of an equation is **symmetric with respect to the y-axis** if the replacement of x with $-x$ results in an equivalent equation.

The graph of an equation is **symmetric with respect to the x-axis** if the replacement of y with $-y$ results in an equivalent equation.

Examples: Of the basic functions in the previous section, graphs of the squaring and absolute value functions are symmetric with respect to the y-axis.

EXAMPLE 3   Testing for Symmetry with Respect to an Axis

Test for symmetry with respect to the x-axis and the y-axis.

(a) $y = x^2 + 4$ **(b)** $x = y^2 - 3$ **(c)** $x^2 + y^2 = 16$ **(d)** $2x + y = 4$

SOLUTION

(a) In $y = x^2 + 4$, replace x with $-x$.

$$y = x^2 + 4$$

Use parentheses around $-x$. $\quad y = (-x)^2 + 4 \quad$ The result is equivalent to the original equation.

$$y = x^2 + 4$$

Thus the graph, shown in **Figure 75,** is symmetric with respect to the y-axis. The y-axis cuts the graph in half, with the halves being mirror images.

Now replace y with $-y$ to test for symmetry with respect to the x-axis.

$$y = x^2 + 4$$

$$-y = x^2 + 4 \quad$$ The result is *not* equivalent to the original equation.

Multiply by -1. $\quad y = -x^2 - 4$

The graph is *not* symmetric with respect to the x-axis. See **Figure 75.**

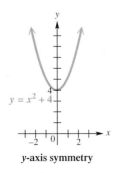

$y = x^2 + 4$

y-axis symmetry

Figure 75

(b) In $x = y^2 - 3$, replace y with $-y$.

$$x = (-y)^2 - 3 = y^2 - 3 \qquad \text{Same as the original equation}$$

The graph is symmetric with respect to the x-axis, as shown in **Figure 76.** It is *not* symmetric with respect to the y-axis.

(c) Substitute $-x$ for x and then $-y$ for y in $x^2 + y^2 = 16$.

$$(-x)^2 + y^2 = 16 \quad \text{and} \quad x^2 + (-y)^2 = 16$$

Both simplify to the original equation,

$$x^2 + y^2 = 16.$$

The graph, a circle of radius 4 centered at the origin, is symmetric with respect to *both* axes. See **Figure 77.**

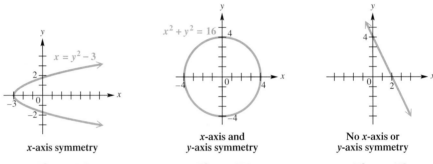

x-axis symmetry	x-axis and y-axis symmetry	No x-axis or y-axis symmetry
Figure 76	**Figure 77**	**Figure 78**

(d) $2x + y = 4$

Replace x with $-x$ and then replace y with $-y$.

$$
\begin{array}{l}
2x + y = 4 \\
2(-x) + y = 4 \\
-2x + y = 4
\end{array}
\quad \text{Not equivalent}
\qquad
\begin{array}{l}
2x + y = 4 \\
2x + (-y) = 4 \\
2x - y = 4
\end{array}
\quad \text{Not equivalent}
$$

See **Figure 78.** ✔ **Now Try Exercise 45.**

Another kind of symmetry occurs when a graph can be rotated $180°$ about the origin, with the result coinciding exactly with the original graph. Symmetry of this type is *symmetry with respect to the origin.* ***In general, for a graph to be symmetric with respect to the origin, the point $(-x, -y)$ is on the graph whenever the point (x, y) is on the graph.***

Figure 79 shows two such graphs.

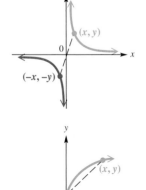

Origin symmetry

Figure 79

Symmetry with Respect to the Origin

The graph of an equation is **symmetric with respect to the origin** if the replacement of both x with $-x$ and y with $-y$ at the same time results in an equivalent equation.

Examples: Of the basic functions in the previous section, graphs of the cubing and cube root functions are symmetric with respect to the origin.

EXAMPLE 4 **Testing for Symmetry with Respect to the Origin**

Determine whether the graph of each equation is symmetric with respect to the origin.

(a) $x^2 + y^2 = 16$

(b) $y = x^3$

SOLUTION

(a) Replace x with $-x$ and y with $-y$.

> Use parentheses around $-x$ and $-y$.

$$x^2 + y^2 = 16$$
$$(-x)^2 + (-y)^2 = 16 \quad \text{Equivalent}$$
$$x^2 + y^2 = 16$$

The graph, which is the circle shown in **Figure 77** in **Example 3(c),** is symmetric with respect to the origin.

(b) Replace x with $-x$ and y with $-y$.

$$y = x^3$$
$$-y = (-x)^3 \quad \text{Equivalent}$$
$$-y = -x^3$$
$$y = x^3$$

$y = x^3$

Origin symmetry

Figure 80

The graph, which is that of the cubing function, is symmetric with respect to the origin and is shown in **Figure 80.**

✔ **Now Try Exercise 49.**

Notice the following important concepts regarding symmetry:

- A graph symmetric with respect to both the x- and y-axes is automatically symmetric with respect to the origin. (See **Figure 77.**)

- A graph symmetric with respect to the origin need *not* be symmetric with respect to either axis. (See **Figure 80.**)

- Of the three types of symmetry—with respect to the x-axis, with respect to the y-axis, and with respect to the origin—a graph possessing any two types must also exhibit the third type of symmetry.

- A graph symmetric with respect to the x-axis does not represent a function. (See **Figures 76 and 77.**)

Summary of Tests for Symmetry

	Symmetry with Respect to:		
	***x*-axis**	***y*-axis**	**Origin**
Equation is unchanged if:	y is replaced with $-y$	x is replaced with $-x$	x is replaced with $-x$ and y is replaced with $-y$
Example:			

Even and Odd Functions   The concepts of symmetry with respect to the y-axis and symmetry with respect to the origin are closely associated with the concepts of *even* and *odd functions*.

Even and Odd Functions

A function f is an **even function** if $f(-x) = f(x)$ for all x in the domain of f. (Its graph is symmetric with respect to the y-axis.)

A function f is an **odd function** if $f(-x) = -f(x)$ for all x in the domain of f. (Its graph is symmetric with respect to the origin.)

EXAMPLE 5   **Determining Whether Functions Are Even, Odd, or Neither**

Determine whether each function defined is *even*, *odd*, or *neither*.

(a) $f(x) = 8x^4 - 3x^2 + 1$ (b) $f(x) = 6x^3 - 9x$ (c) $f(x) = 3x^2 + 5x$

SOLUTION

(a) Replacing x with $-x$ gives the following.

$$f(x) = 8x^4 - 3x^2 + 1$$
$$f(-x) = 8(-x)^4 - 3(-x)^2 + 1 \quad \text{Replace } x \text{ with } -x.$$
$$f(-x) = 8x^4 - 3x^2 + 1 \quad\quad\quad \text{Apply the exponents.}$$
$$f(-x) = f(x) \quad\quad\quad\quad\quad 8x^4 - 3x^2 + 1 = f(x)$$

Because $f(-x) = f(x)$ for each x in the domain of the function, f is even.

(b) $\quad f(x) = 6x^3 - 9x$
$$f(-x) = 6(-x)^3 - 9(-x) \quad \text{Replace } x \text{ with } -x.$$
$$f(-x) = -6x^3 + 9x \quad \boxed{\text{Be careful with signs.}}$$
$$f(-x) = -f(x) \quad\quad\quad -6x^3 + 9x = -(6x^3 - 9x) = -f(x)$$

The function f is odd because $f(-x) = -f(x)$.

(c) $\quad f(x) = 3x^2 + 5x$
$$f(-x) = 3(-x)^2 + 5(-x) \quad \text{Replace } x \text{ with } -x.$$
$$f(-x) = 3x^2 - 5x \quad\quad\quad \text{Simplify.}$$

Because $f(-x) \neq f(x)$ and $f(-x) \neq -f(x)$, the function f is neither even nor odd.

✔ **Now Try Exercises 53, 55, and 57.**

NOTE   Consider a function defined by a polynomial in x.

- If the function has only *even* exponents on x (including the case of a constant where x^0 is understood to have the even exponent 0), it will *always* be an even function.

- Similarly, if only *odd* exponents appear on x, the function will be an odd function.

Translations   The next examples show the results of horizontal and vertical shifts, or **translations,** of the graph of $f(x) = |x|$.

EXAMPLE 6   Translating a Graph Vertically

Graph $g(x) = |x| - 4$.

SOLUTION   Comparing the table shown with **Figure 81,** we see that for corresponding x-values, the y-values of g are each 4 *less* than those for f. The graph of $g(x) = |x| - 4$ is the same as that of $f(x) = |x|$, but translated 4 units down. The lowest point is at $(0, -4)$. The graph is symmetric with respect to the y-axis and is therefore the graph of an even function.

x	$f(x) = |x|$	$g(x) = |x| - 4$
-4	4	0
-1	1	-3
0	0	-4
1	1	-3
4	4	0

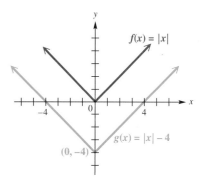

Figure 81

✔ **Now Try Exercise 67.**

The graphs in **Example 6** suggest the following generalization.

Vertical Translations

Given a function g defined by $g(x) = f(x) + c$, where c is a real number:

- For every point (x, y) on the graph of f, there will be a corresponding point $(x, y + c)$ on the graph of g.

- The graph of g will be the same as the graph of f, but translated c units up if c is positive or $|c|$ units down if c is negative.

The graph of g is a **vertical translation** of the graph of f.

Figure 82 shows a graph of a function f and two vertical translations of f. **Figure 83** shows two vertical translations of $y_1 = x^2$ on a TI-84 Plus calculator screen.

Figure 82

y_2 is the graph of $y_1 = x^2$ translated 2 units *up*. y_3 is that of y_1 translated 6 units *down*.

Figure 83

EXAMPLE 7 **Translating a Graph Horizontally**

Graph $g(x) = |x - 4|$.

SOLUTION Comparing the tables of values given with **Figure 84** shows that for corresponding y-values, the x-values of g are each 4 *more* than those for f. The graph of $g(x) = |x - 4|$ is the same as that of $f(x) = |x|$, but translated 4 units to the right. The lowest point is at $(4, 0)$. As suggested by the graphs in **Figure 84,** this graph is symmetric with respect to the line $x = 4$.

x	$f(x) = \lvert x \rvert$	$g(x) = \lvert x - 4 \rvert$
-2	2	6
0	0	4
2	2	2
4	4	0
6	6	2

Figure 84

✔ **Now Try Exercise 65.**

The graphs in **Example 7** suggest the following generalization.

Horizontal Translations

Given a function g defined by $g(x) = f(x - c)$, where c is a real number:

- For every point (x, y) on the graph of f, there will be a corresponding point $(x + c, y)$ on the graph of g.

- The graph of g will be the same as the graph of f, but translated c units to the right if c is positive or $|c|$ units to the left if c is negative.

The graph of g is a **horizontal translation** of the graph of f.

Figure 85 shows a graph of a function f and two horizontal translations of f. **Figure 86** shows two horizontal translations of $y_1 = x^2$ on a TI-84 Plus calculator screen.

Figure 85

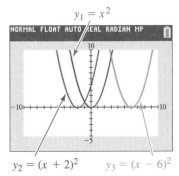

y_2 is the graph of $y_1 = x^2$ translated 2 units to the *left*. y_3 is that of y_1 translated 6 units to the *right*.

Figure 86

Summary of Translations

$(c > 0)$ To Graph:	Shift the Graph of $y = f(x)$ by c Units:
$y = f(x) + c$	up
$y = f(x) - c$	down
$y = f(x + c)$	left
$y = f(x - c)$	right

Vertical and horizontal translations are summarized in the table, where f is a function and c is a positive number.

CAUTION *Errors frequently occur when horizontal shifts are involved.* To determine the direction and magnitude of a horizontal shift, find the value that causes the expression $x - h$ to equal 0, as shown below.

$F(x) = (x - 5)^2$	$F(x) = (x + 5)^2$
Because **+5** causes $x - 5$ to equal 0, the graph of $F(x)$ illustrates a shift of	Because **−5** causes $x + 5$ to equal 0, the graph of $F(x)$ illustrates a shift of
5 units to the right.	**5 units to the left.**

EXAMPLE 8 **Using More Than One Transformation**

Graph each function.

(a) $f(x) = -|x + 3| + 1$ **(b)** $h(x) = |2x - 4|$ **(c)** $g(x) = -\dfrac{1}{2}x^2 + 4$

SOLUTION

(a) To graph $f(x) = -|x + 3| + 1$, the *lowest* point on the graph of $y = |x|$ is translated 3 units to the left and 1 unit up. The graph opens down because of the negative sign in front of the absolute value expression, making the lowest point now the highest point on the graph, as shown in **Figure 87.** The graph is symmetric with respect to the line $x = -3$.

Figure 87

(b) To determine the horizontal translation, factor out 2.

$$h(x) = |2x - 4|$$
$$h(x) = |2(x - 2)| \qquad \text{Factor out 2.}$$
$$h(x) = |2| \cdot |x - 2| \qquad |ab| = |a| \cdot |b|$$
$$h(x) = 2|x - 2| \qquad |2| = 2$$

The graph of h is the graph of $y = |x|$ translated 2 units to the right, and vertically stretched by a factor of 2. Horizontal shrinking gives the same appearance as vertical stretching for this function. See **Figure 88.**

Figure 88

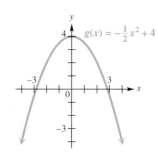

Figure 89

(c) The graph of $g(x) = -\frac{1}{2}x^2 + 4$ has the same shape as that of $y = x^2$, but it is wider (that is, shrunken vertically), reflected across the x-axis because the coefficient $-\frac{1}{2}$ is negative, and then translated 4 units up. See **Figure 89.**

✔ **Now Try Exercises 71, 73, and 81.**

EXAMPLE 9 **Graphing Translations and Reflections of a Given Graph**

A graph of a function $y = f(x)$ is shown in **Figure 90.** Use this graph to sketch each of the following graphs.

(a) $g(x) = f(x) + 3$ **(b)** $h(x) = f(x + 3)$

(c) $k(x) = f(x - 2) + 3$ **(d)** $F(x) = -f(x)$

SOLUTION   In each part, pay close attention to how the plotted points in **Figure 90** are translated or reflected.

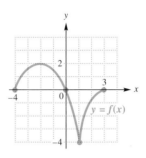

Figure 90

(a) The graph of $g(x) = f(x) + 3$ is the same as the graph in **Figure 90,** translated 3 units up. See **Figure 91(a).**

(b) To obtain the graph of $h(x) = f(x + 3)$, the graph of $y = f(x)$ must be translated 3 units to the left because $x + 3 = 0$ when $x = -3$. See **Figure 91(b).**

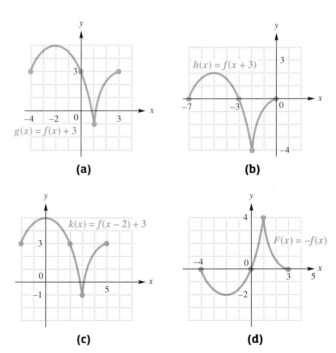

Figure 91

(c) The graph of $k(x) = f(x - 2) + 3$ will look like the graph of $f(x)$ translated 2 units to the right and 3 units up, as shown in **Figure 91(c).**

(d) The graph of $F(x) = -f(x)$ is that of $y = f(x)$ reflected across the x-axis. See **Figure 91(d).**

✔ **Now Try Exercise 87.**

Summary of Graphing Techniques

In the descriptions that follow, assume that $a > 0$, $h > 0$, and $k > 0$. In comparison with the graph of $y = f(x)$:

1. The graph of $y = f(x) + k$ is translated k units up.

2. The graph of $y = f(x) - k$ is translated k units down.

3. The graph of $y = f(x + h)$ is translated h units to the left.

4. The graph of $y = f(x - h)$ is translated h units to the right.

5. The graph of $y = af(x)$ is a vertical stretching of the graph of $y = f(x)$ if $a > 1$. It is a vertical shrinking if $0 < a < 1$.

6. The graph of $y = f(ax)$ is a horizontal stretching of the graph of $y = f(x)$ if $0 < a < 1$. It is a horizontal shrinking if $a > 1$.

7. The graph of $y = -f(x)$ is reflected across the x-axis.

8. The graph of $y = f(-x)$ is reflected across the y-axis.

2.7 Exercises

CONCEPT PREVIEW *Fill in the blank(s) to correctly complete each sentence.*

1. To graph the function $f(x) = x^2 - 3$, shift the graph of $y = x^2$ down _____ units.

2. To graph the function $f(x) = x^2 + 5$, shift the graph of $y = x^2$ up _____ units.

3. The graph of $f(x) = (x + 4)^2$ is obtained by shifting the graph of $y = x^2$ to the _____ 4 units.

4. The graph of $f(x) = (x - 7)^2$ is obtained by shifting the graph of $y = x^2$ to the _____ 7 units.

5. The graph of $f(x) = -\sqrt{x}$ is a reflection of the graph of $y = \sqrt{x}$ across the _____-axis.

6. The graph of $f(x) = \sqrt{-x}$ is a reflection of the graph of $y = \sqrt{x}$ across the _____-axis.

7. To obtain the graph of $f(x) = (x + 2)^3 - 3$, shift the graph of $y = x^3$ to the left _____ units and down _____ units.

8. To obtain the graph of $f(x) = (x - 3)^3 + 6$, shift the graph of $y = x^3$ to the right _____ units and up _____ units.

9. The graph of $f(x) = |-x|$ is the same as the graph of $y = |x|$ because reflecting it across the _____-axis yields the same ordered pairs.

10. The graph of $x = y^2$ is the same as the graph of $x = (-y)^2$ because reflecting it across the _____-axis yields the same ordered pairs.

11. *Concept Check* Match each equation in Column I with a description of its graph from Column II as it relates to the graph of $y = x^2$.

I	II
(a) $y = (x - 7)^2$	**A.** a translation 7 units to the left
(b) $y = x^2 - 7$	**B.** a translation 7 units to the right
(c) $y = 7x^2$	**C.** a translation 7 units up
(d) $y = (x + 7)^2$	**D.** a translation 7 units down
(e) $y = x^2 + 7$	**E.** a vertical stretching by a factor of 7

12. *Concept Check* Match each equation in Column I with a description of its graph from Column II as it relates to the graph of $y = \sqrt[3]{x}$.

I	II
(a) $y = 4\sqrt[3]{x}$	**A.** a translation 4 units to the right
(b) $y = -\sqrt[3]{x}$	**B.** a translation 4 units down
(c) $y = \sqrt[3]{-x}$	**C.** a reflection across the x-axis
(d) $y = \sqrt[3]{x-4}$	**D.** a reflection across the y-axis
(e) $y = \sqrt[3]{x} - 4$	**E.** a vertical stretching by a factor of 4

13. *Concept Check* Match each equation with the sketch of its graph in A–I.

(a) $y = x^2 + 2$ **(b)** $y = x^2 - 2$ **(c)** $y = (x+2)^2$

(d) $y = (x-2)^2$ **(e)** $y = 2x^2$ **(f)** $y = -x^2$

(g) $y = (x-2)^2 + 1$ **(h)** $y = (x+2)^2 + 1$ **(i)** $y = (x+2)^2 - 1$

A.   **B.**   **C.**

D.   **E.**   **F.**

G.   **H.**   **I.**

14. *Concept Check* Match each equation with the sketch of its graph in A–I.

(a) $y = \sqrt{x+3}$ **(b)** $y = \sqrt{x} - 3$ **(c)** $y = \sqrt{x} + 3$

(d) $y = 3\sqrt{x}$ **(e)** $y = -\sqrt{x}$ **(f)** $y = \sqrt{x-3}$

(g) $y = \sqrt{x-3} + 2$ **(h)** $y = \sqrt{x+3} + 2$ **(i)** $y = \sqrt{x-3} - 2$

A.   **B.**   **C.**

D.   **E.**   **F.**

G.

H.

I.

15. *Concept Check* Match each equation with the sketch of its graph in A–I.

(a) $y = |x - 2|$ (b) $y = |x| - 2$ (c) $y = |x| + 2$

(d) $y = 2|x|$ (e) $y = -|x|$ (f) $y = |-x|$

(g) $y = -2|x|$ (h) $y = |x - 2| + 2$ (i) $y = |x + 2| - 2$

A.

B.

C.

D.

E.

F.

G.

H.

I.

16. *Concept Check* Describe how the graph of $f(x) = 2(x + 1)^3 - 6$ compares to the graph of $y = x^3$.

Graph each function. ***See Examples 1 and 2.***

17. $f(x) = 3|x|$ **18.** $f(x) = 4|x|$ **19.** $f(x) = \dfrac{2}{3}|x|$

20. $f(x) = \dfrac{3}{4}|x|$ **21.** $g(x) = 2x^2$ **22.** $g(x) = 3x^2$

23. $g(x) = \dfrac{1}{2}x^2$ **24.** $g(x) = \dfrac{1}{3}x^2$ **25.** $f(x) = -\dfrac{1}{2}x^2$

26. $f(x) = -\dfrac{1}{3}x^2$ **27.** $f(x) = -3|x|$ **28.** $f(x) = -2|x|$

29. $h(x) = \left|-\dfrac{1}{2}x\right|$ **30.** $h(x) = \left|-\dfrac{1}{3}x\right|$ **31.** $h(x) = \sqrt{4x}$

32. $h(x) = \sqrt{9x}$ **33.** $f(x) = -\sqrt{-x}$ **34.** $f(x) = -|-x|$

Concept Check *Suppose the point* (8, 12) *is on the graph of* $y = f(x)$.

35. Find a point on the graph of
    (a) $y = f(x + 4)$ **(b)** $y = f(x) + 4$.

36. Find a point on the graph of
    (a) $y = \frac{1}{4}f(x)$ **(b)** $y = 4f(x)$.

37. Find a point on the graph of
    (a) $y = f(4x)$ **(b)** $y = f\left(\frac{1}{4}x\right)$.

38. Find a point on the graph of the reflection of $y = f(x)$
    (a) across the *x*-axis   **(b)** across the *y*-axis.

Concept Check *Plot each point, and then plot the points that are symmetric to the given point with respect to the **(a)** x-axis, **(b)** y-axis, and **(c)** origin.*

39. $(5, -3)$ **40.** $(-6, 1)$ **41.** $(-4, -2)$ **42.** $(-8, 0)$

43. *Concept Check* The graph of $y = |x - 2|$ is symmetric with respect to a vertical line. What is the equation of that line?

44. *Concept Check* Repeat **Exercise 43** for the graph of $y = -|x + 1|$.

*Without graphing, determine whether each equation has a graph that is symmetric with respect to the x-axis, the y-axis, the origin, or none of these. **See Examples 3 and 4.***

45. $y = x^2 + 5$ **46.** $y = 2x^4 - 3$

47. $x^2 + y^2 = 12$ **48.** $y^2 - x^2 = -6$

49. $y = -4x^3 + x$ **50.** $y = x^3 - x$

51. $y = x^2 - x + 8$ **52.** $y = x + 15$

*Determine whether each function is even, odd, or neither. **See Example 5.***

53. $f(x) = -x^3 + 2x$ **54.** $f(x) = x^5 - 2x^3$

55. $f(x) = 0.5x^4 - 2x^2 + 6$ **56.** $f(x) = 0.75x^2 + |x| + 4$

57. $f(x) = x^3 - x + 9$ **58.** $f(x) = x^4 - 5x + 8$

*Graph each function. **See Examples 6–8 and the Summary of Graphing Techniques box following Example 9.***

59. $f(x) = x^2 - 1$ **60.** $f(x) = x^2 - 2$ **61.** $f(x) = x^2 + 2$

62. $f(x) = x^2 + 3$ **63.** $g(x) = (x - 4)^2$ **64.** $g(x) = (x - 2)^2$

65. $g(x) = (x + 2)^2$ **66.** $g(x) = (x + 3)^2$ **67.** $g(x) = |x| - 1$

68. $g(x) = |x + 3| + 2$ **69.** $h(x) = -(x + 1)^3$ **70.** $h(x) = -(x - 1)^3$

71. $h(x) = 2x^2 - 1$ **72.** $h(x) = 3x^2 - 2$ **73.** $f(x) = 2(x - 2)^2 - 4$

74. $f(x) = -3(x - 2)^2 + 1$ **75.** $f(x) = \sqrt{x + 2}$ **76.** $f(x) = \sqrt{x - 3}$

77. $f(x) = -\sqrt{x}$ **78.** $f(x) = \sqrt{x} - 2$ **79.** $f(x) = 2\sqrt{x} + 1$

80. $f(x) = 3\sqrt{x} - 2$ **81.** $g(x) = \frac{1}{2}x^3 - 4$ **82.** $g(x) = \frac{1}{2}x^3 + 2$

83. $g(x) = (x + 3)^3$ **84.** $f(x) = (x - 2)^3$ **85.** $f(x) = \frac{2}{3}(x - 2)^2$

86. *Concept Check* What is the relationship between the graphs of $f(x) = |x|$ and $g(x) = |-x|$?

Work each problem. ***See Example 9.***

87. Given the graph of $y = g(x)$ in the figure, sketch the graph of each function, and describe how it is obtained from the graph of $y = g(x)$.

(a) $y = g(-x)$

(b) $y = g(x - 2)$

(c) $y = -g(x)$

(d) $y = -g(x) + 2$

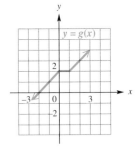

88. Given the graph of $y = f(x)$ in the figure, sketch the graph of each function, and describe how it is obtained from the graph of $y = f(x)$.

(a) $y = -f(x)$

(b) $y = 2f(x)$

(c) $y = f(-x)$

(d) $y = \dfrac{1}{2}f(x)$

Connecting Graphs with Equations *Each of the following graphs is obtained from the graph of $f(x) = |x|$ or $g(x) = \sqrt{x}$ by applying several of the transformations discussed in this section. Describe the transformations and give an equation for the graph.*

89.

90.

91.

92.

93.

94.

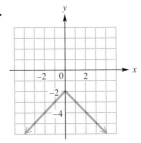

Concept Check *Suppose that for a function f,*

$$f(3) = 6.$$

For the given assumptions, find another function value.

95. The graph of $y = f(x)$ is symmetric with respect to the origin.

96. The graph of $y = f(x)$ is symmetric with respect to the y-axis.

97. The graph of $y = f(x)$ is symmetric with respect to the line $x = 6$.

98. For all x, $f(-x) = f(x)$.

99. For all x, $f(-x) = -f(x)$.

100. f is an odd function.

Work each problem.

101. Find a function $g(x) = ax + b$ whose graph can be obtained by translating the graph of $f(x) = 2x + 5$ up 2 units and 3 units to the left.

102. Find a function $g(x) = ax + b$ whose graph can be obtained by translating the graph of $f(x) = 3 - x$ down 2 units and 3 units to the right.

103. *Concept Check* Complete the left half of the graph of $y = f(x)$ in the figure for each condition.

    (a) $f(-x) = f(x)$ **(b)** $f(-x) = -f(x)$

104. *Concept Check* Complete the right half of the graph of $y = f(x)$ in the figure for each condition.

    (a) f is odd.   **(b)** f is even.

Chapter 2   Quiz (Sections 2.5-2.7)

1. For the line passing through the points $(-3, 5)$ and $(-1, 9)$, find the following.

    (a) the slope-intercept form of its equation   **(b)** its x-intercept

2. Find the slope-intercept form of the equation of the line passing through the point $(-6, 4)$ and perpendicular to the graph of $3x - 2y = 6$.

3. Suppose that P has coordinates $(-8, 5)$. Find the equation of the line through P that is

    (a) vertical   **(b)** horizontal.

4. For each basic function graphed, give the name of the function, the domain, the range, and open intervals over which it is decreasing, increasing, or constant.

(a)

(b)

(c)

5. *(Modeling) Long-Distance Call Charges* A certain long-distance carrier provides service between Podunk and Nowheresville. If x represents the number of minutes for the call, where $x > 0$, then the function

$$f(x) = 0.40[\![x]\!] + 0.75$$

gives the total cost of the call in dollars. Find the cost of a 5.5-min call.

Graph each function.

6. $f(x) = \begin{cases} \sqrt{x} & \text{if } x \geq 0 \\ 2x + 3 & \text{if } x < 0 \end{cases}$ **7.** $f(x) = -x^3 + 1$ **8.** $f(x) = 2|x - 1| + 3$

9. *Connecting Graphs with Equations* The function $g(x)$ graphed here is obtained by stretching, shrinking, reflecting, and/or translating the graph of $f(x) = \sqrt{x}$. Give the equation that defines this function.

10. Determine whether each function is *even*, *odd*, or *neither*.

(a) $f(x) = x^2 - 7$ **(b)** $f(x) = x^3 - x - 1$ **(c)** $f(x) = x^{101} - x^{99}$

2.8 Function Operations and Composition

- Arithmetic Operations on Functions
- The Difference Quotient
- Composition of Functions and Domain

Arithmetic Operations on Functions   **Figure 92** shows the situation for a company that manufactures DVDs. The two lines are the graphs of the linear functions for

$$\text{revenue } R(x) = 168x \quad \text{and} \quad \text{cost } C(x) = 118x + 800,$$

where x is the number of DVDs produced and sold, and x, $R(x)$, and $C(x)$ are given in thousands. When 30,000 (that is, 30 thousand) DVDs are produced and sold, profit is found as follows.

$$P(x) = R(x) - C(x) \qquad \text{Profit function}$$
$$P(30) = R(30) - C(30) \qquad \text{Let } x = 30.$$
$$P(30) = 5040 - 4340 \qquad R(30) = 168(30);\ C(30) = 118(30) + 800$$
$$P(30) = 700 \qquad \text{Subtract.}$$

Thus, the profit from the sale of 30,000 DVDs is $700,000.

The profit function is found by *subtracting* the cost function from the revenue function. New functions can be formed by using other operations as well.

DVD Production

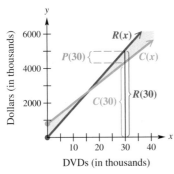

Figure 92

Operations on Functions and Domains

Given two functions f and g, then for all values of x for which both $f(x)$ and $g(x)$ are defined, the functions $f + g$, $f - g$, fg, and $\frac{f}{g}$ are defined as follows.

$$(f + g)(x) = f(x) + g(x) \qquad \text{Sum function}$$

$$(f - g)(x) = f(x) - g(x) \qquad \text{Difference function}$$

$$(fg)(x) = f(x) \cdot g(x) \qquad \text{Product function}$$

$$\left(\frac{f}{g}\right)(x) = \frac{f(x)}{g(x)}, \quad g(x) \neq 0 \qquad \text{Quotient function}$$

The **domains of $f + g$, $f - g$, and fg** include all real numbers in the intersection of the domains of f and g, while the **domain of $\frac{f}{g}$** includes those real numbers in the intersection of the domains of f and g for which $g(x) \neq 0$.

NOTE The condition $g(x) \neq 0$ in the definition of the quotient means that the domain of $\left(\frac{f}{g}\right)(x)$ is restricted to all values of x for which $g(x)$ is not 0. The condition does *not* mean that $g(x)$ is a function that is never 0.

EXAMPLE 1   Using Operations on Functions

Let $f(x) = x^2 + 1$ and $g(x) = 3x + 5$. Find each of the following.

(a) $(f + g)(1)$ **(b)** $(f - g)(-3)$ **(c)** $(fg)(5)$ **(d)** $\left(\dfrac{f}{g}\right)(0)$

SOLUTION

(a) First determine $f(1) = 2$ and $g(1) = 8$. Then use the definition.

$$(f + g)(1)$$
$$= f(1) + g(1) \qquad (f + g)(x) = f(x) + g(x)$$
$$= 2 + 8 \qquad f(1) = 1^2 + 1; g(1) = 3(1) + 5$$
$$= 10 \qquad \text{Add.}$$

(b) $(f - g)(-3)$
$$= f(-3) - g(-3) \qquad (f - g)(x) = f(x) - g(x)$$
$$= 10 - (-4) \qquad f(-3) = (-3)^2 + 1; g(-3) = 3(-3) + 5$$
$$= 14 \qquad \text{Subtract.}$$

(c) $(fg)(5)$
$$= f(5) \cdot g(5) \qquad (fg)(x) = f(x) \cdot g(x)$$
$$= (5^2 + 1)(3 \cdot 5 + 5) \qquad f(x) = x^2 + 1; g(x) = 3x + 5$$
$$= 26 \cdot 20 \qquad f(5) = 26; g(5) = 20$$
$$= 520 \qquad \text{Multiply.}$$

(d) $\left(\dfrac{f}{g}\right)(0)$

$= \dfrac{f(0)}{g(0)}$ $\left(\dfrac{f}{g}\right)(x) = \dfrac{f(x)}{g(x)}$

$= \dfrac{0^2 + 1}{3(0) + 5}$ $f(x) = x^2 + 1$
                 $g(x) = 3x + 5$

$= \dfrac{1}{5}$ Simplify.   ✔ **Now Try Exercises 11, 13, 15, and 17.**

EXAMPLE 2   Using Operations on Functions and Determining Domains

Let $f(x) = 8x - 9$ and $g(x) = \sqrt{2x - 1}$. Find each function in (a)–(d).

(a) $(f + g)(x)$ **(b)** $(f - g)(x)$ **(c)** $(fg)(x)$ **(d)** $\left(\dfrac{f}{g}\right)(x)$

(e) Give the domains of the functions in parts (a)–(d).

SOLUTION

(a) $(f + g)(x)$ **(b)** $(f - g)(x)$

    $= f(x) + g(x)$ $= f(x) - g(x)$

    $= 8x - 9 + \sqrt{2x - 1}$ $= 8x - 9 - \sqrt{2x - 1}$

(c) $(fg)(x)$ **(d)** $\left(\dfrac{f}{g}\right)(x)$

    $= f(x) \cdot g(x)$

    $= (8x - 9)\sqrt{2x - 1}$ $= \dfrac{f(x)}{g(x)}$

                             $= \dfrac{8x - 9}{\sqrt{2x - 1}}$

(e) To find the domains of the functions in parts (a)–(d), we first find the domains of f and g.

The domain of f is the set of all real numbers $(-\infty, \infty)$.

Because g is defined by a square root radical, the radicand must be non-negative (that is, greater than or equal to 0).

$g(x) = \sqrt{2x - 1}$ Rule for $g(x)$

$2x - 1 \geq 0$ $2x - 1$ must be nonnegative.

$x \geq \dfrac{1}{2}$ Add 1 and divide by 2.

Thus, the domain of g is $\left[\dfrac{1}{2}, \infty\right)$.

The domains of $f + g$, $f - g$, and fg are the intersection of the domains of f and g, which is

$(-\infty, \infty) \cap \left[\dfrac{1}{2}, \infty\right) = \left[\dfrac{1}{2}, \infty\right).$ The intersection of two sets is the set of all elements common to both sets.

The domain of $\dfrac{f}{g}$ includes those real numbers in the intersection above for which $g(x) = \sqrt{2x - 1} \neq 0$—that is, the domain of $\dfrac{f}{g}$ is $\left(\dfrac{1}{2}, \infty\right)$.

✔ **Now Try Exercises 19 and 23.**

EXAMPLE 3 Evaluating Combinations of Functions

If possible, use the given representations of functions f and g to evaluate

$$(f + g)(4), \quad (f - g)(-2), \quad (fg)(1), \quad \text{and} \quad \left(\frac{f}{g}\right)(0).$$

(a)

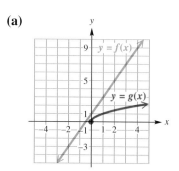

(b)

x	$f(x)$	$g(x)$
-2	-3	undefined
0	1	0
1	3	1
4	9	2

(c) $f(x) = 2x + 1, \quad g(x) = \sqrt{x}$

SOLUTION

(a) From the figure, $f(4) = 9$ and $g(4) = 2$.

$$(f + g)(4)$$

$$= f(4) + g(4) \qquad (f + g)(x) = f(x) + g(x)$$

$$= 9 + 2 \qquad \text{Substitute.}$$

$$= 11 \qquad \text{Add.}$$

For $(f - g)(-2)$, although $f(-2) = -3$, $g(-2)$ is undefined because -2 is not in the domain of g. Thus $(f - g)(-2)$ is undefined.

The domains of f and g both include 1.

$$(fg)(1)$$

$$= f(1) \cdot g(1) \qquad (fg)(x) = f(x) \cdot g(x)$$

$$= 3 \cdot 1 \qquad \text{Substitute.}$$

$$= 3 \qquad \text{Multiply.}$$

The graph of g includes the origin, so $g(0) = 0$. Thus $\left(\frac{f}{g}\right)(0)$ is undefined.

(b) From the table, $f(4) = 9$ and $g(4) = 2$.

$$(f + g)(4)$$

$$= f(4) + g(4) \qquad (f + g)(x) = f(x) + g(x)$$

$$= 9 + 2 \qquad \text{Substitute.}$$

$$= 11 \qquad \text{Add.}$$

In the table, $g(-2)$ is undefined, and thus $(f - g)(-2)$ is also undefined.

$$(fg)(1)$$

$$= f(1) \cdot g(1) \qquad (fg)(x) = f(x) \cdot g(x)$$

$$= 3 \cdot 1 \qquad f(1) = 3 \text{ and } g(1) = 1$$

$$= 3 \qquad \text{Multiply.}$$

The quotient function value $\left(\frac{f}{g}\right)(0)$ is undefined because the denominator, $g(0)$, equals 0.

(c) Using $f(x) = 2x + 1$ and $g(x) = \sqrt{x}$, we can find $(f + g)(4)$ and $(fg)(1)$. Because -2 is not in the domain of g, $(f - g)(-2)$ is not defined.

$$
\begin{aligned}
(f + g)(4) && (fg)(1) \\
= f(4) + g(4) && = f(1) \cdot g(1) \\
= (2 \cdot 4 + 1) + \sqrt{4} && = (2 \cdot 1 + 1) \cdot \sqrt{1} \\
= 9 + 2 && = 3(1) \\
= 11 && = 3
\end{aligned}
$$

$\left(\frac{f}{g}\right)(0)$ is undefined since $g(0) = 0$.   ✔ **Now Try Exercises 33 and 37.**

The Difference Quotient   Suppose a point P lies on the graph of $y = f(x)$ as in **Figure 93,** and suppose h is a positive number. If we let $(x, f(x))$ denote the coordinates of P and $(x + h, f(x + h))$ denote the coordinates of Q, then the line joining P and Q has slope as follows.

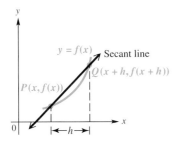

Figure 93

$$
m = \frac{f(x + h) - f(x)}{(x + h) - x} \qquad \text{Slope formula}
$$

$$
m = \frac{f(x + h) - f(x)}{h}, \quad h \neq 0 \qquad \text{Difference quotient}
$$

This boldface expression is the **difference quotient.**

Figure 93 shows the graph of the line PQ (called a **secant line**). As h approaches 0, the slope of this secant line approaches the slope of the line tangent to the curve at P. Important applications of this idea are developed in calculus.

EXAMPLE 4   **Finding the Difference Quotient**

Let $f(x) = 2x^2 - 3x$. Find and simplify the expression for the difference quotient,

$$
\frac{f(x + h) - f(x)}{h}.
$$

SOLUTION   We use a three-step process.

Step 1 Find the first term in the numerator, $f(x + h)$. Replace x in $f(x)$ with $x + h$.

$$
f(x + h)
$$
$$
= 2(x + h)^2 - 3(x + h) \qquad f(x) = 2x^2 - 3x
$$

Step 2 Find the entire numerator, $f(x + h) - f(x)$.

$$
f(x + h) - f(x)
$$
$$
= [2(x + h)^2 - 3(x + h)] - (2x^2 - 3x) \qquad \text{Substitute.}
$$
$$
= 2(x^2 + 2xh + h^2) - 3(x + h) - (2x^2 - 3x) \qquad \text{Square } x + h.
$$
$$
= 2x^2 + 4xh + 2h^2 - 3x - 3h - 2x^2 + 3x \qquad \text{Distributive property}
$$
$$
= 4xh + 2h^2 - 3h \qquad \text{Combine like terms.}
$$

Remember this term when squaring $x + h$

Step 3 Find the difference quotient by dividing by h.

$$\frac{f(x+h) - f(x)}{h}$$

$$= \frac{4xh + 2h^2 - 3h}{h} \qquad \text{Substitute } 4xh + 2h^2 - 3h \text{ for } f(x+h) - f(x), \text{ from Step 2.}$$

$$= \frac{h(4x + 2h - 3)}{h} \qquad \text{Factor out } h.$$

$$= 4x + 2h - 3 \qquad \text{Divide out the common factor.}$$

✔ **Now Try Exercises 45 and 55.**

CAUTION In **Example 4,** notice that the expression $f(x+h)$ is not equivalent to $f(x) + f(h)$. These expressions differ by $4xh$.

$$f(x+h) = 2(x+h)^2 - 3(x+h) = 2x^2 + 4xh + 2h^2 - 3x - 3h$$

$$f(x) + f(h) = (2x^2 - 3x) + (2h^2 - 3h) = 2x^2 - 3x + 2h^2 - 3h$$

In general, for a function f, $f(x+h)$ is *not* equivalent to $f(x) + f(h)$.

Composition of Functions and Domain   The diagram in **Figure 94** shows a function g that assigns to each x in its domain a value $g(x)$. Then another function f assigns to each $g(x)$ in its domain a value $f(g(x))$. This two-step process takes an element x and produces a corresponding element $f(g(x))$.

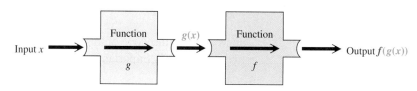

Figure 94

The function with y-values $f(g(x))$ is the *composition* of functions f and g, which is written $\boldsymbol{f \circ g}$ and read **"f of g"** or **"f compose g".**

Composition of Functions and Domain

If f and g are functions, then the **composite function,** or **composition,** of f and g is defined by

$$(f \circ g)(x) = f(g(x)).$$

The **domain of $f \circ g$** is the set of all numbers x in the domain of g such that $g(x)$ is in the domain of f.

As a real-life example of how composite functions occur, consider the following retail situation.

A $40 pair of blue jeans is on sale for 25% off. If we purchase the jeans before noon, they are an additional 10% off. What is the final sale price of the jeans?

We might be tempted to say that the jeans are 35% off and calculate $\$40(0.35) = \$14$, giving a final sale price of

$$\$40 - \$14 = \$26. \qquad \text{Incorrect}$$

$26 *is not correct.* To find the final sale price, we must first find the price after taking 25% off and then take an additional 10% off *that* price. See **Figure 95**.

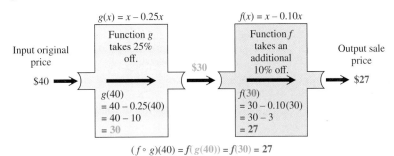

$$(f \circ g)(40) = f(g(40)) = f(30) = 27$$

Figure 95

EXAMPLE 5   **Evaluating Composite Functions**

Let $f(x) = 2x - 1$ and $g(x) = \dfrac{4}{x-1}$.

(a) Find $(f \circ g)(2)$.   **(b)** Find $(g \circ f)(-3)$.

SOLUTION

(a) First find $g(2)$: $g(2) = \dfrac{4}{2-1} = \dfrac{4}{1} = 4$.

Now find $(f \circ g)(2)$.

$$
\begin{aligned}
(f \circ g)(2) & \\
&= f(g(2)) && \text{Definition of composition} \\
&= f(4) && g(2) = 4 \\
&= 2(4) - 1 && \text{Definition of } f \\
&= 7 && \text{Simplify.}
\end{aligned}
$$

(b)

$$
\begin{aligned}
(g \circ f)(-3) & \\
&= g(f(-3)) && \text{Definition of composition} \\
&= g[2(-3) - 1] && f(-3) = 2(-3) - 1 \\
&= g(-7) && \text{Multiply, and then subtract.} \\
&= \dfrac{4}{-7 - 1} && g(x) = \tfrac{4}{x+1} \\
&= -\dfrac{1}{2} && \begin{array}{l}\text{Subtract in the denominator.}\\ \text{Write in lowest terms.}\end{array}
\end{aligned}
$$

✔ **Now Try Exercise 57.**

The screens show how a graphing calculator evaluates the expressions in **Example 5**.

EXAMPLE 6   **Determining Composite Functions and Their Domains**

Given that $f(x) = \sqrt{x}$ and $g(x) = 4x + 2$, find each of the following.

(a) $(f \circ g)(x)$ and its domain   **(b)** $(g \circ f)(x)$ and its domain

SOLUTION

(a)

$$
\begin{aligned}
(f \circ g)(x) & \\
&= f(g(x)) && \text{Definition of composition} \\
&= f(4x + 2) && g(x) = 4x + 2 \\
&= \sqrt{4x + 2} && f(x) = \sqrt{x}
\end{aligned}
$$

The domain and range of g are both the set of all real numbers, $(-\infty, \infty)$. The domain of f is the set of all nonnegative real numbers, $[0, \infty)$. Thus, $g(x)$, which is defined as $4x + 2$, must be greater than or equal to zero.

$$4x + 2 \geq 0 \quad \text{Solve the inequality.}$$

The radicand must be nonnegative.

$$x \geq -\frac{1}{2} \quad \text{Subtract 2. Divide by 4.}$$

Therefore, the domain of $f \circ g$ is $\left[-\frac{1}{2}, \infty\right)$.

(b) $(g \circ f)(x)$

$$= g(f(x)) \quad \text{Definition of composition}$$

$$= g\left(\sqrt{x}\right) \quad f(x) = \sqrt{x}$$

$$= 4\sqrt{x} + 2 \quad g(x) = 4x + 2$$

The domain and range of f are both the set of all nonnegative real numbers, $[0, \infty)$. The domain of g is the set of all real numbers, $(-\infty, \infty)$. Therefore, the domain of $g \circ f$ is $[0, \infty)$.

✔ **Now Try Exercise 75.**

EXAMPLE 7 Determining Composite Functions and Their Domains

Given that $f(x) = \dfrac{6}{x - 3}$ and $g(x) = \dfrac{1}{x}$, find each of the following.

(a) $(f \circ g)(x)$ and its domain   **(b)** $(g \circ f)(x)$ and its domain

SOLUTION

(a) $(f \circ g)(x)$

$$= f(g(x)) \quad \text{By definition}$$

$$= f\left(\frac{1}{x}\right) \quad g(x) = \frac{1}{x}$$

$$= \frac{6}{\dfrac{1}{x} - 3} \quad f(x) = \frac{6}{x-3}$$

$$= \frac{6x}{1 - 3x} \quad \begin{array}{l}\text{Multiply the numerator}\\\text{and denominator by } x.\end{array}$$

The domain of g is the set of all real numbers *except* 0, which makes $g(x)$ undefined. The domain of f is the set of all real numbers *except* 3. The expression for $g(x)$, therefore, cannot equal 3. We determine the value that makes $g(x) = 3$ and *exclude* it from the domain of $f \circ g$.

$$\frac{1}{x} = 3 \quad \text{The solution must be excluded.}$$

$$1 = 3x \quad \text{Multiply by } x.$$

$$x = \frac{1}{3} \quad \text{Divide by 3.}$$

Therefore, the domain of $f \circ g$ is the set of all real numbers *except* 0 and $\frac{1}{3}$, written in interval notation as

$$(-\infty, 0) \cup \left(0, \frac{1}{3}\right) \cup \left(\frac{1}{3}, \infty\right).$$

(b) $(g \circ f)(x)$

$$= g(f(x)) \quad \text{By definition}$$

$$= g\left(\frac{6}{x-3}\right) \quad f(x) = \frac{6}{x-3}$$

$$= \frac{1}{\frac{6}{x-3}} \quad \text{Note that this is meaningless if } x = 3; g(x) = \frac{1}{x}.$$

$$= \frac{x-3}{6} \quad \frac{1}{\frac{a}{b}} = 1 \div \frac{a}{b} = 1 \cdot \frac{b}{a} = \frac{b}{a}$$

The domain of f is the set of all real numbers *except* 3. The domain of g is the set of all real numbers *except* 0. The expression for $f(x)$, which is $\frac{6}{x-3}$, is never zero because the numerator is the nonzero number 6. The domain of $g \circ f$ is the set of all real numbers *except* 3, written $(-\infty, 3) \cup (3, \infty)$.

✔ **Now Try Exercise 87.**

NOTE It often helps to consider the *unsimplified* form of a composition expression when determining the domain in a situation like **Example 7(b).**

LOOKING AHEAD TO CALCULUS

Finding the derivative of a function in calculus is called **differentiation.** To differentiate a composite function such as

$$h(x) = (3x + 2)^4,$$

we interpret $h(x)$ as $(f \circ g)(x)$, where

$$g(x) = 3x + 2 \quad \text{and} \quad f(x) = x^4.$$

The **chain rule** allows us to differentiate composite functions. Notice the use of the composition symbol and function notation in the following, which comes from the chain rule.

If $h(x) = (f \circ g)(x)$, then

$$h'(x) = f'(g(x)) \cdot g'(x).$$

EXAMPLE 8 Showing That $(g \circ f)(x)$ Is Not Equivalent to $(f \circ g)(x)$

Let $f(x) = 4x + 1$ and $g(x) = 2x^2 + 5x$. Show that $(g \circ f)(x) \neq (f \circ g)(x)$. (This is sufficient to prove that this inequality is true in general.)

SOLUTION First, find $(g \circ f)(x)$. Then find $(f \circ g)(x)$.

$$(g \circ f)(x)$$
$$= g(f(x)) \quad \text{By definition}$$
$$= g(4x + 1) \quad f(x) = 4x + 1$$
$$= 2(4x + 1)^2 + 5(4x + 1) \quad g(x) = 2x^2 + 5x$$
$$= 2(16x^2 + 8x + 1) + 20x + 5 \quad \text{Square } 4x + 1 \text{ and apply the distributive property.}$$
$$= 32x^2 + 16x + 2 + 20x + 5 \quad \text{Distributive property}$$
$$= 32x^2 + 36x + 7 \quad \text{Combine like terms.}$$

$$(f \circ g)(x)$$
$$= f(g(x)) \quad \text{By definition}$$
$$= f(2x^2 + 5x) \quad g(x) = 2x^2 + 5x$$
$$= 4(2x^2 + 5x) + 1 \quad f(x) = 4x + 1$$
$$= 8x^2 + 20x + 1 \quad \text{Distributive property}$$

Thus, $(g \circ f)(x) \neq (f \circ g)(x)$.

✔ **Now Try Exercise 91.**

As **Example 8** shows, *it is not always true that $f \circ g = g \circ f$.* One important circumstance in which equality holds occurs when f and g are *inverses* of each other, a concept discussed later in the text.

In calculus it is sometimes necessary to treat a function as a composition of two functions. The next example shows how this can be done.

EXAMPLE 9 Finding Functions That Form a Given Composite

Find functions f and g such that

$$(f \circ g)(x) = (x^2 - 5)^3 - 4(x^2 - 5) + 3.$$

SOLUTION Note the repeated quantity $x^2 - 5$. If we choose $g(x) = x^2 - 5$ and $f(x) = x^3 - 4x + 3$, then we have the following.

$$(f \circ g)(x)$$

$$= f(g(x)) \qquad \text{By definition}$$

$$= f(x^2 - 5) \qquad g(x) = x^2 - 5$$

$$= (x^2 - 5)^3 - 4(x^2 - 5) + 3 \quad \text{Use the rule for } f.$$

There are other pairs of functions f and g that also satisfy these conditions. For instance,

$$f(x) = (x - 5)^3 - 4(x - 5) + 3 \quad \text{and} \quad g(x) = x^2.$$

✔ **Now Try Exercise 99.**

2.8 Exercises

CONCEPT PREVIEW *Without using paper and pencil, evaluate each expression given the following functions.*

$$f(x) = x + 1 \quad \text{and} \quad g(x) = x^2$$

1. $(f + g)(2)$ **2.** $(f - g)(2)$ **3.** $(fg)(2)$

4. $\left(\dfrac{f}{g}\right)(2)$ **5.** $(f \circ g)(2)$ **6.** $(g \circ f)(2)$

CONCEPT PREVIEW *Refer to functions f and g as described in **Exercises 1–6,** and find the following.*

7. domain of f **8.** domain of g **9.** domain of $f + g$ **10.** domain of $\dfrac{f}{g}$

*Let $f(x) = x^2 + 3$ and $g(x) = -2x + 6$. Find each of the following. **See Example 1.***

11. $(f + g)(3)$ **12.** $(f + g)(-5)$ **13.** $(f - g)(-1)$ **14.** $(f - g)(4)$

15. $(fg)(4)$ **16.** $(fg)(-3)$ **17.** $\left(\dfrac{f}{g}\right)(-1)$ **18.** $\left(\dfrac{f}{g}\right)(5)$

*For the pair of functions defined, find $(f + g)(x)$, $(f - g)(x)$, $(fg)(x)$, and $\left(\dfrac{f}{g}\right)(x)$. Give the domain of each. **See Example 2.***

19. $f(x) = 3x + 4$, $g(x) = 2x - 5$ **20.** $f(x) = 6 - 3x$, $g(x) = -4x + 1$

21. $f(x) = 2x^2 - 3x$, $g(x) = x^2 - x + 3$ **22.** $f(x) = 4x^2 + 2x$, $g(x) = x^2 - 3x + 2$

23. $f(x) = \sqrt{4x - 1}$, $g(x) = \dfrac{1}{x}$ **24.** $f(x) = \sqrt{5x - 4}$, $g(x) = -\dfrac{1}{x}$

Associate's Degrees Earned The graph shows the number of associate's degrees earned (in thousands) in the United States from 2004 through 2012.

$M(x)$ gives the number of degrees earned by males.

$F(x)$ gives the number earned by females.

$T(x)$ gives the total number for both groups.

Associate's Degrees Earned

Source: National Center for Education Statistics.

25. Estimate $M(2008)$ and $F(2008)$, and use the results to estimate $T(2008)$.

26. Estimate $M(2012)$ and $F(2012)$, and use the results to estimate $T(2012)$.

27. Use the slopes of the line segments to decide in which period (2004–2008 or 2008–2012) the total number of associate's degrees earned increased more rapidly.

28. *Concept Check* Refer to the graph of Associate's Degrees Earned.

If $2004 \le k \le 2012$, $T(k) = r$, and $F(k) = s$, then $M(k) = $ _____.

Science and Space/Technology Spending The graph shows dollars (in billions) spent for general science and for space/other technologies in selected years.

$G(x)$ represents the dollars spent for general science.

$S(x)$ represents the dollars spent for space and other technologies.

$T(x)$ represents the total expenditures for these two categories.

29. Estimate $(T - S)(2000)$. What does this function represent?

30. Estimate $(T - G)(2010)$. What does this function represent?

31. In which category and which period(s) does spending decrease?

32. In which period does spending for $T(x)$ increase most?

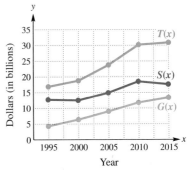

Science and Space Spending

Source: U.S. Office of Management and Budget.

*Use the graph to evaluate each expression. **See Example 3(a).***

33. (a) $(f + g)(2)$ (b) $(f - g)(1)$

(c) $(fg)(0)$ (d) $\left(\dfrac{f}{g}\right)(1)$

34. (a) $(f + g)(0)$ (b) $(f - g)(-1)$

(c) $(fg)(1)$ (d) $\left(\dfrac{f}{g}\right)(2)$

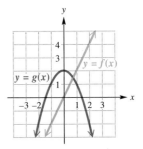

35. **(a)** $(f + g)(-1)$ **(b)** $(f - g)(-2)$ **36.** **(a)** $(f + g)(1)$ **(b)** $(f - g)(0)$

   (c) $(fg)(0)$ **(d)** $\left(\dfrac{f}{g}\right)(2)$ **(c)** $(fg)(-1)$ **(d)** $\left(\dfrac{f}{g}\right)(1)$

Use the table to evaluate each expression in parts (a)–(d), if possible. See Example 3(b).

(a) $(f + g)(2)$ **(b)** $(f - g)(4)$ **(c)** $(fg)(-2)$ **(d)** $\left(\dfrac{f}{g}\right)(0)$

37.

x	$f(x)$	$g(x)$
-2	0	6
0	5	0
2	7	-2
4	10	5

38.

x	$f(x)$	$g(x)$
-2	-4	2
0	8	-1
2	5	4
4	0	0

39. Use the table in **Exercise 37** to complete the following table.

x	$(f + g)(x)$	$(f - g)(x)$	$(fg)(x)$	$\left(\dfrac{f}{g}\right)(x)$
-2				
0				
2				
4				

40. Use the table in **Exercise 38** to complete the following table.

x	$(f + g)(x)$	$(f - g)(x)$	$(fg)(x)$	$\left(\dfrac{f}{g}\right)(x)$
-2				
0				
2				
4				

41. *Concept Check* How is the difference quotient related to slope?

42. *Concept Check* Refer to **Figure 93.** How is the secant line PQ related to the tangent line to a curve at point P?

For each function, find (a) $f(x + h)$, (b) $f(x + h) - f(x)$, and (c) $\dfrac{f(x + h) - f(x)}{h}$.
See Example 4.

43. $f(x) = 2 - x$ **44.** $f(x) = 1 - x$ **45.** $f(x) = 6x + 2$

46. $f(x) = 4x + 11$ **47.** $f(x) = -2x + 5$ **48.** $f(x) = -4x + 2$

49. $f(x) = \dfrac{1}{x}$ **50.** $f(x) = \dfrac{1}{x^2}$ **51.** $f(x) = x^2$

52. $f(x) = -x^2$ **53.** $f(x) = 1 - x^2$ **54.** $f(x) = 1 + 2x^2$

55. $f(x) = x^2 + 3x + 1$ **56.** $f(x) = x^2 - 4x + 2$

*Let $f(x) = 2x - 3$ and $g(x) = -x + 3$. Find each function value. **See Example 5.***

57. $(f \circ g)(4)$ **58.** $(f \circ g)(2)$ **59.** $(f \circ g)(-2)$ **60.** $(g \circ f)(3)$

61. $(g \circ f)(0)$ **62.** $(g \circ f)(-2)$ **63.** $(f \circ f)(2)$ **64.** $(g \circ g)(-2)$

Concept Check The tables give some selected ordered pairs for functions f and g.

x	3	4	6
$f(x)$	1	3	9

x	2	7	1	9
$g(x)$	3	6	9	12

Find each of the following.

65. $(f \circ g)(2)$ **66.** $(f \circ g)(7)$ **67.** $(g \circ f)(3)$

68. $(g \circ f)(6)$ **69.** $(f \circ f)(4)$ **70.** $(g \circ g)(1)$

71. *Concept Check* Why can we not determine $(f \circ g)(1)$ given the information in the tables for **Exercises 65–70?**

72. *Concept Check* Extend the concept of composition of functions to evaluate $(g \circ (f \circ g))(7)$ using the tables for **Exercises 65–70.**

*Given functions f and g, find **(a)** $(f \circ g)(x)$ and its domain, and **(b)** $(g \circ f)(x)$ and its domain. **See Examples 6 and 7.***

73. $f(x) = -6x + 9$, $g(x) = 5x + 7$ **74.** $f(x) = 8x + 12$, $g(x) = 3x - 1$

75. $f(x) = \sqrt{x}$, $g(x) = x + 3$ **76.** $f(x) = \sqrt{x}$, $g(x) = x - 1$

77. $f(x) = x^3$, $g(x) = x^2 + 3x - 1$ **78.** $f(x) = x + 2$, $g(x) = x^4 + x^2 - 4$

79. $f(x) = \sqrt{x - 1}$, $g(x) = 3x$ **80.** $f(x) = \sqrt{x - 2}$, $g(x) = 2x$

81. $f(x) = \dfrac{2}{x}$, $g(x) = x + 1$ **82.** $f(x) = \dfrac{4}{x}$, $g(x) = x + 4$

83. $f(x) = \sqrt{x + 2}$, $g(x) = -\dfrac{1}{x}$ **84.** $f(x) = \sqrt{x + 4}$, $g(x) = -\dfrac{2}{x}$

85. $f(x) = \sqrt{x}$, $g(x) = \dfrac{1}{x + 5}$ **86.** $f(x) = \sqrt{x}$, $g(x) = \dfrac{3}{x + 6}$

87. $f(x) = \dfrac{1}{x - 2}$, $g(x) = \dfrac{1}{x}$ **88.** $f(x) = \dfrac{1}{x + 4}$, $g(x) = -\dfrac{1}{x}$

89. *Concept Check* Fill in the missing entries in the table.

x	$f(x)$	$g(x)$	$g(f(x))$
1	3	2	7
2	1	5	
3	2		

90. *Concept Check* Suppose $f(x)$ is an odd function and $g(x)$ is an even function. Fill in the missing entries in the table.

x	-2	-1	0	1	2
$f(x)$			0	-2	
$g(x)$	0	2	1		
$(f \circ g)(x)$		1	-2		

91. Show that $(f \circ g)(x)$ is not equivalent to $(g \circ f)(x)$ for

$$f(x) = 3x - 2 \quad \text{and} \quad g(x) = 2x - 3.$$

92. Show that for the functions

$$f(x) = x^3 + 7 \quad \text{and} \quad g(x) = \sqrt[3]{x - 7},$$

both $(f \circ g)(x)$ and $(g \circ f)(x)$ equal x.

*For certain pairs of functions f and g, $(f \circ g)(x) = x$ and $(g \circ f)(x) = x$. Show that this is true for each pair in **Exercises 93–96**.*

93. $f(x) = 4x + 2, \quad g(x) = \dfrac{1}{4}(x - 2)$

94. $f(x) = -3x, \quad g(x) = -\dfrac{1}{3}x$

95. $f(x) = \sqrt[3]{5x + 4}, \quad g(x) = \dfrac{1}{5}x^3 - \dfrac{4}{5}$

96. $f(x) = \sqrt[3]{x + 1}, \quad g(x) = x^3 - 1$

*Find functions f and g such that $(f \circ g)(x) = h(x)$. (There are many possible ways to do this.) **See Example 9.***

97. $h(x) = (6x - 2)^2$

98. $h(x) = (11x^2 + 12x)^2$

99. $h(x) = \sqrt{x^2 - 1}$

100. $h(x) = (2x - 3)^3$

101. $h(x) = \sqrt{6x} + 12$

102. $h(x) = \sqrt[3]{2x + 3} - 4$

Solve each problem.

103. *Relationship of Measurement Units* The function $f(x) = 12x$ computes the number of inches in x feet, and the function $g(x) = 5280x$ computes the number of feet in x miles. What is $(f \circ g)(x)$, and what does it compute?

104. The function $f(x) = 3x$ computes the number of feet in x yards, and the function $g(x) = 1760x$ computes the number of yards in x miles. What is $(f \circ g)(x)$, and what does it compute?

105. *Area of an Equilateral Triangle* The area of an equilateral triangle with sides of length x is given by the function $\mathcal{A}(x) = \dfrac{\sqrt{3}}{4}x^2$.

 (a) Find $\mathcal{A}(2x)$, the function representing the area of an equilateral triangle with sides of length twice the original length.

 (b) Find the area of an equilateral triangle with side length 16. Use the formula $\mathcal{A}(2x)$ found in part (a).

106. *Perimeter of a Square* The perimeter x of a square with side of length s is given by the formula $x = 4s$.

 (a) Solve for s in terms of x.

 (b) If y represents the area of this square, write y as a function of the perimeter x.

 (c) Use the composite function of part (b) to find the area of a square with perimeter 6.

107. *Oil Leak* An oil well off the Gulf Coast is leaking, with the leak spreading oil over the water's surface as a circle. At any time *t*, in minutes, after the beginning of the leak, the radius of the circular oil slick on the surface is $r(t) = 4t$ feet. Let $\mathcal{A}(r) = \pi r^2$ represent the area of a circle of radius *r*.

(a) Find $(\mathcal{A} \circ r)(t)$.

(b) Interpret $(\mathcal{A} \circ r)(t)$.

(c) What is the area of the oil slick after 3 min?

108. *Emission of Pollutants* When a thermal inversion layer is over a city (as happens in Los Angeles), pollutants cannot rise vertically but are trapped below the layer and must disperse horizontally. Assume that a factory smokestack begins emitting a pollutant at 8 A.M. Assume that the pollutant disperses horizontally over a circular area. If *t* represents the time, in hours, since the factory began emitting pollutants ($t = 0$ represents 8 A.M.), assume that the radius of the circle of pollutants at time *t* is $r(t) = 2t$ miles. Let $\mathcal{A}(r) = \pi r^2$ represent the area of a circle of radius *r*.

(a) Find $(\mathcal{A} \circ r)(t)$.　　(b) Interpret $(\mathcal{A} \circ r)(t)$.

(c) What is the area of the circular region covered by the layer at noon?

109. *(Modeling) Product of Two Factors* Suppose that *x* represents a real number between 0 and 100. Consider two factors of the following forms: $N(x)$ represents the number that is *x* less than 100 and $G(x)$ represents the number that is 20 greater than the product of 5 and *x*.

(a) Express the factor $N(x)$ in terms of *x*.

(b) Express the factor $G(x)$ in terms of *x*.

(c) Express their product $C(x)$.

(d) Suppose that in an application $C(x)$ represents a cost in dollars and the value of *x* is 20. Evaluate $C(20)$.

110. *Software Author Royalties* A software author invests his royalties in two accounts for 1 yr.

(a) The first account pays 2% simple interest. If he invests *x* dollars in this account, write an expression for y_1 in terms of *x*, where y_1 represents the amount of interest earned.

(b) He invests in a second account $500 more than he invested in the first account. This second account pays 1.5% simple interest. Write an expression for y_2, where y_2 represents the amount of interest earned.

(c) What does $y_1 + y_2$ represent?

(d) How much interest will he receive if $250 is invested in the first account?

111. *Sale Pricing* In the sale room at a clothing store, every item is on sale for half the original price, plus 1 dollar.

(a) Write a function *g* that finds half of *x*.

(b) Write a function *f* that adds 1 to *x*.

(c) Write and simplify the function $(f \circ g)(x)$.

(d) Use the function from part (c) to find the sale price of a shirt at this store that has original price $60.

112. *Area of a Square* The area of a square is x^2 square inches. Suppose that 3 in. is added to one dimension and 1 in. is subtracted from the other dimension. Express the area $\mathcal{A}(x)$ of the resulting rectangle as a product of two functions.

Chapter 2 Test Prep

Key Terms

2.1 ordered pair	**2.3** dependent variable	slope	vertex
origin	independent	average rate of	piecewise-defined
x-axis	variable	change	function
y-axis	relation	mathematical	step function
rectangular (Cartesian)	function	modeling	**2.7** symmetry
coordinate system	input	linear cost function	even function
coordinate plane	output	cost	odd function
(xy-plane)	input-output (function)	fixed cost	vertical
quadrants	machine	revenue function	translation
coordinates	domain	profit function	horizontal
conditional statement	range	**2.5** point-slope form	translation
collinear	increasing function	slope-intercept form	**2.8** difference
graph of an equation	decreasing function	negative reciprocals	quotient
x-intercept	constant function	scatter diagram	secant line
y-intercept	**2.4** linear function	linear regression	composite function
2.2 circle	constant function	zero (of a function)	(composition)
radius	standard form	**2.6** continuous function	
center of a circle	relatively prime	parabola	

New Symbols

(a, b)	ordered pair	m	slope	
$f(x)$	function f evaluated at x (read "f of x" or "f at x")	$[\![x]\!]$	the greatest integer less than or equal to x	
Δx	change in x	$f \circ g$	composite function	
Δy	change in y			

Quick Review

Concepts	Examples

2.1 Rectangular Coordinates and Graphs

Distance Formula

Suppose that $P(x_1, y_1)$ and $R(x_2, y_2)$ are two points in a coordinate plane. The distance between P and R, written $d(P, R)$, is given by the following formula.

$$d(P, R) = \sqrt{(x_2 - x_1)^2 + (y_2 - y_1)^2}$$

Find the distance between the points $P(-1, 4)$ and $R(6, -3)$.

$$d(P, R) = \sqrt{[6 - (-1)]^2 + (-3 - 4)^2}$$
$$= \sqrt{49 + 49}$$
$$= \sqrt{98}$$
$$= 7\sqrt{2} \quad \sqrt{98} = \sqrt{49 \cdot 2} = 7\sqrt{2}$$

Midpoint Formula

The coordinates of the midpoint M of the line segment with endpoints $P(x_1, y_1)$ and $Q(x_2, y_2)$ are given by the following.

$$M = \left(\frac{x_1 + x_2}{2}, \frac{y_1 + y_2}{2} \right)$$

Find the coordinates of the midpoint M of the line segment with endpoints $(-1, 4)$ and $(6, -3)$.

$$M = \left(\frac{-1 + 6}{2}, \frac{4 + (-3)}{2} \right) = \left(\frac{5}{2}, \frac{1}{2} \right)$$

Concepts	**Examples**

2.2 Circles

Center-Radius Form of the Equation of a Circle
The equation of a circle with center (h, k) and radius r is given by the following.

$$(x - h)^2 + (y - k)^2 = r^2$$

General Form of the Equation of a Circle

$$x^2 + y^2 + Dx + Ey + F = 0$$

Find the center-radius form of the equation of the circle with center $(-2, 3)$ and radius 4.

$$[x - (-2)]^2 + (y - 3)^2 = 4^2$$
$$(x + 2)^2 + (y - 3)^2 = 16$$

The general form of the equation of the preceding circle is

$$x^2 + y^2 + 4x - 6y - 3 = 0.$$

2.3 Functions

A **relation** is a set of ordered pairs. A **function** is a relation in which, for each value of the first component of the ordered pairs, there is *exactly one* value of the second component.

The set of first components is the **domain.**

The set of second components is the **range.**

The relation $y = x^2$ defines a function because each choice of a number for x corresponds to one and only one number for y. The domain is $(-\infty, \infty)$, and the range is $[0, \infty)$.

The relation $x = y^2$ does *not* define a function because a number x may correspond to two numbers for y. The domain is $[0, \infty)$, and the range is $(-\infty, \infty)$.

Vertical Line Test
If every vertical line intersects the graph of a relation in no more than one point, then the relation is a function.

Determine whether each graph is that of a function.

A.   **B.**

By the vertical line test, graph A is the graph of a function, but graph B is not.

Increasing, Decreasing, and Constant Functions

Discuss the function in graph A in terms of whether it is increasing, decreasing, or constant.

The function in graph A is decreasing on the open interval $(-\infty, 0)$ and increasing on the open interval $(0, \infty)$.

2.4 Linear Functions

A function f is a **linear function** if, for real numbers a and b,

$$f(x) = ax + b.$$

The graph of a linear function is a line.

The equation

$$y = f(x) = \frac{1}{2}x - 4$$

defines y as a linear function f of x.

Definition of Slope
The slope m of the line through the points (x_1, y_1) and (x_2, y_2) is given by the following.

$$m = \frac{\text{rise}}{\text{run}} = \frac{\Delta y}{\Delta x} = \frac{y_2 - y_1}{x_2 - x_1}, \quad \text{where } \Delta x \neq 0$$

Find the slope of the line through the points $(2, 4)$ and $(-1, 7)$.

$$m = \frac{7 - 4}{-1 - 2} = \frac{3}{-3} = -1$$

Concepts	Examples

2.5 Equations of Lines and Linear Models

Summary of Forms of Linear Equations

Equation	Description
$y = mx + b$	**Slope-Intercept Form** Slope is m. y-intercept is $(0, b)$.
$y - y_1 = m(x - x_1)$	**Point-Slope Form** Slope is m. Line passes through (x_1, y_1).
$Ax + By = C$	**Standard Form** (A, B, and C integers, $A \geq 0$.) Slope is $-\frac{A}{B}$ $\quad$ $(B \neq 0)$. x-intercept is $\left(\frac{C}{A}, 0\right)$ $\quad$ $(A \neq 0)$. y-intercept is $\left(0, \frac{C}{B}\right)$ $\quad$ $(B \neq 0)$.
$y = b$	**Horizontal Line** Slope is 0. y-intercept is $(0, b)$.
$x = a$	**Vertical Line** Slope is undefined. x-intercept is $(a, 0)$.

Consider the following equations.

$$y = 3x + \frac{2}{3} \quad \text{Slope-intercept form}$$

The slope m is 3, and the y-intercept is $\left(0, \frac{2}{3}\right)$.

$$y - 3 = -2(x - 4) \quad \text{Point-slope form}$$

The slope m is -2. The line passes through the point $(4, 3)$.

$$4x + 5y = 7 \quad \text{Standard form with } A = 4, B = 5, C = 7$$

The slope is $m = -\frac{A}{B} = -\frac{4}{5}$.

The x-intercept has $\frac{C}{A} = \frac{7}{4}$ and is $\left(\frac{7}{4}, 0\right)$.

The y-intercept has $\frac{C}{B} = \frac{7}{5}$ and is $\left(0, \frac{7}{5}\right)$.

$$y = -6 \quad \text{Horizontal line}$$

The slope is 0. The y-intercept is $(0, -6)$.

$$x = 3 \quad \text{Vertical line}$$

The slope is undefined. The x-intercept is $(3, 0)$.

2.6 Graphs of Basic Functions

Basic Functions

Identity Function $\quad f(x) = x$

Squaring Function $\quad f(x) = x^2$

Cubing Function $\quad f(x) = x^3$

Square Root Function $\quad f(x) = \sqrt{x}$

Cube Root Function $\quad f(x) = \sqrt[3]{x}$

Absolute Value Function $\quad f(x) = |x|$

Greatest Integer Function $\quad f(x) = [\![x]\!]$

Refer to the function boxes in **Section 2.6.** Graphs of the basic functions are also shown on the back inside cover of the print text.

2.7 Graphing Techniques

Stretching and Shrinking

If $a > 1$, then the graph of $y = af(x)$ is a **vertical stretching** of the graph of $y = f(x)$.

If $0 < a < 1$, then the graph of $y = af(x)$ is a **vertical shrinking** of the graph of $y = f(x)$.

If $0 < a < 1$, then the graph of $y = f(ax)$ is a **horizontal stretching** of the graph of $y = f(x)$.

If $a > 1$, then the graph of $y = f(ax)$ is a **horizontal shrinking** of the graph of $y = f(x)$.

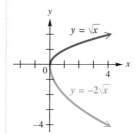

The graph of

$$y = -2\sqrt{x}$$

is the graph of $y = \sqrt{x}$ stretched vertically by a factor of 2 and reflected across the x-axis.

Concepts	Examples

Reflection across an Axis

The graph of $y = -f(x)$ is the same as the graph of $y = f(x)$ reflected across the x-axis.

The graph of $y = f(-x)$ is the same as the graph of $y = f(x)$ reflected across the y-axis.

Symmetry

The graph of an equation is **symmetric with respect to the y-axis** if the replacement of x with $-x$ results in an equivalent equation.

The graph of an equation is **symmetric with respect to the x-axis** if the replacement of y with $-y$ results in an equivalent equation.

The graph of an equation is **symmetric with respect to the origin** if the replacement of both x with $-x$ and y with $-y$ at the same time results in an equivalent equation.

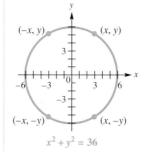

The graph of
$$x^2 + y^2 = 36$$
is symmetric with respect to the y-axis, the x-axis, and the origin.

Translations

Let f be a function and c be a positive number.

To Graph:	Shift the Graph of $y = f(x)$ by c Units:
$y = f(x) + c$	up
$y = f(x) - c$	down
$y = f(x + c)$	left
$y = f(x - c)$	right

The graph of
$$y = (x - 1)^3 + 2$$
is the graph of $y = x^3$ translated 1 unit to the right and 2 units up.

2.8 Function Operations and Composition

Operations on Functions

Given two functions f and g, then for all values of x for which both $f(x)$ and $g(x)$ are defined, the following operations are defined.

$$(f + g)(x) = f(x) + g(x) \qquad \text{Sum function}$$

$$(f - g)(x) = f(x) - g(x) \qquad \text{Difference function}$$

$$(fg)(x) = f(x) \cdot g(x) \qquad \text{Product function}$$

$$\left(\frac{f}{g}\right)(x) = \frac{f(x)}{g(x)}, \quad g(x) \neq 0 \qquad \text{Quotient function}$$

The **domains of $f + g$, $f - g$,** and **fg** include all real numbers in the intersection of the domains of f and g, while the **domain of $\frac{f}{g}$** includes those real numbers in the intersection of the domains of f and g for which $g(x) \neq 0$.

Let $f(x) = 2x - 4$ and $g(x) = \sqrt{x}$.

$$(f + g)(x) = 2x - 4 + \sqrt{x}$$
$$(f - g)(x) = 2x - 4 - \sqrt{x}$$
$$(fg)(x) = (2x - 4)\sqrt{x}$$

The domain is $[0, \infty)$.

$$\left(\frac{f}{g}\right)(x) = \frac{2x - 4}{\sqrt{x}}$$

The domain is $(0, \infty)$.

Difference Quotient

The line joining $P(x, f(x))$ and $Q(x + h, f(x + h))$ has slope

$$m = \frac{f(x + h) - f(x)}{h}, \quad h \neq 0.$$

The boldface expression is the difference quotient.

Refer to **Example 4** in **Section 2.8.**

Concepts	Examples
Composition of Functions If f and g are functions, then the composite function, or composition, of f and g is defined by $$(f \circ g)(x) = f(g(x)).$$ The domain of $f \circ g$ is the set of all x in the domain of g such that $g(x)$ is in the domain of f.	Let $f(x) = 2x - 4$ and $g(x) = \sqrt{x}$. $$(f \circ g)(x) = 2\sqrt{x} - 4$$ The domain is all x such that $x \geq 0$, represented by the interval $[0, \infty)$.

Chapter 2 Review Exercises

Find the distance between each pair of points, and give the coordinates of the midpoint of the line segment joining them.

1. $P(3, -1), Q(-4, 5)$ **2.** $M(-8, 2), N(3, -7)$ **3.** $A(-6, 3), B(-6, 8)$

4. Are the points $(5, 7)$, $(3, 9)$, and $(6, 8)$ the vertices of a right triangle? If so, at what point is the right angle?

5. Determine the coordinates of B for line segment AB, given that A has coordinates $(-6, 10)$ and the coordinates of its midpoint M are $(8, 2)$.

6. Use the distance formula to determine whether the points $(-2, -5)$, $(1, 7)$, and $(3, 15)$ are collinear.

Find the center-radius form of the equation for each circle.

7. center $(-2, 3)$, radius 15   **8.** center $\left(\sqrt{5}, -\sqrt{7}\right)$, radius $\sqrt{3}$

9. center $(-8, 1)$, passing through $(0, 16)$ **10.** center $(3, -6)$, tangent to the x-axis

Connecting Graphs with Equations *Use each graph to determine an equation of the circle. Express in center-radius form.*

11.

$(3, 5)$

12.

$(-2, 3)$

13.
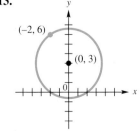
$(-2, 6)$
$(0, 3)$

14.
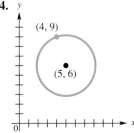
$(4, 9)$
$(5, 6)$

Find the center and radius of each circle.

15. $x^2 + y^2 - 4x + 6y + 12 = 0$

16. $x^2 + y^2 - 6x - 10y + 30 = 0$

17. $2x^2 + 2y^2 + 14x + 6y + 2 = 0$

18. $3x^2 + 3y^2 + 33x - 15y = 0$

For each graph, decide whether y is a function of x. Give the domain and range of each relation.

19.

20.

21.

22.

23.

24.

Decide whether each equation defines y as a function of x.

25. $y = 6 - x^2$

26. $x = \frac{1}{3}y^2$

27. $y = \pm \sqrt{x - 2}$

28. $y = -\dfrac{4}{x}$

Give the domain of each function.

29. $f(x) = -4 + |x|$

30. $f(x) = \dfrac{8 + x}{8 - x}$

31. $f(x) = \sqrt{6 - 3x}$

32. For the function graphed in **Exercise 22**, determine the largest open intervals over which it is **(a)** increasing, **(b)** decreasing, and **(c)** constant.

Let $f(x) = -2x^2 + 3x - 6$. Find each of the following.

33. $f(3)$

34. $f(-0.5)$

35. $f(0)$

36. $f(k)$

Graph each equation.

37. $2x - 5y = 5$

38. $3x + 7y = 14$

39. $2x + 5y = 20$

40. $3y = x$

41. $f(x) = x$

42. $x - 4y = 8$

43. $x = -5$

44. $f(x) = 3$

45. $y + 2 = 0$

46. *Concept Check* The equation of the line that lies along the *x*-axis is _____.

Graph the line satisfying the given conditions.

47. through $(0, 5)$, $m = -\frac{2}{3}$

48. through $(2, -4)$, $m = \frac{3}{4}$

Find the slope of each line, provided that it has a slope.

49. through $(2, -2)$ and $(3, -4)$

50. through $(8, 7)$ and $\left(\frac{1}{2}, -2\right)$

51. through $(0, -7)$ and $(3, -7)$

52. through $(5, 6)$ and $(5, -2)$

53. $11x + 2y = 3$

54. $9x - 4y = 2$

55. $x - 2 = 0$

56. $x - 5y = 0$

Work each problem.

57. *(Modeling) Distance from Home* The graph depicts the distance y that a person driving a car on a straight road is from home after x hours. Interpret the graph. At what speeds did the car travel?

Distance from Home

58. *(Modeling) Job Market* The figure shows the number of jobs gained or lost in a recent period from September to May.

(a) Is this the graph of a function?

(b) In what month were the most jobs lost? the most gained?

(c) What was the largest number of jobs lost? of jobs gained?

(d) Do these data show an upward or a downward trend? If so, which is it?

Job Market Trends

59. *(Modeling) E-Filing Tax Returns* The percent of tax returns filed electronically in 2001 was 30.7%. In 2013, the figure was 82.9%. (*Source:* Internal Revenue Service.)

(a) Use the information given for the years 2001 and 2013, letting $x = 0$ represent 2001, $x = 12$ represent 2013, and y represent the percent of returns filed electronically, to find a linear equation that models the data. Write the equation in slope-intercept form. Interpret the slope of the graph of this equation.

(b) Use the equation from part (a) to approximate the percent of tax returns that were filed electronically in 2009.

60. *Family Income* In 1980 the median family income in the United States was about $21,000 per year. In 2013 it was about $63,800 per year. Find the average annual rate of change of median family income to the nearest dollar over that period. (*Source:* U.S. Census Bureau.)

For each line described, write an equation in (a) slope-intercept form, if possible, and (b) standard form.

61. through $(3, -5)$ with slope -2

62. through $(-2, 4)$ and $(1, 3)$

63. through $(2, -1)$, parallel to $3x - y = 1$

64. x-intercept $(-3, 0)$, y-intercept $(0, 5)$

65. through $(2, -10)$, perpendicular to a line with undefined slope

66. through $(0, 5)$, perpendicular to $8x + 5y = 3$

67. through $(-7, 4)$, perpendicular to $y = 8$

68. through $(3, -5)$, parallel to $y = 4$

Graph each function.

69. $f(x) = |x| - 3$

70. $f(x) = -|x|$

71. $f(x) = -(x + 1)^2 + 3$

72. $f(x) = -\sqrt{x} - 2$

73. $f(x) = [\![x - 3]\!]$

74. $f(x) = 2\sqrt[3]{x + 1} - 2$

75. $f(x) = \begin{cases} -4x + 2 & \text{if } x \le 1 \\ 3x - 5 & \text{if } x > 1 \end{cases}$

76. $f(x) = \begin{cases} x^2 + 3 & \text{if } x < 2 \\ -x + 4 & \text{if } x \geq 2 \end{cases}$ **77.** $f(x) = \begin{cases} |x| & \text{if } x < 3 \\ 6 - x & \text{if } x \geq 3 \end{cases}$

78. *Concept Check* If x represents an integer, then what is the simplest form of the expression $[\![x]\!] + x$?

Concept Check *Decide whether each statement is* true *or* false. *If false, tell why.*

79. The graph of an even function is symmetric with respect to the y-axis.

80. The graph of a nonzero function cannot be symmetric with respect to the x-axis.

81. If (a, b) is on the graph of an even function, then so is $(a, -b)$.

82. The graph of an odd function is symmetric with respect to the origin.

83. The constant function $f(x) = 0$ is both even and odd.

84. If (a, b) is on the graph of an odd function, then so is $(-a, b)$.

Decide whether each equation has a graph that is symmetric with respect to the x-axis, *the* y-axis, *the* origin, *or* none of these.

85. $x + y^2 = 10$ **86.** $5y^2 + 5x^2 = 30$ **87.** $x^2 = y^3$

88. $y^3 = x + 4$ **89.** $6x + y = 4$ **90.** $|y| = -x$

91. $y = 1$ **92.** $|x| = |y|$

93. $x^2 - y^2 = 0$ **94.** $x^2 + (y - 2)^2 = 4$

Describe how the graph of each function can be obtained from the graph of $f(x) = |x|$.

95. $g(x) = -|x|$ **96.** $h(x) = |x| - 2$ **97.** $k(x) = 2|x - 4|$

Let $f(x) = 3x - 4$. *Find an equation for each reflection of the graph of* $f(x)$.

98. across the x-axis   **99.** across the y-axis   **100.** across the origin

101. *Concept Check* The graph of a function f is shown in the figure. Sketch the graph of each function defined as follows.

(a) $y = f(x) + 3$
(b) $y = f(x - 2)$
(c) $y = f(x + 3) - 2$
(d) $y = |f(x)|$

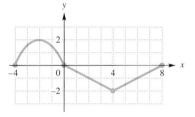

102. *Concept Check* Must the domain of g be a subset of the domain of $f \circ g$?

Let $f(x) = 3x^2 - 4$ *and* $g(x) = x^2 - 3x - 4$. *Find each of the following.*

103. $(fg)(x)$ **104.** $(f - g)(4)$ **105.** $(f + g)(-4)$

106. $(f + g)(2k)$ **107.** $\left(\dfrac{f}{g}\right)(3)$ **108.** $\left(\dfrac{f}{g}\right)(-1)$

109. the domain of $(fg)(x)$ **110.** the domain of $\left(\dfrac{f}{g}\right)(x)$

For each function, find and simplify $\dfrac{f(x + h) - f(x)}{h}$, $h \neq 0$.

111. $f(x) = 2x + 9$ **112.** $f(x) = x^2 - 5x + 3$

Let $f(x) = \sqrt{x - 2}$ and $g(x) = x^2$. Find each of the following, if possible.

113. $(g \circ f)(x)$ **114.** $(f \circ g)(x)$ **115.** $(g \circ f)(3)$

116. $(f \circ g)(-6)$ **117.** $(g \circ f)(-1)$ **118.** the domain of $f \circ g$

Use the table to evaluate each expression, if possible.

119. $(f + g)(1)$

120. $(f - g)(3)$

121. $(fg)(-1)$

122. $\left(\dfrac{f}{g}\right)(0)$

x	$f(x)$	$g(x)$
-1	3	-2
0	5	0
1	7	1
3	9	9

Use the tables for f and g to evaluate each expression.

123. $(g \circ f)(-2)$

124. $(f \circ g)(3)$

x	$f(x)$
-2	1
0	4
2	3
4	2

x	$g(x)$
1	2
2	4
3	-2
4	0

Concept Check The graphs of two functions f and g are shown in the figures.

125. Find $(f \circ g)(2)$.

126. Find $(g \circ f)(3)$.

 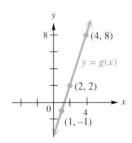

Solve each problem.

127. *Relationship of Measurement Units* There are 36 in. in 1 yd, and there are 1760 yd in 1 mi. Express the number of inches in x miles by forming two functions and then considering their composition.

128. *(Modeling) Perimeter of a Rectangle* Suppose the length of a rectangle is twice its width. Let x represent the width of the rectangle. Write a formula for the perimeter P of the rectangle in terms of x alone. Then use $P(x)$ notation to describe it as a function. What type of function is this?

129. *(Modeling) Volume of a Sphere* The formula for the volume of a sphere is $V(r) = \frac{4}{3}\pi r^3$, where r represents the radius of the sphere. Construct a model function V representing the amount of volume gained when the radius r (in inches) of a sphere is increased by 3 in.

130. *(Modeling) Dimensions of a Cylinder* A cylindrical can makes the most efficient use of materials when its height is the same as the diameter of its top.

  (a) Express the volume V of such a can as a function of the diameter d of its top.

  (b) Express the surface area S of such a can as a function of the diameter d of its top. (*Hint:* The curved side is made from a rectangle whose length is the circumference of the top of the can.)

Chapter 2	Test

1. Match the set described in Column I with the correct interval notation from Column II. Choices in Column II may be used once, more than once, or not at all.

I

(a) Domain of $f(x) = \sqrt{x + 3}$

(b) Range of $f(x) = \sqrt{x - 3}$

(c) Domain of $f(x) = x^2 - 3$

(d) Range of $f(x) = x^2 + 3$

(e) Domain of $f(x) = \sqrt[3]{x - 3}$

(f) Range of $f(x) = \sqrt[3]{x + 3}$

(g) Domain of $f(x) = |x| - 3$

(h) Range of $f(x) = |x + 3|$

(i) Domain of $x = y^2$

(j) Range of $x = y^2$

II

A. $[-3, \infty)$

B. $[3, \infty)$

C. $(-\infty, \infty)$

D. $[0, \infty)$

E. $(-\infty, 3)$

F. $(-\infty, 3]$

G. $(3, \infty)$

H. $(-\infty, 0]$

The graph shows the line that passes through the points $(-2, 1)$ and $(3, 4)$. Refer to it to answer Exercises 2–6.

2. What is the slope of the line?

3. What is the distance between the two points shown?

4. What are the coordinates of the midpoint of the line segment joining the two points?

5. Find the standard form of the equation of the line.

6. Write the linear function $f(x) = ax + b$ that has this line as its graph.

7. *Connecting Graphs with Equations* Use each graph to determine an equation of the circle. Express it in center-radius form.

(a)

(b)

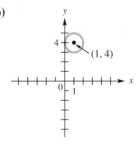

8. Graph the circle with equation $x^2 + y^2 + 4x - 10y + 13 = 0$.

9. In each case, determine whether y is a function of x. Give the domain and range. If it is a function, give the largest open intervals over which it is increasing, decreasing, or constant.

(a)

(b)

10. Suppose point A has coordinates $(5, -3)$. What is the equation of the

    (a) vertical line through A?   **(b)** horizontal line through A?

11. Find the slope-intercept form of the equation of the line passing through $(2, 3)$ and

    (a) parallel to the graph of $y = -3x + 2$

    (b) perpendicular to the graph of $y = -3x + 2$.

12. Consider the graph of the function shown here. Give the open interval(s) over which the function is

    (a) increasing   **(b)** decreasing

    (c) constant   **(d)** continuous.

    (e) What is the domain of this function?

    (f) What is the range of this function?

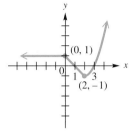

Graph each function.

13. $f(x) = |x - 2| - 1$ **14.** $f(x) = [\![x + 1]\!]$ **15.** $f(x) = \begin{cases} 3 & \text{if } x < -2 \\ 2 - \frac{1}{2}x & \text{if } x \geq -2 \end{cases}$

16. The graph of $y = f(x)$ is shown here. Sketch the graph of each of the following. Use ordered pairs to indicate three points on the graph.

    (a) $y = f(x) + 2$ **(b)** $y = f(x + 2)$

    (c) $y = -f(x)$ **(d)** $y = f(-x)$

    (e) $y = 2f(x)$

17. Describe how the graph of $f(x) = -2\sqrt{x + 2} - 3$ can be obtained from the graph of $y = \sqrt{x}$.

18. Determine whether the graph of $3x^2 - 2y^2 = 3$ is symmetric with respect to the

    (a) x-axis   **(b)** y-axis   **(c)** origin.

19. Let $f(x) = 2x^2 - 3x + 2$ and $g(x) = -2x + 1$. Find each of the following. Simplify the expressions when possible.

    (a) $(f - g)(x)$ **(b)** $\left(\dfrac{f}{g}\right)(x)$

    (c) domain of $\dfrac{f}{g}$ **(d)** $\dfrac{f(x + h) - f(x)}{h}$ $(h \neq 0)$

    (e) $(f + g)(1)$ **(f)** $(fg)(2)$ **(g)** $(f \circ g)(0)$

Let $f(x) = \sqrt{x + 1}$ and $g(x) = 2x - 7$. Find each of the following.

20. $(f \circ g)(x)$ and its domain   **21.** $(g \circ f)(x)$ and its domain

22. *(Modeling) Cost, Revenue, and Profit Analysis* Dotty starts up a small business manufacturing bobble-head figures of famous soccer players. Her initial cost is $3300. Each figure costs $4.50 to manufacture.

    (a) Write a cost function C, where x represents the number of figures manufactured.

    (b) Find the revenue function R if each figure in part (a) sells for $10.50.

    (c) Give the profit function P.

    (d) How many figures must be produced and sold for Dotty to earn a profit?

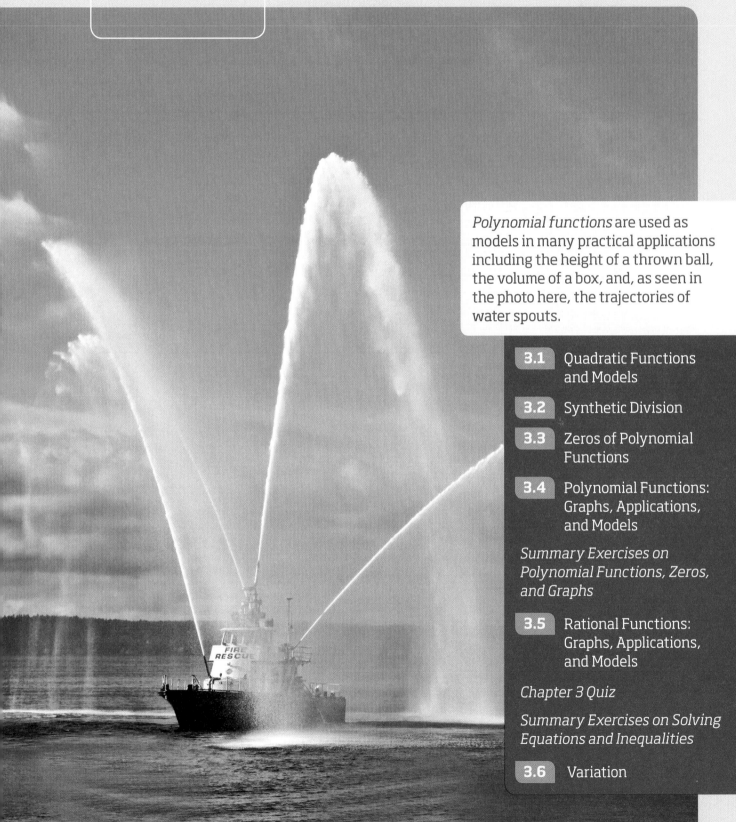

3

Polynomial and Rational Functions

Polynomial functions are used as models in many practical applications including the height of a thrown ball, the volume of a box, and, as seen in the photo here, the trajectories of water spouts.

303

3.1 Quadratic Functions and Models

- Polynomial Functions
- Quadratic Functions
- Graphing Techniques
- Completing the Square
- The Vertex Formula
- Quadratic Models

Polynomial Functions

Polynomial Function

A **polynomial function** f of degree n, where n is a nonnegative integer, is given by

$$f(x) = a_n x^n + a_{n-1} x^{n-1} + \cdots + a_1 x + a_0,$$

where $a_n, a_{n-1}, \ldots, a_1$, and a_0 are complex numbers, with $a_n \neq 0$.

In this chapter we primarily consider polynomial functions having real coefficients. When analyzing a polynomial function, the degree n and the **leading coefficient** a_n are important. These are both given in the **dominating term** $a_n x^n$.

LOOKING AHEAD TO CALCULUS

In calculus, polynomial functions are used to approximate more complicated functions. For example, the trigonometric function $\sin x$ is approximated by the polynomial

$$x - \frac{x^3}{6} + \frac{x^5}{120} - \frac{x^7}{5040}.$$

Polynomial Function	Function Type	Degree n	Leading Coefficient a_n
$f(x) = 2$	Constant	0	2
$f(x) = 5x - 1$	Linear	1	5
$f(x) = 4x^2 - x + 1$	Quadratic	2	4
$f(x) = 2x^3 - \frac{1}{2}x + 5$	Cubic	3	2
$f(x) = x^4 + \sqrt{2}x^3 - 3x^2$	Quartic	4	1

The function $f(x) = 0$ is the **zero polynomial** and has no degree.

Quadratic Functions

Polynomial functions of degree 2 are *quadratic functions*. Again, we are most often concerned with real coefficients.

Quadratic Function

A function f is a **quadratic function** if

$$f(x) = ax^2 + bx + c,$$

where a, b, and c are complex numbers, with $a \neq 0$.

The simplest quadratic function is

$$f(x) = x^2. \quad \text{Squaring function}$$

See **Figure 1.** This graph is a **parabola.** Every quadratic function with real coefficients defined over the real numbers has a graph that is a parabola. The domain of $f(x) = x^2$ is $(-\infty, \infty)$, and the range is $[0, \infty)$. The lowest point on the graph occurs at the origin $(0, 0)$. Thus, the function decreases on the open interval $(-\infty, 0)$ and increases on the open interval $(0, \infty)$. (Remember that these intervals indicate x-values.)

x	$f(x)$
-2	4
-1	1
0	0
1	1
2	4

Figure 1

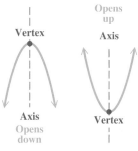

Figure 2

Parabolas are symmetric with respect to a line (the y-axis in **Figure 1**). This line is the **axis of symmetry,** or **axis,** of the parabola. The point where the axis intersects the parabola is the **vertex** of the parabola. As **Figure 2** shows, the vertex of a parabola that opens down is the highest point of the graph, and the vertex of a parabola that opens up is the lowest point of the graph.

Graphing Techniques   Graphing techniques may be applied to the graph of $f(x) = x^2$ to give the graph of a different quadratic function. Compared to the basic graph of $f(x) = x^2$, the graph of $F(x) = a(x - h)^2 + k$, with $a \neq 0$, has the following characteristics.

$$F(x) = a(x - h)^2 + k$$

	Horizontal shift:	Vertical shift:
• Opens up if $a > 0$	• h units right if $h > 0$	• k units up if $k > 0$
• Opens down if $a < 0$	• $|h|$ units left if $h < 0$	• $|k|$ units down if $k < 0$
• Vertically stretched (narrower) if $|a| > 1$		
• Vertically shrunk (wider) if $0 < |a| < 1$		

EXAMPLE 1   **Graphing Quadratic Functions**

Graph each function. Give the domain and range.

(a) $f(x) = x^2 - 4x - 2$ (by plotting points)

(b) $g(x) = -\frac{1}{2}x^2$ $\left(\text{and compare to } y = x^2 \text{ and } y = \frac{1}{2}x^2\right)$

(c) $F(x) = -\frac{1}{2}(x - 4)^2 + 3$ (and compare to the graph in part (b))

SOLUTION

(a) See the table with **Figure 3.** The domain of $f(x) = x^2 - 4x - 2$ is $(-\infty, \infty)$, the range is $[-6, \infty)$, the vertex is the point $(2, -6)$, and the axis has equation $x = 2$. **Figure 4** shows how a graphing calculator displays this graph.

x	$f(x)$
-1	3
0	-2
1	-5
2	-6
3	-5
4	-2
5	3

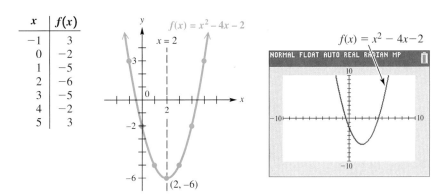

Figure 3   **Figure 4**

(b) Think of $g(x) = -\frac{1}{2}x^2$ as $g(x) = -\left(\frac{1}{2}x^2\right)$. The graph of $y = \frac{1}{2}x^2$ is a wider version of the graph of $y = x^2$, and the graph of $g(x) = -\left(\frac{1}{2}x^2\right)$ is a reflection of the graph of $y = \frac{1}{2}x^2$ across the x-axis. See **Figure 5** on the next page. The vertex is the point $(0, 0)$, and the axis of the parabola is the line $x = 0$ (the y-axis). The domain is $(-\infty, \infty)$, and the range is $(-\infty, 0]$.

Calculator graphs are shown in **Figure 6.**

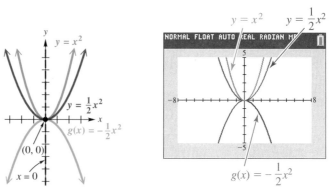

Figure 5   **Figure 6**

(c) Notice that $F(x) = -\frac{1}{2}(x - 4)^2 + 3$ is related to $g(x) = -\frac{1}{2}x^2$ from part (b). The graph of $F(x)$ is the graph of $g(x)$ translated 4 units to the right and 3 units up. See **Figure 7.** The vertex is the point $(4, 3)$, which is also shown in the calculator graph in **Figure 8,** and the axis of the parabola is the line $x = 4$. The domain is $(-\infty, \infty)$, and the range is $(-\infty, 3]$.

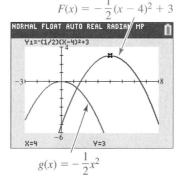

Figure 7   **Figure 8**

✔ **Now Try Exercises 19 and 21.**

Completing the Square   In general, the graph of the quadratic function

$$f(x) = a(x - h)^2 + k \quad (a \neq 0)$$

is a parabola with **vertex (h, k)** and **axis of symmetry $x = h$.** The parabola opens up if a is positive and down if a is negative. With these facts in mind, we *complete the square* to graph the general quadratic function

$$f(x) = ax^2 + bx + c.$$

EXAMPLE 2   **Graphing a Parabola ($a = 1$)**

Graph $f(x) = x^2 - 6x + 7$. Find the largest open intervals over which the function is increasing or decreasing.

SOLUTION   We express $x^2 - 6x + 7$ in the form $(x - h)^2 + k$ by completing the square. In preparation for this, we first write

$$f(x) = (x^2 - 6x \qquad) + 7. \quad \text{Prepare to complete the square.}$$

We must add a number inside the parentheses to obtain a perfect square trinomial. Find this number by taking half the coefficient of x and squaring the result.

$f(x) = x^2 - 6x + 7$
$f(x) = (x - 3)^2 - 2$

This screen shows that the vertex of the graph in **Figure 9** is the point $(3, -2)$. Because it is the *lowest* point on the graph, we direct the calculator to find the *minimum*.

$$\left[\tfrac{1}{2}(-6)\right]^2 = (-3)^2 = 9 \quad \text{Take half the coefficient of } x.$$
$$\text{Square the result.}$$

$f(x) = (x^2 - 6x + 9 - 9) + 7 \quad$ Add and subtract 9. — This is the same as adding 0.

$f(x) = (x^2 - 6x + 9) - 9 + 7 \quad$ Regroup terms.

$f(x) = (x - 3)^2 - 2 \quad$ Factor and simplify.

The vertex of the parabola is the point $(3, -2)$, and the axis is the line $x = 3$. We find additional ordered pairs that satisfy the equation, as shown in the table, and plot and join these points to obtain the graph in **Figure 9**.

x	y	
0	7	← y-intercept
1	2	
3	−2	← Vertex
5	2	Find using symmetry
6	7	about the axis.

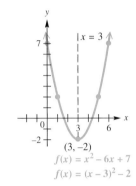

$f(x) = x^2 - 6x + 7$
$f(x) = (x - 3)^2 - 2$

Figure 9

The domain of this function is $(-\infty, \infty)$, and the range is $[-2, \infty)$. Because the lowest point on the graph is the vertex $(3, -2)$, the function is decreasing on $(-\infty, 3)$ and increasing on $(3, \infty)$.

✔ **Now Try Exercise 31.**

NOTE In **Example 2** we added and subtracted 9 *on the same side* of the equation to complete the square. This differs from adding the same number to *each side of the equation*, as is sometimes done in the procedure. We want $f(x)$—that is, y—alone on one side of the equation, so we adjusted that step in the process of completing the square here.

EXAMPLE 3 **Graphing a Parabola ($a \neq 1$)**

Graph $f(x) = -3x^2 - 2x + 1$. Identify the intercepts of the graph.

SOLUTION To complete the square, the coefficient of x^2 must be 1.

$$f(x) = -3\left(x^2 + \frac{2}{3}x \qquad\right) + 1 \qquad \text{Factor } -3 \text{ from the first two terms.}$$

$$f(x) = -3\left(x^2 + \frac{2}{3}x + \frac{1}{9} - \frac{1}{9}\right) + 1 \qquad \left[\tfrac{1}{2}\left(\tfrac{2}{3}\right)\right]^2 = \left(\tfrac{1}{3}\right)^2 = \tfrac{1}{9}, \text{ so add and subtract } \tfrac{1}{9}.$$

$$f(x) = -3\left(x^2 + \frac{2}{3}x + \frac{1}{9}\right) - 3\left(-\frac{1}{9}\right) + 1 \qquad \text{Distributive property}$$

Be careful here.

$$f(x) = -3\left(x + \frac{1}{3}\right)^2 + \frac{4}{3} \qquad \text{Factor and simplify.}$$

The vertex is the point $\left(-\frac{1}{3}, \frac{4}{3}\right)$. The intercepts are good additional points to find. The y-intercept is found by evaluating $f(0)$.

$$f(0) = -3(0)^2 - 2(0) + 1 \qquad \text{Let } x = 0 \text{ in } f(x) = -3x^2 - 2x + 1.$$

$$f(0) = 1 \longleftarrow \text{The } y\text{-intercept is } (0, 1).$$

The x-intercepts are found by setting $f(x)$ equal to 0 and solving for x.

$$0 = -3x^2 - 2x + 1 \qquad \text{Set } f(x) = 0.$$

$$0 = 3x^2 + 2x - 1 \qquad \text{Multiply by } -1.$$

$$0 = (3x - 1)(x + 1) \qquad \text{Factor.}$$

$$x = \frac{1}{3} \quad \text{or} \quad x = -1 \qquad \text{Zero-factor property}$$

Therefore, the x-intercepts are $\left(\frac{1}{3}, 0\right)$ and $(-1, 0)$. The graph is shown in **Figure 10.**

This screen gives the vertex of the graph in **Figure 10** as the point $\left(-\frac{1}{3}, \frac{4}{3}\right)$. (The display shows decimal approximations.) We want the highest point on the graph, so we direct the calculator to find the *maximum*.

$$f(x) = -3x^2 - 2x + 1$$
$$f(x) = -3\left(x + \frac{1}{3}\right)^2 + \frac{4}{3}$$

Figure 10 ✔ **Now Try Exercise 33.**

NOTE It is possible to reverse the process of **Example 3** and write the quadratic function from its graph in **Figure 10** if the vertex and any other point on the graph are known. Because quadratic functions take the form

$$f(x) = a(x - h)^2 + k,$$

we can substitute the x- and y-values of the vertex, $\left(-\frac{1}{3}, \frac{4}{3}\right)$, for h and k.

$$f(x) = a\left[x - \left(-\frac{1}{3}\right)\right]^2 + \frac{4}{3} \qquad \text{Let } h = -\frac{1}{3} \text{ and } k = \frac{4}{3}.$$

$$f(x) = a\left(x + \frac{1}{3}\right)^2 + \frac{4}{3} \qquad \text{Simplify.}$$

We find the value of a by substituting the x- and y-coordinates of any other point on the graph, say $(0, 1)$, into this function and solving for a.

$$1 = a\left(0 + \frac{1}{3}\right)^2 + \frac{4}{3} \qquad \text{Let } x = 0 \text{ and } y = 1.$$

$$1 = a\left(\frac{1}{9}\right) + \frac{4}{3} \qquad \text{Square.}$$

$$-\frac{1}{3} = \frac{1}{9}a \qquad \text{Subtract } \frac{4}{3}.$$

$$a = -3 \qquad \text{Multiply by 9. Interchange sides.}$$

Verify in **Example 3** that the vertex form of the quadratic function is

$$f(x) = -3\left(x + \frac{1}{3}\right)^2 + \frac{4}{3}.$$

Exercises of this type are labeled *Connecting Graphs with Equations*.

The Vertex Formula   We can generalize the earlier work to obtain a formula for the vertex of a parabola.

$$f(x) = ax^2 + bx + c \qquad \text{General quadratic form}$$

$$= a\left(x^2 + \frac{b}{a}x \qquad\right) + c \qquad \text{Factor } a \text{ from the first two terms.}$$

$$= a\left(x^2 + \frac{b}{a}x + \frac{b^2}{4a^2}\right) + c - a\left(\frac{b^2}{4a^2}\right) \qquad \begin{array}{l}\text{Add } \left[\frac{1}{2}\left(\frac{b}{a}\right)\right]^2 = \frac{b^2}{4a^2} \text{ inside the} \\ \text{parentheses. Subtract } a\left(\frac{b^2}{4a^2}\right) \\ \text{outside the parentheses.}\end{array}$$

$$= a\left(x + \frac{b}{2a}\right)^2 + c - \frac{b^2}{4a} \qquad \text{Factor and simplify.}$$

$$f(x) = a\underbrace{\left[x - \left(-\frac{b}{2a}\right)\right]^2}_{h} + \underbrace{\frac{4ac - b^2}{4a}}_{k} \qquad \begin{array}{l}\text{Vertex form of} \\ f(x) = a(x - h)^2 + k\end{array}$$

Thus, the vertex (h, k) can be expressed in terms of a, b, and c. **It is not necessary to memorize the expression for k because it is equal to $f(h) = f\left(-\frac{b}{2a}\right).$**

Graph of a Quadratic Function

The quadratic function $f(x) = ax^2 + bx + c$ can be written as

$$y = f(x) = a(x - h)^2 + k, \quad \text{with } a \neq 0,$$

where $\qquad h = -\dfrac{b}{2a} \quad$ and $\quad k = f(h)$.   Vertex formula

The graph of f has the following characteristics.

1. It is a parabola with vertex (h, k) and the vertical line $x = h$ as axis.

2. It opens up if $a > 0$ and down if $a < 0$.

3. It is wider than the graph of $y = x^2$ if $|a| < 1$ and narrower if $|a| > 1$.

4. The y-intercept is $(0, f(0)) = (0, c)$.

5. The x-intercepts are found by solving the equation $ax^2 + bx + c = 0$.

   - If $b^2 - 4ac > 0$, then the x-intercepts are $\left(\dfrac{-b \pm \sqrt{b^2 - 4ac}}{2a}, 0\right)$.

   - If $b^2 - 4ac = 0$, then the x-intercept is $\left(-\dfrac{b}{2a}, 0\right)$.

   - If $b^2 - 4ac < 0$, then there are no x-intercepts.

EXAMPLE 4   **Using the Vertex Formula**

Find the axis and vertex of the parabola having equation $f(x) = 2x^2 + 4x + 5$.

SOLUTION   The axis of the parabola is the vertical line

$$x = h = -\frac{b}{2a} = -\frac{4}{2(2)} = -1. \qquad \begin{array}{l}\text{Use the vertex formula.} \\ \text{Here } a = 2 \text{ and } b = 4.\end{array}$$

The vertex is $(-1, f(-1))$. Evaluate $f(-1)$.

$$f(-1) = 2(-1)^2 + 4(-1) + 5 = 3$$

The vertex is $(-1, 3)$.   ✔ **Now Try Exercise 31(a).**

Quadratic Models Because the vertex of a vertical parabola is the highest or lowest point on the graph, equations of the form

$$y = ax^2 + bx + c$$

are important in certain problems where we must find the maximum or minimum value of some quantity.

- When $a < 0$, the y-coordinate of the vertex gives the maximum value of y.

- When $a > 0$, the y-coordinate of the vertex gives the minimum value of y.

The x-coordinate of the vertex tells *where* the maximum or minimum value occurs.

If air resistance is neglected, the height s (in feet) of an object projected directly upward from an initial height s_0 feet with initial velocity v_0 feet per second is

$$s(t) = -16t^2 + v_0 t + s_0,$$

where t is the number of seconds after the object is projected. The coefficient of t^2 (that is, -16) is a constant based on the gravitational force of Earth. This constant is different on other surfaces, such as the moon and the other planets.

EXAMPLE 5 **Solving a Problem Involving Projectile Motion**

A ball is projected directly upward from an initial height of 100 ft with an initial velocity of 80 ft per sec.

(a) Give the function that describes the height of the ball in terms of time t.

(b) After how many seconds does the ball reach its maximum height? What is this maximum height?

(c) For what interval of time is the height of the ball greater than 160 ft?

(d) After how many seconds will the ball hit the ground?

ALGEBRAIC SOLUTION

(a) Use the projectile height function.

$$s(t) = -16t^2 + v_0 t + s_0 \quad \text{Let } v_0 = 80 \text{ and}$$
$$s(t) = -16t^2 + 80t + 100 \quad s_0 = 100.$$

(b) The coefficient of t^2 is -16, so the graph of the projectile function is a parabola that opens down. Find the coordinates of the vertex to determine the maximum height and when it occurs. Let $a = -16$ and $b = 80$ in the vertex formula.

$$t = -\frac{b}{2a} = -\frac{80}{2(-16)} = 2.5$$

$$s(t) = -16t^2 + 80t + 100$$

$$s(2.5) = -16(2.5)^2 + 80(2.5) + 100$$

$$s(2.5) = 200$$

Therefore, after 2.5 sec the ball reaches its maximum height of 200 ft.

GRAPHING CALCULATOR SOLUTION

(a) Use the projectile height function as in the algebraic solution, with $v_0 = 80$ and $s_0 = 100$.

$$s(t) = -16t^2 + 80t + 100$$

(b) Using the capabilities of a calculator, we see in **Figure 11** that the vertex coordinates are indeed $(2.5, 200)$.

$y_1 = -16x^2 + 80x + 100$ — Here $x = t$ and $y_1 = s(t)$.

Figure 11

Be careful not to misinterpret the graph in **Figure 11.** *It does not show the path followed by the ball. It defines height as a function of time.*

(c) We must solve the related quadratic *inequality*.

$$-16t^2 + 80t + 100 > 160$$

$$-16t^2 + 80t - 60 > 0 \qquad \text{Subtract 160.}$$

$$4t^2 - 20t + 15 < 0 \qquad \begin{array}{l}\text{Divide by } -4. \text{ Reverse} \\ \text{the inequality symbol.}\end{array}$$

Use the quadratic formula to find the solutions of $4t^2 - 20t + 15 = 0$.

$$t = \frac{-(-20) \pm \sqrt{(-20)^2 - 4(4)(15)}}{2(4)} \quad \begin{array}{l}\text{Here } a = 4, \\ b = -20, \\ \text{and } c = 15.\end{array}$$

$$t = \frac{5 - \sqrt{10}}{2} \approx 0.92 \quad \text{or} \quad t = \frac{5 + \sqrt{10}}{2} \approx 4.08$$

These numbers divide the number line into three intervals:

$$(-\infty, 0.92), \quad (0.92, 4.08), \quad \text{and} \quad (4.08, \infty).$$

Using a test value from each interval shows that $(0.92, 4.08)$ satisfies the *inequality*. The ball is greater than 160 ft above the ground between 0.92 sec and 4.08 sec.

(d) The height is 0 when the ball hits the ground. We use the quadratic formula to find the *positive* solution of the equation

$$-16t^2 + 80t + 100 = 0.$$

Here, $a = -16$, $b = 80$, and $c = 100$.

$$t = \frac{-80 \pm \sqrt{80^2 - 4(-16)(100)}}{2(-16)}$$

$$t \approx -1.04 \quad \text{or} \quad t \approx 6.04$$
$$\text{Reject}$$

The ball hits the ground after about 6.04 sec.

(c) If we graph

$$y_1 = -16x^2 + 80x + 100 \quad \text{and} \quad y_2 = 160,$$

as shown in **Figures 12 and 13,** and locate the two points of intersection, we find that the *x*-coordinates for these points are approximately

$$0.92 \quad \text{and} \quad 4.08.$$

Therefore, between 0.92 sec and 4.08 sec, y_1 is greater than y_2, and the ball is greater than 160 ft above the ground.

Figure 12   **Figure 13**

(d) **Figure 14** shows that the *x*-intercept of the graph of $y = -16x^2 + 80x + 100$ in the given window is approximately $(6.04, 0)$, which means that the ball hits the ground after about 6.04 sec.

Figure 14

✔ **Now Try Exercise 57.**

[icon] **EXAMPLE 6** **Modeling the Number of Hospital Outpatient Visits**

The number of hospital outpatient visits (in millions) for selected years is shown in the table.

Year	Visits	Year	Visits
99	573.5	106	690.4
100	592.7	107	693.5
101	612.0	108	710.0
102	640.5	109	742.0
103	648.6	110	750.4
104	662.1	111	754.5
105	673.7	112	778.0

Source: American Hospital Association.

(a)

$f(x) = -0.16161x^2 + 49.071x - 2695.5$

(b)

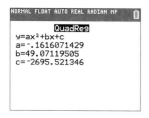

(c)

Figure 15

In the table on the preceding page, 99 represents 1999, 100 represents 2000, and so on, and the number of outpatient visits is given in millions.

(a) Prepare a scatter diagram, and determine a quadratic model for these data.

(b) Use the model from part (a) to predict the number of visits in 2016.

SOLUTION

(a) Linear regression is used to determine linear equations that model data. With a graphing calculator, we can use **quadratic regression** to find quadratic equations that model data.

The scatter diagram in **Figure 15(a)** suggests that a quadratic function with a negative value of a (so the graph opens down) would be a reasonable model for the data. Using quadratic regression, the quadratic function

$$f(x) = -0.16161x^2 + 49.071x - 2695.5$$

approximates the data well. See **Figure 15(b).** The quadratic regression values of a, b, and c are displayed in **Figure 15(c).**

(b) The year 2016 corresponds to $x = 116$. The model predicts that there will be 822 million visits in 2016.

$$f(x) = -0.16161x^2 + 49.071x - 2695.5$$

$$f(116) = -0.16161(116)^2 + 49.071(116) - 2695.5$$

$$f(116) \approx 822 \text{ million}$$

✔ **Now Try Exercise 73.**

3.1 Exercises

CONCEPT PREVIEW *Fill in the blank(s) to correctly complete each sentence.*

1. A polynomial function with leading term $3x^5$ has degree _____.

2. The lowest point on the graph of a parabola that opens up is the _____ of the parabola.

3. The highest point on the graph of a parabola that opens down is the _____ of the parabola.

4. The axis of symmetry of the graph of $f(x) = 2(x + 4)^2 - 6$ has equation $x =$ _____.

5. The vertex of the graph of $f(x) = x^2 + 2x + 4$ has x-coordinate _____.

6. The graph of $f(x) = -2x^2 - 6x + 5$ opens down with y-intercept $(0,$ _____$)$, so it has _____ x-intercept(s).
   (no/one/two)

CONCEPT PREVIEW *Match each equation in Column I with the description of the parabola that is its graph in Column II.*

I	II
7. $y = (x + 4)^2 + 2$	**A.** vertex $(-2, 4)$, opens up
8. $y = (x + 2)^2 + 4$	**B.** vertex $(-2, 4)$, opens down
9. $y = -(x + 4)^2 + 2$	**C.** vertex $(-4, 2)$, opens up
10. $y = -(x + 2)^2 + 4$	**D.** vertex $(-4, 2)$, opens down

*Consider the graph of each quadratic function. Do the following. **See Examples 1–4.***

(a) *Give the domain and range.* **(b)** *Give the coordinates of the vertex.*

(c) *Give the equation of the axis.* **(d)** *Find the y-intercept.*

(e) *Find the x-intercepts.*

11.

12.

13.

14.

Concept Check *Match each function with its graph without actually entering it into a calculator. Then, after completing the exercises, check the answers with a calculator. Use the standard viewing window.*

15. $f(x) = (x - 4)^2 - 3$ **16.** $f(x) = -(x - 4)^2 + 3$

17. $f(x) = (x + 4)^2 - 3$ **18.** $f(x) = -(x + 4)^2 + 3$

A.   **B.**

C.   **D.**

19. Graph the following on the same coordinate system.

    (a) $y = x^2$ **(b)** $y = 3x^2$ **(c)** $y = \dfrac{1}{3}x^2$

    (d) How does the coefficient of x^2 affect the shape of the graph?

20. Graph the following on the same coordinate system.

    (a) $y = x^2$ **(b)** $y = x^2 - 2$ **(c)** $y = x^2 + 2$

    (d) How do the graphs in parts (b) and (c) differ from the graph of $y = x^2$?

21. Graph the following on the same coordinate system.

    (a) $y = (x - 2)^2$ **(b)** $y = (x + 1)^2$ **(c)** $y = (x + 3)^2$

    (d) How do these graphs differ from the graph of $y = x^2$?

22. *Concept Check* A quadratic function $f(x)$ has vertex $(0, 0)$, and all of its intercepts are the same point. What is the general form of its equation?

Graph each quadratic function. Give the (a) vertex, (b) axis, (c) domain, and (d) range. Then determine (e) the largest open interval of the domain over which the function is increasing and (f) the largest open interval over which the function is decreasing. See Examples 1–4.

23. $f(x) = (x - 2)^2$

24. $f(x) = (x + 4)^2$

25. $f(x) = (x + 3)^2 - 4$

26. $f(x) = (x - 5)^2 - 4$

27. $f(x) = -\dfrac{1}{2}(x + 1)^2 - 3$

28. $f(x) = -3(x - 2)^2 + 1$

29. $f(x) = x^2 - 2x + 3$

30. $f(x) = x^2 + 6x + 5$

31. $f(x) = x^2 - 10x + 21$

32. $f(x) = 2x^2 - 4x + 5$

33. $f(x) = -2x^2 - 12x - 16$

34. $f(x) = -3x^2 + 24x - 46$

35. $f(x) = -\dfrac{1}{2}x^2 - 3x - \dfrac{1}{2}$

36. $f(x) = \dfrac{2}{3}x^2 - \dfrac{8}{3}x + \dfrac{5}{3}$

Concept Check The figure shows the graph of a quadratic function $y = f(x)$. Use it to answer each question.

37. What is the minimum value of $f(x)$?

38. For what value of x is $f(x)$ as small as possible?

39. How many real solutions are there to the equation $f(x) = 1$?

40. How many real solutions are there to the equation $f(x) = 4$?

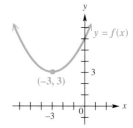

Concept Check Several graphs of the quadratic function

$$f(x) = ax^2 + bx + c$$

are shown below. For the given restrictions on a, b, and c, select the corresponding graph from choices A–F. (Hint: Use the discriminant.)

41. $a < 0;\ b^2 - 4ac = 0$

42. $a > 0;\ b^2 - 4ac < 0$

43. $a < 0;\ b^2 - 4ac < 0$

44. $a < 0;\ b^2 - 4ac > 0$

45. $a > 0;\ b^2 - 4ac > 0$

46. $a > 0;\ b^2 - 4ac = 0$

A.

B.

C.

D.

E.

F.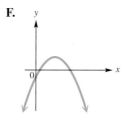

Connecting Graphs with Equations *Find a quadratic function f having the graph shown. (Hint: See the Note following* **Example 3.***)*

47.

48.

49.

50.

(Modeling) *In each scatter diagram, tell whether a linear or a quadratic model is appropriate for the data. If linear, tell whether the slope should be positive or negative. If quadratic, tell whether the leading coefficient of x^2 should be positive or negative.*

51. number of shopping centers as a function of time

52. growth in science centers/museums as a function of time

53. value of U.S. salmon catch as a function of time

54. height of an object projected upward as a function of time

55. Social Security assets as a function of time

56. newborns with AIDS as a function of time

(Modeling) *Solve each problem. Give approximations to the nearest hundredth.* **See Example 5.**

57. *Height of a Toy Rocket* A toy rocket (not internally powered) is launched straight up from the top of a building 50 ft tall at an initial velocity of 200 ft per sec.

  (a) Give the function that describes the height of the rocket in terms of time *t*.

  (b) Determine the time at which the rocket reaches its maximum height and the maximum height in feet.

  (c) For what time interval will the rocket be more than 300 ft above ground level?

  (d) After how many seconds will it hit the ground?

58. *Height of a Projected Rock* A rock is projected directly upward from ground level with an initial velocity of 90 ft per sec.

(a) Give the function that describes the height of the rock in terms of time t.

(b) Determine the time at which the rock reaches its maximum height and the maximum height in feet.

(c) For what time interval will the rock be more than 120 ft above ground level?

(d) After how many seconds will it hit the ground?

59. *Area of a Parking Lot* One campus of Houston Community College has plans to construct a rectangular parking lot on land bordered on one side by a highway. There are 640 ft of fencing available to fence the other three sides. Let x represent the length of each of the two parallel sides of fencing.

(a) Express the length of the remaining side to be fenced in terms of x.

(b) What are the restrictions on x?

(c) Determine a function $\mathscr{A}$ that represents the area of the parking lot in terms of x.

(d) Determine the values of x that will give an area between 30,000 and 40,000 ft².

(e) What dimensions will give a maximum area, and what will this area be?

60. *Area of a Rectangular Region* A farmer wishes to enclose a rectangular region bordering a river with fencing, as shown in the diagram. Suppose that x represents the length of each of the three parallel pieces of fencing. She has 600 ft of fencing available.

(a) What is the length of the remaining piece of fencing in terms of x?

(b) Determine a function $\mathscr{A}$ that represents the total area of the enclosed region. Give any restrictions on x.

(c) What dimensions for the total enclosed region would give an area of 22,500 ft²?

(d) What is the maximum area that can be enclosed?

61. *Volume of a Box* A piece of cardboard is twice as long as it is wide. It is to be made into a box with an open top by cutting 2-in. squares from each corner and folding up the sides. Let x represent the width (in inches) of the original piece of cardboard.

(a) Represent the length of the original piece of cardboard in terms of x.

(b) What will be the dimensions of the bottom rectangular base of the box? Give the restrictions on x.

(c) Determine a function V that represents the volume of the box in terms of x.

(d) For what dimensions of the bottom of the box will the volume be 320 in.³?

(e) Find the values of x if such a box is to have a volume between 400 and 500 in.³.

62. *Volume of a Box* A piece of sheet metal is 2.5 times as long as it is wide. It is to be made into a box with an open top by cutting 3-in. squares from each corner and folding up the sides. Let x represent the width (in inches) of the original piece.

(a) Represent the length of the original piece of sheet metal in terms of x.

(b) What are the restrictions on x?

(c) Determine a function V that represents the volume of the box in terms of x.

(d) For what values of x (that is, original widths) will the volume of the box be between 600 and 800 in.^{3}?

63. *Shooting a Free Throw* If a person shoots a free throw from a position 8 ft above the floor, then the path of the ball may be modeled by the parabola

$$y = \frac{-16x^2}{0.434v^2} + 1.15x + 8,$$

where v is the initial velocity of the ball in feet per second, as illustrated in the figure. (*Source:* Rist, C., "The Physics of Foul Shots," *Discover.*)

(a) If the basketball hoop is 10 ft high and located 15 ft away, what initial velocity v should the basketball have?

(b) What is the maximum height of the basketball?

64. *Shooting a Free Throw* See **Exercise 63.** If a person shoots a free throw from an underhand position 3 ft above the floor, then the path of the ball may be modeled by

$$y = \frac{-16x^2}{0.117v^2} + 2.75x + 3.$$

Repeat parts (a) and (b) from **Exercise 63.** Then compare the paths for the overhand shot and the underhand shot.

65. *Sum and Product of Two Numbers* Find two numbers whose sum is 20 and whose product is the maximum possible value. (*Hint: Let x be one number. Then 20 − x is the other number. Form a quadratic function by multiplying them, and then find the maximum value of the function.*)

66. *Sum and Product of Two Numbers* Find two numbers whose sum is 32 and whose product is the maximum possible value.

67. *Height of an Object* If an object is projected upward from ground level with an initial velocity of 32 ft per sec, then its height in feet after t seconds is given by

$$s(t) = -16t^2 + 32t.$$

Find the number of seconds it will take the object to reach its maximum height. What is this maximum height?

68. *Height of an Object* If an object is projected upward from an initial height of 100 ft with an initial velocity of 64 ft per sec, then its height in feet after t seconds is given by

$$s(t) = -16t^2 + 64t + 100.$$

Find the number of seconds it will take the object to reach its maximum height. What is this maximum height?

(Modeling) *Solve each problem.* ***See Example 6.***

69. *Price of Chocolate Chip Cookies* The average price in dollars of a pound of chocolate chip cookies from 2002 to 2013 is shown in the table.

Year	Price per Pound	Year	Price per Pound
2002	2.59	2008	2.88
2003	2.81	2009	3.17
2004	2.65	2010	3.25
2005	2.67	2011	3.35
2006	2.88	2012	3.62
2007	2.70	2013	3.64

Source: Consumer Price Index.

The data are modeled by the quadratic function

$$f(x) = 0.0095x^2 - 0.0076x + 2.660,$$

where $x = 0$ corresponds to 2002 and $f(x)$ is the price in dollars. If this model continues to apply, what will it predict for the price of a pound of chocolate chip cookies in 2018?

70. *Concentration of Atmospheric CO_2* The quadratic function

$$f(x) = 0.0118x^2 + 0.8633x + 317$$

models the worldwide atmospheric concentration of carbon dioxide in parts per million (ppm) over the period 1960–2013, where $x = 0$ represents the year 1960. If this model continues to apply, what will be the atmospheric CO_2 concentration in 2020? Round to the nearest unit. (*Source*: U.S. Department of Energy.)

71. *Spending on Shoes and Clothing* The total amount spent by Americans on shoes and clothing from 2000 to 2013 can be modeled by

$$f(x) = 0.7714x^2 - 3.693x + 297.9,$$

where $x = 0$ represents 2000 and $f(x)$ is in billions of dollars. Based on this model, in what year did spending on shoes and clothing reach a minimum? (*Source*: Bureau of Economic Analysis.)

72. *Accident Rate* According to data from the National Highway Traffic Safety Administration, the accident rate as a function of the age of the driver in years x can be approximated by the function

$$f(x) = 0.0232x^2 - 2.28x + 60.0, \quad \text{for } 16 \le x \le 85.$$

Find both the age at which the accident rate is a minimum and the minimum rate to the nearest hundredth.

73. *College Enrollment* The table lists total fall enrollments in degree-granting postsecondary colleges in the United States for selected years.

Year	Enrollment (in millions)
2008	19.1
2009	20.4
2010	21.0
2011	21.0
2012	20.6

Source: National Center for Education Statistics.

(a) Plot the data. Let $x = 0$ correspond to the year 2008.

(b) Find a quadratic function $f(x) = ax^2 + bx + c$ that models the data.

(c) Plot the data together with f in the same window. How well does f model enrollment?

(d) Use f to estimate total enrollment in 2013 to the nearest tenth of a million.

74. *Two-Year College Enrollment* The table lists total fall enrollments in degree-granting two-year colleges in the United States for selected years.

(a) Plot the data. Let $x = 0$ correspond to the year 2008.

(b) Find a quadratic function $g(x) = ax^2 + bx + c$ that models the data.

(c) Plot the data together with g in the same window. How well does g model enrollment?

(d) Use g to estimate total enrollment in 2013 to the nearest tenth of a million.

Year	Enrollment (in millions)
2008	6.9
2009	7.5
2010	7.7
2011	7.5
2012	7.2

Source: National Center for Education Statistics.

75. *Foreign-Born Americans* The table lists the percent of the U.S. population that was foreign-born for selected years.

(a) Plot the data. Let $x = 0$ correspond to the year 1930, $x = 10$ correspond to 1940, and so on.

(b) Find a quadratic function $f(x) = a(x - h)^2 + k$ that models the data. Use $(40, 4.7)$ as the vertex and $(20, 6.9)$ as the other point to determine a.

(c) Plot the data together with f in the same window. How well does f model the percent of the U.S. population that is foreign-born?

(d) Use the quadratic regression feature of a graphing calculator to determine the quadratic function g that provides the best fit for the data.

(e) Use functions f and g to predict the percent, to the nearest tenth, of the U.S. population in 2019 that will be foreign-born.

Year	Percent
1930	11.6
1940	8.8
1950	6.9
1960	5.4
1970	4.7
1980	6.2
1990	7.9
2000	11.1
2010	12.4
2012	12.9

Source: U.S. Census Bureau.

76. *Automobile Stopping Distance* Selected values of the stopping distance y, in feet, of a car traveling x miles per hour are given in the table.

(a) Plot the data.

(b) The quadratic function

$$f(x) = 0.056057x^2 + 1.06657x$$

is one model that has been used to approximate stopping distances. Find $f(45)$ to the nearest foot, and interpret this result.

(c) How well does f model the car's stopping distance?

Speed (in mph)	Stopping Distance (in feet)
20	46
30	87
40	140
50	240
60	282
70	371

Source: National Safety Institute Student Workbook.

Concept Check Work each problem.

77. Find a value of c so that $y = x^2 - 10x + c$ has exactly one x-intercept.

78. For what values of a does $y = ax^2 - 8x + 4$ have no x-intercepts?

79. Define the quadratic function f having x-intercepts $(2, 0)$ and $(5, 0)$ and y-intercept $(0, 5)$.

80. Define the quadratic function f having x-intercepts $(1, 0)$ and $(-2, 0)$ and y-intercept $(0, 4)$.

81. The distance between the two points $P(x_1, y_1)$ and $R(x_2, y_2)$ is

$$d(P, R) = \sqrt{(x_1 - x_2)^2 + (y_1 - y_2)^2}. \quad \text{Distance formula}$$

Find the closest point on the line $y = 2x$ to the point $(1, 7)$. (*Hint:* Every point on $y = 2x$ has the form $(x, 2x)$, and the closest point has the minimum distance.)

82. A quadratic equation $f(x) = 0$ has a solution $x = 2$. Its graph has vertex $(5, 3)$. What is the other solution of the equation?

Relating Concepts

For individual or collaborative investigation *(Exercises 83–86)*

A quadratic inequality such as

$$x^2 + 2x - 8 < 0$$

can be solved by first solving the related quadratic equation

$$x^2 + 2x - 8 = 0,$$

identifying intervals determined by the solutions of this equation, and then using a test value from each interval to determine which intervals form the solution set. **Work Exercises 83–86 in order** *to learn a graphical method of solving inequalities.*

83. Graph $f(x) = x^2 + 2x - 8$.

84. The real solutions of $x^2 + 2x - 8 = 0$ are the x-values of the x-intercepts of the graph in **Exercise 83.** These are values of x for which $f(x) = 0$. What are these values? What is the solution set of this equation?

85. The real solutions of $x^2 + 2x - 8 < 0$ are the x-values for which the graph in **Exercise 83** lies *below* the x-axis. These are values of x for which $f(x) < 0$ is true. What interval of x-values represents the solution set of this inequality?

86. The real solutions of $x^2 + 2x - 8 > 0$ are the x-values for which the graph in **Exercise 83** lies *above* the x-axis. These are values of x for which $f(x) > 0$ is true. What intervals of x-values represent the solution set of this inequality?

Use the technique described in **Exercises 83–86** *to solve each inequality. Write the solution set in interval notation.*

87. $x^2 - x - 6 < 0$

88. $x^2 - 9x + 20 < 0$

89. $2x^2 - 9x \geq 18$

90. $3x^2 + x \geq 4$

91. $-x^2 + 4x + 1 \geq 0$

92. $-x^2 + 2x + 6 > 0$

3.2 Synthetic Division

- Synthetic Division
- Remainder Theorem
- Potential Zeros of Polynomial Functions

The outcome of a division problem can be written using multiplication, even when the division involves polynomials. The **division algorithm** illustrates this.

Division Algorithm

Let $f(x)$ and $g(x)$ be polynomials with $g(x)$ of lesser degree than $f(x)$ and $g(x)$ of degree 1 or more. There exist unique polynomials $q(x)$ and $r(x)$ such that

$$f(x) = g(x) \cdot q(x) + r(x),$$

where either $r(x) = 0$ or the degree of $r(x)$ is less than the degree of $g(x)$.

This is the long-division process for the result shown to the right.

$$
\begin{array}{r}
3x - 2 \\
x^2 + 0x - 4 \overline{) 3x^3 - 2x^2 + 0x - 150} \\
\underline{3x^3 + 0x^2 - 12x} \\
-2x^2 + 12x - 150 \\
\underline{-2x^2 - 0x + 8} \\
12x - 158
\end{array}
$$

Consider the result shown here.

$$
\frac{3x^3 - 2x^2 - 150}{x^2 - 4} = 3x - 2 + \frac{12x - 158}{x^2 - 4}.
$$

We can express it using the preceding division algorithm.

$$
\underbrace{3x^3 - 2x^2 - 150}_{\substack{f(x) \\ \text{Dividend} \\ \text{(original polynomial)}}} = \underbrace{(x^2 - 4)}_{\substack{g(x) \\ \text{Divisor}}} \underbrace{(3x - 2)}_{\substack{q(x) \\ \cdot \; \text{Quotient}}} + \underbrace{12x - 158}_{\substack{r(x) \\ + \; \text{Remainder}}}
$$

Synthetic Division   When a given polynomial in x is divided by a first-degree binomial of the form $x - k$, a shortcut method called **synthetic division** may be used. The example on the left below is simplified by omitting all variables and writing only coefficients, with 0 used to represent the coefficient of any missing terms. The coefficient of x in the divisor is always 1 in these divisions, so it too can be omitted. These omissions simplify the problem, as shown on the right.

$$
\begin{array}{r}
3x^2 + 10x + 40 \\
x - 4 \overline{) 3x^3 - 2x^2 + 0x - 150} \\
\underline{3x^3 - 12x^2} \\
10x^2 + 0x \\
\underline{10x^2 - 40x} \\
40x - 150 \\
\underline{40x - 160} \\
10
\end{array}
\qquad
\begin{array}{r}
3 \quad 10 \quad 40 \\
-4 \overline{) 3 \; -2 \quad 0 \; -150} \\
\underline{3 \; -12} \\
10 \quad 0 \\
\underline{10 \; -40} \\
40 \; -150 \\
\underline{40 \; -160} \\
10
\end{array}
$$

The numbers in color that are repetitions of the numbers directly above them can also be omitted, as shown on the left below.

$$
\begin{array}{r}
3 \quad 10 \quad 40 \\
-4 \overline{) 3 \; -2 \quad 0 \; -150} \\
\underline{-12} \\
10 \quad 0 \\
\underline{-40} \\
40 \; -150 \\
\underline{-160} \\
10
\end{array}
\qquad
\begin{array}{r}
3 \quad 10 \quad 40 \\
-4 \overline{) 3 \; -2 \quad 0 \; -150} \\
\underline{-12} \\
10 \\
\underline{-40} \\
40 \\
\underline{-160} \\
10
\end{array}
$$

The numbers in color are again repetitions of those directly above them. They may be omitted, as shown on the right above.

The entire process can now be condensed vertically.

$$
\begin{array}{r}
-4 \overline{) 3 \quad -2 \qquad 0 \quad -150} \\
\underline{-12 \quad -40 \quad -160} \\
3 \quad 10 \qquad 40 \qquad 10
\end{array}
$$

The top row of numbers can be omitted since it duplicates the bottom row if the 3 is brought down.

The rest of the bottom row is obtained by subtracting -12, -40, and -160 from the corresponding terms above them.

To simplify the arithmetic, we replace subtraction in the second row by addition and compensate by changing the -4 at the upper left to its additive inverse, 4.

Additive inverse $\longrightarrow$

$$
\begin{array}{r}
4 \overline{) 3 \qquad -2 \qquad 0 \quad -150} \\
12 \qquad 40 \qquad 160 \quad \longleftarrow \text{Signs changed} \\
\begin{array}{cccc} 3 & 10 & 40 & 10 \\ \downarrow & \downarrow & \downarrow & \downarrow \end{array}
\end{array}
$$

$$
\text{Quotient} \longrightarrow \quad 3x^2 + 10x + 40 + \frac{10}{x - 4} \quad \longleftarrow \text{Remainder}
$$

Synthetic division provides an efficient process for dividing a polynomial in x by a binomial of the form $x - k$. Begin by writing the coefficients of the polynomial in decreasing powers of the variable, using 0 as the coefficient of any missing powers. The number k is written to the left in the same row. The answer is found in the bottom row with the remainder farthest to the right and the coefficients of the quotient on the left when written in order of decreasing degree.

CAUTION *To avoid errors, use **0** as the coefficient for any missing terms, including a missing constant, when setting up the division.*

EXAMPLE 1   Using Synthetic Division

Use synthetic division to perform the division.

$$\frac{5x^3 - 6x^2 - 28x - 2}{x + 2}$$

SOLUTION Express $x + 2$ in the form $x - k$ by writing it as $x - (-2)$.

$\boxed{x + 2 \text{ leads to } -2.}$ $-2\overline{)5 \quad -6 \quad -28 \quad -2}$ ← Coefficients of the polynomial

Bring down the 5, and multiply: $-2(5) = -10$.

$$
\begin{array}{r}
-2\overline{)5 \quad\ -6 \quad -28 \quad -2} \\
\downarrow \quad -10 \\
\hline
5
\end{array}
$$

Add -6 and -10 to obtain -16. Multiply: $-2(-16) = 32$.

$$
\begin{array}{r}
-2\overline{)5 \quad -6 \quad -28 \quad -2} \\
-10 \quad\ 32 \\
\hline
5 \quad -16
\end{array}
$$

Add -28 and 32, obtaining 4. Multiply: $-2(4) = -8$.

$$
\begin{array}{r}
-2\overline{)5 \quad -6 \quad -28 \quad -2} \\
-10 \quad\ 32 \quad -8 \\
\hline
5 \quad -16 \quad\ 4
\end{array}
$$
$\boxed{\text{Add columns. Be careful with signs.}}$

Add -2 and -8 to obtain -10.

$$
\begin{array}{r}
-2\overline{)5 \quad -6 \quad -28 \quad -2} \\
-10 \quad\ 32 \quad -8 \\
\hline
5 \quad -16 \quad\ 4 \quad -10
\end{array}
$$
← Remainder

$\underbrace{\qquad\qquad\qquad}_{\text{Quotient}}$

Because the divisor $x - k$ has degree 1, the degree of the quotient will always be one less than the degree of the polynomial to be divided.

$$\frac{5x^3 - 6x^2 - 28x - 2}{x + 2} = 5x^2 - 16x + 4 + \frac{-10}{x + 2}$$
$\boxed{\text{Remember to add } \frac{\text{remainder}}{\text{divisor}}.}$

✔ **Now Try Exercise 15.**

The result of the division in **Example 1** can be written as

$$5x^3 - 6x^2 - 28x - 2 = (x + 2)(5x^2 - 16x + 4) + (-10). \quad \text{Multiply by } x + 2.$$

The theorem that follows is a generalization of this product form.

Special Case of the Division Algorithm

For any polynomial $f(x)$ and any complex number k, there exists a unique polynomial $q(x)$ and number r such that the following holds.

$$f(x) = (x - k)q(x) + r$$

We can illustrate this connection using the mathematical statement

$$\underbrace{5x^3 - 6x^2 - 28x - 2}_{f(x)} = \underbrace{(x + 2)}_{= \ (x-k)}\underbrace{(5x^2 - 16x + 4)}_{\cdot \ \ \ q(x)} + \underbrace{(-10)}_{+ \ \ \ r}.$$

This form of the division algorithm is used to develop the *remainder theorem*.

Remainder Theorem Suppose $f(x)$ is written as $f(x) = (x - k)q(x) + r$. This equality is true for all complex values of x, so it is true for $x = k$.

$$f(k) = (k - k)q(k) + r, \quad \text{or} \quad f(k) = r \quad \text{\small Replace x with k.}$$

This proves the following **remainder theorem,** which gives a new method of evaluating polynomial functions.

Remainder Theorem

If the polynomial $f(x)$ is divided by $x - k$, then the remainder is equal to $f(k)$.

In **Example 1,** when $f(x) = 5x^3 - 6x^2 - 28x - 2$ was divided by $x + 2$, or $x - (-2)$, the remainder was -10. Substitute -2 for x in $f(x)$.

$$f(-2) = 5(-2)^3 - 6(-2)^2 - 28(-2) - 2$$

$$f(-2) = -40 - 24 + 56 - 2$$

$$f(-2) = -10$$

> Use parentheses around substituted values to avoid errors.

An alternative way to find the value of a polynomial is to use synthetic division. By the remainder theorem, instead of replacing x by -2 to find $f(-2)$, divide $f(x)$ by $x + 2$ as in **Example 1.** Then $f(-2)$ is the remainder, -10.

$$
\begin{array}{r|rrrr}
-2 & 5 & -6 & -28 & -2 \\
   &   & -10 & 32 & -8 \\
\hline
   & 5 & -16 & 4 & -10 \leftarrow f(-2)
\end{array}
$$

EXAMPLE 2 Applying the Remainder Theorem

Let $f(x) = -x^4 + 3x^2 - 4x - 5$. Use the remainder theorem to find $f(-3)$.

SOLUTION Use synthetic division with $k = -3$.

$$
\begin{array}{r|rrrrr}
-3 & -1 & 0 & 3 & -4 & -5 \\
   &   & 3 & -9 & 18 & -42 \\
\hline
   & -1 & 3 & -6 & 14 & -47 \leftarrow \text{Remainder}
\end{array}
$$

> $f(-3)$ is equal to the remainder when dividing by $x + 3$.

By this result, $f(-3) = -47$.

✔ **Now Try Exercise 37.**

Potential Zeros of Polynomial Functions A **zero** of a polynomial function $f(x)$ is a number k such that $f(k) = 0$. ***Real number zeros are the x-values of the x-intercepts of the graph of the function.***

The remainder theorem gives a quick way to decide whether a number k is a zero of a polynomial function $f(x)$, as follows.

1. Use synthetic division to find $f(k)$.

2. If the remainder is 0, then $f(k) = 0$ and k is a zero of $f(x)$. If the remainder is not 0, then k is not a zero of $f(x)$.

A zero of $f(x)$ is a **root,** or **solution,** of the equation $f(x) = 0$.

EXAMPLE 3 **Deciding Whether a Number Is a Zero**

Decide whether the given number k is a zero of $f(x)$.

(a) $f(x) = x^3 - 4x^2 + 9x - 6$; $k = 1$

(b) $f(x) = x^4 + x^2 - 3x + 1$; $k = -1$

(c) $f(x) = x^4 - 2x^3 + 4x^2 + 2x - 5$; $k = 1 + 2i$

SOLUTION

Figure 16

(a) To decide whether 1 is a zero of $f(x) = x^3 - 4x^2 + 9x - 6$, use synthetic division.

$$\text{Proposed zero} \longrightarrow 1)\overline{\begin{array}{rrrr} 1 & -4 & 9 & -6 \end{array}} \longleftarrow f(x) = x^3 - 4x^2 + 9x - 6$$
$$\begin{array}{rrrr} & 1 & -3 & 6 \end{array}$$
$$\begin{array}{rrrr} 1 & -3 & 6 & 0 \end{array} \longleftarrow \text{Remainder}$$

Because the remainder is 0, $f(1) = 0$, and 1 is a zero of the given polynomial function. An x-intercept of the graph of $f(x) = x^3 - 4x^2 + 9x - 6$ is the point $(1, 0)$. The graph in **Figure 16** supports this.

(b) For $f(x) = x^4 + x^2 - 3x + 1$, remember to use 0 as coefficient for the missing x^3-term in the synthetic division.

$$\text{Proposed zero} \longrightarrow -1)\overline{\begin{array}{rrrrr} 1 & 0 & 1 & -3 & 1 \end{array}}$$
$$\begin{array}{rrrrr} & -1 & 1 & -2 & 5 \end{array}$$
$$\begin{array}{rrrrr} 1 & -1 & 2 & -5 & 6 \end{array} \longleftarrow \text{Remainder}$$

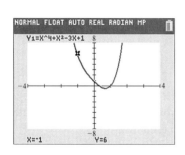

Figure 17

The remainder is not 0, so -1 is not a zero of $f(x) = x^4 + x^2 - 3x + 1$. In fact, $f(-1) = 6$, indicating that $(-1, 6)$ is on the graph of $f(x)$. The graph in **Figure 17** supports this.

(c) Use synthetic division and operations with complex numbers to determine whether $1 + 2i$ is a zero of $f(x) = x^4 - 2x^3 + 4x^2 + 2x - 5$.

$$1 + 2i)\overline{\begin{array}{rrrrr} 1 & -2 & 4 & 2 & -5 \end{array}}$$
$$\begin{array}{rrrrr} & 1 + 2i & -5 & -1 - 2i & 5 \end{array} \quad i^2 = -1$$
$$\begin{array}{rrrrr} 1 & -1 + 2i & -1 & 1 - 2i & 0 \end{array} \longleftarrow \text{Remainder}$$

$(1 + 2i)(-1 + 2i)$
$= -1 + 4i^2$
$= -5$

The remainder is 0, so $1 + 2i$ is a zero of the given polynomial function. Notice that $1 + 2i$ is *not* a real number zero. Therefore, it is not associated with an x-intercept on the graph of $f(x)$.

✔ **Now Try Exercises 49 and 59.**

3.2 Exercises

CONCEPT PREVIEW *Fill in the blank(s) to correctly complete each sentence.*

1. In arithmetic, the result of the division

$$\begin{array}{r} 3 \\ 5\overline{)19} \\ 15 \\ \hline 4 \end{array}$$

can be written

$$19 = 5 \cdot \underline{\hspace{1cm}} + \underline{\hspace{1cm}}.$$

2. In algebra, the result of the division

$$\begin{array}{r} x + 3 \\ x - 1\overline{)x^2 + 2x + 3} \\ \underline{x^2 - x} \\ 3x + 3 \\ \underline{3x - 3} \\ 6 \end{array}$$

can be written $x^2 + 2x + 3 = (x - 1)(\underline{\hspace{1cm}}) + \underline{\hspace{1cm}}.$

3. To perform the division in **Exercise 2** using synthetic division, we begin by writing the following.

$$\underline{\hspace{0.8cm}}\overline{)\underline{\hspace{0.8cm}}\quad 2\quad 3}$$

4. To perform the division

$$x + 2\overline{)x^3 + 4x + 2}$$

using synthetic division, we begin by writing the following.

$$\underline{\hspace{0.8cm}}\overline{)1\quad \underline{\hspace{0.8cm}}\quad 4\quad 2}$$

5. To perform the division

$$x - 3\overline{)x^3 + 6x^2 + 2x}$$

using synthetic division, we begin by writing the following.

$$\underline{\hspace{0.8cm}}\overline{)1\quad \underline{\hspace{0.8cm}}\quad 2\quad \underline{\hspace{0.8cm}}}$$

6. Consider the following function.

$$f(x) = 2x^4 + 6x^3 - 5x^2 + 3x + 8$$

$$f(x) = (x - 2)(2x^3 + 10x^2 + 15x + 33) + 74$$

By inspection, we can state that $f(2) = \underline{\hspace{1cm}}.$

Use synthetic division to perform each division. See Example 1.

7. $\dfrac{x^3 + 3x^2 + 11x + 9}{x + 1}$

8. $\dfrac{x^3 + 7x^2 + 13x + 6}{x + 2}$

9. $\dfrac{5x^4 + 5x^3 + 2x^2 - x - 3}{x + 1}$

10. $\dfrac{2x^4 - x^3 - 7x^2 + 7x - 10}{x - 2}$

11. $\dfrac{x^4 + 4x^3 + 2x^2 + 9x + 4}{x + 4}$

12. $\dfrac{x^4 + 5x^3 + 4x^2 - 3x + 9}{x + 3}$

13. $\dfrac{x^5 + 3x^4 + 2x^3 + 2x^2 + 3x + 1}{x + 2}$

14. $\dfrac{x^6 - 3x^4 + 2x^3 - 6x^2 - 5x + 3}{x + 2}$

15. $\dfrac{-9x^3 + 8x^2 - 7x + 2}{x - 2}$

16. $\dfrac{-11x^4 - 2x^3 - 8x^2 - 4}{x + 1}$

17. $\dfrac{\frac{1}{3}x^3 - \frac{2}{9}x^2 + \frac{2}{27}x - \frac{1}{81}}{x - \frac{1}{3}}$

18. $\dfrac{x^3 + x^2 + \frac{1}{2}x + \frac{1}{8}}{x + \frac{1}{2}}$

19. $\dfrac{x^4 - 3x^3 - 4x^2 + 12x}{x - 2}$

20. $\dfrac{x^4 - x^3 - 5x^2 - 3x}{x + 1}$

21. $\dfrac{x^3 - 1}{x - 1}$

22. $\dfrac{x^4 - 1}{x - 1}$

23. $\dfrac{x^5 + 1}{x + 1}$

24. $\dfrac{x^7 + 1}{x + 1}$

Use synthetic division to divide $f(x)$ by $x - k$ for the given value of k. Then express $f(x)$ in the form $f(x) = (x - k)q(x) + r$.

25. $f(x) = 2x^3 + x^2 + x - 8;\quad k = -1$

26. $f(x) = 2x^3 + 3x^2 - 16x + 10;\quad k = -4$

27. $f(x) = x^3 + 4x^2 + 5x + 2;\quad k = -2$

28. $f(x) = -x^3 + x^2 + 3x - 2;\quad k = 2$

29. $f(x) = 4x^4 - 3x^3 - 20x^2 - x;\quad k = 3$

30. $f(x) = 2x^4 + x^3 - 15x^2 + 3x;\quad k = -3$

31. $f(x) = 3x^4 + 4x^3 - 10x^2 + 15;\quad k = -1$

32. $f(x) = -5x^4 + x^3 + 2x^2 + 3x + 1;\quad k = 1$

For each polynomial function, use the remainder theorem to find $f(k)$. **See Example 2.**

33. $f(x) = x^2 + 5x + 6;\quad k = -2$

34. $f(x) = x^2 - 4x - 5;\quad k = 5$

35. $f(x) = 2x^2 - 3x - 3;\quad k = 2$

36. $f(x) = -x^3 + 8x^2 + 63;\quad k = 4$

37. $f(x) = x^3 - 4x^2 + 2x + 1;\quad k = -1$

38. $f(x) = 2x^3 - 3x^2 - 5x + 4;\quad k = 2$

39. $f(x) = x^2 - 5x + 1;\quad k = 2 + i$

40. $f(x) = x^2 - x + 3;\quad k = 3 - 2i$

41. $f(x) = x^2 + 4;\quad k = 2i$

42. $f(x) = 2x^2 + 10;\quad k = i\sqrt{5}$

43. $f(x) = 2x^5 - 10x^3 - 19x^2 - 50;\quad k = 3$

44. $f(x) = x^4 + 6x^3 + 9x^2 + 3x - 3;\quad k = 4$

45. $f(x) = 6x^4 + x^3 - 8x^2 + 5x + 6;\quad k = \frac{1}{2}$

46. $f(x) = 6x^3 - 31x^2 - 15x;\quad k = -\frac{1}{2}$

Use synthetic division to decide whether the given number k is a zero of the polynomial function. If it is not, give the value of $f(k)$. **See Examples 2 and 3.**

47. $f(x) = x^2 + 2x - 8;\quad k = 2$

48. $f(x) = x^2 + 4x - 5;\quad k = -5$

49. $f(x) = x^3 - 3x^2 + 4x - 4;\quad k = 2$

50. $f(x) = x^3 + 2x^2 - x + 6;\quad k = -3$

51. $f(x) = 2x^3 - 6x^2 - 9x + 4;\quad k = 1$

52. $f(x) = 2x^3 + 9x^2 - 16x + 12;\quad k = 1$

53. $f(x) = x^3 + 7x^2 + 10x;\quad k = 0$

54. $f(x) = 2x^3 - 3x^2 - 5x;\quad k = 0$

55. $f(x) = 5x^4 + 2x^3 - x + 3;\quad k = \frac{2}{5}$

56. $f(x) = 16x^4 + 3x^2 - 2;\quad k = \frac{1}{2}$

57. $f(x) = x^2 - 2x + 2;\quad k = 1 - i$

58. $f(x) = x^2 - 4x + 5;\quad k = 2 - i$

59. $f(x) = x^2 + 3x + 4;\quad k = 2 + i$

60. $f(x) = x^2 - 3x + 5;\quad k = 1 - 2i$

61. $f(x) = 4x^4 + x^2 + 17x + 3;\quad k = -\frac{3}{2}$

62. $f(x) = 3x^4 + 13x^3 - 10x + 8;\quad k = -\frac{4}{3}$

63. $f(x) = x^3 + 3x^2 - x + 1;\quad k = 1 + i$

64. $f(x) = 2x^3 - x^2 + 3x - 5;\quad k = 2 - i$

Relating Concepts

For individual or collaborative investigation *(Exercises 65-74)*

The remainder theorem indicates that when a polynomial $f(x)$ is divided by $x - k$, the remainder is equal to $f(k)$. For

$$f(x) = x^3 - 2x^2 - x + 2,$$

use the remainder theorem to find each of the following. Then determine the coordinates of the corresponding point on the graph of $f(x)$.

65. $f(-2)$ **66.** $f(-1)$ **67.** $f\left(-\dfrac{1}{2}\right)$ **68.** $f(0)$

69. $f(1)$ **70.** $f\left(\dfrac{3}{2}\right)$ **71.** $f(2)$ **72.** $f(3)$

73. Use the results from **Exercises 65–72** to plot eight points on the graph of $f(x)$. Join these points with a smooth curve.

74. Apply the method above to graph $f(x) = -x^3 - x^2 + 2x$. Use x-values $-3, -1, \frac{1}{2}$, and 2 and the fact that $f(0) = 0$.

3.3 Zeros of Polynomial Functions

- Factor Theorem
- Rational Zeros Theorem
- Number of Zeros
- Conjugate Zeros Theorem
- Zeros of a Polynomial Function
- Descartes' Rule of Signs

Factor Theorem   Consider the polynomial function

$$f(x) = x^2 + x - 2,$$

which is written in factored form as

$$f(x) = (x - 1)(x + 2).$$

For this function, $f(1) = 0$ and $f(-2) = 0$, and thus 1 and -2 are zeros of $f(x)$. Notice the special relationship between each linear factor and its corresponding zero. The **factor theorem** summarizes this relationship.

Factor Theorem

For any polynomial function $f(x)$, $x - k$ is a factor of the polynomial if and only if $f(k) = 0$.

EXAMPLE 1   **Deciding Whether $x - k$ Is a Factor**

Determine whether $x - 1$ is a factor of each polynomial.

(a) $f(x) = 2x^4 + 3x^2 - 5x + 7$

(b) $f(x) = 3x^5 - 2x^4 + x^3 - 8x^2 + 5x + 1$

SOLUTION

(a) By the factor theorem, $x - 1$ will be a factor if $f(1) = 0$. Use synthetic division and the remainder theorem to decide.

$$
\begin{array}{r}
1\,\overline{)2\quad 0\quad 3\quad -5\quad 7} \\
\underline{\quad\quad 2\quad 2\quad 5\quad 0} \\
2\quad 2\quad 5\quad\quad 0\quad 7
\end{array}
\longleftarrow f(1) = 7
$$

Use a zero coefficient for the missing term.

The remainder is 7, not 0, so $x - 1$ is not a factor of $2x^4 + 3x^2 - 5x + 7$.

(b)

$$
\begin{array}{r|rrrrrr}
1) & 3 & -2 & 1 & -8 & 5 & 1 \\
& & 3 & 1 & 2 & -6 & -1 \\
\hline
& 3 & 1 & 2 & -6 & -1 & 0
\end{array}
$$

$\longleftarrow f(x) = 3x^5 - 2x^4 + x^3 - 8x^2 + 5x + 1$

$\longleftarrow f(1) = 0$

Because the remainder is 0, $x - 1$ is a factor. Additionally, we can determine from the coefficients in the bottom row that the other factor is

$$3x^4 + 1x^3 + 2x^2 - 6x - 1.$$

Thus, we can express the polynomial in factored form.

$$f(x) = (x - 1)(3x^4 + x^3 + 2x^2 - 6x - 1)$$

✔ **Now Try Exercises 9 and 11.**

We can use the factor theorem to factor a polynomial of greater degree into linear factors of the form $ax - b$.

EXAMPLE 2 **Factoring a Polynomial Given a Zero**

Factor $f(x) = 6x^3 + 19x^2 + 2x - 3$ into linear factors given that -3 is a zero.

SOLUTION Because -3 is a zero of f, $x - (-3) = x + 3$ is a factor.

$$
\begin{array}{r|rrrr}
-3) & 6 & 19 & 2 & -3 \\
& & -18 & -3 & 3 \\
\hline
& 6 & 1 & -1 & 0
\end{array}
$$
Use synthetic division to divide $f(x)$ by $x + 3$.

The quotient is $6x^2 + x - 1$, which is the factor that accompanies $x + 3$.

$$f(x) = (x + 3)(6x^2 + x - 1)$$
$$f(x) = (x + 3)\underbrace{(2x + 1)(3x - 1)}}$$ Factor $6x^2 + x - 1$.

These factors are all linear.

✔ **Now Try Exercise 21.**

LOOKING AHEAD TO CALCULUS
Finding the derivative of a polynomial function is one of the basic skills required in a first calculus course. For the functions

$$f(x) = x^4 - x^2 + 5x - 4,$$
$$g(x) = -x^6 + x^2 - 3x + 4,$$
and $h(x) = 3x^3 - x^2 + 2x - 4,$

the derivatives are

$$f'(x) = 4x^3 - 2x + 5,$$
$$g'(x) = -6x^5 + 2x - 3,$$
and $h'(x) = 9x^2 - 2x + 2.$

Notice the use of the "prime" notation. For example, the derivative of $f(x)$ is denoted $f'(x)$.

Look for the pattern among the exponents and the coefficients. Using this pattern, what is the derivative of

$$F(x) = 4x^4 - 3x^3 + 6x - 4?$$

The answer is at the top of the next page.

Rational Zeros Theorem The **rational zeros theorem** gives a method to determine all possible candidates for rational zeros of a polynomial function with integer coefficients.

Rational Zeros Theorem

If $\frac{p}{q}$ is a rational number written in lowest terms, and if $\frac{p}{q}$ is a zero of f, a polynomial function with integer coefficients, then p is a factor of the constant term and q is a factor of the leading coefficient.

Proof $f\left(\frac{p}{q}\right) = 0$ because $\frac{p}{q}$ is a zero of $f(x)$.

$$a_n\left(\frac{p}{q}\right)^n + a_{n-1}\left(\frac{p}{q}\right)^{n-1} + \cdots + a_1\left(\frac{p}{q}\right) + a_0 = 0$$ Definition of zero of f

$$a_n\left(\frac{p^n}{q^n}\right) + a_{n-1}\left(\frac{p^{n-1}}{q^{n-1}}\right) + \cdots + a_1\left(\frac{p}{q}\right) + a_0 = 0$$ Power rule for exponents

$$a_np^n + a_{n-1}p^{n-1}q + \cdots + a_1pq^{n-1} = -a_0q^n$$ Multiply by q^n. Subtract a_0q^n.

$$p(a_np^{n-1} + a_{n-1}p^{n-2}q + \cdots + a_1q^{n-1}) = -a_0q^n$$ Factor out p.

Answer to Looking Ahead to Calculus:

$$F'(x) = 16x^3 - 9x^2 + 6$$

This result shows that $-a_0 q^n$ equals the product of the two factors p and $(a_n p^{n-1} + \cdots + a_1 q^{n-1})$. For this reason, p must be a factor of $-a_0 q^n$. Because it was assumed that $\frac{p}{q}$ is written in lowest terms, p and q have no common factor other than 1, so p is not a factor of q^n. Thus, p must be a factor of a_0. In a similar way, it can be shown that q is a factor of a_n.

EXAMPLE 3 Using the Rational Zeros Theorem

Consider the polynomial function.

$$f(x) = 6x^4 + 7x^3 - 12x^2 - 3x + 2$$

(a) List all possible rational zeros.

(b) Find all rational zeros and factor $f(x)$ into linear factors.

SOLUTION

(a) For a rational number $\frac{p}{q}$ to be a zero, p must be a factor of $a_0 = 2$, and q must be a factor of $a_4 = 6$. Thus, p can be ± 1 or ± 2, and q can be ± 1, ± 2, ± 3, or ± 6. The possible rational zeros $\frac{p}{q}$ are ± 1, ± 2, $\pm\frac{1}{2}$, $\pm\frac{1}{3}$, $\pm\frac{1}{6}$, and $\pm\frac{2}{3}$.

(b) Use the remainder theorem to show that 1 is a zero.

Use "trial and error" to find zeros.

$$1\overline{)6 \quad 7 \quad -12 \quad -3 \quad 2}$$
$$\underline{ \quad 6 \quad 13 \quad 1 \quad -2}$$
$$6 \quad 13 \quad 1 \quad -2 \quad 0 \leftarrow f(1) = 0$$

The 0 remainder shows that 1 is a zero. The quotient is $6x^3 + 13x^2 + x - 2$.

$$f(x) = (x - 1)(6x^3 + 13x^2 + x - 2) \quad \text{Begin factoring } f(x).$$

Now, use the quotient polynomial and synthetic division to find that -2 is a zero.

$$-2\overline{)6 \quad 13 \quad 1 \quad -2}$$
$$\underline{ \quad -12 \quad -2 \quad 2}$$
$$6 \quad 1 \quad -1 \quad 0 \leftarrow f(-2) = 0$$

The new quotient polynomial is $6x^2 + x - 1$. Therefore, $f(x)$ can now be completely factored as follows.

$$f(x) = (x - 1)(x + 2)(6x^2 + x - 1)$$

$$f(x) = (x - 1)(x + 2)(3x - 1)(2x + 1)$$

Setting $3x - 1 = 0$ and $2x + 1 = 0$ yields the zeros $\frac{1}{3}$ and $-\frac{1}{2}$. In summary, the rational zeros are 1, -2, $\frac{1}{3}$, and $-\frac{1}{2}$. These zeros correspond to the x-intercepts of the graph of $f(x)$ in **Figure 18.** The linear factorization of $f(x)$ is as follows.

$$f(x) = 6x^4 + 7x^3 - 12x^2 - 3x + 2$$

$$f(x) = (x - 1)(x + 2)(3x - 1)(2x + 1)$$

Check by multiplying these factors.

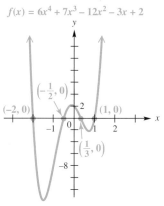

$f(x) = 6x^4 + 7x^3 - 12x^2 - 3x + 2$

Figure 18

✔ **Now Try Exercise 39.**

NOTE   Once we obtained the quadratic factor

$$6x^2 + x - 1$$

in **Example 3,** we were able to complete the work by factoring it directly. Had it not been easily factorable, we could have used the quadratic formula to find the other two zeros (and factors).

CAUTION   *The rational zeros theorem gives only possible rational zeros. It does not tell us whether these rational numbers are actual zeros.* We must rely on other methods to determine whether or not they are indeed zeros. Furthermore, the polynomial must have integer coefficients.

To apply the rational zeros theorem to a polynomial with fractional coefficients, multiply through by the least common denominator of all the fractions. For example, any rational zeros of $p(x)$ defined below will also be rational zeros of $q(x)$.

$$p(x) = x^4 - \frac{1}{6}x^3 + \frac{2}{3}x^2 - \frac{1}{6}x - \frac{1}{3}$$

$$q(x) = 6x^4 - x^3 + 4x^2 - x - 2 \qquad \text{Multiply the terms of } p(x) \text{ by 6.}$$

Number of Zeros   The **fundamental theorem of algebra** says that every function defined by a polynomial of degree 1 or more has a zero, which means that every such polynomial can be factored.

Carl Friedrich Gauss (1777–1855)

The **fundamental theorem of algebra** was first proved by Carl Friedrich Gauss in his doctoral thesis in 1799, when he was 22 years old.

Fundamental Theorem of Algebra

Every function defined by a polynomial of degree 1 or more has at least one complex zero.

From the fundamental theorem, if $f(x)$ is of degree 1 or more, then there is some number k_1 such that $f(k_1) = 0$. By the factor theorem,

$$f(x) = (x - k_1)q_1(x), \quad \text{for some polynomial } q_1(x).$$

If $q_1(x)$ is of degree 1 or more, the fundamental theorem and the factor theorem can be used to factor $q_1(x)$ in the same way. There is some number k_2 such that $q_1(k_2) = 0$, so

$$q_1(x) = (x - k_2)q_2(x)$$

and

$$f(x) = (x - k_1)(x - k_2)q_2(x).$$

Assuming that $f(x)$ has degree n and repeating this process n times gives

$$f(x) = a(x - k_1)(x - k_2) \cdots (x - k_n). \quad \text{a is the leading coefficient.}$$

Each of these factors leads to a zero of $f(x)$, so $f(x)$ has the n zeros $k_1, k_2, k_3, \ldots, k_n$. This result suggests the **number of zeros theorem.**

Number of Zeros Theorem

A function defined by a polynomial of degree n has *at most* n distinct zeros.

For example, a polynomial function of degree 3 has *at most* three distinct zeros but can have as few as one zero. Consider the following polynomial.

$$f(x) = x^3 + 3x^2 + 3x + 1$$

$$f(x) = (x + 1)^3 \qquad \text{Factored form of } f(x)$$

The function f is of degree 3 but has one distinct zero, -1. Actually, the zero -1 occurs *three* times because there are three factors of $x + 1$. The number of times a zero occurs is referred to as the **multiplicity of the zero.**

EXAMPLE 4 Finding a Polynomial Function That Satisfies Given Conditions (Real Zeros)

Find a polynomial function $f(x)$ of degree 3 with real coefficients that satisfies the given conditions.

(a) Zeros of -1, 2, and 4; $f(1) = 3$

(b) -2 is a zero of multiplicity 3; $f(-1) = 4$

SOLUTION

(a) These three zeros give $x - (-1) = x + 1$, $x - 2$, and $x - 4$ as factors of $f(x)$. Because $f(x)$ is to be of degree 3, these are the only possible factors by the number of zeros theorem. Therefore, $f(x)$ has the form

$$f(x) = a(x + 1)(x - 2)(x - 4), \quad \text{for some real number } a.$$

To find a, use the fact that $f(1) = 3$.

$$f(1) = a(1 + 1)(1 - 2)(1 - 4) \qquad \text{Let } x = 1.$$

$$3 = a(2)(-1)(-3) \qquad\qquad f(1) = 3$$

$$3 = 6a \qquad\qquad\qquad \text{Multiply.}$$

$$a = \frac{1}{2} \qquad\qquad\qquad \text{Divide by 6.}$$

Thus, $f(x) = \frac{1}{2}(x + 1)(x - 2)(x - 4), \quad \text{Let } a = \frac{1}{2}.$

or, $f(x) = \frac{1}{2}x^3 - \frac{5}{2}x^2 + x + 4. \qquad \text{Multiply.}$

(b) The polynomial function $f(x)$ has the following form.

$$f(x) = a(x + 2)(x + 2)(x + 2) \qquad \text{Factor theorem}$$

$$f(x) = a(x + 2)^3 \qquad\qquad (x + 2) \text{ is a factor three times.}$$

To find a, use the fact that $f(-1) = 4$.

$$f(-1) = a(-1 + 2)^3 \qquad \text{Let } x = -1.$$

$$4 = a(1)^3 \qquad\qquad f(-1) = 4$$

$$a = 4 \qquad\qquad \text{Solve for } a.$$

Thus, $f(x) = 4(x + 2)^3,$ Remember: $(x + 2)^3 \neq x^3 + 2^3$

or, $f(x) = 4x^3 + 24x^2 + 48x + 32. \qquad \text{Multiply.}$

✔ **Now Try Exercises 53 and 57.**

NOTE In **Example 4(a),** we cannot clear the denominators in $f(x)$ by multiplying each side by 2 because the result would equal $2 \cdot f(x)$, not $f(x)$.

Conjugate Zeros Theorem The following properties of complex conjugates are needed to prove the **conjugate zeros theorem.** We use a simplified notation for conjugates here. If $z = a + bi$, then the conjugate of z is written $\bar{z}$, where $\bar{z} = a - bi$. For example, if $z = -5 + 2i$, then $\bar{z} = -5 - 2i$.

Properties of Conjugates

For any complex numbers c and d, the following properties hold.

$$\overline{c + d} = \bar{c} + \bar{d}, \quad \overline{c \cdot d} = \bar{c} \cdot \bar{d}, \quad \text{and} \quad \overline{c^n} = (\bar{c})^n$$

In general, if z is a zero of a polynomial function with *real* coefficients, then so is $\bar{z}$. For example, the remainder theorem can be used to show that both $2 + i$ and $2 - i$ are zeros of $f(x) = x^3 - x^2 - 7x + 15$.

Conjugate Zeros Theorem

If $f(x)$ defines a polynomial function *having only real coefficients* and if $z = a + bi$ is a zero of $f(x)$, where a and b are real numbers, then

$$\bar{z} = a - bi \text{ is also a zero of } f(x).$$

Proof Start with the polynomial function

$$f(x) = a_n x^n + a_{n-1} x^{n-1} + \cdots + a_1 x + a_0,$$

where all coefficients are real numbers. If the complex number z is a zero of $f(x)$, then we have the following.

$$f(z) = a_n z^n + a_{n-1} z^{n-1} + \cdots + a_1 z + a_0 = 0$$

Take the conjugate of both sides of this equation.

$$\overline{a_n z^n + a_{n-1} z^{n-1} + \cdots + a_1 z + a_0} = \bar{0}$$

$$\overline{a_n z^n} + \overline{a_{n-1} z^{n-1}} + \cdots + \overline{a_1 z} + \overline{a_0} = \bar{0} \qquad \text{Use generalizations of the properties } \overline{c + d} = \bar{c} + \bar{d} \text{ and } \overline{c \cdot d} = \bar{c} \cdot \bar{d}.$$

$$\overline{a_n}\, \overline{z^n} + \overline{a_{n-1}}\, \overline{z^{n-1}} + \cdots + \overline{a_1}\, \overline{z} + \overline{a_0} = \bar{0}$$

$$a_n (\bar{z})^n + a_{n-1} (\bar{z})^{n-1} + \cdots + a_1 (\bar{z}) + a_0 = 0 \qquad \text{Use the property } \overline{c^n} = (\bar{c})^n \text{ and the fact that for any real number } a,$$

$$f(\bar{z}) = 0 \qquad \bar{a} = a.$$

Hence $\bar{z}$ is also a zero of $f(x)$, which completes the proof.

CAUTION *When the conjugate zeros theorem is applied, it is essential that the polynomial have only real coefficients. For example,*

$$f(x) = x - (1 + i)$$

has $1 + i$ as a zero, but the conjugate $1 - i$ is not a zero.

EXAMPLE 5 **Finding a Polynomial Function That Satisfies Given Conditions (Complex Zeros)**

Find a polynomial function $f(x)$ of least degree having only real coefficients and zeros 3 and $2 + i$.

SOLUTION The complex number $2 - i$ must also be a zero, so the polynomial has at least three zeros: 3, $2 + i$, and $2 - i$. For the polynomial to be of least degree, these must be the only zeros. By the factor theorem there must be three factors: $x - 3$, $x - (2 + i)$, and $x - (2 - i)$.

$$f(x) = (x - 3)[x - (2 + i)][x - (2 - i)] \qquad \text{Factor theorem}$$

$$f(x) = (x - 3)(x - 2 - i)(x - 2 + i) \qquad \text{Distribute negative signs.}$$

> The multiplication steps are not shown here.

$$f(x) = (x - 3)(x^2 - 4x + 5) \qquad \text{Multiply and combine like terms; } i^2 = -1.$$

$$f(x) = x^3 - 7x^2 + 17x - 15 \qquad \text{Multiply again.}$$

Any nonzero multiple of $x^3 - 7x^2 + 17x - 15$ also satisfies the given conditions on zeros. The information on zeros given in the problem is not sufficient to give a specific value for the leading coefficient.

✔ **Now Try Exercise 69.**

Zeros of a Polynomial Function The theorem on conjugate zeros helps predict the number of real zeros of polynomial functions with real coefficients.

- A polynomial function with real coefficients of *odd* degree n, where $n \geq 1$, must have at least one real zero (because zeros of the form $a + bi$, where $b \neq 0$, occur in conjugate pairs).

- A polynomial function with real coefficients of *even* degree n may have no real zeros.

EXAMPLE 6 **Finding All Zeros Given One Zero**

Find all zeros of $f(x) = x^4 - 7x^3 + 18x^2 - 22x + 12$, given that $1 - i$ is a zero.

SOLUTION Because the polynomial function has only real coefficients and $1 - i$ is a zero, by the conjugate zeros theorem $1 + i$ is also a zero. To find the remaining zeros, first use synthetic division to divide the original polynomial by $x - (1 - i)$.

$$(1 - i)(-6 - i) = -6 - i + 6i + i^2 = -7 + 5i$$

$$
\begin{array}{r|rrrrr}
1 - i & 1 & -7 & 18 & -22 & 12 \\
 & & 1 - i & -7 + 5i & 16 - 6i & -12 \\
\hline
 & 1 & -6 - i & 11 + 5i & -6 - 6i & 0
\end{array}
$$

By the factor theorem, because $x = 1 - i$ is a zero of $f(x)$, $x - (1 - i)$ is a factor, and $f(x)$ can be written as follows.

$$f(x) = [x - (1 - i)][x^3 + (-6 - i)x^2 + (11 + 5i)x + (-6 - 6i)]$$

We know that $x = 1 + i$ is also a zero of $f(x)$. Continue to use synthetic division and divide the quotient polynomial above by $x - (1 + i)$.

$$
\begin{array}{r|rrrr}
1 + i & 1 & -6 - i & 11 + 5i & -6 - 6i \\
 & & 1 + i & -5 - 5i & 6 + 6i \\
\hline
 & 1 & -5 & 6 & 0
\end{array}
$$

Using the result of the synthetic division, $f(x)$ can be written in the following factored form.

$$f(x) = [x - (1 - i)][x - (1 + i)](x^2 - 5x + 6)$$
$$f(x) = [x - (1 - i)][x - (1 + i)](x - 2)(x - 3)$$

The remaining zeros are 2 and 3. The four zeros are $1 - i$, $1 + i$, 2, and 3.

✔ **Now Try Exercise 35.**

NOTE If we had been unable to factor $x^2 - 5x + 6$ into linear factors, in **Example 6,** we would have used the quadratic formula to solve the equation $x^2 - 5x + 6 = 0$ to find the remaining two zeros of the function.

Descartes' Rule of Signs The following rule helps to determine the number of positive and negative real zeros of a polynomial function. A **variation in sign** is a change from positive to negative or from negative to positive in successive terms of the polynomial when they are written in order of descending powers of the variable. *Missing terms (those with **0** coefficients) are counted as no change in sign and can be ignored.*

Descartes' Rule of Signs

Let $f(x)$ define a polynomial function with real coefficients and a nonzero constant term, with terms in descending powers of x.

(a) The number of positive real zeros of f either equals the number of variations in sign occurring in the coefficients of $f(x)$, or is less than the number of variations by a positive even integer.

(b) The number of negative real zeros of f either equals the number of variations in sign occurring in the coefficients of $f(-x)$, or is less than the number of variations by a positive even integer.

EXAMPLE 7 **Applying Descartes' Rule of Signs**

Determine the different possibilities for the numbers of positive, negative, and nonreal complex zeros of

$$f(x) = x^4 - 6x^3 + 8x^2 + 2x - 1.$$

SOLUTION We first consider the possible number of positive zeros by observing that $f(x)$ has three variations in signs.

$$f(x) = +x^4 - 6x^3 + 8x^2 + 2x - 1$$
$$\quad\quad\quad\;\; 1 \quad\; 2 \quad\quad\quad\; 3$$

Thus, $f(x)$ has either three or one (because $3 - 2 = 1$) positive real zeros.

For negative zeros, consider the variations in signs for $f(-x)$.

$$f(-x) = (-x)^4 - 6(-x)^3 + 8(-x)^2 + 2(-x) - 1$$
$$f(-x) = x^4 + 6x^3 + 8x^2 - 2x - 1$$
$$\quad\quad\quad\quad\quad\quad\quad\quad\quad\quad\quad\; 1$$

There is only one variation in sign, so $f(x)$ has exactly one negative real zero.

Because $f(x)$ is a fourth-degree polynomial function, it must have four complex zeros, some of which may be repeated. Descartes' rule of signs has indicated that exactly one of these zeros is a negative real number.

- One possible combination of the zeros is one negative real zero, three positive real zeros, and no nonreal complex zeros.

- Another possible combination of the zeros is one negative real zero, one positive real zero, and two nonreal complex zeros.

By the conjugate zeros theorem, any possible nonreal complex zeros must occur in conjugate pairs because $f(x)$ has real coefficients. The table below summarizes these possibilities.

Possible Number of Zeros		
Positive	**Negative**	**Nonreal Complex**
3	1	0
1	1	2

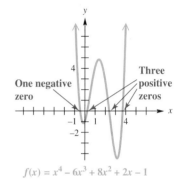

$$f(x) = x^4 - 6x^3 + 8x^2 + 2x - 1$$

Figure 19

The graph of $f(x)$ in **Figure 19** verifies the correct combination of three positive real zeros with one negative real zero, as seen in the first row of the table.*

✔ **Now Try Exercise 79.**

NOTE *Descartes' rule of signs does not identify the multiplicity of the zeros of a function.* For example, if it indicates that a function $f(x)$ has exactly two positive real zeros, then $f(x)$ may have two distinct positive real zeros or one positive real zero of multiplicity 2.

3.3 Exercises

CONCEPT PREVIEW *Determine whether each statement is* true *or* false. *If false, explain why.*

1. Because $x - 1$ is a factor of $f(x) = x^6 - x^4 + 2x^2 - 2$, we can also conclude that $f(1) = 0$.

2. Because $f(1) = 0$ for $f(x) = x^6 - x^4 + 2x^2 - 2$, we can conclude that $x - 1$ is a factor of $f(x)$.

3. For $f(x) = (x + 2)^4(x - 3)$, the number 2 is a zero of multiplicity 4.

4. Because $2 + 3i$ is a zero of $f(x) = x^2 - 4x + 13$, we can conclude that $2 - 3i$ is also a zero.

---

* The authors would like to thank Mary Hill of College of DuPage for her input into **Example 7.**

5. A polynomial function having degree 6 and only real coefficients may have no real zeros.

6. The polynomial function $f(x) = 2x^5 + 3x^4 - 8x^3 - 5x + 6$ has three variations in sign.

7. If $z = 7 - 6i$, then $\bar{z} = -7 + 6i$.

8. The product of a complex number and its conjugate is always a real number.

Use the factor theorem and synthetic division to decide whether the second polynomial is a factor of the first. **See Example 1.**

9. $x^3 - 5x^2 + 3x + 1$; $x - 1$
10. $x^3 + 6x^2 - 2x - 7$; $x + 1$

11. $2x^4 + 5x^3 - 8x^2 + 3x + 13$; $x + 1$
12. $-3x^4 + x^3 - 5x^2 + 2x + 4$; $x - 1$

13. $-x^3 + 3x - 2$; $x + 2$
14. $-2x^3 + x^2 - 63$; $x + 3$

15. $4x^2 + 2x + 54$; $x - 4$
16. $5x^2 - 14x + 10$; $x + 2$

17. $x^3 + 2x^2 + 3$; $x - 1$
18. $2x^3 + x + 2$; $x + 1$

19. $2x^4 + 5x^3 - 2x^2 + 5x + 6$; $x + 3$
20. $5x^4 + 16x^3 - 15x^2 + 8x + 16$; $x + 4$

Factor $f(x)$ into linear factors given that k is a zero. **See Example 2.**

21. $f(x) = 2x^3 - 3x^2 - 17x + 30$; $k = 2$

22. $f(x) = 2x^3 - 3x^2 - 5x + 6$; $k = 1$

23. $f(x) = 6x^3 + 13x^2 - 14x + 3$; $k = -3$

24. $f(x) = 6x^3 + 17x^2 - 63x + 10$; $k = -5$

25. $f(x) = 6x^3 + 25x^2 + 3x - 4$; $k = -4$

26. $f(x) = 8x^3 + 50x^2 + 47x - 15$; $k = -5$

27. $f(x) = x^3 + (7 - 3i)x^2 + (12 - 21i)x - 36i$; $k = 3i$

28. $f(x) = 2x^3 + (3 + 2i)x^2 + (1 + 3i)x + i$; $k = -i$

29. $f(x) = 2x^3 + (3 - 2i)x^2 + (-8 - 5i)x + (3 + 3i)$; $k = 1 + i$

30. $f(x) = 6x^3 + (19 - 6i)x^2 + (16 - 7i)x + (4 - 2i)$; $k = -2 + i$

31. $f(x) = x^4 + 2x^3 - 7x^2 - 20x - 12$; $k = -2$ (multiplicity 2)

32. $f(x) = 2x^4 + x^3 - 9x^2 - 13x - 5$; $k = -1$ (multiplicity 3)

For each polynomial function, one zero is given. Find all other zeros. **See Examples 2 and 6.**

33. $f(x) = x^3 - x^2 - 4x - 6$; 3
34. $f(x) = x^3 + 4x^2 - 5$; 1

35. $f(x) = x^3 - 7x^2 + 17x - 15$; $2 - i$
36. $f(x) = 4x^3 + 6x^2 - 2x - 1$; $\frac{1}{2}$

37. $f(x) = x^4 + 5x^2 + 4$; $-i$
38. $f(x) = x^4 + 26x^2 + 25$; i

For each polynomial function, (a) list all possible rational zeros, (b) find all rational zeros, and (c) factor $f(x)$ into linear factors. **See Example 3.**

39. $f(x) = x^3 - 2x^2 - 13x - 10$
40. $f(x) = x^3 + 5x^2 + 2x - 8$

41. $f(x) = x^3 + 6x^2 - x - 30$
42. $f(x) = x^3 - x^2 - 10x - 8$

43. $f(x) = 6x^3 + 17x^2 - 31x - 12$ **44.** $f(x) = 15x^3 + 61x^2 + 2x - 8$

45. $f(x) = 24x^3 + 40x^2 - 2x - 12$ **46.** $f(x) = 24x^3 + 80x^2 + 82x + 24$

For each polynomial function, find all zeros and their multiplicities.

47. $f(x) = (x - 2)^3(x^2 - 7)$ **48.** $f(x) = (x + 1)^2(x - 1)^3(x^2 - 10)$

49. $f(x) = 3x(x - 2)(x + 3)(x^2 - 1)$ **50.** $f(x) = 5x^2(x^2 - 16)(x + 5)$

51. $f(x) = (x^2 + x - 2)^5\left(x - 1 + \sqrt{3}\right)^2$

52. $f(x) = (2x^2 - 7x + 3)^3\left(x - 2 - \sqrt{5}\right)$

Find a polynomial function $f(x)$ of degree 3 with real coefficients that satisfies the given conditions. See Example 4.

53. Zeros of -3, 1, and 4; $f(2) = 30$ **54.** Zeros of 1, -1, and 0; $f(2) = 3$

55. Zeros of -2, 1, and 0; $f(-1) = -1$ **56.** Zeros of 2, -3, and 5; $f(3) = 6$

57. Zero of -3 having multiplicity 3; $f(3) = 36$

58. Zero of 2 and zero of 4 having multiplicity 2; $f(1) = -18$

59. Zero of 0 and zero of 1 having multiplicity 2; $f(2) = 10$

60. Zero of -4 and zero of 0 having multiplicity 2; $f(-1) = -6$

Find a polynomial function $f(x)$ of least degree having only real coefficients and zeros as given. Assume multiplicity 1 unless otherwise stated. See Examples 4–6.

61. $5 + i$ and $5 - i$ **62.** $7 - 2i$ and $7 + 2i$

63. 0, i, and $1 + i$ **64.** 0, $-i$, and $2 + i$

65. $1 + \sqrt{2}$, $1 - \sqrt{2}$, and 1   **66.** $1 - \sqrt{3}$, $1 + \sqrt{3}$, and 1

67. $2 - i$, 3, and -1 **68.** $3 + 2i$, -1, and 2

69. 2 and $3 + i$ **70.** -1 and $4 - 2i$

71. $1 - \sqrt{2}$, $1 + \sqrt{2}$, and $1 - i$ **72.** $2 + \sqrt{3}$, $2 - \sqrt{3}$, and $2 + 3i$

73. $2 - i$ and $6 - 3i$ **74.** $5 + i$ and $4 - i$

75. 4, $1 - 2i$, and $3 + 4i$ **76.** -1, $5 - i$, and $1 + 4i$

77. $1 + 2i$ and 2 (multiplicity 2)   **78.** $2 + i$ and -3 (multiplicity 2)

Determine the different possibilities for the numbers of positive, negative, and nonreal complex zeros of each function. See Example 7.

79. $f(x) = 2x^3 - 4x^2 + 2x + 7$ **80.** $f(x) = x^3 + 2x^2 + x - 10$

81. $f(x) = 4x^3 - x^2 + 2x - 7$ **82.** $f(x) = 3x^3 + 6x^2 + x + 7$

83. $f(x) = 5x^4 + 3x^2 + 2x - 9$ **84.** $f(x) = 3x^4 + 2x^3 - 8x^2 - 10x - 1$

85. $f(x) = -8x^4 + 3x^3 - 6x^2 + 5x - 7$ **86.** $f(x) = 6x^4 + 2x^3 + 9x^2 + x + 5$

87. $f(x) = x^5 + 3x^4 - x^3 + 2x + 3$ **88.** $f(x) = 2x^5 - x^4 + x^3 - x^2 + x + 5$

89. $f(x) = 2x^5 - 7x^3 + 6x + 8$ **90.** $f(x) = 11x^5 - x^3 + 7x - 5$

91. $f(x) = 5x^6 - 6x^5 + 7x^3 - 4x^2 + x + 2$ **92.** $f(x) = 9x^6 - 7x^4 + 8x^2 + x + 6$

93. $f(x) = 7x^5 + 6x^4 + 2x^3 + 9x^2 + x + 5$

94. $f(x) = -2x^5 + 10x^4 - 6x^3 + 8x^2 - x + 1$

Find all complex zeros of each polynomial function. Give exact values. List multiple zeros as necessary. *

95. $f(x) = x^4 + 2x^3 - 3x^2 + 24x - 180$

96. $f(x) = x^3 - x^2 - 8x + 12$

97. $f(x) = x^4 + x^3 - 9x^2 + 11x - 4$

98. $f(x) = x^3 - 14x + 8$

99. $f(x) = 2x^5 + 11x^4 + 16x^3 + 15x^2 + 36x$

100. $f(x) = 3x^3 - 9x^2 - 31x + 5$

101. $f(x) = x^5 - 6x^4 + 14x^3 - 20x^2 + 24x - 16$

102. $f(x) = 9x^4 + 30x^3 + 241x^2 + 720x + 600$

103. $f(x) = 2x^4 - x^3 + 7x^2 - 4x - 4$

104. $f(x) = 32x^4 - 188x^3 + 261x^2 + 54x - 27$

105. $f(x) = 5x^3 - 9x^2 + 28x + 6$

106. $f(x) = 4x^3 + 3x^2 + 8x + 6$

107. $f(x) = x^4 + 29x^2 + 100$

108. $f(x) = x^4 + 4x^3 + 6x^2 + 4x + 1$

109. $f(x) = x^4 + 2x^2 + 1$

110. $f(x) = x^4 - 8x^3 + 24x^2 - 32x + 16$

111. $f(x) = x^4 - 6x^3 + 7x^2$

112. $f(x) = 4x^4 - 65x^2 + 16$

113. $f(x) = x^4 - 8x^3 + 29x^2 - 66x + 72$

114. $f(x) = 12x^4 - 43x^3 + 50x^2 + 38x - 12$

115. $f(x) = x^6 - 9x^4 - 16x^2 + 144$

116. $f(x) = x^6 - x^5 - 26x^4 + 44x^3 + 91x^2 - 139x + 30$

If c and d are complex numbers, prove each statement. (Hint: Let c = a + bi and d = m + ni and form all the conjugates, the sums, and the products.)

117. $\overline{c + d} = \overline{c} + \overline{d}$

118. $\overline{c \cdot d} = \overline{c} \cdot \overline{d}$

119. $\overline{a} = a$ for any real number a

120. $\overline{c^2} = (\overline{c})^2$

In 1545, a method of solving a cubic equation of the form

$$x^3 + mx = n,$$

developed by Niccolo Tartaglia, was published in the Ars Magna, *a work by Girolamo Cardano. The formula for finding the one real solution of the equation is*

$$x = \sqrt[3]{\frac{n}{2} + \sqrt{\left(\frac{n}{2}\right)^2 + \left(\frac{m}{3}\right)^3}} - \sqrt[3]{\frac{-n}{2} + \sqrt{\left(\frac{n}{2}\right)^2 + \left(\frac{m}{3}\right)^3}}.$$

(Source: Gullberg, J., *Mathematics from the Birth of Numbers,* W.W. Norton & Company.)
Use the formula to solve each equation for the one real solution.

121. $x^3 + 9x = 26$

122. $x^3 + 15x = 124$

---

* The authors would like to thank Aileen Solomon of Trident Technical College for preparing and suggesting the inclusion of **Exercises 95–108.**

3.4 Polynomial Functions: Graphs, Applications, and Models

- Graphs of $f(x) = ax^n$
- Graphs of General Polynomial Functions
- Behavior at Zeros
- Turning Points and End Behavior
- Graphing Techniques
- Intermediate Value and Boundedness Theorems
- Approximations of Real Zeros
- Polynomial Models

Graphs of $f(x) = ax^n$ We can now graph polynomial functions of degree 3 or greater with real number coefficients and domains (because the graphs are in the real number plane). We begin by inspecting the graphs of several functions of the form

$$f(x) = ax^n, \quad \text{with } a = 1.$$

The identity function $f(x) = x$, the squaring function $f(x) = x^2$, and the cubing function $f(x) = x^3$ were graphed earlier using a general point-plotting method.

Each function in **Figure 20** has odd degree and is an odd function exhibiting symmetry about the origin. Each has domain $(-\infty, \infty)$ and range $(-\infty, \infty)$ and is continuous on its entire domain $(-\infty, \infty)$. Additionally, these odd functions are increasing on their entire domain $(-\infty, \infty)$, appearing as though they fall to the left and rise to the right.

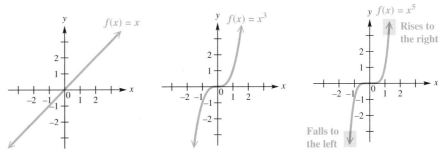

Figure 20

Each function in **Figure 21** has even degree and is an even function exhibiting symmetry about the y-axis. Each has domain $(-\infty, \infty)$ but restricted range $[0, \infty)$. These even functions are also continuous on their entire domain $(-\infty, \infty)$. However, they are decreasing on $(-\infty, 0)$ and increasing on $(0, \infty)$, appearing as though they rise both to the left and to the right.

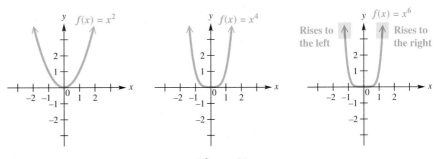

Figure 21

The behaviors in the graphs of these basic polynomial functions as x increases (decreases) without bound also apply to more complicated polynomial functions.

Graphs of General Polynomial Functions   As with quadratic functions, the absolute value of a in $f(x) = ax^n$ determines the width of the graph.

- When $|a| > 1$, the graph is stretched vertically, making it narrower.

- When $0 < |a| < 1$, the graph is shrunk or compressed vertically, making it wider.

Compared to the graph of $f(x) = ax^n$, the following also hold true.

- The graph of $f(x) = -ax^n$ is reflected across the x-axis.

- The graph of $f(x) = ax^n + k$ is translated (shifted) k units up if $k > 0$ and $|k|$ units down if $k < 0$.

- The graph of $f(x) = a(x - h)^n$ is translated h units to the right if $h > 0$ and $|h|$ units to the left if $h < 0$.

- The graph of $f(x) = a(x - h)^n + k$ shows a combination of these translations.

EXAMPLE 1 Examining Vertical and Horizontal Translations

Graph each polynomial function. Determine the largest open intervals of the domain over which each function is increasing or decreasing.

(a) $f(x) = x^5 - 2$ **(b)** $f(x) = (x + 1)^6$ **(c)** $f(x) = -2(x - 1)^3 + 3$

SOLUTION

(a) The graph of $f(x) = x^5 - 2$ will be the same as that of $f(x) = x^5$, but translated 2 units down. See **Figure 22.** This function is increasing on its entire domain $(-\infty, \infty)$.

(b) In $f(x) = (x + 1)^6$, function f has a graph like that of $f(x) = x^6$, but because

$$x + 1 = x - (-1),$$

it is translated 1 unit to the left. See **Figure 23.** This function is decreasing on $(-\infty, -1)$ and increasing on $(-1, \infty)$.

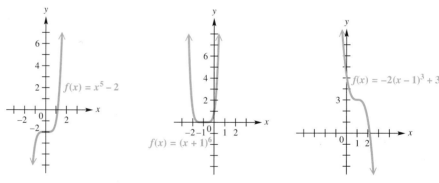

| Figure 22 | Figure 23 | Figure 24 |

(c) The negative sign in -2 causes the graph of

$$f(x) = -2(x - 1)^3 + 3$$

to be reflected across the x-axis when compared with the graph of $f(x) = x^3$. Because $|-2| > 1$, the graph is stretched vertically when compared to the graph of $f(x) = x^3$. As shown in **Figure 24,** the graph is also translated 1 unit to the right and 3 units up. This function is decreasing on its entire domain $(-\infty, \infty)$.

✔ **Now Try Exercises 13, 15, and 19.**

Unless otherwise restricted, the domain of a polynomial function is the set of all real numbers. Polynomial functions are smooth, continuous curves on the interval $(-\infty, \infty)$. The range of a polynomial function of odd degree is also the set of all real numbers.

The graphs in **Figure 25** suggest that for every polynomial function f of odd degree there is at least one real value of x that satisfies $f(x) = 0$. The real zeros correspond to the x-intercepts of the graph and can be determined by inspecting the factored form of each polynomial.

Odd Degree

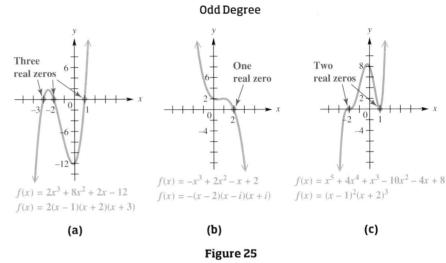

$f(x) = 2x^3 + 8x^2 + 2x - 12$
$f(x) = 2(x - 1)(x + 2)(x + 3)$

(a)

$f(x) = -x^3 + 2x^2 - x + 2$
$f(x) = -(x - 2)(x - i)(x + i)$

(b)

$f(x) = x^5 + 4x^4 + x^3 - 10x^2 - 4x + 8$
$f(x) = (x - 1)^2(x + 2)^3$

(c)

Figure 25

A polynomial function of even degree has a range of the form $(-\infty, k]$ or $[k, \infty)$, for some real number k. **Figure 26** shows two typical graphs.

Even Degree

$f(x) = x^4 - 5x^2 + 4$
$f(x) = (x - 1)(x + 1)(x - 2)(x + 2)$

(a)

$f(x) = -x^6 - x^5 + 4x^4 + 2x^3 - 5x^2 - x + 2$
$f(x) = -(x + 2)(x + 1)^2(x - 1)^3$

(b)

Figure 26

Behavior at Zeros   **Figure 26(b)** shows a sixth-degree polynomial function with three distinct zeros, yet the behavior of the graph at each zero is different. This behavior depends on the multiplicity of the zero as determined by the exponent on the corresponding factor. The factored form of the polynomial function $f(x)$ is

$$-(x + 2)^1(x + 1)^2(x - 1)^3.$$

- $(x + 2)$ is a factor of multiplicity 1. Therefore, the graph crosses the x-axis at $(-2, 0)$.

- $(x + 1)$ is a factor of multiplicity 2. Therefore, the graph is tangent to the x-axis at $(-1, 0)$. This means that it touches the x-axis, then turns and changes behavior from decreasing to increasing similar to that of the squaring function $f(x) = x^2$ at its zero.

- $(x - 1)$ is a factor of multiplicity 3. Therefore, the graph crosses the x-axis **and** is tangent to the x-axis at $(1, 0)$. This causes a change in concavity (that is, how the graph opens upward or downward) at this x-intercept with behavior similar to that of the cubing function $f(x) = x^3$ at its zero.

Figure 27 generalizes the behavior of such graphs at their zeros.

The graph crosses the x-axis at $(c, 0)$ if c is a zero of multiplicity 1.

The graph is tangent to the x-axis at $(c, 0)$ if c is a zero of even multiplicity. The graph bounces, or turns, at c.

The graph crosses **and** is tangent to the x-axis at $(c, 0)$ if c is a zero of odd multiplicity greater than 1. The graph wiggles at c.

Figure 27

Turning Points and End Behavior   The graphs in **Figures 25 and 26** show that polynomial functions often have **turning points** where the function changes from increasing to decreasing or from decreasing to increasing.

> ### Turning Points
>
> A polynomial function of degree n has at most $n - 1$ turning points, with at least one turning point between each pair of successive zeros.

The **end behavior** of a polynomial graph is determined by the *dominating term*—that is, the term of greatest degree. A polynomial of the form

$$f(x) = a_n x^n + a_{n-1} x^{n-1} + \cdots + a_0$$

has the same end behavior as $f(x) = a_n x^n$. For example,

$$f(x) = 2x^3 + 8x^2 + 2x - 12$$

has the same end behavior as $f(x) = 2x^3$. It is large and positive for large positive values of x, while it is large and negative for negative values of x with large absolute value. That is, it rises to the right and falls to the left.

Figure 25(a) shows that as x increases without bound, y does also. For the same graph, as x decreases without bound, y does also.

$$\text{As} \quad x \to \infty, \quad y \to \infty \quad \text{and} \quad \text{as} \quad x \to -\infty, \quad y \to -\infty.$$

LOOKING AHEAD TO CALCULUS

To find the x-coordinates of the two turning points of the graph of

$$f(x) = 2x^3 + 8x^2 + 2x - 12,$$

we can use the "maximum" and "minimum" capabilities of a graphing calculator and determine that, to the nearest thousandth, they are -0.131 and -2.535. In calculus, their exact values can be found by determining the zeros of the derivative function of $f(x)$,

$$f'(x) = 6x^2 + 16x + 2,$$

because the turning points occur precisely where the tangent line has slope 0. Using the quadratic formula would show that the zeros are

$$\frac{-4 \pm \sqrt{13}}{3},$$

which agree with the calculator approximations.

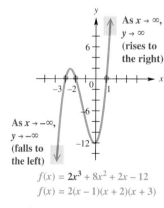

$f(x) = 2x^3 + 8x^2 + 2x - 12$
$f(x) = 2(x - 1)(x + 2)(x + 3)$

$f(x) = -x^3 + 2x^2 - x + 2$
$f(x) = -(x - 2)(x - i)(x + i)$

Figure 25(a) (repeated)   **Figure 25(b) (repeated)**

The graph in **Figure 25(b)** has the same end behavior as $f(x) = -x^3$.

$$\text{As} \quad x \to \infty, \quad y \to -\infty \quad \text{and} \quad \text{as} \quad x \to -\infty, \quad y \to \infty.$$

The graph of a polynomial function with a dominating term of even degree will show end behavior in the same direction. See **Figure 26.**

End Behavior of Graphs of Polynomial Functions

Suppose that ax^n is the dominating term of a polynomial function f of **odd degree.**

1. If $a > 0$, then as $x \to \infty$, $f(x) \to \infty$, and as $x \to -\infty$, $f(x) \to -\infty$. Therefore, the end behavior of the graph is of the type shown in **Figure 28(a).** We symbolize it as $\swarrow\nearrow$.

2. If $a < 0$, then as $x \to \infty$, $f(x) \to -\infty$, and as $x \to -\infty$, $f(x) \to \infty$. Therefore, the end behavior of the graph is of the type shown in **Figure 28(b).** We symbolize it as $\nwarrow\searrow$.

(a)   (b)   (a)   (b)

Figure 28   **Figure 29**

Suppose that ax^n is the dominating term of a polynomial function f of **even degree.**

1. If $a > 0$, then as $|x| \to \infty$, $f(x) \to \infty$. Therefore, the end behavior of the graph is of the type shown in **Figure 29(a).** We symbolize it as $\uparrow\!\smile\!\uparrow$.

2. If $a < 0$, then as $|x| \to \infty$, $f(x) \to -\infty$. Therefore, the end behavior of the graph is of the type shown in **Figure 29(b).** We symbolize it as $\downarrow\!\frown\!\downarrow$.

EXAMPLE 2   **Determining End Behavior**

The graphs of the polynomial functions defined as follows are shown in A–D.

$$f(x) = x^4 - x^2 + 5x - 4, \qquad g(x) = -x^6 + x^2 - 3x - 4,$$
$$h(x) = 3x^3 - x^2 + 2x - 4, \quad \text{and} \quad k(x) = -x^7 + x - 4$$

Based on the discussion of end behavior, match each function with its graph.

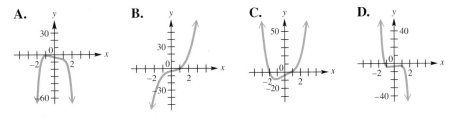

A.   **B.**   **C.**   **D.**

SOLUTION

- Function f has even degree and a dominating term with positive leading coefficient, as in C.

- Function g has even degree and a dominating term with negative leading coefficient, as in A.

- Function h has odd degree and a dominating term with positive coefficient, as in B.

- Function k has odd degree and a dominating term with negative coefficient, as in D.

✔ **Now Try Exercises 21, 23, 25, and 27.**

Graphing Techniques   We have discussed several characteristics of the graphs of polynomial functions that are useful for graphing the function by hand. A **comprehensive graph** of a polynomial function $f(x)$ will show the following characteristics.

- all x-intercepts (indicating the real zeros) and the behavior of the graph at these zeros

- the y-intercept

- the sign of $f(x)$ within the intervals formed by the x-intercepts

- enough of the domain to show the end behavior

In **Examples 3 and 4,** we sketch the graphs of two polynomial functions by hand. We use the following general guidelines.

Graphing a Polynomial Function

Let $f(x) = a_n x^n + a_{n-1} x^{n-1} + \cdots + a_1 x + a_0$, with $a_n \neq 0$, be a polynomial function of degree n. To sketch its graph, follow these steps.

Step 1 Find the real zeros of f. Plot the corresponding x-intercepts.

Step 2 Find $f(0) = a_0$. Plot the corresponding y-intercept.

Step 3 Use end behavior, whether the graph crosses, bounces on, or wiggles through the x-axis at the x-intercepts, and selected points as necessary to complete the graph.

EXAMPLE 3   Graphing a Polynomial Function

Graph $f(x) = 2x^3 + 5x^2 - x - 6$.

SOLUTION

Step 1 The possible rational zeros are ± 1, ± 2, ± 3, ± 6, $\pm \frac{1}{2}$, and $\pm \frac{3}{2}$. Use synthetic division to show that 1 is a zero.

$$
\begin{array}{r}
1)\overline{2 \quad 5 \quad -1 \quad -6} \\
2 \quad\ \ 7 \quad\ \ 6 \\
\hline
2 \quad 7 \quad\ \ 6 \quad\ \ \ 0 \leftarrow f(1) = 0
\end{array}
$$

We use the results of the synthetic division to factor as follows.

$$f(x) = (x - 1)(2x^2 + 7x + 6)$$

$$f(x) = (x - 1)(2x + 3)(x + 2) \quad \text{Factor again.}$$

Set each linear factor equal to 0, and then solve for x to find zeros. The three zeros of f are 1, $-\frac{3}{2}$, and -2. Plot the corresponding x-intercepts. See **Figure 30.**

Step 2 $f(0) = -6$, so plot $(0, -6)$. See **Figure 30.**

Step 3 The dominating term of $f(x)$ is $2x^3$, so the graph will have end behavior similar to that of $f(x) = x^3$. It will rise to the right and fall to the left as ⤴. See **Figure 30.** Each zero of $f(x)$ occurs with multiplicity 1, meaning that the graph of $f(x)$ will cross the x-axis at each of its zeros. Because the graph of a polynomial function has no breaks, gaps, or sudden jumps, we now have sufficient information to sketch the graph of $f(x)$.

Figure 30

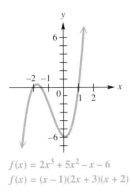

$f(x) = 2x^3 + 5x^2 - x - 6$

$f(x) = (x - 1)(2x + 3)(x + 2)$

Figure 31

Begin sketching at either end of the graph with the appropriate end behavior, and draw a smooth curve that crosses the x-axis at each zero, has a turning point between successive zeros, and passes through the y-intercept as shown in **Figure 31.**

Additional points may be used to verify whether the graph is above or below the x-axis between the zeros and to add detail to the sketch of the graph. The zeros divide the x-axis into four intervals:

$$\left(-\infty, -2\right), \quad \left(-2, -\frac{3}{2}\right), \quad \left(-\frac{3}{2}, 1\right), \quad \text{and} \quad (1, \infty).$$

Select an x-value as a test point in each interval, and substitute it into the equation for $f(x)$ to determine additional points on the graph. A typical selection of test points and the results of the tests are shown in the table.

Interval	Test Point x	Value of $f(x)$	Sign of $f(x)$	Graph Above or Below x-Axis
$(-\infty, -2)$	-3	-12	Negative	Below
$\left(-2, -\frac{3}{2}\right)$	$-\frac{7}{4}$	$\frac{11}{32}$	Positive	Above
$\left(-\frac{3}{2}, 1\right)$	0	-6	Negative	Below
$(1, \infty)$	2	28	Positive	Above

✔ **Now Try Exercise 29.**

EXAMPLE 4 **Graphing a Polynomial Function**

Graph $f(x) = -(x - 1)(x - 3)(x + 2)^2$.

SOLUTION

Step 1 Because the polynomial is given in factored form, the zeros can be determined by inspection. They are 1, 3, and -2. Plot the corresponding x-intercepts of the graph of $f(x)$. See **Figure 32.**

Step 2
$$f(0) = -(0 - 1)(0 - 3)(0 + 2)^2 \quad \text{Find } f(0).$$
$$f(0) = -(-1)(-3)(2)^2 \quad \text{Simplify in parentheses.}$$
$$f(0) = -12 \longleftarrow \text{The } y\text{-intercept is } (0, -12).$$

Plot the y-intercept $(0, -12)$. See **Figure 32.**

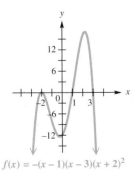

$f(x) = -(x - 1)(x - 3)(x + 2)^2$

Figure 32

Step 3 The dominating term of $f(x)$ can be found by multiplying the factors and identifying the term of greatest degree. Here it is $-(x)(x)(x)^2 = -x^4$, indicating that the end behavior of the graph is ⌢⌣. Because 1 and 3 are zeros of multiplicity 1, the graph will cross the x-axis at these zeros. The graph of $f(x)$ will touch the x-axis at -2 and then turn and change direction because it is a zero of even multiplicity.

Begin at either end of the graph with the appropriate end behavior and draw a smooth curve that crosses the x-axis at 1 and 3 and that touches the x-axis at -2, then turns and changes direction. The graph will also pass through the y-intercept $(0, -12)$. See **Figure 32.**

Using test points within intervals formed by the x-intercepts is a good way to add detail to the graph and verify the accuracy of the sketch. A typical selection of test points is $(-3, -24)$, $(-1, -8)$, $(2, 16)$, and $(4, -108)$.

✔ **Now Try Exercise 33.**

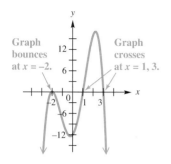

Figure 32 (repeated)

NOTE It is possible to reverse the process of **Example 4** and write the polynomial function from its graph if the zeros and any other point on the graph are known. Suppose that we are asked to find a polynomial function of least degree having the graph shown in **Figure 32** (repeated in the margin). Because the graph crosses the x-axis at 1 and 3 and bounces at -2, we know that the factored form of the function is as follows.

Multiplicity one Multiplicity two

$$f(x) = a(x - 1)^1(x - 3)^1(x + 2)^2$$

Now find the value of a by substituting the x- and y-values of any other point on the graph, say $(0, -12)$, into this function and solving for a.

$$f(x) = a(x - 1)(x - 3)(x + 2)^2$$
$$-12 = a(0 - 1)(0 - 3)(0 + 2)^2 \quad \text{Let } x = 0 \text{ and } y = -12.$$
$$-12 = a(12) \quad \text{Simplify.}$$
$$a = -1 \quad \text{Divide by 12. Interchange sides.}$$

Verify in **Example 4** that the polynomial function is

$$f(x) = -(x - 1)(x - 3)(x + 2)^2.$$

Exercises of this type are labeled *Connecting Graphs with Equations*.

We emphasize the important relationships among the following concepts.

- the x-intercepts of the graph of $y = f(x)$
- the zeros of the function f
- the solutions of the equation $f(x) = 0$
- the factors of $f(x)$

For example, the graph of the function

$$f(x) = 2x^3 + 5x^2 - x - 6 \qquad \text{Example 3}$$
$$f(x) = (x - 1)(2x + 3)(x + 2) \qquad \text{Factored form}$$

has x-intercepts $(1, 0)$, $\left(-\frac{3}{2}, 0\right)$, and $(-2, 0)$ as shown in **Figure 31** on the previous page. Because 1, $-\frac{3}{2}$, and -2 are the x-values where the function is 0, they are the zeros of f. Also, 1, $-\frac{3}{2}$, and -2 are the solutions of the polynomial equation

$$2x^3 + 5x^2 - x - 6 = 0.$$

This discussion is summarized as follows.

Relationships among x-Intercepts, Zeros, Solutions, and Factors

If f is a polynomial function and $(c, 0)$ is an x-intercept of the graph of $y = f(x)$, then

$$c \text{ is a zero of } f, \quad c \text{ is a solution of } f(x) = 0,$$

and $x - c$ is a factor of $f(x)$.

Intermediate Value and Boundedness Theorems   As **Examples 3 and 4** show, one key to graphing a polynomial function is locating its zeros. In the special case where the potential zeros are rational numbers, the zeros are found by the rational zeros theorem.

Occasionally, irrational zeros can be found by inspection. For instance, $f(x) = x^3 - 2$ has the irrational zero $\sqrt[3]{2}$.

The next two theorems apply to the zeros of every polynomial function with real coefficients. The first theorem uses the fact that graphs of polynomial functions are continuous curves. The proof requires advanced methods, so it is not given here. **Figure 33** illustrates the theorem.

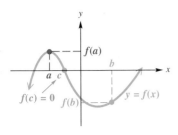

Figure 33

Intermediate Value Theorem

If $f(x)$ is a polynomial function with only *real coefficients,* and if for real numbers a and b the values $f(a)$ and $f(b)$ are opposite in sign, then there exists at least one real zero between a and b.

This theorem helps identify intervals where zeros of polynomial functions are located. If $f(a)$ and $f(b)$ are opposite in sign, then 0 is between $f(a)$ and $f(b)$, and so there must be a number c between a and b where $f(c) = 0$.

EXAMPLE 5   **Locating a Zero**

Use synthetic division and a graph to show that $f(x) = x^3 - 2x^2 - x + 1$ has a real zero between 2 and 3.

ALGEBRAIC SOLUTION

Use synthetic division to find $f(2)$ and $f(3)$.

$$
\begin{array}{r}
2)\overline{1 \quad -2 \quad -1 \quad1} \\
\phantom{2)\overline{1}}\;\; 2 \quad0 \quad -2 \\
\hline
1 \quad0 \quad -1 \quad -1 = f(2)
\end{array}
$$

$$
\begin{array}{r}
3)\overline{1 \quad -2 \quad -1 \quad1} \\
\phantom{3)\overline{1}}\;\; 3 \quad3 \quad6 \\
\hline
1 \quad1 \quad2 \quad7 = f(3)
\end{array}
$$

Because $f(2)$ is negative and $f(3)$ is positive, by the intermediate value theorem there must be a real zero between 2 and 3.

GRAPHING CALCULATOR SOLUTION

The graphing calculator screen in **Figure 34** indicates that this zero is approximately 2.2469796. (Notice that there are two other zeros as well.)

Figure 34

✔ **Now Try Exercise 49.**

CAUTION   *Be careful when interpreting the intermediate value theorem.* If $f(a)$ and $f(b)$ are *not* opposite in sign, it does not necessarily mean that there is no zero between a and b. In **Figure 35**, $f(a)$ and $f(b)$ are both negative, but -3 and -1, which are between a and b, are zeros of $f(x)$.

The intermediate value theorem for polynomials helps limit the search for real zeros to smaller and smaller intervals. In **Example 5,** we used the theorem to verify that there is a real zero between 2 and 3. To locate the zero more accurately, we can use the theorem repeatedly. (Prior to modern-day methods involving calculators and computers, this was done by hand.)

Figure 35

The **boundedness theorem** shows how the bottom row of a synthetic division is used to place upper and lower bounds on possible real zeros of a polynomial function.

> ## Boundedness Theorem
>
> Let $f(x)$ be a polynomial function of degree $n \geq 1$ with *real coefficients* and with a *positive* leading coefficient. Suppose $f(x)$ is divided synthetically by $x - c$.
>
> **(a)** If $c > 0$ and all numbers in the bottom row of the synthetic division are nonnegative, then $f(x)$ has no zero greater than c.
>
> **(b)** If $c < 0$ and the numbers in the bottom row of the synthetic division alternate in sign (with 0 considered positive or negative, as needed), then $f(x)$ has no zero less than c.

Proof   We outline the proof of part (a). The proof for part (b) is similar.

By the division algorithm, if $f(x)$ is divided by $x - c$, then for some $q(x)$ and r,

$$f(x) = (x - c)q(x) + r,$$

where all coefficients of $q(x)$ are nonnegative, $r \geq 0$, and $c > 0$. If $x > c$, then $x - c > 0$. Because $q(x) > 0$ and $r \geq 0$,

$$f(x) = (x - c)q(x) + r > 0.$$

This means that $f(x)$ will never be 0 for $x > c$.

EXAMPLE 6   Using the Boundedness Theorem

Show that the real zeros of $f(x) = 2x^4 - 5x^3 + 3x + 1$ satisfy these conditions.

(a) No real zero is greater than 3.   **(b)** No real zero is less than -1.

SOLUTION

(a) Because $f(x)$ has real coefficients and the leading coefficient, 2, is positive, use the boundedness theorem. Divide $f(x)$ synthetically by $x - 3$.

$$
\begin{array}{r|rrrrr}
3 & 2 & -5 & 0 & 3 & 1 \\
  &   & 6 & 3 & 9 & 36 \\
\hline
  & 2 & 1 & 3 & 12 & 37
\end{array}
$$
←— All are nonnegative.

Here $3 > 0$ and all numbers in the last row of the synthetic division are nonnegative, so $f(x)$ has no real zero greater than 3.

(b) We use the boundedness theorem again and divide $f(x)$ synthetically by $x - (-1)$, or $x + 1$.

$$
\begin{array}{r|rrrrr}
-1 & 2 & -5 & 0 & 3 & 1 \\
   &   & -2 & 7 & -7 & 4 \\
\hline
   & 2 & -7 & 7 & -4 & 5
\end{array}
$$
←— These numbers alternate in sign.

Here $-1 < 0$ and the numbers in the last row alternate in sign, so $f(x)$ has no real zero less than -1.

✔ **Now Try Exercises 57 and 59.**

Approximations of Real Zeros   We can approximate the irrational real zeros of a polynomial function using a graphing calculator.

EXAMPLE 7   Approximating Real Zeros of a Polynomial Function

Approximate the real zeros of $f(x) = x^4 - 6x^3 + 8x^2 + 2x - 1$.

SOLUTION   The dominating term is x^4, so the graph will have end behavior similar to the graph of $f(x) = x^4$, which is positive for all values of x with large absolute values. That is, the end behavior is up at the left and the right, $\smile$. There are at most four real zeros because the polynomial is fourth-degree.

Since $f(0) = -1$, the y-intercept is $(0, -1)$. Because the end behavior is positive on the left and the right, by the intermediate value theorem f has at least one real zero on either side of $x = 0$. To approximate the zeros, we use a graphing calculator. The graph in **Figure 36** shows that there are four real zeros, and the table indicates that they are between

$$-1 \text{ and } 0, \quad 0 \text{ and } 1, \quad 2 \text{ and } 3, \quad \text{and} \quad 3 \text{ and } 4$$

because there is a sign change in $f(x) = y_1$ in each case.

Figure 36   **Figure 37**

Using a calculator, we can find zeros to a great degree of accuracy. **Figure 37** shows that the negative zero is approximately -0.4142136. Similarly, we find that the other three zeros are approximately

$$0.26794919, \quad 2.4142136, \quad \text{and} \quad 3.7320508.$$

✔ **Now Try Exercise 77.**

Polynomial Models

EXAMPLE 8   Examining a Polynomial Model

The table shows the number of transactions, in millions, by users of bank debit cards for selected years.

(a) Using $x = 0$ to represent 1995, $x = 3$ to represent 1998, and so on, use the regression feature of a calculator to determine the quadratic function that best fits the data. Plot the data and the graph.

(b) Repeat part (a) for a cubic function (degree 3).

(c) Repeat part (a) for a quartic function (degree 4).

(d) The **correlation coefficient**, R, is a measure of the strength of the relationship between two variables. The values of R and R^2 are used to determine how well a regression model fits a set of data. The closer the value of R^2 is to 1, the better the fit. Compare R^2 for the three functions found in parts (a)–(c) to decide which function best fits the data.

Year	Transactions (in millions)
1995	829
1998	3765
2000	5290
2004	14,106
2008	28,464
2012	44,351

Source: Statistical Abstract of the United States.

SOLUTION

(a) The best-fitting quadratic function for the data is

$$y = 131.4x^2 + 342.9x + 901.4.$$

The regression coordinates screen and the graph are shown in **Figure 38.**

Figure 38

(b) The best-fitting cubic function is shown in **Figure 39** and is

$$y = -1.606x^3 + 172.1x^2 + 92.33x + 1119.$$

Figure 39

(c) The best-fitting quartic function is shown in **Figure 40** and is

$$y = -1.088x^4 + 34.71x^3 - 195.1x^2 + 1190x + 868.6.$$

Figure 40

(d) With the statistical diagnostics turned on, the value of R^2 is displayed with the regression results on the TI-84 Plus each time that a regression model is executed. By inspecting the R^2 value for each model above, we see that the quartic function provides the best fit because it has the largest R^2 value of 0.9998725617.

✔ **Now Try Exercise 99.**

NOTE In **Example 8(d),** we selected the quartic function as the best model based on the comparison of R^2 values of the models. In practice, however, the best choice of a model should also depend on the set of data being analyzed as well as analysis of its trends and attributes.

3.4 Exercises

CONCEPT PREVIEW *Comprehensive graphs of four polynomial functions are shown in A–D. They represent the graphs of functions defined by these four equations, but not necessarily in the order listed.*

$$y = x^3 - 3x^2 - 6x + 8 \qquad y = x^4 + 7x^3 - 5x^2 - 75x$$

$$y = -x^3 + 9x^2 - 27x + 17 \qquad y = -x^5 + 36x^3 - 22x^2 - 147x - 90$$

Apply the concepts of this section to work each problem.

A.

B.

C.

D.

1. Which one of the graphs is that of $y = x^3 - 3x^2 - 6x + 8$?

2. Which one of the graphs is that of $y = x^4 + 7x^3 - 5x^2 - 75x$?

3. How many real zeros does the function graphed in C have?

4. Which one of C and D is the graph of $y = -x^3 + 9x^2 - 27x + 17$?

5. Which of the graphs cannot be that of a cubic polynomial function?

6. Which one of the graphs is that of a function whose range is *not* $(-\infty, \infty)$?

7. The function $f(x) = x^4 + 7x^3 - 5x^2 - 75x$ has the graph shown in B. Use the graph to factor the polynomial.

8. The function $f(x) = -x^5 + 36x^3 - 22x^2 - 147x - 90$ has the graph shown in D. Use the graph to factor the polynomial.

Graph each function. Determine the largest open intervals of the domain over which each function is (a) increasing or (b) decreasing. See Example 1.

9. $f(x) = 2x^4$

10. $f(x) = \dfrac{1}{4}x^6$

11. $f(x) = -\dfrac{2}{3}x^5$

12. $f(x) = -\dfrac{5}{4}x^5$

13. $f(x) = \dfrac{1}{2}x^3 + 1$

14. $f(x) = -x^4 + 2$

15. $f(x) = -(x + 1)^3 + 1$

16. $f(x) = (x + 2)^3 - 1$

17. $f(x) = (x - 1)^4 + 2$ **18.** $f(x) = \dfrac{1}{3}(x + 3)^4 - 3$

19. $f(x) = \dfrac{1}{2}(x - 2)^2 + 4$ **20.** $f(x) = \dfrac{1}{3}(x + 1)^3 - 3$

Use an end behavior diagram, ⤿, ⤾, ⬊, or ⬈, to describe the end behavior of the graph of each polynomial function. See Example 2.

21. $f(x) = 5x^5 + 2x^3 - 3x + 4$ **22.** $f(x) = -x^3 - 4x^2 + 2x - 1$

23. $f(x) = -4x^3 + 3x^2 - 1$ **24.** $f(x) = 4x^7 - x^5 + x^3 - 1$

25. $f(x) = 9x^6 - 3x^4 + x^2 - 2$ **26.** $f(x) = 10x^6 - x^5 + 2x - 2$

27. $f(x) = 3 + 2x - 4x^2 - 5x^{10}$ **28.** $f(x) = 7 + 2x - 5x^2 - 10x^4$

Graph each polynomial function. Factor first if the polynomial is not in factored form. See Examples 3 and 4.

29. $f(x) = x^3 + 5x^2 + 2x - 8$ **30.** $f(x) = x^3 + 3x^2 - 13x - 15$

31. $f(x) = 2x(x - 3)(x + 2)$ **32.** $f(x) = x(x + 1)(x - 1)$

33. $f(x) = x^2(x - 2)(x + 3)^2$ **34.** $f(x) = x^2(x - 5)(x + 3)(x - 1)$

35. $f(x) = (3x - 1)(x + 2)^2$ **36.** $f(x) = (4x + 3)(x + 2)^2$

37. $f(x) = x^3 + 5x^2 - x - 5$ **38.** $f(x) = x^3 + x^2 - 36x - 36$

39. $f(x) = x^3 - x^2 - 2x$ **40.** $f(x) = 3x^4 + 5x^3 - 2x^2$

41. $f(x) = 2x^3(x^2 - 4)(x - 1)$ **42.** $f(x) = x^2(x - 3)^3(x + 1)$

43. $f(x) = 2x^3 - 5x^2 - x + 6$ **44.** $f(x) = 2x^4 + x^3 - 6x^2 - 7x - 2$

45. $f(x) = 3x^4 - 7x^3 - 6x^2 + 12x + 8$ **46.** $f(x) = x^4 + 3x^3 - 3x^2 - 11x - 6$

Use the intermediate value theorem to show that each polynomial function has a real zero between the numbers given. See Example 5.

47. $f(x) = 2x^2 - 7x + 4;$ 2 and 3

48. $f(x) = 3x^2 - x - 4;$ 1 and 2

49. $f(x) = 2x^3 - 5x^2 - 5x + 7;$ 0 and 1

50. $f(x) = 2x^3 - 9x^2 + x + 20;$ 2 and 2.5

51. $f(x) = 2x^4 - 4x^2 + 4x - 8;$ 1 and 2

52. $f(x) = x^4 - 4x^3 - x + 3;$ 0.5 and 1

53. $f(x) = x^4 + x^3 - 6x^2 - 20x - 16;$ 3.2 and 3.3

54. $f(x) = x^4 - 2x^3 - 2x^2 - 18x + 5;$ 3.7 and 3.8

55. $f(x) = x^4 - 4x^3 - 20x^2 + 32x + 12;$ -1 and 0

56. $f(x) = x^5 + 2x^4 + x^3 + 3;$ -1.8 and -1.7

Show that the real zeros of each polynomial function satisfy the given conditions. See Example 6.

57. $f(x) = x^4 - x^3 + 3x^2 - 8x + 8;$ no real zero greater than 2

58. $f(x) = 2x^5 - x^4 + 2x^3 - 2x^2 + 4x - 4;$ no real zero greater than 1

59. $f(x) = x^4 + x^3 - x^2 + 3;$ no real zero less than -2

60. $f(x) = x^5 + 2x^3 - 2x^2 + 5x + 5$; no real zero less than -1

61. $f(x) = 3x^4 + 2x^3 - 4x^2 + x - 1$; no real zero greater than 1

62. $f(x) = 3x^4 + 2x^3 - 4x^2 + x - 1$; no real zero less than -2

63. $f(x) = x^5 - 3x^3 + x + 2$; no real zero greater than 2

64. $f(x) = x^5 - 3x^3 + x + 2$; no real zero less than -3

Connecting Graphs with Equations *Find a polynomial function f of least degree having the graph shown. (Hint: See the Note following **Example 4**.)*

65.

66.

67.

68.

69.

70.

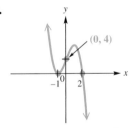

Graph each function in the viewing window specified. Compare the graph to the one shown in the answer section of this text. Then use the graph to find $f(1.25)$.

71. $f(x) = 2x(x - 3)(x + 2)$; window: $[-3, 4]$ by $[-20, 12]$
Compare to **Exercise 31.**

72. $f(x) = x^2(x - 2)(x + 3)^2$; window: $[-4, 3]$ by $[-24, 4]$
Compare to **Exercise 33.**

73. $f(x) = (3x - 1)(x + 2)^2$; window: $[-4, 2]$ by $[-15, 15]$
Compare to **Exercise 35.**

74. $f(x) = x^3 + 5x^2 - x - 5$; window: $[-6, 2]$ by $[-30, 30]$
Compare to **Exercise 37.**

*Approximate the real zero discussed in each specified exercise. See **Example 7**.*

75. Exercise 47   **76. Exercise 49**

77. Exercise 51   **78. Exercise 50**

*For the given polynomial function, approximate each zero as a decimal to the nearest tenth. See **Example 7**.*

79. $f(x) = x^3 + 3x^2 - 2x - 6$ **80.** $f(x) = x^3 - 3x + 3$

81. $f(x) = -2x^4 - x^2 + x + 5$ **82.** $f(x) = -x^4 + 2x^3 + 3x^2 + 6$

Use a graphing calculator to find the coordinates of the turning points of the graph of each polynomial function in the given domain interval. Give answers to the nearest hundredth.

83. $f(x) = 2x^3 - 5x^2 - x + 1;$ $[-1, 0]$

84. $f(x) = x^3 + 4x^2 - 8x - 8;$ $[0.3, 1]$

85. $f(x) = 2x^3 - 5x^2 - x + 1;$ $[1.4, 2]$

86. $f(x) = x^3 - x + 3;$ $[-1, 0]$

87. $f(x) = x^3 + 4x^2 - 8x - 8;$ $[-3.8, -3]$

88. $f(x) = x^4 - 7x^3 + 13x^2 + 6x - 28;$ $[-1, 0]$

Solve each problem.

89. *(Modeling) Social Security Numbers* Your Social Security number (SSN) is unique, and with it you can construct your own personal Social Security polynomial. Let the polynomial function be defined as follows, where a_i represents the ith digit in your SSN:

$$SSN(x) = (x - a_1)(x + a_2)(x - a_3)(x + a_4)(x - a_5) \cdot$$
$$(x + a_6)(x - a_7)(x + a_8)(x - a_9).$$

For example, if the SSN is 539-58-0954, the polynomial function is

$$SSN(x) = (x - 5)(x + 3)(x - 9)(x + 5)(x - 8)(x + 0)(x - 9)(x + 5)(x - 4).$$

A comprehensive graph of this function is shown in **Figure A.** In **Figure B,** we show a screen obtained by zooming in on the positive zeros, as the comprehensive graph does not show the local behavior well in this region. Use a graphing calculator to graph your own "personal polynomial."

Figure A

Figure B

90. A comprehensive graph of

$$f(x) = x^4 - 7x^3 + 18x^2 - 22x + 12$$

is shown in the two screens, along with displays of the two real zeros. Find the two remaining nonreal complex zeros.

(Modeling) *The following exercises are geometric in nature and lead to polynomial models. Solve each problem.*

91. *Volume of a Box* A rectangular piece of cardboard measuring 12 in. by 18 in. is to be made into a box with an open top by cutting equal-size squares from each corner and folding up the sides. Let x represent the length of a side of each such square in inches. Give approximations to the nearest hundredth.

(a) Give the restrictions on x.

(b) Determine a function V that gives the volume of the box as a function of x.

(c) For what value of x will the volume be a maximum? What is this maximum volume? (*Hint:* Use the function of a graphing calculator that allows us to determine a maximum point within a given interval.)

(d) For what values of x will the volume be greater than 80 in.^{3}?

92. *Construction of a Rain Gutter* A piece of rectangular sheet metal is 20 in. wide. It is to be made into a rain gutter by turning up the edges to form parallel sides. Let x represent the length of each of the parallel sides. Give approximations to the nearest hundredth.

(a) Give the restrictions on x.

(b) Determine a function $\mathcal{A}$ that gives the area of a cross section of the gutter.

(c) For what value of x will $\mathcal{A}$ be a maximum (and thus maximize the amount of water that the gutter will hold)? What is this maximum area?

(d) For what values of x will the area of a cross section be less than 40 in.^{2}?

93. *Sides of a Right Triangle* A certain right triangle has area 84 in.^{2}. One leg of the triangle measures 1 in. less than the hypotenuse. Let x represent the length of the hypotenuse.

(a) Express the length of the leg mentioned above in terms of x. Give the domain of x.

(b) Express the length of the other leg in terms of x.

(c) Write an equation based on the information determined thus far. Square both sides and then write the equation with one side as a polynomial with integer coefficients, in descending powers, and the other side equal to 0.

(d) Solve the equation in part (c) graphically. Find the lengths of the three sides of the triangle.

94. *Area of a Rectangle* Find the value of x in the figure that will maximize the area of rectangle *ABCD*. Round to the nearest thousandth.

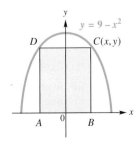

95. *Butane Gas Storage* A storage tank for butane gas is to be built in the shape of a right circular cylinder of altitude 12 ft, with a half sphere attached to each end. If *x* represents the radius of each half sphere, what radius should be used to cause the volume of the tank to be 144π ft^3?

— 12 ft —

96. *Volume of a Box* A standard piece of notebook paper measuring 8.5 in. by 11 in. is to be made into a box with an open top by cutting equal-size squares from each corner and folding up the sides. Let *x* represent the length of a side of each such square in inches. Use the table feature of a graphing calculator to do the following. Round to the nearest hundredth.

(a) Find the maximum volume of the box.

(b) Determine when the volume of the box will be greater than 40 in.^{3}.

97. *Floating Ball* The polynomial function

$$f(x) = \frac{\pi}{3}x^3 - 5\pi x^2 + \frac{500\pi d}{3}$$

can be used to find the depth that a ball 10 cm in diameter sinks in water. The constant *d* is the density of the ball, where the density of water is 1. The smallest *positive* zero of $f(x)$ equals the depth that the ball sinks. Approximate this depth for each material and interpret the results.

(a) A wooden ball with $d = 0.8$ (to the nearest hundredth)

(b) A solid aluminum ball with $d = 2.7$

(c) A spherical water balloon with $d = 1$

98. *Floating Ball* Refer to **Exercise 97.** If a ball has a 20-cm diameter, then the function becomes

$$f(x) = \frac{\pi}{3}x^3 - 10\pi x^2 + \frac{4000\pi d}{3}.$$

This function can be used to determine the depth that the ball sinks in water. Find the depth that this size ball sinks when $d = 0.6$. Round to the nearest hundredth.

(Modeling) Solve each problem. **See Example 8.**

99. *Highway Design* To allow enough distance for cars to pass on two-lane highways, engineers calculate minimum sight distances between curves and hills. The table shows the minimum sight distance *y* in feet for a car traveling at *x* miles per hour.

x (in mph)	20	30	40	50	60	65	70
y (in feet)	810	1090	1480	1840	2140	2310	2490

Source: Haefner, L., *Introduction to Transportation Systems*, Holt, Rinehart and Winston.

(a) Make a scatter diagram of the data.

(b) Use the regression feature of a calculator to find the best-fitting linear function for the data. Graph the function with the data.

(c) Repeat part (b) for a cubic function.

(d) Estimate the minimum sight distance for a car traveling 43 mph using the functions from parts (b) and (c).

(e) By comparing graphs of the functions in parts (b) and (c) with the data, decide which function best fits the given data.

100. *Water Pollution* Copper in high doses can be lethal to aquatic life. The table lists copper concentrations in freshwater mussels after 45 days at various distances downstream from an electroplating plant. The concentration C is measured in micrograms of copper per gram of mussel x kilometers downstream.

x	5	21	37	53	59
C	20	13	9	6	5

Source: Foster, R., and J. Bates, "Use of mussels to monitor point source industrial discharges," *Environ. Sci. Technol.*; Mason, C., *Biology of Freshwater Pollution*, John Wiley & Sons.

(a) Make a scatter diagram of the data.

(b) Use the regression feature of a calculator to find the best-fitting quadratic function for the data. Graph the function with the data.

(c) Repeat part (b) for a cubic function.

(d) By comparing graphs of the functions in parts (b) and (c) with the data, decide which function best fits the given data.

(e) Concentrations above 10 are lethal to mussels. Find the values of x (using the cubic function) for which this is the case.

101. *Government Spending on Health Research and Training* The table lists the annual amount (in billions of dollars) spent by the federal government on health research and training programs over a 10-yr period.

Year	Amount (billions of $)	Year	Amount (billions of $)
2004	27.1	2009	30.6
2005	28.1	2010	34.2
2006	28.8	2011	36.2
2007	29.3	2012	34.5
2008	29.9	2013	32.9

Source: U.S. Office of Management and Budget.

Which one of the following provides the best model for these data, where x represents the year?

A. $f(x) = 0.2(x - 2004)^2 + 27.1$ **B.** $g(x) = (x - 2004) + 27.1$

C. $h(x) = 2.5\sqrt{x - 2004} + 27.1$ **D.** $k(x) = 0.1(x - 2004)^3 + 27.1$

102. *Swing of a Pendulum* Grandfather clocks use pendulums to keep accurate time. The relationship between the length of a pendulum L and the time T for one complete oscillation can be expressed by the equation

$$L = kT^n,$$

where k is a constant and n is a positive integer to be determined. The data in the table were taken for different lengths of pendulums.

L (ft)	T (sec)	L (ft)	T (sec)
1.0	1.11	3.0	1.92
1.5	1.36	3.5	2.08
2.0	1.57	4.0	2.22
2.5	1.76		

(a) As the length of the pendulum increases, what happens to T?

(b) Use the data to approximate k and determine the best value for n.

(c) Using the values of k and n from part (b), predict T for a pendulum having length 5 ft. Round to the nearest hundredth.

(d) If the length L of a pendulum doubles, what happens to the period T?

Relating Concepts

For individual or collaborative investigation *(Exercises 103–108)*

For any function $y = f(x)$, the following hold true.

  (a) The real solutions of $f(x) = 0$ correspond to the x-intercepts of the graph.

  (b) The real solutions of $f(x) < 0$ are the x-values for which the graph lies *below* the x-axis.

  (c) The real solutions of $f(x) > 0$ are the x-values for which the graph lies *above* the x-axis.

In each exercise, a polynomial function $f(x)$ is given in both expanded and factored forms. Graph each function, and solve the equations and inequalities. Give multiplicities of solutions when applicable.

103. $f(x) = x^3 - 3x^2 - 6x + 8$

  $f(x) = (x - 4)(x - 1)(x + 2)$

  (a) $f(x) = 0$ **(b)** $f(x) < 0$

  (c) $f(x) > 0$

104. $f(x) = x^3 + 4x^2 - 11x - 30$

  $f(x) = (x - 3)(x + 2)(x + 5)$

  (a) $f(x) = 0$ **(b)** $f(x) < 0$

  (c) $f(x) > 0$

105. $f(x) = 2x^4 - 9x^3 - 5x^2 + 57x - 45$

  $f(x) = (x - 3)^2(2x + 5)(x - 1)$

  (a) $f(x) = 0$ **(b)** $f(x) < 0$

  (c) $f(x) > 0$

106. $f(x) = 4x^4 + 27x^3 - 42x^2$
  $- 445x - 300$

  $f(x) = (x + 5)^2(4x + 3)(x - 4)$

  (a) $f(x) = 0$ **(b)** $f(x) < 0$

  (c) $f(x) > 0$

107. $f(x) = -x^4 - 4x^3 + 3x^2 + 18x$

  $f(x) = x(2 - x)(x + 3)^2$

  (a) $f(x) = 0$ **(b)** $f(x) \geq 0$

  (c) $f(x) \leq 0$

108. $f(x) = -x^4 + 2x^3 + 8x^2$

  $f(x) = x^2(4 - x)(x + 2)$

  (a) $f(x) = 0$ **(b)** $f(x) \geq 0$

  (c) $f(x) \leq 0$

Summary Exercises on Polynomial Functions, Zeros, and Graphs

We use all of the theorems for finding complex zeros of polynomial functions in the next example.

EXAMPLE Finding All Zeros of a Polynomial Function

Find all zeros of $f(x) = x^4 - 3x^3 + 6x^2 - 12x + 8$.

SOLUTION We consider the number of positive zeros by observing the variations in signs for $f(x)$.

$$f(x) = +x^4 - 3x^3 + 6x^2 - 12x + 8$$

$$\underbrace{\qquad}_{1} \underbrace{\qquad}_{2} \underbrace{\qquad}_{3} \underbrace{\qquad}_{4}$$

Because $f(x)$ has four sign changes, we can use Descartes' rule of signs to determine that there are four, two, or zero positive real zeros. For negative zeros, we consider the variations in signs for $f(-x)$.

$$f(-x) = (-x)^4 - 3(-x)^3 + 6(-x)^2 - 12(-x) + 8$$

$$f(-x) = x^4 + 3x^3 + 6x^2 + 12x + 8$$

Because $f(-x)$ has no sign changes, there are no negative real zeros. The function has degree 4, so it has a maximum of four zeros with possibilities summarized in the table on the next page.

Positive	Negative	Nonreal Complex
4	0	0
2	0	2
0	0	4

We can now use the rational zeros theorem to determine that the possible rational zeros are ± 1, ± 2, ± 4, and ± 8. Based on Descartes' rule of signs, we discard the negative rational zeros from this list and try to find a positive rational zero. We start by using synthetic division to check 4.

$$\text{Proposed zero} \longrightarrow 4)\overline{1 \quad -3 \quad 6 \quad -12 \quad 8}$$
$$4 \quad 4 \quad 40 \quad 112$$
$$\overline{1 \quad 1 \quad 10 \quad 28 \quad 120} \longleftarrow f(4) = 120$$

We find that 4 is not a zero. However, $4 > 0$, and the numbers in the bottom row of the synthetic division are nonnegative. Thus, the boundedness theorem indicates that there are no zeros greater than 4. We can discard 8 as a possible rational zero and use synthetic division to show that 1 and 2 are zeros.

$$1)\overline{1 \quad -3 \quad 6 \quad -12 \quad 8}$$
$$1 \quad -2 \quad 4 \quad -8$$
$$2)\overline{1 \quad -2 \quad 4 \quad -8 \quad 0} \longleftarrow f(1) = 0$$
$$2 \quad 0 \quad 8$$
$$\overline{1 \quad 0 \quad 4 \quad 0} \longleftarrow f(2) = 0$$

The polynomial now factors as

$$f(x) = (x - 1)(x - 2)(x^2 + 4).$$

We find the remaining two zeros using algebra to solve for x in the quadratic factor of the following equation.

$$(x - 1)(x - 2)(x^2 + 4) = 0$$

$x - 1 = 0$ or $x - 2 = 0$ or $x^2 + 4 = 0$ Zero-factor property

$x = 1$ or $x = 2$ or $x^2 = -4$

$x = \pm 2i$ Square root property

The linear factored form of the polynomial is

$$f(x) = (x - 1)(x - 2)(x - 2i)(x + 2i),$$

and the corresponding zeros are 1, 2, $2i$, and $-2i$. ✔ **Now Try Exercise 3.**

EXERCISES

For each polynomial function, complete the following in order.

(a) *Use Descartes' rule of signs to determine the different possibilities for the numbers of positive, negative, and nonreal complex zeros.*

(b) *Use the rational zeros theorem to determine the possible rational zeros.*

(c) *Use synthetic division with the boundedness theorem where appropriate and/or factoring to find the rational zeros, if any.*

(d) *Find all other complex zeros (both real and nonreal), if any.*

1. $f(x) = 6x^3 - 41x^2 + 26x + 24$ **2.** $f(x) = 2x^3 - 5x^2 - 4x + 3$

3. $f(x) = 3x^4 - 5x^3 + 14x^2 - 20x + 8$ **4.** $f(x) = 2x^4 - 3x^3 + 16x^2 - 27x - 18$

5. $f(x) = 6x^4 - 5x^3 - 11x^2 + 10x - 2$ **6.** $f(x) = 5x^4 + 8x^3 - 19x^2 - 24x + 12$

7. $f(x) = x^5 - 6x^4 + 16x^3 - 24x^2 + 16x$ (*Hint:* Factor out x first.)

8. $f(x) = 2x^4 + 8x^3 - 7x^2 - 42x - 9$

9. $f(x) = 8x^4 + 8x^3 - x - 1$ (*Hint:* Factor the polynomial.)

10. $f(x) = 2x^5 + 5x^4 - 9x^3 - 11x^2 + 19x - 6$

For each polynomial function, complete the following in order.

   (a) *Use Descartes' rule of signs to determine the different possibilities for the numbers of positive, negative, and nonreal complex zeros.*

   (b) *Use the rational zeros theorem to determine the possible rational zeros.*

   (c) *Find the rational zeros, if any.*

   (d) *Find all other real zeros, if any.*

   (e) *Find any other complex zeros (that is, zeros that are not real), if any.*

   (f) *Find the x-intercepts of the graph, if any.*

   (g) *Find the y-intercept of the graph.*

   (h) *Use synthetic division to find $f(4)$, and give the coordinates of the corresponding point on the graph.*

   (i) *Determine the end behavior of the graph.*

   (j) *Sketch the graph.*

11. $f(x) = x^4 + 3x^3 - 3x^2 - 11x - 6$

12. $f(x) = -2x^5 + 5x^4 + 34x^3 - 30x^2 - 84x + 45$

13. $f(x) = 2x^5 - 10x^4 + x^3 - 5x^2 - x + 5$

14. $f(x) = 3x^4 - 4x^3 - 22x^2 + 15x + 18$

15. $f(x) = -2x^4 - x^3 + x + 2$

16. $f(x) = 4x^5 + 8x^4 + 9x^3 + 27x^2 + 27x$ (*Hint:* Factor out x first.)

17. $f(x) = 3x^4 - 14x^2 - 5$ (*Hint:* Factor the polynomial.)

18. $f(x) = -x^5 - x^4 + 10x^3 + 10x^2 - 9x - 9$

19. $f(x) = -3x^4 + 22x^3 - 55x^2 + 52x - 12$

20. For the polynomial functions in **Exercises 11–19** that have irrational zeros, find approximations to the nearest thousandth.

3.5 Rational Functions: Graphs, Applications, and Models

- **The Reciprocal Function $f(x) = \frac{1}{x}$**
- **The Function $f(x) = \frac{1}{x^2}$**
- **Asymptotes**
- **Graphing Techniques**
- **Rational Models**

A rational expression is a fraction that is the quotient of two polynomials. A *rational function* is defined by a quotient of two polynomial functions.

Rational Function

A function f of the form

$$f(x) = \frac{p(x)}{q(x)},$$

where $p(x)$ and $q(x)$ are polynomial functions, with $q(x) \neq 0$, is a **rational function.**

$$f(x) = \frac{1}{x}, \quad f(x) = \frac{x + 1}{2x^2 + 5x - 3}, \quad f(x) = \frac{3x^2 - 3x - 6}{x^2 + 8x + 16} \quad \text{Rational functions}$$

Any values of x such that $q(x) = 0$ are excluded from the domain of a rational function, so this type of function often has a **discontinuous graph**—that is, a graph that has one or more breaks in it.

The Reciprocal Function $f(x) = \frac{1}{x}$ The simplest rational function with a variable denominator is the **reciprocal function.**

$$f(x) = \frac{1}{x} \quad \text{Reciprocal function}$$

The domain of this function is the set of all nonzero real numbers. The number 0 cannot be used as a value of x, but it is helpful to find values of $f(x)$ for some values of x very close to 0. We use the table feature of a graphing calculator to do this. The tables in **Figure 41** suggest that $|f(x)|$ increases without bound as x gets closer and closer to 0, which is written in symbols as

$$|f(x)| \to \infty \quad \text{as} \quad x \to 0.$$

(The symbol $x \to 0$ means that x approaches 0, without necessarily ever being equal to 0.) Because x cannot equal 0, the graph of $f(x) = \frac{1}{x}$ will never intersect the vertical line $x = 0$. This line is a **vertical asymptote.**

As x approaches 0 from the left, $y_1 = \frac{1}{x}$ approaches $-\infty$. (-1E-6 means -1×10^{-6}.)

As x approaches 0 from the right, $y_1 = \frac{1}{x}$ approaches ∞.

Figure 41

As $|x|$ increases without bound (written $|x| \to \infty$), the values of $f(x) = \frac{1}{x}$ get closer and closer to 0, as shown in the tables in **Figure 42.** Letting $|x|$ increase without bound causes the graph of $f(x) = \frac{1}{x}$ to move closer and closer to the horizontal line $y = 0$. This line is a **horizontal asymptote.**

As x approaches ∞, $y_1 = \frac{1}{x}$ approaches 0 through positive values.

As x approaches $-\infty$, $y_1 = \frac{1}{x}$ approaches 0 through negative values.

Figure 42

The graph of $f(x) = \frac{1}{x}$ is shown in **Figure 43**.

Reciprocal Function $f(x) = \dfrac{1}{x}$

Domain: $(-\infty, 0) \cup (0, \infty)$ Range: $(-\infty, 0) \cup (0, \infty)$

x	y
-2	$-\frac{1}{2}$
-1	-1
$-\frac{1}{2}$	-2
0	undefined
$\frac{1}{2}$	2
1	1
2	$\frac{1}{2}$

Figure 43

- $f(x) = \frac{1}{x}$ decreases on the open intervals $(-\infty, 0)$ and $(0, \infty)$.

- It is discontinuous at $x = 0$.

- The y-axis is a vertical asymptote, and the x-axis is a horizontal asymptote.

- It is an odd function, and its graph is symmetric with respect to the origin.

The graph of $y = \frac{1}{x}$ can be translated and/or reflected.

EXAMPLE 1 **Graphing a Rational Function**

Graph $y = -\frac{2}{x}$. Give the domain and range and the largest open intervals of the domain over which the function is increasing or decreasing.

SOLUTION The expression $-\frac{2}{x}$ can be written as $-2\left(\frac{1}{x}\right)$ or $2\left(\frac{1}{-x}\right)$, indicating that the graph may be obtained by stretching the graph of $y = \frac{1}{x}$ vertically by a factor of 2 and reflecting it across either the x-axis or the y-axis. The x- and y-axes remain the horizontal and vertical asymptotes. The domain and range are both still $(-\infty, 0) \cup (0, \infty)$. See **Figure 44.**

The graph in **Figure 44** is shown here using a **decimal window.** Using a nondecimal window *may* produce an extraneous vertical line that is not part of the graph.

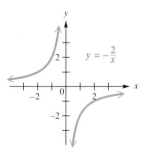

Figure 44

The graph shows that $f(x)$ is increasing on both sides of its vertical asymptote. Thus, it is increasing on $(-\infty, 0)$ and $(0, \infty)$.

✔ **Now Try Exercise 17.**

EXAMPLE 2 **Graphing a Rational Function**

Graph $f(x) = \frac{2}{x+1}$. Give the domain and range and the largest open intervals of the domain over which the function is increasing or decreasing.

ALGEBRAIC SOLUTION

The expression $\frac{2}{x+1}$ can be written as $2\left(\frac{1}{x+1}\right)$, indicating that the graph may be obtained by shifting the graph of $y = \frac{1}{x}$ to the left 1 unit and stretching it vertically by a factor of 2. See **Figure 45**.

The horizontal shift affects the domain, which is now $(-\infty, -1) \cup (-1, \infty)$. The line $x = -1$ is the vertical asymptote, and the line $y = 0$ (the x-axis) remains the horizontal asymptote. The range is still $(-\infty, 0) \cup (0, \infty)$. The graph shows that $f(x)$ is decreasing on both sides of its vertical asymptote. Thus, it is decreasing on $(-\infty, -1)$ and $(-1, \infty)$.

Figure 45

GRAPHING CALCULATOR SOLUTION

When entering this rational function into the function editor of a calculator, make sure that the numerator is 2 and the denominator is the entire expression $(x + 1)$.

The graph of this function has a vertical asymptote at $x = -1$ and a horizontal asymptote at $y = 0$, so it is reasonable to choose a viewing window that contains the locations of both asymptotes as well as enough of the graph to determine its basic characteristics. See **Figure 46.**

Figure 46

✔ **Now Try Exercise 19.**

The Function $f(x) = \frac{1}{x^2}$ The rational function

$$f(x) = \frac{1}{x^2} \quad \text{Rational function}$$

also has domain $(-\infty, 0) \cup (0, \infty)$. We can use the table feature of a graphing calculator to examine values of $f(x)$ for some x-values close to 0. See **Figure 47.**

X	Y1
-1	1
-.1	100
-.01	10000
-.001	1E6
-1E-4	1E8
-1E-5	1E10
-1E-6	1E12

Y1=1/X²

As x approaches 0 from the left, $y_1 = \frac{1}{x^2}$ approaches ∞.

X	Y1
1	1
.1	100
.01	10000
.001	1E6
1E-4	1E8
1E-5	1E10
1E-6	1E12

Y1=1/X²

As x approaches 0 from the right, $y_1 = \frac{1}{x^2}$ approaches ∞.

Figure 47

The tables suggest that $f(x)$ increases without bound as x gets closer and closer to 0. Notice that as x approaches 0 from *either* side, function values are all positive and there is symmetry with respect to the y-axis. Thus, $f(x) \to \infty$ as $x \to 0$. The y-axis ($x = 0$) is the vertical asymptote.

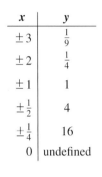

As x approaches ∞, $y_1 = \frac{1}{x^2}$ approaches 0 through positive values.

As x approaches $-\infty$, $y_1 = \frac{1}{x^2}$ approaches 0 through positive values.

Figure 48

As $|x|$ increases without bound, $f(x)$ approaches 0, as suggested by the tables in **Figure 48.** Again, function values are all positive. The x-axis is the horizontal asymptote of the graph.

The graph of $f(x) = \frac{1}{x^2}$ is shown in **Figure 49.**

Rational Function $\quad f(x) = \dfrac{1}{x^2}$

Domain: $(-\infty, 0) \cup (0, \infty)$ Range: $(0, \infty)$

x	y
± 3	$\frac{1}{9}$
± 2	$\frac{1}{4}$
± 1	1
$\pm \frac{1}{2}$	4
$\pm \frac{1}{4}$	16
0	undefined

Figure 49

- $f(x) = \frac{1}{x^2}$ increases on the open interval $(-\infty, 0)$ and decreases on the open interval $(0, \infty)$.

- It is discontinuous at $x = 0$.

- The y-axis is a vertical asymptote, and the x-axis is a horizontal asymptote.

- It is an even function, and its graph is symmetric with respect to the y-axis.

EXAMPLE 3 **Graphing a Rational Function**

Graph $g(x) = \frac{1}{(x+2)^2} - 1$. Give the domain and range and the largest open intervals of the domain over which the function is increasing or decreasing.

SOLUTION   The function $g(x) = \frac{1}{(x+2)^2} - 1$ is equivalent to

$$g(x) = f(x+2) - 1, \quad \text{where} \quad f(x) = \frac{1}{x^2}.$$

This indicates that the graph will be shifted 2 units to the left and 1 unit down. The horizontal shift affects the domain, now $(-\infty, -2) \cup (-2, \infty)$. The vertical shift affects the range, now $(-1, \infty)$.

The vertical asymptote has equation $x = -2$, and the horizontal asymptote has equation $y = -1$. A traditional graph is shown in **Figure 50,** with a calculator graph in **Figure 51.** Both graphs show that this function is increasing on $(-\infty, -2)$ and decreasing on $(-2, \infty)$.

This is the graph of $y = \frac{1}{x^2}$ shifted 2 units to the left and 1 unit down.

Figure 50

✔ **Now Try Exercise 27.**

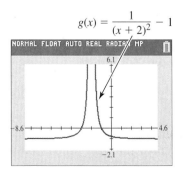

Figure 51

LOOKING AHEAD TO CALCULUS

The rational function

$$f(x) = \frac{2}{x+1}$$

in **Example 2** has a vertical asymptote at $x = -1$. In calculus, the behavior of the graph of this function for values close to -1 is described using **one-sided limits**. As x approaches -1 from the *left*, the function values decrease without bound. This is written

$$\lim_{x \to -1^-} f(x) = -\infty.$$

As x approaches -1 from the *right*, the function values increase without bound. This is written

$$\lim_{x \to -1^+} f(x) = \infty.$$

Asymptotes   The preceding examples suggest the following definitions of vertical and horizontal asymptotes.

Asymptotes

Let $p(x)$ and $q(x)$ define polynomial functions. Consider the rational function $f(x) = \frac{p(x)}{q(x)}$, written in lowest terms, and real numbers a and b.

1. If $|f(x)| \to \infty$ as $x \to a$, then the line $x = a$ is a **vertical asymptote.**

2. If $f(x) \to b$ as $|x| \to \infty$, then the line $y = b$ is a **horizontal asymptote.**

Locating asymptotes is important when graphing rational functions.

- We find *vertical asymptotes* by determining the values of x that make the denominator equal to 0.

- We find *horizontal asymptotes* (and, in some cases, *oblique asymptotes*), by considering what happens to $f(x)$ as $|x| \to \infty$. These asymptotes determine the end behavior of the graph.

Determining Asymptotes

To find the asymptotes of a rational function defined by a rational expression in *lowest terms,* use the following procedures.

1. **Vertical Asymptotes**
   Find any vertical asymptotes by setting the denominator equal to 0 and solving for x. If a is a zero of the denominator, then **the line $x = a$ is a vertical asymptote.**

2. **Other Asymptotes**
   Determine any other asymptotes by considering three possibilities:

   (a) If the numerator has lesser degree than the denominator, then there is a **horizontal asymptote $y = 0$** (the x-axis).

   (b) If the numerator and denominator have the same degree, and the function is of the form

   $$f(x) = \frac{a_n x^n + \cdots + a_0}{b_n x^n + \cdots + b_0}, \quad \text{where} \quad a_n, b_n \neq 0,$$

   then the **horizontal asymptote has equation $y = \frac{a_n}{b_n}$.**

   (c) If the numerator is of degree exactly one more than the denominator, then there will be an **oblique (slanted) asymptote.** To find it, divide the numerator by the denominator and disregard the remainder. Set the rest of the quotient equal to y to obtain the equation of the asymptote.

NOTE   The graph of a rational function may have more than one vertical asymptote, or it may have none at all.

The graph cannot intersect any vertical asymptote. There can be at most one other (nonvertical) asymptote, and the graph may intersect that asymptote. (See Example 7.)

EXAMPLE 4 **Finding Asymptotes of Rational Functions**

Give the equations of any vertical, horizontal, or oblique asymptotes for the graph of each rational function.

(a) $f(x) = \dfrac{x+1}{(2x-1)(x+3)}$ **(b)** $f(x) = \dfrac{2x+1}{x-3}$ **(c)** $f(x) = \dfrac{x^2+1}{x-2}$

SOLUTION

(a) To find the vertical asymptotes, set the denominator equal to 0 and solve.

$$(2x-1)(x+3) = 0$$

$$2x - 1 = 0 \quad \text{or} \quad x + 3 = 0 \qquad \text{Zero-factor property}$$

$$x = \frac{1}{2} \quad \text{or} \quad x = -3 \qquad \text{Solve each equation.}$$

The equations of the vertical asymptotes are $x = \frac{1}{2}$ and $x = -3$.

To find the equation of the horizontal asymptote, begin by multiplying the factors in the denominator.

$$f(x) = \frac{x+1}{(2x-1)(x+3)} = \frac{x+1}{2x^2+5x-3}$$

Now divide each term in the numerator and denominator by x^2. We choose the exponent 2 because it is the greatest power of x in the entire expression.

$$f(x) = \frac{\dfrac{x}{x^2} + \dfrac{1}{x^2}}{\dfrac{2x^2}{x^2} + \dfrac{5x}{x^2} - \dfrac{3}{x^2}} = \frac{\dfrac{1}{x} + \dfrac{1}{x^2}}{2 + \dfrac{5}{x} - \dfrac{3}{x^2}} \qquad \boxed{\text{Stop here. Leave the expression in complex form.}}$$

As $|x|$ increases without bound, the quotients $\frac{1}{x}, \frac{1}{x^2}, \frac{5}{x}$, and $\frac{3}{x^2}$ all approach 0, and the value of $f(x)$ approaches

$$\frac{0+0}{2+0-0} = 0. \qquad \tfrac{0}{2} = 0$$

The line $y = 0$ (that is, the x-axis) is therefore the horizontal asymptote. This supports procedure 2(a) of determining asymptotes on the previous page.

(b) Set the denominator $x - 3$ equal to 0 to find that the vertical asymptote has equation $x = 3$. To find the horizontal asymptote, divide each term in the rational expression by x since the greatest power of x in the expression is 1.

$$f(x) = \frac{2x+1}{x-3} = \frac{\dfrac{2x}{x} + \dfrac{1}{x}}{\dfrac{x}{x} - \dfrac{3}{x}} = \frac{2 + \dfrac{1}{x}}{1 - \dfrac{3}{x}}$$

As $|x|$ increases without bound, the quotients $\frac{1}{x}$ and $\frac{3}{x}$ both approach 0, and the value of $f(x)$ approaches

$$\frac{2+0}{1-0} = 2.$$

The line $y = 2$ is the horizontal asymptote. This supports procedure 2(b) of determining asymptotes on the previous page.

(c) Setting the denominator $x - 2$ equal to 0 shows that the vertical asymptote has equation $x = 2$. If we divide by the greatest power of x as before (x^2 in this case), we see that there is no horizontal asymptote because

$$f(x) = \frac{x^2 + 1}{x - 2} = \frac{\dfrac{x^2}{x^2} + \dfrac{1}{x^2}}{\dfrac{x}{x^2} - \dfrac{2}{x^2}} = \frac{1 + \dfrac{1}{x^2}}{\dfrac{1}{x} - \dfrac{2}{x^2}}$$

does not approach any real number as $|x| \to \infty$, due to the fact that $\frac{1 + 0}{0 - 0} = \frac{1}{0}$ is undefined. This happens whenever the degree of the numerator is greater than the degree of the denominator.

Setup for synthetic division

In such cases, divide the denominator into the numerator to write the expression in another form. We use synthetic division, as shown in the margin. The result enables us to write the function as follows.

$$f(x) = x + 2 + \frac{5}{x - 2}$$

For very large values of $|x|$, $\frac{5}{x-2}$ is close to 0, and the graph approaches the line $y = x + 2$. This line is an **oblique asymptote** (slanted, neither vertical nor horizontal) for the graph of the function. This supports procedure 2(c) of determining asymptotes.

✔ **Now Try Exercises 37, 39, and 41.**

Graphing Techniques   A comprehensive graph of a rational function will show the following characteristics.

- all x- and y-intercepts

- all asymptotes: vertical, horizontal, and/or oblique

- the point at which the graph intersects its nonvertical asymptote (if there is any such point)

- the behavior of the function on each domain interval determined by the vertical asymptotes and x-intercepts

Graphing a Rational Function

Let $f(x) = \frac{p(x)}{q(x)}$ define a function where $p(x)$ and $q(x)$ are polynomial functions and the rational expression is written in lowest terms. To sketch its graph, follow these steps.

Step 1 Find any vertical asymptotes.

Step 2 Find any horizontal or oblique asymptotes.

Step 3 If $q(0) \neq 0$, plot the y-intercept by evaluating $f(0)$.

Step 4 Plot the x-intercepts, if any, by solving $f(x) = 0$. (These will correspond to the zeros of the numerator, $p(x)$.)

Step 5 Determine whether the graph will intersect its nonvertical asymptote $y = b$ or $y = mx + b$ by solving $f(x) = b$ or $f(x) = mx + b$.

Step 6 Plot selected points, as necessary. Choose an x-value in each domain interval determined by the vertical asymptotes and x-intercepts.

Step 7 Complete the sketch.

EXAMPLE 5 **Graphing a Rational Function (x-Axis as Horizontal Asymptote)**

Graph $f(x) = \dfrac{x + 1}{2x^2 + 5x - 3}$.

SOLUTION

Steps 1 and 2 In **Example 4(a),** we found that $2x^2 + 5x - 3 = (2x - 1)(x + 3)$, so the vertical asymptotes have equations $x = \frac{1}{2}$ and $x = -3$, and the horizontal asymptote is the x-axis.

Step 3 The y-intercept is $\left(0, -\frac{1}{3}\right)$, as justified below.

$$f(0) = \frac{0 + 1}{2(0)^2 + 5(0) - 3} = -\frac{1}{3} \quad \text{The y-intercept corresponds to the ratio of the constant terms.}$$

Step 4 The x-intercept is found by solving $f(x) = 0$.

$$\frac{x + 1}{2x^2 + 5x - 3} = 0 \quad \text{Set } f(x) = 0.$$

$$x + 1 = 0 \quad \text{If a rational expression is equal to 0, then its numerator must equal 0.}$$

$$x = -1 \quad \text{The x-intercept is } (-1, 0).$$

Step 5 To determine whether the graph intersects its horizontal asymptote, solve this equation.

$$f(x) = 0 \longleftarrow \text{y-value of horizontal asymptote}$$

The horizontal asymptote is the x-axis, so the solution of $f(x) = 0$ was found in Step 4. The graph intersects its horizontal asymptote at $(-1, 0)$.

Step 6 Plot a point in each of the intervals determined by the x-intercepts and vertical asymptotes, $(-\infty, -3), (-3, -1), \left(-1, \frac{1}{2}\right)$ and $\left(\frac{1}{2}, \infty\right)$, to get an idea of how the graph behaves in each interval.

Interval	Test Point x	Value of $f(x)$	Sign of $f(x)$	Graph Above or Below x-Axis
$(-\infty, -3)$	-4	$-\frac{1}{3}$	Negative	Below
$(-3, -1)$	-2	$\frac{1}{5}$	Positive	Above
$\left(-1, \frac{1}{2}\right)$	0	$-\frac{1}{3}$	Negative	Below
$\left(\frac{1}{2}, \infty\right)$	2	$\frac{1}{5}$	Positive	Above

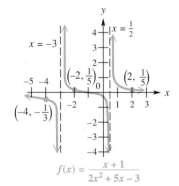

$f(x) = \dfrac{x + 1}{2x^2 + 5x - 3}$

Figure 52

Step 7 Complete the sketch. See **Figure 52.** This function is decreasing on each interval of its domain—that is, on $(-\infty, -3), \left(-3, \frac{1}{2}\right)$ and $\left(\frac{1}{2}, \infty\right)$.

✔ **Now Try Exercise 67.**

EXAMPLE 6 **Graphing a Rational Function (Does Not Intersect Its Horizontal Asymptote)**

Graph $f(x) = \dfrac{2x + 1}{x - 3}$.

SOLUTION

Steps 1 and 2 As determined in **Example 4(b),** the equation of the vertical asymptote is $x = 3$. The horizontal asymptote has equation $y = 2$.

Step 3 $f(0) = -\frac{1}{3}$, so the *y*-intercept is $\left(0, -\frac{1}{3}\right)$.

Step 4   Solve $f(x) = 0$ to find any *x*-intercepts.

$$\frac{2x + 1}{x - 3} = 0 \qquad \text{Set } f(x) = 0.$$

$$2x + 1 = 0 \qquad \begin{array}{l}\text{If a rational expression is equal to 0,}\\ \text{then its numerator must equal 0.}\end{array}$$

$$x = -\frac{1}{2} \qquad \text{The } x\text{-intercept is } \left(-\frac{1}{2}, 0\right).$$

Step 5   The graph does not intersect its horizontal asymptote because $f(x) = 2$ has no solution.

$$\frac{2x + 1}{x - 3} = 2 \qquad \text{Set } f(x) = 2.$$

$$2x + 1 = 2x - 6 \qquad \text{Multiply each side by } x - 3.$$

A false statement results. $\quad 1 = -6 \qquad \text{Subtract } 2x.$

Steps 6 and 7   The points $(-4, 1)$, $\left(1, -\frac{3}{2}\right)$, and $\left(6, \frac{13}{3}\right)$ are on the graph and can be used to complete the sketch of this function, which decreases on every interval of its domain. See **Figure 53**.

✔ **Now Try Exercise 63.**

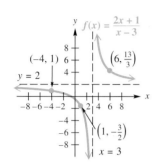

Figure 53

EXAMPLE 7   **Graphing a Rational Function (Intersects Its Horizontal Asymptote)**

Graph $f(x) = \dfrac{3x^2 - 3x - 6}{x^2 + 8x + 16}$.

SOLUTION

Step 1   To find the vertical asymptote(s), solve $x^2 + 8x + 16 = 0$.

$$x^2 + 8x + 16 = 0 \qquad \text{Set the denominator equal to 0.}$$

$$(x + 4)^2 = 0 \qquad \text{Factor.}$$

The numerator is not 0 when $x = -4$. $\qquad x = -4 \qquad \text{Zero-factor property}$

The vertical asymptote has equation $x = -4$.

Step 2   We divide all terms by x^2 and consider the behavior of each term as $|x|$ increases without bound to get the equation of the horizontal asymptote,

$$y = \frac{3}{1}, \quad \begin{array}{l}\longleftarrow \text{ Leading coefficient of numerator}\\ \longleftarrow \text{ Leading coefficient of denominator}\end{array} \quad \text{or} \quad y = 3.$$

Step 3 $f(0) = -\frac{3}{8}$, so the *y*-intercept is $\left(0, -\frac{3}{8}\right)$.

Step 4   Solve $f(x) = 0$ to find any *x*-intercepts.

$$\frac{3x^2 - 3x - 6}{x^2 + 8x + 16} = 0 \qquad \text{Set } f(x) = 0.$$

$$3x^2 - 3x - 6 = 0 \qquad \text{Set the numerator equal to 0.}$$

$$x^2 - x - 2 = 0 \qquad \text{Divide by 3.}$$

$$(x - 2)(x + 1) = 0 \qquad \text{Factor.}$$

$$x = 2 \quad \text{or} \quad x = -1 \qquad \text{Zero-factor property}$$

The *x*-intercepts are $(-1, 0)$ and $(2, 0)$.

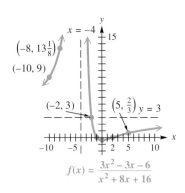

Figure 54

Step 5 We set $f(x) = 3$ and solve to locate the point where the graph intersects the horizontal asymptote.

$$\frac{3x^2 - 3x - 6}{x^2 + 8x + 16} = 3 \qquad \text{Set } f(x) = 3.$$

$$3x^2 - 3x - 6 = 3x^2 + 24x + 48 \qquad \text{Multiply each side by } x^2 + 8x + 16.$$

$$-27x = 54 \qquad \text{Subtract } 3x^2 \text{ and } 24x. \text{ Add 6.}$$

$$x = -2 \qquad \text{Divide by } -27.$$

The graph intersects its horizontal asymptote at $(-2, 3)$.

Steps 6 Some other points that lie on the graph are $(-10, 9)$, $\left(-8, 13\frac{1}{8}\right)$, and
and 7 $\left(5, \frac{2}{3}\right)$. These are used to complete the graph, as shown in **Figure 54.**

✔ **Now Try Exercise 83.**

Notice the behavior of the graph of the function in **Figure 54** near the line $x = -4$. As $x \to -4$ from either side, $f(x) \to \infty$.

If we examine the behavior of the graph of the function in **Figure 53** (on the previous page) near the line $x = 3$, we find that $f(x) \to -\infty$ as x approaches 3 from the left, while $f(x) \to \infty$ as x approaches 3 from the right. The behavior of the graph of a rational function near a vertical asymptote $x = a$ partially depends on the exponent on $x - a$ in the denominator.

LOOKING AHEAD TO CALCULUS

The rational function

$$f(x) = \frac{2x + 1}{x - 3},$$

seen in **Example 6**, has horizontal asymptote $y = 2$. In calculus, the behavior of the graph of this function as x approaches $-\infty$ and as x approaches ∞ is described using **limits at infinity**. As x approaches $-\infty$, $f(x)$ approaches 2. This is written

$$\lim_{x \to -\infty} f(x) = 2.$$

As x approaches ∞, $f(x)$ approaches 2. This is written

$$\lim_{x \to \infty} f(x) = 2.$$

Behavior of Graphs of Rational Functions near Vertical Asymptotes

Suppose that $f(x)$ is a rational expression in lowest terms. If n is the largest positive integer such that $(x - a)^n$ is a factor of the denominator of $f(x)$, then the graph will behave in the manner illustrated.

We have observed that the behavior of the graph of a polynomial function near its zeros is dependent on the multiplicity of the zero. The same statement can be made for rational functions.

Suppose that $f(x)$ is defined by a rational expression in lowest terms. If n is the greatest positive integer such that $(x - c)^n$ is a factor of the numerator of $f(x)$, then the graph will behave in the manner illustrated.

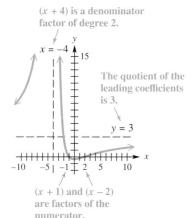

(x + 4) is a denominator factor of degree 2.

$x = -4$

The quotient of the leading coefficients is 3.

$y = 3$

(x + 1) and (x − 2) are factors of the numerator.

Figure 54 (repeated)

NOTE Suppose that we are asked to reverse the process of **Example 7** and find the equation of a rational function having the graph shown in **Figure 54** (repeated in the margin). Because the graph crosses the x-axis at its x-intercepts $(-1, 0)$ and $(2, 0)$, the numerator must have factors

$$(x + 1) \quad \text{and} \quad (x - 2),$$

each of degree 1.

The behavior of the graph at its vertical asymptote $x = -4$ suggests that there is a factor of $(x + 4)$ of even degree in the denominator. The horizontal asymptote at $y = 3$ indicates that the numerator and denominator have the same degree (both 2) and that the ratio of leading coefficients is 3.

Verify in **Example 7** that the rational function is

$$f(x) = \frac{3(x + 1)(x - 2)}{(x + 4)^2}$$

$$f(x) = \frac{3x^2 - 3x - 6}{x^2 + 8x + 16}.$$ Multiply the factors in the numerator. Square in the denominator.

Exercises of this type are labeled *Connecting Graphs with Equations*.

EXAMPLE 8 **Graphing a Rational Function with an Oblique Asymptote**

Graph $f(x) = \dfrac{x^2 + 1}{x - 2}$.

SOLUTION As shown in **Example 4(c)**, the vertical asymptote has equation $x = 2$, and the graph has an oblique asymptote with equation $y = x + 2$. The y-intercept is $\left(0, -\frac{1}{2}\right)$, and the graph has no x-intercepts because the numerator, $x^2 + 1$, has no real zeros. The graph does not intersect its oblique asymptote because the following has no solution.

$$\frac{x^2 + 1}{x - 2} = x + 2$$ Set the expressions defining the function and the oblique asymptote equal.

$$x^2 + 1 = x^2 - 4$$ Multiply each side by $x - 2$.

$$1 = -4$$ False

Using the y-intercept, asymptotes, the points $\left(4, \frac{17}{2}\right)$ and $\left(-1, -\frac{2}{3}\right)$, and the general behavior of the graph near its asymptotes leads to the graph in **Figure 55**.

Figure 55   ✔ **Now Try Exercise 87.**

LOOKING AHEAD TO CALCULUS

Different types of discontinuity are discussed in calculus. The function in **Example 9**,

$$f(x) = \frac{x^2 - 4}{x - 2},$$

is said to have a **removable discontinuity** at $x = 2$, because the discontinuity can be removed by redefining f at 2. The function in **Example 8**,

$$f(x) = \frac{x^2 + 1}{x - 2},$$

has **infinite discontinuity** at $x = 2$, as indicated by the vertical asymptote there. The greatest integer function has **jump discontinuities** because the function values "jump" from one value to another for integer domain values.

A rational function that is not in lowest terms often has a **point of discontinuity** in its graph. Such a point is sometimes called a *hole*.

EXAMPLE 9 **Graphing a Rational Function Defined by an Expression That Is Not in Lowest Terms**

Graph $f(x) = \dfrac{x^2 - 4}{x - 2}$.

ALGEBRAIC SOLUTION

The domain of this function cannot include 2. The expression $\dfrac{x^2 - 4}{x - 2}$ should be written in lowest terms.

$$f(x) = \frac{x^2 - 4}{x - 2} \quad \boxed{\text{Factor and then divide.}}$$

$$f(x) = \frac{(x + 2)(x - 2)}{x - 2} \quad \text{Factor.}$$

$$f(x) = x + 2, \quad x \neq 2$$

Therefore, the graph of this function will be the same as the graph of $y = x + 2$ (a straight line), with the exception of the point with x-value 2. A hole appears in the graph at $(2, 4)$. See **Figure 56.**

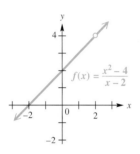

$f(x) = \dfrac{x^2 - 4}{x - 2}$

Figure 56

GRAPHING CALCULATOR SOLUTION

If we set the window of a graphing calculator so that an x-value of 2 is displayed, then we can see that the calculator cannot determine a value for y. We define

$$y_1 = \frac{x^2 - 4}{x - 2}$$

and graph it in such a window, as in **Figure 57.** The error message in the table further supports the existence of a discontinuity at $x = 2$. (For the table, $y_2 = x + 2$.)

Figure 57

Notice the visible discontinuity at $x = 2$ in the graph for the chosen window. (If the standard viewing window is chosen, the discontinuity is not visible.)

✔ **Now Try Exercise 91.**

Rational Models Rational functions have a variety of applications.

EXAMPLE 10 **Modeling Traffic Intensity with a Rational Function**

Vehicles arrive randomly at a parking ramp at an average rate of 2.6 vehicles per minute. The parking attendant can admit 3.2 vehicles per minute. However, since arrivals are random, lines form at various times. (*Source:* Mannering, F. and W. Kilareski, *Principles of Highway Engineering and Traffic Analysis,* 2nd ed., John Wiley & Sons.)

(a) The **traffic intensity** x is defined as the ratio of the average arrival rate to the average admittance rate. Determine x for this parking ramp.

(b) The average number of vehicles waiting in line to enter the ramp is given by

$$f(x) = \frac{x^2}{2(1 - x)},$$

where $0 \leq x < 1$ is the traffic intensity. Graph $f(x)$ and compute $f(0.8125)$ for this parking ramp.

(c) What happens to the number of vehicles waiting as the traffic intensity approaches 1?

SOLUTION

(a) The average arrival rate is 2.6 vehicles per minute and the average admittance rate is 3.2 vehicles per minute, so

$$x = \frac{2.6}{3.2} = 0.8125.$$

(b) A calculator graph of f is shown in **Figure 58.**

$$f(0.8125) = \frac{0.8125^2}{2(1 - 0.8125)} \approx 1.76 \text{ vehicles}$$

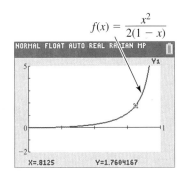

$$f(x) = \frac{x^2}{2(1 - x)}$$

Figure 58

(c) From the graph we see that as x approaches 1, $y = f(x)$ gets very large. Thus, the average number of waiting vehicles gets very large. This is what we would expect.

✔ **Now Try Exercise 113.**

3.5 Exercises

CONCEPT PREVIEW *Provide a short answer to each question.*

1. What is the domain of the function $f(x) = \dfrac{1}{x}$? What is its range?

2. What is the domain of the function $f(x) = \dfrac{1}{x^2}$? What is its range?

3. What is the largest open interval of the domain over which the function $f(x) = \dfrac{1}{x}$ increases? decreases? is constant?

4. What is the largest open interval of the domain over which the function $f(x) = \dfrac{1}{x^2}$ increases? decreases? is constant?

5. What is the equation of the vertical asymptote of the graph of $y = \dfrac{1}{x - 3} + 2$? Of the horizontal asymptote?

6. What is the equation of the vertical asymptote of the graph of $y = \dfrac{1}{(x + 2)^2} - 4$? Of the horizontal asymptote?

7. Is $f(x) = \dfrac{1}{x^2}$ an even or an odd function? What symmetry does its graph exhibit?

8. Is $f(x) = \dfrac{1}{x}$ an even or an odd function? What symmetry does its graph exhibit?

Concept Check Use the graphs of the rational functions in choices A–D to answer each question. There may be more than one correct choice.

A.

B.

C.

D.

9. Which choices have domain $(-\infty, 3) \cup (3, \infty)$?

10. Which choices have range $(-\infty, 3) \cup (3, \infty)$?

11. Which choices have range $(-\infty, 0) \cup (0, \infty)$?

12. Which choices have range $(0, \infty)$?

13. If f represents the function, only one choice has a single solution to the equation $f(x) = 3$. Which one is it?

14. Which choices have domain $(-\infty, 0) \cup (0, 3) \cup (3, \infty)$?

15. Which choices have the x-axis as a horizontal asymptote?

16. Which choices are symmetric with respect to a vertical line?

*Explain how the graph of each function can be obtained from the graph of $y = \frac{1}{x}$ or $y = \frac{1}{x^2}$. Then graph f and give the (**a**) domain and (**b**) range. Determine the largest open intervals of the domain over which the function is (**c**) increasing or (**d**) decreasing. See Examples 1–3.*

17. $f(x) = \dfrac{2}{x}$

18. $f(x) = -\dfrac{3}{x}$

19. $f(x) = \dfrac{1}{x + 2}$

20. $f(x) = \dfrac{1}{x - 3}$

21. $f(x) = \dfrac{1}{x} + 1$

22. $f(x) = \dfrac{1}{x} - 2$

23. $f(x) = -\dfrac{2}{x^2}$

24. $f(x) = \dfrac{1}{x^2} + 3$

25. $f(x) = \dfrac{1}{(x - 3)^2}$

26. $f(x) = \dfrac{-2}{(x - 3)^2}$

27. $f(x) = \dfrac{-1}{(x + 2)^2} - 3$

28. $f(x) = \dfrac{-1}{(x - 4)^2} + 2$

Concept Check *Match the rational function in Column I with the appropriate description in Column II. Choices in Column II can be used only once.*

I   **II**

29. $f(x) = \dfrac{x + 7}{x + 1}$ **A.** The x-intercept is $(-3, 0)$.

30. $f(x) = \dfrac{x + 10}{x + 2}$ **B.** The y-intercept is $(0, 5)$.

31. $f(x) = \dfrac{1}{x + 4}$ **C.** The horizontal asymptote is $y = 4$.

32. $f(x) = \dfrac{-3}{x^2}$ **D.** The vertical asymptote is $x = -1$.

33. $f(x) = \dfrac{x^2 - 16}{x + 4}$ **E.** There is a hole in its graph at $(-4, -8)$.

34. $f(x) = \dfrac{4x + 3}{x - 7}$ **F.** The graph has an oblique asymptote.

35. $f(x) = \dfrac{x^2 + 3x + 4}{x - 5}$ **G.** The x-axis is its horizontal asymptote, and the y-axis is not its vertical asymptote.

36. $f(x) = \dfrac{x + 3}{x - 6}$ **H.** The x-axis is its horizontal asymptote, and the y-axis is its vertical asymptote.

Give the equations of any vertical, horizontal, or oblique asymptotes for the graph of each rational function. *See Example 4.*

37. $f(x) = \dfrac{3}{x - 5}$ **38.** $f(x) = \dfrac{-6}{x + 9}$ **39.** $f(x) = \dfrac{4 - 3x}{2x + 1}$

40. $f(x) = \dfrac{2x + 6}{x - 4}$ **41.** $f(x) = \dfrac{x^2 - 1}{x + 3}$ **42.** $f(x) = \dfrac{x^2 + 4}{x - 1}$

43. $f(x) = \dfrac{x^2 - 2x - 3}{2x^2 - x - 10}$ **44.** $f(x) = \dfrac{3x^2 - 6x - 24}{5x^2 - 26x + 5}$

45. $f(x) = \dfrac{x^2 + 1}{x^2 + 9}$ **46.** $f(x) = \dfrac{4x^2 + 25}{x^2 + 9}$

Concept Check *Work each problem.*

47. Let f be the function whose graph is obtained by translating the graph of $y = \frac{1}{x}$ to the right 3 units and up 2 units.

   (a) Write an equation for $f(x)$ as a quotient of two polynomials.

   (b) Determine the zero(s) of f.

   (c) Identify the asymptotes of the graph of $f(x)$.

48. Repeat **Exercise 47** if f is the function whose graph is obtained by translating the graph of $y = -\frac{1}{x^2}$ to the left 3 units and up 1 unit.

49. After the numerator is divided by the denominator,

$$f(x) = \frac{x^5 + x^4 + x^2 + 1}{x^4 + 1} \quad \text{becomes} \quad f(x) = x + 1 + \frac{x^2 - x}{x^4 + 1}.$$

   (a) What is the oblique asymptote of the graph of the function?

   (b) Where does the graph of the function intersect its asymptote?

   (c) As $x \to \infty$, does the graph of the function approach its asymptote from above or below?

50. Choices A–D below show the four ways in which the graph of a rational function can approach the vertical line $x = 2$ as an asymptote. Identify the graph of each rational function defined in parts (a)–(d).

(a) $f(x) = \dfrac{1}{(x-2)^2}$ **(b)** $f(x) = \dfrac{1}{x-2}$ **(c)** $f(x) = \dfrac{-1}{x-2}$ **(d)** $f(x) = \dfrac{-1}{(x-2)^2}$

A.

B.

C.

D.

51. Which function has a graph that does not have a vertical asymptote?

A. $f(x) = \dfrac{1}{x^2 + 2}$ **B.** $f(x) = \dfrac{1}{x^2 - 2}$ **C.** $f(x) = \dfrac{3}{x^2}$ **D.** $f(x) = \dfrac{2x + 1}{x - 8}$

52. Which function has a graph that does not have a horizontal asymptote?

A. $f(x) = \dfrac{2x - 7}{x + 3}$ **B.** $f(x) = \dfrac{3x}{x^2 - 9}$

C. $f(x) = \dfrac{x^2 - 9}{x + 3}$ **D.** $f(x) = \dfrac{x + 5}{(x + 2)(x - 3)}$

Identify any vertical, horizontal, or oblique asymptotes in the graph of $y = f(x)$. State the domain of f.

53.

54.

55.

56.

57.

58.

59.

60.

Graph each rational function. ***See Examples 5–9.***

61. $f(x) = \dfrac{x+1}{x-4}$

62. $f(x) = \dfrac{x-5}{x+3}$

63. $f(x) = \dfrac{x+2}{x-3}$

64. $f(x) = \dfrac{x-3}{x+4}$

65. $f(x) = \dfrac{4-2x}{8-x}$

66. $f(x) = \dfrac{6-3x}{4-x}$

67. $f(x) = \dfrac{3x}{x^2-x-2}$

68. $f(x) = \dfrac{2x+1}{x^2+6x+8}$

69. $f(x) = \dfrac{5x}{x^2-1}$

70. $f(x) = \dfrac{x}{4-x^2}$

71. $f(x) = \dfrac{(x+6)(x-2)}{(x+3)(x-4)}$

72. $f(x) = \dfrac{(x+3)(x-5)}{(x+1)(x-4)}$

73. $f(x) = \dfrac{3x^2+3x-6}{x^2-x-12}$

74. $f(x) = \dfrac{4x^2+4x-24}{x^2-3x-10}$

75. $f(x) = \dfrac{9x^2-1}{x^2-4}$

76. $f(x) = \dfrac{16x^2-9}{x^2-9}$

77. $f(x) = \dfrac{(x-3)(x+1)}{(x-1)^2}$

78. $f(x) = \dfrac{x(x-2)}{(x+3)^2}$

79. $f(x) = \dfrac{x}{x^2-9}$

80. $f(x) = \dfrac{-5}{2x+4}$

81. $f(x) = \dfrac{1}{x^2+1}$

82. $f(x) = \dfrac{(x-5)(x-2)}{x^2+9}$

83. $f(x) = \dfrac{(x+4)^2}{(x-1)(x+5)}$

84. $f(x) = \dfrac{(x+1)^2}{(x+2)(x-3)}$

85. $f(x) = \dfrac{20+6x-2x^2}{8+6x-2x^2}$

86. $f(x) = \dfrac{18+6x-4x^2}{4+6x+2x^2}$

87. $f(x) = \dfrac{x^2+1}{x+3}$

88. $f(x) = \dfrac{2x^2+3}{x-4}$

89. $f(x) = \dfrac{x^2+2x}{2x-1}$

90. $f(x) = \dfrac{x^2-x}{x+2}$

91. $f(x) = \dfrac{x^2-9}{x+3}$

92. $f(x) = \dfrac{x^2-16}{x+4}$

93. $f(x) = \dfrac{2x^2-5x-2}{x-2}$

94. $f(x) = \dfrac{x^2-5}{x-3}$

95. $f(x) = \dfrac{x^2-1}{x^2-4x+3}$

96. $f(x) = \dfrac{x^2-4}{x^2+3x+2}$

97. $f(x) = \dfrac{(x^2-9)(2+x)}{(x^2-4)(3+x)}$

98. $f(x) = \dfrac{(x^2-16)(3+x)}{(x^2-9)(4+x)}$

99. $f(x) = \dfrac{x^4-20x^2+64}{x^4-10x^2+9}$

100. $f(x) = \dfrac{x^4-5x^2+4}{x^4-24x^2+108}$

Connecting Graphs with Equations *Find a rational function f having the graph shown.* *(Hint: See the note preceding **Example 8**.)*

101.

102.

103.

104.

105.

106.

Concept Check *Find a rational function f having a graph with the given features.*

107. x-intercepts: $(-1, 0)$ and $(3, 0)$
y-intercept: $(0, -3)$
vertical asymptote: $x = 1$
horizontal asymptote: $y = 1$

108. x-intercepts: $(1, 0)$ and $(3, 0)$
y-intercept: none
vertical asymptotes: $x = 0$ and $x = 2$
horizontal asymptote: $y = 1$

Use a graphing calculator to graph the rational function in each specified exercise. Then use the graph to find $f(1.25)$.

109. Exercise 61   **110. Exercise 67**   **111. Exercise 89**   **112. Exercise 91**

(Modeling) *Solve each problem.* **See Example 10.**

113. *Traffic Intensity* Let the average number of vehicles arriving at the gate of an amusement park per minute be equal to k, and let the average number of vehicles admitted by the park attendants be equal to r. Then the average waiting time T (in minutes) for each vehicle arriving at the park is given by the rational function

$$T(r) = \frac{2r - k}{2r^2 - 2kr},$$

where $r > k$. (*Source:* Mannering, F., and W. Kilareski, *Principles of Highway Engineering and Traffic Analysis*, 2nd ed., John Wiley & Sons.)

(a) It is known from experience that on Saturday afternoon $k = 25$. Use graphing to estimate the admittance rate r that is necessary to keep the average waiting time T for each vehicle to 30 sec.

(b) If one park attendant can serve 5.3 vehicles per minute, how many park attendants will be needed to keep the average wait to 30 sec?

114. *Waiting in Line* Queuing theory (also known as **waiting-line theory**) investigates the problem of providing adequate service economically to customers waiting in line. Suppose customers arrive at a fast-food service window at the rate of 9 people per hour. With reasonable assumptions, the average time (in hours) that a customer will wait in line before being served is modeled by

$$f(x) = \frac{9}{x(x - 9)},$$

where x is the average number of people served per hour. A graph of $f(x)$ for $x > 9$ is shown in the figure on the next page.

(a) Why is the function meaningless if the average number of people served per hour is less than 9?

Suppose the average time to serve a customer is 5 min.

(b) How many customers can be served in an hour?

(c) How many minutes will a customer have to wait in line (on the average)?

Average Waiting Time

(d) Suppose we want to halve the average waiting time to 7.5 min $\left(\frac{1}{8} \text{hr}\right)$. How fast must an employee work to serve a customer (on the average)? (*Hint:* Let $f(x) = \frac{1}{8}$ and solve the equation for x. Convert the answer to minutes and round to the nearest hundredth.) How might this reduction in serving time be accomplished?

115. ***Braking Distance*** Braking distance for automobiles traveling at x miles per hour, where $20 \le x \le 70$, can be modeled by the rational function

$$d(x) = \frac{8710x^2 - 69{,}400x + 470{,}000}{1.08x^2 - 324x + 82{,}200}.$$

(*Source:* Mannering, F., and W. Kilareski, *Principles of Highway Engineering and Traffic Analysis*, 2nd ed., John Wiley & Sons.)

(a) Use graphing to estimate x to the nearest unit when $d(x) = 300$.

(b) Complete the table for each value of x.

(c) If a car doubles its speed, does the braking distance double or more than double? Explain.

(d) Suppose that the automobile braking distance doubled whenever the speed doubled. What type of relationship would exist between the braking distance and the speed?

x	$d(x)$	x	$d(x)$
20		50	
25		55	
30		60	
35		65	
40		70	
45			

116. ***Braking Distance*** The **grade** x of a hill is a measure of its steepness. For example, if a road rises 10 ft for every 100 ft of horizontal distance, then it has an uphill grade of

$$x = \frac{10}{100}, \quad \text{or} \quad 10\%.$$

Grades are typically kept quite small—usually less than 10%. The braking distance D for a car traveling at 50 mph on a wet, uphill grade is given by

$$D(x) = \frac{2500}{30(0.3 + x)}.$$

(*Source:* Haefner, L., *Introduction to Transportation Systems*, Holt, Rinehart and Winston.)

(a) Evaluate $D(0.05)$ and interpret the result.

(b) Describe what happens to braking distance as the hill becomes steeper. Does this agree with your driving experience?

(c) Estimate the grade associated with a braking distance of 220 ft.

117. *Tax Revenue* Economist Arthur Laffer has been a center of controversy because of his **Laffer curve**, an idealized version of which is shown here.

Tax rate percent

According to this curve, increasing a tax rate, say from x_1 percent to x_2 percent on the graph, can actually lead to a decrease in government revenue. All economists agree on the endpoints, 0 revenue at tax rates of both 0% and 100%, but there is much disagreement on the location of the rate x_1 that produces maximum revenue. Suppose an economist studying the Laffer curve produces the rational function

$$R(x) = \frac{80x - 8000}{x - 110},$$

where $R(x)$ is government revenue in tens of millions of dollars for a tax rate of x percent, with the function valid for $55 \le x \le 100$. Find the revenue for the following tax rates. Round to the nearest tenth if necessary.

(a) 55%   (b) 60%   (c) 70%   (d) 90%   (e) 100%

118. *Tax Revenue* **See Exercise 117.** Suppose an economist determines that

$$R(x) = \frac{60x - 6000}{x - 120},$$

where $y = R(x)$ is government revenue in tens of millions of dollars for a tax rate of x percent, with $y = R(x)$ valid for $50 \le x \le 100$. Find the revenue for each tax rate. Round to the nearest tenth if necessary.

(a) 50%   (b) 60%   (c) 80%   (d) 100%

Relating Concepts

For individual or collaborative investigation *(Exercises 119–128)*

Consider the following "monster" rational function.

$$f(x) = \frac{x^4 - 3x^3 - 21x^2 + 43x + 60}{x^4 - 6x^3 + x^2 + 24x - 20}$$

Analyzing this function will synthesize many of the concepts of this and earlier sections. **Work Exercises 119–128 in order.**

119. Find the equation of the horizontal asymptote.

120. Given that -4 and -1 are zeros of the numerator, factor the numerator completely.

121. **(a)** Given that 1 and 2 are zeros of the denominator, factor the denominator completely.

(b) Write the entire quotient for f so that the numerator and the denominator are in factored form.

122. **(a)** What is the common factor in the numerator and the denominator?

(b) For what value of x will there be a point of discontinuity (i.e., a hole)?

123. What are the x-intercepts of the graph of f?

124. What is the y-intercept of the graph of f?

125. Find the equations of the vertical asymptotes.

126. Determine the point or points of intersection of the graph of f with its horizontal asymptote.

127. Sketch the graph of f.

128. Use the graph of f to solve each inequality.

(a) $f(x) < 0$ (b) $f(x) > 0$

Chapter 3   Quiz (Sections 3.1-3.5)

1. Graph each quadratic function. Give the vertex, axis, domain, range, and largest open intervals of the domain over which the function is increasing or decreasing.

(a) $f(x) = -2(x + 3)^2 - 1$ (b) $f(x) = 2x^2 - 8x + 3$

2. *(Modeling) Height of a Projected Object* A ball is projected directly upward from an initial height of 200 ft with an initial velocity of 64 ft per sec.

(a) Use the function $s(t) = -16t^2 + v_0t + s_0$ to describe the height of the ball in terms of time t.

(b) For what interval of time is the height of the ball greater than 240 ft? Round to the nearest hundredth.

Use synthetic division to decide whether the given number k is a zero of the polynomial function. If it is not, give the value of $f(k)$.

3. $f(x) = 2x^4 + x^3 - 3x + 4;\ k = 2$ **4.** $f(x) = x^2 - 4x + 5;\ k = 2 + i$

5. Find a polynomial function f of least degree having only real coefficients with zeros $-2, 3,$ and $3 - i$.

Graph each polynomial function. Factor first if the polynomial is not in factored form.

6. $f(x) = x(x - 2)^3(x + 2)^2$

7. $f(x) = 2x^4 - 9x^3 - 5x^2 + 57x - 45$

8. $f(x) = -4x^5 + 16x^4 + 13x^3 - 76x^2 - 3x + 18$

Graph each rational function.

9. $f(x) = \dfrac{3x + 1}{x^2 + 7x + 10}$ **10.** $f(x) = \dfrac{x^2 + 2x + 1}{x - 1}$

Summary Exercises on Solving Equations and Inequalities

A rational inequality can be solved by rewriting it so that 0 is on one side. Then we determine the values that cause either the numerator or the denominator to *equal* 0, and by using a test value from each interval determined by these values, we can find the solution set.

We now solve rational inequalities by inspecting the graph of a related function. The graphs can be obtained using technology or the steps for graphing a rational function.

EXAMPLE   Solving a Rational Inequality

Solve the inequality.

$$\frac{1-x}{x+4} \geq 0$$

SOLUTION Graph the related rational function

$$y = \frac{1-x}{x+4}.$$

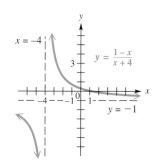

The real solutions of $\frac{1-x}{x+4} \geq 0$ are the x-values for which the graph lies above or on the x-axis. This is true for all x to the right of the vertical asymptote at $x = -4$, up to and including the x-intercept at $(1, 0)$. Therefore, the solution set of the inequality is $(-4, 1]$.

By inspecting the graph of the related function, we can also determine that the solution set of $\frac{1-x}{x+4} < 0$ is $(-\infty, -4) \cup (1, \infty)$ and that the solution set of the equation $\frac{1-x}{x+4} = 0$ is $\{1\}$, the x-value of the x-intercept. (This graphical method may be used to solve other equations and inequalities including those defined by polynomials.)

✔ **Now Try Exercise 19.**

EXERCISES

Concept Check *Use the graph of the function to solve each equation or inequality.*

1. (a) $f(x) > 0$ **(b)** $f(x) \leq 0$ **2. (a)** $f(x) < 0$ **(b)** $f(x) > 0$

3. (a) $f(x) = 0$ **(b)** $f(x) > 0$ **4. (a)** $f(x) = 0$ **(b)** $f(x) \leq 0$

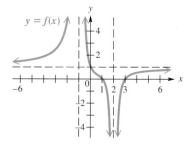

Solve each equation.

5. $\dfrac{5x+8}{-2} = 2x - 10$

6. $\dfrac{1}{5}x + 0.25x = \dfrac{1}{2}x - 1$

7. $(x-5)^{-4} - 13(x-5)^{-2} = -36$

8. $x = 13\sqrt{x} - 40$

9. $\sqrt{2x-5} - \sqrt{x-3} = 1$

10. $3 = \sqrt{x+2} + \sqrt{x-1}$

11. $x^{2/3} + \dfrac{1}{2} = \dfrac{3}{4}$

12. $27 - (x-4)^{3/2} = 0$

Sketch the graph of an appropriate function and then use the graph to solve each equation or inequality. (Note: First determine whether the expression is an equation or an inequality. When appropriate, use the steps for graphing polynomial functions or rational functions.)

13. $25x^2 - 20x + 4 > 0$

14. $3x + 4 \leq x^2$

15. $x^4 - 2x^3 - 3x^2 + 4x + 4 = 0$

16. $x^3 + 5x^2 + 3x - 9 \geq 0$

17. $-x^4 - x^3 + 12x^2 = 0$

18. $-4x^4 + 13x^2 - 3 > 0$

19. $\dfrac{2x^2 - 13x + 15}{x^2 - 3x} \geq 0$

20. $\dfrac{x^2 + 3x - 1}{x + 1} > 3$

21. $\dfrac{x - 1}{(x - 3)^2} = 0$

22. $\dfrac{x}{x^2 - 4} > 0$

3.6 Variation

- **Direct Variation**
- **Inverse Variation**
- **Combined and Joint Variation**

Direct Variation To apply mathematics we often need to express relationships between quantities. For example,

- In chemistry, the ideal gas law describes how temperature, pressure, and volume are related.

- In physics, various formulas in optics describe the relationship between the focal length of a lens and the size of an image.

When one quantity is a constant multiple of another quantity, the two quantities are said to *vary directly*. For example, if you work for an hourly wage of $10, then

$$[\text{pay}] = 10 \cdot [\text{hours worked}].$$

Doubling the hours doubles the pay. Tripling the hours triples the pay, and so on. This is stated more precisely as follows.

Direct Variation

y **varies directly** as *x*, or *y* is **directly proportional** to *x*, if there exists a nonzero real number *k*, called the **constant of variation,** such that for all *x*,

$$y = kx.$$

The direct variation equation $y = kx$ defines a linear function, where the constant of variation k is the slope of the line. For $k > 0$,

- As the value of x *increases*, the value of y *increases*.

- As the value of x *decreases*, the value of y *decreases*.

When used to describe a direct variation relationship, the phrase "directly proportional" is sometimes abbreviated to just "proportional."

The steps involved in solving a variation problem are summarized on the next page.

Solving a Variation Problem

Step 1 Write the general relationship among the variables as an equation. Use the constant k.

Step 2 Substitute given values of the variables and find the value of k.

Step 3 Substitute this value of k into the equation from Step 1, obtaining a specific formula.

Step 4 Substitute the remaining values and solve for the required unknown.

EXAMPLE 1 **Solving a Direct Variation Problem**

The area of a rectangle varies directly as its length. If the area is 50 m² when the length is 10 m, find the area when the length is 25 m. (See **Figure 59.**)

SOLUTION

Step 1 The area varies directly as the length, so

$$\mathcal{A} = kL,$$

where $\mathcal{A}$ represents the area of the rectangle, L is the length, and k is a nonzero constant.

Step 2 Because $\mathcal{A} = 50$ when $L = 10$, we can solve the equation $\mathcal{A} = kL$ for k.

$$50 = 10k \quad \text{Substitute for } \mathcal{A} \text{ and L.}$$
$$k = 5 \quad \text{Divide by 10. Interchange sides.}$$

Step 3 Using this value of k, we can express the relationship between the area and the length as follows.

$$\mathcal{A} = 5L \quad \text{Direct variation equation}$$

Step 4 To find the area when the length is 25, we replace L with 25.

$$\mathcal{A} = 5L$$
$$\mathcal{A} = 5(25) \quad \text{Substitute for L.}$$
$$\mathcal{A} = 125 \quad \text{Multiply.}$$

The area of the rectangle is 125 m² when the length is 25 m.

✔ **Now Try Exercise 27.**

Figure 59 (area diagram: $\mathcal{A} = 50$ m², 10 m; $\mathcal{A} = ?$, 25 m)

Sometimes y varies as a power of x. If n is a positive integer greater than or equal to 2, then y is a greater-power polynomial function of x.

Direct Variation as nth Power

Let n be a positive real number. Then y **varies directly as the nth power** of x, or y is **directly proportional to the nth power** of x, if for all x there exists a nonzero real number k such that

$$y = kx^n.$$

For example, the area of a square of side x is given by the formula $\mathcal{A} = x^2$, so the area varies directly as the square of the length of a side. Here $k = 1$.

Inverse Variation   Another type of variation is *inverse variation*. With inverse variation, where $k > 0$, as the value of one variable increases, the value of the other decreases. This relationship can be expressed as a rational function.

Inverse Variation as *n*th Power

Let n be a positive real number. Then y **varies inversely as the *n*th power** of x, or y is **inversely proportional to the *n*th power** of x, if for all x there exists a nonzero real number k such that

$$y = \frac{k}{x^n}.$$

If $n = 1$, then $y = \frac{k}{x}$, and y **varies inversely** as x.

EXAMPLE 2   Solving an Inverse Variation Problem

In a certain manufacturing process, the cost of producing a single item varies inversely as the square of the number of items produced. If 100 items are produced, each costs \$2. Find the cost per item if 400 items are produced.

SOLUTION

Step 1 Let x represent the number of items produced and y represent the cost per item. Then, for some nonzero constant k, the following holds.

$$y = \frac{k}{x^2} \qquad \text{\textit{y} varies inversely as the square of \textit{x}.}$$

Step 2 $$2 = \frac{k}{100^2} \qquad \text{Substitute; } y = 2 \text{ when } x = 100.$$

$$k = 20{,}000 \qquad \text{Solve for } k.$$

Step 3 The relationship between x and y is $y = \dfrac{20{,}000}{x^2}$.

Step 4 When 400 items are produced, the cost per item is found as follows.

$$y = \frac{20{,}000}{x^2} = \frac{20{,}000}{400^2} = 0.125$$

The cost per item is \$0.125, or 12.5 cents.

✔ **Now Try Exercise 37.**

Combined and Joint Variation   In **combined variation**, one variable depends on more than one other variable. Specifically, when a variable depends on the *product* of two or more other variables, it is referred to as *joint variation*.

Joint Variation

Let m and n be real numbers. Then y **varies jointly** as the *n*th power of x and the *m*th power of z if for all x and z, there exists a nonzero real number k such that

$$y = kx^n z^m.$$

CAUTION Note that *and* in the expression "*y* varies jointly as *x* and *z*" translates as the product $y = kxz$. The word "and" does not indicate addition here.

EXAMPLE 3 Solving a Joint Variation Problem

The area of a triangle varies jointly as the lengths of the base and the height. A triangle with base 10 ft and height 4 ft has area 20 ft². Find the area of a triangle with base 3 ft and height 8 ft. (See **Figure 60**.)

SOLUTION

Step 1 Let $\mathcal{A}$ represent the area, b the base, and h the height of the triangle. Then, for some number k,

$$\mathcal{A} = kbh. \quad \text{\small $\mathcal{A}$ varies jointly as b and h.}$$

Step 2 $\mathcal{A}$ is 20 when b is 10 and h is 4, so substitute and solve for k.

$$20 = k(10)(4) \quad \text{\small Substitute for $\mathcal{A}$, b, and h.}$$

$$\frac{1}{2} = k \quad \text{\small Solve for k.}$$

Step 3 The relationship among the variables is the familiar formula for the area of a triangle,

$$\mathcal{A} = \frac{1}{2}bh.$$

Step 4 To find $\mathcal{A}$ when $b = 3$ ft and $h = 8$ ft, substitute into the formula.

$$\mathcal{A} = \frac{1}{2}(3)(8) = 12 \text{ ft}^2 \qquad ✔ \text{ Now Try Exercise 39.}$$

4 ft
10 ft
Area = 20 ft²

8 ft

3 ft
$\mathcal{A} = ?$

Figure 60

EXAMPLE 4 Solving a Combined Variation Problem

The number of vibrations per second (the pitch) of a steel guitar string varies directly as the square root of the tension and inversely as the length of the string. If the number of vibrations per second is 50 when the tension is 225 newtons and the length is 0.60 m, find the number of vibrations per second when the tension is 196 newtons and the length is 0.65 m.

SOLUTION

Step 1 Let n represent the number of vibrations per second, T represent the tension, and L represent the length of the string. Then, from the information in the problem, write the variation equation.

$$n = \frac{k\sqrt{T}}{L} \quad \text{\small n varies directly as the square root of T and inversely as L.}$$

Step 2 Substitute the given values for n, T, and L and solve for k.

$$50 = \frac{k\sqrt{225}}{0.60} \quad \text{\small Let $n = 50$, $T = 225$, $L = 0.60$.}$$

$$30 = k\sqrt{225} \quad \text{\small Multiply by 0.60.}$$

$$30 = 15k \quad \text{\small $\sqrt{225} = 15$}$$

$$k = 2 \quad \text{\small Divide by 15. Interchange sides.}$$

Step 3 Substitute for k to find the relationship among the variables.

$$n = \frac{2\sqrt{T}}{L}$$

Step 4 Now use the second set of values for T and L to find n.

$$n = \frac{2\sqrt{196}}{0.65} \approx 43 \qquad \text{Let } T = 196, L = 0.65.$$

The number of vibrations per second is approximately 43.

✔ **Now Try Exercise 43.**

3.6 Exercises

CONCEPT PREVIEW *Fill in the blank(s) to correctly complete each sentence, or answer the question as appropriate.*

1. For $k > 0$, if y varies directly as x, then when x increases, y _____, and when x decreases, y _____.

2. For $k > 0$, if y varies inversely as x, then when x increases, y _____, and when x decreases, y _____.

3. In the equation $y = 6x$, y varies directly as x. When $x = 5$, $y = 30$. What is the value of y when $x = 10$?

4. In the equation $y = \frac{12}{x}$, y varies inversely as x. When $x = 3$, $y = 4$. What is the value of y when $x = 6$?

5. Consider the two ordered pairs (x, y) from **Exercise 3.** Divide the y-value by the x-value. What is the result in each case?

6. Consider the two ordered pairs (x, y) from **Exercise 4.** Multiply the y-value by the x-value. What is the result in each case?

Solve each problem. **See Examples 1–4.**

7. If y varies directly as x, and $y = 20$ when $x = 4$, find y when $x = -6$.

8. If y varies directly as x, and $y = 9$ when $x = 30$, find y when $x = 40$.

9. If m varies jointly as x and y, and $m = 10$ when $x = 2$ and $y = 14$, find m when $x = 21$ and $y = 8$.

10. If m varies jointly as z and p, and $m = 10$ when $z = 2$ and $p = 7.5$, find m when $z = 6$ and $p = 9$.

11. If y varies inversely as x, and $y = 10$ when $x = 3$, find y when $x = 20$.

12. If y varies inversely as x, and $y = 20$ when $x = \frac{1}{4}$, find y when $x = 15$.

13. Suppose r varies directly as the square of m, and inversely as s. If $r = 12$ when $m = 6$ and $s = 4$, find r when $m = 6$ and $s = 20$.

14. Suppose p varies directly as the square of z, and inversely as r. If $p = \frac{32}{5}$ when $z = 4$ and $r = 10$, find p when $z = 3$ and $r = 32$.

15. Let a be directly proportional to m and n^2, and inversely proportional to y^3. If $a = 9$ when $m = 4$, $n = 9$, and $y = 3$, find a when $m = 6$, $n = 2$, and $y = 5$.

16. Let y vary directly as x, and inversely as m^2 and r^2. If $y = \frac{5}{3}$ when $x = 1$, $m = 2$, and $r = 3$, find y when $x = 3$, $m = 1$, and $r = 8$.

Concept Check *Match each statement with its corresponding graph in choices A–D. In each case, k > 0.*

17. y varies directly as x. $(y = kx)$

18. y varies inversely as x. $\left(y = \frac{k}{x}\right)$

19. y varies directly as the second power of x. $(y = kx^2)$

20. x varies directly as the second power of y. $(x = ky^2)$

A.   **B.**   **C.**   **D.**

Concept Check *Write each formula as an English phrase using the word* varies *or* proportional.

21. $C = 2\pi r$, where C is the circumference of a circle of radius r

22. $d = \frac{1}{5}s$, where d is the approximate distance (in miles) from a storm, and s is the number of seconds between seeing lightning and hearing thunder

23. $r = \frac{d}{t}$, where r is the speed when traveling d miles in t hours

24. $d = \dfrac{1}{4\pi nr^2}$, where d is the distance a gas atom of radius r travels between collisions, and n is the number of atoms per unit volume

25. $s = kx^3$, where s is the strength of a muscle that has length x

26. $f = \dfrac{mv^2}{r}$, where f is the centripetal force of an object of mass m moving along a circle of radius r at velocity v

Solve each problem. **See Examples 1–4.**

27. *Circumference of a Circle* The circumference of a circle varies directly as the radius. A circle with radius 7 in. has circumference 43.96 in. Find the circumference of the circle if the radius changes to 11 in.

28. *Pressure Exerted by a Liquid* The pressure exerted by a certain liquid at a given point varies directly as the depth of the point beneath the surface of the liquid. The pressure at 10 ft is 50 pounds per square inch (psi). What is the pressure at 15 ft?

29. *Resistance of a Wire* The resistance in ohms of a platinum wire temperature sensor varies directly as the temperature in kelvins (K). If the resistance is 646 ohms at a temperature of 190 K, find the resistance at a temperature of 250 K.

30. *Weight on the Moon* The weight of an object on Earth is directly proportional to the weight of that same object on the moon. A 200-lb astronaut would weigh 32 lb on the moon. How much would a 50-lb dog weigh on the moon?

31. *Distance to the Horizon* The distance that a person can see to the horizon on a clear day from a point above the surface of Earth varies directly as the square root of the height at that point. If a person 144 m above the surface of Earth can see 18 km to the horizon, how far can a person see to the horizon from a point 64 m above the surface?

32. *Water Emptied by a Pipe* The amount of water emptied by a pipe varies directly as the square of the diameter of the pipe. For a certain constant water flow, a pipe emptying into a canal will allow 200 gal of water to escape in an hour. The diameter of the pipe is 6 in. How much water would a 12-in. pipe empty into the canal in an hour, assuming the same water flow?

33. *Hooke's Law for a Spring* Hooke's law for an elastic spring states that the distance a spring stretches varies directly as the force applied. If a force of 15 lb stretches a certain spring 8 in., how much will a force of 30 lb stretch the spring?

34. *Current in a Circuit* The current in a simple electrical circuit varies inversely as the resistance. If the current is 50 amps when the resistance is 10 ohms, find the current if the resistance is 5 ohms.

35. *Speed of a Pulley* The speed of a pulley varies inversely as its diameter. One kind of pulley, with diameter 3 in., turns at 150 revolutions per minute. Find the speed of a similar pulley with diameter 5 in.

36. *Weight of an Object* The weight of an object varies inversely as the square of its distance from the center of Earth. If an object 8000 mi from the center of Earth weighs 90 lb, find its weight when it is 12,000 mi from the center of Earth.

37. *Current Flow* In electric current flow, it is found that the resistance offered by a fixed length of wire of a given material varies inversely as the square of the diameter of the wire. If a wire 0.01 in. in diameter has a resistance of 0.4 ohm, what is the resistance of a wire of the same length and material with diameter 0.03 in., to the nearest ten-thousandth of an ohm?

38. *Illumination* The illumination produced by a light source varies inversely as the square of the distance from the source. The illumination of a light source at 5 m is 70 candelas. What is the illumination 12 m from the source?

39. *Simple Interest* Simple interest varies jointly as principal and time. If $1000 invested for 2 yr earned $70, find the amount of interest earned by $5000 for 5 yr.

40. *Volume of a Gas* The volume of a gas varies inversely as the pressure and directly as the temperature in kelvins (K). If a certain gas occupies a volume of 1.3 L at 300 K and a pressure of 18 newtons, find the volume at 340 K and a pressure of 24 newtons.

41. *Force of Wind* The force of the wind blowing on a vertical surface varies jointly as the area of the surface and the square of the velocity. If a wind of 40 mph exerts a force of 50 lb on a surface of $\frac{1}{2}$ ft^2, how much force will a wind of 80 mph place on a surface of 2 ft^2?

42. *Volume of a Cylinder* The volume of a right circular cylinder is jointly proportional to the square of the radius of the circular base and to the height. If the volume is 300 cm^3 when the height is 10.62 cm and the radius is 3 cm, find the volume, to the nearest tenth, of a cylinder with radius 4 cm and height 15.92 cm.

43. *Sports Arena Construction* The roof of a new sports arena rests on round concrete pillars. The maximum load a cylindrical column of circular cross section can hold varies directly as the fourth power of the diameter and inversely as the square of the height. The arena has 9-m-tall columns that are 1 m in diameter and will support a load of 8 metric tons. How many metric tons will be supported by a column 12 m high and $\frac{2}{3}$ m in diameter?

Load = 8 metric tons

44. *Sports Arena Construction* The sports arena in **Exercise 43** requires a horizontal beam 16 m long, 24 cm wide, and 8 cm high. The maximum load of such a horizontal beam that is supported at both ends varies directly as the width of the beam and the square of its height and inversely as the length between supports. If a beam of the same material 8 m long, 12 cm wide, and 15 cm high can support a maximum of 400 kg, what is the maximum load the beam in the arena will support?

45. *Period of a Pendulum* The period of a pendulum varies directly as the square root of the length of the pendulum and inversely as the square root of the acceleration due to gravity. Find the period when the length is 121 cm and the acceleration due to gravity is 980 cm per second squared, if the period is 6π seconds when the length is 289 cm and the acceleration due to gravity is 980 cm per second squared.

46. *Skidding Car* The force needed to keep a car from skidding on a curve varies inversely as the radius r of the curve and jointly as the weight of the car and the square of the speed. It takes 3000 lb of force to keep a 2000-lb car from skidding on a curve of radius 500 ft at 30 mph. What force will keep the same car from skidding on a curve of radius 800 ft at 60 mph?

47. *Body Mass Index* The federal government has developed the **body mass index** (BMI) to determine ideal weights. A person's BMI is directly proportional to his or her weight in pounds and inversely proportional to the square of his or her height in inches. (A BMI of 19 to 25 corresponds to a healthy weight.) A 6-foot-tall person weighing 177 lb has BMI 24. Find the BMI (to the nearest whole number) of a person whose weight is 130 lb and whose height is 66 in.

48. *Poiseuille's Law* According to Poiseuille's law, the resistance to flow of a blood vessel, R, is directly proportional to the length, l, and inversely proportional to the fourth power of the radius, r. If $R = 25$ when $l = 12$ and $r = 0.2$, find R, to the nearest hundredth, as r increases to 0.3, while l is unchanged.

49. *Stefan-Boltzmann Law* The Stefan-Boltzmann law says that the radiation of heat R from an object is directly proportional to the fourth power of the kelvin temperature of the object. For a certain object, $R = 213.73$ at room temperature (293 K). Find R, to the nearest hundredth, if the temperature increases to 335 K.

50. *Nuclear Bomb Detonation* Suppose the effects of detonating a nuclear bomb will be felt over a distance from the point of detonation that is directly proportional to the cube root of the yield of the bomb. Suppose a 100-kiloton bomb has certain effects to a radius of 3 km from the point of detonation. Find the distance to the nearest tenth that the effects would be felt for a 1500-kiloton bomb.

51. *Malnutrition Measure* A measure of malnutrition, called the **pelidisi**, varies directly as the cube root of a person's weight in grams and inversely as the person's sitting height in centimeters. A person with a pelidisi below 100 is considered undernourished, while a pelidisi greater than 100 indicates overfeeding. A person who weighs 48,820 g with a sitting height of 78.7 cm has a pelidisi of 100. Find the pelidisi (to the nearest whole number) of a person whose weight is 54,430 g and whose sitting height is 88.9 cm. Is this individual undernourished or overfed?

Weight: 48,820 g   Weight: 54,430 g

52. *Photography* Variation occurs in a formula from photography. In

$$L = \frac{25F^2}{st},$$

the luminance, L, varies directly as the square of the F-stop, F, and inversely as the product of the film ASA number, s, and the shutter speed, t.

(a) What would an appropriate F-stop be for 200 ASA film and a shutter speed of $\frac{1}{250}$ sec when 500 footcandles of light is available?

(b) If 125 footcandles of light is available and an F-stop of 2 is used with 200 ASA film, what shutter speed should be used?

Concept Check *Work each problem.*

53. What happens to y if y varies inversely as x, and x is doubled?

54. What happens to y if y varies directly as x, and x is halved?

55. Suppose y is directly proportional to x, and x is replaced by $\frac{1}{3}x$. What happens to y?

56. Suppose y is inversely proportional to x, and x is tripled. What happens to y?

57. Suppose p varies directly as r^3 and inversely as t^2. If r is halved and t is doubled, what happens to p?

58. Suppose m varies directly as p^2 and q^4. If p doubles and q triples, what happens to m?

Chapter 3 Test Prep

Key Terms

3.1 polynomial function
leading coefficient
dominating term
zero polynomial
quadratic function
parabola
axis of symmetry (axis)
vertex
quadratic regression

3.2 synthetic division
zero of a polynomial
function
root (or solution) of an
equation
3.3 multiplicity of a zero
3.4 turning points
end behavior

3.5 rational function
discontinuous graph
vertical asymptote
horizontal asymptote
oblique asymptote
point of discontinuity
(hole)

3.6 varies directly
(directly
proportional to)
constant of variation
varies inversely
(inversely
proportional to)
combined variation
varies jointly

New Symbols

$\bar{z}$ conjugate of $z = a + bi$

      end behavior diagrams

$|f(x)| \to \infty$ absolute value of $f(x)$ increases
without bound

$x \to a$ x approaches a

Quick Review

Concepts	Examples

3.1 Quadratic Functions and Models

1. The graph of

$$f(x) = a(x - h)^2 + k, \quad \text{with } a \neq 0,$$

is a parabola with vertex at (h, k) and the vertical line $x = h$ as axis.

2. The graph opens up if $a > 0$ and down if $a < 0$.

3. The graph is wider than the graph of $f(x) = x^2$ if $|a| < 1$ and narrower if $|a| > 1$.

Vertex Formula
The vertex of the graph of $f(x) = ax^2 + bx + c$, with $a \neq 0$, may be found by completing the square or using the vertex formula.

$$\left(-\frac{b}{2a}, f\left(-\frac{b}{2a}\right)\right) \quad \text{Vertex}$$

Graphing a Quadratic Function $f(x) = ax^2 + bx + c$

Step 1 Find the vertex either by using the vertex formula or by completing the square. Plot the vertex.

Step 2 Plot the y-intercept by evaluating $f(0)$.

Step 3 Plot any x-intercepts by solving $f(x) = 0$.

Step 4 Plot any additional points as needed, using symmetry about the axis.

The graph opens up if $a > 0$ and down if $a < 0$.

Graph $f(x) = -(x + 3)^2 + 1$.

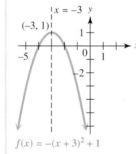

$$f(x) = -(x + 3)^2 + 1$$

The graph opens down because $a < 0$. It is the graph of $y = -x^2$ shifted 3 units left and 1 unit up, so the vertex is $(-3, 1)$, with axis $x = -3$. The domain is $(-\infty, \infty)$, and the range is $(-\infty, 1]$. The function is increasing on $(-\infty, -3)$ and decreasing on $(-3, \infty)$.

Graph $f(x) = x^2 + 4x + 3$. The vertex of the graph is

$$\left(-\frac{b}{2a}, f\left(-\frac{b}{2a}\right)\right) = (-2, -1). \quad a = 1, b = 4, c = 3$$

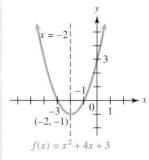

$$f(x) = x^2 + 4x + 3$$

The graph opens up because $a > 0$. $f(0) = 3$, so the y-intercept is $(0, 3)$. The solutions of $x^2 + 4x + 3 = 0$ are -1 and -3, which correspond to the x-intercepts. The domain is $(-\infty, \infty)$, and the range is $[-1, \infty)$. The function is decreasing on $(-\infty, -2)$ and increasing on $(-2, \infty)$.

Concepts	Examples

3.2 Synthetic Division

Division Algorithm

Let $f(x)$ and $g(x)$ be polynomials with $g(x)$ of lesser degree than $f(x)$ and $g(x)$ of degree 1 or more. There exist unique polynomials $q(x)$ and $r(x)$ such that

$$f(x) = g(x) \cdot q(x) + r(x),$$

where either $r(x) = 0$ or the degree of $r(x)$ is less than the degree of $g(x)$.

Synthetic Division

Synthetic division is a shortcut method for dividing a polynomial by a binomial of the form $x - k$.

Remainder Theorem

If the polynomial $f(x)$ is divided by $x - k$, the remainder is $f(k)$.

Use synthetic division to divide

$$f(x) = 2x^3 - 3x + 2 \quad \text{by} \quad x - 1,$$

and write the result as $f(x) = g(x) \cdot q(x) + r(x)$.

$$
\begin{array}{r|rrrr}
1) & 2 & 0 & -3 & 2 \\
 & & 2 & 2 & -1 \\
\hline
 & 2 & 2 & -1 & 1
\end{array}
$$

$$\underbrace{\text{Coefficients of the quotient}} \qquad \underbrace{\text{Remainder}}$$

$$2x^3 - 3x + 2 = \underbrace{(x - 1)}_{f(x) \;=\; g(x)}\underbrace{(2x^2 + 2x - 1)}_{\cdot \quad q(x)} + \underbrace{1}_{+ \quad r(x)}$$

By the result above, for $f(x) = 2x^3 - 3x + 2$,

$$f(1) = 1.$$

3.3 Zeros of Polynomial Functions

Factor Theorem

For any polynomial function $f(x)$, $x - k$ is a factor of the polynomial if and only if $f(k) = 0$.

Rational Zeros Theorem

If $\frac{p}{q}$ is a rational number written in lowest terms, and if $\frac{p}{q}$ is a zero of f, a polynomial function with integer coefficients, then p is a factor of the constant term and q is a factor of the leading coefficient.

Fundamental Theorem of Algebra

Every function defined by a polynomial of degree 1 or more has at least one complex zero.

Number of Zeros Theorem

A function defined by a polynomial of degree n has at most n distinct zeros.

Conjugate Zeros Theorem

If $f(x)$ defines a polynomial function *having only real coefficients* and if $z = a + bi$ is a zero of $f(x)$, where a and b are real numbers, then the conjugate

$$\bar{z} = a - bi \text{ is also a zero of } f(x).$$

For the polynomial functions

$$f(x) = x^3 + x + 2 \quad \text{and} \quad g(x) = x^3 - 1,$$

$f(-1) = 0$. Therefore, $x - (-1)$, or $x + 1$, is a factor of $f(x)$. Because $x - 1$ is a factor of $g(x)$, $g(1) = 0$.

The only rational numbers that can possibly be zeros of

$$f(x) = 2x^3 - 9x^2 - 4x - 5$$

are ± 1, ± 5, $\pm\frac{1}{2}$, and $\pm\frac{5}{2}$. By synthetic division, it can be shown that the only rational zero of $f(x)$ is 5.

$$
\begin{array}{r|rrrr}
5) & 2 & -9 & -4 & -5 \\
 & & 10 & 5 & 5 \\
\hline
 & 2 & 1 & 1 & 0 \leftarrow f(5)
\end{array}
$$

$f(x) = x^3 + x + 2$ has at least one and at most three distinct zeros.

$1 + 2i$ is a zero of

$$f(x) = x^3 - 5x^2 + 11x - 15,$$

and therefore its conjugate $1 - 2i$ is also a zero.

Concepts	Examples

Descartes' Rule of Signs

Let $f(x)$ define a polynomial function with real coefficients and a nonzero constant term, with terms in descending powers of x.

(a) The number of positive real zeros of f either equals the number of variations in sign occurring in the coefficients of $f(x)$ or is less than the number of variations by a positive even integer.

(b) The number of negative real zeros of f either equals the number of variations in sign occurring in the coefficients of $f(-x)$ or is less than the number of variations by a positive even integer.

There are three sign changes for

$$f(x) = +3x^3 - 2x^2 + x - 4,$$

so there will be three or one positive real zeros. Because

$$f(-x) = -3x^3 - 2x^2 - x - 4$$

has no sign changes, there will be no negative real zeros. The table shows the possibilities for the numbers of positive, negative, and nonreal complex zeros.

Positive	Negative	Nonreal Complex
3	0	0
1	0	2

3.4 Polynomial Functions: Graphs, Applications, and Models

Graphing Using Translations

The graph of the function

$$f(x) = a(x - h)^n + k$$

can be found by considering the effects of the constants a, h, and k on the graph of $f(x) = ax^n$.

- When $|a| > 1$, the graph is stretched vertically.

- When $0 < |a| < 1$, the graph is shrunk vertically.

- When $a < 0$, the graph is reflected across the x-axis.

- The graph is translated h units right if $h > 0$ and $|h|$ units left if $h < 0$.

- The graph is translated k units up if $k > 0$ and $|k|$ units down if $k < 0$.

Graph $f(x) = -(x + 2)^4 + 1$.

$f(x) = -(x + 2)^4 + 1$

The negative sign causes the graph to be reflected across the x-axis compared to the graph of $f(x) = x^4$. The graph is translated 2 units to the left and 1 unit up. The function is increasing on $(-\infty, -2)$ and decreasing on $(-2, \infty)$.

Multiplicity of a Zero

The behavior of the graph of a polynomial function $f(x)$ near a zero depends on the multiplicity of the zero. If $(x - c)^n$ is a factor of $f(x)$, then the graph will behave in the following manner.

- For $n = 1$, the graph will cross the x-axis at $(c, 0)$.

- For n even, the graph will bounce, or turn, at $(c, 0)$.

- For n an odd integer greater than 1, the graph will wiggle through the x-axis at $(c, 0)$.

Determine the behavior of f near its zeros, and graph.

$$f(x) = (x - 1)(x - 3)^2(x + 1)^3$$

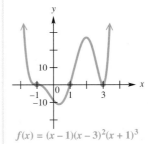

$f(x) = (x - 1)(x - 3)^2(x + 1)^3$

The graph will cross the x-axis at $x = 1$, bounce at $x = 3$, and wiggle through the x-axis at $x = -1$. Since the dominating term is x^6, the end behavior is $\smile$. The y-intercept is $(0, -9)$ because $f(0) = -9$.

Concepts	Examples

Turning Points

A polynomial function of degree n has at most $n - 1$ turning points, with at least one turning point between each pair of successive zeros.

The graph of

$$f(x) = 4x^5 - 2x^3 + 3x^2 + x - 10$$

has at most four turning points (because $5 - 1 = 4$).

End Behavior

The end behavior of the graph of a polynomial function $f(x)$ is determined by the dominating term, or term of greatest degree. If ax^n is the dominating term of $f(x)$, then the end behavior is as follows.

The end behavior of

$$f(x) = 3x^5 + 2x^2 + 7$$

is ⤴ .

The end behavior of

$$f(x) = -x^4 - 3x^3 + 2x - 9$$

is ⤵ ⤵.

Graphing Polynomial Functions

To graph a polynomial function f, first find the x-intercepts and y-intercept.

Then use end behavior, whether the graph crosses, bounces on, or wiggles through the x-axis at the x-intercepts, and selected points as necessary to complete the graph.

Graph $f(x) = (x + 2)(x - 1)(x + 3)$.

$f(x) = (x + 2)(x - 1)(x + 3)$

The x-intercepts correspond to the zeros of f, which are -2, 1, and -3. Because $f(0) = 2(-1)(3) = -6$, the y-intercept is $(0, -6)$. The dominating term is $x(x)(x)$, or x^3, so the end behavior is ⤴. Begin at either end of the graph with the correct end behavior, and draw a smooth curve that crosses the x-axis at each zero, has a turning point between successive zeros, and passes through the y-intercept.

Intermediate Value Theorem

If $f(x)$ is a polynomial function with only *real coefficients*, and if for real numbers a and b the values of $f(a)$ and $f(b)$ are opposite in sign, then there exists at least one real zero between a and b.

For the polynomial function

$$f(x) = -x^4 + 2x^3 + 3x^2 + 6,$$

$$f(3.1) = 2.0599 \quad \text{and} \quad f(3.2) = -2.6016.$$

Because $f(3.1) > 0$ and $f(3.2) < 0$, there exists at least one real zero between 3.1 and 3.2.

Boundedness Theorem

Let $f(x)$ be a polynomial function of degree $n \geq 1$ with *real coefficients* and with a *positive* leading coefficient. Suppose $f(x)$ is divided synthetically by $x - c$.

(a) If $c > 0$ and all numbers in the bottom row of the synthetic division are nonnegative, then $f(x)$ has no zero greater than c.

(b) If $c < 0$ and the numbers in the bottom row of the synthetic division alternate in sign (with 0 considered positive or negative, as needed), then $f(x)$ has no zero less than c.

Show that $f(x) = x^3 - x^2 - 8x + 12$ has no zero greater than 4 and no zero less than -4.

$$
\begin{array}{r|rrrr}
4) & 1 & -1 & -8 & 12 \\
   &   & 4 & 12 & 16 \\
\hline
   & 1 & 3 & 4 & 28
\end{array}
$$
⟵ All signs positive

$$
\begin{array}{r|rrrr}
-4) & 1 & -1 & -8 & 12 \\
    &   & -4 & 20 & -48 \\
\hline
    & 1 & -5 & 12 & -36
\end{array}
$$
⟵ Alternating signs

Concepts	Examples

3.5 Rational Functions: Graphs, Applications, and Models

Graphing Rational Functions

To graph a rational function in lowest terms, find the asymptotes and intercepts. Determine whether the graph intersects its nonvertical asymptote. Plot selected points, as necessary, to complete the sketch.

Graph $f(x) = \dfrac{x^2 - 1}{(x + 3)(x - 2)}$.

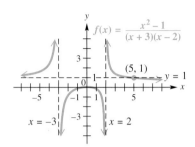

Point of Discontinuity

A rational function that is not in lowest terms often has a hole, or point of discontinuity, in its graph.

Graph $f(x) = \dfrac{x^2 - 1}{x + 1}$.

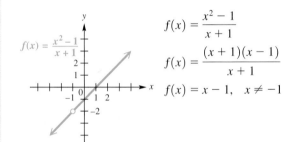

$$f(x) = \frac{x^2 - 1}{x + 1}$$

$$f(x) = \frac{(x + 1)(x - 1)}{x + 1}$$

$$f(x) = x - 1, \quad x \neq -1$$

The graph is that of $y = x - 1$, with a hole at $(-1, -2)$.

3.6 Variation

Direct Variation

y varies directly as the nth power of x if for all x there exists a nonzero real number k such that

$$y = kx^n.$$

The area of a circle varies directly as the square of the radius.

$$\mathcal{A} = kr^2 \quad (k = \pi)$$

Inverse Variation

y varies inversely as the nth power of x if for all x there exists a nonzero real number k such that

$$y = \frac{k}{x^n}.$$

Pressure of a gas varies inversely as volume.

$$P = \frac{k}{V}$$

Joint Variation

For real numbers m and n, y varies jointly as the nth power of x and the mth power of z if for all x and z, there exists a nonzero real number k such that

$$y = kx^n z^m.$$

The area of a triangle varies jointly as its base and its height.

$$\mathcal{A} = kbh \quad \left(k = \tfrac{1}{2}\right)$$

Chapter 3   **Review Exercises**

Graph each quadratic function. Give the vertex, axis, x-intercepts, y-intercept, domain, range, and largest open intervals of the domain over which each function is increasing or decreasing.

1. $f(x) = 3(x + 4)^2 - 5$

2. $f(x) = -\dfrac{2}{3}(x - 6)^2 + 7$

3. $f(x) = -3x^2 - 12x - 1$

4. $f(x) = 4x^2 - 4x + 3$

Concept Check *Consider the function*

$$f(x) = a(x - h)^2 + k, \quad \text{for } a > 0.$$

5. What are the coordinates of the lowest point of its graph?

6. What is the *y*-intercept of its graph?

7. Under what conditions will its graph have one or more *x*-intercepts? For these conditions, express the *x*-intercept(s) in terms of *a*, *h*, and *k*.

8. If *a* is positive, what is the least value of $ax^2 + bx + c$ in terms of *a*, *b*, and *c*?

(Modeling) *Solve each problem.*

9. *Area of a Rectangle* Use a quadratic function to find the dimensions of the rectangular region of maximum area that can be enclosed with 180 m of fencing, if no fencing is needed along one side of the region.

10. *Height of a Projectile* A projectile is fired vertically upward, and its height *s(t)* in feet after *t* seconds is given by the function

$$s(t) = -16t^2 + 800t + 600.$$

(a) From what height was the projectile fired?

(b) After how many seconds will it reach its maximum height?

(c) What is the maximum height it will reach?

(d) Between what two times (in seconds, to the nearest tenth) will it be more than 5000 ft above the ground?

(e) After how many seconds, to the nearest tenth, will the projectile hit the ground?

11. *Food Bank Volunteers* During the course of a year, the number of volunteers available to run a food bank each month is modeled by *V(x)*, where

$$V(x) = 2x^2 - 32x + 150$$

between the months of January and August. Here *x* is time in months, with $x = 1$ representing January. From August to December, *V(x)* is modeled by

$$V(x) = 31x - 226.$$

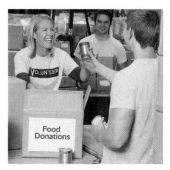

Find the number of volunteers in each of the following months.

(a) January   **(b)** May   **(c)** August   **(d)** October   **(e)** December

(f) Sketch a graph of $y = V(x)$ for January through December. In what month are the fewest volunteers available?

12. *Concentration of Atmospheric CO_2* In 1990, the International Panel on Climate Change (IPCC) stated that if current trends of burning fossil fuel and deforestation were to continue, then future amounts of atmospheric carbon dioxide in parts per million (ppm) would increase, as shown in the table.

Year	Carbon Dioxide
1990	353
2000	375
2075	590
2175	1090
2275	2000

Source: IPCC.

(a) Let $x = 0$ represent 1990, $x = 10$ represent 2000, and so on. Find a function of the form

$$f(x) = a(x - h)^2 + k$$

that models the data. Use $(0, 353)$ as the vertex and $(285, 2000)$ as another point to determine a.

(b) Use the function to predict the amount of carbon dioxide in 2300. Round to the nearest unit.

Consider the function $f(x) = -2.64x^2 + 5.47x + 3.54$.

13. Use the discriminant to explain how to determine the number of x-intercepts the graph of $f(x)$ will have before graphing it on a calculator.

14. Graph the function in the standard viewing window of a calculator, and use the calculator to solve the equation $f(x) = 0$. Express solutions as approximations to the nearest hundredth.

15. Use the answer to **Exercise 14** and the graph of f to solve the following. Give approximations to the nearest hundredth.

    (a) $f(x) > 0$ (b) $f(x) < 0$

16. Use the capabilities of a calculator to find the coordinates of the vertex of the graph. Express coordinates to the nearest hundredth.

Use synthetic division to perform each division.

17. $\dfrac{x^3 + x^2 - 11x - 10}{x - 3}$

18. $\dfrac{3x^3 + 8x^2 + 5x + 10}{x + 2}$

19. $\dfrac{2x^3 - x + 6}{x + 4}$

20. $\dfrac{3x^3 + 6x^2 - 8x + 3}{x + 3}$

Use synthetic division to divide $f(x)$ by $x - k$ for the given value of k. Then express $f(x)$ in the form $f(x) = (x - k)q(x) + r$.

21. $f(x) = 5x^3 - 3x^2 + 2x - 6$; $k = 2$

22. $f(x) = -3x^3 + 5x - 6$; $k = -1$

Use synthetic division to find $f(2)$.

23. $f(x) = -x^3 + 5x^2 - 7x + 1$

24. $f(x) = 2x^3 - 3x^2 + 7x - 12$

25. $f(x) = 5x^4 - 12x^2 + 2x - 8$

26. $f(x) = x^5 + 4x^2 - 2x - 4$

Use synthetic division to determine whether k is a zero of the function.

27. $f(x) = x^3 + 2x^2 + 3x + 2$; $k = -1$

28. $f(x) = 2x^3 + 5x^2 + 30$; $k = -4$

29. *Concept Check* If $f(x)$ is a polynomial function with real coefficients, and if $7 + 2i$ is a zero of the function, then what other complex number must also be a zero?

30. *Concept Check* Suppose the polynomial function f has a zero at $x = -3$. Which of the following statements *must* be true?

    A. $(3, 0)$ is an x-intercept of the graph of f.

    B. $(0, 3)$ is a y-intercept of the graph of f.

    C. $x - 3$ is a factor of $f(x)$.

    D. $f(-3) = 0$

Find a polynomial function $f(x)$ of least degree with real coefficients having zeros as given.

31. $-1, 4, 7$

32. $8, 2, 3$

33. $\sqrt{3}, -\sqrt{3}, 2, 3$

34. $-2 + \sqrt{5}, -2 - \sqrt{5}, -2, 1$

35. $2, 4, -i$

36. $0, 5, 1 + 2i$

Find all rational zeros of each function.

37. $f(x) = 2x^3 - 9x^2 - 6x + 5$

38. $f(x) = 8x^4 - 14x^3 - 29x^2 - 4x + 3$

Show that each polynomial function has a real zero as described in parts (a) and (b). In Exercises 39 and 40, also work part (c).

39. $f(x) = 3x^3 - 8x^2 + x + 2$

    (a) between -1 and 0 **(b)** between 2 and 3

    (c) Find the zero in part (b) to three decimal places.

40. $f(x) = 4x^3 - 37x^2 + 50x + 60$

    (a) between 2 and 3   **(b)** between 7 and 8

    (c) Find the zero in part (b) to three decimal places.

41. $f(x) = 6x^4 + 13x^3 - 11x^2 - 3x + 5$

    (a) no zero greater than 1   **(b)** no zero less than -3

Solve each problem.

42. Use Descartes' rule of signs to determine the different possibilities for the numbers of positive, negative, and nonreal complex zeros of

$$f(x) = x^3 + 3x^2 - 4x - 2.$$

43. Is $x + 1$ a factor of $f(x) = x^3 + 2x^2 + 3x + 2$?

44. Find a polynomial function f with real coefficients of degree 4 with 3, 1, and $-1 + 3i$ as zeros, and $f(2) = -36$.

45. Find a polynomial function f of degree 3 with -2, 1, and 4 as zeros, and $f(2) = 16$.

46. Find all zeros of $f(x) = x^4 - 3x^3 - 8x^2 + 22x - 24$, given that $1 + i$ is a zero.

47. Find all zeros of $f(x) = 2x^4 - x^3 + 7x^2 - 4x - 4$, given that 1 and $-2i$ are zeros.

48. Find a value of k such that $x - 4$ is a factor of $f(x) = x^3 - 2x^2 + kx + 4$.

49. Find a value of k such that when the polynomial $x^3 - 3x^2 + kx - 4$ is divided by $x - 2$, the remainder is 5.

50. Give the maximum number of turning points of the graph of each function.

    (a) $f(x) = x^5 - 9x^2$ **(b)** $f(x) = 4x^3 - 6x^2 + 2$

51. *Concept Check* Give an example of a cubic polynomial function having exactly one real zero, and then sketch its graph.

52. *Concept Check* Give an example of a fourth-degree polynomial function having exactly two distinct real zeros, and then sketch its graph.

53. *Concept Check* If the dominating term of a polynomial function is $10x^7$, what can we conclude about each of the following features of the graph of the function?

    (a) domain   **(b)** range   **(c)** end behavior   **(d)** number of zeros

    (e) number of turning points

54. *Concept Check* Repeat **Exercise 53** for a polynomial function with dominating term $-9x^6$.

Graph each polynomial function.

55. $f(x) = (x - 2)^2(x + 3)$

56. $f(x) = -2x^3 + 7x^2 - 2x - 3$

57. $f(x) = 2x^3 + x^2 - x$

58. $f(x) = x^4 - 3x^2 + 2$

59. $f(x) = x^4 + x^3 - 3x^2 - 4x - 4$

60. $f(x) = -2x^4 + 7x^3 - 4x^2 - 4x$

Concept Check For each polynomial function, identify its graph from choices A–F.

61. $f(x) = (x - 2)^2(x - 5)$

62. $f(x) = -(x - 2)^2(x - 5)$

63. $f(x) = (x - 2)^2(x - 5)^2$

64. $f(x) = (x - 2)(x - 5)$

65. $f(x) = -(x - 2)(x - 5)$

66. $f(x) = -(x - 2)^2(x - 5)^2$

A.

B.

C.

D.

E.

F.

 Graph each polynomial function in the viewing window specified. Then approximate the real zeros to as many decimal places as the calculator will provide.

67. $f(x) = x^3 - 8x^2 + 2x + 5$;   window: $[-10, 10]$ by $[-60, 60]$

68. $f(x) = x^4 - 4x^3 - 5x^2 + 14x - 15$;   window: $[-10, 10]$ by $[-60, 60]$

Solve each problem.

69. *(Modeling) Medicare Beneficiary Spending* Out-of-pocket spending projections for a typical Medicare beneficiary as a share of his or her income are given in the table. Let $x = 0$ represent 1990, so $x = 8$ represents 1998. Use a graphing calculator to do the following.

Year	Percent of Income
1998	18.6
2000	19.3
2005	21.7
2010	24.7
2015	27.5
2020	28.3
2025	28.6

Source: Urban Institute's Analysis of Medicare Trustees' Report.

(a) Graph the data points.

(b) Find a quadratic function to model the data.

(c) Find a cubic function to model the data.

(d) Graph each function in the same viewing window as the data points.

(e) Compare the two functions. Which is a better fit for the data?

70. *Dimensions of a Cube* After a 2-in. slice is cut off the top of a cube, the resulting solid has a volume of 32 in.³. Find the dimensions of the original cube.

71. ***Dimensions of a Box*** The width of a rectangular box is three times its height, and its length is 11 in. more than its height. Find the dimensions of the box if its volume is 720 in.³.

72. The function $f(x) = \frac{1}{x}$ is negative at $x = -1$ and positive at $x = 1$ but has no zero between -1 and 1. Explain why this does not contradict the intermediate value theorem.

Graph each rational function.

73. $f(x) = \dfrac{4}{x - 1}$

74. $f(x) = \dfrac{4x - 2}{3x + 1}$

75. $f(x) = \dfrac{6x}{x^2 + x - 2}$

76. $f(x) = \dfrac{2x}{x^2 - 1}$

77. $f(x) = \dfrac{x^2 + 4}{x + 2}$

78. $f(x) = \dfrac{x^2 - 1}{x}$

79. $f(x) = \dfrac{-2}{x^2 + 1}$

80. $f(x) = \dfrac{4x^2 - 9}{2x + 3}$

Solve each problem.

81. ***Concept Check*** Work each of the following.

  (a) Sketch the graph of a function that does not intersect its horizontal asymptote $y = 1$, has the line $x = 3$ as a vertical asymptote, and has x-intercepts $(2, 0)$ and $(4, 0)$.

  (b) Find an equation for a possible corresponding rational function.

82. ***Concept Check*** Work each of the following.

  (a) Sketch the graph of a function that is never negative and has the lines $x = -1$ and $x = 1$ as vertical asymptotes, the x-axis as a horizontal asymptote, and the origin as an x-intercept.

  (b) Find an equation for a possible corresponding rational function.

83. ***Connecting Graphs with Equations*** Find a rational function f having the graph shown.

84. ***Concept Check*** The rational function

$$f(x) = \frac{x^3 + 7x^2 - 25x - 175}{x^3 + 3x^2 - 25x - 75}$$

has two holes and one vertical asymptote.

  (a) What are the x-values of the holes?

  (b) What is the equation of the vertical asymptote?

(Modeling) Solve each problem.

85. *Environmental Pollution* In situations involving environmental pollution, a cost-benefit model expresses cost as a function of the percentage of pollutant removed from the environment. Suppose a cost-benefit model is expressed as

$$C(x) = \frac{6.7x}{100 - x},$$

where $C(x)$ is cost in thousands of dollars of removing x percent of a pollutant.

(a) Graph the function in the window $[0, 100]$ by $[0, 100]$.

(b) How much would it cost to remove 95% of the pollutant? Round to the nearest tenth.

86. *Antique-Car Competition* Antique-car owners often enter their cars in a **concours d'elegance** in which a maximum of 100 points can be awarded to a particular car based on its attractiveness. The function

$$C(x) = \frac{10x}{49(101 - x)}$$

models the cost, in thousands of dollars, of restoring a car so that it will win x points.

(a) Graph the function in the window $[0, 101]$ by $[0, 10]$.

(b) How much would an owner expect to pay to restore a car in order to earn 95 points? Round to the nearest tenth.

Solve each problem.

87. If x varies directly as y, and $x = 20$ when $y = 14$, find y when $x = 50$.

88. If x varies directly as y, and $x = 12$ when $y = 4$, find x when $y = 12$.

89. If t varies inversely as s, and $t = 3$ when $s = 5$, find s when $t = 20$.

90. If z varies inversely as w, and $z = 10$ when $w = \frac{1}{2}$, find z when $w = 10$.

91. f varies jointly as g^2 and h, and $f = 50$ when $g = 5$ and $h = 4$. Find f when $g = 3$ and $h = 6$.

92. p varies jointly as q and r^2, and $p = 100$ when $q = 2$ and $r = 3$. Find p when $q = 5$ and $r = 2$.

93. *Power of a Windmill* The power a windmill obtains from the wind varies directly as the cube of the wind velocity. If a wind of 10 km per hr produces 10,000 units of power, how much power is produced by a wind of 15 km per hr?

94. *Pressure in a Liquid* The pressure on a point in a liquid is directly proportional to the distance from the surface to the point. In a certain liquid, the pressure at a depth of 4 m is 60 kg per m². Find the pressure at a depth of 10 m.

Chapter 3   Test

1. Graph the quadratic function $f(x) = -2x^2 + 6x - 3$. Give the intercepts, vertex, axis, domain, range, and the largest open intervals of the domain over which the function is increasing or decreasing.

2. *(Modeling) Height of a Projectile* A small rocket is fired directly upward, and its height s in feet after t seconds is given by the function

$$s(t) = -16t^2 + 88t + 48.$$

(a) Determine the time at which the rocket reaches its maximum height.

(b) Determine the maximum height.

(c) Between what two times (in seconds, to the nearest tenth) will the rocket be more than 100 ft above ground level?

(d) After how many seconds will the rocket hit the ground?

Use synthetic division to perform each division.

3. $\dfrac{3x^3 + 4x^2 - 9x + 6}{x + 2}$

4. $\dfrac{2x^3 - 11x^2 + 25}{x - 5}$

5. Use synthetic division to determine $f(5)$ for

$$f(x) = 2x^3 - 9x^2 + 4x + 8.$$

6. Use the factor theorem to determine whether the polynomial $x - 3$ is a factor of

$$6x^4 - 11x^3 - 35x^2 + 34x + 24.$$

If it is, what is the other factor? If it is not, explain why.

7. Given that -2 is a zero, find all zeros of

$$f(x) = x^3 + 8x^2 + 25x + 26.$$

8. Find a fourth degree polynomial function f having only real coefficients, -1, 2, and i as zeros, and $f(3) = 80$.

9. Why can't the polynomial function $f(x) = x^4 + 8x^2 + 12$ have any real zeros?

10. Consider the polynomial function

$$f(x) = x^3 - 5x^2 + 2x + 7.$$

(a) Use the intermediate value theorem to show that f has a zero between 1 and 2.

(b) Use Descartes' rule of signs to determine the different possibilities for the numbers of positive, negative, and nonreal complex zeros.

(c) Use a graphing calculator to find all real zeros to as many decimal places as the calculator will give.

11. Graph the polynomial functions

$$f(x) = x^4 \quad \text{and} \quad g(x) = -2(x + 5)^4 + 3$$

on the same axes. How can the graph of g be obtained by a transformation of the graph of f?

12. Use end behavior to determine which one of the following graphs is that of $f(x) = -x^7 + x - 4$.

A.

B.

C.

D.

Graph each polynomial function.

13. $f(x) = x^3 - 5x^2 + 3x + 9$

14. $f(x) = 2x^2(x - 2)^2$

15. $f(x) = -x^3 - 4x^2 + 11x + 30$

16. *Connecting Graphs with Equations* Find a cubic polynomial function f having the graph shown.

17. *(Modeling) Oil Pressure* The pressure of oil in a reservoir tends to drop with time. Engineers found that the change in pressure is modeled by

$$f(t) - 1.06t^3 - 24.6t^2 + 180t,$$

for t (in years) in the interval $[0, 15]$.

(a) What was the change after 2 yr?

(b) For what time periods, to the nearest tenth of a year, is the amount of change in pressure increasing? decreasing? Use a graph to decide.

Graph each rational function.

18. $f(x) = \dfrac{3x - 1}{x - 2}$

19. $f(x) = \dfrac{x^2 - 1}{x^2 - 9}$

20. Consider the rational function $f(x) = \dfrac{2x^2 + x - 6}{x - 1}$.

(a) Determine the equation of the oblique asymptote.

(b) Determine the x-intercepts.

(c) Determine the y-intercept.

(d) Determine the equation of the vertical asymptote.

(e) Sketch the graph.

21. If y varies directly as the square root of x, and $y = 12$ when $x = 4$, find y when $x = 100$.

22. *Weight on and above Earth* The weight w of an object varies inversely as the square of the distance d between the object and the center of Earth. If a man weighs 90 kg on the surface of Earth, how much would he weigh 800 km above the surface? (*Hint:* The radius of Earth is about 6400 km.)

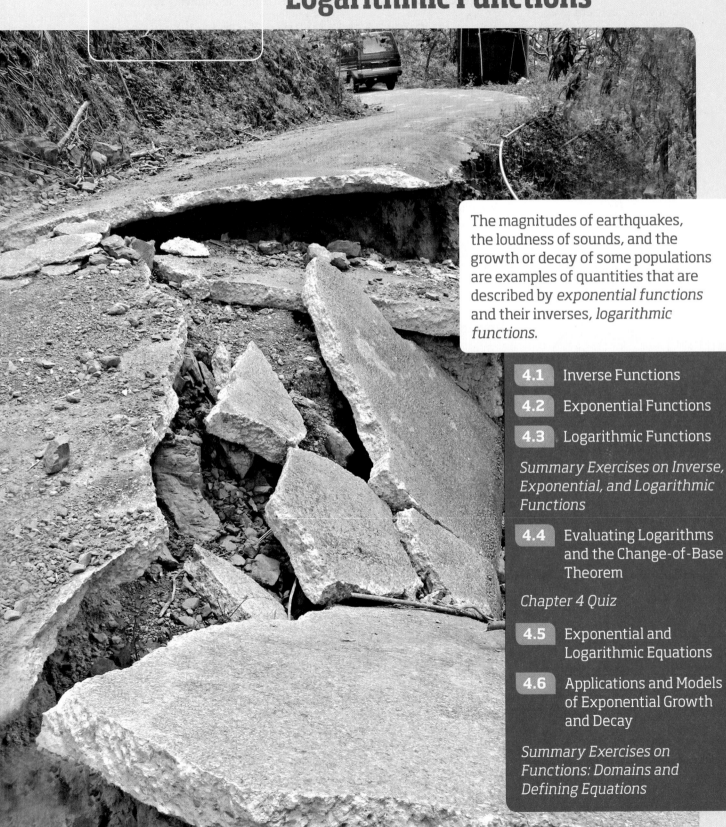

4 Inverse, Exponential, and Logarithmic Functions

The magnitudes of earthquakes, the loudness of sounds, and the growth or decay of some populations are examples of quantities that are described by *exponential functions* and their inverses, *logarithmic functions.*

4.1 Inverse Functions

- One-to-One Functions
- Inverse Functions
- Equations of Inverses
- An Application of Inverse Functions to Cryptography

One-to-One Functions   Suppose we define the following function F.

$$F = \{(-2, 2), (-1, 1), (0, 0), (1, 3), (2, 5)\}$$

(We have defined F so that each *second* component is used only once.) We can form another set of ordered pairs from F by interchanging the x- and y-values of each pair in F. We call this set G.

$$G = \{(2, -2), (1, -1), (0, 0), (3, 1), (5, 2)\}$$

G is the *inverse* of F. Function F was defined with each *second* component used only once, so set G will also be a function. (Each *first* component must be used only once.) In order for a function to have an inverse that is also a function, it must exhibit this one-to-one relationship.

> *In a one-to-one function, each x-value corresponds to only one y-value, and each y-value corresponds to only one x-value.*

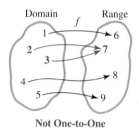

Not One-to-One

Figure 1

The function f shown in **Figure 1** is not one-to-one because the y-value 7 corresponds to *two* x-values, 2 and 3. That is, the ordered pairs $(2, 7)$ and $(3, 7)$ both belong to the function. The function f in **Figure 2** is one-to-one.

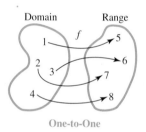

One-to-One

Figure 2

One-to-One Function

A function f is a **one-to-one function** if, for elements a and b in the domain of f,

$$a \neq b \quad \text{implies} \quad f(a) \neq f(b).$$

That is, different values of the domain correspond to different values of the range.

Using the concept of the *contrapositive* from the study of logic, the boldface statement in the preceding box is equivalent to

$$f(a) = f(b) \quad \text{implies} \quad a = b.$$

This means that if two range values are equal, then their corresponding domain values are equal. We use this statement to show that a function f is one-to-one in **Example 1(a).**

EXAMPLE 1   Deciding Whether Functions Are One-to-One

Determine whether each function is one-to-one.

(a) $f(x) = -4x + 12$ **(b)** $f(x) = \sqrt{25 - x^2}$

SOLUTION

(a) We can determine that the function $f(x) = -4x + 12$ is one-to-one by showing that $f(a) = f(b)$ leads to the result $a = b$.

$$f(a) = f(b)$$

$$-4a + 12 = -4b + 12 \qquad \text{\small $f(x) = -4x + 12$}$$

$$-4a = -4b \qquad \qquad \text{\small Subtract 12.}$$

$$a = b \qquad \qquad \text{\small Divide by -4.}$$

By the definition, $f(x) = -4x + 12$ is one-to-one.

(b) We can determine that the function $f(x) = \sqrt{25 - x^2}$ is not one-to-one by showing that *different* values of the domain correspond to the *same* value of the range. If we choose $a = 3$ and $b = -3$, then $3 \neq -3$, but

$$f(3) = \sqrt{25 - 3^2} = \sqrt{25 - 9} = \sqrt{16} = 4$$

and $\qquad f(-3) = \sqrt{25 - (-3)^2} = \sqrt{25 - 9} = 4.$

Here, even though $3 \neq -3$, $f(3) = f(-3) = 4$. By the definition, f is *not* a one-to-one function.

✔ **Now Try Exercises 17 and 19.**

As illustrated in **Example 1(b),** a way to show that a function is *not* one-to-one is to produce a pair of different domain elements that lead to the same function value. There is a useful graphical test for this, the **horizontal line test.**

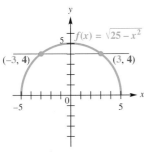

Figure 3

Horizontal Line Test

A function is one-to-one if every horizontal line intersects the graph of the function at most once.

NOTE In **Example 1(b),** the graph of the function is a semicircle, as shown in **Figure 3.** Because there is at least one horizontal line that intersects the graph in more than one point, this function is not one-to-one.

EXAMPLE 2 Using the Horizontal Line Test

Determine whether each graph is the graph of a one-to-one function.

(a)

(b)

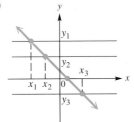

SOLUTION

(a) Each point where the horizontal line intersects the graph has the same value of y but a different value of x. Because more than one different value of x (here three) lead to the same value of y, the function is not one-to-one.

(b) Every horizontal line will intersect the graph at exactly one point, so this function is one-to-one.

✔ **Now Try Exercises 11 and 13.**

The function graphed in **Example 2(b)** decreases on its entire domain.

In general, a function that is either increasing or decreasing on its entire domain, such as $f(x) = -x$, $g(x) = x^3$, and $h(x) = \frac{1}{x}$, must be one-to-one.

> **Tests to Determine Whether a Function Is One-to-One**
>
> **1.** Show that $f(a) = f(b)$ implies $a = b$. This means that f is one-to-one. **(See Example 1(a).)**
>
> **2.** In a one-to-one function, every y-value corresponds to no more than one x-value. To show that a function is not one-to-one, find at least two x-values that produce the same y-value. **(See Example 1(b).)**
>
> **3.** Sketch the graph and use the horizontal line test. **(See Example 2.)**
>
> **4.** If the function either increases or decreases on its entire domain, then it is one-to-one. A sketch is helpful here, too. **(See Example 2(b).)**

Inverse Functions   Certain pairs of one-to-one functions "undo" each other. For example, consider the functions

$$g(x) = 8x + 5 \quad \text{and} \quad f(x) = \frac{1}{8}x - \frac{5}{8}.$$

We choose an arbitrary element from the domain of g, say 10. Evaluate $g(10)$.

$$g(x) = 8x + 5 \qquad \text{Given function}$$
$$g(10) = 8 \cdot 10 + 5 \qquad \text{Let } x = 10.$$
$$g(10) = 85 \qquad \text{Multiply and then add.}$$

Now, we evaluate $f(85)$.

$$f(x) = \frac{1}{8}x - \frac{5}{8} \qquad \text{Given function}$$

$$f(85) = \frac{1}{8}(85) - \frac{5}{8} \qquad \text{Let } x = 85.$$

$$f(85) = \frac{85}{8} - \frac{5}{8} \qquad \text{Multiply.}$$

$$f(85) = 10 \qquad \text{Subtract and then divide.}$$

Starting with 10, we "applied" function g and then "applied" function f to the result, which returned the number 10. See **Figure 4.**

These functions contain inverse operations that "undo" each other.

Figure 4

As further examples, confirm the following.

$$g(3) = 29 \quad \text{and} \quad f(29) = 3$$
$$g(-5) = -35 \quad \text{and} \quad f(-35) = -5$$
$$g(2) = 21 \quad \text{and} \quad f(21) = 2$$
$$f(2) = -\frac{3}{8} \quad \text{and} \quad g\left(-\frac{3}{8}\right) = 2$$

In particular, for the pair of functions $g(x) = 8x + 5$ and $f(x) = \frac{1}{8}x - \frac{5}{8}$,

$$f(g(2)) = 2 \quad \text{and} \quad g(f(2)) = 2.$$

In fact, for *any* value of x,

$$f(g(x)) = x \quad \text{and} \quad g(f(x)) = x.$$

Using the notation for composition of functions, these two equations can be written as follows.

$$(f \circ g)(x) = x \quad \text{and} \quad (g \circ f)(x) = x \qquad \text{The result is the identity function.}$$

Because the compositions of f and g yield the *identity* function, they are *inverses* of each other.

Inverse Function

Let f be a one-to-one function. Then g is the **inverse function** of f if

$$(\boldsymbol{f \circ g})(\boldsymbol{x}) = \boldsymbol{x} \quad \text{for every } x \text{ in the domain of } g,$$

and $\qquad (\boldsymbol{g \circ f})(\boldsymbol{x}) = \boldsymbol{x} \quad \text{for every } x \text{ in the domain of } f.$

The condition that f is one-to-one in the definition of inverse function is essential. Otherwise, g will not define a function.

EXAMPLE 3 Determining Whether Two Functions Are Inverses

Let functions f and g be defined respectively by

$$f(x) = x^3 - 1 \quad \text{and} \quad g(x) = \sqrt[3]{x + 1}.$$

Is g the inverse function of f?

SOLUTION As shown in **Figure 5**, the horizontal line test applied to the graph indicates that f is one-to-one, so the function has an inverse. Because it is one-to-one, we now find $(f \circ g)(x)$ and $(g \circ f)(x)$.

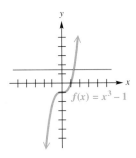

Figure 5

$$\begin{aligned}(f \circ g)(x) \\ &= f(g(x)) \\ &= \left(\sqrt[3]{x + 1}\right)^3 - 1 \\ &= x + 1 - 1 \\ &= x\end{aligned} \qquad \begin{aligned}(g \circ f)(x) \\ &= g(f(x)) \\ &= \sqrt[3]{(x^3 - 1) + 1} \\ &= \sqrt[3]{x^3} \\ &= x\end{aligned}$$

Since $(f \circ g)(x) = x$ and $(g \circ f)(x) = x$, function g is the inverse of function f.

✔ **Now Try Exercise 41.**

A special notation is used for inverse functions: If g is the inverse of a function f, then g is written as $\boldsymbol{f^{-1}}$ (read "**f-inverse**").

$$f(x) = x^3 - 1 \quad \text{has inverse} \quad f^{-1}(x) = \sqrt[3]{x + 1}. \qquad \text{See Example 3.}$$

CAUTION *Do not confuse the* -1 *in* f^{-1} *with a negative exponent.* The symbol $f^{-1}(x)$ represents the inverse function of f, *not* $\frac{1}{f(x)}$.

By the definition of inverse function, the domain of f is the range of f^{-1}, and the range of f is the domain of f^{-1}. See **Figure 6**.

Figure 6

EXAMPLE 4 **Finding Inverses of One-to-One Functions**

Find the inverse of each function that is one-to-one.

(a) $F = \{(-2, 1), (-1, 0), (0, 1), (1, 2), (2, 2)\}$

(b) $G = \{(3, 1), (0, 2), (2, 3), (4, 0)\}$

(c) The table in the margin shows the number of hurricanes recorded in the North Atlantic during the years 2009–2013. Let f be the function defined in the table, with the years forming the domain and the numbers of hurricanes forming the range.

Year	Number of Hurricanes
2009	3
2010	12
2011	7
2012	10
2013	2

Source: www.wunderground.com

SOLUTION

(a) Each x-value in F corresponds to just one y-value. However, the y-value 2 corresponds to two x-values, 1 and 2. Also, the y-value 1 corresponds to both -2 and 0. Because at least one y-value corresponds to more than one x-value, F is not one-to-one and does not have an inverse.

(b) Every x-value in G corresponds to only one y-value, and every y-value corresponds to only one x-value, so G is a one-to-one function. The inverse function is found by interchanging the x- and y-values in each ordered pair.

$$G^{-1} = \{(1, 3), (2, 0), (3, 2), (0, 4)\}$$

Notice how the domain and range of G become the range and domain, respectively, of G^{-1}.

(c) Each x-value in f corresponds to only one y-value, and each y-value corresponds to only one x-value, so f is a one-to-one function. The inverse function is found by interchanging the x- and y-values in the table.

$$f^{-1}(x) = \{(3, 2009), (12, 2010), (7, 2011), (10, 2012), (2, 2013)\}$$

The domain and range of f become the range and domain of f^{-1}.

✔ **Now Try Exercises 37, 51, and 53.**

Equations of Inverses The inverse of a one-to-one function is found by interchanging the x- and y-values of each of its ordered pairs. The equation of the inverse of a function defined by $y = f(x)$ is found in the same way.

Finding the Equation of the Inverse of $y = f(x)$

For a one-to-one function f defined by an equation $y = f(x)$, find the defining equation of the inverse as follows. (If necessary, replace $f(x)$ with y first. Any restrictions on x and y should be considered.)

Step 1 Interchange x and y.

Step 2 Solve for y.

Step 3 Replace y with $f^{-1}(x)$.

EXAMPLE 5 Finding Equations of Inverses

Determine whether each equation defines a one-to-one function. If so, find the equation of the inverse.

(a) $f(x) = 2x + 5$ **(b)** $y = x^2 + 2$ **(c)** $f(x) = (x - 2)^3$

SOLUTION

(a) The graph of $y = 2x + 5$ is a nonhorizontal line, so by the horizontal line test, f is a one-to-one function. Find the equation of the inverse as follows.

$$f(x) = 2x + 5 \quad \text{Given function}$$

$$y = 2x + 5 \quad \text{Let } y = f(x).$$

Step 1 $\quad x = 2y + 5 \quad$ Interchange x and y.

Step 2 $\quad x - 5 = 2y \quad$ Subtract 5.

$$y = \frac{x - 5}{2} \quad \begin{array}{l} \text{Divide by 2.} \\ \text{Rewrite.} \end{array} \Bigg\} \text{ Solve for } y.$$

Step 3 $\quad f^{-1}(x) = \dfrac{1}{2}x - \dfrac{5}{2} \quad \begin{array}{l} \text{Replace } y \text{ with } f^{-1}(x). \\ \frac{a-b}{c} = \left(\frac{1}{c}\right)a - \frac{b}{c} \end{array}$

Thus, the equation $f^{-1}(x) = \frac{x-5}{2} = \frac{1}{2}x - \frac{5}{2}$ represents a linear function. In the function $y = 2x + 5$, the value of y is found by starting with a value of x, multiplying by 2, and adding 5.

The equation $f^{-1}(x) = \frac{x-5}{2}$ for the inverse *subtracts* 5 and then *divides* by 2. An inverse is used to "undo" what a function does to the variable x.

(b) The equation $y = x^2 + 2$ has a parabola opening up as its graph, so some horizontal lines will intersect the graph at two points. For example, both $x = 3$ and $x = -3$ correspond to $y = 11$. Because of the presence of the x^2-term, there are many pairs of x-values that correspond to the same y-value. This means that the function defined by $y = x^2 + 2$ is not one-to-one and does not have an inverse.

Proceeding with the steps for finding the equation of an inverse leads to

$$y = x^2 + 2$$

$$x = y^2 + 2 \quad \text{Interchange } x \text{ and } y.$$

$$x - 2 = y^2 \quad \text{Solve for } y.$$

Remember *both* roots. $\quad \pm \sqrt{x - 2} = y. \quad$ Square root property

The last equation shows that there are two y-values for each choice of x greater than 2, indicating that this is not a function.

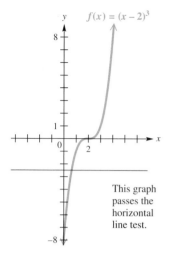

Figure 7

This graph passes the horizontal line test.

(c) **Figure 7** shows that the horizontal line test assures us that this horizontal translation of the graph of the cubing function is one-to-one.

$$f(x) = (x - 2)^3 \qquad \text{Given function}$$

$$y = (x - 2)^3 \qquad \text{Replace } f(x) \text{ with } y.$$

Step 1 $\qquad x = (y - 2)^3 \qquad \text{Interchange } x \text{ and } y.$

Step 2 $\qquad \sqrt[3]{x} = \sqrt[3]{(y - 2)^3} \qquad \text{Take the cube root on each side.}$

$$\sqrt[3]{x} = y - 2 \qquad \qquad \sqrt[3]{a^3} = a \qquad \left.\vphantom{\begin{array}{c}1\\1\\1\end{array}}\right\} \text{Solve for } y.$$

$$\sqrt[3]{x} + 2 = y \qquad \qquad \text{Add 2.}$$

Step 3 $\quad f^{-1}(x) = \sqrt[3]{x} + 2 \qquad \text{Replace } y \text{ with } f^{-1}(x). \text{ Rewrite.}$

✔ **Now Try Exercises 59(a), 63(a), and 65(a).**

EXAMPLE 6 **Finding the Equation of the Inverse of a Rational Function**

The following rational function is one-to-one. Find its inverse.

$$f(x) = \frac{2x + 3}{x - 4}, \quad x \neq 4$$

SOLUTION $\qquad f(x) = \dfrac{2x + 3}{x - 4}, \quad x \neq 4 \qquad \text{Given function}$

$$y = \frac{2x + 3}{x - 4} \qquad \qquad \text{Replace } f(x) \text{ with } y.$$

Step 1 $\qquad x = \dfrac{2y + 3}{y - 4}, \quad y \neq 4 \qquad \text{Interchange } x \text{ and } y.$

Step 2 $\qquad x(y - 4) = 2y + 3 \qquad \qquad \text{Multiply by } y - 4.$

$$xy - 4x = 2y + 3 \qquad \qquad \text{Distributive property}$$

Pay close attention here. $\quad xy - 2y = 4x + 3 \qquad \qquad \text{Add } 4x \text{ and } -2y.$

$$y(x - 2) = 4x + 3 \qquad \qquad \text{Factor out } y.$$

$$y = \frac{4x + 3}{x - 2}, \quad x \neq 2 \qquad \text{Divide by } x - 2.$$

Solve for y.

In the final line, we give the condition $x \neq 2$. (Note that 2 is not in the *range* of f, so it is not in the domain of f^{-1}.)

Step 3 $\qquad f^{-1}(x) = \dfrac{4x + 3}{x - 2}, \quad x \neq 2 \qquad \text{Replace } y \text{ with } f^{-1}(x).$

✔ **Now Try Exercise 71(a).**

One way to graph the inverse of a function f whose equation is known follows.

Step 1 Find some ordered pairs that are on the graph of f.

Step 2 Interchange x and y to find ordered pairs that are on the graph of f^{-1}.

Step 3 Plot those points, and sketch the graph of f^{-1} through them.

Another way is to select points on the graph of f and use symmetry to find corresponding points on the graph of f^{-1}.

For example, suppose the point (a, b) shown in **Figure 8** is on the graph of a one-to-one function f. Then the point (b, a) is on the graph of f^{-1}. The line segment connecting (a, b) and (b, a) is perpendicular to, and cut in half by, the line $y = x$. The points (a, b) and (b, a) are "mirror images" of each other with respect to $y = x$.

Thus, we can find the graph of f^{-1} from the graph of f by locating the mirror image of each point in f with respect to the line $y = x$.

Figure 8

EXAMPLE 7 Graphing f^{-1} Given the Graph of f

In each set of axes in **Figure 9,** the graph of a one-to-one function f is shown in blue. Graph f^{-1} in red.

SOLUTION In **Figure 9,** the graphs of two functions f shown in blue are given with their inverses shown in red. In each case, the graph of f^{-1} is a reflection of the graph of f with respect to the line $y = x$.

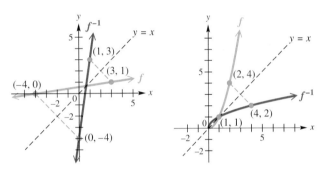

Figure 9

✔ **Now Try Exercises 77 and 81.**

EXAMPLE 8 Finding the Inverse of a Function (Restricted Domain)

Let $f(x) = \sqrt{x + 5}, \quad x \geq -5$. Find $f^{-1}(x)$.

SOLUTION The domain of f is restricted to the interval $[-5, \infty)$. Function f is one-to-one because it is an increasing function and thus has an inverse function. Now we find the equation of the inverse.

$$f(x) = \sqrt{x + 5}, \quad x \geq -5 \quad \text{Given function}$$

$$y = \sqrt{x + 5}, \quad x \geq -5 \quad \text{Replace } f(x) \text{ with } y.$$

Step 1 $\quad x = \sqrt{y + 5}, \quad y \geq -5 \quad \text{Interchange } x \text{ and } y.$

Step 2 $\quad x^2 = \left(\sqrt{y + 5}\right)^2 \qquad \text{Square each side.}$

$$x^2 = y + 5 \qquad \left(\sqrt{a}\right)^2 = a \text{ for } a \geq 0 \quad \Big\} \quad \text{Solve for } y.$$

$$y = x^2 - 5 \qquad \text{Subtract 5. Rewrite.}$$

However, we cannot define $f^{-1}(x)$ as $x^2 - 5$. The domain of f is $[-5, \infty)$, and its range is $[0, \infty)$. The range of f is the domain of f^{-1}, so f^{-1} must be defined as follows.

Step 3 $\qquad\qquad\qquad f^{-1}(x) = x^2 - 5, \quad x \geq 0$

As a check, the range of f^{-1}, $[-5, \infty)$, is the domain of f.

Graphs of f and f^{-1} are shown in **Figures 10 and 11.** The line $y = x$ is included on the graphs to show that the graphs of f and f^{-1} are mirror images with respect to this line.

Figure 10

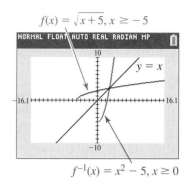

Figure 11

✔ **Now Try Exercise 75.**

Important Facts about Inverses

1. If f is one-to-one, then f^{-1} exists.

2. The domain of f is the range of f^{-1}, and the range of f is the domain of f^{-1}.

3. If the point (a, b) lies on the graph of f, then (b, a) lies on the graph of f^{-1}. The graphs of f and f^{-1} are reflections of each other across the line $y = x$.

4. To find the equation for f^{-1}, replace $f(x)$ with y, interchange x and y, and solve for y. This gives $f^{-1}(x)$.

Despite the fact that $y = x^2$ is not one-to-one, the calculator will draw its "inverse," $x = y^2$.

Figure 12

Some graphing calculators have the capability of "drawing" the reflection of a graph across the line $y = x$. This feature does not require that the function be one-to-one, however, so the resulting figure may not be the graph of a function. See **Figure 12.** *It is necessary to understand the mathematics to interpret results correctly.* ∎

An Application of Inverse Functions to Cryptography A one-to-one function and its inverse can be used to make information secure. The function is used to encode a message, and its inverse is used to decode the coded message. In practice, complicated functions are used.

EXAMPLE 9 **Using Functions to Encode and Decode a Message**

Use the one-to-one function $f(x) = 3x + 1$ and the following numerical values assigned to each letter of the alphabet to encode and decode the message BE MY FACEBOOK FRIEND.

A	1	**H**	8	**O**	15	**V**	22
B	2	**I**	9	**P**	16	**W**	23
C	3	**J**	10	**Q**	17	**X**	24
D	4	**K**	11	**R**	18	**Y**	25
E	5	**L**	12	**S**	19	**Z**	26
F	6	**M**	13	**T**	20		
G	7	**N**	14	**U**	21		

A	1	N	14
B	2	O	15
C	3	P	16
D	4	Q	17
E	5	R	18
F	6	S	19
G	7	T	20
H	8	U	21
I	9	V	22
J	10	W	23
K	11	X	24
L	12	Y	25
M	13	Z	26

SOLUTION The message BE MY FACEBOOK FRIEND would be encoded as

$$7 \quad 16 \quad 40 \quad 76 \quad 19 \quad 4 \quad 10 \quad 16 \quad 7$$
$$46 \quad 46 \quad 34 \quad 19 \quad 55 \quad 28 \quad 16 \quad 43 \quad 13$$

because

B corresponds to 2   and $f(2) = 3(2) + 1 = 7,$

E corresponds to 5   and $f(5) = 3(5) + 1 = 16,$ and so on.

Using the inverse $f^{-1}(x) = \frac{1}{3}x - \frac{1}{3}$ to decode yields

$$f^{-1}(7) = \frac{1}{3}(7) - \frac{1}{3} = 2, \quad \text{which corresponds to B,}$$

$$f^{-1}(16) = \frac{1}{3}(16) - \frac{1}{3} = 5, \quad \text{which corresponds to E,} \quad \text{and so on.}$$

✔ **Now Try Exercise 97.**

4.1 Exercises

CONCEPT PREVIEW *Determine whether the function represented in each table is one-to-one.*

1. The table shows the number of registered passenger cars in the United States for the years 2008–2012.

Year	Registered Passenger Cars (in thousands)
2008	137,080
2009	134,880
2010	139,892
2011	125,657
2012	111,290

Source: U.S. Federal Highway Administration.

2. The table gives the number of representatives currently in Congress from each of five New England states.

State	Number of Representatives
Connecticut	5
Maine	2
Massachusetts	9
New Hampshire	2
Vermont	1

Source: www.house.gov

CONCEPT PREVIEW *Fill in the blank(s) to correctly complete each sentence.*

3. For a function to have an inverse, it must be _____.

4. If two functions f and g are inverses, then $(f \circ g)(x) = $ _____ and _____ $= x$.

5. The domain of f is equal to the _____ of f^{-1}, and the range of f is equal to the _____ of f^{-1}.

6. If the point (a, b) lies on the graph of f, and f has an inverse, then the point _____ lies on the graph of f^{-1}.

7. If $f(x) = x^3$, then $f^{-1}(x) = $ _____.

8. If a function f has an inverse, then the graph of f^{-1} may be obtained by reflecting the graph of f across the line with equation _____.

9. If a function f has an inverse and $f(-3) = 6$, then $f^{-1}(6) = $ _____.

10. If $f(-4) = 16$ and $f(4) = 16$, then f _____ have an inverse because
_____. $\underset{\text{(does/does not)}}{}$

Determine whether each function graphed or defined is one-to-one. ***See Examples 1 and 2.***

11.

12.

13.

14.

15.

16.

17. $y = 2x - 8$ **18.** $y = 4x + 20$ **19.** $y = \sqrt{36 - x^2}$

20. $y = -\sqrt{100 - x^2}$ **21.** $y = 2x^3 - 1$ **22.** $y = 3x^3 - 6$

23. $y = \dfrac{-1}{x + 2}$ **24.** $y = \dfrac{4}{x - 8}$ **25.** $y = 2(x + 1)^2 - 6$

26. $y = -3(x - 6)^2 + 8$ **27.** $y = \sqrt[3]{x + 1} - 3$ **28.** $y = -\sqrt[3]{x + 2} - 8$

Concept Check *Answer each question.*

29. Can a constant function, such as $f(x) = 3$, defined over the set of real numbers, be one-to-one?

30. Can a polynomial function of even degree defined over the set of real numbers have an inverse?

Concept Check *An everyday activity is described. Keeping in mind that an inverse operation "undoes" what an operation does, describe each inverse activity.*

31. tying your shoelaces   **32.** starting a car

33. entering a room   **34.** climbing the stairs

35. screwing in a light bulb   **36.** filling a cup

Determine whether the given functions are inverses. **See Example 4.**

37.

x	f(x)
3	−4
2	−6
5	8
1	9
4	3

x	g(x)
−4	3
−6	2
8	5
9	1
3	4

38.

x	f(x)
−2	−8
−1	−1
0	0
1	1
2	8

x	g(x)
8	−2
1	−1
0	0
−1	1
−8	2

39. $f = \{(2, 5), (3, 5), (4, 5)\}$; $g = \{(5, 2)\}$

40. $f = \{(1, 1), (3, 3), (5, 5)\}$; $g = \{(1, 1), (3, 3), (5, 5)\}$

Use the definition of inverses to determine whether f and g are inverses. **See Example 3.**

41. $f(x) = 2x + 4$, $g(x) = \dfrac{1}{2}x - 2$ **42.** $f(x) = 3x + 9$, $g(x) = \dfrac{1}{3}x - 3$

43. $f(x) = -3x + 12$, $g(x) = -\dfrac{1}{3}x - 12$ **44.** $f(x) = -4x + 2$, $g(x) = -\dfrac{1}{4}x - 2$

45. $f(x) = \dfrac{x + 1}{x - 2}$, $g(x) = \dfrac{2x + 1}{x - 1}$ **46.** $f(x) = \dfrac{x - 3}{x + 4}$, $g(x) = \dfrac{4x + 3}{1 - x}$

47. $f(x) = \dfrac{2}{x + 6}$, $g(x) = \dfrac{6x + 2}{x}$ **48.** $f(x) = \dfrac{-1}{x + 1}$, $g(x) = \dfrac{1 - x}{x}$

49. $f(x) = x^2 + 3$, $x \geq 0$; $g(x) = \sqrt{x - 3}$, $x \geq 3$

50. $f(x) = \sqrt{x + 8}$, $x \geq -8$; $g(x) = x^2 - 8$, $x \geq 0$

Find the inverse of each function that is one-to-one. **See Example 4.**

51. $\{(-3, 6), (2, 1), (5, 8)\}$ **52.** $\left\{(3, -1), (5, 0), (0, 5), \left(4, \dfrac{2}{3}\right)\right\}$

53. $\{(1, -3), (2, -7), (4, -3), (5, -5)\}$ **54.** $\{(6, -8), (3, -4), (0, -8), (5, -4)\}$

Determine whether each pair of functions graphed are inverses. **See Example 7.**

55.

56.

57.

58.

*For each function that is one-to-one, (**a**) write an equation for the inverse function, (**b**) graph f and f^{-1} on the same axes, and (**c**) give the domain and range of both f and f^{-1}. If the function is not one-to-one, say so.* **See Examples 5–8.**

59. $f(x) = 3x - 4$

60. $f(x) = 4x - 5$

61. $f(x) = -4x + 3$

62. $f(x) = -6x - 8$

63. $f(x) = x^3 + 1$

64. $f(x) = -x^3 - 2$

65. $f(x) = x^2 + 8$

66. $f(x) = -x^2 + 2$

67. $f(x) = \dfrac{1}{x}, \quad x \neq 0$

68. $f(x) = \dfrac{4}{x}, \quad x \neq 0$

69. $f(x) = \dfrac{1}{x - 3}, \quad x \neq 3$

70. $f(x) = \dfrac{1}{x + 2}, \quad x \neq -2$

71. $f(x) = \dfrac{x + 1}{x - 3}, \quad x \neq 3$

72. $f(x) = \dfrac{x + 2}{x - 1}, \quad x \neq 1$

73. $f(x) = \dfrac{2x + 6}{x - 3}, \quad x \neq 3$

74. $f(x) = \dfrac{-3x + 12}{x - 6}, \quad x \neq 6$

75. $f(x) = \sqrt{x + 6}, \quad x \geq -6$

76. $f(x) = -\sqrt{x^2 - 16}, \quad x \geq 4$

Graph the inverse of each one-to-one function. **See Example 7.**

77.

78.

79.

80.

81.

82.

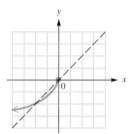

Concept Check *The graph of a function f is shown in the figure. Use the graph to find each value.*

83. $f^{-1}(4)$

84. $f^{-1}(2)$

85. $f^{-1}(0)$

86. $f^{-1}(-2)$

87. $f^{-1}(-3)$

88. $f^{-1}(-4)$

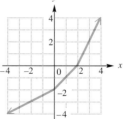

Concept Check *Answer each of the following.*

89. Suppose $f(x)$ is the number of cars that can be built for x dollars. What does $f^{-1}(1000)$ represent?

90. Suppose $f(r)$ is the volume (in cubic inches) of a sphere of radius r inches. What does $f^{-1}(5)$ represent?

91. If a line has slope a, what is the slope of its reflection across the line $y = x$?

92. For a one-to-one function f, find $(f^{-1} \circ f)(2)$, where $f(2) = 3$.

Use a graphing calculator to graph each function defined as follows, using the given viewing window. Use the graph to decide which functions are one-to-one. If a function is one-to-one, give the equation of its inverse.

93. $f(x) = 6x^3 + 11x^2 - 6$;

$[-3, 2]$ by $[-10, 10]$

94. $f(x) = x^4 - 5x^2$;

$[-3, 3]$ by $[-8, 8]$

95. $f(x) = \dfrac{x - 5}{x + 3}$, $x \neq -3$;

$[-8, 8]$ by $[-6, 8]$

96. $f(x) = \dfrac{-x}{x - 4}$, $x \neq 4$;

$[-1, 8]$ by $[-6, 6]$

Use the following alphabet coding assignment to work each problem. **See Example 9.**

A	1	**H**	8	**O**	15	**V**	22
B	2	**I**	9	**P**	16	**W**	23
C	3	**J**	10	**Q**	17	**X**	24
D	4	**K**	11	**R**	18	**Y**	25
E	5	**L**	12	**S**	19	**Z**	26
F	6	**M**	13	**T**	20		
G	7	**N**	14	**U**	21		

97. The function $f(x) = 3x - 2$ was used to encode a message as

37   25   19   61   13   34   22   1   55   1   52   52   25   64   13   10.

Find the inverse function and determine the message.

98. The function $f(x) = 2x - 9$ was used to encode a message as

−5   9   5   5   9   27   15   29   −1   21   19   31   −3   27   41.

Find the inverse function and determine the message.

99. Encode the message SEND HELP, using the one-to-one function

$$f(x) = x^3 - 1.$$

Give the inverse function that the decoder will need when the message is received.

100. Encode the message SAILOR BEWARE, using the one-to-one function

$$f(x) = (x + 1)^3.$$

Give the inverse function that the decoder will need when the message is received.

4.2   Exponential Functions

- Exponents and Properties
- Exponential Functions
- Exponential Equations
- Compound Interest
- The Number e and Continuous Compounding
- Exponential Models

Exponents and Properties   Recall the definition of $a^{m/n}$: If a is a real number, m is an integer, n is a positive integer, and $\sqrt[n]{a}$ is a real number, then

$$a^{m/n} = \left(\sqrt[n]{a} \right)^m.$$

For example, $16^{3/4} = \left(\sqrt[4]{16} \right)^3 = 2^3 = 8,$

$$27^{-1/3} = \frac{1}{27^{1/3}} = \frac{1}{\sqrt[3]{27}} = \frac{1}{3}, \quad \text{and} \quad 64^{-1/2} = \frac{1}{64^{1/2}} = \frac{1}{\sqrt{64}} = \frac{1}{8}.$$

In this section, we extend the definition of a^r to include all *real* (not just rational) values of the exponent r. Consider the graphs of $y = 2^x$ for different domains in **Figure 13.**

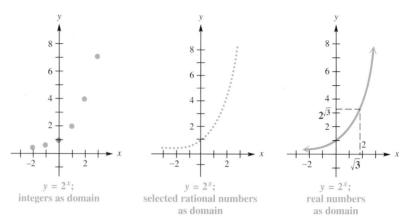

$y = 2^x$;
integers as domain

$y = 2^x$;
selected rational numbers
as domain

$y = 2^x$;
real numbers
as domain

Figure 13

The equations that use just integers or selected rational numbers as domain in **Figure 13** leave holes in the graphs. In order for the graph to be continuous, we must extend the domain to include irrational numbers such as $\sqrt{3}$. We might evaluate $2^{\sqrt{3}}$ by approximating the exponent with the rational numbers 1.7, 1.73, 1.732, and so on. Because these values approach the value of $\sqrt{3}$ more and more closely, it is reasonable that $2^{\sqrt{3}}$ should be approximated more and more closely by the numbers $2^{1.7}$, $2^{1.73}$, $2^{1.732}$, and so on. These expressions can be evaluated using rational exponents as follows.

$$2^{1.7} = 2^{17/10} = \left(\sqrt[10]{2}\right)^{17} \approx 3.249009585$$

Because any irrational number may be approximated more and more closely using rational numbers, we can extend the definition of a^r to include all real number exponents and apply all previous theorems for exponents. In addition to the rules for exponents presented earlier, we use several new properties in this chapter.

Additional Properties of Exponents

For any real number $a > 0$, $a \neq 1$, the following statements hold.

Property	Description
(a) a^x is a unique real number for all real numbers x.	$y = a^x$ can be considered a function $f(x) = a^x$ with domain $(-\infty, \infty)$.
(b) $a^b = a^c$ if and only if $b = c$.	The function $f(x) = a^x$ is one-to-one.
(c) If $a > 1$ and $m < n$, then $a^m < a^n$.	*Example:* $2^3 < 2^4$ $(a > 1)$ Increasing the exponent leads to a *greater* number. The function $f(x) = 2^x$ is an *increasing* function.
(d) If $0 < a < 1$ and $m < n$, then $a^m > a^n$.	*Example:* $\left(\frac{1}{2}\right)^2 > \left(\frac{1}{2}\right)^3$ $(0 < a < 1)$ Increasing the exponent leads to a *lesser* number. The function $f(x) = \left(\frac{1}{2}\right)^x$ is a *decreasing* function.

Exponential Functions We now define a function $f(x) = a^x$ whose domain is the set of all real numbers. Notice how the independent variable x appears in the exponent in this function. In earlier chapters, this was not the case.

Exponential Function

If $a > 0$ and $a \neq 1$, then the **exponential function with base a** is

$$f(x) = a^x.$$

NOTE The restrictions on a in the definition of an exponential function are important. Consider the outcome of breaking each restriction.

If $a < 0$, say $a = -2$, and we let $x = \frac{1}{2}$, then $f\left(\frac{1}{2}\right) = (-2)^{1/2} = \sqrt{-2}$, which is *not* a real number.

If $a = 1$, then the function becomes the constant function $f(x) = 1^x = 1$, which is *not* an exponential function.

EXAMPLE 1 Evaluating an Exponential Function

For $f(x) = 2^x$, find each of the following.

(a) $f(-1)$ **(b)** $f(3)$ **(c)** $f\left(\dfrac{5}{2}\right)$ **(d)** $f(4.92)$

SOLUTION

(a) $f(-1) = 2^{-1} = \dfrac{1}{2}$ Replace x with -1.   **(b)** $f(3) = 2^3 = 8$

(c) $f\left(\dfrac{5}{2}\right) = 2^{5/2} = (2^5)^{1/2} = 32^{1/2} = \sqrt{32} = \sqrt{16 \cdot 2} = 4\sqrt{2}$

(d) $f(4.92) = 2^{4.92} \approx 30.2738447$ Use a calculator.

✔ **Now Try Exercises 13, 19, and 23.**

We repeat the final graph of $y = 2^x$ (with real numbers as domain) from **Figure 13** and summarize important details of the function $f(x) = 2^x$ here.

- The y-intercept is $(0, 1)$.

- Because $2^x > 0$ for all x and $2^x \to 0$ as $x \to -\infty$, the x-axis is a horizontal asymptote.

- As the graph suggests, the domain of the function is $(-\infty, \infty)$ and the range is $(0, \infty)$.

- The function is increasing on its entire domain. Therefore, it is one-to-one.

These observations lead to the following generalizations about the graphs of exponential functions.

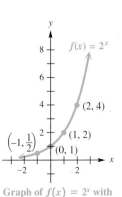

Graph of $f(x) = 2^x$ with domain $(-\infty, \infty)$

Figure 13 (repeated)

Exponential Function $f(x) = a^x$

Domain: $(-\infty, \infty)$ Range: $(0, \infty)$

For $f(x) = 2^x$:

x	$f(x)$
-2	$\frac{1}{4}$
-1	$\frac{1}{2}$
0	1
1	2
2	4
3	8

$f(x) = a^x, a > 1$

This is the general behavior seen on a calculator graph for **any base a, for $a > 1$.**

Figure 14

- $f(x) = a^x$, **for $a > 1$,** is increasing and continuous on its entire domain, $(-\infty, \infty)$.

- The x-axis is a horizontal asymptote as $x \to -\infty$.

- The graph passes through the points $\left(-1, \frac{1}{a}\right)$, $(0, 1)$, and $(1, a)$.

For $f(x) = \left(\frac{1}{2}\right)^x$:

x	$f(x)$
-3	8
-2	4
-1	2
0	1
1	$\frac{1}{2}$
2	$\frac{1}{4}$

$f(x) = a^x, 0 < a < 1$

This is the general behavior seen on a calculator graph for **any base a, for $0 < a < 1$.**

Figure 15

- $f(x) = a^x$, **for $0 < a < 1$,** is decreasing and continuous on its entire domain, $(-\infty, \infty)$.

- The x-axis is a horizontal asymptote as $x \to \infty$.

- The graph passes through the points $\left(-1, \frac{1}{a}\right)$, $(0, 1)$, and $(1, a)$.

Recall that the graph of $y = f(-x)$ is the graph of $y = f(x)$ reflected across the y-axis. Thus, we have the following.

$$\text{If } f(x) = 2^x, \quad \text{then} \quad f(-x) = 2^{-x} = 2^{-1 \cdot x} = \left(2^{-1}\right)^x = \left(\frac{1}{2}\right)^x.$$

This is supported by the graphs in **Figures 14 and 15.**

The graph of $f(x) = 2^x$ is typical of graphs of $f(x) = a^x$ where $a > 1$. For larger values of a, the graphs rise more steeply, but the general shape is similar to the graph in **Figure 14.** When $0 < a < 1$, the graph decreases in a manner similar to the graph of $f(x) = \left(\frac{1}{2}\right)^x$ in **Figure 15.**

In **Figure 16,** the graphs of several typical exponential functions illustrate these facts.

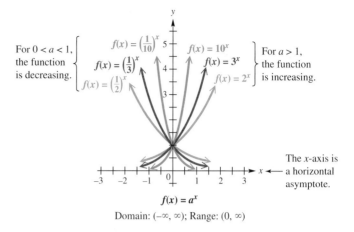

For $0 < a < 1$, the function is decreasing. $\left\{ \begin{array}{l} f(x) = \left(\frac{1}{10}\right)^x \\ f(x) = \left(\frac{1}{3}\right)^x \\ f(x) = \left(\frac{1}{2}\right)^x \end{array} \right.$

$f(x) = 10^x$
$f(x) = 3^x$
$f(x) = 2^x$ $\left. \begin{array}{l} \\ \\ \\ \end{array} \right\}$ For $a > 1$, the function is increasing.

The x-axis is a horizontal asymptote.

$f(x) = a^x$

Domain: $(-\infty, \infty)$; Range: $(0, \infty)$

Figure 16

In summary, the graph of a function of the form $f(x) = a^x$ has the following features.

Characteristics of the Graph of $f(x) = a^x$

1. The points $\left(-1, \frac{1}{a}\right)$, $(0, 1)$, and $(1, a)$ are on the graph.
2. If $a > 1$, then f is an increasing function.
   If $0 < a < 1$, then f is a decreasing function.
3. The x-axis is a horizontal asymptote.
4. The domain is $(-\infty, \infty)$, and the range is $(0, \infty)$.

EXAMPLE 2 **Graphing an Exponential Function**

Graph $f(x) = \left(\frac{1}{5}\right)^x$. Give the domain and range.

SOLUTION The y-intercept is $(0, 1)$, and the x-axis is a horizontal asymptote. Plot a few ordered pairs, and draw a smooth curve through them as shown in **Figure 17.**

x	$f(x)$
-2	25
-1	5
0	1
1	$\frac{1}{5}$
2	$\frac{1}{25}$

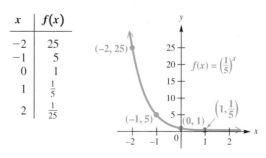

$f(x) = \left(\frac{1}{5}\right)^x$

Figure 17

This function has domain $(-\infty, \infty)$, range $(0, \infty)$, and is one-to-one. It is decreasing on its entire domain.

✔ **Now Try Exercise 29.**

EXAMPLE 3 **Graphing Reflections and Translations**

Graph each function. Show the graph of $y = 2^x$ for comparison. Give the domain and range.

(a) $f(x) = -2^x$ **(b)** $f(x) = 2^{x+3}$ **(c)** $f(x) = 2^{x-2} - 1$

SOLUTION In each graph, we show in particular how the point $(0, 1)$ on the graph of $y = 2^x$ has been translated.

(a) The graph of $f(x) = -2^x$ is that of $f(x) = 2^x$ reflected across the x-axis. See **Figure 18.** The domain is $(-\infty, \infty)$, and the range is $(-\infty, 0)$.

(b) The graph of $f(x) = 2^{x+3}$ is the graph of $f(x) = 2^x$ translated 3 units to the left, as shown in **Figure 19.** The domain is $(-\infty, \infty)$, and the range is $(0, \infty)$.

(c) The graph of $f(x) = 2^{x-2} - 1$ is that of $f(x) = 2^x$ translated 2 units to the right and 1 unit down. See **Figure 20.** The domain is $(-\infty, \infty)$, and the range is $(-1, \infty)$.

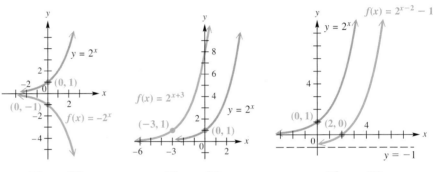

Figure 18   Figure 19   Figure 20

✔ **Now Try Exercises 39, 41, and 47.**

Exponential Equations Because the graph of $f(x) = a^x$ is that of a one-to-one function, to solve $a^{x_1} = a^{x_2}$, we need only show that $x_1 = x_2$. This property is used to solve an **exponential equation,** which is an equation with a variable as exponent.

EXAMPLE 4 **Solving an Exponential Equation**

Solve $\left(\frac{1}{3}\right)^x = 81$.

SOLUTION *Write each side of the equation using a common base.*

$$\left(\frac{1}{3}\right)^x = 81$$

$$(3^{-1})^x = 81 \qquad \text{Definition of negative exponent}$$

$$3^{-x} = 81 \qquad (a^m)^n = a^{mn}$$

$$3^{-x} = 3^4 \qquad \text{Write 81 as a power of 3.}$$

$$-x = 4 \qquad \text{Set exponents equal (Property (b) given earlier).}$$

$$x = -4 \qquad \text{Multiply by } -1.$$

Check by substituting -4 for x in the original equation. The solution set is $\{-4\}$.

✔ **Now Try Exercise 73.**

The x-intercept of the graph of $y = \left(\frac{1}{3}\right)^x - 81$ can be used to verify the solution in **Example 4.**

> **EXAMPLE 5** Solving an Exponential Equation

Solve $2^{x+4} = 8^{x-6}$.

SOLUTION Write each side of the equation using a common base.

$$2^{x+4} = 8^{x-6}$$

$$2^{x+4} = \left(2^3\right)^{x-6} \qquad \text{Write 8 as a power of 2.}$$

$$2^{x+4} = 2^{3x-18} \qquad (a^m)^n = a^{mn}$$

$$x + 4 = 3x - 18 \qquad \text{Set exponents equal (Property (b)).}$$

$$-2x = -22 \qquad \text{Subtract } 3x \text{ and } 4.$$

$$x = 11 \qquad \text{Divide by } -2.$$

Check by substituting 11 for x in the original equation. The solution set is $\{11\}$.

✔ **Now Try Exercise 81.**

Later in this chapter, we describe a general method for solving exponential equations where the approach used in **Examples 4 and 5** is not possible. For instance, the above method could not be used to solve an equation like

$$7^x = 12$$

because it is not easy to express both sides as exponential expressions with the same base.

In **Example 6,** we solve an equation that has the variable as the base of an exponential expression.

> **EXAMPLE 6** Solving an Equation with a Fractional Exponent

Solve $x^{4/3} = 81$.

SOLUTION Notice that the variable is in the base rather than in the exponent.

$$x^{4/3} = 81$$

$$\left(\sqrt[3]{x}\right)^4 = 81 \qquad \text{Radical notation for } a^{m/n}$$

$$\sqrt[3]{x} = \pm 3 \qquad \begin{array}{l}\text{Take fourth roots on each side.}\\ \text{Remember to use } \pm.\end{array}$$

$$x = \pm 27 \qquad \text{Cube each side.}$$

Check *both* solutions in the original equation. Both check, so the solution set is $\{\pm 27\}$.

Alternative Method There may be more than one way to solve an exponential equation, as shown here.

$$x^{4/3} = 81$$

$$(x^{4/3})^3 = 81^3 \qquad \text{Cube each side.}$$

$$x^4 = \left(3^4\right)^3 \qquad \text{Write 81 as } 3^4.$$

$$x^4 = 3^{12} \qquad (a^m)^n = a^{mn}$$

$$x = \pm \sqrt[4]{3^{12}} \qquad \text{Take fourth roots on each side.}$$

$$x = \pm 3^3 \qquad \text{Simplify the radical.}$$

$$x = \pm 27 \qquad \text{Apply the exponent.}$$

The same solution set, $\{\pm 27\}$, results. ✔ **Now Try Exercise 83.**

Compound Interest   Recall the formula for simple interest, $I = Prt$, where P is principal (amount deposited), r is annual rate of interest expressed as a decimal, and t is time in years that the principal earns interest. Suppose $t = 1$ yr. Then at the end of the year, the amount has grown to the following.

$$P + Pr = P(1 + r) \qquad \text{Original principal plus interest}$$

If this balance earns interest at the same interest rate for another year, the balance at the end of *that* year will increase as follows.

$$[P(1 + r)] + [P(1 + r)]r = [P(1 + r)](1 + r) \qquad \text{Factor.}$$
$$= P(1 + r)^2 \qquad a \cdot a = a^2$$

After the third year, the balance will grow in a similar pattern.

$$[P(1 + r)^2] + [P(1 + r)^2]r = [P(1 + r)^2](1 + r) \qquad \text{Factor.}$$
$$= P(1 + r)^3 \qquad a^2 \cdot a = a^3$$

Continuing in this way produces a formula for interest compounded annually.

$$A = P(1 + r)^t$$

The general formula for compound interest can be derived in the same way.

Compound Interest

If P dollars are deposited in an account paying an annual rate of interest r compounded (paid) n times per year, then after t years the account will contain A dollars, according to the following formula.

$$A = P\left(1 + \frac{r}{n}\right)^{tn}$$

EXAMPLE 7   **Using the Compound Interest Formula**

Suppose \$1000 is deposited in an account paying 4% interest per year compounded quarterly (four times per year).

(a) Find the amount in the account after 10 yr with no withdrawals.

(b) How much interest is earned over the 10-yr period?

SOLUTION

(a)
$$A = P\left(1 + \frac{r}{n}\right)^{tn} \qquad \text{Compound interest formula}$$

$$A = 1000\left(1 + \frac{0.04}{4}\right)^{10(4)} \qquad \text{Let } P = 1000, r = 0.04, n = 4, \text{ and } t = 10.$$

$$A = 1000(1 + 0.01)^{40} \qquad \text{Simplify.}$$

$$A = 1488.86 \qquad \text{Round to the nearest cent.}$$

Thus, \$1488.86 is in the account after 10 yr.

(b) The interest earned for that period is

$$\$1488.86 - \$1000 = \$488.86.$$

✔ Now Try Exercise 97(a).

In the formula for compound interest

$$A = P\left(1 + \frac{r}{n}\right)^{tn},$$

A is sometimes called the **future value** and P the **present value.** A is also called the **compound amount** and is the balance *after* interest has been earned.

EXAMPLE 8 Finding Present Value

Becky must pay a lump sum of $6000 in 5 yr.

(a) What amount deposited today (present value) at 3.1% compounded annually will grow to $6000 in 5 yr?

(b) If only $5000 is available to deposit now, what annual interest rate is necessary for the money to increase to $6000 in 5 yr?

SOLUTION

(a)

$$A = P\left(1 + \frac{r}{n}\right)^{tn} \qquad \text{Compound interest formula}$$

$$6000 = P\left(1 + \frac{0.031}{1}\right)^{5(1)} \qquad \text{Let } A = 6000, r = 0.031, n = 1, \text{ and } t = 5.$$

$$6000 = P(1.031)^5 \qquad \text{Simplify.}$$

$$P = \frac{6000}{(1.031)^5} \qquad \text{Divide by } (1.031)^5 \text{ to solve for } P.$$

$$P \approx 5150.60 \qquad \text{Use a calculator.}$$

If Becky leaves $5150.60 for 5 yr in an account paying 3.1% compounded annually, she will have $6000 when she needs it. Thus, $5150.60 is the present value of $6000 if interest of 3.1% is compounded annually for 5 yr.

(b)

$$A = P\left(1 + \frac{r}{n}\right)^{tn} \qquad \text{Compound interest formula}$$

$$6000 = 5000(1 + r)^5 \qquad \text{Let } A = 6000, P = 5000, n = 1, \text{ and } t = 5.$$

$$\frac{6}{5} = (1 + r)^5 \qquad \text{Divide by 5000.}$$

$$\left(\frac{6}{5}\right)^{1/5} = 1 + r \qquad \text{Take the fifth root on each side.}$$

$$\left(\frac{6}{5}\right)^{1/5} - 1 = r \qquad \text{Subtract 1.}$$

$$r \approx 0.0371 \qquad \text{Use a calculator.}$$

An interest rate of 3.71% will produce enough interest to increase the $5000 to $6000 by the end of 5 yr.

✔ **Now Try Exercises 99 and 103.**

CAUTION   When performing the computations in problems like those in **Examples 7 and 8,** do not round off during intermediate steps. Keep all calculator digits and round at the end of the process.

n	$\left(1 + \dfrac{1}{n}\right)^n$ (rounded)
1	2
2	2.25
5	2.48832
10	2.59374
100	2.70481
1000	2.71692
10,000	2.71815
1,000,000	2.71828

The Number e and Continuous Compounding The more often interest is compounded within a given time period, the more interest will be earned. Surprisingly, however, there is a limit on the amount of interest, no matter how often it is compounded.

Suppose that \$1 is invested at 100% interest per year, compounded n times per year. Then the interest rate (in decimal form) is 1.00, and the interest rate per period is $\frac{1}{n}$. According to the formula (with $P = 1$), the compound amount at the end of 1 yr will be

$$A = \left(1 + \frac{1}{n}\right)^n.$$

A calculator gives the results in the margin for various values of n. The table suggests that as n increases, the value of $\left(1 + \frac{1}{n}\right)^n$ gets closer and closer to some fixed number. This is indeed the case. This fixed number is called e. **(In mathematics, e is a real number and not a variable.)**

Value of e

$$e \approx 2.718281828459045$$

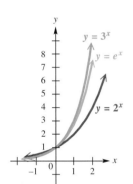

Figure 21

Figure 21 shows graphs of the functions

$$y = 2^x, \quad y = 3^x, \quad \text{and} \quad y = e^x.$$

Because $2 < e < 3$, the graph of $y = e^x$ lies "between" the other two graphs.

As mentioned above, the amount of interest earned increases with the frequency of compounding, but the value of the expression $\left(1 + \frac{1}{n}\right)^n$ approaches e as n gets larger. Consequently, the formula for compound interest approaches a limit as well, called the compound amount from **continuous compounding.**

Continuous Compounding

If P dollars are deposited at a rate of interest r compounded continuously for t years, then the compound amount A in dollars on deposit is given by the following formula.

$$A = Pe^{rt}$$

EXAMPLE 9 Solving a Continuous Compounding Problem

Suppose \$5000 is deposited in an account paying 3% interest compounded continuously for 5 yr. Find the total amount on deposit at the end of 5 yr.

SOLUTION

$A = Pe^{rt}$	Continuous compounding formula
$A = 5000e^{0.03(5)}$	Let $P = 5000$, $r = 0.03$, and $t = 5$.
$A = 5000e^{0.15}$	Multiply exponents.
$A \approx 5809.17$ or \$5809.17	Use a calculator.

Check that daily compounding would have produced a compound amount about \$0.03 less.

✔ **Now Try Exercise 97(b).**

> **EXAMPLE 10** **Comparing Interest Earned as Compounding Is More Frequent**

In **Example 7,** we found that $1000 invested at 4% compounded quarterly for 10 yr grew to $1488.86. Compare this same investment compounded annually, semiannually, monthly, daily, and continuously.

SOLUTION Substitute 0.04 for r, 10 for t, and the appropriate number of compounding periods for n into the formulas

$$A = P\left(1 + \frac{r}{n}\right)^{tn} \quad \text{Compound interest formula}$$

and

$$A = Pe^{rt}. \quad \text{Continuous compounding formula}$$

The results for amounts of $1 and $1000 are given in the table.

Compounded	$1	$1000
Annually	$(1 + 0.04)^{10} \approx 1.48024$	$1480.24
Semiannually	$\left(1 + \dfrac{0.04}{2}\right)^{10(2)} \approx 1.48595$	$1485.95
Quarterly	$\left(1 + \dfrac{0.04}{4}\right)^{10(4)} \approx 1.48886$	$1488.86
Monthly	$\left(1 + \dfrac{0.04}{12}\right)^{10(12)} \approx 1.49083$	$1490.83
Daily	$\left(1 + \dfrac{0.04}{365}\right)^{10(365)} \approx 1.49179$	$1491.79
Continuously	$e^{10(0.04)} \approx 1.49182$	$1491.82

Comparing the results for a $1000 investment, we notice the following.

- Compounding semiannually rather than annually increases the value of the account after 10 yr by $5.71.

- Quarterly compounding grows to $2.91 more than semiannual compounding after 10 yr.

- Daily compounding yields only $0.96 more than monthly compounding.

- Continuous compounding yields only $0.03 more than daily compounding.

Each increase in compounding frequency earns less additional interest.

✔ **Now Try Exercise 105.**

LOOKING AHEAD TO CALCULUS

In calculus, the derivative allows us to determine the slope of a tangent line to the graph of a function. For the function

$$f(x) = e^x,$$

the derivative is the function f itself:

$$f'(x) = e^x.$$

Therefore, in calculus the exponential function with base e is much easier to work with than exponential functions having other bases.

Exponential Models The number e is important as the base of an exponential function in many practical applications. In situations involving growth or decay of a quantity, the amount or number present at time t often can be closely modeled by a function of the form

$$y = y_0 e^{kt},$$

where y_0 is the amount or number present at time $t = 0$ and k is a constant.

Exponential functions are used to model the growth of microorganisms in a culture, the growth of certain populations, and the decay of radioactive material.

⊞ **EXAMPLE 11** **Using Data to Model Exponential Growth**

Data from recent years indicate that future amounts of carbon dioxide in the atmosphere may grow according to the table. Amounts are given in parts per million.

(a) Make a scatter diagram of the data. Do the carbon dioxide levels appear to grow exponentially?

(b) One model for the data is the function

$$y = 0.001942e^{0.00609x},$$

where x is the year and $1990 \leq x \leq 2275$. Use a graph of this model to estimate when future levels of carbon dioxide will double and triple over the preindustrial level of 280 ppm.

Year	Carbon Dioxide (ppm)
1990	353
2000	375
2075	590
2175	1090
2275	2000

Source: International Panel on Climate Change (IPCC).

SOLUTION

(a) We show a calculator graph for the data in **Figure 22(a).** The data appear to resemble the graph of an increasing exponential function.

(a)

(b)

Figure 22

(b) A graph of $y = 0.001942e^{0.00609x}$ in **Figure 22(b)** shows that it is very close to the data points. We graph $y_2 = 2 \cdot 280 = 560$ in **Figure 23(a)** and $y_2 = 3 \cdot 280 = 840$ in **Figure 23(b)** on the same coordinate axes as the given function, and we use the calculator to find the intersection points.

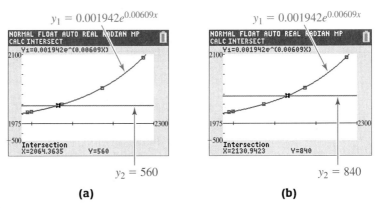

$y_1 = 0.001942e^{0.00609x}$

$y_2 = 560$

(a)

$y_1 = 0.001942e^{0.00609x}$

$y_2 = 840$

(b)

Figure 23

The graph of the function intersects the horizontal lines at x-values of approximately 2064.4 and 2130.9. According to this model, carbon dioxide levels will have doubled during 2064 and tripled by 2131.

✔ **Now Try Exercise 107.**

(a)

$y_1 = 0.001923(1.006109)^x$

(b)

Figure 24

⊞ Graphing calculators are capable of fitting exponential curves to scatter diagrams like the one found in **Example 11.** The TI-84 Plus displays another (different) equation in **Figure 24(a)** for the atmospheric carbon dioxide example, approximated as follows.

$$y = 0.001923(1.006109)^x$$

This calculator form differs from the model in **Example 11. Figure 24(b)** shows the data points and the graph of this exponential regression equation. ∎

4.2 Exercises

CONCEPT PREVIEW *Fill in the blank(s) to correctly complete each sentence.*

1. If $f(x) = 4^x$, then $f(2) =$ _____ and $f(-2) =$ _____.

2. If $a > 1$, then the graph of $f(x) = a^x$ _____ from left to right.
 (rises/falls)

3. If $0 < a < 1$, then the graph of $f(x) = a^x$ _____ from left to right.
 (rises/falls)

4. The domain of $f(x) = 4^x$ is _____ and the range is _____ .

5. The graph of $f(x) = 8^x$ passes through the points $(-1, \text{\_\_})$, $(0, \text{\_\_})$, and $(1, \text{\_\_})$.

6. The graph of $f(x) = -\left(\frac{1}{3}\right)^{x+4} - 5$ is that of $f(x) = \left(\frac{1}{3}\right)^x$ reflected across the
_____-axis, translated _____ units to the left and _____ units down.

CONCEPT PREVIEW *Solve each equation. Round answers to the nearest hundredth as needed.*

7. $\left(\dfrac{1}{4}\right)^x = 64$

8. $x^{2/3} = 36$

9. $A = 2000\left(1 + \dfrac{0.03}{4}\right)^{8(4)}$

10. $10{,}000 = 5000(1 + r)^{25}$

For $f(x) = 3^x$ and $g(x) = \left(\frac{1}{4}\right)^x$, find each of the following. Round answers to the nearest thousandth as needed. **See Example 1.**

11. $f(2)$	**12.** $f(3)$	**13.** $f(-2)$	**14.** $f(-3)$
15. $g(2)$	**16.** $g(3)$	**17.** $g(-2)$	**18.** $g(-3)$
19. $f\left(\dfrac{3}{2}\right)$	**20.** $f\left(-\dfrac{5}{2}\right)$	**21.** $g\left(\dfrac{3}{2}\right)$	**22.** $g\left(-\dfrac{5}{2}\right)$
23. $f(2.34)$	**24.** $f(-1.68)$	**25.** $g(-1.68)$	**26.** $g(2.34)$

Graph each function. **See Example 2.**

27. $f(x) = 3^x$

28. $f(x) = 4^x$

29. $f(x) = \left(\dfrac{1}{3}\right)^x$

30. $f(x) = \left(\dfrac{1}{4}\right)^x$

31. $f(x) = \left(\dfrac{3}{2}\right)^x$

32. $f(x) = \left(\dfrac{5}{3}\right)^x$

33. $f(x) = \left(\dfrac{1}{10}\right)^{-x}$

34. $f(x) = \left(\dfrac{1}{6}\right)^{-x}$

35. $f(x) = 4^{-x}$

36. $f(x) = 10^{-x}$

37. $f(x) = 2^{|x|}$

38. $f(x) = 2^{-|x|}$

Graph each function. Give the domain and range. **See Example 3.**

39. $f(x) = 2^x + 1$

40. $f(x) = 2^x - 4$

41. $f(x) = 2^{x+1}$

42. $f(x) = 2^{x-4}$

43. $f(x) = -2^{x+2}$

44. $f(x) = -2^{x-3}$

45. $f(x) = 2^{-x}$

46. $f(x) = -2^{-x}$

47. $f(x) = 2^{x-1} + 2$

48. $f(x) = 2^{x+3} + 1$

49. $f(x) = 2^{x+2} - 4$

50. $f(x) = 2^{x-3} - 1$

*Graph each function. Give the domain and range. **See Example 3.***

51. $f(x) = \left(\dfrac{1}{3}\right)^x - 2$

52. $f(x) = \left(\dfrac{1}{3}\right)^x + 4$

53. $f(x) = \left(\dfrac{1}{3}\right)^{x+2}$

54. $f(x) = \left(\dfrac{1}{3}\right)^{x-4}$

55. $f(x) = \left(\dfrac{1}{3}\right)^{-x+1}$

56. $f(x) = \left(\dfrac{1}{3}\right)^{-x-2}$

57. $f(x) = \left(\dfrac{1}{3}\right)^{-x}$

58. $f(x) = -\left(\dfrac{1}{3}\right)^{-x}$

59. $f(x) = \left(\dfrac{1}{3}\right)^{x-2} + 2$

60. $f(x) = \left(\dfrac{1}{3}\right)^{x-1} + 3$

61. $f(x) = \left(\dfrac{1}{3}\right)^{x+2} - 1$

62. $f(x) = \left(\dfrac{1}{3}\right)^{x+3} - 2$

Connecting Graphs with Equations *Write an equation for the graph given. Each represents an exponential function f with base 2 or 3, translated and/or reflected.*

63.

64.

65.

66.

67.

68.

69.

70.

*Solve each equation. **See Examples 4–6.***

71. $4^x = 2$

72. $125^x = 5$

73. $\left(\dfrac{5}{2}\right)^x = \dfrac{4}{25}$

74. $\left(\dfrac{2}{3}\right)^x = \dfrac{9}{4}$

75. $2^{3-2x} = 8$

76. $5^{2+2x} = 25$

77. $e^{4x-1} = (e^2)^x$

78. $e^{3-x} = (e^3)^{-x}$

79. $27^{4x} = 9^{x+1}$

80. $32^{2x} = 16^{x-1}$

81. $4^{x-2} = 2^{3x+3}$

82. $2^{6-3x} = 8^{x+1}$

83. $x^{2/3} = 4$

84. $x^{2/5} = 16$

85. $x^{5/2} = 32$

86. $x^{3/2} = 27$

87. $x^{-6} = \dfrac{1}{64}$

88. $x^{-4} = \dfrac{1}{256}$

89. $x^{5/3} = -243$

90. $x^{7/5} = -128$

91. $\left(\dfrac{1}{e}\right)^{-x} = \left(\dfrac{1}{e^2}\right)^{x+1}$

92. $e^{x-1} = \left(\dfrac{1}{e^4}\right)^{x+1}$

93. $\left(\sqrt{2}\right)^{x+4} = 4^x$

94. $\left(\sqrt[3]{5}\right)^{-x} = \left(\dfrac{1}{5}\right)^{x+2}$

95. $\dfrac{1}{27} = x^{-3}$

96. $\dfrac{1}{32} = x^{-5}$

Solve each problem. ***See Examples 7–9.***

97. *Future Value* Find the future value and interest earned if $8906.54 is invested for 9 yr at 3% compounded

    (a) semiannually   **(b)** continuously.

98. *Future Value* Find the future value and interest earned if $56,780 is invested at 2.8% compounded

    (a) quarterly for 23 quarters   **(b)** continuously for 15 yr.

99. *Present Value* Find the present value that will grow to $25,000 if interest is 3.2% compounded quarterly for 11 quarters.

100. *Present Value* Find the present value that will grow to $45,000 if interest is 3.6% compounded monthly for 1 yr.

101. *Present Value* Find the present value that will grow to $5000 if interest is 3.5% compounded quarterly for 10 yr.

102. *Interest Rate* Find the required annual interest rate to the nearest tenth of a percent for $65,000 to grow to $65,783.91 if interest is compounded monthly for 6 months.

103. *Interest Rate* Find the required annual interest rate to the nearest tenth of a percent for $1200 to grow to $1500 if interest is compounded quarterly for 9 yr.

104. *Interest Rate* Find the required annual interest rate to the nearest tenth of a percent for $5000 to grow to $6200 if interest is compounded quarterly for 8 yr.

Solve each problem. ***See Example 10.***

105. *Comparing Loans* Bank A is lending money at 6.4% interest compounded annually. The rate at Bank B is 6.3% compounded monthly, and the rate at Bank C is 6.35% compounded quarterly. At which bank will we pay the *least* interest?

106. *Future Value* Suppose $10,000 is invested at an annual rate of 2.4% for 10 yr. Find the future value if interest is compounded as follows.

    (a) annually   **(b)** quarterly   **(c)** monthly   **(d)** daily (365 days)

(Modeling) *Solve each problem.* ***See Example 11.***

107. *Atmospheric Pressure* The atmospheric pressure (in millibars) at a given altitude (in meters) is shown in the table.

Altitude	Pressure	Altitude	Pressure
0	1013	6000	472
1000	899	7000	411
2000	795	8000	357
3000	701	9000	308
4000	617	10,000	265
5000	541		

Source: Miller, A. and J. Thompson, *Elements of Meteorology,* Fourth Edition, Charles E. Merrill Publishing Company, Columbus, Ohio.

    (a) Use a graphing calculator to make a scatter diagram of the data for atmospheric pressure P at altitude x.

    (b) Would a linear or an exponential function fit the data better?

    (c) The following function approximates the data.

$$P(x) = 1013e^{-0.0001341x}$$

    Use a graphing calculator to graph P and the data on the same coordinate axes.

    (d) Use P to predict the pressures at 1500 m and 11,000 m, and compare them to the actual values of 846 millibars and 227 millibars, respectively.

108. *World Population Growth* World population in millions closely fits the exponential function

$$f(x) = 6084e^{0.0120x},$$

where x is the number of years since 2000. (*Source:* U.S. Census Bureau.)

(a) The world population was about 6853 million in 2010. How closely does the function approximate this value?

(b) Use this model to predict world population in 2020 and 2030.

109. *Deer Population* The exponential growth of the deer population in Massachusetts can be approximated using the model

$$f(x) = 50,000(1 + 0.06)^x,$$

where 50,000 is the initial deer population and 0.06 is the rate of growth. $f(x)$ is the total population after x years have passed. Find each value to the nearest thousand.

(a) Predict the total population after 4 yr.

(b) If the initial population was 30,000 and the growth rate was 0.12, how many deer would be present after 3 yr?

(c) How many additional deer can we expect in 5 yr if the initial population is 45,000 and the current growth rate is 0.08?

110. *Employee Training* A person learning certain skills involving repetition tends to learn quickly at first. Then learning tapers off and skill acquisition approaches some upper limit. Suppose the number of symbols per minute that a person using a keyboard can type is given by

$$f(t) = 250 - 120(2.8)^{-0.5t},$$

where t is the number of months the operator has been in training. Find each value to the nearest whole number.

(a) $f(2)$ **(b)** $f(4)$ **(c)** $f(10)$

(d) What happens to the number of symbols per minute after several months of training?

Use a graphing calculator to find the solution set of each equation. Approximate the solution(s) to the nearest tenth.

111. $5e^{3x} = 75$ **112.** $6^{-x} = 1 - x$ **113.** $3x + 2 = 4^x$ **114.** $x = 2^x$

115. A function of the form $f(x) = x^r$, where r is a constant, is a **power function.** Discuss the difference between an exponential function and a power function.

116. *Concept Check* If $f(x) = a^x$ and $f(3) = 27$, determine each function value.

     (a) $f(1)$ **(b)** $f(-1)$ **(c)** $f(2)$ **(d)** $f(0)$

Concept Check Give an equation of the form $f(x) = a^x$ to define the exponential function whose graph contains the given point.

117. $(3, 8)$ **118.** $(3, 125)$ **119.** $(-3, 64)$ **120.** $(-2, 36)$

Concept Check Use properties of exponents to write each function in the form $f(t) = ka^t$, where k is a constant. (Hint: Recall that $a^{x+y} = a^x \cdot a^y$.)

121. $f(t) = 3^{2t+3}$ **122.** $f(t) = 2^{3t+2}$ **123.** $f(t) = \left(\dfrac{1}{3}\right)^{1-2t}$ **124.** $f(t) = \left(\dfrac{1}{2}\right)^{1-2t}$

In calculus, the following can be shown.

$$e^x = 1 + x + \frac{x^2}{2 \cdot 1} + \frac{x^3}{3 \cdot 2 \cdot 1} + \frac{x^4}{4 \cdot 3 \cdot 2 \cdot 1} + \frac{x^5}{5 \cdot 4 \cdot 3 \cdot 2 \cdot 1} + \cdots$$

Using more terms, one can obtain a more accurate approximation for e^x.

125. Use the terms shown, and replace x with 1 to approximate $e^1 = e$ to three decimal places. Check the result with a calculator.

126. Use the terms shown, and replace x with -0.05 to approximate $e^{-0.05}$ to four decimal places. Check the result with a calculator.

Relating Concepts

For individual or collaborative investigation *(Exercises 127–132)*

Consider $f(x) = a^x$, where $a > 1$. **Work these exercises in order.**

127. Is f a one-to-one function? If so, what kind of related function exists for f?

128. If f has an inverse function f^{-1}, sketch f and f^{-1} on the same set of axes.

129. If f^{-1} exists, find an equation for $y = f^{-1}(x)$. (You need not solve for y.)

130. If $a = 10$, what is the equation for $y = f^{-1}(x)$? (You need not solve for y.)

131. If $a = e$, what is the equation for $y = f^{-1}(x)$? (You need not solve for y.)

132. If the point (p, q) is on the graph of f, then the point _____ is on the graph of f^{-1}.

4.3 Logarithmic Functions

- **Logarithms**
- **Logarithmic Equations**
- **Logarithmic Functions**
- **Properties of Logarithms**

Logarithms The previous section dealt with exponential functions of the form $y = a^x$ for all positive values of a, where $a \neq 1$. The horizontal line test shows that exponential functions are one-to-one and thus have inverse functions. The equation defining the inverse of a function is found by interchanging x and y in the equation that defines the function. Starting with $y = a^x$ and interchanging x and y yields

$$x = a^y.$$

Here y is the exponent to which a must be raised in order to obtain x. We call this exponent a **logarithm,** symbolized by the abbreviation "**log.**" The expression $\log_a x$ represents the logarithm in this discussion. The number a is the **base** of the logarithm, and x is the **argument** of the expression. It is read "**logarithm with base a of x,**" or "**logarithm of x with base a,**" or "**base a logarithm of x.**"

Logarithm

For all real numbers y and all positive numbers a and x, where $a \neq 1$,

$$y = \log_a x \quad \text{is equivalent to} \quad x = a^y.$$

The expression $\log_a x$ represents the exponent to which the base a must be raised in order to obtain x.

EXAMPLE 1 Writing Equivalent Logarithmic and Exponential Forms

The table shows several pairs of equivalent statements, written in both logarithmic and exponential forms.

SOLUTION

Logarithmic Form	Exponential Form
$\log_2 8 = 3$	$2^3 = 8$
$\log_{1/2} 16 = -4$	$\left(\frac{1}{2}\right)^{-4} = 16$
$\log_{10} 100{,}000 = 5$	$10^5 = 100{,}000$
$\log_3 \frac{1}{81} = -4$	$3^{-4} = \frac{1}{81}$
$\log_5 5 = 1$	$5^1 = 5$
$\log_{3/4} 1 = 0$	$\left(\frac{3}{4}\right)^0 = 1$

To remember the relationships among a, x, and y in the two equivalent forms $y = \log_a x$ and $x = a^y$, refer to these diagrams.

A logarithm is an exponent.

Exponent
↓
Logarithmic form: $y = \log_a x$
↑
Base

Exponent
↓
Exponential form: $a^y = x$
↑
Base

✔ **Now Try Exercises 11, 13, 15, and 17.**

Logarithmic Equations The definition of logarithm can be used to solve a **logarithmic equation,** which is an equation with a logarithm in at least one term.

EXAMPLE 2 Solving Logarithmic Equations

Solve each equation.

(a) $\log_x \dfrac{8}{27} = 3$ **(b)** $\log_4 x = \dfrac{5}{2}$ **(c)** $\log_{49} \sqrt[3]{7} = x$

SOLUTION Many logarithmic equations can be solved by first writing the equation in exponential form.

(a)
$$\log_x \frac{8}{27} = 3$$

$$x^3 = \frac{8}{27} \qquad \text{Write in exponential form.}$$

$$x^3 = \left(\frac{2}{3}\right)^3 \qquad \tfrac{8}{27} = \left(\tfrac{2}{3}\right)^3$$

$$x = \frac{2}{3} \qquad \text{Take cube roots.}$$

CHECK $\qquad \log_x \dfrac{8}{27} = 3 \qquad$ Original equation

$$\log_{2/3} \frac{8}{27} \overset{?}{=} 3 \qquad \text{Let } x = \tfrac{2}{3}.$$

$$\left(\frac{2}{3}\right)^3 \overset{?}{=} \frac{8}{27} \qquad \text{Write in exponential form.}$$

$$\frac{8}{27} = \frac{8}{27} \quad \checkmark \quad \text{True}$$

The solution set is $\left\{\dfrac{2}{3}\right\}$.

(b)

$$\log_4 x = \frac{5}{2}$$

$4^{5/2} = x$ Write in exponential form.

$(4^{1/2})^5 = x$ $a^{mn} = (a^m)^n$

$2^5 = x$ $4^{1/2} = (2^2)^{1/2} = 2$

$32 = x$ Apply the exponent.

CHECK $\log_4 32 \stackrel{?}{=} \frac{5}{2}$ Let $x = 32$.

$4^{5/2} \stackrel{?}{=} 32$

$2^5 \stackrel{?}{=} 32$ $4^{5/2} = \left(\sqrt{4}\right)^5 = 2^5$

$32 = 32$ ✓   True

The solution set is $\{32\}$.

(c) $\log_{49} \sqrt[3]{7} = x$

$49^x = \sqrt[3]{7}$ Write in exponential form.

$(7^2)^x = 7^{1/3}$ Write with the same base.

$7^{2x} = 7^{1/3}$ Power rule for exponents

$2x = \frac{1}{3}$ Set exponents equal.

$x = \frac{1}{6}$ Divide by 2.

A check shows that the solution set is $\left\{\frac{1}{6}\right\}$.

✔ **Now Try Exercises 19, 29, and 35.**

Logarithmic Functions   We define the logarithmic function with base a.

Logarithmic Function

If $a > 0$, $a \neq 1$, and $x > 0$, then the **logarithmic function with base a** is

$$f(x) = \log_a x.$$

Exponential and logarithmic functions are inverses of each other. To show this, we use the three steps for finding the inverse of a function.

$f(x) = 2^x$ Exponential function with base 2

$y = 2^x$ Let $y = f(x)$.

Step 1 $x = 2^y$ Interchange x and y.

Step 2 $y = \log_2 x$ Solve for y by writing in equivalent logarithmic form.

Step 3 $f^{-1}(x) = \log_2 x$ Replace y with $f^{-1}(x)$.

The graph of $f(x) = 2^x$ has the x-axis as horizontal asymptote and is shown in red in **Figure 25.** Its inverse, $f^{-1}(x) = \log_2 x$, has the y-axis as vertical asymptote and is shown in blue. The graphs are reflections of each other across the line $y = x$. As a result, their domains and ranges are interchanged.

x	$f(x) = 2^x$
-2	$\frac{1}{4}$
-1	$\frac{1}{2}$
0	1
1	2
2	4

x	$f^{-1}(x) = \log_2 x$
$\frac{1}{4}$	-2
$\frac{1}{2}$	-1
1	0
2	1
4	2

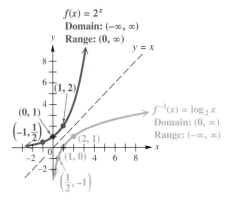

Figure 25

The domain of an exponential function is the set of all real numbers, so the range of a logarithmic function also will be the set of all real numbers. In the same way, both the range of an exponential function and the domain of a logarithmic function are the set of all positive real numbers.

Thus, logarithms can be found for positive numbers only.

Logarithmic Function $f(x) = \log_a x$

Domain: $(0, \infty)$ Range: $(-\infty, \infty)$

For $f(x) = \log_2 x$:

x	$f(x)$
$\frac{1}{4}$	-2
$\frac{1}{2}$	-1
1	0
2	1
4	2
8	3

This is the general behavior seen on a calculator graph for **any base a, for $a > 1$.**

Figure 26

- $f(x) = \log_a x$, **for $a > 1$,** is increasing and continuous on its entire domain, $(0, \infty)$.

- The y-axis is a vertical asymptote as $x \rightarrow 0$ from the right.

- The graph passes through the points $\left(\frac{1}{a}, -1\right)$, $(1, 0)$, and $(a, 1)$.

For $f(x) = \log_{1/2} x$:

x	$f(x)$
$\frac{1}{4}$	2
$\frac{1}{2}$	1
1	0
2	-1
4	-2
8	-3

This is the general behavior seen on a calculator graph for **any base a, for $0 < a < 1$.**

Figure 27

- $f(x) = \log_a x$, **for $0 < a < 1$,** is decreasing and continuous on its entire domain, $(0, \infty)$.

- The y-axis is a vertical asymptote as $x \rightarrow 0$ from the right.

- The graph passes through the points $\left(\frac{1}{a}, -1\right)$, $(1, 0)$, and $(a, 1)$.

Calculator graphs of logarithmic functions sometimes do not give an accurate picture of the behavior of the graphs near the vertical asymptotes. While it may seem as if the graph has an endpoint, this is not the case. The resolution of the calculator screen is not precise enough to indicate that the graph approaches the vertical asymptote as the value of x gets closer to it. Do not draw incorrect conclusions just because the calculator does not show this behavior. ∎

The graphs in **Figures 26 and 27** and the information with them suggest the following generalizations about the graphs of logarithmic functions of the form $f(x) = \log_a x$.

Characteristics of the Graph of $f(x) = \log_a x$

1. The points $\left(\frac{1}{a}, -1\right)$, $(1, 0)$, and $(a, 1)$ are on the graph.

2. If $a > 1$, then f is an increasing function.

 If $0 < a < 1$, then f is a decreasing function.

3. The y-axis is a vertical asymptote.

4. The domain is $(0, \infty)$, and the range is $(-\infty, \infty)$.

EXAMPLE 3 Graphing Logarithmic Functions

Graph each function.

(a) $f(x) = \log_{1/2} x$ **(b)** $f(x) = \log_3 x$

SOLUTION

(a) One approach is to first graph $y = \left(\frac{1}{2}\right)^x$, which defines the inverse function of f, by plotting points. Some ordered pairs are given in the table with the graph shown in red in **Figure 28.**

The graph of $f(x) = \log_{1/2} x$ is the reflection of the graph of $y = \left(\frac{1}{2}\right)^x$ across the line $y = x$. The ordered pairs for $y = \log_{1/2} x$ are found by interchanging the x- and y-values in the ordered pairs for $y = \left(\frac{1}{2}\right)^x$. See the graph in blue in **Figure 28.**

x	$y = \left(\frac{1}{2}\right)^x$
-2	4
-1	2
0	1
1	$\frac{1}{2}$
2	$\frac{1}{4}$
4	$\frac{1}{16}$

x	$f(x) = \log_{1/2} x$
4	-2
2	-1
1	0
$\frac{1}{2}$	1
$\frac{1}{4}$	2
$\frac{1}{16}$	4

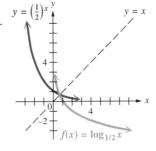

Figure 28

x	$f(x) = \log_3 x$
$\frac{1}{3}$	-1
1	0
3	1
9	2

Think: $x = 3^y$

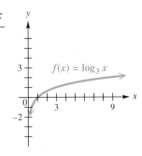

Figure 29

(b) Another way to graph a logarithmic function is to write $f(x) = y = \log_3 x$ in exponential form as $x = 3^y$, and then select y-values and calculate corresponding x-values. Several selected ordered pairs are shown in the table for the graph in **Figure 29.**

✔ **Now Try Exercise 55.**

CAUTION If we write a logarithmic function in exponential form in order to graph it, as in **Example 3(b),** we start *first* with *y*-values to calculate corresponding *x*-values. *Be careful to write the values in the ordered pairs in the correct order.*

More general logarithmic functions can be obtained by forming the composition of $f(x) = \log_a x$ with a function $g(x)$. For example, if $f(x) = \log_2 x$ and $g(x) = x - 1$, then

$$(f \circ g)(x) = f(g(x)) = \log_2 (x - 1).$$

The next example shows how to graph such functions.

EXAMPLE 4 **Graphing Translated Logarithmic Functions**

Graph each function. Give the domain and range.

(a) $f(x) = \log_2 (x - 1)$ **(b)** $f(x) = (\log_3 x) - 1$

(c) $f(x) = \log_4 (x + 2) + 1$

SOLUTION

(a) The graph of $f(x) = \log_2 (x - 1)$ is the graph of $g(x) = \log_2 x$ translated 1 unit to the right. The vertical asymptote has equation $x = 1$. Because logarithms can be found only for positive numbers, we solve $x - 1 > 0$ to find the domain, $(1, \infty)$. To determine ordered pairs to plot, use the equivalent exponential form of the equation $y = \log_2 (x - 1)$.

$$y = \log_2 (x - 1)$$
$$x - 1 = 2^y \qquad \text{Write in exponential form.}$$
$$x = 2^y + 1 \qquad \text{Add 1.}$$

We first choose values for *y* and then calculate each of the corresponding *x*-values. The range is $(-\infty, \infty)$. See **Figure 30.**

Figure 30 **Figure 31**

(b) The function $f(x) = (\log_3 x) - 1$ has the same graph as $g(x) = \log_3 x$ translated 1 unit down. We find ordered pairs to plot by writing the equation $y = (\log_3 x) - 1$ in exponential form.

$$y = (\log_3 x) - 1$$
$$y + 1 = \log_3 x \qquad \text{Add 1.}$$
$$x = 3^{y+1} \qquad \text{Write in exponential form.}$$

Again, choose *y*-values and calculate the corresponding *x*-values. The graph is shown in **Figure 31.** The domain is $(0, \infty)$, and the range is $(-\infty, \infty)$.

(c) The graph of $f(x) = \log_4(x + 2) + 1$ is obtained by shifting the graph of $y = \log_4 x$ to the left 2 units and up 1 unit. The domain is found by solving

$$x + 2 > 0,$$

which yields $(-2, \infty)$. The vertical asymptote has been shifted to the left 2 units as well, and it has equation $x = -2$. The range is unaffected by the vertical shift and remains $(-\infty, \infty)$. See **Figure 32.**

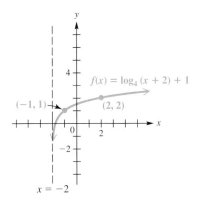

Figure 32

✔ **Now Try Exercises 43, 47, and 61.**

NOTE   If we are given a graph such as the one in **Figure 31** and asked to find its equation, we could reason as follows: The point $(1, 0)$ on the basic logarithmic graph has been shifted *down* 1 unit, and the point $(3, 0)$ on the given graph is 1 unit lower than $(3, 1)$, which is on the graph of $y = \log_3 x$. Thus, the equation will be

$$y = (\log_3 x) - 1.$$

Properties of Logarithms   The properties of logarithms enable us to change the form of logarithmic statements so that products can be converted to sums, quotients can be converted to differences, and powers can be converted to products.

Properties of Logarithms

For $x > 0$, $y > 0$, $a > 0$, $a \neq 1$, and any real number r, the following properties hold.

Property	Description
Product Property $\log_a xy = \log_a x + \log_a y$	The logarithm of the product of two numbers is equal to the sum of the logarithms of the numbers.
Quotient Property $\log_a \dfrac{x}{y} = \log_a x - \log_a y$	The logarithm of the quotient of two numbers is equal to the difference between the logarithms of the numbers.
Power Property $\log_a x^r = r \log_a x$	The logarithm of a number raised to a power is equal to the exponent multiplied by the logarithm of the number.
Logarithm of 1 $\log_a 1 = 0$	The base a logarithm of 1 is 0.
Base a Logarithm of a $\log_a a = 1$	The base a logarithm of a is 1.

Proof   To prove the product property, let $m = \log_a x$ and $n = \log_a y$.

$$\log_a x = m \quad \text{means} \quad a^m = x \qquad \text{Write in exponential form.}$$
$$\log_a y = n \quad \text{means} \quad a^n = y$$

Now consider the product xy.

$$xy = a^m \cdot a^n \qquad x = a^m \text{ and } y = a^n; \text{ Substitute.}$$

$$xy = a^{m+n} \qquad \text{Product rule for exponents}$$

$$\log_a xy = m + n \qquad \text{Write in logarithmic form.}$$

$$\log_a xy = \log_a x + \log_a y \qquad \text{Substitute.}$$

The last statement is the result we wished to prove. The quotient and power properties are proved similarly and are left as exercises.

EXAMPLE 5 Using Properties of Logarithms

Use the properties of logarithms to rewrite each expression. Assume all variables represent positive real numbers, with $a \neq 1$ and $b \neq 1$.

(a) $\log_6 (7 \cdot 9)$ **(b)** $\log_9 \dfrac{15}{7}$ **(c)** $\log_5 \sqrt{8}$

(d) $\log_a \sqrt[3]{m^2}$ **(e)** $\log_a \dfrac{mnq}{p^2 t^4}$ **(f)** $\log_b \sqrt[n]{\dfrac{x^3 y^5}{z^m}}$

SOLUTION

(a) $\log_6 (7 \cdot 9)$

$\qquad = \log_6 7 + \log_6 9 \qquad$ Product property

(b) $\log_9 \dfrac{15}{7}$

$\qquad = \log_9 15 - \log_9 7 \qquad$ Quotient property

(c) $\log_5 \sqrt{8}$

$\qquad = \log_5 (8^{1/2}) \qquad \sqrt{a} = a^{1/2}$

$\qquad = \dfrac{1}{2} \log_5 8 \qquad$ Power property

(d) $\log_a \sqrt[3]{m^2}$

$\qquad = \log_a m^{2/3} \qquad \sqrt[n]{a^m} = a^{m/n}$

$\qquad = \dfrac{2}{3} \log_a m \qquad$ Power property

(e) $\log_a \dfrac{mnq}{p^2 t^4}$

> Use parentheses to avoid errors.

$\qquad = \log_a m + \log_a n + \log_a q - (\log_a p^2 + \log_a t^4) \qquad$ Product and quotient properties

$\qquad = \log_a m + \log_a n + \log_a q - (2 \log_a p + 4 \log_a t) \qquad$ Power property

$\qquad = \log_a m + \log_a n + \log_a q - 2 \log_a p - 4 \log_a t \qquad$ Distributive property

> Be careful with signs.

(f) $\log_b \sqrt[n]{\dfrac{x^3 y^5}{z^m}}$

$\qquad = \log_b \left(\dfrac{x^3 y^5}{z^m} \right)^{1/n} \qquad \sqrt[n]{a} = a^{1/n}$

$\qquad = \dfrac{1}{n} \log_b \dfrac{x^3 y^5}{z^m} \qquad$ Power property

$\qquad = \dfrac{1}{n} (\log_b x^3 + \log_b y^5 - \log_b z^m) \qquad$ Product and quotient properties

$\qquad = \dfrac{1}{n} (3 \log_b x + 5 \log_b y - m \log_b z) \qquad$ Power property

$\qquad = \dfrac{3}{n} \log_b x + \dfrac{5}{n} \log_b y - \dfrac{m}{n} \log_b z \qquad$ Distributive property

✔ **Now Try Exercises 71, 73, and 77.**

EXAMPLE 6 **Using Properties of Logarithms**

Write each expression as a single logarithm with coefficient 1. Assume all variables represent positive real numbers, with $a \neq 1$ and $b \neq 1$.

(a) $\log_3 (x + 2) + \log_3 x - \log_3 2$ **(b)** $2 \log_a m - 3 \log_a n$

(c) $\frac{1}{2} \log_b m + \frac{3}{2} \log_b 2n - \log_b m^2 n$

SOLUTION

(a) $\log_3 (x + 2) + \log_3 x - \log_3 2$

$= \log_3 \dfrac{(x + 2)x}{2}$ Product and quotient properties

(b) $2 \log_a m - 3 \log_a n$

$= \log_a m^2 - \log_a n^3$ Power property

$= \log_a \dfrac{m^2}{n^3}$ Quotient property

(c) $\frac{1}{2} \log_b m + \frac{3}{2} \log_b 2n - \log_b m^2 n$

$= \log_b m^{1/2} + \log_b (2n)^{3/2} - \log_b m^2 n$ Power property

$= \log_b \dfrac{m^{1/2}(2n)^{3/2}}{m^2 n}$ [Use parentheses around 2n.]   Product and quotient properties

$= \log_b \dfrac{2^{3/2} n^{1/2}}{m^{3/2}}$ Rules for exponents

$= \log_b \left(\dfrac{2^3 n}{m^3} \right)^{1/2}$ Rules for exponents

$= \log_b \sqrt{\dfrac{8n}{m^3}}$ Definition of $a^{1/n}$

✔ **Now Try Exercises 83, 87, and 89.**

CAUTION   *There is no property of logarithms to rewrite a logarithm of a sum or difference.* That is why, in **Example 6(a)**,

$$\log_3 (x + 2) \quad \text{cannot be written as} \quad \log_3 x + \log_3 2.$$

The distributive property does not apply here because $\log_3 (x + y)$ is one term. The abbreviation "log" is a function name, *not* a factor.

Napier's Rods

The search for ways to make calculations easier has been a long, ongoing process. Machines built by Charles Babbage and Blaise Pascal, a system of "rods" used by John Napier, and slide rules were the forerunners of today's calculators and computers. The invention of logarithms by John Napier in the 16th century was a great breakthrough in the search for easier calculation methods.

Source: IBM Corporate Archives.

EXAMPLE 7 **Using Properties of Logarithms with Numerical Values**

Given that $\log_{10} 2 \approx 0.3010$, find each logarithm without using a calculator.

(a) $\log_{10} 4$ **(b)** $\log_{10} 5$

SOLUTION

(a) $\log_{10} 4$

$= \log_{10} 2^2$

$= 2 \log_{10} 2$

$\approx 2(0.3010)$

≈ 0.6020

(b) $\log_{10} 5$

$= \log_{10} \dfrac{10}{2}$

$= \log_{10} 10 - \log_{10} 2$

$\approx 1 - 0.3010$

≈ 0.6990

✔ **Now Try Exercises 93 and 95.**

> **NOTE** The values in **Example 7** are approximations of logarithms, so the final digit may differ from the actual 4-decimal-place approximation after properties of logarithms are applied.

Recall that for inverse functions f and g, $(f \circ g)(x) = (g \circ f)(x) = x$. We can use this property with exponential and logarithmic functions to state two more properties. If $f(x) = a^x$ and $g(x) = \log_a x$, then

$$(f \circ g)(x) = a^{\log_a x} = x \quad \text{and} \quad (g \circ f)(x) = \log_a(a^x) = x.$$

Theorem on Inverses

For $a > 0$, $a \neq 1$, the following properties hold.

$$a^{\log_a x} = x \ (\textbf{for } x > 0) \quad \text{and} \quad \log_a a^x = x$$

Examples: $7^{\log_7 10} = 10$, $\log_5 5^3 = 3$, and $\log_r r^{k+1} = k + 1$

The second statement in the theorem will be useful when we solve logarithmic and exponential equations.

4.3 Exercises

CONCEPT PREVIEW *Match the logarithm in Column I with its value in Column II. Remember that* $\log_a x$ *is the exponent to which a must be raised in order to obtain x.*

I	II		I	II
1. (a) $\log_2 16$	**A.** 0		**2. (a)** $\log_3 81$	**A.** -2
(b) $\log_3 1$	**B.** $\dfrac{1}{2}$		**(b)** $\log_3 \dfrac{1}{3}$	**B.** -1
(c) $\log_{10} 0.1$	**C.** 4		**(c)** $\log_{10} 0.01$	**C.** 0
(d) $\log_2 \sqrt{2}$	**D.** -3		**(d)** $\log_6 \sqrt{6}$	**D.** $\dfrac{1}{2}$
(e) $\log_e \dfrac{1}{e^2}$	**E.** -1		**(e)** $\log_e 1$	**E.** $\dfrac{9}{2}$
(f) $\log_{1/2} 8$	**F.** -2		**(f)** $\log_3 27^{3/2}$	**F.** 4

CONCEPT PREVIEW *Write each equivalent form.*

3. Write $\log_2 8 = 3$ in exponential form.   **4.** Write $10^3 = 1000$ in logarithmic form.

CONCEPT PREVIEW *Solve each logarithmic equation.*

5. $\log_x \dfrac{16}{81} = 2$ **6.** $\log_{36} \sqrt[3]{6} = x$

CONCEPT PREVIEW *Sketch the graph of each function. Give the domain and range.*

7. $f(x) = \log_5 x$ **8.** $g(x) = \log_{1/5} x$

CONCEPT PREVIEW *Use the properties of logarithms to rewrite each expression. Assume all variables represent positive real numbers.*

9. $\log_{10} \dfrac{2x}{7}$

10. $3 \log_4 x - 5 \log_4 y$

If the statement is in exponential form, write it in an equivalent logarithmic form. If the statement is in logarithmic form, write it in exponential form. **See Example 1.**

11. $3^4 = 81$

12. $2^5 = 32$

13. $\left(\dfrac{2}{3}\right)^{-3} = \dfrac{27}{8}$

14. $10^{-4} = 0.0001$

15. $\log_6 36 = 2$

16. $\log_5 5 = 1$

17. $\log_{\sqrt{3}} 81 = 8$

18. $\log_4 \dfrac{1}{64} = -3$

Solve each equation. **See Example 2.**

19. $x = \log_5 \dfrac{1}{625}$

20. $x = \log_3 \dfrac{1}{81}$

21. $\log_x \dfrac{1}{32} = 5$

22. $\log_x \dfrac{27}{64} = 3$

23. $x = \log_8 \sqrt[4]{8}$

24. $x = \log_7 \sqrt[5]{7}$

25. $x = 3^{\log_3 8}$

26. $x = 12^{\log_{12} 5}$

27. $x = 2^{\log_2 9}$

28. $x = 8^{\log_8 11}$

29. $\log_x 25 = -2$

30. $\log_x 16 = -2$

31. $\log_4 x = 3$

32. $\log_2 x = 3$

33. $x = \log_4 \sqrt[3]{16}$

34. $x = \log_5 \sqrt[4]{25}$

35. $\log_9 x = \dfrac{5}{2}$

36. $\log_4 x = \dfrac{7}{2}$

37. $\log_{1/2}(x + 3) = -4$

38. $\log_{1/3}(x + 6) = -2$

39. $\log_{(x+3)} 6 = 1$

40. $\log_{(x-4)} 19 = 1$

41. $3x - 15 = \log_x 1 \quad (x > 0, x \neq 1)$

42. $4x - 24 = \log_x 1 \quad (x > 0, x \neq 1)$

Graph each function. Give the domain and range. **See Example 4.**

43. $f(x) = (\log_2 x) + 3$

44. $f(x) = \log_2(x + 3)$

45. $f(x) = |\log_2(x + 3)|$

Graph each function. Give the domain and range. **See Example 4.**

46. $f(x) = (\log_{1/2} x) - 2$

47. $f(x) = \log_{1/2}(x - 2)$

48. $f(x) = |\log_{1/2}(x - 2)|$

Concept Check *In Exercises 49–54, match the function with its graph from choices A–F.*

49. $f(x) = \log_2 x$

50. $f(x) = \log_2 2x$

51. $f(x) = \log_2 \dfrac{1}{x}$

52. $f(x) = \log_2\left(\dfrac{1}{2}x\right)$

53. $f(x) = \log_2(x - 1)$

54. $f(x) = \log_2(-x)$

A.

B.

C.

D.

E.

F.

Graph each function. ***See Examples 3 and 4.***

55. $f(x) = \log_5 x$

56. $f(x) = \log_{10} x$

57. $f(x) = \log_5 (x + 1)$

58. $f(x) = \log_6 (x - 2)$

59. $f(x) = \log_{1/2} (1 - x)$

60. $f(x) = \log_{1/3} (3 - x)$

61. $f(x) = \log_3 (x - 1) + 2$

62. $f(x) = \log_2 (x + 2) - 3$

63. $f(x) = \log_{1/2} (x + 3) - 2$

64. *Concept Check* To graph the function $f(x) = -\log_5 (x - 7) - 4$, reflect the graph of $y = \log_5 x$ across the _____-axis, then shift the graph _____ units to the right and _____ units down.

Connecting Graphs with Equations *Write an equation for the graph given. Each represents a logarithmic function f with base 2 or 3, translated and/or reflected.* ***See the Note following Example 4.***

65.

66.

67.

68.

69.

70.

Use the properties of logarithms to rewrite each expression. Simplify the result if possible. Assume all variables represent positive real numbers. ***See Example 5.***

71. $\log_2 \dfrac{6x}{y}$

72. $\log_3 \dfrac{4p}{q}$

73. $\log_5 \dfrac{5\sqrt{7}}{3}$

74. $\log_2 \dfrac{2\sqrt{3}}{5}$

75. $\log_4 (2x + 5y)$

76. $\log_6 (7m + 3q)$

77. $\log_2 \sqrt{\dfrac{5r^3}{z^5}}$

78. $\log_3 \sqrt[3]{\dfrac{m^5 n^4}{t^2}}$

79. $\log_2 \dfrac{ab}{cd}$

80. $\log_2 \dfrac{xy}{tqr}$

81. $\log_3 \dfrac{\sqrt{x} \cdot \sqrt[3]{y}}{w^2 \sqrt{z}}$

82. $\log_4 \dfrac{\sqrt[3]{a} \cdot \sqrt[4]{b}}{\sqrt{c} \cdot \sqrt[3]{d^2}}$

Write each expression as a single logarithm with coefficient 1. Assume all variables represent positive real numbers, with $a \neq 1$ and $b \neq 1$. ***See Example 6.***

83. $\log_a x + \log_a y - \log_a m$

84. $\log_b k + \log_b m - \log_b a$

85. $\log_a m - \log_a n - \log_a t$

86. $\log_b p - \log_b q - \log_b r$

87. $\dfrac{1}{3} \log_b x^4 y^5 - \dfrac{3}{4} \log_b x^2 y$

88. $\dfrac{1}{2} \log_a p^3 q^4 - \dfrac{2}{3} \log_a p^4 q^3$

89. $2 \log_a (z + 1) + \log_a (3z + 2)$

90. $5 \log_a (z + 7) + \log_a (2z + 9)$

91. $-\dfrac{2}{3} \log_5 5m^2 + \dfrac{1}{2} \log_5 25m^2$

92. $-\dfrac{3}{4} \log_3 16p^4 - \dfrac{2}{3} \log_3 8p^3$

Given that $\log_{10} 2 \approx 0.3010$ and $\log_{10} 3 \approx 0.4771$, find each logarithm without using a calculator. See Example 7.

93. $\log_{10} 6$ **94.** $\log_{10} 12$ **95.** $\log_{10} \dfrac{3}{2}$ **96.** $\log_{10} \dfrac{2}{9}$

97. $\log_{10} \dfrac{9}{4}$ **98.** $\log_{10} \dfrac{20}{27}$ **99.** $\log_{10} \sqrt{30}$ **100.** $\log_{10} 36^{1/3}$

Solve each problem.

101. *(Modeling) Interest Rates of Treasury Securities* The table gives interest rates for various U.S. Treasury Securities on January 2, 2015.

(a) Make a scatter diagram of the data.

(b) Which type of function will model this data best: linear, exponential, or logarithmic?

Time	Yield
3-month	0.02%
6-month	0.10%
2-year	0.66%
5-year	1.61%
10-year	2.11%
30-year	2.60%

*Source:*www.federal reserve.gov

102. *Concept Check* Use the graph to estimate each logarithm.

(a) $\log_3 0.3$ (b) $\log_3 0.8$

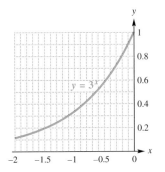

103. *Concept Check* Suppose $f(x) = \log_a x$ and $f(3) = 2$. Determine each function value.

(a) $f\left(\dfrac{1}{9}\right)$ (b) $f(27)$ (c) $f(9)$ (d) $f\left(\dfrac{\sqrt{3}}{3}\right)$

104. Use properties of logarithms to evaluate each expression.

(a) $100^{\log_{10} 3}$ (b) $\log_{10}(0.01)^3$ (c) $\log_{10}(0.0001)^5$ (d) $1000^{\log_{10} 5}$

105. Using the compound interest formula $A = P\left(1 + \dfrac{r}{n}\right)^{tn}$, show that the amount of time required for a deposit to double is

$$\frac{1}{\log_2\left(1 + \dfrac{r}{n}\right)^n}.$$

106. *Concept Check* If $(5, 4)$ is on the graph of the logarithmic function with base a, which of the following statements is true:

$$5 = \log_a 4 \quad \text{or} \quad 4 = \log_a 5?$$

Use a graphing calculator to find the solution set of each equation. Give solutions to the nearest hundredth.

107. $\log_{10} x = x - 2$ **108.** $2^{-x} = \log_{10} x$

109. Prove the quotient property of logarithms: $\log_a \dfrac{x}{y} = \log_a x - \log_a y.$

110. Prove the power property of logarithms: $\log_a x^r = r \log_a x.$

Summary Exercises on Inverse, Exponential, and Logarithmic Functions

The following exercises are designed to help solidify your understanding of inverse, exponential, and logarithmic functions from **Sections 4.1–4.3.**

Determine whether the functions in each pair are inverses of each other.

1. $f(x) = 3x - 4$, $\quad g(x) = \dfrac{1}{3}x + \dfrac{4}{3}$

2. $f(x) = 8 - 5x$, $\quad g(x) = 8 + \dfrac{1}{5}x$

3. $f(x) = 1 + \log_2 x$, $\quad g(x) = 2^{x-1}$

4. $f(x) = 3^{x/5} - 2$, $\quad g(x) = 5 \log_3 (x + 2)$

Determine whether each function is one-to-one. If it is, then sketch the graph of its inverse function.

5.

6.

7.

8.

In Exercises 9–12, match each function with its graph from choices A–D.

9. $y = \log_3 (x + 2)$

10. $y = 5 - 2^x$

11. $y = \log_2 (5 - x)$

12. $y = 3^x - 2$

A.

B.

C.

D.

13. The functions in **Exercises 9–12** form two pairs of inverse functions. Determine which functions are inverses of each other.

14. Determine the inverse of the function $f(x) = \log_5 x$. (*Hint:* Replace $f(x)$ with y, and write in exponential form.)

For each function that is one-to-one, write an equation for the inverse function. Give the domain and range of both f and f⁻¹. If the function is not one-to-one, say so.

15. $f(x) = 3x - 6$

16. $f(x) = 2(x + 1)^3$

17. $f(x) = 3x^2$

18. $f(x) = \dfrac{2x - 1}{5 - 3x}$

19. $f(x) = \sqrt[3]{5 - x^4}$

20. $f(x) = \sqrt{x^2 - 9}, \quad x \geq 3$

Write an equivalent statement in logarithmic form.

21. $\left(\dfrac{1}{10}\right)^{-3} = 1000$

22. $a^b = c$

23. $\left(\sqrt{3}\right)^4 = 9$

24. $4^{-3/2} = \dfrac{1}{8}$

25. $2^x = 32$

26. $27^{4/3} = 81$

Solve each equation.

27. $3x = 7^{\log_7 6}$

28. $x = \log_{10} 0.001$

29. $x = \log_6 \dfrac{1}{216}$

30. $\log_x 5 = \dfrac{1}{2}$

31. $\log_{10} 0.01 = x$

32. $\log_x 3 = -1$

33. $\log_x 1 = 0$

34. $x = \log_2 \sqrt{8}$

35. $\log_x \sqrt[3]{5} = \dfrac{1}{3}$

36. $\log_{1/3} x = -5$

37. $\log_{10}\left(\log_2 2^{10}\right) = x$

38. $x = \log_{4/5} \dfrac{25}{16}$

39. $2x - 1 = \log_6 6^x$

40. $x = \sqrt{\log_{1/2} \dfrac{1}{16}}$

41. $2^x = \log_2 16$

42. $\log_3 x = -2$

43. $\left(\dfrac{1}{3}\right)^{x+1} = 9^x$

44. $5^{2x-6} = 25^{x-3}$

4.4 Evaluating Logarithms and the Change-of-Base Theorem

- Common Logarithms
- Applications and Models with Common Logarithms
- Natural Logarithms
- Applications and Models with Natural Logarithms
- Logarithms with Other Bases

Common Logarithms   Two of the most important bases for logarithms are 10 and e. Base 10 logarithms are **common logarithms.** The common logarithm of x is written **log x,** where the base is understood to be 10.

Common Logarithm

For all positive numbers x,

$$\log x = \log_{10} x.$$

A calculator with a log key can be used to find the base 10 logarithm of any positive number.

EXAMPLE 1 **Evaluating Common Logarithms with a Calculator**

Use a calculator to find the values of

$$\log 1000, \quad \log 142, \quad \text{and} \quad \log 0.005832.$$

SOLUTION **Figure 33** shows that the exact value of log 1000 is 3 (because $10^3 = 1000$), and that

$$\log 142 \approx 2.152288344$$

and $\log 0.005832 \approx -2.234182485.$

Most common logarithms that appear in calculations are approximations, as seen in the second and third displays.

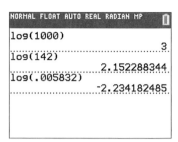

Figure 33

☑ Now Try Exercises 11, 15, and 17.

For a > 1, base a logarithms of numbers between 0 and 1 are always negative, and base a logarithms of numbers greater than 1 are always positive.

Applications and Models with Common Logarithms   In chemistry, the **pH** of a solution is defined as

$$pH = -\log[H_3O^+],$$

where $[H_3O^+]$ is the hydronium ion concentration in moles* per liter. The pH value is a measure of the acidity or alkalinity of a solution. Pure water has pH 7.0, substances with pH values greater than 7.0 are alkaline, and substances with pH values less than 7.0 are acidic. See **Figure 34.** It is customary to round pH values to the nearest tenth.

Figure 34

EXAMPLE 2 **Finding pH**

(a) Find the pH of a solution with $[H_3O^+] = 2.5 \times 10^{-4}$.

(b) Find the hydronium ion concentration of a solution with pH = 7.1.

SOLUTION

(a) $pH = -\log[H_3O^+]$

$pH = -\log(2.5 \times 10^{-4})$ Substitute $[H_3O^+] = 2.5 \times 10^{-4}$.

$pH = -(\log 2.5 + \log 10^{-4})$ Product property

$pH = -(0.3979 - 4)$ $\log 10^{-4} = -4$

$pH = -0.3979 + 4$ Distributive property

$pH \approx 3.6$ Add.

---

*A *mole* is the amount of a substance that contains the same number of molecules as the number of atoms in exactly 12 grams of carbon-12.

(b)
$$\text{pH} = -\log\left[H_3O^+\right]$$
$$7.1 = -\log\left[H_3O^+\right] \qquad \text{Substitute pH} = 7.1.$$
$$-7.1 = \log\left[H_3O^+\right] \qquad \text{Multiply by } -1.$$
$$\left[H_3O^+\right] = 10^{-7.1} \qquad \text{Write in exponential form.}$$
$$\left[H_3O^+\right] \approx 7.9 \times 10^{-8} \qquad \text{Evaluate } 10^{-7.1} \text{ with a calculator.}$$

✔ **Now Try Exercises 29 and 33.**

NOTE In the fourth line of the solution in **Example 2(a)**, we use the equality symbol, $=$, rather than the approximate equality symbol, $\approx$, when replacing log 2.5 with 0.3979. This is often done for convenience, despite the fact that most logarithms used in applications are indeed approximations.

EXAMPLE 3 **Using pH in an Application**

Wetlands are classified as *bogs, fens, marshes,* and *swamps* based on pH values. A pH value between 6.0 and 7.5 indicates that the wetland is a "rich fen." When the pH is between 3.0 and 6.0, it is a "poor fen," and if the pH falls to 3.0 or less, the wetland is a "bog." (*Source:* R. Mohlenbrock, "Summerby Swamp, Michigan," *Natural History.*)

Suppose that the hydronium ion concentration of a sample of water from a wetland is 6.3×10^{-5}. How would this wetland be classified?

SOLUTION
$$\text{pH} = -\log\left[H_3O^+\right] \qquad \text{Definition of pH}$$
$$\text{pH} = -\log(6.3 \times 10^{-5}) \qquad \text{Substitute for } \left[H_3O^+\right].$$
$$\text{pH} = -(\log 6.3 + \log 10^{-5}) \qquad \text{Product property}$$
$$\text{pH} = -\log 6.3 - (-5) \qquad \text{Distributive property; } \log 10^n = n$$
$$\text{pH} = -\log 6.3 + 5 \qquad \text{Definition of subtraction}$$
$$\text{pH} \approx 4.2 \qquad \text{Use a calculator.}$$

The pH is between 3.0 and 6.0, so the wetland is a poor fen.

✔ **Now Try Exercise 37.**

EXAMPLE 4 **Measuring the Loudness of Sound**

The loudness of sounds is measured in **decibels.** We first assign an intensity of I_0 to a very faint **threshold sound.** If a particular sound has intensity I, then the decibel rating d of this louder sound is given by the following formula.

$$d = 10 \log \frac{I}{I_0}$$

Find the decibel rating d of a sound with intensity $10,000I_0$.

SOLUTION
$$d = 10 \log \frac{10,000I_0}{I_0} \qquad \text{Let } I = 10,000I_0.$$
$$d = 10 \log 10,000 \qquad \frac{I_0}{I_0} = 1$$
$$d = 10(4) \qquad \log 10,000 = \log 10^4 = 4$$
$$d = 40 \qquad \text{Multiply.}$$

The sound has a decibel rating of 40. ✔ **Now Try Exercise 63.**

Natural Logarithms In most practical applications of logarithms, the irrational number e is used as the base. Logarithms with base e are **natural logarithms** because they occur in the life sciences and economics in natural situations that involve growth and decay. The base e logarithm of x is written **ln x** (read **"el-en x"**). *The expression* **ln x** *represents the exponent to which e must be raised in order to obtain x.*

Figure 35

Natural Logarithm

For all positive numbers x,

$$\ln x = \log_e x.$$

A graph of the natural logarithmic function $f(x) = \ln x$ is given in **Figure 35.**

EXAMPLE 5 **Evaluating Natural Logarithms with a Calculator**

Use a calculator to find the values of

$$\ln e^3, \quad \ln 142, \quad \text{and} \quad \ln 0.005832.$$

SOLUTION **Figure 36** shows that the exact value of $\ln e^3$ is 3, and that

$$\ln 142 \approx 4.955827058$$

and

$$\ln 0.005832 \approx -5.144395284.$$

Figure 36

✔ **Now Try Exercises 45, 51, and 53.**

Figure 37

Figure 37 illustrates that **ln x is the exponent to which e must be raised in order to obtain x.**

Applications and Models with Natural Logarithms

EXAMPLE 6 **Measuring the Age of Rocks**

Geologists sometimes measure the age of rocks by using "atomic clocks." By measuring the amounts of argon-40 and potassium-40 in a rock, it is possible to find the age t of the specimen in years with the formula

$$t = (1.26 \times 10^9)\frac{\ln\left(1 + 8.33\left(\frac{A}{K}\right)\right)}{\ln 2},$$

where A and K are the numbers of atoms of argon-40 and potassium-40, respectively, in the specimen.

(a) How old is a rock in which $A = 0$ and $K > 0$?

(b) The ratio $\frac{A}{K}$ for a sample of granite from New Hampshire is 0.212. How old is the sample?

LOOKING AHEAD TO CALCULUS

The natural logarithmic function $f(x) = \ln x$ and the reciprocal function $g(x) = \frac{1}{x}$ have an important relationship in calculus. The derivative of the natural logarithmic function is the reciprocal function. Using **Leibniz notation** (named after one of the co-inventors of calculus), we write this fact as $\frac{d}{dx}(\ln x) = \frac{1}{x}$.

SOLUTION

(a) If $A = 0$, then $\frac{A}{K} = 0$ and the equation is as follows.

$$t = (1.26 \times 10^9) \frac{\ln\left(1 + 8.33\left(\frac{A}{K}\right)\right)}{\ln 2} \qquad \text{Given formula}$$

$$t = (1.26 \times 10^9) \frac{\ln 1}{\ln 2} \qquad \frac{A}{K} = 0, \text{ so } \ln(1 + 0) = \ln 1$$

$$t = (1.26 \times 10^9)(0) \qquad \ln 1 = 0$$

$$t = 0$$

The rock is new (0 yr old).

(b) Because $\frac{A}{K} = 0.212$, we have the following.

$$t = (1.26 \times 10^9) \frac{\ln(1 + 8.33(0.212))}{\ln 2} \qquad \text{Substitute.}$$

$$t \approx 1.85 \times 10^9 \qquad \text{Use a calculator.}$$

The granite is about 1.85 billion yr old. ✔ **Now Try Exercise 77.**

> **EXAMPLE 7** **Modeling Global Temperature Increase**

Carbon dioxide in the atmosphere traps heat from the sun. The additional solar radiation trapped by carbon dioxide is **radiative forcing.** It is measured in watts per square meter (w/m^2). In 1896 the Swedish scientist Svante Arrhenius modeled radiative forcing R caused by additional atmospheric carbon dioxide, using the logarithmic equation

$$R = k \ln \frac{C}{C_0},$$

where C_0 is the preindustrial amount of carbon dioxide, C is the current carbon dioxide level, and k is a constant. Arrhenius determined that $10 \le k \le 16$ when $C = 2C_0$. (*Source:* Clime, W., *The Economics of Global Warming*, Institute for International Economics, Washington, D.C.)

(a) Let $C = 2C_0$. Is the relationship between R and k linear or logarithmic?

(b) The average global temperature increase T (in °F) is given by $T(R) = 1.03R$. Write T as a function of k.

SOLUTION

(a) If $C = 2C_0$, then $\frac{C}{C_0} = 2$, so $R = k \ln 2$ is a linear relation, because $\ln 2$ is a constant.

(b) $$T(R) = 1.03R$$

$$T(k) = 1.03k \ln \frac{C}{C_0} \qquad \text{Use the given expression for } R.$$

✔ **Now Try Exercise 75.**

Logarithms with Other Bases   We can use a calculator to find the values of either natural logarithms (base e) or common logarithms (base 10). However, sometimes we must use logarithms with other bases. The change-of-base theorem can be used to convert logarithms from one base to another.

Change-of-Base Theorem

For any positive real numbers x, a, and b, where $a \neq 1$ and $b \neq 1$, the following holds.

$$\log_a x = \frac{\log_b x}{\log_b a}$$

Proof Let

$$y = \log_a x.$$

Then

$a^y = x$	Write in exponential form.
$\log_b a^y = \log_b x$	Take the base b logarithm on each side.
$y \log_b a = \log_b x$	Power property
$y = \dfrac{\log_b x}{\log_b a}$	Divide each side by $\log_b a$.
$\log_a x = \dfrac{\log_b x}{\log_b a}.$	Substitute $\log_a x$ for y.

Any positive number other than 1 can be used for base b in the change-of-base theorem, but usually the only practical bases are e and 10 since most calculators give logarithms for these two bases only.

Using the change-of-base theorem, we can graph an equation such as $y = \log_2 x$ by directing the calculator to graph $y = \frac{\log x}{\log 2}$, or, equivalently, $y = \frac{\ln x}{\ln 2}$. ∎

EXAMPLE 8 **Using the Change-of-Base Theorem**

Use the change-of-base theorem to find an approximation to four decimal places for each logarithm.

(a) $\log_5 17$ **(b)** $\log_2 0.1$

SOLUTION

(a) We use natural logarithms to approximate this logarithm. Because $\log_5 5 = 1$ and $\log_5 25 = 2$, we can estimate $\log_5 17$ to be a number between 1 and 2.

$$\log_5 17 = \frac{\ln 17}{\ln 5} \approx 1.7604 \quad \text{Check: } 5^{1.7604} \approx 17$$

The first two entries in **Figure 38(a)** show that the results are the same whether natural or common logarithms are used.

(a)

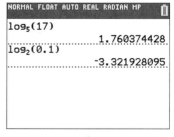

(b)

Figure 38

(b) We use common logarithms for this approximation.

$$\log_2 0.1 = \frac{\log 0.1}{\log 2} \approx -3.3219 \quad \boxed{\text{Check: } 2^{-3.3219} \approx 0.1}$$

The last two entries in **Figure 38(a)** show that the results are the same whether natural or common logarithms are used.

Some calculators, such as the TI-84 Plus, evaluate these logarithms directly without using the change-of-base theorem. See **Figure 38(b).**

✔ **Now Try Exercises 79 and 81.**

EXAMPLE 9   Modeling Diversity of Species

One measure of the diversity of the species in an ecological community is modeled by the formula

$$H = -[P_1 \log_2 P_1 + P_2 \log_2 P_2 + \cdots + P_n \log_2 P_n],$$

where $P_1, P_2, \ldots, P_n$ are the proportions of a sample that belong to each of n species found in the sample. (*Source:* Ludwig, J., and J. Reynolds, *Statistical Ecology: A Primer on Methods and Computing,* © 1988, John Wiley & Sons, NY.)

Find the measure of diversity in a community with two species where there are 90 of one species and 10 of the other.

SOLUTION   There are 100 members in the community, so $P_1 = \frac{90}{100} = 0.9$ and $P_2 = \frac{10}{100} = 0.1$.

$$H = -[0.9 \log_2 0.9 + 0.1 \log_2 0.1] \quad \text{Substitute for } P_1 \text{ and } P_2.$$

In **Example 8(b),** we found that $\log_2 0.1 \approx -3.32$. Now we find $\log_2 0.9$.

$$\log_2 0.9 = \frac{\log 0.9}{\log 2} \approx -0.152 \quad \text{Change-of-base theorem}$$

Now evaluate H.

$$H = -[0.9 \log_2 0.9 + 0.1 \log_2 0.1]$$
$$H \approx -[0.9(-0.152) + 0.1(-3.32)] \quad \text{Substitute approximate values.}$$
$$H \approx 0.469 \quad\quad\quad\quad\quad\quad\quad\quad\quad\quad \text{Simplify.}$$

Verify that $H \approx 0.971$ if there are 60 of one species and 40 of the other. As the proportions of n species get closer to $\frac{1}{n}$ each, the measure of diversity increases to a maximum of $\log_2 n$.

✔ **Now Try Exercise 73.**

We saw previously that graphing calculators are capable of fitting exponential curves to data that suggest such behavior. The same is true for logarithmic curves. For example, during the early 2000s on one particular day, interest rates for various U.S. Treasury Securities were as shown in the table.

Time	3-mo	6-mo	2-yr	5-yr	10-yr	30-yr
Yield	0.83%	0.91%	1.35%	2.46%	3.54%	4.58%

Source: U.S. Treasury.

Figure 39 shows how a calculator gives the best-fitting natural logarithmic curve for the data, as well as the data points and the graph of this curve. ∎

$y = 1.479 + 0.809 \ln x$

Figure 39

4.4 Exercises

CONCEPT PREVIEW *Answer each of the following.*

1. For the exponential function $f(x) = a^x$, where $a > 1$, is the function increasing or decreasing over its entire domain?

2. For the logarithmic function $g(x) = \log_a x$, where $a > 1$, is the function increasing or decreasing over its entire domain?

3. If $f(x) = 5^x$, what is the rule for $f^{-1}(x)$?

4. What is the name given to the exponent to which 4 must be raised to obtain 11?

5. A base e logarithm is called a(n) _____ logarithm, and a base 10 logarithm is called a(n) _____ logarithm.

6. How is $\log_3 12$ written in terms of natural logarithms using the change-of-base theorem?

7. Why is $\log_2 0$ undefined?

8. Between what two consecutive integers must $\log_2 12$ lie?

9. The graph of $y = \log x$ shows a point on the graph. Write the logarithmic equation associated with that point.

10. The graph of $y = \ln x$ shows a point on the graph. Write the logarithmic equation associated with that point.

*Find each value. If applicable, give an approximation to four decimal places. **See Example 1.***

11. $\log 10^{12}$	**12.** $\log 10^7$	**13.** $\log 0.1$	**14.** $\log 0.01$
15. $\log 63$	**16.** $\log 94$	**17.** $\log 0.0022$	**18.** $\log 0.0055$
19. $\log(387 \times 23)$	**20.** $\log(296 \times 12)$	**21.** $\log \dfrac{518}{342}$	**22.** $\log \dfrac{643}{287}$
23. $\log 387 + \log 23$		**24.** $\log 296 + \log 12$	
25. $\log 518 - \log 342$		**26.** $\log 643 - \log 287$	

Answer each question.

27. Why is the result in **Exercise 23** the same as that in **Exercise 19**?

28. Why is the result in **Exercise 25** the same as that in **Exercise 21**?

For each substance, find the pH from the given hydronium ion concentration. See Example 2(a).

29. grapefruit, 6.3×10^{-4}

30. limes, 1.6×10^{-2}

31. crackers, 3.9×10^{-9}

32. sodium hydroxide (lye), 3.2×10^{-14}

Find the $[H_3O^+]$ for each substance with the given pH. See Example 2(b).

33. soda pop, 2.7

34. wine, 3.4

35. beer, 4.8

36. drinking water, 6.5

Suppose that water from a wetland area is sampled and found to have the given hydronium ion concentration. Determine whether the wetland is a rich fen, a poor fen, or a bog. See Example 3.

37. 2.49×10^{-5}

38. 6.22×10^{-5}

39. 2.49×10^{-2}

40. 3.14×10^{-2}

41. 2.49×10^{-7}

42. 5.86×10^{-7}

Solve each problem.

43. Use a calculator to find an approximation for each logarithm.

(a) log 398.4   (b) log 39.84   (c) log 3.984

(d) From the answers to parts (a)–(c), make a conjecture concerning the decimal values in the approximations of common logarithms of numbers greater than 1 that have the same digits.

44. Given that log 25 ≈ 1.3979, log 250 ≈ 2.3979, and log 2500 ≈ 3.3979, make a conjecture for an approximation of log 25,000. Why does this pattern continue?

Find each value. If applicable, give an approximation to four decimal places. See Example 5.

45. $\ln e^{1.6}$

46. $\ln e^{5.8}$

47. $\ln \dfrac{1}{e^2}$

48. $\ln \dfrac{1}{e^4}$

49. $\ln \sqrt{e}$

50. $\ln \sqrt[3]{e}$

51. $\ln 28$

52. $\ln 39$

53. $\ln 0.00013$

54. $\ln 0.0077$

55. $\ln (27 \times 943)$

56. $\ln (33 \times 568)$

57. $\ln \dfrac{98}{13}$

58. $\ln \dfrac{84}{17}$

59. $\ln 27 + \ln 943$

60. $\ln 33 + \ln 568$

61. $\ln 98 - \ln 13$

62. $\ln 84 - \ln 17$

Solve each problem. See Examples 4, 6, 7, and 9.

63. *Decibel Levels* Find the decibel ratings of sounds having the following intensities.

(a) $100I_0$ (b) $1000I_0$ (c) $100,000I_0$ (d) $1,000,000I_0$

(e) If the intensity of a sound is doubled, by how much is the decibel rating increased? Round to the nearest whole number.

64. *Decibel Levels* Find the decibel ratings of the following sounds, having intensities as given. Round each answer to the nearest whole number.

(a) whisper, $115I_0$ (b) busy street, $9,500,000I_0$

(c) heavy truck, 20 m away, $1,200,000,000I_0$

(d) rock music, $895,000,000,000I_0$

(e) jetliner at takeoff, $109,000,000,000,000I_0$

65. *Earthquake Intensity* The magnitude of an earthquake, measured on the Richter scale, is $\log_{10} \frac{I}{I_0}$, where I is the amplitude registered on a seismograph 100 km from the epicenter of the earthquake, and I_0 is the amplitude of an earthquake of a certain (small) size. Find the Richter scale ratings for earthquakes having the following amplitudes.

   (a) $1000 I_0$ (b) $1,000,000 I_0$ (c) $100,000,000 I_0$

66. *Earthquake Intensity* On December 26, 2004, an earthquake struck in the Indian Ocean with a magnitude of 9.1 on the Richter scale. The resulting tsunami killed an estimated 229,900 people in several countries. Express this reading in terms of I_0 to the nearest hundred thousand.

67. *Earthquake Intensity* On February 27, 2010, a massive earthquake struck Chile with a magnitude of 8.8 on the Richter scale. Express this reading in terms of I_0 to the nearest hundred thousand.

68. *Earthquake Intensity Comparison* Compare the answers to **Exercises 66 and 67.** How many times greater was the force of the 2004 earthquake than that of the 2010 earthquake?

69. *(Modeling) Bachelor's Degrees in Psychology* The table gives the number of bachelor's degrees in psychology (in thousands) earned at U.S. colleges and universities for selected years from 1980 through 2012. Suppose x represents the number of years since 1950. Thus, 1980 is represented by 30, 1990 is represented by 40, and so on.

Year	Degrees Earned (in thousands)
1980	42.1
1990	54.0
2000	74.2
2010	97.2
2011	100.9
2012	109.0

Source: National Center for Education Statistics.

The following function is a logarithmic model for the data.

$$f(x) = -273 + 90.6 \ln x$$

Use this function to estimate the number of bachelor's degrees in psychology earned in the year 2016 to the nearest tenth thousand. What assumption must we make to estimate the number of degrees in years beyond 2012?

70. *(Modeling) Domestic Leisure Travel* The bar graph shows numbers of leisure trips within the United States (in millions of person-trips of 50 or more miles one-way) over the years 2009–2014. The function

   $$f(t) = 1458 + 95.42 \ln t, \quad t \geq 1,$$

   where t represents the number of years since 2008 and $f(t)$ is the number of person-trips, in millions, approximates the curve reasonably well.

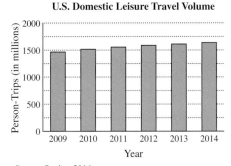

U.S. Domestic Leisure Travel Volume

Source: Statista 2014.

Use the function to approximate the number of person-trips in 2012 to the nearest million. How does this approximation compare to the actual number of 1588 million?

71. *(Modeling) Diversity of Species* The number of species $S(n)$ in a sample is given by

$$S(n) = a \ln \left(1 + \frac{n}{a} \right),$$

where n is the number of individuals in the sample, and a is a constant that indicates the diversity of species in the community. If $a = 0.36$, find $S(n)$ for each value of n. (*Hint:* $S(n)$ must be a whole number.)

(a) 100 (b) 200 (c) 150 (d) 10

72. *(Modeling) Diversity of Species* In **Exercise 71,** find $S(n)$ if a changes to 0.88. Use the following values of n.

(a) 50 (b) 100 (c) 250

73. *(Modeling) Diversity of Species* Suppose a sample of a small community shows two species with 50 individuals each. Find the measure of diversity H.

74. *(Modeling) Diversity of Species* A virgin forest in northwestern Pennsylvania has 4 species of large trees with the following proportions of each:

hemlock, 0.521; beech, 0.324; birch, 0.081; maple, 0.074.

Find the measure of diversity H to the nearest thousandth.

75. *(Modeling) Global Temperature Increase* In **Example 7,** we expressed the average global temperature increase T (in °F) as

$$T(k) = 1.03k \ln \frac{C}{C_0},$$

where C_0 is the preindustrial amount of carbon dioxide, C is the current carbon dioxide level, and k is a constant. Arrhenius determined that $10 \le k \le 16$ when C was double the value C_0. Use $T(k)$ to find the range of the rise in global temperature T (rounded to the nearest degree) that Arrhenius predicted. (*Source:* Clime, W., *The Economics of Global Warming,* Institute for International Economics, Washington, D.C.)

76. *(Modeling) Global Temperature Increase* (Refer to **Exercise 75.**) According to one study by the IPCC, future increases in average global temperatures (in °F) can be modeled by

$$T(C) = 6.489 \ln \frac{C}{280},$$

where C is the concentration of atmospheric carbon dioxide (in ppm). C can be modeled by the function

$$C(x) = 353(1.006)^{x-1990},$$

where x is the year. (*Source:* International Panel on Climate Change (IPCC).)

(a) Write T as a function of x.

(b) Using a graphing calculator, graph $C(x)$ and $T(x)$ on the interval [1990, 2275] using different coordinate axes. Describe the graph of each function. How are C and T related?

(c) Approximate the slope of the graph of T. What does this slope represent?

(d) Use graphing to estimate x and $C(x)$ when $T(x) = 10°F$.

77. *Age of Rocks* Use the formula of **Example 6** to estimate the age of a rock sample having $\frac{A}{K} = 0.103$. Give the answer in billions of years, rounded to the nearest hundredth.

78. *(Modeling) Planets' Distances from the Sun and Periods of Revolution* The table contains the planets' average distances D from the sun and their periods P of revolution around the sun in years. The distances have been normalized so that Earth is one unit away from the sun. For example, since Jupiter's distance is 5.2, its distance from the sun is 5.2 times farther than Earth's.

Planet	D	P
Mercury	0.39	0.24
Venus	0.72	0.62
Earth	1	1
Mars	1.52	1.89
Jupiter	5.2	11.9
Saturn	9.54	29.5
Uranus	19.2	84.0
Neptune	30.1	164.8

Source: Ronan, C., *The Natural History of the Universe*, MacMillan Publishing Co., New York.

(a) Using a graphing calculator, make a scatter diagram by plotting the point $(\ln D, \ln P)$ for each planet on the xy-coordinate axes. Do the data points appear to be linear?

(b) Determine a linear equation that models the data points. Graph the line and the data on the same coordinate axes.

(c) Use this linear model to predict the period of Pluto if its distance is 39.5. Compare the answer to the actual value of 248.5 yr.

Use the change-of-base theorem to find an approximation to four decimal places for each logarithm. See Example 8.

79. $\log_2 5$ **80.** $\log_2 9$ **81.** $\log_8 0.59$ **82.** $\log_8 0.71$

83. $\log_{1/2} 3$ **84.** $\log_{1/3} 2$ **85.** $\log_\pi e$ **86.** $\log_\pi \sqrt{2}$

87. $\log_{\sqrt{13}} 12$ **88.** $\log_{\sqrt{19}} 5$ **89.** $\log_{0.32} 5$ **90.** $\log_{0.91} 8$

Let $u = \ln a$ and $v = \ln b$. Write each expression in terms of u and v without using the ln *function.*

91. $\ln\left(b^4\sqrt{a}\right)$ **92.** $\ln\dfrac{a^3}{b^2}$ **93.** $\ln\sqrt{\dfrac{a^3}{b^5}}$ **94.** $\ln\left(\sqrt[3]{a} \cdot b^4\right)$

Concept Check Use the various properties of exponential and logarithmic functions to evaluate the expressions in parts (a)–(c).

95. Given $g(x) = e^x$, find   (a) $g(\ln 4)$ (b) $g(\ln 5^2)$ (c) $g\left(\ln \frac{1}{e}\right)$.

96. Given $f(x) = 3^x$, find   (a) $f(\log_3 2)$ (b) $f(\log_3 (\ln 3))$ (c) $f(\log_3 (2 \ln 3))$.

97. Given $f(x) = \ln x$, find   (a) $f(e^6)$ (b) $f(e^{\ln 3})$ (c) $f(e^{2 \ln 3})$.

98. Given $f(x) = \log_2 x$, find   (a) $f(2^7)$ (b) $f(2^{\log_2 2})$ (c) $f(2^{2 \log_2 2})$.

Work each problem.

99. *Concept Check* Which of the following is equivalent to $2 \ln (3x)$ for $x > 0$?

A. $\ln 9 + \ln x$ **B.** $\ln 6x$ **C.** $\ln 6 + \ln x$ **D.** $\ln 9x^2$

100. *Concept Check* Which of the following is equivalent to $\ln (4x) - \ln (2x)$ for $x > 0$?

A. $2 \ln x$ **B.** $\ln 2x$ **C.** $\dfrac{\ln 4x}{\ln 2x}$ **D.** $\ln 2$

101. The function $f(x) = \ln |x|$ plays a prominent role in calculus. Find its domain, its range, and the symmetries of its graph.

102. Consider the function $f(x) = \log_3 |x|$.

(a) What is the domain of this function?

(b) Use a graphing calculator to graph $f(x) = \log_3 |x|$ in the window $[-4, 4]$ by $[-4, 4]$.

(c) How might one easily misinterpret the domain of the function by merely observing the calculator graph?

Use properties of logarithms to rewrite each function, and describe how the graph of the given function compares to the graph of $g(x) = \ln x$.

103. $f(x) = \ln(e^2 x)$ **104.** $f(x) = \ln \dfrac{x}{e}$ **105.** $f(x) = \ln \dfrac{x}{e^2}$

Chapter 4   Quiz (Sections 4.1–4.4)

1. For the one-to-one function $f(x) = \sqrt[3]{3x - 6}$, find $f^{-1}(x)$.

2. Solve $4^{2x+1} = 8^{3x-6}$.

3. Graph $f(x) = -3^x$. Give the domain and range.

4. Graph $f(x) = \log_4(x + 2)$. Give the domain and range.

5. *Future Value* Suppose that $15,000 is deposited in a bank certificate of deposit at an annual rate of 2.7% for 8 yr. Find the future value if interest is compounded as follows.

    (a) annually   **(b)** quarterly   **(c)** monthly   **(d)** daily (365 days)

6. Use a calculator to evaluate each logarithm to four decimal places.

    (a) log 34.56   **(b)** ln 34.56

7. What is the meaning of the expression $\log_6 25$?

8. Solve each equation.

    (a) $x = 3^{\log_3 4}$ **(b)** $\log_x 25 = 2$ **(c)** $\log_4 x = -2$

9. Assuming all variables represent positive real numbers, use properties of logarithms to rewrite

$$\log_3 \frac{\sqrt{x} \cdot y}{pq^4}.$$

10. Given $\log_b 9 = 3.1699$ and $\log_b 5 = 2.3219$, find the value of $\log_b 225$.

11. Find the value of $\log_3 40$ to four decimal places.

12. If $f(x) = 4^x$, what is the value of $f(\log_4 12)$?

4.5   Exponential and Logarithmic Equations

- Exponential Equations
- Logarithmic Equations
- Applications and Models

Exponential Equations   We solved exponential equations in earlier sections. General methods for solving these equations depend on the property below, which follows from the fact that logarithmic functions are one-to-one.

Property of Logarithms

If $x > 0$, $y > 0$, $a > 0$, and $a \neq 1$, then the following holds.

$$x = y \quad \text{is equivalent to} \quad \log_a x = \log_a y.$$

EXAMPLE 1 Solving an Exponential Equation

Solve $7^x = 12$. Give the solution to the nearest thousandth.

SOLUTION The properties of exponents cannot be used to solve this equation, so we apply the preceding property of logarithms. While any appropriate base b can be used, the best practical base is base 10 or base e. We choose base e (natural) logarithms here.

As seen in the display at the bottom of the screen, when rounded to three decimal places, the solution of $7^x - 12 = 0$ agrees with that found in **Example 1.**

$$7^x = 12$$

$$\ln 7^x = \ln 12 \qquad \text{Property of logarithms}$$

$$x \ln 7 = \ln 12 \qquad \text{Power property}$$

$$\boxed{\text{This is } exact.} \qquad x = \frac{\ln 12}{\ln 7} \qquad \text{Divide by } \ln 7.$$

$$x \approx 1.277 \qquad \text{Use a calculator.}$$

The solution set is $\{1.277\}$. $\boxed{\text{This is } approximate.}$

✔ **Now Try Exercise 11.**

CAUTION Do not confuse a quotient like $\frac{\ln 12}{\ln 7}$ in **Example 1** with $\ln \frac{12}{7}$, which can be written as $\ln 12 - \ln 7$. *We cannot change the quotient of two logarithms to a difference of logarithms.*

$$\frac{\ln 12}{\ln 7} \neq \ln \frac{12}{7}$$

EXAMPLE 2 Solving an Exponential Equation

Solve $3^{2x-1} = 0.4^{x+2}$. Give the solution to the nearest thousandth.

SOLUTION
$$3^{2x-1} = 0.4^{x+2}$$

$$\ln 3^{2x-1} = \ln 0.4^{x+2} \qquad \text{Take the natural logarithm on each side.}$$

$$(2x - 1) \ln 3 = (x + 2) \ln 0.4 \qquad \text{Power property}$$

$$2x \ln 3 - \ln 3 = x \ln 0.4 + 2 \ln 0.4 \qquad \text{Distributive property}$$

$$2x \ln 3 - x \ln 0.4 = 2 \ln 0.4 + \ln 3 \qquad \begin{array}{l}\text{Write so that the terms with } x \text{ are}\\\text{on one side.}\end{array}$$

$$x(2 \ln 3 - \ln 0.4) = 2 \ln 0.4 + \ln 3 \qquad \text{Factor out } x.$$

$$x = \frac{2 \ln 0.4 + \ln 3}{2 \ln 3 - \ln 0.4} \qquad \text{Divide by } 2 \ln 3 - \ln 0.4.$$

$$x = \frac{\ln 0.4^2 + \ln 3}{\ln 3^2 - \ln 0.4} \qquad \text{Power property}$$

$$x = \frac{\ln 0.16 + \ln 3}{\ln 9 - \ln 0.4} \qquad \text{Apply the exponents.}$$

$$\boxed{\text{This is } exact.} \qquad x = \frac{\ln 0.48}{\ln 22.5} \qquad \text{Product and quotient properties}$$

$$x \approx -0.236 \qquad \text{Use a calculator.}$$

The solution set is $\{-0.236\}$. $\boxed{\text{This is } approximate.}$

✔ **Now Try Exercise 19.**

This screen supports the solution found in **Example 2.**

> **EXAMPLE 3** Solving Base e Exponential Equations

Solve each equation. Give solutions to the nearest thousandth.

(a) $e^{x^2} = 200$ **(b)** $e^{2x+1} \cdot e^{-4x} = 3e$

SOLUTION

(a)

$$e^{x^2} = 200$$

$$\ln e^{x^2} = \ln 200 \qquad \text{Take the natural logarithm on each side.}$$

$$x^2 = \ln 200 \qquad \ln e^{x^2} = x^2$$

> Remember both roots.

$$x = \pm \sqrt{\ln 200} \qquad \text{Square root property}$$

$$x \approx \pm 2.302 \qquad \text{Use a calculator.}$$

The solution set is $\{\pm 2.302\}$.

(b)

$$e^{2x+1} \cdot e^{-4x} = 3e$$

$$e^{-2x+1} = 3e \qquad a^m \cdot a^n = a^{m+n}$$

$$e^{-2x} = 3 \qquad \text{Divide by } e; \frac{e^{-2x+1}}{e^1} = e^{-2x+1-1} = e^{-2x}.$$

$$\ln e^{-2x} = \ln 3 \qquad \text{Take the natural logarithm on each side.}$$

$$-2x \ln e = \ln 3 \qquad \text{Power property}$$

$$-2x = \ln 3 \qquad \ln e = 1$$

$$x = -\frac{1}{2} \ln 3 \qquad \text{Multiply by } -\frac{1}{2}.$$

$$x \approx -0.549 \qquad \text{Use a calculator.}$$

The solution set is $\{-0.549\}$.

✔ **Now Try Exercises 21 and 23.**

> **EXAMPLE 4** Solving an Exponential Equation (Quadratic in Form)

Solve $e^{2x} - 4e^x + 3 = 0$. Give exact value(s) for x.

SOLUTION If we substitute $u = e^x$, we notice that the equation is quadratic in form.

$$e^{2x} - 4e^x + 3 = 0$$

$$(e^x)^2 - 4e^x + 3 = 0 \qquad a^{mn} = (a^n)^m$$

$$u^2 - 4u + 3 = 0 \qquad \text{Let } u = e^x.$$

$$(u - 1)(u - 3) = 0 \qquad \text{Factor.}$$

$$u - 1 = 0 \quad \text{or} \quad u - 3 = 0 \qquad \text{Zero-factor property}$$

$$u = 1 \quad \text{or} \quad u = 3 \qquad \text{Solve for } u.$$

$$e^x = 1 \quad \text{or} \quad e^x = 3 \qquad \text{Substitute } e^x \text{ for } u.$$

$$\ln e^x = \ln 1 \quad \text{or} \quad \ln e^x = \ln 3 \qquad \begin{array}{l}\text{Take the natural logarithm} \\ \text{on each side.}\end{array}$$

$$x = 0 \quad \text{or} \quad x = \ln 3 \qquad \ln e^x = x; \ln 1 = 0$$

Both values check, so the solution set is $\{0, \ln 3\}$.

✔ **Now Try Exercise 35.**

Logarithmic Equations   The following equations involve logarithms of variable expressions.

EXAMPLE 5   Solving Logarithmic Equations

Solve each equation. Give exact values.

(a) $7 \ln x = 28$ $\qquad\qquad\qquad$ **(b)** $\log_2(x^3 - 19) = 3$

SOLUTION

(a) $\qquad\qquad 7 \ln x = 28$

$\qquad\qquad\quad \log_e x = 4 \qquad$ $\ln x = \log_e x$; Divide by 7.

$\qquad\qquad\qquad\quad x = e^4 \qquad$ Write in exponential form.

The solution set is $\{e^4\}$.

(b) $\qquad\qquad \log_2(x^3 - 19) = 3$

$\qquad\qquad\qquad x^3 - 19 = 2^3 \qquad$ Write in exponential form.

$\qquad\qquad\qquad x^3 - 19 = 8 \qquad$ Apply the exponent.

$\qquad\qquad\qquad\qquad x^3 = 27 \qquad$ Add 19.

$\qquad\qquad\qquad\qquad x = \sqrt[3]{27} \qquad$ Take cube roots.

$\qquad\qquad\qquad\qquad x = 3 \qquad$ $\sqrt[3]{27} = 3$

The solution set is $\{3\}$.

✔ **Now Try Exercises 41 and 49.**

EXAMPLE 6   Solving a Logarithmic Equation

Solve $\log(x + 6) - \log(x + 2) = \log x$. Give exact value(s).

SOLUTION   Recall that logarithms are defined only for nonnegative numbers.

$$\log(x + 6) - \log(x + 2) = \log x$$

$\qquad\qquad \log \dfrac{x + 6}{x + 2} = \log x \qquad$ Quotient property

$\qquad\qquad\quad \dfrac{x + 6}{x + 2} = x \qquad$ Property of logarithms

$\qquad\qquad\quad x + 6 = x(x + 2) \qquad$ Multiply by $x + 2$.

$\qquad\qquad\quad x + 6 = x^2 + 2x \qquad$ Distributive property

$\qquad\quad x^2 + x - 6 = 0 \qquad$ Standard form

$\qquad (x + 3)(x - 2) = 0 \qquad$ Factor.

$x + 3 = 0 \quad$ or $\quad x - 2 = 0 \qquad$ Zero-factor property

$x = -3 \quad$ or $\qquad x = 2 \qquad$ Solve for x.

The proposed negative solution (-3) is not in the domain of $\log x$ in the original equation, so the only valid solution is the positive number 2. The solution set is $\{2\}$.

✔ **Now Try Exercise 69.**

CAUTION Recall that the domain of $y = \log_a x$ is $(0, \infty)$. *For this reason, it is always necessary to check that proposed solutions of a logarithmic equation result in logarithms of positive numbers in the original equation.*

EXAMPLE 7 **Solving a Logarithmic Equation**

Solve $\log_2 [(3x - 7)(x - 4)] = 3$. Give exact value(s).

SOLUTION

$$\log_2 [(3x - 7)(x - 4)] = 3$$

$$(3x - 7)(x - 4) = 2^3 \qquad \text{Write in exponential form.}$$

$$3x^2 - 19x + 28 = 8 \qquad \text{Multiply. Apply the exponent.}$$

$$3x^2 - 19x + 20 = 0 \qquad \text{Standard form}$$

$$(3x - 4)(x - 5) = 0 \qquad \text{Factor.}$$

$$3x - 4 = 0 \quad \text{or} \quad x - 5 = 0 \qquad \text{Zero-factor property}$$

$$x = \frac{4}{3} \quad \text{or} \qquad x = 5 \qquad \text{Solve for } x.$$

A check is necessary to be sure that the argument of the logarithm in the given equation is positive. In both cases, the product $(3x - 7)(x - 4)$ leads to 8, and $\log_2 8 = 3$ is true. The solution set is $\left\{\frac{4}{3}, 5\right\}$.

✔ **Now Try Exercise 53.**

EXAMPLE 8 **Solving a Logarithmic Equation**

Solve $\log(3x + 2) + \log(x - 1) = 1$. Give exact value(s).

SOLUTION

$$\log(3x + 2) + \log(x - 1) = 1$$

$$\log_{10}[(3x + 2)(x - 1)] = 1 \qquad \log x = \log_{10} x; \text{ product property}$$

$$(3x + 2)(x - 1) = 10^1 \qquad \text{Write in exponential form.}$$

$$3x^2 - x - 2 = 10 \qquad \text{Multiply; } 10^1 = 10.$$

$$3x^2 - x - 12 = 0 \qquad \text{Subtract 10.}$$

$$x = \frac{-b \pm \sqrt{b^2 - 4ac}}{2a}$$

Quadratic formula

$$x = \frac{-(-1) \pm \sqrt{(-1)^2 - 4(3)(-12)}}{2(3)}$$

Substitute $a = 3, b = -1, c = -12$.

The two proposed solutions are

$$\frac{1 - \sqrt{145}}{6} \quad \text{and} \quad \frac{1 + \sqrt{145}}{6}.$$

The first proposed solution, $\frac{1 - \sqrt{145}}{6}$, is negative. Substituting for x in $\log(x - 1)$ results in a negative argument, which is not allowed. Therefore, this solution must be rejected.

The second proposed solution, $\frac{1 + \sqrt{145}}{6}$, is positive. Substituting it for x in $\log(3x + 2)$ results in a positive argument. Substituting it for x in $\log(x + 1)$ also results in a positive argument. Both are necessary conditions. Therefore, the solution set is $\left\{\frac{1 + \sqrt{145}}{6}\right\}$.

✔ **Now Try Exercise 77.**

NOTE We could have replaced 1 with $\log_{10} 10$ in **Example 8** by first writing

$$\log(3x + 2) + \log(x - 1) = 1 \qquad \text{Equation from \textbf{Example 8}}$$
$$\log_{10}[(3x + 2)(x - 1)] = \log_{10} 10 \quad \text{Substitute.}$$
$$(3x + 2)(x - 1) = 10, \qquad \text{Property of logarithms}$$

and then continuing as shown on the preceding page.

EXAMPLE 9 Solving a Base e Logarithmic Equation

Solve $\ln e^{\ln x} - \ln(x - 3) = \ln 2$. Give exact value(s).

SOLUTION This logarithmic equation differs from those in **Examples 7 and 8** because the expression on the right side involves a logarithm.

$$\ln e^{\ln x} - \ln(x - 3) = \ln 2$$
$$\ln x - \ln(x - 3) = \ln 2 \qquad e^{\ln x} = x$$
$$\ln \frac{x}{x - 3} = \ln 2 \qquad \text{Quotient property}$$
$$\frac{x}{x - 3} = 2 \qquad \text{Property of logarithms}$$
$$x = 2(x - 3) \quad \text{Multiply by } x - 3.$$
$$x = 2x - 6 \qquad \text{Distributive property}$$
$$x = 6 \qquad \text{Solve for } x.$$

Check that the solution set is $\{6\}$.

✔ **Now Try Exercise 79.**

Solving an Exponential or Logarithmic Equation

To solve an exponential or logarithmic equation, change the given equation into one of the following forms, where a and b are real numbers, $a > 0$ and $a \neq 1$, and follow the guidelines.

1. $a^{f(x)} = b$

   Solve by taking logarithms on each side.

2. $\log_a f(x) = b$

   Solve by changing to exponential form $a^b = f(x)$.

3. $\log_a f(x) = \log_a g(x)$

   The given equation is equivalent to the equation $f(x) = g(x)$. Solve algebraically.

4. In a more complicated equation, such as

   $$e^{2x+1} \cdot e^{-4x} = 3e, \quad \text{See \textbf{Example 3(b)}.}$$

   it may be necessary to first solve for $a^{f(x)}$ or $\log_a f(x)$ and then solve the resulting equation using one of the methods given above.

5. Check that each proposed solution is in the domain.

Applications and Models

EXAMPLE 10 **Applying an Exponential Equation to the Strength of a Habit**

The strength of a habit is a function of the number of times the habit is repeated. If N is the number of repetitions and H is the strength of the habit, then, according to psychologist C.L. Hull,

$$H = 1000(1 - e^{-kN}),$$

where k is a constant. Solve this equation for k.

SOLUTION

$H = 1000(1 - e^{-kN})$ *First solve for e^{-kN}.*

$$\dfrac{H}{1000} = 1 - e^{-kN} \qquad \text{Divide by 1000.}$$

$$\dfrac{H}{1000} - 1 = -e^{-kN} \qquad \text{Subtract 1.}$$

$$e^{-kN} = 1 - \dfrac{H}{1000} \qquad \text{Multiply by } -1 \text{ and rewrite.}$$

Now solve for k.

$$\ln e^{-kN} = \ln\left(1 - \dfrac{H}{1000}\right) \qquad \text{Take the natural logarithm on each side.}$$

$$-kN = \ln\left(1 - \dfrac{H}{1000}\right) \qquad \ln e^x = x$$

$$k = -\dfrac{1}{N}\ln\left(1 - \dfrac{H}{1000}\right) \qquad \text{Multiply by } -\tfrac{1}{N}.$$

With the final equation, if one pair of values for H and N is known, k can be found, and the equation can then be used to find either H or N for given values of the other variable.

✔ **Now Try Exercise 91.**

EXAMPLE 11 **Modeling PC Tablet Sales in the U.S.**

The table gives U.S. tablet sales (in millions) for several years. The data can be modeled by the function

$$f(t) = 20.57 \ln t + 10.58, \quad t \geq 1,$$

where t is the number of years after 2009.

Year	Sales (in millions)
2010	10.3
2011	24.1
2012	35.1
2013	39.8
2014	42.1

Source: Forrester Research.

(a) Use the function to estimate the number of tablets sold in the United States in 2015.

(b) If this trend continues, approximately when will annual sales reach 60 million?

SOLUTION

(a) The year 2015 is represented by $t = 2015 - 2009 = 6$.

$$f(t) = 20.57 \ln t + 10.58 \qquad \text{Given function}$$

$$f(6) = 20.57 \ln 6 + 10.58 \qquad \text{Let } t = 6.$$

$$f(6) \approx 47.4 \qquad \text{Use a calculator.}$$

Based on this model, 47.4 million tablets were sold in 2015.

(b) Replace $f(t)$ with 60 and solve for t.

$$f(t) = 20.57 \ln t + 10.58 \quad \text{Given function}$$

$$60 = 20.57 \ln t + 10.58 \quad \text{Let } f(t) = 60.$$

$$49.42 = 20.57 \ln t \quad \text{Subtract 10.58.}$$

$$\ln t = \frac{49.42}{20.57} \quad \text{Divide by 20.57 and rewrite.}$$

$$t = e^{49.42/20.57} \quad \text{Write in exponential form.}$$

$$t \approx 11.05 \quad \text{Use a calculator.}$$

Adding 11 to 2009 gives the year 2020. Based on this model, annual sales will reach 60 million in 2020.

✔ **Now Try Exercise 111.**

4.5 Exercises

CONCEPT PREVIEW *Match each equation in Column I with the best first step for solving it in Column II.*

I	II
1. $10^x = 150$	**A.** Use the product rule for exponents.
2. $e^{2x-1} = 24$	**B.** Take the common logarithm on each side.
3. $\log_4(x^2 - 10) = 2$	**C.** Write the sum of logarithms as the logarithm of a product.
4. $e^{2x} \cdot e^x = 2e$	**D.** Let $u = e^x$ and write the equation in quadratic form.
5. $2e^{2x} - 5e^x - 3 = 0$	**E.** Change to exponential form.
6. $\log(2x - 1) + \log(x + 4) = 1$	**F.** Take the natural logarithm on each side.

CONCEPT PREVIEW *An exponential equation such as*

$$5^x = 9$$

can be solved for its exact solution using the meaning of logarithm and the change-of-base theorem. Because x is the exponent to which 5 must be raised in order to obtain 9, the exact solution is

$$\log_5 9, \quad \text{or} \quad \frac{\log 9}{\log 5}, \quad \text{or} \quad \frac{\ln 9}{\ln 5}.$$

For each equation, give the exact solution in three forms similar to the forms above.

7. $7^x = 19$ **8.** $3^x = 10$ **9.** $\left(\frac{1}{2}\right)^x = 12$ **10.** $\left(\frac{1}{3}\right)^x = 4$

Solve each equation. In Exercises 11–34, give irrational solutions as decimals correct to the nearest thousandth. In Exercises 35–40, give solutions in exact form. **See Examples 1–4.**

11. $3^x = 7$ **12.** $5^x = 13$ **13.** $\left(\frac{1}{2}\right)^x = 5$

14. $\left(\frac{1}{3}\right)^x = 6$ **15.** $0.8^x = 4$ **16.** $0.6^x = 3$

17. $4^{x-1} = 3^{2x}$ **18.** $2^{x+3} = 5^{2x}$ **19.** $6^{x+1} = 4^{2x-1}$

20. $3^{x-4} = 7^{2x+5}$ **21.** $e^{x^2} = 100$ **22.** $e^{x^4} = 1000$

23. $e^{3x-7} \cdot e^{-2x} = 4e$ **24.** $e^{1-3x} \cdot e^{5x} = 2e$ **25.** $\left(\dfrac{1}{3}\right)^x = -3$

26. $\left(\dfrac{1}{9}\right)^x = -9$ **27.** $0.05(1.15)^x = 5$ **28.** $1.2(0.9)^x = 0.6$

29. $3(2)^{x-2} + 1 = 100$ **30.** $5(1.2)^{3x-2} + 1 = 7$ **31.** $2(1.05)^x + 3 = 10$

32. $3(1.4)^x - 4 = 60$ **33.** $5(1.015)^{x-1980} = 8$ **34.** $6(1.024)^{x-1900} = 9$

35. $e^{2x} - 6e^x + 8 = 0$ **36.** $e^{2x} - 8e^x + 15 = 0$ **37.** $2e^{2x} + e^x = 6$

38. $3e^{2x} + 2e^x = 1$ **39.** $5^{2x} + 3(5^x) = 28$ **40.** $3^{2x} - 12(3^x) = -35$

Solve each equation. Give solutions in exact form. ***See Examples 5–9.***

41. $5 \ln x = 10$ **42.** $3 \ln x = 9$

43. $\ln 4x = 1.5$ **44.** $\ln 2x = 5$

45. $\log(2 - x) = 0.5$ **46.** $\log(3 - x) = 0.75$

47. $\log_6(2x + 4) = 2$ **48.** $\log_5(8 - 3x) = 3$

49. $\log_4(x^3 + 37) = 3$ **50.** $\log_7(x^3 + 65) = 0$

51. $\ln x + \ln x^2 = 3$ **52.** $\log x + \log x^2 = 3$

53. $\log_3[(x + 5)(x - 3)] = 2$ **54.** $\log_4[(3x + 8)(x - 6)] = 3$

55. $\log_2[(2x + 8)(x + 4)] = 5$ **56.** $\log_5[(3x + 5)(x + 1)] = 1$

57. $\log x + \log(x + 15) = 2$ **58.** $\log x + \log(2x + 1) = 1$

59. $\log(x + 25) = \log(x + 10) + \log 4$ **60.** $\log(3x + 5) - \log(2x + 4) = 0$

61. $\log(x - 10) - \log(x - 6) = \log 2$ **62.** $\log(x^2 - 9) - \log(x - 3) = \log 5$

63. $\ln(7 - x) + \ln(1 - x) = \ln(25 - x)$ **64.** $\ln(3 - x) + \ln(5 - x) = \ln(50 - 6x)$

65. $\log_8(x + 2) + \log_8(x + 4) = \log_8 8$ **66.** $\log_2(5x - 6) - \log_2(x + 1) = \log_2 3$

67. $\log_2(x^2 - 100) - \log_2(x + 10) = 1$ **68.** $\log_2(x - 2) + \log_2(x - 1) = 1$

69. $\log x + \log(x - 21) = \log 100$ **70.** $\log x + \log(3x - 13) = \log 10$

71. $\log(9x + 5) = 3 + \log(x + 2)$ **72.** $\log(11x + 9) = 3 + \log(x + 3)$

73. $\ln(4x - 2) - \ln 4 = -\ln(x - 2)$ **74.** $\ln(5 + 4x) - \ln(3 + x) = \ln 3$

75. $\log_5(x + 2) + \log_5(x - 2) = 1$ **76.** $\log_2(x - 7) + \log_2 x = 3$

77. $\log_2(2x - 3) + \log_2(x + 1) = 1$ **78.** $\log_5(3x + 2) + \log_5(x - 1) = 1$

79. $\ln e^x - 2 \ln e = \ln e^4$ **80.** $\ln e^x - \ln e^3 = \ln e^3$

81. $\log_2(\log_2 x) = 1$ **82.** $\log x = \sqrt{\log x}$

83. $\log x^2 = (\log x)^2$ **84.** $\log_2 \sqrt{2x^2} = \dfrac{3}{2}$

85. *Concept Check* Consider the following statement: "We must reject any negative proposed solution when we solve an equation involving logarithms." Is this correct? Why or why not?

86. *Concept Check* What values of x could not possibly be solutions of the following equation?

$$\log_a(4x - 7) + \log_a(x^2 + 4) = 0$$

Solve each equation for the indicated variable. Use logarithms with the appropriate bases. **See Example 10.**

87. $p = a + \dfrac{k}{\ln x}$,   for x

88. $r = p - k \ln t$,   for t

89. $T = T_0 + (T_1 - T_0)10^{-kt}$,   for t

90. $A = \dfrac{Pr}{1 - (1 + r)^{-n}}$,   for n

91. $I = \dfrac{E}{R}(1 - e^{-Rt/2})$,   for t

92. $y = \dfrac{K}{1 + ae^{-bx}}$,   for b

93. $y = A + B(1 - e^{-Cx})$,   for x

94. $m = 6 - 2.5 \log \dfrac{M}{M_0}$,   for M

95. $\log A = \log B - C \log x$,   for A

96. $d = 10 \log \dfrac{I}{I_0}$,   for I

97. $A = P\left(1 + \dfrac{r}{n}\right)^{tn}$,   for t

98. $D = 160 + 10 \log x$,   for x

To solve each problem, refer to the formulas for compound interest.

$$A = P\left(1 + \frac{r}{n}\right)^{tn} \quad \text{and} \quad A = Pe^{rt}$$

99. *Compound Amount* If $10,000 is invested in an account at 3% annual interest compounded quarterly, how much will be in the account in 5 yr if no money is withdrawn?

100. *Compound Amount* If $5000 is invested in an account at 4% annual interest compounded continuously, how much will be in the account in 8 yr if no money is withdrawn?

101. *Investment Time* Kurt wants to buy a $30,000 truck. He has saved $27,000. Find the number of years (to the nearest tenth) it will take for his $27,000 to grow to $30,000 at 4% interest compounded quarterly.

102. *Investment Time* Find t to the nearest hundredth of a year if $1786 becomes $2063 at 2.6%, with interest compounded monthly.

103. *Interest Rate* Find the interest rate to the nearest hundredth of a percent that will produce $2500, if $2000 is left at interest compounded semiannually for 8.5 yr.

104. *Interest Rate* At what interest rate, to the nearest hundredth of a percent, will $16,000 grow to $20,000 if invested for 7.25 yr and interest is compounded quarterly?

(Modeling) Solve each application. **See Example 11.**

105. In the central Sierra Nevada (a mountain range in California), the percent of moisture that falls as snow rather than rain is approximated reasonably well by

$$f(x) = 86.3 \ln x - 680,$$

where x is the altitude in feet and $f(x)$ is the percent of moisture that falls as snow. Find the percent of moisture, to the nearest tenth, that falls as snow at each altitude.

  (a) 3000 ft   **(b)** 4000 ft   **(c)** 7000 ft

106. Northwest Creations finds that its total sales in dollars, $T(x)$, from the distribution of x thousand catalogues is approximated by

$$T(x) = 5000 \log(x + 1).$$

Find the total sales, to the nearest dollar, resulting from the distribution of each number of catalogues.

  (a) 5000   **(b)** 24,000   **(c)** 49,000

107. *Average Annual Public University Costs* The table shows the cost of a year's tuition, room and board, and fees at 4-year public colleges for the years 2006–2014. Letting y represent the cost in dollars and x the number of years since 2006, the function

$$f(x) = 13,017(1.05)^x$$

models the data quite well. According to this function, in what year will the 2006 cost be doubled?

Year	Average Annual Cost
2006	$12,837
2007	$13,558
2008	$14,372
2009	$15,235
2010	$16,178
2011	$17,156
2012	$17,817
2013	$18,383
2014	$18,943

Source: The College Board, *Annual Survey of Colleges.*

108. *Race Speed* At the World Championship races held at Rome's Olympic Stadium in 1987, American sprinter Carl Lewis ran the 100-m race in 9.86 sec. His speed in meters per second after t seconds is closely modeled by the function

$$f(t) = 11.65(1 - e^{-t/1.27}).$$

(*Source:* Banks, Robert B., *Towing Icebergs, Falling Dominoes, and Other Adventures in Applied Mathematics,* Princeton University Press.)

(a) How fast, to the nearest hundredth, was he running as he crossed the finish line?

(b) After how many seconds, to the nearest hundredth, was he running at the rate of 10 m per sec?

109. *Women in Labor Force* The percent of women in the U.S. civilian labor force can be modeled fairly well by the function

$$f(x) = \frac{67.21}{1 + 1.081e^{-x/24.71}},$$

where x represents the number of years since 1950. (*Source: Monthly Labor Review,* U.S. Bureau of Labor Statistics.)

(a) What percent, to the nearest whole number, of U.S. women were in the civilian labor force in 2014?

(b) In what year were 55% of U.S. women in the civilian labor force?

110. *Height of the Eiffel Tower* One side of the Eiffel Tower in Paris has a shape that can be approximated by the graph of the function

$$f(x) = -301 \ln \frac{x}{207}, \quad x > 0,$$

where x and $f(x)$ are both measured in feet. (*Source:* Banks, Robert B., *Towing Icebergs, Falling Dominoes, and Other Adventures in Applied Mathematics,* Princeton University Press.)

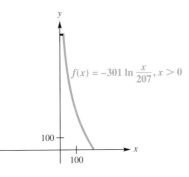

(a) Why does the shape of the left side of the graph of the Eiffel Tower have the formula given by $f(-x)$?

(b) The short horizontal segment at the top of the figure has length 7.8744 ft. How tall, to the nearest foot, is the Eiffel Tower?

(c) How far from the center of the tower is the point on the right side that is 500 ft above the ground? Round to the nearest foot.

111. *CO_2 Emissions Tax* One action that government could take to reduce carbon emissions into the atmosphere is to levy a tax on fossil fuel. This tax would be based on the amount of carbon dioxide emitted into the air when the fuel is burned. The **cost-benefit equation**

$$\ln(1 - P) = -0.0034 - 0.0053x$$

models the approximate relationship between a tax of x dollars per ton of carbon and the corresponding percent reduction P (in decimal form) of emissions of carbon dioxide. (*Source:* Nordhause, W., "To Slow or Not to Slow: The Economics of the Greenhouse Effect," Yale University, New Haven, Connecticut.)

(a) Write P as a function of x.

(b) Graph P for $0 \le x \le 1000$. Discuss the benefit of continuing to raise taxes on carbon.

(c) Determine P, to the nearest tenth, when $x = \$60$. Interpret this result.

(d) What value of x will give a 50% reduction in carbon emissions?

112. *Radiative Forcing* Radiative forcing, R, measures the influence of carbon dioxide in altering the additional solar radiation trapped in Earth's atmosphere. The International Panel on Climate Change (IPCC) in 1990 estimated k to be 6.3 in the radiative forcing equation

$$R = k \ln \frac{C}{C_0},$$

where C_0 is the preindustrial amount of carbon dioxide and C is the current level. (*Source:* Clime, W., *The Economics of Global Warming,* Institute for International Economics, Washington, D.C.)

(a) Use the equation $R = 6.3 \ln \frac{C}{C_0}$ to determine the radiative forcing R (in watts per square meter to the nearest tenth) expected by the IPCC if the carbon dioxide level in the atmosphere doubles from its preindustrial level.

(b) Determine the global temperature increase T, to the nearest tenth, that the IPCC predicted would occur if atmospheric carbon dioxide levels were to double, given $T(R) = 1.03R$.

Find $f^{-1}(x)$, and give the domain and range.

113. $f(x) = e^{x-5}$ **114.** $f(x) = e^x + 10$ **115.** $f(x) = e^{x+1} - 4$

116. $f(x) = \ln(x + 2)$ **117.** $f(x) = 2 \ln 3x$ **118.** $f(x) = \ln(x - 1) + 6$

Use a graphing calculator to solve each equation. Give irrational solutions correct to the nearest hundredth.

119. $e^x + \ln x = 5$ **120.** $e^x - \ln(x + 1) = 3$ **121.** $2e^x + 1 = 3e^{-x}$

122. $e^x + 6e^{-x} = 5$ **123.** $\log x = x^2 - 8x + 14$ **124.** $\ln x = -\sqrt[3]{x + 3}$

125. Find the **error** in the following **"proof"** that $2 < 1$.

$$\frac{1}{9} < \frac{1}{3} \qquad \text{True statement}$$

$$\left(\frac{1}{3}\right)^2 < \frac{1}{3} \qquad \text{Rewrite the left side.}$$

$$\log\left(\frac{1}{3}\right)^2 < \log \frac{1}{3} \qquad \text{Take the logarithm on each side.}$$

$$2 \log \frac{1}{3} < 1 \log \frac{1}{3} \qquad \text{Property of logarithms; identity property}$$

$$2 < 1 \qquad \text{Divide each side by } \log \tfrac{1}{3}.$$

4.6 Applications and Models of Exponential Growth and Decay

- The Exponential Growth or Decay Function
- Growth Function Models
- Decay Function Models

LOOKING AHEAD TO CALCULUS

The exponential growth and decay function formulas are studied in calculus in conjunction with the topic known as **differential equations.**

The Exponential Growth or Decay Function　In many situations in ecology, biology, economics, and the social sciences, a quantity changes at a rate proportional to the amount present. The amount present at time t is a special function of t called an **exponential growth or decay function.**

Exponential Growth or Decay Function

Let y_0 be the amount or number present at time $t = 0$. Then, under certain conditions, the amount y present at any time t is modeled by

$$y = y_0 e^{kt}, \quad \text{where } k \text{ is a constant.}$$

The constant k determines the type of function.

- When $k > 0$, the function describes growth. Examples of exponential growth include compound interest and atmospheric carbon dioxide.

- When $k < 0$, the function describes decay. One example of exponential decay is radioactive decay.

Growth Function Models　The amount of time it takes for a quantity that grows exponentially to become twice its initial amount is its **doubling time.**

Year	Carbon Dioxide (ppm)
1990	353
2000	375
2075	590
2175	1090
2275	2000

Source: International Panel on Climate Change (IPCC).

EXAMPLE 1　**Determining a Function to Model Exponential Growth**

Earlier in this chapter, we discussed the growth of atmospheric carbon dioxide over time using a function based on the data from the table. Now we determine such a function from the data.

(a) Find an exponential function that gives the amount of carbon dioxide y in year x.

(b) Estimate the year when future levels of carbon dioxide will be double the preindustrial level of 280 ppm.

SOLUTION

(a) The data points exhibit exponential growth, so the equation will take the form

$$y = y_0 e^{kx}.$$

We must find the values of y_0 and k. The data begin with the year 1990, so to simplify our work we let 1990 correspond to $x = 0$, 1991 correspond to $x = 1$, and so on. Here y_0 is the initial amount and $y_0 = 353$ in 1990 when $x = 0$. Thus the equation is

$$y = 353e^{kx}. \quad \text{Let } y_0 = 353.$$

From the last pair of values in the table, we know that in 2275 the carbon dioxide level is expected to be 2000 ppm. The year 2275 corresponds to $2275 - 1990 = 285$. Substitute 2000 for y and 285 for x, and solve for k.

$$y = 353e^{kx} \qquad \text{Solve for } k.$$

$$2000 = 353e^{k(285)} \qquad \text{Substitute 2000 for } y \text{ and 285 for } x.$$

$$\frac{2000}{353} = e^{285k} \qquad \text{Divide by 353.}$$

$$\ln \frac{2000}{353} = \ln e^{285k} \qquad \text{Take the natural logarithm on each side.}$$

$$\ln \frac{2000}{353} = 285k \qquad \ln e^x = x, \text{ for all } x.$$

$$k = \frac{1}{285} \cdot \ln \frac{2000}{353} \qquad \text{Multiply by } \tfrac{1}{285} \text{ and rewrite.}$$

$$k \approx 0.00609 \qquad \text{Use a calculator.}$$

A function that models the data is

$$y = 353e^{0.00609x}.$$

(b)

$$y = 353e^{0.00609x} \qquad \text{Solve the model from part (a) for the year } x.$$

$$560 = 353e^{0.00609x} \qquad \text{To double the level 280, let } y = 2(280) = 560.$$

$$\frac{560}{353} = e^{0.00609x} \qquad \text{Divide by 353.}$$

$$\ln \frac{560}{353} = \ln e^{0.00609x} \qquad \text{Take the natural logarithm on each side.}$$

$$\ln \frac{560}{353} = 0.00609x \qquad \ln e^x = x, \text{ for all } x.$$

$$x = \frac{1}{0.00609} \cdot \ln \frac{560}{353} \qquad \text{Multiply by } \tfrac{1}{0.00609} \text{ and rewrite.}$$

$$x \approx 75.8 \qquad \text{Use a calculator.}$$

Since $x = 0$ corresponds to 1990, the preindustrial carbon dioxide level will double in the 75th year after 1990, or during 2065, according to this model.

✔ **Now Try Exercise 43.**

EXAMPLE 2 Finding Doubling Time for Money

How long will it take for money in an account that accrues interest at a rate of 3%, compounded continuously, to double?

SOLUTION

$$A = Pe^{rt} \qquad \text{Continuous compounding formula}$$

$$2P = Pe^{0.03t} \qquad \text{Let } A = 2P \text{ and } r = 0.03.$$

$$2 = e^{0.03t} \qquad \text{Divide by } P.$$

$$\ln 2 = \ln e^{0.03t} \qquad \text{Take the natural logarithm on each side.}$$

$$\ln 2 = 0.03t \qquad \ln e^x = x$$

$$\frac{\ln 2}{0.03} = t \qquad \text{Divide by 0.03.}$$

$$23.10 \approx t \qquad \text{Use a calculator.}$$

It will take about 23 yr for the amount to double. ✔ **Now Try Exercise 31.**

EXAMPLE 3 **Using an Exponential Function to Model Population Growth**

According to the U.S. Census Bureau, the world population reached 6 billion people during 1999 and was growing exponentially. By the end of 2010, the population had grown to 6.947 billion. The projected world population (in billions of people) t years after 2010 is given by the function

$$f(t) = 6.947e^{0.00745t}.$$

(a) Based on this model, what will the world population be in 2025?

(b) If this trend continues, approximately when will the world population reach 9 billion?

SOLUTION

(a) Since $t = 0$ represents the year 2010, in 2025, t would be $2025 - 2010 = 15$ yr. We must find $f(t)$ when t is 15.

$$f(t) = 6.947e^{0.00745t} \qquad \text{Given function}$$
$$f(15) = 6.947e^{0.00745(15)} \qquad \text{Let } t = 15.$$
$$f(15) \approx 7.768 \qquad \text{Use a calculator.}$$

The population will be 7.768 billion at the end of 2025.

(b)
$$f(t) = 6.947e^{0.00745t} \qquad \text{Given function}$$
$$9 = 6.947e^{0.00745t} \qquad \text{Let } f(t) = 9.$$

$$\frac{9}{6.947} = e^{0.00745t} \qquad \text{Divide by 6.947.}$$

$$\ln \frac{9}{6.947} = \ln e^{0.00745t} \qquad \text{Take the natural logarithm on each side.}$$

$$\ln \frac{9}{6.947} = 0.00745t \qquad \ln e^x = x, \text{ for all } x.$$

$$t = \frac{\ln \frac{9}{6.947}}{0.00745} \qquad \text{Divide by 0.00745 and rewrite.}$$

$$t \approx 34.8 \qquad \text{Use a calculator.}$$

Thus, 34.8 yr after 2010, during the year 2044, world population will reach 9 billion.

✔ **Now Try Exercise 39.**

Decay Function Models **Half-life** is the amount of time it takes for a quantity that decays exponentially to become half its initial amount.

NOTE In **Example 4** on the next page, the initial amount of substance is given as 600 g. Because half-life is constant over the lifetime of a decaying quantity, starting with any initial amount, y_0, and substituting $\frac{1}{2}y_0$ for y in $y = y_0 e^{kt}$ would allow the common factor y_0 to be divided out. The rest of the work would be the same.

EXAMPLE 4 Determining an Exponential Function to Model Radioactive Decay

Suppose 600 g of a radioactive substance are present initially and 3 yr later only 300 g remain.

(a) Determine an exponential function that models this decay.

(b) How much of the substance will be present after 6 yr?

SOLUTION

(a) We use the given values to find k in the exponential equation

$$y = y_0 e^{kt}.$$

Because the initial amount is 600 g, $y_0 = 600$, which gives $y = 600e^{kt}$. The initial amount (600 g) decays to half that amount (300 g) in 3 yr, so its half-life is 3 yr. Now we solve this exponential equation for k.

$y = 600e^{kt}$	Let $y_0 = 600$.
$300 = 600e^{3k}$	Let $y = 300$ and $t = 3$.
$0.5 = e^{3k}$	Divide by 600.
$\ln 0.5 = \ln e^{3k}$	Take the natural logarithm on each side.
$\ln 0.5 = 3k$	$\ln e^x = x$, for all x.
$\dfrac{\ln 0.5}{3} = k$	Divide by 3.
$k \approx -0.231$	Use a calculator.

A function that models the situation is

$$y = 600e^{-0.231t}.$$

(b) To find the amount present after 6 yr, let $t = 6$.

$y = 600e^{-0.231t}$	Model from part (a)
$y = 600e^{-0.231(6)}$	Let $t = 6$.
$y = 600e^{-1.386}$	Multiply.
$y \approx 150$	Use a calculator.

After 6 yr, 150 g of the substance will remain. ✔ **Now Try Exercise 19.**

EXAMPLE 5 Solving a Carbon Dating Problem

Carbon-14, also known as radiocarbon, is a radioactive form of carbon that is found in all living plants and animals. After a plant or animal dies, the radiocarbon disintegrates. Scientists can determine the age of the remains by comparing the amount of radiocarbon with the amount present in living plants and animals. This technique is called **carbon dating.** The amount of radiocarbon present after t years is given by

$$y = y_0 e^{-0.0001216t},$$

where y_0 is the amount present in living plants and animals.

(a) Find the half-life of carbon-14.

(b) Charcoal from an ancient fire pit on Java contained $\frac{1}{4}$ the carbon-14 of a living sample of the same size. Estimate the age of the charcoal.

SOLUTION

(a) If y_0 is the amount of radiocarbon present in a living thing, then $\frac{1}{2}y_0$ is half this initial amount. We substitute and solve the given equation for t.

$$y = y_0 e^{-0.0001216t} \qquad \text{Given equation}$$

$$\frac{1}{2}y_0 = y_0 e^{-0.0001216t} \qquad \text{Let } y = \tfrac{1}{2}y_0.$$

$$\frac{1}{2} = e^{-0.0001216t} \qquad \text{Divide by } y_0.$$

$$\ln \frac{1}{2} = \ln e^{-0.0001216t} \qquad \text{Take the natural logarithm on each side.}$$

$$\ln \frac{1}{2} = -0.0001216t \qquad \ln e^x = x, \text{ for all } x.$$

$$\frac{\ln \frac{1}{2}}{-0.0001216} = t \qquad \text{Divide by } -0.0001216.$$

$$5700 \approx t \qquad \text{Use a calculator.}$$

The half-life is 5700 yr.

(b) Solve again for t, this time letting the amount $y = \frac{1}{4}y_0$.

$$y = y_0 e^{-0.0001216t} \qquad \text{Given equation}$$

$$\frac{1}{4}y_0 = y_0 e^{-0.0001216t} \qquad \text{Let } y = \tfrac{1}{4}y_0.$$

$$\frac{1}{4} = e^{-0.0001216t} \qquad \text{Divide by } y_0.$$

$$\ln \frac{1}{4} = \ln e^{-0.0001216t} \qquad \text{Take the natural logarithm on each side.}$$

$$\frac{\ln \frac{1}{4}}{-0.0001216} = t \qquad \ln e^x = x; \text{ Divide by } -0.0001216.$$

$$t \approx 11{,}400 \qquad \text{Use a calculator.}$$

The charcoal is 11,400 yr old. ✔ **Now Try Exercise 23.**

EXAMPLE 6 **Modeling Newton's Law of Cooling**

Newton's law of cooling says that the rate at which a body cools is proportional to the difference in temperature between the body and the environment around it. The temperature $f(t)$ of the body at time t in appropriate units after being introduced into an environment having constant temperature T_0 is

$$f(t) = T_0 + Ce^{-kt}, \quad \text{where } C \text{ and } k \text{ are constants.}$$

A pot of coffee with a temperature of 100°C is set down in a room with a temperature of 20°C. The coffee cools to 60°C after 1 hr.

(a) Write an equation to model the data.

(b) Find the temperature after half an hour.

(c) How long will it take for the coffee to cool to 50°C?

SOLUTION

(a) We must find values for C and k in the given formula. As given, when $t = 0$, $T_0 = 20$, and the temperature of the coffee is $f(0) = 100$.

$$f(t) = T_0 + Ce^{-kt} \quad \text{Given function}$$

$$100 = 20 + Ce^{-0k} \quad \text{Let } t = 0, f(0) = 100, \text{ and } T_0 = 20.$$

$$100 = 20 + C \quad e^0 = 1$$

$$80 = C \quad \text{Subtract 20.}$$

The following function models the data.

$$f(t) = 20 + 80e^{-kt} \quad \text{Let } T_0 = 20 \text{ and } C = 80.$$

The coffee cools to 60°C after 1 hr, so when $t = 1$, $f(1) = 60$.

$$f(t) = 20 + 80e^{-kt} \quad \text{Above function with } T_0 = 20 \text{ and } C = 80$$

$$60 = 20 + 80e^{-1k} \quad \text{Let } t = 1 \text{ and } f(1) = 60.$$

$$40 = 80e^{-k} \quad \text{Subtract 20.}$$

$$\frac{1}{2} = e^{-k} \quad \text{Divide by 80.}$$

$$\ln \frac{1}{2} = \ln e^{-k} \quad \text{Take the natural logarithm on each side.}$$

$$\ln \frac{1}{2} = -k \quad \ln e^x = x, \text{ for all } x.$$

$$k \approx 0.693 \quad \text{Multiply by } -1, \text{ rewrite, and use a calculator.}$$

Thus, the model is $f(t) = 20 + 80e^{-0.693t}$.

(b) To find the temperature after $\frac{1}{2}$ hr, let $t = \frac{1}{2}$ in the model from part (a).

$$f(t) = 20 + 80e^{-0.693t} \quad \text{Model from part (a)}$$

$$f\left(\frac{1}{2}\right) = 20 + 80e^{(-0.693)(1/2)} \quad \text{Let } t = \frac{1}{2}.$$

$$f\left(\frac{1}{2}\right) \approx 76.6°\text{C} \quad \text{Use a calculator.}$$

(c) To find how long it will take for the coffee to cool to 50°C, let $f(t) = 50$.

$$f(t) = 20 + 80e^{-0.693t} \quad \text{Model from part (a)}$$

$$50 = 20 + 80e^{-0.693t} \quad \text{Let } f(t) = 50.$$

$$30 = 80e^{-0.693t} \quad \text{Subtract 20.}$$

$$\frac{3}{8} = e^{-0.693t} \quad \text{Divide by 80.}$$

$$\ln \frac{3}{8} = \ln e^{-0.693t} \quad \text{Take the natural logarithm on each side.}$$

$$\ln \frac{3}{8} = -0.693t \quad \ln e^x = x, \text{ for all } x.$$

$$t = \frac{\ln \frac{3}{8}}{-0.693} \quad \text{Divide by } -0.693 \text{ and rewrite.}$$

$$t \approx 1.415 \text{ hr, } \quad \text{or} \quad \text{about 1 hr, 25 min} \quad \checkmark \text{ Now Try Exercise 27.}$$

4.6 Exercises

CONCEPT PREVIEW *Population Growth* A population is increasing according to the exponential function

$$y = 2e^{0.02x},$$

where *y* is in millions and *x* is the number of years. Match each question in Column I with the correct procedure in Column II to answer the question.

I	**II**
1. How long will it take for the population to triple?	**A.** Evaluate $y = 2e^{0.02(1/3)}$.
2. When will the population reach 3 million?	**B.** Solve $2e^{0.02x} = 6$.
3. How large will the population be in 3 yr?	**C.** Evaluate $y = 2e^{0.02(3)}$.
4. How large will the population be in 4 months?	**D.** Solve $2e^{0.02x} = 3$.

CONCEPT PREVIEW *Radioactive Decay* Strontium-90 decays according to the exponential function

$$y = y_0 e^{-0.0241t},$$

where *t* is time in years. Match each question in Column I with the correct procedure in Column II to answer the question.

I	**II**
5. If the initial amount of Strontium-90 is 200 g, how much will remain after 10 yr?	**A.** Solve $0.75y_0 = y_0 e^{-0.0241t}$.
6. If the initial amount of Strontium-90 is 200 g, how much will remain after 20 yr?	**B.** Evaluate $y = 200e^{-0.0241(10)}$.
7. What is the half-life of Strontium-90?	**C.** Solve $\frac{1}{2}y_0 = y_0 e^{-0.0241t}$.
8. How long will it take for any amount of Strontium-90 to decay to 75% of its initial amount?	**D.** Evaluate $y = 200e^{-0.0241(20)}$.

(Modeling) The exercises in this set are grouped according to discipline. They involve exponential or logarithmic models. **See Examples 1–6.**

Physical Sciences *(Exercises 9–28)*

An initial amount of a radioactive substance y_0 is given, along with information about the amount remaining after a given time t in appropriate units. For an equation of the form $y = y_0 e^{kt}$ that models the situation, give the exact value of k in terms of natural logarithms.

9. $y_0 = 60$ g; After 3 hr, 20 g remain.   **10.** $y_0 = 30$ g; After 6 hr, 10 g remain.

11. $y_0 = 10$ mg; The half-life is 100 days.   **12.** $y_0 = 20$ mg; The half-life is 200 days.

13. $y_0 = 2.4$ lb; After 2 yr, 0.6 lb remains.   **14.** $y_0 = 8.1$ kg; After 4 yr, 0.9 kg remains.

Solve each problem.

15. *Decay of Lead* A sample of 500 g of radioactive lead-210 decays to polonium-210 according to the function

$$A(t) = 500e^{-0.032t},$$

where *t* is time in years. Find the amount of radioactive lead remaining after

(a) 4 yr,   **(b)** 8 yr,   **(c)** 20 yr.   **(d)** Find the half-life.

16. *Decay of Plutonium* Repeat **Exercise 15** for 500 g of plutonium-241, which decays according to the function $A(t) = A_0 e^{-0.053t}$, where t is time in years.

17. *Decay of Radium* Find the half-life of radium-226, which decays according to the function $A(t) = A_0 e^{-0.00043t}$, where t is time in years.

18. *Decay of Tritium* Find the half-life of tritium, a radioactive isotope of hydrogen, which decays according to the function $A(t) = A_0 e^{-0.056t}$, where t is time in years.

19. *Radioactive Decay* If 12 g of a radioactive substance are present initially and 4 yr later only 6.0 g remain, how much of the substance will be present after 7 yr?

20. *Radioactive Decay* If 1 g of strontium-90 is present initially, and 2 yr later 0.95 g remains, how much strontium-90 will be present after 5 yr?

21. *Decay of Iodine* How long will it take any quantity of iodine-131 to decay to 25% of its initial amount, knowing that it decays according to the exponential function $A(t) = A_0 e^{-0.087t}$, where t is time in days?

22. *Magnitude of a Star* The magnitude M of a star is modeled by

$$M = 6 - \frac{5}{2} \log \frac{I}{I_0},$$

where I_0 is the intensity of a just-visible star and I is the actual intensity of the star being measured. The dimmest stars are of magnitude 6, and the brightest are of magnitude 1. Determine the ratio of light intensities between a star of magnitude 1 and a star of magnitude 3.

23. *Carbon-14 Dating* Suppose an Egyptian mummy is discovered in which the amount of carbon-14 present is only about one-third the amount found in living human beings. How long ago did the Egyptian die?

24. *Carbon-14 Dating* A sample from a refuse deposit near the Strait of Magellan had 60% of the carbon-14 of a contemporary sample. How old was the sample?

25. *Carbon-14 Dating* Paint from the Lascaux caves of France contains 15% of the normal amount of carbon-14. Estimate the age of the paintings.

26. *Dissolving a Chemical* The amount of a chemical that will dissolve in a solution increases exponentially as the (Celsius) temperature t is increased according to the model

$$A(t) = 10e^{0.0095t}.$$

At what temperature will 15 g dissolve?

27. *Newton's Law of Cooling* Boiling water, at 100°C, is placed in a freezer at 0°C. The temperature of the water is 50°C after 24 min. Find the temperature of the water to the nearest hundredth after 96 min. (*Hint:* Change minutes to hours.)

28. *Newton's Law of Cooling* A piece of metal is heated to 300°C and then placed in a cooling liquid at 50°C. After 4 min, the metal has cooled to 175°C. Find its temperature to the nearest hundredth after 12 min. (*Hint:* Change minutes to hours.)

Finance (Exercises 29–34)

29. *Comparing Investments* Russ, who is self-employed, wants to invest $60,000 in a pension plan. One investment offers 3% compounded quarterly. Another offers 2.75% compounded continuously.

(a) Which investment will earn more interest in 5 yr?

(b) How much more will the better plan earn?

30. *Growth of an Account* If Russ (see **Exercise 29**) chooses the plan with continuous compounding, how long will it take for his $60,000 to grow to $70,000?

31. *Doubling Time* Find the doubling time of an investment earning 2.5% interest if interest is compounded continuously.

32. *Doubling Time* If interest is compounded continuously and the interest rate is tripled, what effect will this have on the time required for an investment to double?

33. *Growth of an Account* How long will it take an investment to triple if interest is compounded continuously at 3%?

34. *Growth of an Account* Use the Table feature of a graphing calculator to find how long it will take $1500 invested at 2.75% compounded daily to triple in value. Zoom in on the solution by systematically decreasing the increment for x. Find the answer to the nearest day. (Find the answer to the nearest day by eventually letting the increment of x equal $\frac{1}{365}$. The decimal part of the solution can be multiplied by 365 to determine the number of days greater than the nearest year. For example, if the solution is determined to be 16.2027 yr, then multiply 0.2027 by 365 to get 73.9855. The solution is then, to the nearest day, 16 yr, 74 days.) Confirm the answer algebraically.

Social Sciences (Exercises 35–44)

35. *Legislative Turnover* The turnover of legislators is a problem of interest to political scientists. It was found that one model of legislative turnover in a particular body was

$$M(t) = 434e^{-0.08t},$$

where $M(t)$ represents the number of continuously serving members at time t. Here, $t = 0$ represents 1965, $t = 1$ represents 1966, and so on. Use this model to approximate the number of continuously serving members in each year.

(a) 1969   **(b)** 1973   **(c)** 1979

36. *Legislative Turnover* Use the model in **Exercise 35** to determine the year in which the number of continuously serving members was 338.

37. *Population Growth* In 2000 India's population reached 1 billion, and it is projected to be 1.4 billion in 2025. (*Source:* U.S. Census Bureau.)

(a) Find values for P_0 and a so that $P(x) = P_0 a^{x-2000}$ models the population of India in year x. Round a to five decimal places.

(b) Predict India's population in 2020 to the nearest tenth of a billion.

(c) In what year is India's population expected to reach 1.5 billion?

38. *Population Decline* A midwestern city finds its residents moving to the suburbs. Its population is declining according to the function

$$P(t) = P_0 e^{-0.04t},$$

where t is time measured in years and P_0 is the population at time $t = 0$. Assume that $P_0 = 1,000,000$.

(a) Find the population at time $t = 1$ to the nearest thousand.

(b) How long, to the nearest tenth of a year, will it take for the population to decline to 750,000?

(c) How long, to the nearest tenth of a year, will it take for the population to decline to half the initial number?

39. *Health Care Spending* Out-of-pocket spending in the United States for health care increased between 2008 and 2012. The function

$$f(x) = 7446e^{0.0305x}$$

models average annual expenditures per household, in dollars. In this model, x represents the year, where $x = 0$ corresponds to 2008. (*Source:* U.S. Bureau of Labor Statistics.)

(a) Estimate out-of-pocket household spending on health care in 2012 to the nearest dollar.

(b) In what year did spending reach $7915 per household?

40. *Recreational Expenditures* Personal consumption expenditures for recreation in billions of dollars in the United States during the years 2000–2013 can be approximated by the function

$$A(t) = 632.37e^{0.0351t},$$

where $t = 0$ corresponds to the year 2000. Based on this model, how much were personal consumption expenditures in 2013 to the nearest billion? (*Source:* U.S. Bureau of Economic Analysis.)

41. *Housing Costs* Average annual per-household spending on housing over the years 2000–2012 is approximated by

$$H = 12{,}744e^{0.0264t},$$

where t is the number of years since 2000. Find H to the nearest dollar for each year. (*Source:* U.S. Bureau of Labor Statistics.)

(a) 2005   **(b)** 2009   **(c)** 2012

42. *Evolution of Language* The number of years, n, since two independently evolving languages split off from a common ancestral language is approximated by

$$n \approx -7600 \log r,$$

where r is the proportion of words from the ancestral language common to both languages. Find each of the following to the nearest year.

(a) Find n if $r = 0.9$.   **(b)** Find n if $r = 0.3$.

(c) How many years have elapsed since the split if half of the words of the ancestral language are common to both languages?

43. *School District Growth* Student enrollment in the Wentzville School District, one of the fastest-growing school districts in the state of Missouri, has projected growth as shown in the graph.

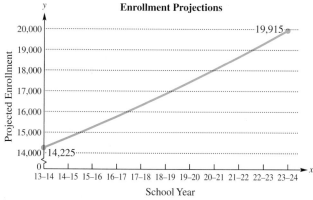

Source: Wentzville School District.

(a) Use the model $y = y_0 e^{kx}$ to find an exponential function that gives the projected enrollment y in school year x. Let the school year 2013–14 correspond to $x = 0$, 2014–15 correspond to $x = 1$, and so on, and use the two points indicated on the graph.

(b) Estimate the school year for which projected enrollment will be 21,500 students.

44. *YouTube Views* The number of views of a YouTube video increases after the number of hours posted as shown in the table.

Hour	Number of Views
20	100
25	517
30	2015
35	10,248

(a) Use the model $y = y_0 e^{kx}$ to find an exponential function that gives projected number of views y after number of hours x. Let hour 20 correspond to $x = 0$, hour 25 correspond to $x = 5$, and so on, and use the first and last data values given in the table.

(b) Estimate the number of views after 50 hr.

Life Sciences (Exercises 45–50)

45. *Spread of Disease* During an epidemic, the number of people who have never had the disease and who are not immune (they are *susceptible*) decreases exponentially according to the function

$$f(t) = 15{,}000e^{-0.05t},$$

where t is time in days. Find the number of susceptible people at each time.

(a) at the beginning of the epidemic   (b) after 10 days   (c) after 3 weeks

46. *Spread of Disease* Refer to **Exercise 45** and determine how long it will take, to the nearest day, for the initial number of people susceptible to decrease to half its amount.

47. *Growth of Bacteria* The growth of bacteria makes it necessary to time-date some food products so that they will be sold and consumed before the bacteria count is too high. Suppose for a certain product the number of bacteria present is given by

$$f(t) = 500e^{0.1t},$$

where t is time in days and the value of $f(t)$ is in millions. Find the number of bacteria, in millions, present at each time.

(a) 2 days   (b) 4 days   (c) 1 week

48. *Growth of Bacteria* How long will it take the bacteria population in **Exercise 47** to double? Round the answer to the nearest tenth.

49. *Medication Effectiveness* Drug effectiveness decreases over time. If, each hour, a drug is only 90% as effective as the previous hour, at some point the patient will not be receiving enough medication and must receive another dose. If the initial dose was 200 mg and the drug was administered 3 hr ago, the expression $200(0.90)^3$, which equals 145.8, represents the amount of effective medication still in the system. (The exponent is equal to the number of hours since the drug was administered.)

The amount of medication still available in the system is given by the function

$$f(t) = 200(0.90)^t.$$

In this model, t is in hours and $f(t)$ is in milligrams. How long will it take for this initial dose to reach the dangerously low level of 50 mg? Round the answer to the nearest tenth.

50. *Population Size* Many environmental situations place effective limits on the growth of the number of an organism in an area. Many such limited-growth situations are described by the **logistic function**

$$G(x) = \frac{MG_0}{G_0 + (M - G_0)e^{-kMx}},$$

where G_0 is the initial number present, M is the maximum possible size of the population, and k is a positive constant. The screens illustrate a typical logistic function calculation and graph.

Assume that $G_0 = 100$, $M = 2500$, $k = 0.0004$, and $x =$ time in decades (10-yr periods).

(a) Use a calculator to graph the function, using $0 \le x \le 8$ and $0 \le y \le 2500$.

(b) Estimate the value of $G(2)$ from the graph. Then evaluate $G(2)$ algebraically to find the population after 20 yr.

(c) Find the x-coordinate of the intersection of the curve with the horizontal line $y = 1000$ to estimate the number of decades required for the population to reach 1000. Then solve $G(x) = 1000$ algebraically to obtain the exact value of x.

Economics (Exercises 51–56)

51. *Consumer Price Index* The U.S. Consumer Price Index for the years 1990–2013 is approximated by

$$A(t) = 100e^{0.0264t},$$

where t represents the number of years after 1990. (Since $A(16)$ is about 153, the amount of goods that could be purchased for $100 in 1990 cost about $153 in 2006.) Use the function to determine the year in which costs will be 125% higher than in 1990. (*Source:* U.S. Bureau of Labor Statistics.)

52. *Product Sales* Sales of a product, under relatively stable market conditions but in the absence of promotional activities such as advertising, tend to decline at a constant yearly rate. This rate of sales decline varies considerably from product to product, but it seems to remain the same for any particular product. The sales decline can be expressed by the function

$$S(t) = S_0 e^{-at},$$

where $S(t)$ is the rate of sales at time t measured in years, S_0 is the rate of sales at time $t = 0$, and a is the sales decay constant.

(a) Suppose the sales decay constant for a particular product is $a = 0.10$. Let $S_0 = 50{,}000$ and find $S(1)$ and $S(3)$ to the nearest thousand.

(b) Find $S(2)$ and $S(10)$ to the nearest thousand if $S_0 = 80{,}000$ and $a = 0.05$.

53. *Product Sales* Use the sales decline function given in **Exercise 52.** If $a = 0.10$, $S_0 = 50{,}000$, and t is time measured in years, find the number of years it will take for sales to fall to half the initial sales. Round the answer to the nearest tenth.

54. *Cost of Bread* Assume the cost of a loaf of bread is $4. With continuous compounding, find the number of years, to the nearest tenth, it would take for the cost to triple at an annual inflation rate of 4%.

55. *Electricity Consumption* Suppose that in a certain area the consumption of electricity has increased at a continuous rate of 6% per year. If it continued to increase at this rate, find the number of years, to the nearest tenth, before twice as much electricity would be needed.

56. *Electricity Consumption* Suppose a conservation campaign, together with higher rates, caused demand for electricity to increase at only 2% per year. (See **Exercise 55.**) Find the number of years, to the nearest tenth, before twice as much electricity would be needed.

(Modeling) Solve each problem that uses a logistic function.

57. *Heart Disease* As age increases, so does the likelihood of coronary heart disease (CHD). The fraction of people x years old with some CHD is modeled by

$$f(x) = \frac{0.9}{1 + 271e^{-0.122x}}.$$

(*Source:* Hosmer, D., and S. Lemeshow, *Applied Logistic Regression,* John Wiley and Sons.)

(a) Evaluate $f(25)$ and $f(65)$ to the nearest hundredth. Interpret the results.

(b) At what age, to the nearest year, does this likelihood equal 50%?

58. *Tree Growth* The height of a certain tree in feet after x years is modeled by

$$f(x) = \frac{50}{1 + 47.5e^{-0.22x}}.$$

(a) Make a table for f starting at $x = 10$, and incrementing by 10. What appears to be the maximum height of the tree?

(b) Graph f and identify the horizontal asymptote. Explain its significance.

(c) After how many years was the tree 30 ft tall? Round to the nearest tenth.

Summary Exercises on Functions: Domains and Defining Equations

Finding the Domain of a Function: A Summary   To find the domain of a function, given the equation that defines the function, remember that the value of x input into the equation must yield a real number for y when the function is evaluated. For the functions studied so far in this book, there are three cases to consider when determining domains.

Guidelines for Domain Restrictions

1. No input value can lead to 0 in a denominator, because division by 0 is undefined.

2. No input value can lead to an even root of a negative number, because this situation does not yield a real number.

3. No input value can lead to the logarithm of a negative number or 0, because this situation does not yield a real number.

Unless otherwise specified, we determine domains as follows.

- The domain of a **polynomial function** is the set of all real numbers.

- The domain of an **absolute value function** is the set of all real numbers for which the expression inside the absolute value bars (the argument) is defined.

- If a **function is defined by a rational expression,** the domain is the set of all real numbers for which the denominator is not zero.

- The domain of a **function defined by a radical with *even* root index** is the set of all real numbers that make the radicand greater than or equal to zero.

  If the root index is *odd*, the domain is the set of all real numbers for which the radicand is itself a real number.

- For an **exponential function** with constant base, the domain is the set of all real numbers for which the exponent is a real number.

- For a **logarithmic function,** the domain is the set of all real numbers that make the argument of the logarithm greater than zero.

Determining Whether an Equation Defines *y* as a Function of *x*

For y to be a function of x, it is necessary that every input value of x in the domain leads to one and only one value of y.

To determine whether an equation such as

$$x - y^3 = 0 \quad \text{or} \quad x - y^2 = 0$$

represents a function, solve the equation for *y*. In the first equation above, doing so leads to

$$y = \sqrt[3]{x}.$$

Notice that every value of *x* in the domain (that is, all real numbers) leads to one and only one value of *y*. So in the first equation, we can write *y* as a function of *x*. However, in the second equation above, solving for *y* leads to

$$y = \pm\sqrt{x}.$$

If we let *x* = 4, for example, we get two values of *y*: −2 and 2. Thus, in the second equation, we cannot write *y* as a function of *x*.

EXERCISES

Find the domain of each function. Write answers using interval notation.

1. $f(x) = 3x - 6$

2. $f(x) = \sqrt{2x - 7}$

3. $f(x) = |x + 4|$

4. $f(x) = \dfrac{x + 2}{x - 6}$

5. $f(x) = \dfrac{-2}{x^2 + 7}$

6. $f(x) = \sqrt{x^2 - 9}$

7. $f(x) = \dfrac{x^2 + 7}{x^2 - 9}$

8. $f(x) = \sqrt[3]{x^3 + 7x - 4}$

9. $f(x) = \log_5(16 - x^2)$

10. $f(x) = \log \dfrac{x + 7}{x - 3}$

11. $f(x) = \sqrt{x^2 - 7x - 8}$

12. $f(x) = 2^{1/x}$

13. $f(x) = \dfrac{1}{2x^2 - x + 7}$

14. $f(x) = \dfrac{x^2 - 25}{x + 5}$

15. $f(x) = \sqrt{x^3 - 1}$

16. $f(x) = \ln|x^2 - 5|$ **17.** $f(x) = e^{x^2 + x + 4}$ **18.** $f(x) = \dfrac{x^3 - 1}{x^2 - 1}$

19. $f(x) = \sqrt{\dfrac{-1}{x^3 - 1}}$ **20.** $f(x) = \sqrt[3]{\dfrac{1}{x^3 - 8}}$

21. $f(x) = \ln(x^2 + 1)$ **22.** $f(x) = \sqrt{(x-3)(x+2)(x-4)}$

23. $f(x) = \log\left(\dfrac{x+2}{x-3}\right)^2$ **24.** $f(x) = \sqrt[12]{(4-x)^2(x+3)}$

25. $f(x) = e^{|1/x|}$ **26.** $f(x) = \dfrac{1}{|x^2 - 7|}$

27. $f(x) = x^{100} - x^{50} + x^2 + 5$ **28.** $f(x) = \sqrt{-x^2 - 9}$

29. $f(x) = \sqrt[4]{16 - x^4}$ **30.** $f(x) = \sqrt[3]{16 - x^4}$

31. $f(x) = \sqrt{\dfrac{x^2 - 2x - 63}{x^2 + x - 12}}$ **32.** $f(x) = \sqrt[5]{5 - x}$

33. $f(x) = \left|\sqrt{5 - x}\right|$ **34.** $f(x) = \sqrt{\dfrac{-1}{x - 3}}$

35. $f(x) = \log\left|\dfrac{1}{4 - x}\right|$ **36.** $f(x) = 6^{x^2 - 9}$

37. $f(x) = 6^{\sqrt{x^2 - 25}}$ **38.** $f(x) = 6^{\sqrt[3]{x^2 - 25}}$

39. $f(x) = \ln\left(\dfrac{-3}{(x+2)(x-6)}\right)$ **40.** $f(x) = \dfrac{-2}{\log x}$

Determine which one of the choices (A, B, C, or D) is an equation in which y can be written as a function of x.

41. A. $3x + 2y = 6$ **B.** $x = \sqrt{|y|}$ **C.** $x = |y + 3|$ **D.** $x^2 + y^2 = 9$

42. A. $3x^2 + 2y^2 = 36$ **B.** $x^2 + y - 2 = 0$ **C.** $x - |y| = 0$ **D.** $x = y^2 - 4$

43. A. $x = \sqrt{y^2}$ **B.** $x = \log y^2$ **C.** $x^3 + y^3 = 5$ **D.** $x = \dfrac{1}{y^2 + 3}$

44. A. $\dfrac{x^2}{4} + \dfrac{y^2}{4} = 1$ **B.** $x = 5y^2 - 3$ **C.** $\dfrac{x^2}{4} - \dfrac{y^2}{9} = 1$ **D.** $x = 10^y$

45. A. $x = \dfrac{2 - y}{y + 3}$ **B.** $x = \ln(y + 1)^2$ **C.** $\sqrt{x} = |y + 1|$ **D.** $\sqrt[4]{x} = y^2$

46. A. $e^{y^2} = x$ **B.** $e^{y+2} = x$ **C.** $e^{|y|} = x$ **D.** $10^{|y+2|} = x$

47. A. $x^2 = \dfrac{1}{y^2}$ **B.** $x + 2 = \dfrac{1}{y^2}$ **C.** $3x = \dfrac{1}{y^4}$ **D.** $2x = \dfrac{1}{y^3}$

48. A. $|x| = |y|$ **B.** $x = |y^2|$ **C.** $x = \dfrac{1}{y}$ **D.** $x^4 + y^4 = 81$

49. A. $\dfrac{x^2}{4} - \dfrac{y^2}{9} = 1$ **B.** $\dfrac{y^2}{4} - \dfrac{x^2}{9} = 1$ **C.** $\dfrac{x}{4} - \dfrac{y}{9} = 0$ **D.** $\dfrac{x^2}{4} - \dfrac{y^2}{9} = 0$

50. A. $y^2 - \sqrt{(x+2)^2} = 0$ **B.** $y - \sqrt{(x+2)^2} = 0$

C. $y^6 - \sqrt{(x+1)^2} = 0$ **D.** $y^4 - \sqrt{x^2} = 0$

Chapter 4 Test Prep

Key Terms

4.1 one-to-one function
inverse function
4.2 exponential function
exponential equation
compound interest

future value
present value
compound amount
continuous
  compounding

4.3 logarithm
base
argument
logarithmic equation
logarithmic function

4.4 common logarithm
pH
natural logarithm
4.6 doubling time
half-life

New Symbols

$f^{-1}(x)$ inverse of $f(x)$

e a constant, approximately 2.718281828459045

$\log_a x$ logarithm of x with the base a

$\log x$ common (base 10) logarithm of x

$\ln x$ natural (base e) logarithm of x

Quick Review

Concepts	Examples

4.1   Inverse Functions

One-to-One Function

In a one-to-one function, each x-value corresponds to only one y-value, and each y-value corresponds to only one x-value.

A function f is one-to-one if, for elements a and b in the domain of f,

$$a \neq b \quad \text{implies} \quad f(a) \neq f(b).$$

The function $y = f(x) = x^2$ is not one-to-one, because $y = 16$, for example, corresponds to both $x = 4$ and $x = -4$.

Horizontal Line Test

A function is one-to-one if every horizontal line intersects the graph of the function at most once.

The graph of $f(x) = 2x - 1$ is a straight line with slope 2. f is a one-to-one function by the horizontal line test.

Inverse Functions

Let f be a one-to-one function. Then g is the inverse function of f if

$$(f \circ g)(x) = x \quad \text{for every } x \text{ in the domain of } g$$

and

$$(g \circ f)(x) = x \quad \text{for every } x \text{ in the domain of } f.$$

To find $g(x)$, interchange x and y in $y = f(x)$, solve for y, and replace y with $g(x)$, which is $f^{-1}(x)$.

Find the inverse of f.

$$f(x) = 2x - 1 \qquad \text{Given function}$$
$$y = 2x - 1 \qquad \text{Let } y = f(x).$$
$$x = 2y - 1 \qquad \text{Interchange } x \text{ and } y.$$
$$y = \frac{x + 1}{2} \qquad \text{Solve for } y.$$
$$f^{-1}(x) = \frac{x + 1}{2} \qquad \text{Replace } y \text{ with } f^{-1}(x).$$
$$f^{-1}(x) = \frac{1}{2}x + \frac{1}{2} \qquad \frac{x+1}{2} = \frac{x}{2} + \frac{1}{2} = \frac{1}{2}x + \frac{1}{2}$$

Concepts	Examples

4.2 Exponential Functions

Additional Properties of Exponents
For any real number $a > 0$, $a \neq 1$, the following hold true.

(a) a^x is a unique real number for all real numbers x.

(b) $a^b = a^c$ if and only if $b = c$.

(c) If $a > 1$ and $m < n$, then $a^m < a^n$.

(d) If $0 < a < 1$ and $m < n$, then $a^m > a^n$.

(a) 2^x is a unique real number for all real numbers x.

(b) $2^x = 2^3$ if and only if $x = 3$.

(c) $2^5 < 2^{10}$, because $2 > 1$ and $5 < 10$.

(d) $\left(\frac{1}{2}\right)^5 > \left(\frac{1}{2}\right)^{10}$ because $0 < \frac{1}{2} < 1$ and $5 < 10$.

Exponential Function
If $a > 0$ and $a \neq 1$, then the exponential function with base a is $\boldsymbol{f(x) = a^x}$.

$f(x) = 3^x$ is the exponential function with base 3.

Graph of $\boldsymbol{f(x) = a^x}$

1. The points $\left(-1, \frac{1}{a}\right)$, $(0, 1)$, and $(1, a)$ are on the graph.

2. If $a > 1$, then f is an increasing function.
If $0 < a < 1$, then f is a decreasing function.

3. The x-axis is a horizontal asymptote.

4. The domain is $(-\infty, \infty)$, and the range is $(0, \infty)$.

x	y
-1	$\frac{1}{3}$
0	1
1	3

4.3 Logarithmic Functions

Logarithm
For all real numbers y and all positive numbers a and x, where $a \neq 1$, $\boldsymbol{y = \log_a x}$ is equivalent to $\boldsymbol{x = a^y}$.

$\log_3 81 = 4$ is equivalent to $3^4 = 81$.

Logarithmic Function
If $a > 0$, $a \neq 1$, and $x > 0$, then the logarithmic function with base a is $\boldsymbol{f(x) = \log_a x}$.

$f(x) = \log_3 x$ is the logarithmic function with base 3.

Graph of $\boldsymbol{f(x) = \log_a x}$

1. The points $\left(\frac{1}{a}, -1\right)$, $(1, 0)$, and $(a, 1)$ are on the graph.

2. If $a > 1$, then f is an increasing function.
If $0 < a < 1$, then f is a decreasing function.

3. The y-axis is a vertical asymptote.

4. The domain is $(0, \infty)$, and the range is $(-\infty, \infty)$.

x	y
$\frac{1}{3}$	-1
1	0
3	1

Properties of Logarithms
For $x > 0$, $y > 0$, $a > 0$, $a \neq 1$, and any real number r, the following properties hold.

$\log_a xy = \log_a x + \log_a y$ Product property

$\log_a \dfrac{x}{y} = \log_a x - \log_a y$ Quotient property

$\log_a x^r = r \log_a x$ Power property

$\log_a 1 = 0$ Logarithm of 1

$\log_a a = 1$ Base a logarithm of a

$\log_2(3 \cdot 5) = \log_2 3 + \log_2 5$

$\log_2 \dfrac{3}{5} = \log_2 3 - \log_2 5$

$\log_6 3^5 = 5 \log_6 3$

$\log_{10} 1 = 0$

$\log_{10} 10 = 1$

Theorem on Inverses
For $a > 0$ and $a \neq 1$, the following properties hold.

$$a^{\log_a x} = x \;\; (x > 0) \quad \text{and} \quad \log_a a^x = x$$

$2^{\log_2 5} = 5 \quad \text{and} \quad \log_2 2^5 = 5$

Concepts	Examples

4.4 Evaluating Logarithms and the Change-of-Base Theorem

Common and Natural Logarithms
For all positive numbers x, base 10 logarithms and base e logarithms are written as follows.

$$\log x = \log_{10} x \quad \text{Common logarithm}$$

$$\ln x = \log_e x \quad \text{Natural logarithm}$$

Approximate $\log 0.045$ and $\ln 247.1$.

$$\log 0.045 \approx -1.3468$$
$$\ln 247.1 \approx 5.5098$$
Use a calculator.

Change-of-Base Theorem
For any positive real numbers x, a, and b, where $a \neq 1$ and $b \neq 1$, the following holds.

$$\log_a x = \frac{\log_b x}{\log_b a}$$

Approximate $\log_8 7$.

$$\log_8 7 = \frac{\log 7}{\log 8} = \frac{\ln 7}{\ln 8} \approx 0.9358 \quad \text{Use a calculator.}$$

4.5 Exponential and Logarithmic Equations

Property of Logarithms
If $x > 0$, $y > 0$, $a > 0$, and $a \neq 1$, then the following holds.

$$x = y \quad \text{is equivalent to} \quad \log_a x = \log_a y.$$

Solve.

$$e^{5x} = 10$$
$$\ln e^{5x} = \ln 10 \quad \text{Take natural logarithms.}$$
$$5x = \ln 10 \quad \ln e^x = x, \text{ for all } x.$$
$$x = \frac{\ln 10}{5} \quad \text{Divide by 5.}$$
$$x \approx 0.461 \quad \text{Use a calculator.}$$

The solution set can be written with the exact value, $\left\{\frac{\ln 10}{5}\right\}$, or with the approximate value, $\{0.461\}$.

$$\log_2 (x^2 - 3) = \log_2 6$$
$$x^2 - 3 = 6 \quad \text{Property of logarithms}$$
$$x^2 = 9 \quad \text{Add 3.}$$
$$x = \pm 3 \quad \text{Take square roots.}$$

Both values check, so the solution set is $\{\pm 3\}$.

4.6 Applications and Models of Exponential Growth and Decay

Exponential Growth or Decay Function
Let y_0 be the amount or number present at time $t = 0$. Then, under certain conditions, the amount present at any time t is modeled by

$$y = y_0 e^{kt}, \quad \text{where } k \text{ is a constant.}$$

The formula for continuous compounding,

$$A = Pe^{rt},$$

is an example of exponential growth. Here, A is the compound amount if P dollars are invested at an annual interest rate r for t years.

If $P = \$200$, $r = 3\%$, and $t = 5$ yr, find A.

$$A = Pe^{rt}$$
$$A = 200e^{0.03(5)} \quad \text{Substitute.}$$
$$A \approx \$232.37 \quad \text{Use a calculator.}$$

Chapter 4   Review Exercises

Determine whether each function as graphed or defined is one-to-one.

1.

2.

3.

4. $y = x^3 + 1$

5. $y = (x + 3)^2$

6. $y = \sqrt{3x^2 + 2}$

Find the inverse of each function that is one-to-one.

7. $f(x) = x^3 - 3$

8. $f(x) = \sqrt{25 - x^2}$

Concept Check   *Work each problem.*

9. Suppose $f(t)$ is the amount an investment will grow to t years after 2004. What does $f^{-1}(\$50{,}000)$ represent?

10. The graphs of two functions are shown. Based on their graphs, are these functions inverses?

11. To have an inverse, a function must be a(n) _____ function.

12. *True* or *false*?   The x-coordinate of the x-intercept of the graph of $y = f(x)$ is the y-coordinate of the y-intercept of the graph of $y = f^{-1}(x)$.

Match each equation with the figure that most closely resembles its graph.

13. $y = \log_{0.3} x$ **14.** $y = e^x$ **15.** $y = \ln x$ **16.** $y = 0.3^x$

A.
    B.
    C.
    D.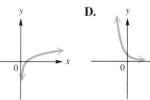

Write each equation in logarithmic form.

17. $2^5 = 32$

18. $100^{1/2} = 10$

19. $\left(\dfrac{3}{4}\right)^{-1} = \dfrac{4}{3}$

20. Graph $f(x) = \left(\dfrac{1}{5}\right)^{x+2} - 1$. Give the domain and range.

Write each equation in exponential form.

21. $\log 1000 = 3$

22. $\log_9 27 = \dfrac{3}{2}$

23. $\ln \sqrt{e} = \dfrac{1}{2}$

24. *Concept Check* What is the base of the logarithmic function whose graph contains the point $(81, 4)$?

25. *Concept Check* What is the base of the exponential function whose graph contains the point $\left(-4, \frac{1}{16}\right)$?

Use properties of logarithms to rewrite each expression. Simplify the result if possible. Assume all variables represent positive real numbers.

26. $\log_5\left(x^2 y^4 \sqrt[5]{m^3 p}\right)$ **27.** $\log_3 \dfrac{mn}{5r}$ **28.** $\log_7\left(7k + 5r^2\right)$

Use a calculator to find an approximation to four decimal places for each logarithm.

29. $\log 0.0411$ **30.** $\log 45.6$ **31.** $\ln 144{,}000$

32. $\ln 470$ **33.** $\log_{2/3} \dfrac{5}{8}$ **34.** $\log_3 769$

Solve each equation. Unless otherwise specified, give irrational solutions as decimals correct to the nearest thousandth.

35. $16^{x+4} = 8^{3x-2}$ **36.** $4^x = 12$ **37.** $3^{2x-5} = 13$

38. $2^{x+3} = 5^x$ **39.** $6^{x+3} = 4^x$ **40.** $e^{x-1} = 4$

41. $e^{2-x} = 12$ **42.** $2e^{5x+2} = 8$ **43.** $10e^{3x-7} = 5$

44. $5^{x+2} = 2^{2x-1}$ **45.** $6^{x-3} = 3^{4x+1}$ **46.** $e^{8x} \cdot e^{2x} = e^{20}$

47. $e^{6x} \cdot e^x = e^{21}$ **48.** $100(1.02)^{x/4} = 200$ **49.** $2e^{2x} - 5e^x - 3 = 0$
(Give exact form.)

50. $\left(\dfrac{1}{2}\right)^x + 2 = 0$ **51.** $4(1.06)^x + 2 = 8$

52. *Concept Check* Which one or more of the following choices is the solution set of $5^x = 9$?

     A. $\{\log_5 9\}$ **B.** $\{\log_9 5\}$ **C.** $\left\{\dfrac{\log 9}{\log 5}\right\}$ **D.** $\left\{\dfrac{\ln 9}{\ln 5}\right\}$

Solve each equation. Give solutions in exact form.

53. $3\ln x = 13$ **54.** $\ln 5x = 16$

55. $\log(2x + 7) = 0.25$ **56.** $\ln x + \ln x^3 = 12$

57. $\log_2(x^3 + 5) = 5$ **58.** $\log_3(x^2 - 9) = 3$

59. $\log_4\left[(3x + 1)(x - 4)\right] = 2$ **60.** $\ln e^{\ln x} - \ln(x - 4) = \ln 3$

61. $\log x + \log(13 - 3x) = 1$ **62.** $\log_7(3x + 2) - \log_7(x - 2) = 1$

63. $\ln(6x) - \ln(x + 1) = \ln 4$ **64.** $\log_{16}\sqrt{x + 1} = \dfrac{1}{4}$

65. $\ln\left[\ln e^{-x}\right] = \ln 3$ **66.** $S = a \ln\left(1 + \dfrac{n}{a}\right)$,   for n

67. $d = 10 \log \dfrac{I}{I_0}$,   for I_0 **68.** $D = 200 + 100 \log x$,   for x

69. Use a graphing calculator to solve the equation $e^x = 4 - \ln x$. Give solution(s) to the nearest thousandth.

Solve each problem.

70. *(Modeling) Decibel Levels* Decibel rating of the loudness of a sound is modeled by

$$d = 10 \log \frac{I}{I_0},$$

where I is the intensity of a particular sound, and I_0 is the intensity of a very faint threshold sound. A few years ago, there was a controversy about a proposed government limit on factory noise. One group wanted a maximum of 89 decibels, while another group wanted 86. Find the percent by which the 89-decibel intensity exceeds that for 86 decibels.

71. *Earthquake Intensity* The magnitude of an earthquake, measured on the Richter scale, is $\log \frac{I}{I_0}$, where I is the amplitude registered on a seismograph 100 km from the epicenter of the earthquake, and I_0 is the amplitude of an earthquake of a certain (small) size. On August 24, 2014, the Napa Valley in California was shaken by an earthquake that measured 6.0 on the Richter scale.

   (a) Express this reading in terms of I_0.

   (b) On April 1, 2014, a quake measuring 8.2 on the Richter scale struck off the coast of Chile. It was the largest earthquake in 2014. Express the magnitude of an 8.2 reading in terms of I_0 to the nearest hundred thousand.

   (c) How much greater than the force of the 6.0 earthquake was the force of the earthquake that measured 8.2?

72. *Earthquake Intensity* The San Francisco earthquake of 1906 had a Richter scale rating of 8.3.

   (a) Express the magnitude of this earthquake in terms of I_0 to the nearest hundred thousand.

   (b) In 1989, the San Francisco region experienced an earthquake with a Richter scale rating of 7.1. Express the magnitude of this earthquake in terms of I_0 to the nearest hundred thousand.

   (c) Compare the magnitudes of the two San Francisco earthquakes discussed in parts (a) and (b).

73. *Interest Rate* What annual interest rate, to the nearest tenth, will produce $4700 if $3500 is left at interest compounded annually for 10 yr?

74. *Growth of an Account* Find the number of years (to the nearest tenth) needed for $48,000 to become $53,647 at 2.8% interest compounded semiannually.

75. *Growth of an Account* Manuel deposits $10,000 for 12 yr in an account paying 3% interest compounded annually. He then puts this total amount on deposit in another account paying 4% interest compounded semiannually for another 9 yr. Find the total amount on deposit after the entire 21-yr period.

76. *Growth of an Account* Anne deposits $12,000 for 8 yr in an account paying 2.5% interest compounded annually. She then leaves the money alone with no further deposits at 3% interest compounded annually for an additional 6 yr. Find the total amount on deposit after the entire 14-yr period.

77. *Cost from Inflation* Suppose the inflation rate is 4%. Use the formula for continuous compounding to find the number of years, to the nearest tenth, for a $1 item to cost $2.

78. *(Modeling) Drug Level in the Bloodstream* After a medical drug is injected directly into the bloodstream, it is gradually eliminated from the body. Graph the following functions on the interval $[0, 10]$. Use $[0, 500]$ for the range of $A(t)$. Determine the function that best models the amount $A(t)$ (in milligrams) of a drug remaining in the body after t hours if 350 mg were initially injected.

   (a) $A(t) = t^2 - t + 350$ **(b)** $A(t) = 350 \log(t + 1)$

   (c) $A(t) = 350(0.75)^t$ **(d)** $A(t) = 100(0.95)^t$

79. *(Modeling) Chicago Cubs' Payroll* The table shows the total payroll (in millions of dollars) of the Chicago Cubs baseball team for the years 2010–2014.

Year	Total Payroll (millions of dollars)
2010	145.4
2011	134.3
2012	111.0
2013	107.4
2014	92.7

Source: www.baseballprospectus.com/compensation

Letting $f(x)$ represent the total payroll and x represent the number of years since 2010, we find that the function

$$f(x) = 146.02e^{-0.112x}$$

models the data quite well. According to this function, when will the total payroll halve its 2010 value?

80. *(Modeling) Transistors on Computer Chips* Computing power has increased dramatically as a result of the ability to place an increasing number of transistors on a single processor chip. The table lists the number of transistors on some popular computer chips made by Intel.

Year	Chip	Transistors
1989	486DX	1,200,000
1994	Pentium	3,300,000
2000	Pentium 4	42,000,000
2006	Core 2 Duo	291,000,000
2008	Core 2 Quad	820,000,000
2010	Core (2nd gen.)	1,160,000,000
2012	Core (3rd gen.)	1,400,000,000

Source: Intel.

(a) Make a scatter diagram of the data. Let the x-axis represent the year, where $x = 0$ corresponds to 1989, and let the y-axis represent the number of transistors.

(b) Decide whether a linear, a logarithmic, or an exponential function best describes the data.

(c) Determine a function f that approximates these data. Plot f and the data on the same coordinate axes.

(d) Assuming that this trend continues, use f to estimate the number of transistors on a chip, to the nearest million, in the year 2016.

81. *Financial Planning* The traditional IRA (individual retirement account) is a common tax-deferred saving plan in the United States. Earned income deposited into an IRA is not taxed in the current year, and no taxes are incurred on the interest paid in subsequent years. However, when the money is withdrawn from the account after age $59\frac{1}{2}$, taxes must be paid on the entire amount withdrawn.

Suppose we deposited $5000 of earned income into an IRA, we can earn an annual interest rate of 4%, and we are in a 25% tax bracket. (*Note:* Interest rates and tax brackets are subject to change over time, but some assumptions must be made to evaluate the investment.) Also, suppose that we deposit the $5000 at age 25 and withdraw it at age 60, and that interest is compounded continuously.

(a) How much money will remain after we pay the taxes at age 60?

(b) Suppose that instead of depositing the money into an IRA, we pay taxes on the money and the annual interest. How much money will we have at age 60? (*Note:* We effectively start with \$3750 (75% of \$5000), and the money earns 3% (75% of 4%) interest after taxes.)

(c) To the nearest dollar, how much additional money will we earn with the IRA?

(d) Suppose we pay taxes on the original \$5000 but are then able to earn 4% in a tax-free investment. Compare the balance at age 60 with the IRA balance.

82. Consider $f(x) = \log_4 (2x^2 - x)$.

(a) Use the change-of-base theorem with base e to write $\log_4 (2x^2 - x)$ in a suitable form to graph with a calculator.

(b) Graph the function using a graphing calculator. Use the window $[-2.5, 2.5]$ by $[-5, 2.5]$.

(c) What are the x-intercepts?

(d) Give the equations of the vertical asymptotes.

(e) Why is there no y-intercept?

Chapter 4 Test

1. Consider the function $f(x) = \sqrt[3]{2x - 7}$.

(a) What are the domain and range of f?

(b) Explain why f^{-1} exists.

(c) Write an equation for $f^{-1}(x)$.

(d) What are the domain and range of f^{-1}?

(e) Graph both f and f^{-1}. How are the two graphs related with respect to the line $y = x$?

2. Match each equation with its graph.

(a) $y = \log_{1/3} x$ (b) $y = e^x$ (c) $y = \ln x$ (d) $y = \left(\dfrac{1}{3}\right)^x$

A.   **B.**   **C.**   **D.**

3. Solve $\left(\dfrac{1}{8}\right)^{2x-3} = 16^{x+1}$.

4. (a) Write $4^{3/2} = 8$ in logarithmic form.

(b) Write $\log_8 4 = \dfrac{2}{3}$ in exponential form.

5. Graph $f(x) = \left(\dfrac{1}{2}\right)^x$ and $g(x) = \log_{1/2} x$ on the same axes. What is their relationship?

6. Use properties of logarithms to rewrite the expression. Assume all variables represent positive real numbers.

$$\log_7 \frac{x^2 \sqrt[4]{y}}{z^3}$$

Use a calculator to find an approximation to four decimal places for each logarithm.

7. log 2388   **8.** ln 2388   **9.** $\log_9 13$

10. Solve $x^{2/3} = 25$.

Solve each equation. Give irrational solutions as decimals correct to the nearest thousandth.

11. $12^x = 1$ **12.** $9^x = 4$ **13.** $16^{2x+1} = 8^{3x}$

14. $2^{x+1} = 3^{x-4}$ **15.** $e^{0.4x} = 4^{x-2}$

16. $2e^{2x} - 5e^x + 3 = 0$ (Give both exact and approximate values.)

Solve each equation. Give solutions in exact form.

17. $\log_x \dfrac{9}{16} = 2$ **18.** $\log_2 [(x - 4)(x - 2)] = 3$

19. $\log_2 x + \log_2 (x + 2) = 3$ **20.** $\ln x - 4 \ln 3 = \ln \dfrac{1}{5}x$

21. $\log_3 (x + 1) - \log_3 (x - 3) = 2$

22. A friend is taking another mathematics course and says, "I have no idea what an expression like $\log_5 27$ really means." Write an explanation of what it means, and tell how we can find an approximation for it with a calculator.

Solve each problem.

23. *(Modeling) Skydiver Fall Speed* A skydiver in free fall travels at a speed modeled by

$$v(t) = 176(1 - e^{-0.18t})$$

feet per second after t seconds. How long, to the nearest second, will it take for the skydiver to attain a speed of 147 ft per sec (100 mph)?

24. *Growth of an Account* How many years, to the nearest tenth, will be needed for $5000 to increase to $18,000 at 3.0% annual interest compounded **(a)** monthly **(b)** continuously?

25. *Tripling Time* For *any* amount of money invested at 2.8% annual interest compounded continuously, how long, to the nearest tenth of a year, will it take to triple?

26. *(Modeling) Radioactive Decay* The amount of a certain radioactive material, in grams, present after t days is modeled by

$$A(t) = 600e^{-0.05t}.$$

(a) Find the amount present after 12 days, to the nearest tenth of a gram.

(b) Find the half-life of the material, to the nearest tenth of a day.

5

Systems and Matrices

These two linear trails of jet exhaust crossing each other illustrate the concept of two distinct, nonparallel lines intersecting in a single point–a geometric interpretation of a *system of linear equations* having a single solution.

5.1 Systems of Linear Equations

- Linear Systems
- Substitution Method
- Elimination Method
- Special Systems
- Application of Systems of Equations
- Linear Systems with Three Unknowns (Variables)
- Application of Systems to Model Data

Linear Systems The definition of a linear equation can be extended to more than one variable. Any equation of the form

$$a_1 x_1 + a_2 x_2 + \cdots + a_n x_n = b,$$

for real numbers $a_1, a_2, \ldots, a_n$ (all nonzero) and b, is a **linear equation,** or a **first-degree equation, in n unknowns.**

A set of equations considered simultaneously is a **system of equations.** The **solutions** of a system of equations must satisfy every equation in the system. If all the equations in a system are linear, the system is a **system of linear equations,** or a **linear system.**

The solution set of a linear equation in two unknowns (or variables) is an infinite set of ordered pairs. The graph of such an equation is a straight line, so there are three possibilities for the number of elements in the solution set of a system of two linear equations in two unknowns. See **Figure 1.** The possible graphs of a linear system in two unknowns arc as follows.

1. **The graphs intersect at exactly one point,** which gives the (single) ordered-pair solution of the system. The **system is consistent** and the **equations are independent.** See **Figure 1(a).**

2. **The graphs are parallel lines,** so there is no solution and the solution set is $\varnothing$. The **system is inconsistent** and the **equations are independent.** See **Figure 1(b).**

3. **The graphs are the same line,** and there are an infinite number of solutions. The **system is consistent** and the **equations are dependent.** See **Figure 1(c).**

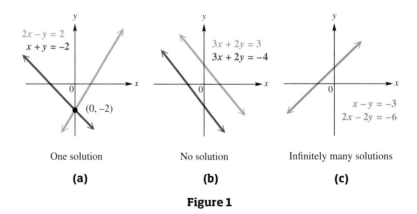

One solution
(a)

No solution
(b)

Infinitely many solutions
(c)

Figure 1

Using graphs to find the solution set of a linear system in two unknowns provides a good visual perspective, but may be inefficient when the solution set contains non-integer values. Thus, we introduce two algebraic methods for solving systems with two unknowns: *substitution* and *elimination.*

Substitution Method In a system of two equations with two variables, the **substitution method** involves using one equation to find an expression for one variable in terms of the other, and then substituting this expression into the other equation of the system.

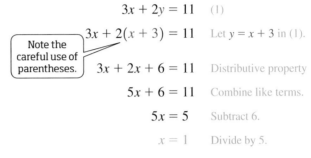

EXAMPLE 1 **Solving a System (Substitution Method)**

Solve the system.

$$3x + 2y = 11 \quad (1)$$
$$-x + y = 3 \quad (2)$$

SOLUTION Begin by solving one of the equations for one of the variables. We solve equation (2) for y.

$$-x + y = 3 \quad (2)$$
$$y = x + 3 \quad \text{Add } x. \quad (3)$$

Now replace y with $x + 3$ in equation (1), and solve for x.

$$3x + 2y = 11 \quad (1)$$
$$3x + 2(x + 3) = 11 \quad \text{Let } y = x + 3 \text{ in (1).}$$
$$3x + 2x + 6 = 11 \quad \text{Distributive property}$$
$$5x + 6 = 11 \quad \text{Combine like terms.}$$
$$5x = 5 \quad \text{Subtract 6.}$$
$$x = 1 \quad \text{Divide by 5.}$$

Note the careful use of parentheses.

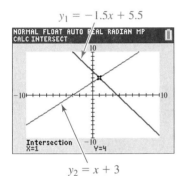

$y_1 = -1.5x + 5.5$

$y_2 = x + 3$

To solve the system in **Example 1** graphically, solve both equations for y.

$3x + 2y = 11$ leads to

$$y_1 = -1.5x + 5.5.$$

$-x + y = 3$ leads to

$$y_2 = x + 3.$$

Graph both y_1 and y_2 in the standard window to find that their point of intersection is $(1, 4)$.

Replace x with 1 in equation (3) to obtain

$$y = x + 3 = 1 + 3 = 4.$$

The solution of the system is the ordered pair $(1, 4)$. ***Check this solution in both equations (1) and (2).***

CHECK
$$3x + 2y = 11 \quad (1) \qquad -x + y = 3 \quad (2)$$
$$3(1) + 2(4) \stackrel{?}{=} 11 \qquad -1 + 4 \stackrel{?}{=} 3$$
$$11 = 11 \checkmark \quad \text{True} \qquad 3 = 3 \checkmark \quad \text{True}$$

True statements result when the solution is substituted in both equations, confirming that the solution set is $\{(1, 4)\}$.

✔ **Now Try Exercise 7.**

Elimination Method The **elimination method** for solving a system of two equations uses multiplication and addition to eliminate a variable from one equation. To eliminate a variable, the coefficients of that variable in the two equations must be additive inverses. We use properties of algebra to change the system to an **equivalent system,** one with the same solution set.

The three transformations that produce an equivalent system are listed here.

Transformations of a Linear System

1. Interchange any two equations of the system.

2. Multiply or divide any equation of the system by a nonzero real number.

3. Replace any equation of the system by the sum of that equation and a multiple of another equation in the system.

EXAMPLE 2 Solving a System (Elimination Method)

Solve the system.

$$3x - 4y = 1 \quad \text{(1)}$$

$$2x + 3y = 12 \quad \text{(2)}$$

SOLUTION One way to eliminate a variable is to use the second transformation and multiply each side of equation (2) by -3, to obtain an equivalent system.

$$3x - 4y = 1 \quad \text{(1)}$$

$$-6x - 9y = -36 \quad \text{Multiply (2) by } -3. \quad \text{(3)}$$

Now multiply each side of equation (1) by 2, and use the third transformation to add the result to equation (3), eliminating x. Solve the result for y.

$$6x - 8y = 2 \qquad \text{Multiply (1) by 2.}$$

$$\underline{-6x - 9y = -36} \qquad \text{(3)}$$

$$-17y = -34 \qquad \text{Add.}$$

$$y = 2 \qquad \text{Solve for } y.$$

Substitute 2 for y in either of the original equations and solve for x.

$$3x - 4y = 1 \qquad \text{(1)}$$

$$3x - 4(2) = 1 \qquad \text{Let } y = 2 \text{ in (1).}$$

$$3x - 8 = 1 \qquad \text{Multiply.}$$

$$3x = 9 \qquad \text{Add 8.}$$

$$x = 3 \qquad \text{Divide by 3.}$$

Write the x-value first.

A check shows that $(3, 2)$ satisfies both equations (1) and (2). Therefore, the solution set is $\{(3, 2)\}$. The graph in **Figure 2** confirms this.

✔ **Now Try Exercise 21.**

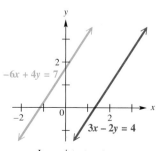

Consistent system
with solution set $\{(3, 2)\}$

Figure 2

Special Systems The systems in **Examples 1 and 2** were both consistent, having a single solution. This is not always the case.

EXAMPLE 3 Solving an Inconsistent System

Solve the system.

$$3x - 2y = 4 \quad \text{(1)}$$

$$-6x + 4y = 7 \quad \text{(2)}$$

SOLUTION To eliminate the variable x, multiply each side of equation (1) by 2, and add the result to equation (2).

$$6x - 4y = 8 \qquad \text{Multiply (1) by 2.}$$

$$\underline{-6x + 4y = 7} \qquad \text{(2)}$$

$$0 = 15 \qquad \text{False}$$

Since $0 = 15$ is false, the system is inconsistent and has no solution. As suggested by **Figure 3**, this means that the graphs of the equations of the system never intersect. (The lines are parallel.) The solution set is $\varnothing$.

Inconsistent system
with solution set $\varnothing$

Figure 3

✔ **Now Try Exercise 31.**

EXAMPLE 4 **Solving a System with Infinitely Many Solutions**

Solve the system.

$$8x - 2y = -4 \quad (1)$$
$$-4x + y = 2 \quad (2)$$

ALGEBRAIC SOLUTION

Divide each side of equation (1) by 2, and add the result to equation (2).

$$\begin{array}{ll} 4x - y = -2 & \text{Divide (1) by 2.} \\ \underline{-4x + y = 2} & (2) \\ 0 = 0 & \text{True} \end{array}$$

The result, $0 = 0$, is a true statement, which indicates that the equations are equivalent. Any ordered pair (x, y) that satisfies either equation will satisfy the system. Solve equation (2) for y.

$$-4x + y = 2 \quad\quad (2)$$
$$y = 4x + 2$$

The solutions of the system can be written in the form of a set of ordered pairs $(x, 4x + 2)$, for any real number x. Some ordered pairs in the solution set are $(0, 4 \cdot 0 + 2)$, or $(0, 2)$, and $(1, 4 \cdot 1 + 2)$, or $(1, 6)$, as well as $(3, 14)$, and $(-2, -6)$.

As shown in **Figure 4,** the equations of the original system are dependent and lead to the same straight-line graph. Using this method, the solution set can be written $\{(x, 4x + 2)\}$.

Consistent system with
infinitely many solutions

Figure 4

GRAPHING CALCULATOR SOLUTION

Solving the equations for y gives

$$y_1 = 4x + 2 \quad (1)$$

and

$$y_2 = 4x + 2. \quad (2)$$

When written in this form, we can immediately determine that the equations are identical. Each has slope 4 and y-intercept $(0, 2)$.

As expected, the graphs coincide. See the top screen in **Figure 5.** The table indicates that

$$y_1 = y_2 \text{ for selected values of } x,$$

providing another way to show that the two equations lead to the same graph.

Figure 5

Refer to the algebraic solution to see how the solution set can be written using an arbitrary variable.

✔ **Now Try Exercise 33.**

NOTE In the algebraic solution for **Example 4,** we wrote the solution set with the variable x arbitrary. We could write the solution set with y arbitrary.

$$\left\{ \left(\frac{y - 2}{4}, y \right) \right\} \quad \text{Solve } -4x + y = 2 \text{ for } x.$$

By selecting values for y and solving for x in this ordered pair, we can find individual solutions. Verify again that $(0, 2)$ is a solution by letting $y = 2$ and solving for x to obtain $\frac{2 - 2}{4} = 0$.

Application of Systems of Equations   Many applied problems involve more than one unknown quantity. Although some problems with two unknowns can be solved using just one variable, it is often easier to use two variables.

To solve a problem with two unknowns, we must write two equations that relate the unknown quantities. The system formed by the pair of equations can then be solved using the methods of this chapter. The following steps, based on the six-step problem-solving method introduced earlier, give a strategy for solving such applied problems.

Solving an Applied Problem by Writing a System of Equations

Step 1 **Read** the problem carefully until you understand what is given and what is to be found.

Step 2 **Assign variables** to represent the unknown values, using diagrams or tables as needed. *Write down* what each variable represents.

Step 3 **Write a system of equations** that relates the unknowns.

Step 4 **Solve** the system of equations.

Step 5 **State the answer** to the problem. Does it seem reasonable?

Step 6 **Check** the answer in the words of the original problem.

EXAMPLE 5   **Using a Linear System to Solve an Application**

Salaries for the same position can vary depending on the location. In 2015, the average of the median salaries for the position of Accountant I in San Diego, California, and Salt Lake City, Utah, was $47,449.50. The median salary in San Diego, however, exceeded the median salary in Salt Lake City by $5333. Determine the median salary for the Accountant I position in San Diego and in Salt Lake City. (*Source:* www.salary.com)

SOLUTION

Step 1 **Read** the problem. We must find the median salary of the Accountant I position in San Diego and in Salt Lake City.

Step 2 **Assign variables.** Let x represent the median salary of the Accountant I position in San Diego and y represent the median salary for the same position in Salt Lake City.

Step 3 **Write a system of equations.** Since the average of the two medians for the Accountant I position in San Diego and Salt Lake City was $47,449.50, one equation is as follows.

$$\frac{x + y}{2} = 47{,}449.50$$

Multiply each side of this equation by 2 to clear the fraction and obtain an equivalent equation.

$$x + y = 94{,}899 \quad (1)$$

The median salary in San Diego exceeded the median salary in Salt Lake City by $5333. Thus, $x - y = 5333$, which gives the following system of equations.

$$x + y = 94{,}899 \quad (1)$$
$$x - y = 5333 \quad (2)$$

Step 4 **Solve** the system. To eliminate y, add the two equations.

$$x + y = 94{,}899 \quad \text{(1)}$$

$$\underline{x - y = 5333} \quad \text{(2)}$$

$$2x = 100{,}232 \quad \text{Add.}$$

$$x = 50{,}116 \quad \text{Solve for } x.$$

To find y, substitute 50,116 for x in equation (2).

$$x - y = 5333 \quad \text{(2)}$$

$$50{,}116 - y = 5333 \quad \text{Let } x = 50{,}116.$$

$$-y = -44{,}783 \quad \text{Subtract 50,116.}$$

$$y = 44{,}783 \quad \text{Multiply by } -1.$$

Step 5 **State the answer.** The median salary for the position of Accountant I was $50,116 in San Diego and $44,783 in Salt Lake City.

Step 6 **Check.** The average of $50,116 and $44,783 is

$$\frac{\$50{,}116 + \$44{,}783}{2} = \$47{,}449.50.$$

Also, $\quad \$50{,}116 - \$44{,}783 = \$5333, \quad$ as required.

✔ **Now Try Exercise 101.**

Linear Systems with Three Unknowns (Variables) We have seen that the graph of a linear equation in two unknowns is a straight line. The graph of a linear equation in three unknowns requires a three-dimensional coordinate system. The three number lines are placed at right angles. The graph of a linear equation in three unknowns is a plane. Some possible intersections of planes representing three equations in three variables are shown in **Figure 6.**

In two dimensions we customarily label the axes x and y. When working in three dimensions they are usually labeled x, y, and z.

A single solution

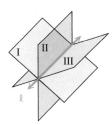

Points of a line in common

Points of a line in common

All points in common

No points in common

No points in common

No points in common

No points in common

Figure 6

Solving a Linear System with Three Unknowns

Step 1 Eliminate a variable from any two of the equations.

Step 2 Eliminate the *same variable* from a different pair of equations.

Step 3 Eliminate a second variable using the resulting two equations in two variables to obtain an equation with just one variable whose value we can now determine.

Step 4 Find the values of the remaining variables by substitution. Write the solution of the system as an **ordered triple.**

EXAMPLE 6 Solving a System of Three Equations with Three Variables

Solve the system.

$$3x + 9y + 6z = 3 \quad (1)$$
$$2x + y - z = 2 \quad (2)$$
$$x + y + z = 2 \quad (3)$$

SOLUTION

Step 1 Eliminate z by adding equations (2) and (3).

$$3x + 2y = 4 \quad (4)$$

Step 2 To eliminate z from another pair of equations, multiply each side of equation (2) by 6 and add the result to equation (1).

> Make sure equation (5) has the *same* two variables as equation (4).

$$
\begin{array}{ll}
12x + 6y - 6z = 12 & \text{Multiply (2) by 6.} \\
\underline{3x + 9y + 6z = 3} & (1) \\
15x + 15y \phantom{{}- 6z} = 15 & (5)
\end{array}
$$

Step 3 To eliminate x from equations (4) and (5), multiply each side of equation (4) by -5 and add the result to equation (5). Solve the resulting equation for y.

$$
\begin{array}{ll}
-15x - 10y = -20 & \text{Multiply (4) by } -5. \\
\underline{15x + 15y = 15} & (5) \\
5y = -5 & \text{Add.} \\
y = -1 & \text{Divide by 5.}
\end{array}
$$

Step 4 Use $y = -1$ to find x from equation (4) by substitution.

$$
\begin{array}{ll}
3x + 2y = 4 & (4) \\
3x + 2(-1) = 4 & \text{Let } y = -1. \\
x = 2 & \text{Solve for } x.
\end{array}
$$

Substitute 2 for x and -1 for y in equation (3) to find z.

$$
\begin{array}{ll}
x + y + z = 2 & (3) \\
2 + (-1) + z = 2 & \text{Let } x = 2, y = -1. \\
z = 1 & \text{Solve for } z.
\end{array}
$$

> Write the values of x, y, and z in the correct order.

Verify that the ordered triple $(2, -1, 1)$ satisfies all three equations in the *original* system. The solution set is $\{(2, -1, 1)\}$.

✔ **Now Try Exercise 47.**

EXAMPLE 7 Solving a System of Two Equations with Three Variables

Solve the system.

$$x + 2y + z = 4 \quad (1)$$
$$3x - y - 4z = -9 \quad (2)$$

SOLUTION Geometrically, the solution is the intersection of the two planes given by equations (1) and (2). The intersection of two different nonparallel planes is a line. Thus there will be an infinite number of ordered triples in the solution set, representing the points on the line of intersection.

To eliminate x, multiply both sides of equation (1) by -3 and add the result to equation (2). (Either y or z could have been eliminated instead.)

$$\begin{array}{ll} -3x - 6y - 3z = -12 & \text{Multiply (1) by } -3. \\ \underline{3x - y - 4z = -9} & (2) \\ -7y - 7z = -21 & (3) \\ -7z = 7y - 21 & \text{Add } 7y. \\ z = -y + 3 & \text{Divide } each \text{ term by } -7. \end{array}$$

Solve this equation for z.

This gives z in terms of y. Express x also in terms of y by solving equation (1) for x and substituting $-y + 3$ for z in the result.

$$\begin{array}{ll} x + 2y + z = 4 & (1) \\ x = -2y - z + 4 & \text{Solve for } x. \\ x = -2y - (-y + 3) + 4 & \text{Substitute } (-y + 3) \text{ for } z. \\ x = -y + 1 & \text{Simplify.} \end{array}$$

Use parentheses around $-y + 3$.

The system has an infinite number of solutions. For any value of y, the value of z is $-y + 3$ and the value of x is $-y + 1$. For example, if $y = 1$, then $x = -1 + 1 = 0$ and $z = -1 + 3 = 2$, giving the solution $(0, 1, 2)$. Verify that another solution is $(-1, 2, 1)$.

With y arbitrary, the solution set is of the form $\{(-y + 1, y, -y + 3)\}$.

✔ **Now Try Exercise 59.**

NOTE Had we solved equation (3) in Example 7 for y instead of z, the solution would have had a different form but would have led to the same set of solutions.

$$\{(z - 2, -z + 3, z)\} \quad \text{Solution set with } z \text{ arbitrary}$$

By choosing $z = 2$, one solution would be $(0, 1, 2)$, which was found above.

Application of Systems to Model Data Applications with three unknowns usually require solving a system of three equations. If we know three points on the graph, we can find the equation of a parabola in the form

$$y = ax^2 + bx + c$$

by solving a system of three equations with three variables.

EXAMPLE 8 **Using Modeling to Find an Equation through Three Points**

Find an equation of the parabola $y = ax^2 + bx + c$ that passes through the points $(2, 4)$, $(-1, 1)$, and $(-2, 5)$.

SOLUTION The three ordered pairs represent points that lie on the graph of the given equation $y = ax^2 + bx + c$, so they must all satisfy the equation. Substituting each ordered pair into the equation gives three equations with three unknowns.

$$4 = a(2)^2 + b(2) + c, \quad \text{or} \quad 4 = 4a + 2b + c \quad (1)$$

$$1 = a(-1)^2 + b(-1) + c, \quad \text{or} \quad 1 = a - b + c \quad (2)$$

$$5 = a(-2)^2 + b(-2) + c, \quad \text{or} \quad 5 = 4a - 2b + c \quad (3)$$

To solve this system, first eliminate c using equations (1) and (2).

$$\begin{array}{ll} 4 = 4a + 2b + c & (1) \\ \underline{-1 = -a + b - c} & \text{Multiply (2) by } -1. \\ 3 = 3a + 3b & (4) \end{array}$$

Now, use equations (2) and (3) to eliminate the same unknown, c.

Equation (5) must have the *same* two unknowns as equation (4).

$$\begin{array}{ll} 1 = a - b + c & (2) \\ \underline{-5 = -4a + 2b - c} & \text{Multiply (3) by } -1. \\ -4 = -3a + b & (5) \end{array}$$

Solve the system of equations (4) and (5) in two unknowns by eliminating a.

$$\begin{array}{ll} 3 = 3a + 3b & (4) \\ \underline{-4 = -3a + b} & (5) \\ -1 = 4b & \text{Add.} \end{array}$$

$$-\frac{1}{4} = b \qquad \text{Divide by 4.}$$

Find a by substituting $-\frac{1}{4}$ for b in equation (4).

$$1 = a + b \qquad \text{Equation (4) divided by 3}$$

$$1 = a + \left(-\frac{1}{4}\right) \qquad \text{Let } b = -\frac{1}{4}.$$

$$\frac{5}{4} = a \qquad \text{Add } \frac{1}{4}.$$

Finally, find c by substituting $a = \frac{5}{4}$ and $b = -\frac{1}{4}$ in equation (2).

$$1 = a - b + c \qquad (2)$$

$$1 = \frac{5}{4} - \left(-\frac{1}{4}\right) + c \qquad \text{Let } a = \frac{5}{4}, b = -\frac{1}{4}.$$

$$1 = \frac{6}{4} + c \qquad \text{Add.}$$

$$-\frac{1}{2} = c \qquad \text{Subtract } \frac{6}{4}.$$

The required equation is $y = \frac{5}{4}x^2 - \frac{1}{4}x - \frac{1}{2}$, or $y = 1.25x^2 - 0.25x - 0.5$.

This graph/table screen shows that the points $(2, 4)$, $(-1, 1)$, and $(-2, 5)$ lie on the graph of $y_1 = 1.25x^2 - 0.25x - 0.5$. This supports the result of **Example 8**.

✔ **Now Try Exercise 79.**

> **EXAMPLE 9** **Solving an Application Using a System of Three Equations**

An animal feed is made from three ingredients: corn, soybeans, and cottonseed. One unit of each ingredient provides units of protein, fat, and fiber as shown in the table. How many units of each ingredient should be used to make a feed that contains 22 units of protein, 28 units of fat, and 18 units of fiber?

	Corn	Soybeans	Cottonseed	Total
Protein	0.25	0.4	0.2	22
Fat	0.4	0.2	0.3	28
Fiber	0.3	0.2	0.1	18

SOLUTION

Step 1 **Read** the problem. We must determine the number of units of corn, soybeans, and cottonseed.

Step 2 **Assign variables.** Let x represent the number of units of corn, y the number of units of soybeans, and z the number of units of cottonseed.

Step 3 **Write a system of equations.** The total amount of protein is to be 22 units, so we use the first row of the table to write equation (1).

$$0.25x + 0.4y + 0.2z = 22 \quad (1)$$

We use the second row of the table to obtain 28 units of fat.

$$0.4x + 0.2y + 0.3z = 28 \quad (2)$$

Finally, we use the third row of the table to obtain 18 units of fiber.

$$0.3x + 0.2y + 0.1z = 18 \quad (3)$$

Multiply equation (1) on each side by 100, and equations (2) and (3) by 10, to obtain an equivalent system.

$$25x + 40y + 20z = 2200 \quad (4)$$
$$4x + 2y + 3z = 280 \quad (5)$$
$$3x + 2y + z = 180 \quad (6)$$

Eliminate the decimal points in equations (1), (2), and (3) by multiplying each equation by an appropriate power of 10.

Step 4 **Solve** the system. Using the methods described earlier in this section, we find the following.

$$x = 40, \quad y = 15, \quad \text{and} \quad z = 30$$

Step 5 **State the answer.** The feed should contain 40 units of corn, 15 units of soybeans, and 30 units of cottonseed.

Step 6 **Check.** Show that the ordered triple $(40, 15, 30)$ satisfies the system formed by equations (1), (2), and (3).

✔ **Now Try Exercise 107.**

NOTE Notice how the table in **Example 9** is used to set up the equations of the system. The coefficients in each equation are read from left to right. This idea is extended in the next section, where we introduce the solution of systems by matrices.

5.1 Exercises

CONCEPT PREVIEW *Fill in the blank(s) to correctly complete each sentence.*

1. The solution set of the following system is $\{(1, \underline{\quad})\}$.

$$-2x + 5y = 18$$
$$x + y = 5$$

2. The solution set of the following system is $\{(\underline{\quad}, 0)\}$.

$$6x + y = -18$$
$$13x + y = -39$$

3. One way of solving the following system by elimination is to multiply equation (2) by the integer _____ to eliminate the y-terms by direct addition.

$$14x + 11y = 80 \quad (1)$$
$$2x + y = 19 \quad (2)$$

4. To solve the system

$$3x + y = 4 \quad (1)$$
$$7x + 8y = -2 \quad (2)$$

by substitution, it is easiest to begin by solving equation (1) for the variable _____ and then substituting into equation (2), because no fractions will appear in the algebraic work.

5. If a system of linear equations in two variables has two graphs that coincide, there is/are _____ solutions to the system.
  (one/no/infinitely many)

6. If a system of linear equations in two variables has two graphs that are parallel lines, there is/are _____ solutions to the system.
  (one/no/infinitely many)

Solve each system by substitution. ***See Example 1.***

7. $4x + 3y = -13$
  $-x + y = 5$

8. $3x + 4y = 4$
  $x - y = 13$

9. $x - 5y = 8$
  $x = 6y$

10. $6x - y = 5$
  $y = 11x$

11. $8x - 10y = -22$
  $3x + y = 6$

12. $4x - 5y = -11$
  $2x + y = 5$

13. $7x - y = -10$
  $3y - x = 10$

14. $4x + 5y = 7$
  $9y = 31 + 2x$

15. $-2x = 6y + 18$
  $-29 = 5y - 3x$

16. $3x - 7y = 15$
  $3x + 7y = 15$

17. $3y = 5x + 6$
  $x + y = 2$

18. $4y = 2x - 4$
  $x - y = 4$

Solve each system by elimination. In systems with fractions, first clear denominators.
See Example 2.

19. $3x - y = -4$
  $x + 3y = 12$

20. $4x + y = -23$
  $x - 2y = -17$

21. $2x - 3y = -7$
  $5x + 4y = 17$

22. $4x + 3y = -1$
  $2x + 5y = 3$

23. $5x + 7y = 6$
  $10x - 3y = 46$

24. $12x - 5y = 9$
  $3x - 8y = -18$

25. $6x + 7y + 2 = 0$
$7x - 6y - 26 = 0$

26. $5x + 4y + 2 = 0$
$4x - 5y - 23 = 0$

27. $\dfrac{x}{2} + \dfrac{y}{3} = 4$
$\dfrac{3x}{2} + \dfrac{3y}{2} = 15$

28. $\dfrac{3x}{2} + \dfrac{y}{2} = -2$
$\dfrac{x}{2} + \dfrac{y}{2} = 0$

29. $\dfrac{2x - 1}{3} + \dfrac{y + 2}{4} = 4$
$\dfrac{x + 3}{2} - \dfrac{x - y}{3} = 3$

30. $\dfrac{x + 6}{5} + \dfrac{2y - x}{10} = 1$
$\dfrac{x + 2}{4} + \dfrac{3y + 2}{5} = -3$

Solve each system of equations. State whether it is an inconsistent system *or has* infinitely many solutions. *If the system has infinitely many solutions, write the solution set with y arbitrary.* **See Examples 3 and 4.**

31. $9x - 5y = 1$
$-18x + 10y = 1$

32. $3x + 2y = 5$
$6x + 4y = 8$

33. $4x - y = 9$
$-8x + 2y = -18$

34. $3x + 5y = -2$
$9x + 15y = -6$

35. $5x - 5y - 3 = 0$
$x - y - 12 = 0$

36. $2x - 3y - 7 = 0$
$-4x + 6y - 14 = 0$

37. $7x + 2y = 6$
$14x + 4y = 12$

38. $2x - 8y = 4$
$x - 4y = 2$

39. $2x - 6y = 0$
$-7x + 21y = 10$

40. *Concept Check* Which screen gives the correct graphical solution of the system? (*Hint:* Solve for y first in each equation and use the slope-intercept forms to answer the question.)

$$4x - 5y = -11$$
$$2x + y = 5$$

A.

B.

C.

D.

Connecting Graphs with Equations Determine the system of equations illustrated in each graph. Write equations in standard form.

41.

42.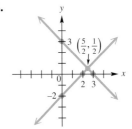

⊟ *Use a graphing calculator to solve each system. Express solutions with approximations to the nearest thousandth.*

43. $\dfrac{11}{3}x + y = 0.5$

$0.6x - y = 3$

44. $\sqrt{3}x - y = 5$

$100x + y = 9$

45. $\sqrt{7}x + \sqrt{2}y = 3$

$\sqrt{6}x - \quad y = \sqrt{3}$

46. $0.2x + \sqrt{2}y = 1$

$\sqrt{5}x + 0.7y = 1$

Solve each system. See Example 6.

47. $x + y + z = 2$
$2x + y - z = 5$
$x - y + z = -2$

48. $2x + y + z = 9$
$-x - y + z = 1$
$3x - y + z = 9$

49. $x + 3y + 4z = 14$
$2x - 3y + 2z = 10$
$3x - y + z = 9$

50. $4x - y + 3z = -2$
$3x + 5y - z = 15$
$-2x + y + 4z = 14$

51. $x + 4y - z = 6$
$2x - y + z = 3$
$3x + 2y + 3z = 16$

52. $4x - 3y + z = 9$
$3x + 2y - 2z = 4$
$x - y + 3z = 5$

53. $x - 3y - 2z = -3$
$3x + 2y - z = 12$
$-x - y + 4z = 3$

54. $x + y + z = 3$
$3x - 3y - 4z = -1$
$x + y + 3z = 11$

55. $2x + 6y - z = 6$
$4x - 3y + 5z = -5$
$6x + 9y - 2z = 11$

56. $8x - 3y + 6z = -2$
$4x + 9y + 4z = 18$
$12x - 3y + 8z = -2$

57. $2x - 3y + 2z - 3 = 0$
$4x + 8y + z - 2 = 0$
$-x - 7y + 3z - 14 = 0$

58. $-x + 2y - z - 1 = 0$
$-x - y - z + 2 = 0$
$x - y + 2z - 2 = 0$

Solve each system in terms of the arbitrary variable z. See Example 7.

59. $x - 2y + 3z = 6$
$2x - y + 2z = 5$

60. $3x - 2y + z = 15$
$x + 4y - z = 11$

61. $5x - 4y + z = 9$
$y + z = 15$

62. $3x - 5y - 4z = -7$
$y - z = -13$

63. $3x + 4y - z = 13$
$x + y + 2z = 15$

64. $x - y + z = -6$
$4x + y + z = 7$

Solve each system. State whether it is an inconsistent system *or has* infinitely many solutions. *If the system has infinitely many solutions, write the solution set with z arbitrary. See Examples 3, 4, 6, and 7.*

65. $3x + 5y - z = -2$
$4x - y + 2z = 1$
$-6x - 10y + 2z = 0$

66. $3x + y + 3z = 1$
$x + 2y - z = 2$
$2x - y + 4z = 4$

67. $5x - 4y + z = 0$
$x + y = 0$
$-10x + 8y - 2z = 0$

68. $2x + y - 3z = 0$
$4x + 2y - 6z = 0$
$x - y + z = 0$

Solve each system. (Hint: In Exercises 69–72, let $\dfrac{1}{x} = t$ and $\dfrac{1}{y} = u$.)

69. $\dfrac{2}{x} + \dfrac{1}{y} = \dfrac{3}{2}$

$\dfrac{3}{x} - \dfrac{1}{y} = 1$

70. $\dfrac{1}{x} + \dfrac{3}{y} = \dfrac{16}{5}$

$\dfrac{5}{x} + \dfrac{4}{y} = 5$

71. $\dfrac{2}{x} + \dfrac{1}{y} = 11$

$\dfrac{3}{x} - \dfrac{5}{y} = 10$

72. $\dfrac{2}{x} + \dfrac{3}{y} = 18$

$\dfrac{4}{x} - \dfrac{5}{y} = -8$

73. $\dfrac{2}{x} + \dfrac{3}{y} - \dfrac{2}{z} = -1$

$\dfrac{8}{x} - \dfrac{12}{y} + \dfrac{5}{z} = 5$

$\dfrac{6}{x} + \dfrac{3}{y} - \dfrac{1}{z} = 1$

74. $-\dfrac{5}{x} + \dfrac{4}{y} + \dfrac{3}{z} = 2$

$\dfrac{10}{x} + \dfrac{3}{y} - \dfrac{6}{z} = 7$

$\dfrac{5}{x} + \dfrac{2}{y} - \dfrac{9}{z} = 6$

75. *Concept Check* For what value(s) of k will the following system of linear equations have no solution? infinitely many solutions?

$$x - 2y = 3$$
$$-2x + 4y = k$$

76. *Concept Check* Consider the linear equation in three variables

$$x + y + z = 4.$$

Find a pair of linear equations in three variables that, when considered together with the given equation, form a system having **(a)** exactly one solution, **(b)** no solution, **(c)** infinitely many solutions.

(Modeling) Use a system of equations to solve each problem. **See Example 8.**

77. Find an equation of the line $y = ax + b$ that passes through the points $(-2, 1)$ and $(-1, -2)$.

78. Find an equation of the line $y = ax + b$ that passes through the points $(3, -4)$ and $(-1, 4)$.

79. Find an equation of the parabola $y = ax^2 + bx + c$ that passes through the points $(2, 3)$, $(-1, 0)$, and $(-2, 2)$.

80. Find an equation of the parabola $y = ax^2 + bx + c$ that passes through the points $(-2, 4)$, $(2, 2)$, and $(4, 9)$.

81. *Connecting Graphs with Equations* Use a system to find an equation of the line through the given points.

82. *Connecting Graphs with Equations* Use a system to find an equation of the parabola through the given points.

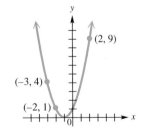

83. *Connecting Graphs with Equations* Find an equation of the parabola. Three views of the same curve are given.

84. *(Modeling)* The table was generated using a function

$$y_1 = ax^2 + bx + c.$$

Use any three points from the table to find an equation for y_1.

```
NORMAL FLOAT AUTO REAL RADIAN MP
PRESS + FOR ∆Tbl
  X    | Y1
 -3    | 5.48
 -2    | 2.9
 -1    | 1.26
  0    | .56
  1    | .8
  2    | 1.98
  3    | 4.1
  4    | 7.16
  5    | 11.16
  6    | 16.1
  7    | 21.98
X= -3
```

(Modeling) *Given three noncollinear points, there is one and only one circle that passes through them. Knowing that the equation of a circle may be written in the form*

$$x^2 + y^2 + ax + by + c = 0,$$

find an equation of the circle passing through the given points.

85. $(-1, 3)$, $(6, 2)$, and $(-2, -4)$ 　　**86.** $(-1, 5)$, $(6, 6)$, and $(7, -1)$

87. $(2, 1)$, $(-1, 0)$, and $(3, 3)$ 　　**88.** $(-5, 0)$, $(2, -1)$, and $(4, 3)$

89. *Connecting Graphs with Equations* 　　**90.** *Connecting Graphs with Equations*

 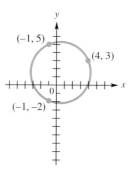

(Modeling) *Use the method of **Example 8** to work each problem.*

91. *Atmospheric Carbon Dioxide* Carbon dioxide concentrations (in parts per million) have been measured directly from the atmosphere since 1960. This concentration has increased quadratically. The table lists readings for three years.

(a) If the quadratic relationship between the carbon dioxide concentration C and the year t is expressed as

$$C = at^2 + bt + c,$$

where $t = 0$ corresponds to 1960, use a system of linear equations to determine the constants a, b, and c, and give the equation.

(b) Predict when the amount of carbon dioxide in the atmosphere will be double its 1960 level.

Year	CO$_2$
1960	317
1980	339
2013	396

Source: U.S. Department of Energy; Carbon Dioxide Information Analysis Center.

92. *Aircraft Speed and Altitude* For certain aircraft there exists a quadratic relationship between an airplane's maximum speed S (in knots) and its ceiling C, or highest altitude possible (in thousands of feet). The table lists three such airplanes.

Airplane	Max Speed (S)	Ceiling (C)
Hawkeye	320	33
Corsair	600	40
Tomcat	1283	50

Source: Sanders, D., *Statistics: A First Course,* Sixth Edition, McGraw Hill.

(a) If the quadratic relationship between C and S is written as

$$C = aS^2 + bS + c,$$

use a system of linear equations to determine the constants a, b and c, and give the equation.

(b) A new aircraft of this type has a ceiling of 45,000 ft. Predict its top speed to the nearest knot.

Changes in Population *The graph shows the populations of the New Orleans, LA, and the Jacksonville, FL, metropolitan areas over the years 2004–2013.*

Population of Metropolitan Areas

Source: U.S. Census Bureau.

93. In what years was the population of the Jacksonville metropolitan area greater than that of the New Orleans metropolitan area?

94. At the time when the populations of the two metropolitan areas were equal, what was the approximate population of each area? Round to the nearest hundredth million.

95. Express the solution of the system as an ordered pair to the nearest tenth of a year and the nearest hundredth million.

96. Use the terms *increasing, decreasing,* and *constant* to describe the trends for the population of the New Orleans metropolitan area.

97. If equations of the form $y = f(t)$ were determined that modeled either of the two graphs, then the variable t would represent _____ and the variable y would represent _____ .

98. Why is each graph that of a function?

Solve each problem. ***See Examples 5 and 9.***

99. *Unknown Numbers* The sum of two numbers is 47, and the difference between the numbers is 1. Find the numbers.

100. *Costs of Goats and Sheep* At the Berger ranch, 6 goats and 5 sheep sell for $305, while 2 goats and 9 sheep sell for $285. Find the cost of a single goat and of a single sheep.

101. *Fan Cost Index* The Fan Cost Index (FCI) is a measure of how much it will cost a family of four to attend a professional sports event. In 2014, the FCI prices for Major League Baseball and the National Football League averaged $345.53. The FCI for baseball was $266.13 less than that for football. What were the FCIs for these sports? (*Source:* Team Marketing Report.)

102. *Money Denominations* A cashier has a total of 30 bills, made up of ones, fives, and twenties. The number of twenties is 9 more than the number of ones. The total value of the money is $351. How many of each denomination of bill are there?

103. *Mixing Water* A sparkling-water distributor wants to make up 300 gal of sparkling water to sell for $6.00 per gallon. She wishes to mix three grades of water selling for $9.00, $3.00, and $4.50 per gallon, respectively. She must use twice as much of the $4.50 water as of the $3.00 water. How many gallons of each should she use?

104. *Mixing Glue* A glue company needs to make some glue that it can sell for $120 per barrel. It wants to use 150 barrels of glue worth $100 per barrel, along with some glue worth $150 per barrel and some glue worth $190 per barrel. It must use the same number of barrels of $150 and $190 glue. How much of the $150 and $190 glue will be needed? How many barrels of $120 glue will be produced?

105. *Triangle Dimensions* The perimeter of a triangle is 59 in. The longest side is 11 in. longer than the medium side, and the medium side is 3 in. longer than the shortest side. Find the length of each side of the triangle.

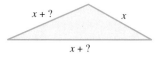

106. *Triangle Dimensions* The sum of the measures of the angles of any triangle is 180°. In a certain triangle, the largest angle measures 55° less than twice the medium angle, and the smallest angle measures 25° less than the medium angle. Find the measures of all three angles.

107. *Investment Decisions* Patrick wins $200,000 in the Louisiana state lottery. He invests part of the money in real estate with an annual return of 3% and another part in a money market account at 2.5% interest. He invests the rest, which amounts to $80,000 less than the sum of the other two parts, in certificates of deposit that pay 1.5%. If the total annual interest on the money is $4900, how much was invested at each rate?

	Amount Invested	Rate (as a decimal)	Annual Interest
Real Estate		0.03	
Money Market		0.025	
CDs		0.015	

108. *Investment Decisions* Jane invests $40,000 received as an inheritance in three parts. With one part she buys mutual funds that offer a return of 2% per year. The second part, which amounts to twice the first, is used to buy government bonds paying 2.5% per year. She puts the rest of the money into a savings account that pays 1.25% annual interest. During the first year, the total interest is $825. How much did she invest at each rate?

	Amount Invested	Rate (as a decimal)	Annual Interest
Mutual Funds		0.02	
Government Bonds		0.025	
Savings Account		0.0125	

109. Solve the system of equations (4), (5), and (6) from **Example 9.**

$$25x + 40y + 20z = 2200 \quad (4)$$
$$4x + 2y + 3z = 280 \quad (5)$$
$$3x + 2y + z = 180 \quad (6)$$

110. Check the solution in **Exercise 109,** showing that it satisfies all three equations of the system.

111. *Blending Coffee Beans* Three varieties of coffee—Arabian Mocha Sanani, Organic Shade Grown Mexico, and Guatemala Antigua—are combined and roasted, yielding a 50-lb batch of coffee beans. Twice as many pounds of Guatemala Antigua, which retails for $10.19 per lb, are needed as of Arabian Mocha Sanani, which retails for $15.99 per lb. Organic Shade Grown Mexico retails for $12.99 per lb. How many pounds, to the nearest hundredth, of each coffee should be used in a blend that sells for $12.37 per lb?

112. *Blending Coffee Beans* Rework **Exercise 111** if Guatemala Antigua retails for $12.49 per lb instead of $10.19 per lb. Does the answer seem reasonable?

Relating Concepts

For individual or collaborative investigation *(Exercises 113–118)*

Supply and Demand In many applications of economics, as the price of an item goes up, demand for the item goes down and supply of the item goes up. The price where supply and demand are equal is the **equilibrium price,** and the resulting supply or demand is the **equilibrium supply** or **equilibrium demand.**

Suppose the supply of a product is related to its price by the equation

$$p = \frac{2}{3}q,$$

where p is in dollars and q is supply in appropriate units. (Here, q stands for quantity.) Furthermore, suppose demand and price for the same product are related by

$$p = -\frac{1}{3}q + 18,$$

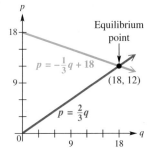

where p is price and q is demand. The system formed by these two equations has solution $(18, 12)$, as seen in the graph. Use this information to **work Exercises 113–118 in order.**

113. Suppose the demand and price for a certain model of electric can opener are related by $p = 16 - \frac{5}{4}q$, where p is price, in dollars, and q is demand, in appropriate units. Find the price when the demand is at each level.

(a) 0 units   (b) 4 units   (c) 8 units

114. Find the demand for the electric can opener at each price.

(a) $6   (b) $11   (c) $16

115. Graph $p = 16 - \frac{5}{4}q$.

116. Suppose the price and supply of the can opener are related by $p = \frac{3}{4}q$, where q represents the supply and p the price. Find the supply at each price.

(a) $0   (b) $10   (c) $20

117. Graph $p = \frac{3}{4}q$ on the same axes used for **Exercise 115.**

118. Use the result of **Exercise 117** to find the equilibrium price and the equilibrium demand.

5.2 Matrix Solution of Linear Systems

- **The Gauss-Jordan Method**
- **Special Systems**

$$\begin{bmatrix} 2 & 3 & 7 \\ 5 & -1 & 10 \end{bmatrix}$$ Matrix

Systems of linear equations occur in many practical situations, and as a result, computer methods have been developed for efficiently solving linear systems. Computer solutions of linear systems depend on the idea of a **matrix** (plural **matrices**), a rectangular array of numbers enclosed in brackets. Each number is an **element** of the matrix.

The Gauss-Jordan Method   In this section, we develop a method for solving linear systems using matrices. We start with a system and write the coefficients of the variables and the constants as an **augmented matrix** of the system.

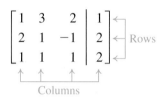

Linear system of equations

$$x + 3y + 2z = 1$$
$$2x + y - z = 2 \quad \text{can be written as}$$
$$x + y + z = 2$$

Augmented matrix

The vertical line, which is optional, separates the coefficients from the constants. Because this matrix has 3 rows (horizontal) and 4 columns (vertical), we say its **dimension*** is 3×4 (read "three by four"). *The number of rows is always given first.* To refer to a number in the matrix, use its row and column numbers. For example, the number 3 is in the first row, second column.

We can treat the rows of this matrix just like the equations of the corresponding system of linear equations. Because an augmented matrix is nothing more than a shorthand form of a system, any transformation of the matrix that results in an equivalent system of equations can be performed.

Matrix Row Transformations

For any augmented matrix of a system of linear equations, the following row transformations will result in the matrix of an equivalent system.

1. Interchange any two rows.

2. Multiply or divide the elements of any row by a nonzero real number.

3. Replace any row of the matrix by the sum of the elements of that row and a multiple of the elements of another row.

These transformations are restatements in matrix form of the transformations of systems discussed in the previous section. From now on, when referring to the third transformation, we will abbreviate "a multiple of the elements of a row" as "a multiple of a row."

Before matrices can be used to solve a linear system, the system must be arranged in the proper form, with variable terms on the left side of the equation and constant terms on the right. The variable terms must be in the same order in each of the equations.

---

*Other terms used to describe the dimension of a matrix are *order* and *size*.

The **Gauss-Jordan method** is a systematic technique for applying matrix row transformations in an attempt to reduce a matrix to **diagonal form,** with 1s along the diagonal, from which the solutions are easily obtained.

$$\begin{bmatrix} 1 & 0 & | & a \\ 0 & 1 & | & b \end{bmatrix} \quad \text{or} \quad \begin{bmatrix} 1 & 0 & 0 & | & a \\ 0 & 1 & 0 & | & b \\ 0 & 0 & 1 & | & c \end{bmatrix} \qquad \text{Diagonal form, or reduced-row echelon form}$$

This form is also called **reduced-row echelon form.**

Using the Gauss-Jordan Method to Transform a Matrix into Diagonal Form

Step 1 Obtain 1 as the first element of the first column.

Step 2 Use the first row to transform the remaining entries in the first column to 0.

Step 3 Obtain 1 as the second entry in the second column.

Step 4 Use the second row to transform the remaining entries in the second column to 0.

Step 5 Continue in this manner as far as possible.

NOTE *The Gauss-Jordan method proceeds column by column, from left to right.* In each column, we work to obtain 1 in the appropriate diagonal location, and then use it to transform the remaining elements in that column to 0s. When we are working with a particular column, no row operation should undo the form of a preceding column.

EXAMPLE 1   Using the Gauss-Jordan Method

Solve the system.

$$3x - 4y = 1$$
$$5x + 2y = 19$$

SOLUTION   Both equations are in the same form, with variable terms in the same order on the left, and constant terms on the right.

$$\begin{bmatrix} 3 & -4 & | & 1 \\ 5 & 2 & | & 19 \end{bmatrix} \qquad \text{Write the augmented matrix.}$$

The goal is to transform the augmented matrix into one in which the value of the variables will be easy to see. That is, because each of the first two columns in the matrix represents the coefficients of one variable, the augmented matrix should be transformed so that it is of the following form.

This form is our goal. $\rightarrow \begin{bmatrix} 1 & 0 & | & k \\ 0 & 1 & | & j \end{bmatrix}$ Here k and j are real numbers.

In this form, the matrix can be rewritten as a linear system.

$$x = k$$
$$y = j$$

It is best to work in columns, beginning in each column with the element that is to become **1.** In the augmented matrix

$$\begin{bmatrix} 3 & -4 & | & 1 \\ 5 & 2 & | & 19 \end{bmatrix},$$

3 is in the first row, first column position. Use transformation 2, multiplying each entry in the first row by $\frac{1}{3}$ (abbreviated $\frac{1}{3}$ R1) to obtain 1 in this position.

$$\begin{bmatrix} 1 & -\frac{4}{3} & | & \frac{1}{3} \\ 5 & 2 & | & 19 \end{bmatrix} \quad \frac{1}{3} \text{R1}$$

Introduce 0 in the second row, first column by multiplying each element of the first row by -5 and adding the result to the corresponding element in the second row, using transformation 3.

$$\begin{bmatrix} 1 & -\frac{4}{3} & | & \frac{1}{3} \\ 0 & \frac{26}{3} & | & \frac{52}{3} \end{bmatrix} \quad -5\text{R1} + \text{R2}$$

Obtain 1 in the second row, second column by multiplying each element of the second row by $\frac{3}{26}$, using transformation 2.

$$\begin{bmatrix} 1 & -\frac{4}{3} & | & \frac{1}{3} \\ 0 & 1 & | & 2 \end{bmatrix} \quad \frac{3}{26} \text{R2}$$

Finally, obtain 0 in the first row, second column by multiplying each element of the second row by $\frac{4}{3}$ and adding the result to the corresponding element in the first row.

$$\begin{bmatrix} 1 & 0 & | & 3 \\ 0 & 1 & | & 2 \end{bmatrix} \quad \frac{4}{3} \text{R2} + \text{R1}$$

This last matrix corresponds to the system

$$x = 3$$
$$y = 2,$$

which indicates the solution $(3, 2)$. We can read this solution directly from the third column of the final matrix.

CHECK   Substitute the solution in both equations of the *original* system.

$3x - 4y = 1$	$5x + 2y = 19$
$3(3) - 4(2) \overset{?}{=} 1$	$5(3) + 2(2) \overset{?}{=} 19$
$9 - 8 \overset{?}{=} 1$	$15 + 4 \overset{?}{=} 19$
$1 = 1$ ✓ True	$19 = 19$ ✓ True

True statements result, so the solution set is $\{(3, 2)\}$.

✔ **Now Try Exercise 23.**

NOTE   Using row operations to write a matrix in diagonal form requires effective use of the inverse properties of addition and multiplication.

A linear system with three equations is solved in a similar way. Row transformations are used to introduce 1s down the diagonal from left to right and 0s above and below each 1.

EXAMPLE 2 **Using the Gauss-Jordan Method**

Solve the system.

$$x - y + 5z = -6$$
$$3x + 3y - z = 10$$
$$x + 3y + 2z = 5$$

SOLUTION
$$\begin{bmatrix} 1 & -1 & 5 & | & -6 \\ 3 & 3 & -1 & | & 10 \\ 1 & 3 & 2 & | & 5 \end{bmatrix}$$ Write the augmented matrix.

There is already a 1 in the first row, first column. Introduce 0 in the second row of the first column by multiplying each element in the first row by -3 and adding the result to the corresponding element in the second row.

$$\begin{bmatrix} 1 & -1 & 5 & | & -6 \\ 0 & 6 & -16 & | & 28 \\ 1 & 3 & 2 & | & 5 \end{bmatrix}$$ $-3R1 + R2$

To change the third element in the first column to 0, multiply each element of the first row by -1. Add the result to the corresponding element of the third row.

$$\begin{bmatrix} 1 & -1 & 5 & | & -6 \\ 0 & 6 & -16 & | & 28 \\ 0 & 4 & -3 & | & 11 \end{bmatrix}$$ $-1R1 + R3$

Use the same procedure to transform the second and third columns. Obtain 1 in the appropriate position of each column by multiplying the elements of the row by the reciprocal of the number in that position.

$$\begin{bmatrix} 1 & -1 & 5 & | & -6 \\ 0 & 1 & -\frac{8}{3} & | & \frac{14}{3} \\ 0 & 4 & -3 & | & 11 \end{bmatrix}$$ $\frac{1}{6}R2$

$$\begin{bmatrix} 1 & 0 & \frac{7}{3} & | & -\frac{4}{3} \\ 0 & 1 & -\frac{8}{3} & | & \frac{14}{3} \\ 0 & 4 & -3 & | & 11 \end{bmatrix}$$ $R2 + R1$

$$\begin{bmatrix} 1 & 0 & \frac{7}{3} & | & -\frac{4}{3} \\ 0 & 1 & -\frac{8}{3} & | & \frac{14}{3} \\ 0 & 0 & \frac{23}{3} & | & -\frac{23}{3} \end{bmatrix}$$ $-4R2 + R3$

$$\begin{bmatrix} 1 & 0 & \frac{7}{3} & | & -\frac{4}{3} \\ 0 & 1 & -\frac{8}{3} & | & \frac{14}{3} \\ 0 & 0 & 1 & | & -1 \end{bmatrix}$$ $\frac{3}{23}R3$

$$\begin{bmatrix} 1 & 0 & 0 & | & 1 \\ 0 & 1 & -\frac{8}{3} & | & \frac{14}{3} \\ 0 & 0 & 1 & | & -1 \end{bmatrix}$$ $-\frac{7}{3}R3 + R1$

$$\begin{bmatrix} 1 & 0 & 0 & | & 1 \\ 0 & 1 & 0 & | & 2 \\ 0 & 0 & 1 & | & -1 \end{bmatrix}$$ $\frac{8}{3}R3 + R2$

$$\begin{bmatrix} 1 & 0 & 0 & | & 1 \\ 0 & 1 & 0 & | & 2 \\ 0 & 0 & 1 & | & -1 \end{bmatrix}$$

Final Matrix

The linear system associated with this final matrix is

$$x = 1$$
$$y = 2$$
$$z = -1.$$

The solution set is $\{(1, 2, -1)\}$. Check the solution in the original system.

✔ **Now Try Exercise 31.**

The TI-84 Plus graphing calculator is able to perform row operations. See **Figure 7(a).** The screen in **Figure 7(b)** shows typical entries for the matrix in the second step of the solution in **Example 2.** The entire Gauss-Jordan method can be carried out in one step with the rref (reduced-row echelon form) command, as shown in **Figure 7(c).**

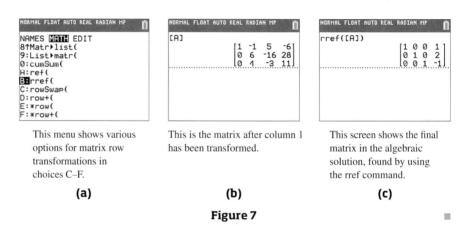

This menu shows various options for matrix row transformations in choices C–F.

(a)

This is the matrix after column 1 has been transformed.

(b)

This screen shows the final matrix in the algebraic solution, found by using the rref command.

(c)

Figure 7

Special Systems The next two examples show how to recognize inconsistent systems or systems with infinitely many solutions when solving such systems using row transformations.

EXAMPLE 3 Solving an Inconsistent System

Use the Gauss-Jordan method to solve the system.

$$x + y = 2$$
$$2x + 2y = 5$$

SOLUTION

$$\begin{bmatrix} 1 & 1 & | & 2 \\ 2 & 2 & | & 5 \end{bmatrix} \quad \text{Write the augmented matrix.}$$

$$\begin{bmatrix} 1 & 1 & | & 2 \\ 0 & 0 & | & 1 \end{bmatrix} \quad -2R1 + R2$$

The next step would be to introduce 1 in the second row, second column. Because of the 0 there, this is impossible. The second row corresponds to

$$0x + 0y = 1$$

which is false for all pairs of x and y, so the system has no solution. The system is inconsistent, and the solution set is $\varnothing$.

✔ **Now Try Exercise 27.**

EXAMPLE 4 **Solving a System with Infinitely Many Solutions**

Use the Gauss-Jordan method to solve the system. Write the solution set with z arbitrary.

$$2x - 5y + 3z = 1$$
$$x - 2y - 2z = 8$$

SOLUTION Recall from the previous section that a system with two equations in three variables usually has an infinite number of solutions. We can use the Gauss-Jordan method to give the solution with z arbitrary.

$$\begin{bmatrix} 2 & -5 & 3 & | & 1 \\ 1 & -2 & -2 & | & 8 \end{bmatrix} \quad \text{Write the augmented matrix.}$$

$$\begin{bmatrix} 1 & -2 & -2 & | & 8 \\ 2 & -5 & 3 & | & 1 \end{bmatrix} \quad \begin{array}{l} \text{Interchange rows to obtain 1 in the} \\ \text{first row, first column position.} \end{array}$$

$$\begin{bmatrix} 1 & -2 & -2 & | & 8 \\ 0 & -1 & 7 & | & -15 \end{bmatrix} \quad -2R1 + R2$$

$$\begin{bmatrix} 1 & -2 & -2 & | & 8 \\ 0 & 1 & -7 & | & 15 \end{bmatrix} \quad -1R2$$

$$\begin{bmatrix} 1 & 0 & -16 & | & 38 \\ 0 & 1 & -7 & | & 15 \end{bmatrix} \quad 2R2 + R1$$

It is not possible to go further with the Gauss-Jordan method. The equations that correspond to the final matrix are

$$x - 16z = 38 \quad \text{and} \quad y - 7z = 15.$$

Solve these equations for x and y, respectively.

$x - 16z = 38$		$y - 7z = 15$	
$x = 16z + 38$	Add $16z$.	$y = 7z + 15$	Add $7z$.

The solution set, written with z arbitrary, is $\{(16z + 38, 7z + 15, z)\}$.

✔ **Now Try Exercise 43.**

Summary of Possible Cases

When matrix methods are used to solve a system of linear equations and the resulting matrix is written in diagonal form (or as close as possible to diagonal form), there are three possible cases.

1. If the number of rows with nonzero elements to the left of the vertical line is equal to the number of variables in the system, then the system has a single solution. **See Examples 1 and 2.**

2. If one of the rows has the form $\begin{bmatrix} 0 & 0 & \cdots & 0 & | & a \end{bmatrix}$ with $a \neq 0$, then the system has no solution. **See Example 3.**

3. If there are fewer rows in the matrix containing nonzero elements than the number of variables, then the system has either no solution or infinitely many solutions. If there are infinitely many solutions, give the solutions in terms of one or more arbitrary variables. **See Example 4.**

5.2 Exercises

CONCEPT PREVIEW *Answer each question.*

1. How many rows and how many columns does this matrix have? What is its dimension?

$$\begin{bmatrix} -2 & 5 & 8 & 0 \\ 1 & 13 & -6 & 9 \end{bmatrix}$$

2. What is the element in the second row, first column of the matrix in **Exercise 1?**

3. What is the augmented matrix of the following system?

$$-3x + 5y = 2$$
$$6x + 2y = 7$$

4. By what number must the first row of the augmented matrix of **Exercise 3** be multiplied so that when it is added to the second row, the element in the second row, first column becomes 0?

5. What is the augmented matrix of the following system?

$$3x + 2y = 5$$
$$-9x + 6z = 1$$
$$-8y + z = 4$$

6. By what number must the first row of the augmented matrix of **Exercise 5** be multiplied so that when it is added to the second row, the element in the second row, first column becomes 0?

Use the given row transformation to change each matrix as indicated. See Example 1.

7. $\begin{bmatrix} 1 & 4 \\ 4 & 7 \end{bmatrix}$; $\begin{array}{l} -4 \text{ times row } 1 \\ \text{added to row } 2 \end{array}$

8. $\begin{bmatrix} 1 & -4 \\ 7 & 0 \end{bmatrix}$; $\begin{array}{l} -7 \text{ times row } 1 \\ \text{added to row } 2 \end{array}$

9. $\begin{bmatrix} 1 & 5 & 6 \\ -2 & 3 & -1 \\ 4 & 7 & 0 \end{bmatrix}$; $\begin{array}{l} 2 \text{ times row } 1 \\ \text{added to row } 2 \end{array}$

10. $\begin{bmatrix} 1 & 5 & 6 \\ -4 & -1 & 2 \\ 3 & 7 & 1 \end{bmatrix}$; $\begin{array}{l} 4 \text{ times row } 1 \\ \text{added to row } 2 \end{array}$

Concept Check Write the augmented matrix for each system and give its dimension. Do not solve.

11. $2x + 3y = 11$
    $x + 2y = 8$

12. $3x + 5y = -13$
    $2x + 3y = -9$

13. $2x + y + z - 3 = 0$
    $3x - 4y + 2z + 7 = 0$
    $x + y + z - 2 = 0$

14. $4x - 2y + 3z - 4 = 0$
    $3x + 5y + z - 7 = 0$
    $5x - y + 4z - 7 = 0$

Concept Check Write the system of equations associated with each augmented matrix. Do not solve.

15. $\begin{bmatrix} 3 & 2 & 1 & | & 1 \\ 0 & 2 & 4 & | & 22 \\ -1 & -2 & 3 & | & 15 \end{bmatrix}$

16. $\begin{bmatrix} 2 & 1 & 3 & | & 12 \\ 4 & -3 & 0 & | & 10 \\ 5 & 0 & -4 & | & -11 \end{bmatrix}$

17. $\begin{bmatrix} 1 & 0 & 0 & | & 2 \\ 0 & 1 & 0 & | & 3 \\ 0 & 0 & 1 & | & -2 \end{bmatrix}$

18. $\begin{bmatrix} 1 & 0 & 0 & | & 4 \\ 0 & 1 & 0 & | & 2 \\ 0 & 0 & 1 & | & 3 \end{bmatrix}$

19.

20.

Use the Gauss-Jordan method to solve each system of equations. For systems in two variables with infinitely many solutions, give the solution with y arbitrary. For systems in three variables with infinitely many solutions, give the solution with z arbitrary. **See Examples 1–4.**

21. $x + y = 5$
$\quad\ x - y = -1$

22. $x + 2y = 5$
$\quad\ 2x + y = -2$

23. $3x + 2y = -9$
$\quad\ 2x - 5y = -6$

24. $2x - 5y = 10$
$\quad\ 3x + y = 15$

25. $6x - 3y - 4 = 0$
$\quad\ 3x + 6y - 7 = 0$

26. $2x - 3y - 10 = 0$
$\quad\ 2x + 2y - 5 = 0$

27. $2x - y = 6$
$\quad\ 4x - 2y = 0$

28. $3x - 2y = 1$
$\quad\ 6x - 4y = -1$

29. $\dfrac{3}{8}x - \dfrac{1}{2}y = \dfrac{7}{8}$
$\quad\ -6x + 8y = -14$

30. $\dfrac{1}{2}x + \dfrac{3}{5}y = \dfrac{1}{4}$
$\quad\ 10x + 12y = 5$

31. $x + y - 5z = -18$
$\quad\ 3x - 3y + z = 6$
$\quad\ x + 3y - 2z = -13$

32. $-x + 2y + 6z = 2$
$\quad\ 3x + 2y + 6z = 6$
$\quad\ x + 4y - 3z = 1$

33. $x + y - z = 6$
$\quad\ 2x - y + z = -9$
$\quad\ x - 2y + 3z = 1$

34. $x + 3y - 6z = 7$
$\quad\ 2x - y + z = 1$
$\quad\ x + 2y + 2z = -1$

35. $x - z = -3$
$\quad\ y + z = 9$
$\quad\ x + z = 7$

36. $-x + y = -1$
$\quad\ y - z = 6$
$\quad\ x + z = -1$

37. $y = -2x - 2z + 1$
$\quad\ x = -2y - z + 2$
$\quad\ z = x - y$

38. $x = -y + 1$
$\quad\ z = 2x$
$\quad\ y = -2z - 2$

39. $2x - y + 3z = 0$
$\quad\ x + 2y - z = 5$
$\quad\ 2y + z = 1$

40. $4x + 2y - 3z = 6$
$\quad\ x - 4y + z = -4$
$\quad\ -x + 2z = 2$

41. $3x + 5y - z + 2 = 0$
$\quad\ 4x - y + 2z - 1 = 0$
$\quad\ -6x - 10y + 2z = 0$

42. $3x + y + 3z - 1 = 0$
$\quad\ x + 2y - z - 2 = 0$
$\quad\ 2x - y + 4z - 4 = 0$

43. $x - 8y + z = 4$
$\quad\ 3x - y + 2z = -1$

44. $5x - 3y + z = 1$
$\quad\ 2x + y - z = 4$

45. $x - y + 2z + w = 4$
$\quad\ y + z = 3$
$\quad\ z - w = 2$
$\quad\ x - y = 0$

46. $x + 2y + z - 3w = 7$
$\quad\ y + z = 0$
$\quad\ x - w = 4$
$\quad\ -x + y = -3$

47. $x + 3y - 2z - w = 9$
$\quad\ 4x + y + z + 2w = 2$
$\quad\ -3x - y + z - w = -5$
$\quad\ x - y - 3z - 2w = 2$

48. $2x + y - z + 3w = 0$
$\quad\ 3x - 2y + z - 4w = -24$
$\quad\ x + y - z + w = 2$
$\quad\ x - y + 2z - 5w = -16$

Solve each system using a graphing calculator capable of performing row operations. Give solutions with values correct to the nearest thousandth.

49. $0.3x + 2.7y - \sqrt{2}z = 3$

$\sqrt{7}x - 20y + 12z = -2$

$4x + \sqrt{3}y - 1.2z = \dfrac{3}{4}$

50. $\sqrt{5}x - 1.2y + z = -3$

$\dfrac{1}{2}x - 3y + 4z = \dfrac{4}{3}$

$4x + 7y - 9z = \sqrt{2}$

Graph each system of three equations together on the same axes, and determine the number of solutions (exactly one, none, or infinitely many). If there is exactly one solution, estimate the solution. Then confirm the answer by solving the system using the Gauss-Jordan method.

51. $2x + 3y = 5$

$-3x + 5y = 22$

$2x + y = -1$

52. $3x - 2y = 3$

$-2x + 4y = 14$

$x + y = 11$

For each equation, determine the constants A and B that make the equation an identity. (Hint: Combine terms on the right, and set coefficients of corresponding terms in the numerators equal.)

53. $\dfrac{1}{(x-1)(x+1)} = \dfrac{A}{x-1} + \dfrac{B}{x+1}$

54. $\dfrac{x+4}{x^2} = \dfrac{A}{x} + \dfrac{B}{x^2}$

55. $\dfrac{x}{(x-a)(x+a)} = \dfrac{A}{x-a} + \dfrac{B}{x+a}$

56. $\dfrac{2x}{(x+2)(x-1)} = \dfrac{A}{x+2} + \dfrac{B}{x-1}$

Solve each problem using the Gauss-Jordan method.

57. Daily Wages Dan is a building contractor. If he hires 7 day laborers and 2 concrete finishers, his payroll for the day is $1384. If he hires 1 day laborer and 5 concrete finishers, his daily cost is $952. Find the daily wage for each type of worker.

58. Mixing Nuts At the Everglades Nut Company, 5 lb of peanuts and 6 lb of cashews cost $33.60, while 3 lb of peanuts and 7 lb of cashews cost $32.40. Find the cost of a single pound of peanuts and a single pound of cashews.

59. Unknown Numbers Find three numbers whose sum is 20, if the first number is three times the difference between the second and the third, and the second number is two more than twice the third.

60. Car Sales Quota To meet a sales quota, a car salesperson must sell 24 new cars, consisting of small, medium, and large cars. She must sell 3 more small cars than medium cars, and the same number of medium cars as large cars. How many of each size must she sell?

61. Mixing Acid Solutions A chemist has two prepared acid solutions, one of which is 2% acid by volume, the other 7% acid. How many cubic centimeters of each should the chemist mix together to obtain 40 cm³ of a 3.2% acid solution?

62. Borrowing Money A small company took out three loans totaling $25,000. The company was able to borrow some of the money at 4% interest. It borrowed $2000 more than one-half the amount of the 4% loan at 6%, and the rest at 5%. The total annual interest was $1220. How much did the company borrow at each rate?

63. Investing Money An investor deposited some money at 1.5% annual interest, and two equal but larger amounts at 2.2% and 2.4%. The total amount invested was $25,000, and the total annual interest earned was $535. How much was invested at each rate?

64. *Investing Money* An investor deposited some money at 1.75% annual interest, some at 2.25%, and twice as much as the sum of the first two at 2.5%. The total amount invested was $30,000, and the total annual interest earned was $710. How much was invested at each rate?

65. *Planning a Diet* In a special diet for a hospital patient, the total amount per meal of food groups A, B, and C must equal 400 g. The diet should include one-third as much of group A as of group B. The sum of the amounts of group A and group C should equal twice the amount of group B. How many grams of each food group should be included? (Give answers to the nearest tenth.)

66. *Planning a Diet* In **Exercise 65,** suppose that, in addition to the conditions given there, foods A and B cost $0.02 per gram, food C costs $0.03 per gram, and a meal must cost $8. Is a solution possible?

(Modeling) Age Distribution in the United States *Use matrices to solve each problem. Let $x = 0$ represent 2015 and $x = 35$ represent 2050. Round values to four decimal places as necessary and percents to the nearest tenth.*

67. In 2015, 14.8% of the population was 65 or older. By 2050, this percent is expected to be 20.9%. The percent of the population aged 25–39 in 2015 was 20.0%. That age group is expected to include 10.3% of the population in 2050. (*Source:* U.S. Census Bureau.)

   (a) Assuming these population changes are linear, use the data for the 65-or-older age group to write a linear equation. Then do the same for the 25–39 age group.

   (b) Solve the system of linear equations from part (a). In what year will the two age groups include the same percent of the population? What is that percent?

   (c) Does the answer to part (b) suggest that the *number* of people in the U.S. population aged 25–39 is decreasing? Explain.

68. In 2015, 19.7% of the U.S. population was aged 40–54. This percent is expected to decrease to 18.7% in 2050. (*Source:* U.S. Census Bureau.)

   (a) Write a linear equation representing this population change.

   (b) Solve the system containing the equation from part (a) and the equation from **Exercise 67** for the 65-or-older age group. Give the year in which these two age groups will include the same percent of the population. What is that percent?

(Modeling) *Solve each problem using matrices.*

69. *Athlete's Weight and Height* The relationship between a professional basketball player's height H (in inches) and weight W (in pounds) was modeled using two different samples of players. The resulting equations that modeled the two samples were

$$W = 7.46H - 374$$

and $\qquad\qquad W = 7.93H - 405.$

   (a) Use each equation to predict the weight of a 6 ft 11 in. professional basketball player to the nearest pound.

   (b) According to each model, what change in weight, to the nearest hundredth pound, is associated with a 1-in. increase in height?

   (c) Determine the weight and height, to the nearest unit, where the two models agree.

70. *Traffic Congestion* At rush hours, substantial traffic congestion is encountered at the traffic intersections shown in the figure. (All streets are one-way.) The city wishes to improve the signals at these corners to speed the flow of traffic. The traffic engineers first gather data. As the figure shows, 700 cars per hour come down M Street to intersection A, and 300 cars per hour come to intersection A on 10th Street. A total of x_1 of these cars leave A on M Street, while x_4 cars leave A on 10th Street. The number of cars entering A must equal the number leaving, which suggests the following equation.

$$x_1 + x_4 = 700 + 300$$
$$x_1 + x_4 = 1000$$

For intersection B, x_1 cars enter B on M Street, and x_2 cars enter B on 11th Street. As the figure shows, 900 cars leave B on 11th Street, while 200 leave on M Street, which leads to the following equation.

$$x_1 + x_2 = 900 + 200$$
$$x_1 + x_2 = 1100$$

At intersection C, 400 cars enter on N Street and 300 on 11th Street, while x_2 cars leave on 11th Street and x_3 cars leave on N Street.

$$x_2 + x_3 = 400 + 300$$
$$x_2 + x_3 = 700$$

Finally, intersection D has x_3 cars entering on N Street and x_4 cars entering on 10th Street. There are 400 cars leaving D on 10th Street and 200 leaving on N Street.

(a) Set up an equation for intersection D.

(b) Use the four equations to write an augmented matrix, and then transform it so that 1s are on the diagonal and 0s are below. This is **triangular form.**

(c) Since there is a row of all 0s, the system of equations does not have a unique solution. Write three equations, corresponding to the three nonzero rows of the matrix. Solve each of the equations for x_4.

(d) One of the equations should have been $x_4 = 1000 - x_1$. What is the greatest possible value of x_1 so that x_4 is not negative?

(e) Another equation should have been $x_4 = x_2 - 100$. Find the least possible value of x_2 so that x_4 is not negative.

(f) Find the greatest possible values of x_3 and x_4 so that neither variable is negative.

(g) Use the results of parts (a)–(f) to give a solution for the problem in which all the equations are satisfied and all variables are nonnegative. Is the solution unique?

(Modeling) Number of Calculations When computers are programmed to solve large linear systems involved in applications like designing aircraft or electrical circuits, they frequently use an algorithm that is similar to the Gauss-Jordan method presented in this section. Solving a linear system with n equations and n variables requires the computer to perform a total of

$$T(n) = \frac{2}{3}n^3 + \frac{3}{2}n^2 - \frac{7}{6}n$$

arithmetic calculations, including additions, subtractions, multiplications, and divisions. Use this model to solve each problem. (*Source:* Burden, R. and J. Faires, *Numerical Analysis*, Sixth Edition, Brooks/Cole Publishing Company.)

71. Compute T for the following values of n. Write the results in a table.

$$n = 3, 6, 10, 29, 100, 200, 400, 1000, 5000, 10{,}000, 100{,}000$$

72. In 1940, John Atanasoff, a physicist from Iowa State University, wanted to solve a 29×29 linear system of equations. How many arithmetic operations would this have required? Is this too many to do by hand? (Atanasoff's work led to the invention of the first fully electronic digital computer.) (*Source: The Gazette.*)

Atanasoff-Berry Computer

73. If the number of equations and variables is doubled, does the number of arithmetic operations double?

74. Suppose that a supercomputer can execute up to 60 billion arithmetic operations per second. How many hours would be required to solve a linear system with 100,000 variables?

Relating Concepts

For individual or collaborative investigation *(Exercises 75-78)*

(Modeling) *Number of Fawns* *To model spring fawn count F from adult pronghorn population A, precipitation P, and severity of the winter W, environmentalists have used the equation*

$$F = a + bA + cP + dW,$$

where a, b, c, and d are constants that must be determined before using the equation. (Winter severity is scaled between 1 and 5, with 1 being mild and 5 being severe.) **Work Exercises 75–78 in order.** (*Source:* Brase, C. and C. Brase, *Understandable Statistics,* D.C. Heath and Company; Bureau of Land Management.)

75. Substitute the values for F, A, P, and W from the table for Years 1–4 into the equation

$$F = a + bA + cP + dW$$

and obtain four linear equations involving a, b, c, and d.

Year	Fawns	Adults	Precip. (in inches)	Winter Severity
1	239	871	11.5	3
2	234	847	12.2	2
3	192	685	10.6	5
4	343	969	14.2	1
5	?	960	12.6	3

76. Write an augmented matrix representing the system in **Exercise 75,** and solve for a, b, c, and d. Round coefficients to three decimal places.

77. Write the equation for F using the values found in **Exercise 76** for the coefficients.

78. Use the information in the table to predict the spring fawn count in Year 5. (Compare this with the actual count of 320.)

5.3 Determinant Solution of Linear Systems

- Determinants
- Cofactors
- $n \times n$ Determinants
- Determinant Theorems
- Cramer's Rule

Determinants   Every $n \times n$ matrix A is associated with a real number called the **determinant** of A, written $|A|$. In this section we show how to evaluate determinants of square matrices, providing mathematical justification as we proceed. Modern graphing calculators are programmed to evaluate determinants in their matrix menus.

The determinant of a 2×2 matrix is defined as follows.

Determinant of a 2 × 2 Matrix

If $A = \begin{bmatrix} a_{11} & a_{12} \\ a_{21} & a_{22} \end{bmatrix}$, then $|A| = \begin{vmatrix} a_{11} & a_{12} \\ a_{21} & a_{22} \end{vmatrix} = a_{11}a_{22} - a_{21}a_{12}.$

NOTE   *Matrices are enclosed with square brackets, while determinants are denoted with vertical bars.* A matrix is an *array* of numbers, but its determinant is a *single* number.

The arrows in the diagram below indicate which products to find when evaluating a 2×2 determinant.

$$\begin{vmatrix} a_{11} & a_{12} \\ a_{21} & a_{22} \end{vmatrix}$$

EXAMPLE 1   **Evaluating a 2 × 2 Determinant**

Let $A = \begin{bmatrix} -3 & 4 \\ 6 & 8 \end{bmatrix}$. Find $|A|$.

ALGEBRAIC SOLUTION

Use the definition with

$$a_{11} = -3, \quad a_{12} = 4, \quad a_{21} = 6, \quad a_{22} = 8.$$

$$|A| = \begin{vmatrix} -3 & 4 \\ 6 & 8 \end{vmatrix}$$

$$|A| = \underset{\substack{\uparrow \quad \uparrow \\ a_{11} \quad a_{22}}}{-3 \cdot 8} \; - \; \underset{\substack{\uparrow \quad \uparrow \\ a_{21} \quad a_{12}}}{6 \cdot 4}$$

$$= -24 - 24 \quad \text{Multiply.}$$

$$= -48 \quad \text{Subtract.}$$

GRAPHING CALCULATOR SOLUTION

We can define a matrix and then use the capability of a graphing calculator to find the determinant of the matrix. In the screen in **Figure 8,** the symbol det($[A]$) represents the determinant of $[A]$.

Figure 8

✔ **Now Try Exercise 7.**

Determinant of a 3 × 3 Matrix

If $A = \begin{bmatrix} a_{11} & a_{12} & a_{13} \\ a_{21} & a_{22} & a_{23} \\ a_{31} & a_{32} & a_{33} \end{bmatrix}$, then the determinant of A, symbolized $|A|$, is defined as follows.

$$|A| = \begin{vmatrix} a_{11} & a_{12} & a_{13} \\ a_{21} & a_{22} & a_{23} \\ a_{31} & a_{32} & a_{33} \end{vmatrix} = (a_{11}a_{22}a_{33} + a_{12}a_{23}a_{31} + a_{13}a_{21}a_{32}) - (a_{31}a_{22}a_{13} + a_{32}a_{23}a_{11} + a_{33}a_{21}a_{12})$$

The terms on the right side of the equation in the definition of $|A|$ can be rearranged to obtain the following.

$$\begin{vmatrix} a_{11} & a_{12} & a_{13} \\ a_{21} & a_{22} & a_{23} \\ a_{31} & a_{32} & a_{33} \end{vmatrix} = a_{11}(a_{22}a_{33} - a_{32}a_{23}) - a_{21}(a_{12}a_{33} - a_{32}a_{13}) + a_{31}(a_{12}a_{23} - a_{22}a_{13})$$

Each quantity in parentheses represents the determinant of a 2 × 2 matrix that is the part of the 3 × 3 matrix remaining when the row and column of the multiplier are eliminated, as shown below.

$$a_{11}(a_{22}a_{33} - a_{32}a_{23}) \quad \begin{bmatrix} a_{11} & a_{12} & a_{13} \\ a_{21} & a_{22} & a_{23} \\ a_{31} & a_{32} & a_{33} \end{bmatrix}$$

$$a_{21}(a_{12}a_{33} - a_{32}a_{13}) \quad \begin{bmatrix} a_{11} & a_{12} & a_{13} \\ a_{21} & a_{22} & a_{23} \\ a_{31} & a_{32} & a_{33} \end{bmatrix}$$

$$a_{31}(a_{12}a_{23} - a_{22}a_{13}) \quad \begin{bmatrix} a_{11} & a_{12} & a_{13} \\ a_{21} & a_{22} & a_{23} \\ a_{31} & a_{32} & a_{33} \end{bmatrix}$$

Cofactors The determinant of each 2 × 2 matrix above is the **minor** of the associated element in the 3 × 3 matrix. The symbol M_{ij} represents the minor that results when row i and column j are eliminated. The following list gives some of the minors from the matrix above.

Element	Minor	Element	Minor
a_{11}	$M_{11} = \begin{vmatrix} a_{22} & a_{23} \\ a_{32} & a_{33} \end{vmatrix}$	a_{22}	$M_{22} = \begin{vmatrix} a_{11} & a_{13} \\ a_{31} & a_{33} \end{vmatrix}$
a_{21}	$M_{21} = \begin{vmatrix} a_{12} & a_{13} \\ a_{32} & a_{33} \end{vmatrix}$	a_{23}	$M_{23} = \begin{vmatrix} a_{11} & a_{12} \\ a_{31} & a_{32} \end{vmatrix}$
a_{31}	$M_{31} = \begin{vmatrix} a_{12} & a_{13} \\ a_{22} & a_{23} \end{vmatrix}$	a_{33}	$M_{33} = \begin{vmatrix} a_{11} & a_{12} \\ a_{21} & a_{22} \end{vmatrix}$

In a 4×4 matrix, the minors are determinants of 3×3 matrices. Similarly, an $n \times n$ matrix has minors that are determinants of $(n-1) \times (n-1)$ matrices.

To find the determinant of a 3×3 or larger matrix, first choose any row or column. Then the minor of each element in that row or column must be multiplied by $+1$ or -1, depending on whether the sum of the row number and column number is even or odd. The product of a minor and the number $+1$ or -1 is a *cofactor*.

Cofactor

Let M_{ij} be the minor for element a_{ij} in an $n \times n$ matrix. The **cofactor** of a_{ij}, written A_{ij}, is defined as follows.

$$A_{ij} = (-1)^{i+j} \cdot M_{ij}$$

EXAMPLE 2   Finding Cofactors of Elements

Find the cofactor of each of the following elements of the given matrix.

$$\begin{bmatrix} 6 & 2 & 4 \\ 8 & 9 & 3 \\ 1 & 2 & 0 \end{bmatrix}$$

(a) 6 **(b)** 3 **(c)** 8

SOLUTION

(a) The element 6 is in the first row, first column of the matrix, so $i = 1$ and $j = 1$. $M_{11} = \begin{vmatrix} 9 & 3 \\ 2 & 0 \end{vmatrix} = -6$. The cofactor is

$$(-1)^{1+1}(-6) = 1(-6) = -6.$$

(b) Here $i = 2$ and $j = 3$, so $M_{23} = \begin{vmatrix} 6 & 2 \\ 1 & 2 \end{vmatrix} = 10$. The cofactor is

$$(-1)^{2+3}(10) = -1(10) = -10.$$

(c) We have $i = 2$ and $j = 1$, and $M_{21} = \begin{vmatrix} 2 & 4 \\ 2 & 0 \end{vmatrix} = -8$. The cofactor is

$$(-1)^{2+1}(-8) = -1(-8) = 8.$$

✔ **Now Try Exercise 17.**

$n \times n$ Determinants   The determinant of a 3×3 or larger matrix is found as follows.

Finding the Determinant of a Matrix

Multiply each element in any row or column of the matrix by its cofactor. The sum of these products gives the value of the determinant.

The process of forming this sum of products is called **expansion by a given row or column.**

EXAMPLE 3 Evaluating a 3 × 3 Determinant

Evaluate $\begin{vmatrix} 2 & -3 & -2 \\ -1 & -4 & -3 \\ -1 & 0 & 2 \end{vmatrix}$, expanding by the second column.

SOLUTION First find the minor of each element in the second column.

$$M_{12} = \begin{vmatrix} -1 & -3 \\ -1 & 2 \end{vmatrix} = -1(2) - (-1)(-3) = -5$$

> Use parentheses, and keep track of all negative signs to avoid errors.

$$M_{22} = \begin{vmatrix} 2 & -2 \\ -1 & 2 \end{vmatrix} = 2(2) - (-1)(-2) = 2$$

$$M_{32} = \begin{vmatrix} 2 & -2 \\ -1 & -3 \end{vmatrix} = 2(-3) - (-1)(-2) = -8$$

Now find the cofactor of each element of these minors.

$$A_{12} = (-1)^{1+2} \cdot M_{12} = (-1)^3 \cdot (-5) = -1(-5) = 5$$

$$A_{22} = (-1)^{2+2} \cdot M_{22} = (-1)^4 \cdot 2 = 1 \cdot 2 = 2$$

$$A_{32} = (-1)^{3+2} \cdot M_{32} = (-1)^5 \cdot (-8) = -1(-8) = 8$$

Find the determinant by multiplying each cofactor by its corresponding element in the matrix and finding the sum of these products.

$$\begin{vmatrix} 2 & -3 & -2 \\ -1 & -4 & -3 \\ -1 & 0 & 2 \end{vmatrix} = a_{12} \cdot A_{12} + a_{22} \cdot A_{22} + a_{32} \cdot A_{32}$$

$$= -3(5) + (-4)2 + 0(8)$$

$$= -15 + (-8) + 0$$

$$= -23 \qquad \text{✔ Now Try Exercise 21.}$$

```
NORMAL FLOAT AUTO REAL RADIAN MP
[B]
                    [2  -3 -2]
                    [-1 -4 -3]
                    [-1  0  2]
det([B])
                              -23
```

The matrix in **Example 3** can be entered as [B] and its determinant found using a graphing calculator, as shown in the screen above.

In **Example 3**, we would have found the same answer using any row or column of the matrix. One reason we used column 2 is that it contains a 0 element, so it was not really necessary to calculate M_{32} and A_{32}.

Instead of calculating $(-1)^{i+j}$ for a given element, we can use the sign checkerboard shown below. The signs alternate for each row and column, beginning with $+$ in the first row, first column position. If we expand a 3 × 3 matrix about row 3, for example, the first minor would have a $+$ sign associated with it, the second minor a $-$ sign, and the third minor a $+$ sign.

*Sign array
for 3 × 3 matrices*

$$+ \quad - \quad +$$
$$- \quad + \quad -$$
$$+ \quad - \quad +$$

This sign array can be extended for determinants of larger matrices.

Determinant Theorems   The following theorems are true for square matrices of any dimension and can be used to simplify finding determinants.

Determinant Theorems

1. If every element in a row (or column) of matrix A is 0, then $|A| = 0$.

2. If the rows of matrix A are the corresponding columns of matrix B, then $|B| = |A|$.

3. If any two rows (or columns) of matrix A are interchanged to form matrix B, then $|B| = -|A|$.

4. Suppose matrix B is formed by multiplying every element of a row (or column) of matrix A by the real number k. Then $|B| = k \cdot |A|$.

5. If two rows (or columns) of matrix A are identical, then $|A| = 0$.

6. Changing a row (or column) of a matrix by adding to it a constant times another row (or column) does not change the determinant of the matrix.

7. If matrix A is in triangular form, having only zeros either above or below the main diagonal, then $|A|$ is the product of the elements on the main diagonal of A.

EXAMPLE 4   **Using the Determinant Theorems**

Use the determinant theorems to evaluate each determinant.

(a) $\begin{vmatrix} -2 & 4 & 2 \\ 6 & 7 & 3 \\ 0 & 16 & 8 \end{vmatrix}$

(b) $\begin{vmatrix} 3 & -7 & 4 & 10 \\ 0 & 1 & 8 & 3 \\ 0 & 0 & -5 & 2 \\ 0 & 0 & 0 & 6 \end{vmatrix}$

SOLUTION

(a) Use determinant theorem 6 to obtain a 0 in the second row of the first column. Multiply each element in the first row by 3, and add the result to the corresponding element in the second row.

$$\begin{vmatrix} -2 & 4 & 2 \\ 0 & 19 & 9 \\ 0 & 16 & 8 \end{vmatrix} \quad \text{3R1 + R2}$$

Now, find the determinant by expanding by column 1.

$$-2(-1)^{1+1} \begin{vmatrix} 19 & 9 \\ 16 & 8 \end{vmatrix} = -2(1)(8) = -16 \quad 19(8) - 16(9) = 152 - 144 = 8$$

(b) Use determinant theorem 7 to find the determinant of this triangular matrix by multiplying the elements on the main diagonal.

$$\begin{vmatrix} 3 & -7 & 4 & 10 \\ 0 & 1 & 8 & 3 \\ 0 & 0 & -5 & 2 \\ 0 & 0 & 0 & 6 \end{vmatrix} = 3(1)(-5)(6) = -90$$

✔ **Now Try Exercises 51 and 53.**

Cramer's Rule The elimination method can be used to develop a process for solving a linear system in two unknowns using determinants. Consider the following system.

$$a_1x + b_1y = c_1 \quad \text{(1)}$$
$$a_2x + b_2y = c_2 \quad \text{(2)}$$

The variable y in this system of equations can be eliminated by using multiplication to create coefficients that are additive inverses and by adding the two equations.

$$a_1b_2x + b_1b_2y = c_1b_2 \qquad \text{Multiply (1) by } b_2.$$
$$\underline{-a_2b_1x - b_1b_2y = -c_2b_1} \qquad \text{Multiply (2) by } -b_1.$$
$$(a_1b_2 - a_2b_1)x \qquad = c_1b_2 - c_2b_1 \qquad \text{Add.}$$

$$x = \frac{c_1b_2 - c_2b_1}{a_1b_2 - a_2b_1}, \quad \text{if } a_1b_2 - a_2b_1 \neq 0$$

Similarly, the variable x can be eliminated.

$$-a_1a_2x - a_2b_1y = -a_2c_1 \qquad \text{Multiply (1) by } -a_2.$$
$$\underline{a_1a_2x + a_1b_2y = a_1c_2} \qquad \text{Multiply (2) by } a_1.$$
$$(a_1b_2 - a_2b_1)y = a_1c_2 - a_2c_1 \qquad \text{Add.}$$

$$y = \frac{a_1c_2 - a_2c_1}{a_1b_2 - a_2b_1}, \quad \text{if } a_1b_2 - a_2b_1 \neq 0$$

Both numerators and the common denominator of these values for x and y can be written as determinants.

$$c_1b_2 - c_2b_1 = \begin{vmatrix} c_1 & b_1 \\ c_2 & b_2 \end{vmatrix}, \quad a_1c_2 - a_2c_1 = \begin{vmatrix} a_1 & c_1 \\ a_2 & c_2 \end{vmatrix}, \quad \text{and} \quad a_1b_2 - a_2b_1 = \begin{vmatrix} a_1 & b_1 \\ a_2 & b_2 \end{vmatrix}$$

The solutions for x and y can be written using these determinants.

$$x = \frac{\begin{vmatrix} c_1 & b_1 \\ c_2 & b_2 \end{vmatrix}}{\begin{vmatrix} a_1 & b_1 \\ a_2 & b_2 \end{vmatrix}} \quad \text{and} \quad y = \frac{\begin{vmatrix} a_1 & c_1 \\ a_2 & c_2 \end{vmatrix}}{\begin{vmatrix} a_1 & b_1 \\ a_2 & b_2 \end{vmatrix}}, \quad \text{if} \quad \begin{vmatrix} a_1 & b_1 \\ a_2 & b_2 \end{vmatrix} \neq 0.$$

We denote the three determinants in the solution as follows.

$$\begin{vmatrix} a_1 & b_1 \\ a_2 & b_2 \end{vmatrix} = D, \quad \begin{vmatrix} c_1 & b_1 \\ c_2 & b_2 \end{vmatrix} = D_x, \quad \text{and} \quad \begin{vmatrix} a_1 & c_1 \\ a_2 & c_2 \end{vmatrix} = D_y$$

NOTE The elements of D are the four coefficients of the variables in the given system. The elements of D_x are obtained by replacing the coefficients of x in D by the respective constants, and the elements of D_y are obtained by replacing the coefficients of y in D by the respective constants.

These results are summarized as **Cramer's rule.**

Cramer's Rule for Two Equations in Two Variables

Given the system

$$a_1 x + b_1 y = c_1$$
$$a_2 x + b_2 y = c_2,$$

if $D \neq 0$, then the system has the unique solution

$$x = \frac{D_x}{D} \quad \text{and} \quad y = \frac{D_y}{D},$$

where $\quad D = \begin{vmatrix} a_1 & b_1 \\ a_2 & b_2 \end{vmatrix}, \quad D_x = \begin{vmatrix} c_1 & b_1 \\ c_2 & b_2 \end{vmatrix}, \quad \text{and} \quad D_y = \begin{vmatrix} a_1 & c_1 \\ a_2 & c_2 \end{vmatrix}.$

CAUTION Evaluate D first. **If $D = 0$, then Cramer's rule does not apply.** The system is inconsistent or has infinitely many solutions.

Because graphing calculators can evaluate determinants, they can also be used to apply Cramer's rule to solve a system of linear equations. The screens above support the result in **Example 5**.

EXAMPLE 5 **Applying Cramer's Rule to a 2 × 2 System**

Use Cramer's rule to solve the system of equations.

$$5x + 7y = -1$$
$$6x + 8y = 1$$

SOLUTION First find D. If $D \neq 0$, then find D_x and D_y.

$$D = \begin{vmatrix} 5 & 7 \\ 6 & 8 \end{vmatrix} = 5(8) - 6(7) = -2$$

$$D_x = \begin{vmatrix} -1 & 7 \\ 1 & 8 \end{vmatrix} = -1(8) - 1(7) = -15$$

$$D_y = \begin{vmatrix} 5 & -1 \\ 6 & 1 \end{vmatrix} = 5(1) - 6(-1) = 11$$

$$x = \frac{D_x}{D} = \frac{-15}{-2} = \frac{15}{2} \quad \text{and} \quad y = \frac{D_y}{D} = \frac{11}{-2} = -\frac{11}{2} \qquad \text{Cramer's rule}$$

The solution set is $\left\{ \left(\frac{15}{2}, -\frac{11}{2} \right) \right\}$. Verify by substituting in the given system.

✔ **Now Try Exercise 65.**

General Form of Cramer's Rule

Let an $n \times n$ system have linear equations of the following form.

$$a_1 x_1 + a_2 x_2 + a_3 x_3 + \cdots + a_n x_n = b$$

Define D as the determinant of the $n \times n$ matrix of all coefficients of the variables. Define D_{x_1} as the determinant obtained from D by replacing the entries in column 1 of D with the constants of the system. Define D_{x_i} as the determinant obtained from D by replacing the entries in column i with the constants of the system. If $D \neq 0$, then the unique solution of the system is

$$x_1 = \frac{D_{x_1}}{D}, \quad x_2 = \frac{D_{x_2}}{D}, \quad x_3 = \frac{D_{x_3}}{D}, \quad \cdots, \quad x_n = \frac{D_{x_n}}{D}.$$

EXAMPLE 6 **Applying Cramer's Rule to a 3 × 3 System**

Use Cramer's rule to solve the system of equations.

$$x + y - z + 2 = 0$$
$$2x - y + z + 5 = 0$$
$$x - 2y + 3z - 4 = 0$$

SOLUTION

$$\begin{aligned} x + y - z &= -2 \\ 2x - y + z &= -5 \\ x - 2y + 3z &= 4 \end{aligned}$$

Rewrite each equation in the form $ax + by + cz + \cdots = k$.

Verify the required determinants.

$$D = \begin{vmatrix} 1 & 1 & -1 \\ 2 & -1 & 1 \\ 1 & -2 & 3 \end{vmatrix} = -3, \qquad D_x = \begin{vmatrix} -2 & 1 & -1 \\ -5 & -1 & 1 \\ 4 & -2 & 3 \end{vmatrix} = 7,$$

$$D_y = \begin{vmatrix} 1 & -2 & -1 \\ 2 & -5 & 1 \\ 1 & 4 & 3 \end{vmatrix} = -22, \qquad D_z = \begin{vmatrix} 1 & 1 & -2 \\ 2 & -1 & -5 \\ 1 & -2 & 4 \end{vmatrix} = -21$$

$$x = \frac{D_x}{D} = \frac{7}{-3} = -\frac{7}{3}, \quad y = \frac{D_y}{D} = \frac{-22}{-3} = \frac{22}{3}, \quad z = \frac{D_z}{D} = \frac{-21}{-3} = 7$$

The solution set is $\left\{ \left(-\frac{7}{3}, \frac{22}{3}, 7 \right) \right\}$.

✔ **Now Try Exercise 81.**

CAUTION As shown in **Example 6,** each equation in the system must be written in the form $ax + by + cz + \cdots = k$ before Cramer's rule is used.

EXAMPLE 7 **Showing That Cramer's Rule Does Not Apply**

Show that Cramer's rule does not apply to the following system.

$$2x - 3y + 4z = 10$$
$$6x - 9y + 12z = 24$$
$$x + 2y - 3z = 5$$

SOLUTION We need to show that $D = 0$. Expand about column 1.

$$D = \begin{vmatrix} 2 & -3 & 4 \\ 6 & -9 & 12 \\ 1 & 2 & -3 \end{vmatrix} = 2 \begin{vmatrix} -9 & 12 \\ 2 & -3 \end{vmatrix} - 6 \begin{vmatrix} -3 & 4 \\ 2 & -3 \end{vmatrix} + 1 \begin{vmatrix} -3 & 4 \\ -9 & 12 \end{vmatrix}$$

$$= 2(3) - 6(1) + 1(0)$$

$$= 0$$

Because $D = 0$, Cramer's rule does not apply.

✔ **Now Try Exercise 79.**

NOTE When $D = 0$, *the system is either inconsistent or has infinitely many solutions.* Use the elimination method to tell which is the case. Verify that the system in **Example 7** is inconsistent, and thus the solution set is $\varnothing$.

5.3 Exercises

CONCEPT PREVIEW *Answer each question.*

1. What is the value of $\begin{vmatrix} 4 & 0 \\ -2 & 0 \end{vmatrix}$?

2. What is the value of $\begin{vmatrix} 4 & 4 \\ -2 & -2 \end{vmatrix}$?

3. What expression in x represents $\begin{vmatrix} x & 4 \\ 3 & x \end{vmatrix}$?

4. What expression in x represents $\begin{vmatrix} 4 & 3 \\ x & x \end{vmatrix}$?

5. What is the value of x if $\begin{vmatrix} x & 0 \\ 0 & x \end{vmatrix} = 9$?

6. What is the value of x if $\begin{vmatrix} 0 & x \\ x & 0 \end{vmatrix} = -4$?

Evaluate each determinant. See Example 1.

7. $\begin{vmatrix} -5 & 9 \\ 4 & -1 \end{vmatrix}$

8. $\begin{vmatrix} -1 & 3 \\ -2 & 9 \end{vmatrix}$

9. $\begin{vmatrix} -1 & -2 \\ 5 & 3 \end{vmatrix}$

10. $\begin{vmatrix} 6 & -4 \\ 0 & -1 \end{vmatrix}$

11. $\begin{vmatrix} 9 & 3 \\ -3 & -1 \end{vmatrix}$

12. $\begin{vmatrix} 0 & 2 \\ 1 & 5 \end{vmatrix}$

13. $\begin{vmatrix} 3 & 4 \\ 5 & -2 \end{vmatrix}$

14. $\begin{vmatrix} -9 & 7 \\ 2 & 6 \end{vmatrix}$

15. $\begin{vmatrix} -7 & 0 \\ 3 & 0 \end{vmatrix}$

16. *Concept Check* Refer to **Exercise 11.** Make a conjecture about the value of the determinant of a matrix in which one row is a multiple of another row.

Find the cofactor of each element in the second row of each matrix. See Example 2.

17. $\begin{bmatrix} -2 & 0 & 1 \\ 1 & 2 & 0 \\ 4 & 2 & 1 \end{bmatrix}$

18. $\begin{bmatrix} 1 & -1 & 2 \\ 1 & 0 & 2 \\ 0 & -3 & 1 \end{bmatrix}$

19. $\begin{bmatrix} 1 & 2 & -1 \\ 2 & 3 & -2 \\ -1 & 4 & 1 \end{bmatrix}$

20. $\begin{bmatrix} 2 & -1 & 4 \\ 3 & 0 & 1 \\ -2 & 1 & 4 \end{bmatrix}$

Evaluate each determinant. See Example 3.

21. $\begin{vmatrix} 4 & -7 & 8 \\ 2 & 1 & 3 \\ -6 & 3 & 0 \end{vmatrix}$

22. $\begin{vmatrix} 8 & -2 & -4 \\ 7 & 0 & 3 \\ 5 & -1 & 2 \end{vmatrix}$

23. $\begin{vmatrix} 1 & 2 & 0 \\ -1 & 2 & -1 \\ 0 & 1 & 4 \end{vmatrix}$

24. $\begin{vmatrix} 2 & 1 & -1 \\ 4 & 7 & -2 \\ 2 & 4 & 0 \end{vmatrix}$

25. $\begin{vmatrix} 10 & 2 & 1 \\ -1 & 4 & 3 \\ -3 & 8 & 10 \end{vmatrix}$

26. $\begin{vmatrix} 7 & -1 & 1 \\ 1 & -7 & 2 \\ -2 & 1 & 1 \end{vmatrix}$

27. $\begin{vmatrix} 1 & -2 & 3 \\ 0 & 0 & 0 \\ 1 & 10 & -12 \end{vmatrix}$

28. $\begin{vmatrix} 2 & 3 & 0 \\ 1 & 9 & 0 \\ -1 & -2 & 0 \end{vmatrix}$

29. $\begin{vmatrix} 3 & 3 & -1 \\ 2 & 6 & 0 \\ -6 & -6 & 2 \end{vmatrix}$

30. $\begin{vmatrix} 5 & -3 & 2 \\ -5 & 3 & -2 \\ 1 & 0 & 1 \end{vmatrix}$

31. $\begin{vmatrix} 1 & 0 & 0 \\ 0 & 1 & 0 \\ 0 & 0 & 1 \end{vmatrix}$

32. $\begin{vmatrix} 1 & 0 & 0 \\ 0 & -1 & 0 \\ 1 & 0 & 1 \end{vmatrix}$

33. $\begin{vmatrix} -2 & 0 & 1 \\ 0 & 1 & 0 \\ 0 & 0 & -1 \end{vmatrix}$ **34.** $\begin{vmatrix} 0 & 0 & -1 \\ -1 & 0 & 1 \\ 0 & -1 & 0 \end{vmatrix}$ **35.** $\begin{vmatrix} \sqrt{2} & 4 & 0 \\ 1 & -\sqrt{5} & 7 \\ -5 & \sqrt{5} & 1 \end{vmatrix}$

36. $\begin{vmatrix} \sqrt{3} & 1 & 0 \\ \sqrt{7} & 4 & -1 \\ 5 & 0 & -\sqrt{7} \end{vmatrix}$ **37.** $\begin{vmatrix} 0.4 & -0.8 & 0.6 \\ 0.3 & 0.9 & 0.7 \\ 3.1 & 4.1 & -2.8 \end{vmatrix}$ **38.** $\begin{vmatrix} -0.3 & -0.1 & 0.9 \\ 2.5 & 4.9 & -3.2 \\ -0.1 & 0.4 & 0.8 \end{vmatrix}$

Use the determinant theorems and the fact that $\begin{vmatrix} 1 & 2 & 3 \\ 4 & 5 & 6 \\ 7 & 9 & 10 \end{vmatrix} = 3$ *to evaluate each determinant.*

39. $\begin{vmatrix} 4 & 5 & 6 \\ 1 & 2 & 3 \\ 7 & 9 & 10 \end{vmatrix}$ **40.** $\begin{vmatrix} 3 & 2 & 1 \\ 6 & 5 & 4 \\ 10 & 9 & 7 \end{vmatrix}$ **41.** $\begin{vmatrix} 5 & 10 & 15 \\ 4 & 5 & 6 \\ 7 & 9 & 10 \end{vmatrix}$

42. $\begin{vmatrix} 1 & 20 & 3 \\ 4 & 50 & 6 \\ 7 & 90 & 10 \end{vmatrix}$ **43.** $\begin{vmatrix} 1 & 2 & 3 \\ 4 & 5 & 6 \\ 8 & 11 & 13 \end{vmatrix}$ **44.** $\begin{vmatrix} 1 & 2 & 0 \\ 4 & 5 & -6 \\ 7 & 9 & -11 \end{vmatrix}$

Use the determinant theorems to evaluate each determinant. **See Example 4.**

45. $\begin{vmatrix} 1 & 0 & 0 \\ 1 & 0 & 1 \\ 3 & 0 & 0 \end{vmatrix}$ **46.** $\begin{vmatrix} -1 & 2 & 4 \\ 4 & -8 & -16 \\ 0 & 0 & 0 \end{vmatrix}$ **47.** $\begin{vmatrix} 6 & 8 & -12 \\ -1 & 0 & 2 \\ 4 & 0 & -8 \end{vmatrix}$

48. $\begin{vmatrix} 4 & 8 & 0 \\ -1 & -2 & 1 \\ 2 & 4 & 3 \end{vmatrix}$ **49.** $\begin{vmatrix} -4 & 1 & 4 \\ 2 & 0 & 1 \\ 0 & 2 & 4 \end{vmatrix}$ **50.** $\begin{vmatrix} 6 & 3 & 2 \\ 1 & 0 & 2 \\ 5 & 7 & 3 \end{vmatrix}$

51. $\begin{vmatrix} 0 & 1 & -3 \\ 7 & 5 & 2 \\ 1 & -2 & 6 \end{vmatrix}$ **52.** $\begin{vmatrix} 7 & 9 & -3 \\ 7 & -6 & 2 \\ 8 & 1 & 0 \end{vmatrix}$ **53.** $\begin{vmatrix} 1 & 6 & 7 \\ 0 & 6 & 7 \\ 0 & 0 & 9 \end{vmatrix}$

54. $\begin{vmatrix} 7 & 0 & 0 \\ 1 & 6 & 0 \\ 4 & 2 & 4 \end{vmatrix}$ **55.** $\begin{vmatrix} 2 & -1 & 3 \\ 6 & 4 & 10 \\ 4 & 5 & 7 \end{vmatrix}$ **56.** $\begin{vmatrix} 9 & 1 & 7 \\ 12 & 5 & 2 \\ 11 & 4 & 3 \end{vmatrix}$

57. $\begin{vmatrix} -1 & 0 & 2 & 3 \\ 5 & 4 & -3 & 7 \\ 8 & 2 & 9 & -5 \\ 4 & 4 & -1 & 10 \end{vmatrix}$ **58.** $\begin{vmatrix} 5 & 1 & 4 & 2 \\ 4 & -3 & 7 & -4 \\ 5 & 8 & -3 & 6 \\ 9 & 9 & 0 & 8 \end{vmatrix}$

59. $\begin{vmatrix} 4 & 0 & 0 & 2 \\ -1 & 0 & 3 & 0 \\ 2 & 4 & 0 & 1 \\ 0 & 0 & 1 & 2 \end{vmatrix}$ **60.** $\begin{vmatrix} -2 & 0 & 4 & 2 \\ 3 & 6 & 0 & 4 \\ 0 & 0 & 0 & 3 \\ 9 & 0 & 2 & -1 \end{vmatrix}$

61. $\begin{vmatrix} 3 & -6 & 5 & -1 \\ 0 & 2 & -1 & 3 \\ -6 & 4 & 2 & 0 \\ -7 & 3 & 1 & 1 \end{vmatrix}$ **62.** $\begin{vmatrix} 4 & 5 & -1 & -1 \\ 2 & -3 & 1 & 0 \\ -5 & 1 & 3 & 9 \\ 0 & -2 & 1 & 5 \end{vmatrix}$

*Use Cramer's rule to solve each system of equations. If D = 0, then use another method to determine the solution set. **See Examples 5–7.***

63. $x + y = 4$
$2x - y = 2$

64. $3x + 2y = -4$
$2x - y = -5$

65. $4x + 3y = -7$
$2x + 3y = -11$

66. $4x - y = 0$
$2x + 3y = 14$

67. $5x + 4y = 10$
$3x - 7y = 6$

68. $3x + 2y = -4$
$5x - y = 2$

69. $1.5x + 3y = 5$
$2x + 4y = 3$

70. $12x + 8y = 3$
$1.5x + y = 0.9$

71. $3x + 2y = 4$
$6x + 4y = 8$

72. $4x + 3y = 9$
$12x + 9y = 27$

73. $\frac{1}{2}x + \frac{1}{3}y = 2$
$\frac{3}{2}x - \frac{1}{2}y = -12$

74. $-\frac{3}{4}x + \frac{2}{3}y = 16$
$\frac{5}{2}x + \frac{1}{2}y = -37$

75. $2x - y + 4z = -2$
$3x + 2y - z = -3$
$x + 4y + 2z = 17$

76. $x + y + z = 4$
$2x - y + 3z = 4$
$4x + 2y - z = -15$

77. $x + 2y + 3z = 4$
$4x + 3y + 2z = 1$
$-x - 2y - 3z = 0$

78. $2x - y + 3z = 1$
$-2x + y - 3z = 2$
$5x - y + z = 2$

79. $-2x - 2y + 3z = 4$
$5x + 7y - z = 2$
$2x + 2y - 3z = -4$

80. $3x - 2y + 4z = 1$
$4x + y - 5z = 2$
$-6x + 4y - 8z = -2$

81. $4x - 3y + z + 1 = 0$
$5x + 7y + 2z + 2 = 0$
$3x - 5y - z - 1 = 0$

82. $2x - 3y + z - 8 = 0$
$-x - 5y + z + 4 = 0$
$3x - 5y + 2z - 12 = 0$

83. $5x - y = -4$
$3x + 2z = 4$
$4y + 3z = 22$

84. $3x + 5y = -7$
$2x + 7z = 2$
$4y + 3z = -8$

85. $x + 2y = 10$
$3x + 4z = 7$
$-y - z = 1$

86. $5x - 2y = 3$
$4y + z = 8$
$x + 2z = 4$

(Modeling) Solve each problem.

87. *Roof Trusses* The simplest type of roof truss is a triangle. The truss shown in the figure is used to frame roofs of small buildings. If a 100-pound force is applied at the peak of the truss, then the forces or weights W_1 and W_2 exerted parallel to each rafter of the truss are determined by the following linear system of equations.

$$\frac{\sqrt{3}}{2}(W_1 + W_2) = 100$$

$$W_1 - W_2 = 0$$

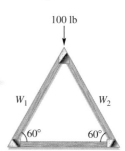

Solve the system to find W_1 and W_2. (*Source:* Hibbeler, R., *Structural Analysis,* Fourth Edition. Copyright © 1999. Reprinted by permission of Pearson Education, Inc., New York, NY.)

88. Roof Trusses **(Refer to Exercise 87.)** Use the following system of equations to determine the forces or weights W_1 and W_2 exerted on each rafter for the truss shown in the figure.

$$W_1 + \sqrt{2}W_2 = 300$$
$$\sqrt{3}W_1 - \sqrt{2}W_2 = 0$$

Area of a Triangle A triangle with vertices at (x_1, y_1), (x_2, y_2), and (x_3, y_3), as shown in the figure, has area equal to the absolute value of D, where

$$D = \frac{1}{2} \begin{vmatrix} x_1 & y_1 & 1 \\ x_2 & y_2 & 1 \\ x_3 & y_3 & 1 \end{vmatrix}.$$

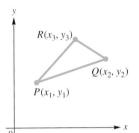

Find the area of each triangle having vertices at P, Q, and R.

89. $P(0,0)$, $Q(0,2)$, $R(1,4)$ **90.** $P(0,1)$, $Q(2,0)$, $R(1,5)$

91. $P(2,5)$, $Q(-1,3)$, $R(4,0)$ **92.** $P(2,-2)$, $Q(0,0)$, $R(-3,-4)$

93. Area of a Triangle Find the area of a triangular lot whose vertices have the following coordinates in feet. Round the answer to the nearest tenth of a foot.

$$(101.3, 52.7), \quad (117.2, 253.9), \quad \text{and} \quad (313.1, 301.6)$$

(*Source:* Al-Khafaji, A. and J. Tooley, *Numerical Methods in Engineering Practice,* Holt, Rinehart, and Winston.)

94. Let $A = \begin{bmatrix} a_{11} & a_{12} & a_{13} \\ a_{21} & a_{22} & a_{23} \\ a_{31} & a_{32} & a_{33} \end{bmatrix}$. Find $|A|$ by expansion about row 3 of the matrix. Show

that the result is really equal to $|A|$ as given in the definition of the determinant of a 3×3 matrix at the beginning of this section.

To solve a **determinant equation** such as

$$\begin{vmatrix} 6 & 4 \\ -2 & x \end{vmatrix} = 2,$$

expand the determinant to obtain

$$6x - (-2)(4) = 2.$$

Then solve to obtain the solution set $\{-1\}$. Use this method to solve each equation.

95. $\begin{vmatrix} 5 & x \\ -3 & 2 \end{vmatrix} = 6$ **96.** $\begin{vmatrix} -0.5 & 2 \\ x & x \end{vmatrix} = 0$ **97.** $\begin{vmatrix} x & 3 \\ x & x \end{vmatrix} = 4$

98. $\begin{vmatrix} 2x & x \\ 11 & x \end{vmatrix} = 6$ **99.** $\begin{vmatrix} -2 & 0 & 1 \\ -1 & 3 & x \\ 5 & -2 & 0 \end{vmatrix} = 3$ **100.** $\begin{vmatrix} 4 & 3 & 0 \\ 2 & 0 & 1 \\ -3 & x & -1 \end{vmatrix} = 5$

101. $\begin{vmatrix} 5 & 3x & -3 \\ 0 & 2 & -1 \\ 4 & -1 & x \end{vmatrix} = -7$ **102.** $\begin{vmatrix} 2x & 1 & -1 \\ 0 & 4 & x \\ 3 & 0 & 2 \end{vmatrix} = x$ **103.** $\begin{vmatrix} x & 0 & -1 \\ 2 & -3 & x \\ x & 0 & 7 \end{vmatrix} = 12$

104. Concept Check Write the sign array representing $(-1)^{i+j}$ for each element of a 4×4 matrix.

Solve each system for x and y using Cramer's rule. Assume a and b are nonzero constants.

105. $bx + y = a^2$
$ax + y = b^2$

106. $ax + by = \dfrac{b}{a}$
$x + y = \dfrac{1}{b}$

107. $b^2x + a^2y = b^2$
$ax + by = a$

108. $x + by = b$
$ax + y = a$

109. Use Cramer's rule to find the solution set if $a, b, c, d, e,$ and f are consecutive integers.

$$ax + by = c$$
$$dx + ey = f$$

110. In the following system, $a, b, c, \ldots, l$ are consecutive integers. Express the solution set in terms of z.

$$ax + by + cz = d$$
$$ex + fy + gz = h$$
$$ix + jy + kz = l$$

Relating Concepts

For individual or collaborative investigation *(Exercises 111–114)*

The determinant of a 3×3 matrix A is defined as follows.

$$\text{If } A = \begin{bmatrix} a_{11} & a_{12} & a_{13} \\ a_{21} & a_{22} & a_{23} \\ a_{31} & a_{32} & a_{33} \end{bmatrix}, \text{ then } |A| = \begin{vmatrix} a_{11} & a_{12} & a_{13} \\ a_{21} & a_{22} & a_{23} \\ a_{31} & a_{32} & a_{33} \end{vmatrix}$$

$$= (a_{11}a_{22}a_{33} + a_{12}a_{23}a_{31} + a_{13}a_{21}a_{32})$$
$$- (a_{31}a_{22}a_{13} + a_{32}a_{23}a_{11} + a_{33}a_{21}a_{12}).$$

Work these exercises in order.

111. The determinant of a 3×3 matrix can also be found using the method of "diagonals."

Step 1 Rewrite columns 1 and 2 of matrix A to the right of matrix A.

Step 2 Identify the diagonals d_1 through d_6 and multiply their elements.

Step 3 Find the sum of the products from d_1, d_2, and d_3.

Step 4 Subtract the sum of the products from d_4, d_5, and d_6 from that sum:

$$(d_1 + d_2 + d_3) - (d_4 + d_5 + d_6).$$

Verify that this method produces the same results as the previous method given.

$d_4 \quad d_5 \quad d_6 \qquad d_1 \quad d_2 \quad d_3$ Each d is a product.

112. Evaluate the determinant $\begin{vmatrix} 1 & 3 & 2 \\ 0 & 2 & 6 \\ 7 & 1 & 5 \end{vmatrix}$ using the method of "diagonals."

113. See **Exercise 112.** Evaluate the determinant by expanding about column 1 and using the method of cofactors. Do these methods give the same determinant for 3×3 matrices?

114. *Concept Check* Does the method of evaluating a determinant using "diagonals" extend to 4×4 matrices?

5.4 Partial Fractions

- Decomposition of Rational Expressions
- Distinct Linear Factors
- Repeated Linear Factors
- Distinct Linear and Quadratic Factors
- Repeated Quadratic Factors

Decomposition of Rational Expressions The sums of rational expressions are found by combining two or more rational expressions into one rational expression. Here, the reverse process is considered:

Given one rational expression, express it as the sum of two or more rational expressions.

A special type of sum involving rational expressions is a **partial fraction decomposition**—each term in the sum is a **partial fraction.**

Add rational expressions →

$$\frac{2}{x+1} + \frac{3}{x} = \frac{5x+3}{x(x+1)}$$

← Partial fraction decomposition

LOOKING AHEAD TO CALCULUS

In calculus, partial fraction decomposition provides a powerful technique for determining integrals of rational functions.

Partial Fraction Decomposition of $\frac{f(x)}{g(x)}$

To form a partial fraction decomposition of a rational expression, follow these steps.

Step 1 If $\frac{f(x)}{g(x)}$ is not a proper fraction (a fraction with the numerator of lesser degree than the denominator), divide $f(x)$ by $g(x)$. For example,

$$\frac{x^4 - 3x^3 + x^2 + 5x}{x^2 + 3} = x^2 - 3x - 2 + \frac{14x + 6}{x^2 + 3}.$$

Then apply the following steps to the remainder, which is a proper fraction.

Step 2 Factor the denominator $g(x)$ completely into factors of the form $(ax + b)^m$ or $(cx^2 + dx + e)^n$, where $cx^2 + dx + e$ is not factorable and m and n are positive integers.

Step 3 **(a)** For each distinct linear factor $(ax + b)$, the decomposition must include the term $\frac{A}{ax + b}$.

(b) For each repeated linear factor $(ax + b)^m$, the decomposition must include the terms

$$\frac{A_1}{ax + b} + \frac{A_2}{(ax + b)^2} + \cdots + \frac{A_m}{(ax + b)^m}.$$

Step 4 **(a)** For each distinct quadratic factor $(cx^2 + dx + e)$, the decomposition must include the term $\frac{Bx + C}{cx^2 + dx + e}$.

(b) For each repeated quadratic factor $(cx^2 + dx + e)^n$, the decomposition must include the terms

$$\frac{B_1x + C_1}{cx^2 + dx + e} + \frac{B_2x + C_2}{(cx^2 + dx + e)^2} + \cdots + \frac{B_nx + C_n}{(cx^2 + dx + e)^n}.$$

Step 5 Use algebraic techniques to solve for the constants in the numerators of the decomposition.

To find the constants in Step 5, the goal is to form a system of equations with as many equations as there are unknowns in the numerators.

Distinct Linear Factors

> **EXAMPLE 1** **Finding a Partial Fraction Decomposition**

Find the partial fraction decomposition of $\dfrac{2x^4 - 8x^2 + 5x - 2}{x^3 - 4x}$.

SOLUTION The given fraction is not a proper fraction—the numerator has greater degree than the denominator. Perform the division (Step 1).

$$
\begin{array}{r}
2x \\
x^3 - 4x{\overline{\smash{\big)}\,2x^4 - 8x^2 + 5x - 2}} \\
\underline{2x^4 - 8x^2} \\
5x - 2
\end{array}
$$

The quotient is $\dfrac{2x^4 - 8x^2 + 5x - 2}{x^3 - 4x} = 2x + \dfrac{5x - 2}{x^3 - 4x}$.

Now, work with the remainder fraction. Factor the denominator (Step 2) as

$$x^3 - 4x = x(x + 2)(x - 2).$$

The factors are distinct linear factors (Step 3(a)). Write the decomposition as

$$\frac{5x - 2}{x^3 - 4x} = \frac{A}{x} + \frac{B}{x + 2} + \frac{C}{x - 2}, \quad (1)$$

where A, B, and C are constants that need to be found (Step 5). Multiply each side of equation (1) by $x(x + 2)(x - 2)$ to obtain

$$5x - 2 = A(x + 2)(x - 2) + Bx(x - 2) + Cx(x + 2). \quad (2)$$

Equation (1) is an identity because each side represents the same rational expression. Thus, equation (2) is also an identity. Equation (1) holds for all values of x except 0, -2, and 2. However, equation (2) holds for all values of x.

We can solve for A by letting $x = 0$ in equation (2).

$$5x - 2 = A(x + 2)(x - 2) + Bx(x - 2) + Cx(x + 2) \quad (2)$$

$$5(0) - 2 = A(0 + 2)(0 - 2) + B(0)(0 - 2) + C(0)(0 + 2) \quad \text{Let } x = 0.$$

$$-2 = -4A \quad \text{Simplify each term.}$$

$$A = \frac{1}{2} \quad \text{Divide by } -4.$$

Similarly, letting $x = -2$ in equation (2) enables us to solve for B.

$$5(-2) - 2 = A(-2 + 2)(-2 - 2) + B(-2)(-2 - 2) + C(-2)(-2 + 2)$$
$$\text{Let } x = -2 \text{ in (2).}$$

$$-12 = 8B \quad \text{Simplify each term.}$$

$$B = -\frac{3}{2} \quad \text{Divide by 8.}$$

Letting $x = 2$ gives the following for C.

$$5(2) - 2 = A(2 + 2)(2 - 2) + B(2)(2 - 2) + C(2)(2 + 2) \quad \text{Let } x = 2 \text{ in (2).}$$

$$8 = 8C \quad \text{Simplify each term.}$$

$$C = 1 \quad \text{Divide by 8.}$$

The remainder rational expression can be written as the following sum of partial fractions. Use $A = \frac{1}{2}$, $B = -\frac{3}{2}$, and $C = 1$ in equation (1).

$$\frac{5x - 2}{x^3 - 4x} = \frac{1}{2x} + \frac{-3}{2(x + 2)} + \frac{1}{x - 2} \qquad \frac{\frac{1}{2}}{x} = \frac{1}{2x} \text{ and } \frac{-\frac{3}{2}}{x + 2} = \frac{-3}{2(x + 2)}$$

The given rational expression can now be written as follows.

$$\frac{2x^4 - 8x^2 + 5x - 2}{x^3 - 4x} = 2x + \frac{1}{2x} + \frac{-3}{2(x + 2)} + \frac{1}{x - 2}$$

Check this result by combining the terms on the right.

✔ **Now Try Exercise 29.**

Repeated Linear Factors

EXAMPLE 2 **Finding a Partial Fraction Decomposition**

Find the partial fraction decomposition of $\dfrac{2x}{(x - 1)^3}$.

SOLUTION This is a proper fraction. The denominator is already factored with repeated linear factors. Write the decomposition as shown using Step 3(b).

$$\frac{2x}{(x - 1)^3} = \frac{A}{x - 1} + \frac{B}{(x - 1)^2} + \frac{C}{(x - 1)^3}$$

Clear the denominators by multiplying each side of this equation by $(x - 1)^3$.

$$2x = A(x - 1)^2 + B(x - 1) + C$$

Substituting 1 for x leads to $C = 2$.

$$2x = A(x - 1)^2 + B(x - 1) + 2 \quad (1)$$

The only root has been substituted, and values for A and B still need to be found. However, *any* number can be substituted for x. For example, when we choose $x = -1$ (because it is easy to substitute), equation (1) becomes the following.

$$2(-1) = A(-1 - 1)^2 + B(-1 - 1) + 2 \qquad \text{Let } x = -1 \text{ in (1).}$$
$$-2 = 4A - 2B + 2 \qquad\qquad \text{Simplify each term.}$$
$$-4 = 4A - 2B \qquad\qquad\quad \text{Subtract 2.}$$
$$-2 = 2A - B \qquad\qquad\quad \text{Divide by 2.} \quad (2)$$

Substituting 0 for x in equation (1) gives another equation in A and B.

$$2(0) = A(0 - 1)^2 + B(0 - 1) + 2 \qquad \text{Let } x = 0 \text{ in (1).}$$
$$0 = A - B + 2 \qquad\qquad\qquad \text{Simplify each term.}$$
$$2 = -A + B \qquad\qquad\qquad\; \text{Subtract 2. Multiply by } -1. \quad (3)$$

$$-2 = 2A - B \quad (2)$$
$$\underline{2 = -A + B \quad (3)}$$
$$0 = A \qquad\qquad \text{Add.}$$

If $A = 0$, then using equation (3),

$$2 = 0 + B$$
$$2 = B.$$

Now, solve the system of equations (2) and (3) as shown in the margin to find $A = 0$ and $B = 2$. Substitute these values for A and B and 2 for C.

$$\frac{2x}{(x - 1)^3} = \frac{2}{(x - 1)^2} + \frac{2}{(x - 1)^3} \qquad \text{Partial fraction decomposition}$$

We needed three substitutions because there were three constants to evaluate: A, B, and C. To check this result, we could combine the terms on the right.

✔ **Now Try Exercise 17.**

Distinct Linear and Quadratic Factors

EXAMPLE 3 **Finding a Partial Fraction Decomposition**

Find the partial fraction decomposition of $\dfrac{x^2 + 3x - 1}{(x + 1)(x^2 + 2)}$.

SOLUTION The denominator $(x + 1)(x^2 + 2)$ has distinct linear and quadratic factors, where neither is repeated. Because $x^2 + 2$ cannot be factored, it is irreducible. The partial fraction decomposition is of the following form.

$$\frac{x^2 + 3x - 1}{(x + 1)(x^2 + 2)} = \frac{A}{x + 1} + \frac{Bx + C}{x^2 + 2}$$

Multiply each side by $(x + 1)(x^2 + 2)$.

$$x^2 + 3x - 1 = A(x^2 + 2) + (Bx + C)(x + 1) \quad (1)$$

First, substitute -1 for x.

> Use parentheses around substituted values to avoid errors.

$$(-1)^2 + 3(-1) - 1 = A[(-1)^2 + 2] + [B(-1) + C](-1 + 1)$$

$$-3 = 3A$$

$$A = -1$$

Replace A with -1 in equation (1) and substitute any value for x. Let $x = 0$.

$$0^2 + 3(0) - 1 = -1(0^2 + 2) + (B \cdot 0 + C)(0 + 1)$$

$$-1 = -2 + C$$

$$C = 1$$

Now, letting $A = -1$ and $C = 1$, substitute again in equation (1), using another value for x. Let $x = 1$.

$$1^2 + 3(1) - 1 = -1(1^2 + 2) + [B(1) + 1](1 + 1)$$

$$3 = -3 + (B + 1)(2)$$

$$6 = 2B + 2$$

$$B = 2$$

Use $A = -1$, $B = 2$, and $C = 1$ to find the partial fraction decomposition.

$$\frac{x^2 + 3x - 1}{(x + 1)(x^2 + 2)} = \frac{-1}{x + 1} + \frac{2x + 1}{x^2 + 2}$$
  Check by combining the terms on the right.

✔ **Now Try Exercise 25.**

For fractions with denominators that have quadratic factors, an alternative method is often more convenient. A system of equations is formed by equating coefficients of like terms on each side of the partial fraction decomposition. For instance, in **Example 3,** equation (1) was

$$x^2 + 3x - 1 = A(x^2 + 2) + (Bx + C)(x + 1). \quad (1)$$

Multiply on the right and collect like terms.

$$x^2 + 3x - 1 = Ax^2 + 2A + Bx^2 + Bx + Cx + C$$

$$1x^2 + 3x - 1 = (A + B)x^2 + (B + C)x + (C + 2A)$$

Now, equate the coefficients of like powers of x to obtain three equations.

$$1 = A + B$$

$$3 = B + C$$

$$-1 = C + 2A$$

Solving this system for A, B, and C gives the partial fraction decomposition.

Repeated Quadratic Factors

EXAMPLE 4 Finding a Partial Fraction Decomposition

Find the partial fraction decomposition of $\dfrac{2x}{(x^2 + 1)^2(x - 1)}$.

SOLUTION This expression has both a linear factor and a repeated quadratic factor. Use Steps 3(a) and 4(b) from the box at the beginning of this section.

$$\frac{2x}{(x^2 + 1)^2(x - 1)} = \frac{Ax + B}{x^2 + 1} + \frac{Cx + D}{(x^2 + 1)^2} + \frac{E}{x - 1}$$

Multiply each side by $(x^2 + 1)^2(x - 1)$.

$$2x = (Ax + B)(x^2 + 1)(x - 1) + (Cx + D)(x - 1) + E(x^2 + 1)^2 \quad \text{(1)}$$

If $x = 1$, then equation (1) reduces to $2 = 4E$, or $E = \frac{1}{2}$. Substitute $\frac{1}{2}$ for E in equation (1), and expand and combine like terms on the right.

$$2x = Ax^4 - Ax^3 + Ax^2 - Ax + Bx^3 - Bx^2 + Bx - B$$

$$+ Cx^2 - Cx + Dx - D + \frac{1}{2}x^4 + x^2 + \frac{1}{2}$$

$$2x = \left(A + \frac{1}{2}\right)x^4 + (-A + B)x^3 + (A - B + C + 1)x^2$$

$$+ (-A + B - C + D)x + \left(-B - D + \frac{1}{2}\right) \quad \text{(2)}$$

To obtain additional equations involving the unknowns, equate the coefficients of like powers of x on the two sides of equation (2). Setting corresponding coefficients of x^4 equal, $0 = A + \frac{1}{2}$, or $A = -\frac{1}{2}$. From the corresponding coefficients of x^3, $0 = -A + B$. Because $A = -\frac{1}{2}$, it follows that $B = -\frac{1}{2}$.

Using the coefficients of x^2, $0 = A - B + C + 1$. Since $A = -\frac{1}{2}$ and $B = -\frac{1}{2}$, it follows that $C = -1$. From the coefficients of x, $2 = -A + B - C + D$. Substituting for A, B, and C gives $D = 1$. With $A = -\frac{1}{2}$, $B = -\frac{1}{2}$, $C = -1$, $D = 1$, and $E = \frac{1}{2}$, the given fraction has partial fraction decomposition as follows.

$$\frac{2x}{(x^2 + 1)^2(x - 1)} = \frac{-\frac{1}{2}x - \frac{1}{2}}{x^2 + 1} + \frac{-1x + 1}{(x^2 + 1)^2} + \frac{\frac{1}{2}}{x - 1} \quad \begin{array}{l}\text{Substitute for}\\ A, B, C, D, \text{and } E.\end{array}$$

$$\frac{2x}{(x^2 + 1)^2(x - 1)} = \frac{-(x + 1)}{2(x^2 + 1)} + \frac{-x + 1}{(x^2 + 1)^2} + \frac{1}{2(x - 1)} \quad \begin{array}{l}\text{Simplify complex}\\ \text{fractions.}\end{array}$$

✔ **Now Try Exercise 31.**

In summary, to solve for the constants in the numerators of a partial fraction decomposition, use either of the following methods or a combination of the two.

Techniques for Decomposition into Partial Fractions

Method 1 **For Linear Factors**

Step 1 Multiply each side of the resulting rational equation by the common denominator.

Step 2 Substitute the zero of each factor in the resulting equation. For repeated linear factors, substitute as many other numbers as necessary to find all the constants in the numerators. The number of substitutions required will equal the number of constants $A, B, \ldots$.

Method 2 **For Quadratic Factors**

Step 1 Multiply each side of the resulting rational equation by the common denominator.

Step 2 Collect like terms on the right side of the equation.

Step 3 Equate the coefficients of like terms to form a system of equations.

Step 4 Solve the system to find the constants in the numerators.

5.4 Exercises

CONCEPT PREVIEW *Answer each question.*

1. By what expression should we multiply each side of

$$\frac{5}{3x(2x+1)} = \frac{A}{3x} + \frac{B}{2x+1}$$

so that there are no fractions in the equation?

2. In **Exercise 1,** after clearing fractions to decompose, the equation

$$A(2x+1) + B(3x) = 5$$

results. If we let $x = 0$, what is the value of A?

3. By what expression should we multiply each side of

$$\frac{3x-2}{(x+4)(3x^2+1)} = \frac{A}{x+4} + \frac{Bx+C}{3x^2+1}$$

so that there are no fractions in the equation?

4. In **Exercise 3,** after clearing fractions to decompose, the equation

$$3x - 2 = A(3x^2+1) + (Bx+C)(x+4)$$

results. If we let $x = -4$, what is the value of A?

5. By what expression should we multiply each side of

$$\frac{3x-1}{x(2x^2+1)^2} = \frac{A}{x} + \frac{Bx+C}{2x^2+1} + \frac{Dx+E}{(2x^2+1)^2}$$

so that there are no fractions in the equation?

6. In **Exercise 5,** after clearing fractions to decompose, the equation

$$3x - 1 = A(2x^2 + 1)^2 + (Bx + C)(x)(2x^2 + 1) + (Dx + E)(x)$$

results. If we let $x = 0$, what is the value of A?

Find the partial fraction decomposition for each rational expression. ***See Examples 1–4.***

7. $\dfrac{5}{3x(2x + 1)}$

8. $\dfrac{3x - 1}{x(x + 1)}$

9. $\dfrac{4x + 2}{(x + 2)(2x - 1)}$

10. $\dfrac{x + 2}{(x + 1)(x - 1)}$

11. $\dfrac{x}{x^2 + 4x - 5}$

12. $\dfrac{5x - 3}{x^2 - 2x - 3}$

13. $\dfrac{4}{x(1 - x)}$

14. $\dfrac{9}{x(x - 3)}$

15. $\dfrac{4x^2 - x - 15}{x(x + 1)(x - 1)}$

16. $\dfrac{3}{(x + 1)(x + 3)}$

17. $\dfrac{2x + 1}{(x + 2)^3}$

18. $\dfrac{2x}{(x + 1)(x + 2)^2}$

19. $\dfrac{x^2}{x^2 + 2x + 1}$

20. $\dfrac{2}{x^2(x + 3)}$

21. $\dfrac{x^3 + 4}{9x^3 - 4x}$

22. $\dfrac{x^3 + 2}{x^3 - 3x^2 + 2x}$

23. $\dfrac{-3}{x^2(x^2 + 5)}$

24. $\dfrac{1}{x^2(x^2 - 2)}$

25. $\dfrac{3x - 2}{(x + 4)(3x^2 + 1)}$

26. $\dfrac{2x + 1}{(x + 1)(x^2 + 2)}$

27. $\dfrac{1}{x(2x + 1)(3x^2 + 4)}$

28. $\dfrac{3}{x(x + 1)(x^2 + 1)}$

29. $\dfrac{2x^5 + 3x^4 - 3x^3 - 2x^2 + x}{2x^2 + 5x + 2}$

30. $\dfrac{6x^5 + 7x^4 - x^2 + 2x}{3x^2 + 2x - 1}$

31. $\dfrac{3x - 1}{x(2x^2 + 1)^2}$

32. $\dfrac{x^4 + 1}{x(x^2 + 1)^2}$

33. $\dfrac{-x^4 - 8x^2 + 3x - 10}{(x + 2)(x^2 + 4)^2}$

34. $\dfrac{3x^4 + x^3 + 5x^2 - x + 4}{(x - 1)(x^2 + 1)^2}$

35. $\dfrac{5x^5 + 10x^4 - 15x^3 + 4x^2 + 13x - 9}{x^3 + 2x^2 - 3x}$

36. $\dfrac{3x^6 + 3x^4 + 3x}{x^4 + x^2}$

37. $\dfrac{x^2}{x^4 - 1}$

38. $\dfrac{-2x^2 - 24}{x^4 - 16}$

39. $\dfrac{4x^2 - 3x - 4}{x^3 + x^2 - 2x}$

40. $\dfrac{2x + 4}{x^3 - 2x^2}$

Chapter 5 · Quiz (Sections 5.1–5.4)

Solve each system, using the method indicated, if possible.

1. (Substitution)

$$2x + y = -4$$
$$-x + 2y = 2$$

2. (Substitution)

$$5x + 10y = 10$$
$$x + 2y = 2$$

3. (Elimination)

$$x - y = 6$$
$$x - y = 4$$

4. (Elimination)

$$2x - 3y = 18$$
$$5x + 2y = 7$$

5. (Gauss-Jordan)

$$3x + 5y = -5$$
$$-2x + 3y = 16$$

6. (Cramer's rule)

$$5x + 2y = -3$$
$$4x - 3y = -30$$

7. (Elimination)

$$x + y + z = 1$$
$$-x + y + z = 5$$
$$y + 2z = 5$$

8. (Gauss-Jordan)

$$2x + 4y + 4z = 4$$
$$x + 3y + z = 4$$
$$-x + 3y + 2z = -1$$

9. (Cramer's rule)

$$7x + y - z = 4$$
$$2x - 3y + z = 2$$
$$-6x + 9y - 3z = -6$$

Solve each problem.

10. *Spending on Food* In 2013, the amount spent by a typical American household on food was about $6602. For every $10 spent on food away from home, about $15 was spent on food at home. Find the amount of household spending on food in each category. (*Source:* U.S. Bureau of Labor Statistics.)

11. *Investments* A sum of $5000 is invested in three accounts that pay 2%, 3%, and 4% interest rates. The amount of money invested in the account paying 4% equals the total amount of money invested in the other two accounts, and the total annual interest from all three investments is $165. Find the amount invested at each rate.

12. Let $A = \begin{bmatrix} -5 & 4 \\ 2 & -1 \end{bmatrix}$. Find $|A|$.

13. Evaluate $\begin{vmatrix} -1 & 2 & 4 \\ -3 & -2 & -3 \\ 2 & -1 & 5 \end{vmatrix}$. Use determinant theorems if desired.

Find the partial fraction decomposition for each rational expression.

14. $\dfrac{10x + 13}{x^2 - x - 20}$

15. $\dfrac{2x^2 - 15x - 32}{(x - 1)(x^2 + 6x + 8)}$

5.5 Nonlinear Systems of Equations

- Nonlinear Systems with Real Solutions
- Nonlinear Systems with Nonreal Complex Solutions
- An Application of Nonlinear Systems

Nonlinear Systems with Real Solutions   A system of equations in which at least one equation is *not* linear is a **nonlinear system.**

$$x^2 - y = 4 \quad (1) \qquad x^2 + y^2 = 16 \quad (1)$$
$$x + y = -2 \quad (2) \qquad |x| + y = 4 \quad (2)$$

Nonlinear systems

The substitution method works well for solving many such systems, particularly when one of the equations is linear, as in the next example.

EXAMPLE 1   Solving a Nonlinear System (Substitution Method)

Solve the system.

$$x^2 - y = 4 \quad (1)$$
$$x + y = -2 \quad (2)$$

ALGEBRAIC SOLUTION

When one of the equations in a nonlinear system is linear, it is usually best to begin by solving the linear equation for one of the variables.

$$y = -2 - x \quad \text{Solve equation (2) for } y.$$

Substitute this result for y in equation (1).

$x^2 - y = 4$	(1)
$x^2 - (-2 - x) = 4$	Let $y = -2 - x$.
$x^2 + 2 + x = 4$	Distributive property
$x^2 + x - 2 = 0$	Standard form
$(x + 2)(x - 1) = 0$	Factor.
$x + 2 = 0 \quad \text{or} \quad x - 1 = 0$	Zero-factor property
$x = -2 \quad \text{or} \qquad x = 1$	Solve each equation.

Substituting -2 for x in equation (2) gives $y = 0$. If $x = 1$, then $y = -3$. The solution set of the given system is $\{(-2, 0), (1, -3)\}$. A graph of the system is shown in **Figure 9.**

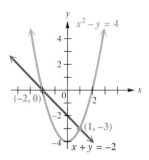

Figure 9

GRAPHING CALCULATOR SOLUTION

Solve each equation for y and graph them in the same viewing window. We obtain

$$y_1 = x^2 - 4 \quad \text{and} \quad y_2 = -x - 2.$$

The screens in **Figure 10,** which indicate that the points of intersection are

$$(-2, 0) \quad \text{and} \quad (1, -3),$$

support the solution found algebraically.

Figure 10

✔ **Now Try Exercise 15.**

CAUTION If we had solved for x in equation (2) to begin the algebraic solution in **Example 1,** we would have found $y = 0$ or $y = -3$. Substituting $y = 0$ into equation (1) gives $x^2 = 4$, so $x = 2$ or $x = -2$, leading to the ordered pairs $(2, 0)$ and $(-2, 0)$. The ordered pair $(2, 0)$ does not satisfy equation (2), however. *This illustrates the necessity of checking by substituting all proposed solutions into each equation of the system.*

Visualizing the types of graphs involved in a nonlinear system helps predict the possible numbers of ordered pairs of real numbers that may be in the solution set of the system. For example, a line and a parabola may have 0, 1, or 2 points of intersection, as shown in **Figure 11.**

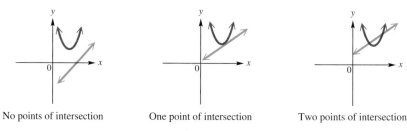

No points of intersection   One point of intersection   Two points of intersection

Figure 11

Nonlinear systems where both variables are squared in both equations are best solved by elimination, as shown in the next example.

EXAMPLE 2 Solving a Nonlinear System (Elimination Method)

Solve the system.

$$x^2 + y^2 = 4 \quad (1)$$
$$2x^2 - y^2 = 8 \quad (2)$$

SOLUTION The graph of equation (1) is a circle, and, as we will see in later work, the graph of equation (2) is a *hyperbola.* These graphs may intersect in 0, 1, 2, 3, or 4 points. We add to eliminate y^2.

$$
\begin{array}{rl}
x^2 + y^2 = 4 & (1) \\
\underline{2x^2 - y^2 = 8} & (2) \\
3x^2 = 12 & \text{Add.} \\
x^2 = 4 & \text{Divide by 3.} \\
x = \pm 2 & \text{Square root property}
\end{array}
$$

Remember to find *both* square roots.

To solve the system in **Example 2** graphically, solve equation (1) for y to obtain

$$y = \pm\sqrt{4 - x^2}.$$

Similarly, equation (2) yields

$$y = \pm\sqrt{-8 + 2x^2}.$$

Graph these *four* functions to find the two solutions. The solution $(-2, 0)$ is indicated in the screen above.

Find y by substituting the values of x in either equation (1) or equation (2).

$$
\begin{array}{ll}
x^2 + y^2 = 4 \quad (1) & x^2 + y^2 = 4 \quad (1) \\
2^2 + y^2 = 4 \quad \text{Let } x = 2. & (-2)^2 + y^2 = 4 \quad \text{Let } x = -2. \\
y^2 = 0 & y^2 = 0 \\
y = 0 & y = 0
\end{array}
$$

The proposed solutions are $(2, 0)$ and $(-2, 0)$. These satisfy both equations, confirming that the solution set is $\{(2, 0), (-2, 0)\}$.

✔ **Now Try Exercise 23.**

> **NOTE** The elimination method works with the system in **Example 2** because the system can be thought of as a system of linear equations where the variables are x^2 and y^2. To see this, substitute u for x^2 and v for y^2. The resulting system is linear in u and v.

Sometimes a combination of the elimination method and the substitution method is effective in solving a system, as illustrated in **Example 3.**

EXAMPLE 3 Solving a Nonlinear System (Combination of Methods)

Solve the system.

$$x^2 + 3xy + y^2 = 22 \quad (1)$$
$$x^2 - xy + y^2 = 6 \quad (2)$$

SOLUTION Begin as with the elimination method.

$$
\begin{array}{ll}
x^2 + 3xy + y^2 = 22 & (1) \\
\underline{-x^2 + xy - y^2 = -6} & \text{Multiply (2) by } -1. \\
\quad\quad 4xy \quad\quad = 16 & \text{Add.} \quad (3) \\
\end{array}
$$

$$y = \frac{4}{x} \qquad \text{Solve for } y \ (x \neq 0). \quad (4)$$

Now substitute $\frac{4}{x}$ for y in either equation (1) or equation (2). We use equation (2).

$$x^2 - xy + y^2 = 6 \qquad\qquad (2)$$

$$x^2 - x\left(\frac{4}{x}\right) + \left(\frac{4}{x}\right)^2 = 6 \qquad \text{Let } y = \frac{4}{x}.$$

$$x^2 - 4 + \frac{16}{x^2} = 6 \qquad \text{Multiply and square.}$$

$$x^4 - 4x^2 + 16 = 6x^2 \qquad \text{Multiply by } x^2 \text{ to clear fractions.}$$

> This equation is quadratic in form.

$$x^4 - 10x^2 + 16 = 0 \qquad \text{Subtract } 6x^2.$$

$$(x^2 - 2)(x^2 - 8) = 0 \qquad \text{Factor.}$$

$$
\begin{array}{lll}
x^2 - 2 = 0 & \text{or} \quad x^2 - 8 = 0 & \text{Zero-factor property} \\
x^2 = 2 & \text{or} \quad\quad x^2 = 8 & \text{Solve each equation.} \\
\end{array}
$$

> For each equation, include *both* square roots.

$$
\begin{array}{lll}
x = \pm\sqrt{2} & \text{or} \quad\quad x = \pm 2\sqrt{2} & \text{Square root property;} \\
& & \pm\sqrt{8} = \pm\sqrt{4}\cdot\sqrt{2} = \pm 2\sqrt{2}
\end{array}
$$

Substitute these x-values into equation (4) to find corresponding values of y.

Let $x = \sqrt{2}$ in (4).	Let $x = -\sqrt{2}$ in (4).	Let $x = 2\sqrt{2}$ in (4).	Let $x = -2\sqrt{2}$ in (4).
$y = \dfrac{4}{\sqrt{2}} = 2\sqrt{2}$	$y = \dfrac{4}{-\sqrt{2}} = -2\sqrt{2}$	$y = \dfrac{4}{2\sqrt{2}} = \sqrt{2}$	$y = \dfrac{4}{-2\sqrt{2}} = -\sqrt{2}$

The solution set of the system is

$$\left\{ \left(\sqrt{2}, 2\sqrt{2}\right), \left(-\sqrt{2}, -2\sqrt{2}\right), \left(2\sqrt{2}, \sqrt{2}\right), \left(-2\sqrt{2}, -\sqrt{2}\right) \right\}.$$

Verify these solutions by substitution in the original system.

✔ **Now Try Exercise 41.**

EXAMPLE 4 **Solving a Nonlinear System (Absolute Value Equation)**

Solve the system.

$$x^2 + y^2 = 16 \quad (1)$$

$$|x| + y = 4 \quad (2)$$

SOLUTION Use the substitution method. Begin by solving equation (2) for $|x|$.

$$|x| = 4 - y \quad (3)$$

In equation (1), the first term is x^2, which is the same as $|x|^2$. Therefore, we substitute $4 - y$ for x in equation (1).

$$x^2 + y^2 = 16 \quad (1)$$

$$(4 - y)^2 + y^2 = 16 \quad \text{Let } x = 4 - y.$$

$$(16 - 8y + y^2) + y^2 = 16 \quad \text{Square the binomial.}$$

> Remember the middle term.

$$2y^2 - 8y = 0 \quad \text{Combine like terms.}$$

$$y^2 - 4y = 0 \quad \text{Divide by 2.}$$

$$y(y - 4) = 0 \quad \text{Factor.}$$

$$y = 0 \quad \text{or} \quad y - 4 = 0 \quad \text{Zero-factor property}$$

$$y = 4 \quad \text{Add 4.}$$

To solve for the corresponding values of x, use either equation (1) or (2).

$x^2 + y^2 = 16 \quad (1)$	$x^2 + y^2 = 16 \quad (1)$
$x^2 + 0^2 = 16 \quad \text{Let } y = 0.$	$x^2 + 4^2 = 16 \quad \text{Let } y = 4.$
$x^2 = 16$	$x^2 = 0$
$x = \pm 4$	$x = 0$

The solution set, $\{(4, 0), (-4, 0), (0, 4)\}$, includes the points of intersection shown in **Figure 12.** Check the solutions in the original system.

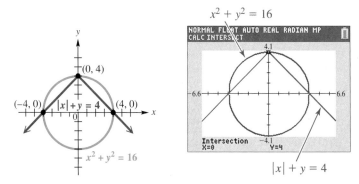

Figure 12   ✔ **Now Try Exercise 45.**

NOTE After solving for y in **Example 4,** the corresponding values of x can be found using equation (2) instead of equation (1).

$|x| + y = 4 \quad (2)$	$|x| + y = 4 \quad (2)$
$|x| + 0 = 4 \quad \text{Let } y = 0.$	$|x| + 4 = 4 \quad \text{Let } y = 4.$
$|x| = 4$	$|x| = 0$
$x = \pm 4$	$x = 0$

Nonlinear Systems with Nonreal Complex Solutions

EXAMPLE 5 Solving a Nonlinear System (Nonreal Complex Solutions)

Solve the system.

$$x^2 + y^2 = 5 \quad (1)$$
$$4x^2 + 3y^2 = 11 \quad (2)$$

SOLUTION Begin by eliminating a variable.

$$\begin{aligned} -3x^2 - 3y^2 &= -15 \qquad \text{Multiply (1) by } -3. \\ \underline{4x^2 + 3y^2 = 11} \qquad &(2) \\ x^2 = -4 \qquad &\text{Add.} \end{aligned}$$

$$x = \pm\sqrt{-4} \qquad \text{Square root property}$$
$$x = \pm 2i \qquad \sqrt{-4} = i\sqrt{4} = 2i$$

To find the corresponding values of y, substitute into equation (1).

The graphs of the two equations in **Example 5** do not intersect as seen here. The graphs are obtained by graphing

$$y = \pm\sqrt{5 - x^2}$$

and

$$y = \pm\sqrt{\frac{11 - 4x^2}{3}}.$$

$$\begin{aligned} x^2 + y^2 &= 5 \qquad (1) \\ (2i)^2 + y^2 &= 5 \qquad \text{Let } x = 2i. \\ \boxed{i^2 = -1}\;{-}4 + y^2 &= 5 \\ y^2 &= 9 \\ y &= \pm 3 \end{aligned}$$

$$\begin{aligned} x^2 + y^2 &= 5 \qquad (1) \\ (-2i)^2 + y^2 &= 5 \qquad \text{Let } x = -2i. \\ -4 + y^2 &= 5 \\ y^2 &= 9 \\ y &= \pm 3 \end{aligned}$$

Checking the proposed solutions confirms the following solution set.

$$\{(2i, 3), (2i, -3), (-2i, 3), (-2i, -3)\}$$

Note that solutions with nonreal complex number components do not appear as intersection points on the graph of the system.

✔ **Now Try Exercise 43.**

An Application of Nonlinear Systems

EXAMPLE 6 Using a Nonlinear System to Find Box Dimensions

A box with an open top has a square base and four sides of equal height. The volume of the box is 75 in.^{3}, and the surface area is 85 in.^{2}. Find the dimensions of the box.

SOLUTION

Step 1 **Read** the problem. We must find the box width, length, and height.

Step 2 **Assign variables.** Let x represent the length and width of the square base, and let y represent the height. See **Figure 13.**

Step 3 **Write a system of equations.** Use the formula for the volume of a box, $V = LWH$, to write one equation using the given volume, 75 in.^{3}.

$$x^2y = 75 \qquad \text{Volume formula}$$

The surface consists of the base, whose area is x^2, and four sides, each having area xy. The total surface area of 85 in.^{2} is used to write a second equation.

$$x^2 + 4xy = 85 \qquad \text{Sum of areas of base and sides}$$

Figure 13

The two equations form this system.

$$x^2 y = 75 \quad (1)$$

$$x^2 + 4xy = 85 \quad (2)$$

Step 4 **Solve** the system. We solve equation (1) for y to obtain $y = \frac{75}{x^2}$.

$$x^2 + 4xy = 85 \qquad (2)$$

$$x^2 + 4x\left(\frac{75}{x^2}\right) = 85 \qquad \text{Let } y = \frac{75}{x^2}.$$

$$x^2 + \frac{300}{x} = 85 \qquad \text{Multiply.}$$

$$x^3 + 300 = 85x \qquad \text{Multiply by } x, \text{ where } x \neq 0.$$

$$x^3 - 85x + 300 = 0 \qquad \text{Subtract } 85x.$$

$$
\begin{array}{r}
5 \overline{)1 \quad 0 \quad -85 \quad 300} \\
\underline{5 \quad 25 \quad -300} \\
1 \quad 5 \quad -60 \quad 0
\end{array}
$$

Coefficients of a
quadratic polynomial factor

We are restricted to positive values for x, and considering the nature of the problem, any solution should be relatively small. By the rational zeros theorem, factors of 300 are the only possible rational solutions. Using synthetic division, as shown in the margin, we see that 5 is a solution. Therefore, one value of x is 5, and $y = \frac{75}{5^2} = 3$. We must now solve

$$x^2 + 5x - 60 = 0$$

for any other possible positive solutions. Use the quadratic formula to find the positive solution.

$$x = \frac{-5 + \sqrt{5^2 - 4(1)(-60)}}{2(1)} \approx 5.639 \qquad \begin{array}{l}\text{Quadratic formula with}\\ a = 1, b = 5, c = -60\end{array}$$

This value of x leads to $y \approx 2.359$.

Step 5 **State the answer.** There are two possible answers.

First answer: length = width = 5 in.; height = 3 in.

Second answer: length = width $\approx$ 5.639 in.; height $\approx$ 2.359 in.

Step 6 **Check.** See **Exercise 69.**   ✔ **Now Try Exercises 67 and 69.**

5.5 Exercises

CONCEPT PREVIEW *Answer each of the following. When appropriate, fill in the blank to correctly complete the sentence.*

1. The following nonlinear system has two solutions, one of which is $(3, \underline{\quad})$.

$$x + y = 7$$
$$x^2 + y^2 = 25$$

2. The following nonlinear system has two solutions with real components, one of which is $(2, \underline{\quad})$.

$$y = x^2 + 6$$
$$x^2 - y^2 = -96$$

3. The following nonlinear system has two solutions, one of which is $(\underline{\quad}, 3)$.

$$2x + y = 1$$
$$x^2 + y^2 = 10$$

4. Refer to the system in **Exercise 2.** The other solution with real components has x-value -2. What is the y-value of this solution?

5. If we want to solve the following nonlinear system by substitution and we decide to solve equation (2) for y, what will be the resulting equation when the substitution is made into equation (1)?

$$x^2 + y = 2 \quad (1)$$
$$x - y = 0 \quad (2)$$

6. If we want to solve the following nonlinear system by eliminating the y^2 terms, by what number should we multiply equation (2)?

$$x^2 + 3y^2 = 4 \quad (1)$$
$$x^2 - y^2 = 0 \quad (2)$$

Concept Check A nonlinear system is given, along with the graphs of both equations in the system. Verify that the points of intersection specified on the graph are solutions of the system by substituting directly into both equations.

7. $x^2 = y - 1$
$\quad y = 3x + 5$

8. $2x^2 = 3y + 23$
$\quad y = 2x - 5$

9. $\quad x^2 + y^2 = 5$
$\quad -3x + 4y = 2$

10. $x + y = -3$
$\quad x^2 + y^2 = 45$

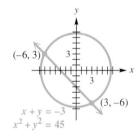

11. $y = 3x^2$
$\quad x^2 + y^2 = 10$

12. $y = -\dfrac{4}{9}x^2$
$\quad x^2 + y^2 = 25$

Concept Check Answer each question.

13. In **Example 1,** we solved the following system. How can we tell, before doing any work, that this system cannot have more than two solutions?

$$x^2 - y = 4$$
$$x + y = -2$$

14. In **Example 5,** there were four solutions to the system, but there were no points of intersection of the graphs. If a nonlinear system has nonreal complex numbers as components of its solutions, will they appear as intersection points of the graphs?

Solve each nonlinear system of equations. Give all solutions, including those with nonreal complex components. **See Examples 1–5.**

15. $x^2 - y = 0$
$x + y = 2$

16. $x^2 + y = 2$
$x - y = 0$

17. $y = x^2 - 2x + 1$
$x - 3y = -1$

18. $y = x^2 + 6x + 9$
$x + 2y = -2$

19. $y = x^2 + 4x$
$2x - y = -8$

20. $y = 6x + x^2$
$4x - y = -3$

21. $3x^2 + 2y^2 = 5$
$x - y = -2$

22. $x^2 + y^2 = 5$
$-3x + 4y = 2$

23. $x^2 + y^2 = 8$
$x^2 - y^2 = 0$

24. $x^2 + y^2 = 10$
$2x^2 - y^2 = 17$

25. $5x^2 - y^2 = 0$
$3x^2 + 4y^2 = 0$

26. $x^2 + y^2 = 0$
$2x^2 - 3y^2 = 0$

27. $3x^2 + y^2 = 3$
$4x^2 + 5y^2 = 26$

28. $x^2 + 2y^2 = 9$
$x^2 + y^2 = 25$

29. $2x^2 + 3y^2 = 5$
$3x^2 - 4y^2 = -1$

30. $3x^2 + 5y^2 = 17$
$2x^2 - 3y^2 = 5$

31. $2x^2 + 2y^2 = 20$
$4x^2 + 4y^2 = 30$

32. $x^2 + y^2 = 4$
$5x^2 + 5y^2 = 28$

33. $2x^2 - 3y^2 = 12$
$6x^2 + 5y^2 = 36$

34. $5x^2 - 2y^2 = 25$
$10x^2 + y^2 = 50$

35. $xy = -15$
$4x + 3y = 3$

36. $xy = 8$
$3x + 2y = -16$

37. $2xy + 1 = 0$
$x + 16y = 2$

38. $-5xy + 2 = 0$
$x - 15y = 5$

39. $3x^2 - y^2 = 11$
$xy = 12$

40. $5x^2 - 2y^2 = 6$
$xy = 2$

41. $x^2 - xy + y^2 = 5$
$2x^2 + xy - y^2 = 10$

42. $3x^2 + 2xy - y^2 = 9$
$x^2 - xy + y^2 = 9$

43. $x^2 + 2xy - y^2 = 14$
$x^2 - y^2 = -16$

44. $x^2 + 3xy - y^2 = 12$
$x^2 - y^2 = -12$

45. $x^2 + y^2 = 25$
$|x| - y = 5$

46. $x^2 + y^2 = 9$
$|x| + y = 3$

47. $x = |y|$
$x^2 + y^2 = 18$

48. $2x + |y| = 4$
$x^2 + y^2 = 5$

49. $2x^2 - y^2 = 4$
$|x| = |y|$

50. $x^2 + y^2 = 9$
$|x| = |y|$

Many nonlinear systems cannot be solved algebraically, so graphical analysis is the only way to determine the solutions of such systems. Use a graphing calculator to solve each nonlinear system. Give x- and y-coordinates to the nearest hundredth.

51. $y = \log(x + 5)$
$y = x^2$

52. $y = 5^x$
$xy = 1$

53. $y = e^{x+1}$
$2x + y = 3$

54. $y = \sqrt[3]{x - 4}$
$x^2 + y^2 = 6$

Solve each problem using a system of equations in two variables. **See Example 6.**

55. *Unknown Numbers* Find two numbers whose sum is -17 and whose product is 42.

56. *Unknown Numbers* Find two numbers whose sum is -10 and whose squares differ by 20.

57. *Unknown Numbers* Find two numbers whose squares have a sum of 100 and a difference of 28.

58. *Unknown Numbers* Find two numbers whose squares have a sum of 194 and a difference of 144.

59. *Unknown Numbers* Find two numbers whose ratio is 9 to 2 and whose product is 162.

60. *Unknown Numbers* Find two numbers whose ratio is 4 to 3 and are such that the sum of their squares is 100.

61. *Triangle Dimensions* The longest side of a right triangle is 13 m in length. One of the other sides is 7 m longer than the shortest side. Find the lengths of the two shorter sides of the triangle.

62. *Triangle Dimensions* The longest side of a right triangle is 29 ft in length. One of the other two sides is 1 ft longer than the shortest side. Find the lengths of the two shorter sides of the triangle.

Answer each question.

63. Does the straight line $3x - 2y = 9$ intersect the circle $x^2 + y^2 = 25$? (*Hint:* To find out, solve the system formed by these two equations.)

64. For what value(s) of b will the line $x + 2y = b$ touch the circle $x^2 + y^2 = 9$ in only one point?

65. A line passes through the points of intersection of the graphs of $y = x^2$ and $x^2 + y^2 = 90$. What is the equation of this line?

66. Suppose we are given the equations of two circles that are known to intersect in exactly two points. How would we find the equation of the only chord common to these circles?

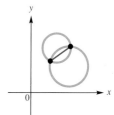

Solve each problem.

67. *Dimensions of a Box* A box with an open top has a square base and four sides of equal height. The volume of the box is 360 ft³. If the surface area is 276 ft², find the dimensions of the box. (Round answers to the nearest thousandth, if necessary.)

68. *Dimensions of a Cylinder* Find the radius and height (to the nearest thousandth) of an open-ended cylinder with volume 50 in.³ and lateral surface area 65 in.².

69. *Checking Answers* Check the two answers in **Example 6.**

70. *(Modeling) Equilibrium Demand and Price* The supply and demand equations for a certain commodity are given.

$$\text{supply: } p = \frac{2000}{2000 - q} \quad \text{and} \quad \text{demand: } p = \frac{7000 - 3q}{2q}$$

(a) Find the equilibrium demand.

(b) Find the equilibrium price (in dollars).

71. *(Modeling) Equilibrium Demand and Price* The supply and demand equations for a certain commodity are given.

$$\text{supply: } p = \sqrt{0.1q + 9} - 2 \quad \text{and} \quad \text{demand: } p = \sqrt{25 - 0.1q}$$

(a) Find the equilibrium demand. **(b)** Find the equilibrium price (in dollars).

72. *(Modeling) Circuit Gain* In electronics, circuit gain is modeled by

$$G = \frac{Bt}{R + R_t},$$

where R is the value of a resistor, t is temperature, R_t is the value of R at temperature t, and B is a constant. The sensitivity of the circuit to temperature is modeled by

$$S = \frac{BR}{(R + R_t)^2}.$$

If $B = 3.7$ and t is 90 K (kelvins), find the values of R and R_t that will result in the values $G = 0.4$ and $S = 0.001$. Round answers to the nearest whole number.

73. *(Modeling) Revenue for Public Colleges* The percents of revenue for public colleges from state sources and tuition are modeled in the accompanying graph.

(a) Interpret this graph. How are the sources of funding for public colleges changing with time?

(b) During what time period was the revenue from state sources increasing?

(c) Use the graph to estimate the year and the percent when the amounts from both sources were equal.

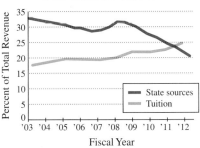

Public College Revenue from Tuition and State Sources

Source: Government Accountability Office.

74. *(Modeling) Revenue for Public Colleges* The following equations model the percents of revenue from both sources in **Exercise 73.** Use the equations to determine the year and percent when the amounts from both sources were equal.

$$S = 32(0.9637)^{x-2003} \quad \text{State sources}$$

$$T = 17(1.0438)^{x-2003} \quad \text{Tuition}$$

Relating Concepts

For individual or collaborative investigation *(Exercises 75–80)*

Consider the following nonlinear system.

$$y = |x - 1|$$
$$y = x^2 - 4$$

Work Exercises 75–80 in order, *to see how concepts of graphing are related to the solutions of this system.*

75. How is the graph of $y = |x - 1|$ obtained by transforming the graph of $y = |x|$?

76. How is the graph of $y = x^2 - 4$ obtained by transforming the graph of $y = x^2$?

77. Use the definition of absolute value to write $y = |x - 1|$ as a piecewise-defined function.

78. Write two quadratic equations that will be used to solve the system. (*Hint:* Set both parts of the piecewise-defined function in **Exercise 77** equal to $x^2 - 4$.)

79. Use the quadratic formula to solve both equations from **Exercise 78.** Pay close attention to the restriction on x.

80. Use the values of x found in **Exercise 79** to find the solution set of the system.

Summary Exercises on Systems of Equations

This chapter has introduced methods for solving systems of equations, including substitution and elimination, and matrix methods such as the Gauss-Jordan method and Cramer's rule. Use each method at least once when solving the systems below. Include solutions with nonreal complex number components. For systems with infinitely many solutions, write the solution set using an arbitrary variable.

1. $2x + 5y = 4$
$3x - 2y = -13$

2. $x - 3y = 7$
$-3x + 4y = -1$

3. $2x^2 + y^2 = 5$
$3x^2 + 2y^2 = 10$

4. $2x - 3y = -2$
$x + y = -16$
$3x - 2y + z = 7$

5. $6x - y = 5$
$xy = 4$

6. $4x + 2z = -12$
$x + y - 3z = 13$
$-3x + y - 2z = 13$

7. $x + 2y + z = 5$
$y + 3z = 9$

8. $x - 4 = 3y$
$x^2 + y^2 = 8$

9. $3x + 6y - 9z = 1$
$2x + 4y - 6z = 1$
$3x + 4y + 5z = 0$

10. $x + 2y + z = 0$
$x + 2y - z = 6$
$2x - y = -9$

11. $x^2 + y^2 = 4$
$y = x + 6$

12. $x + 5y = -23$
$4y - 3z = -29$
$2x + 5z = 19$

13. $y + 1 = x^2 + 2x$
$y + 2x = 4$

14. $3x + 6z = -3$
$y - z = 3$
$2x + 4z = -1$

15. $2x + 3y + 4z = 3$
$-4x + 2y - 6z = 2$
$4x + 3z = 0$

16. $\dfrac{3}{x} + \dfrac{4}{y} = 4$
$\dfrac{1}{x} + \dfrac{2}{y} = \dfrac{2}{3}$

17. $-5x + 2y + z = 5$
$-3x - 2y - z = 3$
$-x + 6y = 1$

18. $x + 5y + 3z = 9$
$2x + 9y + 7z = 5$

19. $2x^2 + y^2 = 9$
$3x - 2y = -6$

20. $2x - 4y - 6 = 0$
$-x + 2y + 3 = 0$

21. $x + y - z = 0$
$2y - z = 1$
$2x + 3y - 4z = -4$

22. $2y = 3x - x^2$
$x + 2y = 12$

23. $2x - 3y = -2$
$x + y - 4z = -16$
$3x - 2y + z = 7$

24. $x - y + 3z = 3$
$-2x + 3y - 11z = -4$
$x - 2y + 8z = 6$

25. $x^2 + 3y^2 = 28$
$y - x = -2$

26. $3x - y = -2$
$y + 5z = -4$
$-2x + 3y - z = -8$

27. $2x^2 + 3y^2 = 20$
$x^2 + 4y^2 = 5$

28. $x + y + z = -1$
$2x + 3y + 2z = 3$
$2x + y + 2z = -7$

29. $x + 2z = 9$
$y + z = 1$
$3x - 2y = 9$

30. $x^2 - y^2 = 15$
$x - 2y = 2$

31. $-x + y = -1$
$x + z = 4$
$6x - 3y + 2z = 10$

32. $2x - y - 2z = -1$
$-x + 2y + 13z = 12$
$3x + 9z = 6$

33. $xy = -3$
$x + y = -2$

34. $-3x + 2z = 1$
$4x + y - 2z = -6$
$x + y + 4z = 3$

35. $y = x^2 + 6x + 9$
$x + y = 3$

36. $5x - 2z = 8$
$4y + 3z = -9$
$\dfrac{1}{2}x + \dfrac{2}{3}y = -1$

5.6 Systems of Inequalities and Linear Programming

- Linear Inequalities in Two Variables
- Systems of Inequalities
- Linear Programming

Linear Inequalities in Two Variables A line divides a plane into three sets of points: the points of the line itself and the points belonging to the two regions determined by the line. Each of these two regions is a **half-plane.** In **Figure 14,** line r divides the plane into three different sets of points: line r, half-plane P, and half-plane Q. The points on r belong neither to P nor to Q. Line r is the **boundary** of each half-plane.

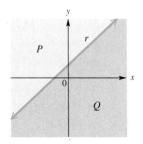

Figure 14

Linear Inequality in Two Variables

A **linear inequality in two variables** is an inequality of the form

$$Ax + By \le C,$$

where A, B, and C are real numbers, with A and B not both equal to 0. (The symbol $\le$ could be replaced with $\ge$, $<$, or $>$.)

The graph of a linear inequality is a half-plane, perhaps with its boundary.

EXAMPLE 1 Graphing a Linear Inequality

Graph $3x - 2y \le 6$.

SOLUTION First graph the boundary, $3x - 2y = 6$, as shown in **Figure 15.** Because the points of the line $3x - 2y = 6$ satisfy $3x - 2y \le 6$, this line is part of the solution set. To decide which half-plane (the one above the line $3x - 2y = 6$ or the one below the line) is part of the solution set, solve the original inequality for y.

$$3x - 2y \le 6$$

$$-2y \le -3x + 6 \quad \text{Subtract } 3x.$$

> Reverse the inequality symbol when dividing by a negative number.

$$y \ge \frac{3}{2}x - 3 \quad \begin{array}{l}\text{Divide by } -2, \text{ and}\\ \text{change } \le \text{ to } \ge.\end{array}$$

For a particular value of x, the inequality will be satisfied by all values of y that are *greater than or equal to* $\frac{3}{2}x - 3$. Thus, the solution set contains the half-plane *above* the line, as shown in **Figure 16.**

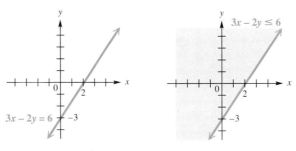

Figure 15   **Figure 16**

Coordinates for x and y from the solution set (the shaded region) satisfy the original inequality, while coordinates outside the solution set do not.

✔ **Now Try Exercise 11.**

CAUTION *A linear inequality must be in slope-intercept form (solved for y) to determine, from the presence of a < symbol or a > symbol, whether to shade the lower or the upper half-plane.* In **Figure 16,** the upper half-plane is shaded, even though the inequality is $3x - 2y \leq 6$ (with a < symbol) in standard form. Only when we write the inequality as

$$y \geq \frac{3}{2}x - 3 \quad \text{Slope-intercept form}$$

does the > symbol indicate to shade the upper half-plane.

EXAMPLE 2   Graphing a Linear Inequality

Graph $x + 4y < 4$.

SOLUTION   The boundary of the graph is the straight line $x + 4y = 4$. Because points on this line do *not* satisfy $x + 4y < 4$, it is customary to make the line dashed, as in **Figure 17.**

Figure 17

To decide which half-plane represents the solution set, solve for y.

$$x + 4y < 4$$
$$4y < -x + 4 \quad \text{Subtract } x.$$
$$y < -\frac{1}{4}x + 1 \quad \text{Divide by 4.}$$

To graph the inequality in **Example 2** using a graphing calculator, solve for y, and then direct the calculator to shade below the boundary line, $y = -\frac{1}{4}x + 1$.

Because y is *less than* $-\frac{1}{4}x + 1$, the graph of the solution set is the half-plane *below* the boundary, as shown in **Figure 17.**

As an alternative method for deciding which half-plane to shade, or as a check, we can choose a test point not on the boundary line and substitute into the inequality. The point $(0, 0)$ is a good choice if it does not lie on the boundary because the substitution is easily done.

$$x + 4y < 4 \quad \text{Use the original inequality.}$$
$$0 + 4(0) \overset{?}{<} 4 \quad \text{Use } (0, 0) \text{ as a test point.}$$
$$0 < 4 \quad \text{True}$$

The test point $(0, 0)$ lies below the boundary, so all points that satisfy the inequality must also lie below the boundary. This agrees with the result above.

✔ **Now Try Exercise 15.**

An inequality containing < or > is a **strict inequality** and does not include the boundary in its solution set. This is indicated with a dashed boundary, as shown in **Example 2.** A **nonstrict inequality** contains $\leq$ or $\geq$ and does include its boundary in the solution set. This is indicated with a solid boundary, as shown in **Example 1.**

Graphing an Inequality in Two Variables

Method 1 If the inequality is or can be solved for y, then the following hold.

- The graph of $y < f(x)$ consists of all the points that are *below* the graph of $y = f(x)$.

- The graph of $y > f(x)$ consists of all the points that are *above* the graph of $y = f(x)$.

Method 2 If the inequality is not or cannot be solved for y, then choose a test point not on the boundary.

- If the test point satisfies the inequality, then the graph includes all points on the *same* side of the boundary as the test point.

- If the test point does not satisfy the inequality, then the graph includes all points on the *other* side of the boundary.

For either method, use a solid boundary for a nonstrict inequality ($\leq$ or $\geq$) or a dashed boundary for a strict inequality ($<$ or $>$).

Systems of Inequalities The solution set of a **system of inequalities** is the intersection of the solution sets of all the inequalities in the system. We find this intersection by graphing the solution sets of all inequalities on the same coordinate axes and identifying, by shading, the region common to all graphs.

EXAMPLE 3 **Graphing Systems of Inequalities**

Graph the solution set of each system.

(a) $x > 6 - 2y$
$x^2 < 2y$

(b) $|x| \leq 3$
$y \leq 0$
$y \geq |x| + 1$

The region shaded twice is the solution set of the system in **Example 3(a).**

SOLUTION

(a) **Figures 18(a) and (b)** show the graphs of $x > 6 - 2y$ and $x^2 < 2y$. The methods presented earlier in this chapter can be used to show that the boundaries intersect at the points $(2, 2)$ and $\left(-3, \frac{9}{2}\right)$.

The solution set of the system is shown in **Figure 18(c).** The points on the boundaries of $x > 6 - 2y$ and $x^2 < 2y$ do not belong to the graph of the solution set, so the boundaries are dashed.

(a)

(b)

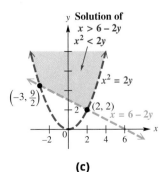

(c)

Figure 18

(b) Writing $|x| \leq 3$ as $-3 \leq x \leq 3$ shows that this inequality is satisfied by points in the region between and including

$$x = -3 \quad \text{and} \quad x = 3.$$

See **Figure 19(a).** The set of points that satisfies $y \leq 0$ includes the points below or on the x-axis. See **Figure 19(b).**

   Graph $y = |x| + 1$ and use a test point to verify that the solutions of $y \geq |x| + 1$ are on or above the boundary. See **Figure 19(c).** Because the solution sets of $y \leq 0$ and $y \geq |x| + 1$ shown in **Figures 19(b) and (c)** have no points in common, ***the solution set of the system is $\varnothing$.***

The solution set of the system is $\varnothing$ because there are no points common to all three regions simultaneously.

        (a)   **(b)**   **(c)**

Figure 19

✔ Now Try Exercises 43, 49, and 51.

NOTE   Although we gave three graphs in the solutions of **Example 3,** in practice we usually give only a final graph showing the solution set of the system. This would be an empty rectangular coordinate system in **Example 3(b).**

Linear Programming   One important application of mathematics to business and social science is **linear programming.** Linear programming is used to find an optimum value—for example, minimum cost or maximum profit. Procedures for solving linear programming problems were developed in 1947 by George Dantzig while he was working on a problem of allocating supplies for the Air Force in a way that minimized total cost.

   To solve a linear programming problem in general, use the following steps. (The italicized terms are defined in **Example 4.**)

Solving a Linear Programming Problem

Step 1 Write all necessary *constraints* and the *objective function.*

Step 2 Graph the *region of feasible solutions.*

Step 3 Identify all *vertices* (*corner points*).

Step 4 Find the value of the *objective function* at each *vertex.*

Step 5 The solution is given by the *vertex* producing the optimum value of the *objective function.*

In this procedure, Step 5 is an application of the following theorem.

Fundamental Theorem of Linear Programming

If an optimal value for a linear programming problem exists, then it occurs at a vertex of the region of feasible solutions.

EXAMPLE 4 Maximizing Rescue Efforts

Earthquake victims in China need medical supplies and bottled water. Each medical kit measures 1 ft^3 and weighs 10 lb. Each container of water is also 1 ft^3 and weighs 20 lb. The plane can carry only 80,000 lb with a total volume of 6000 ft^3. Each medical kit will aid 6 people, and each container of water will serve 10 people.

How many of each should be sent in order to maximize the number of victims aided? What is this maximum number of victims?

SOLUTION

Step 1 We translate the statements of the problem into symbols as follows.

Let x = the number of medical kits to be sent,

and y = the number of containers of water to be sent.

Because negative values of x and y are not valid for this problem, these two inequalities must be satisfied.

$$x \geq 0$$

$$y \geq 0$$

Each medical kit and each container of water will occupy 1 ft^3 of space, and there is a maximum of 6000 ft^3 available.

$$1x + 1y \leq 6000$$

$$x + y \leq 6000$$

Each medical kit weighs 10 lb, and each water container weighs 20 lb. The total weight cannot exceed 80,000 lb.

$$10x + 20y \leq 80,000$$

$$x + 2y \leq 8000 \qquad \text{Divide by 10.}$$

The four inequalities in color form a system of linear inequalities.

$$x \geq 0$$

$$y \geq 0$$

$$x + y \leq 6000$$

$$x + 2y \leq 8000$$

These are the **constraints** on the variables in this application.

Because each medical kit will aid 6 victims and each container of water will serve 10 victims, the total number of victims served is represented by the following **objective function.**

Number of victims served = $6x + 10y$ Multiply the number of items by the number of victims served and add.

Step 2 The maximum number of victims served, subject to these constraints, is found by sketching the graph of the solution set of the system. See **Figure 20.** The only feasible values of x and y are those that satisfy all constraints. These values correspond to points that lie on the boundary or in the shaded region, which is the **region of feasible solutions.**

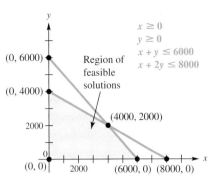

Figure 20

Step 3 The problem may now be stated as follows: Find values of x and y in the region of feasible solutions as shown in **Figure 20** that will produce the maximum possible value of $6x + 10y$. It can be shown that any optimum value (maximum or minimum) will always occur at a **vertex** (or **corner point)** of the region of feasible solutions. The vertices are

$$(0, 0), \quad (0, 4000), \quad (4000, 2000), \quad \text{and} \quad (6000, 0).$$

Step 4 To locate the point (x, y) that gives the maximum value, substitute the coordinates of the vertices into the objective function. See the table below. Find the number of victims served that corresponds to each coordinate pair.

Points are from **Figure 20**.

Point	**Number of Victims Served** $= 6x + 10y$	
$(0, 0)$	$6(0) + 10(0) = 0$	
$(0, 4000)$	$6(0) + 10(4000) = 40{,}000$	
$(4000, 2000)$	$6(4000) + 10(2000) = 44{,}000$	44,000 is the maximum number.
$(6000, 0)$	$6(6000) + 10(0) = 36{,}000$	

Step 5 The vertex $(4000, 2000)$ gives the maximum number. The maximum number of victims served is 44,000, when 4000 medical kits and 2000 containers of water are sent.

✔ **Now Try Exercise 85.**

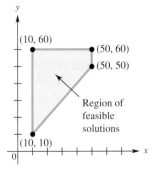

Figure 21

To justify the procedure used in a linear programming problem, suppose that we wish to determine the values of x and y that maximize the objective function

$$30x + 70y$$

with the constraints

$$10 \le x \le 50, \quad 10 \le y \le 60, \quad \text{and} \quad y \ge x.$$

Figure 21 shows this region of feasible solutions.

To locate the point (x, y) that gives the maximum objective function value, add to the graph of **Figure 21** lines corresponding to arbitrarily chosen values of 0, 1000, 3000, and 7000.

$$30x + 70y = 0$$

$$30x + 70y = 1000$$

$$30x + 70y = 3000$$

$$30x + 70y = 7000$$

Figure 22(a) shows the region of feasible solutions, together with these lines. The lines are parallel, and the higher the line, the greater the profit. The line $30x + 70y = 7000$ yields the greatest profit but does not contain any points of the region of feasible solutions. To find the feasible solution of greatest profit, lower the line $30x + 70y = 7000$ until it contains a feasible solution—that is, until it just touches the region of feasible solutions. This occurs at point A, a vertex of the region. The desired maximum value is 5700. See **Figure 22(b)**.

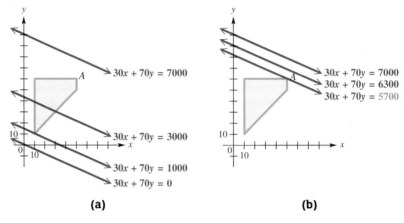

(a)   (b)

Figure 22

EXAMPLE 5   **Minimizing Cost**

Robin takes multivitamins each day. She wants at least 16 units of vitamin A, at least 5 units of vitamin B_1, and at least 20 units of vitamin C. Capsules, costing $0.10 each, contain 8 units of A, 1 of B_1, and 2 of C. Chewable tablets, costing $0.20 each, contain 2 units of A, 1 of B_1, and 7 of C.

How many of each should she take each day to minimize her cost and yet fulfill her daily requirements?

SOLUTION

Step 1 Let x represent the number of capsules to take each day, and let y represent the number of chewable tablets to take. Then the cost in *pennies* per day is

$$\text{cost} = 10x + 20y. \quad \text{Objective function}$$

Robin takes x of the $0.10 capsules and y of the $0.20 chewable tablets, and she gets 8 units of vitamin A from each capsule and 2 units of vitamin A from each tablet. Altogether she gets $8x + 2y$ units of A per day. She wants at least 16 units, which gives the following inequality for A.

$$8x + 2y \geq 16$$

Each capsule and each tablet supplies 1 unit of vitamin B_1. Robin wants at least 5 units per day, so the inequality for B is

$$x + y \geq 5.$$

For vitamin C, the inequality is

$$2x + 7y \geq 20.$$

Because Robin cannot take negative numbers of multivitamins,

$$x \geq 0 \quad \text{and} \quad y \geq 0.$$

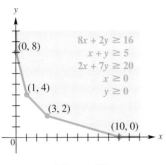

Figure 23

Step 2 **Figure 23** shows the region of feasible solutions formed by the inequalities

$$8x + 2y \geq 16, \quad x + y \geq 5, \quad 2x + 7y \geq 20, \quad x \geq 0, \quad \text{and} \quad y \geq 0.$$

Step 3 The vertices are $(0, 8)$, $(1, 4)$, $(3, 2)$, and $(10, 0)$.

Steps 4 and 5 See the table. The minimum cost of $0.70 occurs at $(3, 2)$.

Points are from **Figure 23**.

Point	Cost $= 10x + 20y$
$(0, 8)$	$10(0) + 20(8) = 160$
$(1, 4)$	$10(1) + 20(4) = 90$
$(3, 2)$	$10(3) + 20(2) = 70$
$(10, 0)$	$10(10) + 20(0) = 100$

70 cents, or $0.70, is the minimum cost.

Robin's best choice is to take 3 capsules and 2 chewable tablets each day, for a total cost of $0.70 per day. She receives just the minimum amounts of vitamins B_1 and C, and an excess of vitamin A.

✔ **Now Try Exercises 79 and 89.**

5.6 Exercises

CONCEPT PREVIEW *Match each system of inequalities with the correct graph from choices A–D.*

1. $x \geq 5$
$y \leq -3$

2. $x \leq 5$
$y \geq -3$

3. $x > 5$
$y < -3$

4. $x < 5$
$y > -3$

A.

B.

C.

D.

CONCEPT PREVIEW *Fill in the blank(s) to correctly complete the sentence, or answer the question.*

5. The test point $(0, 0)$ _____ satisfy the inequality $-3x - 4y \geq 12$.
(does / does not)

6. Any point that lies on the graph of $-3x - 4y = 12$ _____ lie on the graph of $-3x - 4y > 12$.
(does / does not)

7. What are the coordinates of the point of intersection of the boundary lines in the following system?

$$x \geq 2$$
$$y \leq -5$$

8. Does the point $(3, -8)$ satisfy the system in **Exercise 7**?

9. The graph of $4x - 7y < 28$ has a _____ boundary line.
   (solid/dashed)

10. When the inequality in **Exercise 9** is solved for y, the result is _____ , and the points _____ the boundary line are shaded.
    (above/below)

Graph each inequality. See Examples 1–3.

11. $x + 2y \le 6$	12. $x - y \ge 2$	13. $2x + 3y \ge 4$
14. $4y - 3x \le 5$	15. $3x - 5y > 6$	16. $x < 3 + 2y$
17. $5x < 4y - 2$	18. $2x > 3 - 4y$	19. $x \le 3$
20. $y \le -2$	21. $y < 3x^2 + 2$	22. $y \le x^2 - 4$
23. $y > (x - 1)^2 + 2$	24. $y > 2(x + 3)^2 - 1$	25. $x^2 + (y + 3)^2 \le 16$
26. $(x - 4)^2 + y^2 \ge 9$	27. $y > 2^x + 1$	28. $y \le \log(x - 1) - 2$

Concept Check Work each problem.

29. For $Ax + By \ge C$, if $B > 0$, would the region above or below the line be shaded?

30. For $Ax + By \ge C$, if $B < 0$, would the region above or below the line be shaded?

31. Which one of the following is a description of the graph of the inequality

$$(x - 5)^2 + (y - 2)^2 < 4?$$

   A. the region inside a circle with center $(-5, -2)$ and radius 2

   B. the region inside a circle with center $(5, 2)$ and radius 2

   C. the region inside a circle with center $(-5, -2)$ and radius 4

   D. the region outside a circle with center $(5, 2)$ and radius 4

32. Write the inequality that represents the region inside a circle with center $(-5, -2)$ and radius 4.

Concept Check *Match each inequality with the appropriate calculator graph in A–D. Do not use a calculator.*

33. $y \le 3x - 6$ 34. $y \ge 3x - 6$ 35. $y \le -3x - 6$ 36. $y \ge -3x - 6$

A.

B.

C.

D.

*Graph the solution set of each system of inequalities. **See Example 3.***

37. $x + y \geq 0$
$2x - y \geq 3$

38. $x + y \leq 4$
$x - 2y \geq 6$

39. $2x + y > 2$
$x - 3y < 6$

40. $4x + 3y < 12$
$y + 4x > -4$

41. $3x + 5y \leq 15$
$x - 3y \geq 9$

42. $y \leq x$
$x^2 + y^2 < 1$

43. $4x - 3y \leq 12$
$y \leq x^2$

44. $y \leq -x^2$
$y \geq x^2 - 6$

45. $x + 2y \leq 4$
$y \geq x^2 - 1$

46. $x + y \leq 9$
$x \leq -y^2$

47. $y \leq (x + 2)^2$
$y \geq -2x^2$

48. $x - y < 1$
$-1 < y < 1$

49. $x + y \leq 36$
$-4 \leq x \leq 4$

50. $y \geq (x - 2)^2 + 3$
$y \leq -(x - 1)^2 + 6$

51. $y \geq x^2 + 4x + 4$
$y < -x^2$

52. $x \geq 0$
$x + y \leq 4$
$2x + y \leq 5$

53. $3x - 2y \geq 6$
$x + y \leq -5$
$y \leq 4$

54. $-2 < x < 3$
$-1 \leq y \leq 5$
$2x + y < 6$

55. $-2 < x < 2$
$y > 1$
$x - y > 0$

56. $x + y \leq 4$
$x - y \leq 5$
$4x + y \leq -4$

57. $x \leq 4$
$x \geq 0$
$y \geq 0$
$x + 2y \geq 2$

58. $2y + x \geq -5$
$y \leq 3 + x$
$x \leq 0$
$y \leq 0$

59. $2x + 3y \leq 12$
$2x + 3y > -6$
$3x + y < 4$
$x \geq 0$
$y \geq 0$

60. $y \geq 3^x$
$y \geq 2$

61. $y \leq \left(\dfrac{1}{2}\right)^x$
$y \geq 4$

62. $\ln x - y \geq 1$
$x^2 - 2x - y \leq 1$

63. $y \leq \log x$
$y \geq |x - 2|$

64. $e^{-x} - y \leq 1$
$x - 2y \geq 4$

65. $y > x^3 + 1$
$y \geq -1$

66. $y \leq x^3 - x$
$y > -3$

Use the shading capabilities of a graphing calculator to graph each inequality or system of inequalities.

67. $3x + 2y \geq 6$

68. $y \leq x^2 + 5$

69. $x + y \geq 2$
$x + y \leq 6$

70. $y \geq |x + 2|$
$y \leq 6$

Connecting Graphs with Equations *Determine the system of inequalities illustrated in each graph. Write each inequality in standard form.*

71.

72.

73.

74.

75. *Concept Check* Write a system of inequalities for which the graph is the region in the first quadrant inside and including the circle with radius 2 centered at the origin, and above (not including) the line that passes through the points $(0, -1)$ and $(2, 2)$.

76. *Cost of Vitamins* The figure shows the region of feasible solutions for the vitamin problem of **Example 5** and the straight-line graph of all combinations of capsules and chewable tablets for which the cost is $0.40.

Region of feasible solutions

$y = -\frac{1}{2}x + 2$

(a) The cost function is $10x + 20y$. Give the linear equation (in slope-intercept form) of the line of constant cost c.

(b) As c increases, does the line of constant cost move up or down?

(c) By inspection, find the vertex of the region of feasible solutions that gives the optimal value.

The graphs show regions of feasible solutions. Find the maximum and minimum values of each objective function. **See Examples 4 and 5.**

77. objective function $= 3x + 5y$ **78.** objective function $= 6x + y$

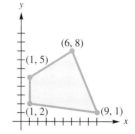

Find the maximum and minimum values of each objective function over the region of feasible solutions shown at the right. **See Examples 4 and 5.**

79. objective function $= 3x + 5y$

80. objective function $= 5x + 5y$

81. objective function $= 10y$

82. objective function $= 3x - y$

Write a system of inequalities for each problem, and then graph the region of feasible solutions of the system. **See Examples 4 and 5.**

83. *Vitamin Requirements* Jane must supplement her daily diet with at least 6000 USP units of vitamin A, at least 195 mg of vitamin C, and at least 600 USP units of vitamin D. She finds that Mason's Pharmacy carries Brand X and Brand Y vitamins. Each Brand X pill contains 3000 USP units of A, 45 mg of C, and 75 USP units of D, while each Brand Y pill contains 1000 USP units of A, 50 mg of C, and 200 USP units of D.

84. *Shipping Requirements* The California Almond Growers have 2400 boxes of almonds to be shipped from their plant in Sacramento to Des Moines and San Antonio. The Des Moines market needs at least 1000 boxes, while the San Antonio market must have at least 800 boxes.

Solve each problem. See Examples 4 and 5.

85. *Aid to Disaster Victims* An agency wants to ship food and clothing to tsunami victims in Japan. Commercial carriers have volunteered to transport the packages, provided they fit in the available cargo space. Each 20-ft^3 box of food weighs 40 lb and each 30-ft^3 box of clothing weighs 10 lb. The total weight cannot exceed 16,000 lb, and the total volume must be at most 18,000 ft^3. Each carton of food will feed 10 victims, and each carton of clothing will help 8 victims.

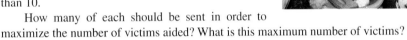

How many cartons of food and clothing should be sent to maximize the number of people assisted? What is the maximum number assisted?

86. *Aid to Disaster Victims* Refer to **Example 4.** Suppose that each medical kit aids 2 victims rather than 6, and each container of water serves 5 victims rather than 10.

How many of each should be sent in order to maximize the number of victims aided? What is this maximum number of victims?

87. *Storage Capacity* An office manager wants to buy some filing cabinets. Cabinet A costs $10 each, requires 6 ft^2 of floor space, and holds 8 ft^3 of files. Cabinet B costs $20 each, requires 8 ft^2 of floor space, and holds 12 ft^3 of files. He can spend no more than $140, and there is room for no more than 72 ft^2 of cabinets.

To maximize storage capacity within the limits imposed by funds and space, how many of each type of cabinet should he buy?

88. *Gasoline Revenues* The manufacturing process requires that oil refineries manufacture at least 2 gal of gasoline for each gallon of fuel oil. To meet the winter demand for fuel oil, at least 3 million gal per day must be produced. The demand for gasoline is no more than 6.4 million gal per day.

If the price of gasoline is $2.90 per gal and the price of fuel oil is $2.50 per gal, how much of each should be produced to maximize revenue?

89. *Diet Requirements* Theo requires two food supplements, I and II. He can get these supplements from two different products, A and B, as shown in the table. He must include at least 15 g of each supplement in his daily diet.

If product A costs $0.25 per serving and product B costs $0.40 per serving, how can he satisfy his requirements most economically?

Supplement (g/serving)	I	II
Product A	3	2
Product B	2	4

90. *Profit from Televisions* The GL company makes televisions. It produces a 32-inch screen that sells for $100 profit and a 48-inch screen that sells for $150 profit. On the assembly line, the 32-inch screen requires 3 hr, and the 48-inch screen takes 5 hr. The cabinet shop spends 1 hr on the cabinet for the 32-inch screen and 3 hr on the cabinet for the 48-inch screen. Both models require 2 hr of time for testing and packing. On one production run, the company has available 3900 work hours on the assembly line, 2100 work hours in the cabinet shop, and 2200 work hours in the testing and packing department.

How many of each model should it produce to maximize profit? What is the maximum profit?

5.7 Properties of Matrices

- Basic Definitions
- Matrix Addition
- Special Matrices
- Matrix Subtraction
- Scalar Multiplication
- Matrix Multiplication
- An Application of Matrix Algebra

Basic Definitions In this section and the next, we discuss algebraic properties of matrices. It is customary to use capital letters to name matrices and subscript notation to name elements of a matrix, as in the following matrix A.

$$A = \begin{bmatrix} a_{11} & a_{12} & a_{13} & \cdots & a_{1n} \\ a_{21} & a_{22} & a_{23} & \cdots & a_{2n} \\ a_{31} & a_{32} & a_{33} & \cdots & a_{3n} \\ \vdots & \vdots & \vdots & & \vdots \\ a_{m1} & a_{m2} & a_{m3} & \cdots & a_{mn} \end{bmatrix}$$

The first row, first column element is a_{11} (read "a-sub-one-one"); the second row, third column element is a_{23}; and in general, the ith row, jth column element is a_{ij}.

An $n \times n$ matrix is a **square matrix of order** n because the number of rows is equal to the number of columns. A matrix with just one row is a **row matrix,** and a matrix with just one column is a **column matrix.**

Two matrices are equal if they have the same dimension and if corresponding elements, position by position, are equal. Using this definition, the matrices

$$\begin{bmatrix} 2 & 1 \\ 3 & -5 \end{bmatrix} \quad \text{and} \quad \begin{bmatrix} 1 & 2 \\ -5 & 3 \end{bmatrix} \quad \text{are } not \text{ equal}$$

(even though they contain the same elements and have the same dimension), because at least one pair of corresponding elements differ.

EXAMPLE 1 Finding Values to Make Two Matrices Equal

Find the values of the variables for which each statement is true, if possible.

(a) $\begin{bmatrix} 2 & 1 \\ p & q \end{bmatrix} = \begin{bmatrix} x & y \\ -1 & 0 \end{bmatrix}$ **(b)** $\begin{bmatrix} x \\ y \end{bmatrix} = \begin{bmatrix} 1 \\ 4 \\ 0 \end{bmatrix}$

SOLUTION

(a) From the definition of equality given above, the only way that the statement can be true is if $2 = x$, $1 = y$, $p = -1$, and $q = 0$.

(b) This statement can never be true because the two matrices have different dimensions. (One is 2×1 and the other is 3×1.)

✔ **Now Try Exercises 13 and 19.**

Matrix Addition Addition of matrices is defined as follows.

Addition of Matrices

To add two matrices of the same dimension, add corresponding elements. *Only matrices of the same dimension can be added.*

It can be shown that matrix addition satisfies the commutative, associative, closure, identity, and inverse properties. **(See Exercises 91 and 92.)**

EXAMPLE 2 **Adding Matrices**

Find each sum, if possible.

(a) $\begin{bmatrix} 5 & -6 \\ 8 & 9 \end{bmatrix} + \begin{bmatrix} -4 & 6 \\ 8 & -3 \end{bmatrix}$

(b) $\begin{bmatrix} 2 \\ 5 \\ 8 \end{bmatrix} + \begin{bmatrix} -6 \\ 3 \\ 12 \end{bmatrix}$

(c) $A + B$, if $A = \begin{bmatrix} 5 & 8 \\ 6 & 2 \end{bmatrix}$ and $B = \begin{bmatrix} 3 & 9 & 1 \\ 4 & 2 & 5 \end{bmatrix}$

ALGEBRAIC SOLUTION

(a) $\begin{bmatrix} 5 & -6 \\ 8 & 9 \end{bmatrix} + \begin{bmatrix} -4 & 6 \\ 8 & -3 \end{bmatrix}$

$= \begin{bmatrix} 5 + (-4) & -6 + 6 \\ 8 + 8 & 9 + (-3) \end{bmatrix}$

$= \begin{bmatrix} 1 & 0 \\ 16 & 6 \end{bmatrix}$

(b) $\begin{bmatrix} 2 \\ 5 \\ 8 \end{bmatrix} + \begin{bmatrix} -6 \\ 3 \\ 12 \end{bmatrix} = \begin{bmatrix} -4 \\ 8 \\ 20 \end{bmatrix}$

(c) The matrices

$$A = \begin{bmatrix} 5 & 8 \\ 6 & 2 \end{bmatrix}$$

and $B = \begin{bmatrix} 3 & 9 & 1 \\ 4 & 2 & 5 \end{bmatrix}$

have different dimensions, so A and B cannot be added. The sum $A + B$ does not exist.

GRAPHING CALCULATOR SOLUTION

(a) **Figure 24** shows the sum of matrices A and B.

Figure 24

(b) The screen in **Figure 25** shows how the sum of two column matrices entered directly on the home screen is displayed.

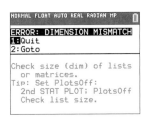

Figure 25   **Figure 26**

(c) A graphing calculator such as the TI-84 Plus will return an ERROR message if it is directed to perform an operation on matrices that is not possible due to dimension mismatch. See **Figure 26.**

✔ **Now Try Exercises 25, 27, and 29.**

Special Matrices   A matrix containing only zero elements is a **zero matrix.** A zero matrix can be written with any dimension.

$$O = \begin{bmatrix} 0 & 0 & 0 \end{bmatrix}$$ 1×3 zero matrix $$O = \begin{bmatrix} 0 & 0 & 0 \\ 0 & 0 & 0 \end{bmatrix}$$ 2×3 zero matrix

By the additive inverse property, each real number has an additive inverse: If a is a real number, then there is a real number $-a$ such that

$$a + (-a) = 0 \quad \text{and} \quad -a + a = 0.$$

Given matrix A, there is a matrix $-A$ such that $A + (-A) = O$. The matrix $-A$ has as elements the additive inverses of the elements of A. (Remember, each element of A is a real number and therefore has an additive inverse.)

Example: If $A = \begin{bmatrix} -5 & 2 & -1 \\ 3 & 4 & -6 \end{bmatrix}$, then $-A = \begin{bmatrix} 5 & -2 & 1 \\ -3 & -4 & 6 \end{bmatrix}$.

CHECK   Confirm that $A + (-A)$ equals the zero matrix, O.

$$A + (-A) = \begin{bmatrix} -5 & 2 & -1 \\ 3 & 4 & -6 \end{bmatrix} + \begin{bmatrix} 5 & -2 & 1 \\ -3 & -4 & 6 \end{bmatrix} = \begin{bmatrix} 0 & 0 & 0 \\ 0 & 0 & 0 \end{bmatrix} = O \checkmark$$

Matrix $-A$ is the **additive inverse,** or **negative,** of matrix A. Every matrix has an additive inverse.

Matrix Subtraction   The real number b is subtracted from the real number a, written $a - b$, by adding a and the additive inverse of b.

$$a - b = a + (-b) \quad \text{Real number subtraction}$$

The same definition applies to subtraction of matrices.

Subtraction of Matrices

If A and B are two matrices of the same dimension, then the following holds.

$$A - B = A + (-B)$$

In practice, the difference of two matrices of the same dimension is found by subtracting corresponding elements.

EXAMPLE 3   Subtracting Matrices

Find each difference, if possible.

(a) $\begin{bmatrix} -5 & 6 \\ 2 & 4 \end{bmatrix} - \begin{bmatrix} -3 & 2 \\ 5 & -8 \end{bmatrix}$ (b) $\begin{bmatrix} 8 & 6 & -4 \end{bmatrix} - \begin{bmatrix} 3 & 5 & -8 \end{bmatrix}$

(c) $A - B$, if $A = \begin{bmatrix} -2 & 5 \\ 0 & 1 \end{bmatrix}$ and $B = \begin{bmatrix} 3 \\ 5 \end{bmatrix}$

SOLUTION

(a) $\begin{bmatrix} -5 & 6 \\ 2 & 4 \end{bmatrix} - \begin{bmatrix} -3 & 2 \\ 5 & -8 \end{bmatrix}$

$= \begin{bmatrix} -5 - (-3) & 6 - 2 \\ 2 - 5 & 4 - (-8) \end{bmatrix}$ Subtract corresponding entries.

$= \begin{bmatrix} -2 & 4 \\ -3 & 12 \end{bmatrix}$ Simplify.

HISTORY	
[C]	
	$\begin{bmatrix} -5 & 6 \\ 2 & 4 \end{bmatrix}$
[D]	
	$\begin{bmatrix} -3 & 2 \\ 5 & -8 \end{bmatrix}$
[C]-[D]	
	$\begin{bmatrix} -2 & 4 \\ -3 & 12 \end{bmatrix}$

This screen supports the result in
Example 3(a).

(b) $\begin{bmatrix} 8 & 6 & -4 \end{bmatrix} - \begin{bmatrix} 3 & 5 & -8 \end{bmatrix}$

$= \begin{bmatrix} 5 & 1 & 4 \end{bmatrix}$ Subtract corresponding entries.

(c) $A = \begin{bmatrix} -2 & 5 \\ 0 & 1 \end{bmatrix}$ and $B = \begin{bmatrix} 3 \\ 5 \end{bmatrix}$

These matrices have different dimensions and cannot be subtracted, so the difference $A - B$ does not exist.

☑ **Now Try Exercises 31, 33, and 35.**

Scalar Multiplication   In work with matrices, a real number is called a **scalar** to distinguish it from a matrix.

> *The product of a scalar k and a matrix X is the matrix kX, each of whose elements is k times the corresponding element of X.*

NORMAL FLOAT AUTO REAL RADIAN MP

[A]

$\begin{bmatrix} 2 & \text{-}3 \\ 0 & 4 \end{bmatrix}$

5[A]

$\begin{bmatrix} 10 & \text{-}15 \\ 0 & 20 \end{bmatrix}$

NORMAL FLOAT AUTO REAL RADIAN MP

[B]

$\begin{bmatrix} 20 & 36 \\ 12 & \text{-}16 \end{bmatrix}$

(3/4)[B]

$\begin{bmatrix} 15 & 27 \\ 9 & \text{-}12 \end{bmatrix}$

These screens support the results in **Example 4.**

EXAMPLE 4   **Multiplying Matrices by Scalars**

Find each product.

(a) $5 \begin{bmatrix} 2 & -3 \\ 0 & 4 \end{bmatrix}$

(b) $\dfrac{3}{4} \begin{bmatrix} 20 & 36 \\ 12 & -16 \end{bmatrix}$

SOLUTION

(a) $5 \begin{bmatrix} 2 & -3 \\ 0 & 4 \end{bmatrix}$

$= \begin{bmatrix} 5(2) & 5(-3) \\ 5(0) & 5(4) \end{bmatrix}$

Multiply each element of the matrix by the scalar 5.

$= \begin{bmatrix} 10 & -15 \\ 0 & 20 \end{bmatrix}$

(b) $\dfrac{3}{4} \begin{bmatrix} 20 & 36 \\ 12 & -16 \end{bmatrix}$

$= \begin{bmatrix} \frac{3}{4}(20) & \frac{3}{4}(36) \\ \frac{3}{4}(12) & \frac{3}{4}(-16) \end{bmatrix}$

Multiply each element of the matrix by the scalar $\frac{3}{4}$.

$= \begin{bmatrix} 15 & 27 \\ 9 & -12 \end{bmatrix}$

✔ **Now Try Exercises 41 and 43.**

The proofs of the following properties of scalar multiplication are left for **Exercises 95–98.**

> **Properties of Scalar Multiplication**
>
> Let A and B be matrices of the same dimension, and let c and d be scalars. Then these properties hold.
>
> $$(c + d)A = cA + dA \qquad (cA)d = (cd)A$$
> $$c(A + B) = cA + cB \qquad (cd)A = c(dA)$$

Matrix Multiplication   We have seen how to multiply a real number (scalar) and a matrix. The product of two matrices can also be found. To illustrate, we multiply

$$A = \begin{bmatrix} -3 & 4 & 2 \\ 5 & 0 & 4 \end{bmatrix} \quad \text{and} \quad B = \begin{bmatrix} -6 & 4 \\ 2 & 3 \\ 3 & -2 \end{bmatrix}.$$

First locate *row* 1 of A and *column* 1 of B, which are shown shaded below.

$$A = \begin{bmatrix} -3 & 4 & 2 \\ 5 & 0 & 4 \end{bmatrix} \quad B = \begin{bmatrix} -6 & 4 \\ 2 & 3 \\ 3 & -2 \end{bmatrix}$$

Multiply corresponding elements, and find the sum of the products.

$$-3(-6) + 4(2) + 2(3) = 32$$

This result is the element for row 1, column 1 of the product matrix.

Now use *row* 1 of A and *column* 2 of B to determine the element in row 1, column 2 of the product matrix.

$$\begin{bmatrix} -3 & 4 & 2 \\ 5 & 0 & 4 \end{bmatrix} \begin{bmatrix} -6 & 4 \\ 2 & 3 \\ 3 & -2 \end{bmatrix} \qquad -3(4) + 4(3) + 2(-2) = -4$$

Next, use *row* 2 of A and *column* 1 of B. This will give the row 2, column 1 element of the product matrix.

$$\begin{bmatrix} -3 & 4 & 2 \\ 5 & 0 & 4 \end{bmatrix} \begin{bmatrix} -6 & 4 \\ 2 & 3 \\ 3 & -2 \end{bmatrix} \qquad 5(-6) + 0(2) + 4(3) = -18$$

Finally, use *row* 2 of A and *column* 2 of B to find the element for row 2, column 2 of the product matrix.

$$\begin{bmatrix} -3 & 4 & 2 \\ 5 & 0 & 4 \end{bmatrix} \begin{bmatrix} -6 & 4 \\ 2 & 3 \\ 3 & -2 \end{bmatrix} \qquad 5(4) + 0(3) + 4(-2) = 12$$

The product matrix can be written using the four elements just found.

$$\begin{bmatrix} -3 & 4 & 2 \\ 5 & 0 & 4 \end{bmatrix} \begin{bmatrix} -6 & 4 \\ 2 & 3 \\ 3 & -2 \end{bmatrix} = \begin{bmatrix} 32 & -4 \\ -18 & 12 \end{bmatrix} \qquad \text{Product } AB$$

NOTE As seen here, the product of a 2×3 matrix and a 3×2 matrix is a 2×2 matrix. The dimension of a product matrix AB is given by the number of rows of A and the number of columns of B, respectively.

By definition, the product AB of an $m \times n$ matrix A and an $n \times p$ matrix B is found as follows.

To find the ith row, jth column element of AB, multiply each element in the ith row of A by the corresponding element in the jth column of B. (Note the shaded areas in the matrices below.) The sum of these products will give the row i, column j element of AB.

$$A = \begin{bmatrix} a_{11} & a_{12} & a_{13} & \cdots & a_{1n} \\ a_{21} & a_{22} & a_{23} & \cdots & a_{2n} \\ \vdots & & & & \\ a_{i1} & a_{i2} & a_{i3} & \cdots & a_{in} \\ \vdots & & & & \\ a_{m1} & a_{m2} & a_{m3} & \cdots & a_{mn} \end{bmatrix} \qquad B = \begin{bmatrix} b_{11} & b_{12} & \cdots & b_{1j} & \cdots & b_{1p} \\ b_{21} & b_{22} & \cdots & b_{2j} & \cdots & b_{2p} \\ \vdots & & & & & \\ b_{n1} & b_{n2} & \cdots & b_{nj} & \cdots & b_{np} \end{bmatrix}$$

Matrix Multiplication

The number of columns of an $m \times n$ matrix A is the same as the number of rows of an $n \times p$ matrix B (i.e., both n). The element c_{ij} of the product matrix $C = AB$ is found as follows.

$$c_{ij} = a_{i1}b_{1j} + a_{i2}b_{2j} + \cdots + a_{in}b_{nj}$$

Matrix AB will be an $m \times p$ matrix.

EXAMPLE 5 **Deciding Whether Two Matrices Can Be Multiplied**

Suppose A is a 3×2 matrix, while B is a 2×4 matrix.

(a) Can the product AB be calculated?

(b) If AB can be calculated, what is its dimension?

(c) Can BA be calculated?

(d) If BA can be calculated, what is its dimension?

SOLUTION

(a) The following diagram shows that AB can be calculated because the number of columns of A is equal to the number of rows of B. (Both are 2.)

Matrix A Matrix B

3×2 2×4

Must match

Dimension of AB

3×4

(b) As indicated in the diagram above, the product AB is a 3×4 matrix.

(c) The diagram below shows that BA cannot be calculated.

Matrix B Matrix A

2×4 3×2

Different

(d) The product BA cannot be calculated because B has 4 columns and A has only 3 rows.

✔ **Now Try Exercises 49 and 51.**

EXAMPLE 6 **Multiplying Matrices**

Let $A = \begin{bmatrix} 1 & -3 \\ 7 & 2 \end{bmatrix}$ and $B = \begin{bmatrix} 1 & 0 & -1 & 2 \\ 3 & 1 & 4 & -1 \end{bmatrix}$. Find each product, if possible.

(a) AB **(b)** BA

SOLUTION

(a) First decide whether AB can be found.

$AB = \begin{bmatrix} 1 & -3 \\ 7 & 2 \end{bmatrix} \begin{bmatrix} 1 & 0 & -1 & 2 \\ 3 & 1 & 4 & -1 \end{bmatrix}$ A is 2×2 and B is 2×4, so the product will be a 2×4 matrix.

$= \begin{bmatrix} 1(1) + (-3)3 & 1(0) + (-3)1 & 1(-1) + (-3)4 & 1(2) + (-3)(-1) \\ 7(1) + 2(3) & 7(0) + 2(1) & 7(-1) + 2(4) & 7(2) + 2(-1) \end{bmatrix}$

Use the definition of matrix multiplication.

$= \begin{bmatrix} -8 & -3 & -13 & 5 \\ 13 & 2 & 1 & 12 \end{bmatrix}$ Perform the operations.

(b) B is a 2×4 matrix, and A is a 2×2 matrix, so the number of columns of B (here 4) does not equal the number of rows of A (here 2). Therefore, the product BA cannot be calculated.

✔ **Now Try Exercises 69 and 73.**

The three screens here support the results of the matrix multiplication in **Example 6.** The final screen indicates that the product BA cannot be found.

EXAMPLE 7 Multiplying Square Matrices in Different Orders

Let $A = \begin{bmatrix} 1 & 3 \\ -2 & 5 \end{bmatrix}$ and $B = \begin{bmatrix} -2 & 7 \\ 0 & 2 \end{bmatrix}$. Find each product.

(a) AB **(b)** BA

SOLUTION

(a) $AB = \begin{bmatrix} 1 & 3 \\ -2 & 5 \end{bmatrix} \begin{bmatrix} -2 & 7 \\ 0 & 2 \end{bmatrix}$

$= \begin{bmatrix} 1(-2) + 3(0) & 1(7) + 3(2) \\ -2(-2) + 5(0) & -2(7) + 5(2) \end{bmatrix}$ Multiply elements of each row of A by elements of each column of B.

$= \begin{bmatrix} -2 & 13 \\ 4 & -4 \end{bmatrix}$

(b) $BA = \begin{bmatrix} -2 & 7 \\ 0 & 2 \end{bmatrix} \begin{bmatrix} 1 & 3 \\ -2 & 5 \end{bmatrix}$

$= \begin{bmatrix} -2(1) + 7(-2) & -2(3) + 7(5) \\ 0(1) + 2(-2) & 0(3) + 2(5) \end{bmatrix}$ Multiply elements of each row of B by elements of each column of A.

$= \begin{bmatrix} -16 & 29 \\ -4 & 10 \end{bmatrix}$

Note that $AB \neq BA$. ✔ **Now Try Exercise 79.**

When multiplying matrices, it is important to pay special attention to the dimensions of the matrices as well as the order in which they are to be multiplied. **Examples 5 and 6** showed that the order in which two matrices are to be multiplied may determine whether their product can be found. **Example 7** showed that even when both products AB and BA can be found, they may not be equal.

Noncommutativity of Matrix Multiplication

In general, if A and B are matrices, then

$$AB \neq BA.$$

Matrix multiplication is not commutative.

Matrix multiplication does satisfy the associative and distributive properties.

Properties of Matrix Multiplication

If A, B, and C are matrices such that all the following products and sums exist, then these properties hold.

$$(AB)C = A(BC), \quad A(B + C) = AB + AC, \quad (B + C)A = BA + CA$$

For proofs of the first two results for the special cases when A, B, and C are square matrices, see **Exercises 93 and 94.** The identity and inverse properties for matrix multiplication are discussed later.

An Application of Matrix Algebra

EXAMPLE 8 **Using Matrix Multiplication to Model Plans for a Subdivision**

A contractor builds three kinds of houses, models A, B, and C, with a choice of two styles, colonial or ranch. Matrix P below shows the number of each kind of house the contractor is planning to build for a new 100-home subdivision. The amounts for each of the main materials used depend on the style of the house. These amounts are shown in matrix Q, while matrix R gives the cost in dollars for each kind of material. Concrete is measured here in cubic yards, lumber in 1000 board feet, brick in 1000s, and shingles in 100 square feet.

$$
\begin{array}{c}
\\
\text{Model A} \\
\text{Model B} \\
\text{Model C}
\end{array}
\begin{array}{c}
\text{Colonial \ Ranch} \\
\begin{bmatrix}
0 & 30 \\
10 & 20 \\
20 & 20
\end{bmatrix} = P
\end{array}
$$

$$
\begin{array}{cc}
& \text{Concrete \ Lumber \ Brick \ Shingles} \\
\begin{array}{c} \text{Colonial} \\ \text{Ranch} \end{array} &
\begin{bmatrix}
10 & 2 & 0 & 2 \\
50 & 1 & 20 & 2
\end{bmatrix} = Q
\end{array}
\qquad
\begin{array}{c}
\\
\text{Concrete} \\
\text{Lumber} \\
\text{Brick} \\
\text{Shingles}
\end{array}
\begin{array}{c}
\text{Cost per Unit} \\
\begin{bmatrix}
20 \\
180 \\
60 \\
25
\end{bmatrix} = R
\end{array}
$$

(a) What is the total cost of materials for all houses of each model?

(b) How much of each of the four kinds of material must be ordered?

(c) What is the total cost of the materials?

SOLUTION

(a) To find the materials cost for each model, first find matrix PQ, which will show the total amount of each material needed for all houses of each model.

$$
PQ =
\begin{bmatrix}
0 & 30 \\
10 & 20 \\
20 & 20
\end{bmatrix}
\begin{bmatrix}
10 & 2 & 0 & 2 \\
50 & 1 & 20 & 2
\end{bmatrix}
=
\begin{array}{c}
\text{Concrete \ Lumber \ Brick \ Shingles} \\
\begin{bmatrix}
1500 & 30 & 600 & 60 \\
1100 & 40 & 400 & 60 \\
1200 & 60 & 400 & 80
\end{bmatrix}
\begin{array}{c}
\text{Model A} \\
\text{Model B} \\
\text{Model C}
\end{array}
\end{array}
$$

Multiplying PQ and the cost matrix R gives the total cost of materials for each model.

$$
(PQ)R =
\begin{bmatrix}
1500 & 30 & 600 & 60 \\
1100 & 40 & 400 & 60 \\
1200 & 60 & 400 & 80
\end{bmatrix}
\begin{bmatrix}
20 \\
180 \\
60 \\
25
\end{bmatrix}
=
\begin{array}{c}
\text{Cost} \\
\begin{bmatrix}
72{,}900 \\
54{,}700 \\
60{,}800
\end{bmatrix}
\begin{array}{c}
\text{Model A} \\
\text{Model B} \\
\text{Model C}
\end{array}
\end{array}
$$

(b) To find how much of each kind of material to order, refer to the columns of matrix PQ. The sums of the elements of the columns will give a matrix whose elements represent the total amounts of all materials needed for the subdivision. Call this matrix T, and write it as a row matrix.

$$
T = \begin{bmatrix} 3800 & 130 & 1400 & 200 \end{bmatrix}
$$

(c) The total cost of all the materials is given by the product of matrix T, the total amounts matrix, and matrix R, the cost matrix. To multiply these matrices and obtain a 1×1 matrix, representing the total cost, requires multiplying a 1×4 matrix and a 4×1 matrix. This is why in part (b) a row matrix was written rather than a column matrix.

The total materials cost is given by TR, so

$$TR = \begin{bmatrix} 3800 & 130 & 1400 & 200 \end{bmatrix} \begin{bmatrix} 20 \\ 180 \\ 60 \\ 25 \end{bmatrix} = \begin{bmatrix} 188{,}400 \end{bmatrix}.$$

The total cost of materials is \$188,400. This total may also be found by summing the elements of the column matrix $(PQ)R$.

✔ **Now Try Exercise 85.**

5.7 Exercises

CONCEPT PREVIEW *Fill in the blank(s) to correctly complete each sentence.*

1. For the following statement to be true, the value of x must be _____, and the value of y must be _____.

$$\begin{bmatrix} 3 & -6 \\ 5 & 1 \end{bmatrix} = \begin{bmatrix} x+1 & -6 \\ 5 & y+1 \end{bmatrix}$$

2. For the following sum to be true, we must have $w = $ _____, $x = $ _____, $y = $ _____, and $z = $ _____.

$$\begin{bmatrix} 0 & 5 \\ -3 & 10 \end{bmatrix} + \begin{bmatrix} 1 & 2 \\ 3 & 4 \end{bmatrix} = \begin{bmatrix} w & x \\ y & z \end{bmatrix}$$

3. For the following difference to be true, we must have $w = $ _____, $x = $ _____, $y = $ _____, and $z = $ _____.

$$\begin{bmatrix} 7 & 2 \\ 5 & 12 \end{bmatrix} - \begin{bmatrix} 1 & 3 \\ 6 & 0 \end{bmatrix} = \begin{bmatrix} w & x \\ y & z \end{bmatrix}$$

4. For the following scalar product to be true, we must have $w = $ _____, $x = $ _____, $y = $ _____, and $z = $ _____.

$$-2\begin{bmatrix} 5 & -1 \\ 6 & 0 \end{bmatrix} = \begin{bmatrix} w & x \\ y & z \end{bmatrix}$$

5. If the dimension of matrix A is 3×2 and the dimension of matrix B is 2×6, then the dimension of AB is _____.

6. For the following matrix product to be true, we must have $x = $ _____.

$$\begin{bmatrix} 2 & 1 \\ -4 & 3 \end{bmatrix}\begin{bmatrix} 3 & -1 \\ 2 & 7 \end{bmatrix} = \begin{bmatrix} x & 5 \\ -6 & 25 \end{bmatrix}$$

*Concept Check Find the dimension of each matrix. Identify any square, column, or row matrices. **See the discussion preceding Example 1.***

7. $\begin{bmatrix} -4 & 8 \\ 2 & 3 \end{bmatrix}$
    8. $\begin{bmatrix} -9 & 6 & 2 \\ 4 & 1 & 8 \end{bmatrix}$
    9. $\begin{bmatrix} -6 & 8 & 0 & 0 \\ 4 & 1 & 9 & 2 \\ 3 & -5 & 7 & 1 \end{bmatrix}$

10. $\begin{bmatrix} 8 & -2 & 4 & 6 & 3 \end{bmatrix}$ **11.** $\begin{bmatrix} 2 \\ 4 \end{bmatrix}$ **12.** $\begin{bmatrix} -9 \end{bmatrix}$

Find the values of the variables for which each statement is true, if possible. **See Examples 1 and 2.**

13. $\begin{bmatrix} -3 & a \\ b & 5 \end{bmatrix} = \begin{bmatrix} c & 0 \\ 4 & d \end{bmatrix}$

14. $\begin{bmatrix} w & x \\ 8 & -12 \end{bmatrix} = \begin{bmatrix} 9 & 17 \\ y & z \end{bmatrix}$

15. $\begin{bmatrix} x+2 & y-6 \\ z-3 & w+5 \end{bmatrix} = \begin{bmatrix} -2 & 8 \\ 0 & 3 \end{bmatrix}$

16. $\begin{bmatrix} 6 & a+3 \\ b+2 & 9 \end{bmatrix} = \begin{bmatrix} c-3 & 4 \\ -2 & d-4 \end{bmatrix}$

17. $\begin{bmatrix} x & y & z \end{bmatrix} = \begin{bmatrix} 21 & 6 \end{bmatrix}$

18. $\begin{bmatrix} p \\ q \\ r \end{bmatrix} = \begin{bmatrix} 3 \\ -9 \end{bmatrix}$

19. $\begin{bmatrix} 0 & 5 & x \\ -1 & 3 & y+2 \\ 4 & 1 & z \end{bmatrix} = \begin{bmatrix} 0 & w & 6 \\ -1 & 3 & 0 \\ 4 & 1 & 8 \end{bmatrix}$

20. $\begin{bmatrix} 5 & x-4 & 9 \\ 2 & -3 & 8 \\ 6 & 0 & 5 \end{bmatrix} = \begin{bmatrix} y+3 & 2 & 9 \\ z+4 & -3 & 8 \\ 6 & 0 & w \end{bmatrix}$

21. $\begin{bmatrix} -7+z & 4r & 8s \\ 6p & 2 & 5 \end{bmatrix} + \begin{bmatrix} -9 & 8r & 3 \\ 2 & 5 & 4 \end{bmatrix} = \begin{bmatrix} 2 & 36 & 27 \\ 20 & 7 & 12a \end{bmatrix}$

22. $\begin{bmatrix} a+2 & 3z+1 & 5m \\ 8k & 0 & 3 \end{bmatrix} + \begin{bmatrix} 3a & 2z & 5m \\ 2k & 5 & 6 \end{bmatrix} = \begin{bmatrix} 10 & -14 & 80 \\ 10 & 5 & 9 \end{bmatrix}$

23. Your friend missed the lecture on adding matrices. Explain to him how to add two matrices.

24. Explain how to subtract two matrices.

Find each sum or difference, if possible. **See Examples 2 and 3.**

25. $\begin{bmatrix} -4 & 3 \\ 12 & -6 \end{bmatrix} + \begin{bmatrix} 2 & -8 \\ 5 & 10 \end{bmatrix}$

26. $\begin{bmatrix} 9 & 4 \\ -8 & 2 \end{bmatrix} + \begin{bmatrix} -3 & 2 \\ -4 & 7 \end{bmatrix}$

27. $\begin{bmatrix} 6 & -9 & 2 \\ 4 & 1 & 3 \end{bmatrix} + \begin{bmatrix} -8 & 2 & 5 \\ 6 & -3 & 4 \end{bmatrix}$

28. $\begin{bmatrix} 4 & -3 \\ 7 & 2 \\ -6 & 8 \end{bmatrix} + \begin{bmatrix} 9 & -10 \\ 0 & 5 \\ -1 & 6 \end{bmatrix}$

29. $\begin{bmatrix} 2 & 4 & 6 \end{bmatrix} + \begin{bmatrix} -2 \\ -4 \\ -6 \end{bmatrix}$

30. $\begin{bmatrix} 3 \\ 1 \\ 0 \end{bmatrix} + \begin{bmatrix} 2 \\ -6 \end{bmatrix}$

31. $\begin{bmatrix} -6 & 8 \\ 0 & 0 \end{bmatrix} - \begin{bmatrix} 0 & 0 \\ -4 & -2 \end{bmatrix}$

32. $\begin{bmatrix} 11 & 0 \\ -4 & 0 \end{bmatrix} - \begin{bmatrix} 0 & 12 \\ 0 & -14 \end{bmatrix}$

33. $\begin{bmatrix} 12 \\ -1 \\ 3 \end{bmatrix} - \begin{bmatrix} 8 \\ 4 \\ -1 \end{bmatrix}$

34. $\begin{bmatrix} 10 & -4 & 6 \end{bmatrix} - \begin{bmatrix} -2 & 5 & 3 \end{bmatrix}$

35. $\begin{bmatrix} -4 & 3 \end{bmatrix} - \begin{bmatrix} 5 & 8 & 2 \end{bmatrix}$

36. $\begin{bmatrix} 4 & 6 \end{bmatrix} - \begin{bmatrix} 2 \\ 3 \end{bmatrix}$

37. $\begin{bmatrix} \sqrt{3} & -4 \\ 2 & -\sqrt{5} \\ -8 & \sqrt{8} \end{bmatrix} - \begin{bmatrix} 2\sqrt{3} & 9 \\ -2 & \sqrt{5} \\ -7 & 3\sqrt{2} \end{bmatrix}$

38. $\begin{bmatrix} 2 & \sqrt{7} \\ 3\sqrt{28} & -6 \end{bmatrix} - \begin{bmatrix} -1 & 5\sqrt{7} \\ 2\sqrt{7} & 2 \end{bmatrix}$

39. $\begin{bmatrix} 3x+y & -2y \\ x+2y & 3y \end{bmatrix} + \begin{bmatrix} 2x & 3y \\ 5x & x \end{bmatrix}$

40. $\begin{bmatrix} 4k-8y \\ 6z-3x \\ 2k+5a \\ -4m+2n \end{bmatrix} - \begin{bmatrix} 5k+6y \\ 2z+5x \\ 4k+6a \\ 4m-2n \end{bmatrix}$

Let $A = \begin{bmatrix} -2 & 4 \\ 0 & 3 \end{bmatrix}$ and $B = \begin{bmatrix} -6 & 2 \\ 4 & 0 \end{bmatrix}$. Find each of the following. **See Examples 2–4.**

41. $2A$ **42.** $-3B$ **43.** $\dfrac{3}{2}B$ **44.** $-\dfrac{3}{2}A$

45. $2A - 3B$ **46.** $-2A + 4B$ **47.** $-A + \dfrac{1}{2}B$ **48.** $\dfrac{3}{4}A - B$

Suppose that matrix A has dimension 2×3, B has dimension 3×5, and C has dimension 5×2. Decide whether the given product can be calculated. If it can, determine its dimension. **See Example 5.**

49. AB **50.** CA **51.** BA **52.** AC **53.** BC **54.** CB

Find each product, if possible. **See Examples 5–7.**

55. $\begin{bmatrix} 1 & 2 \\ 3 & 4 \end{bmatrix}\begin{bmatrix} -1 \\ 7 \end{bmatrix}$ **56.** $\begin{bmatrix} -1 & 5 \\ 7 & 0 \end{bmatrix}\begin{bmatrix} 6 \\ 2 \end{bmatrix}$

57. $\begin{bmatrix} 3 & -4 & 1 \\ 5 & 0 & 2 \end{bmatrix}\begin{bmatrix} -1 \\ 4 \\ 2 \end{bmatrix}$ **58.** $\begin{bmatrix} -6 & 3 & 5 \\ 2 & 9 & 1 \end{bmatrix}\begin{bmatrix} -2 \\ 0 \\ 3 \end{bmatrix}$

59. $\begin{bmatrix} \sqrt{2} & \sqrt{2} & -\sqrt{18} \\ \sqrt{3} & \sqrt{27} & 0 \end{bmatrix}\begin{bmatrix} 8 & -10 \\ 9 & 12 \\ 0 & 2 \end{bmatrix}$ **60.** $\begin{bmatrix} -9 & 2 & 1 \\ 3 & 0 & 0 \end{bmatrix}\begin{bmatrix} \sqrt{5} \\ \sqrt{20} \\ -2\sqrt{5} \end{bmatrix}$

61. $\begin{bmatrix} \sqrt{3} & 1 \\ 2\sqrt{5} & 3\sqrt{2} \end{bmatrix}\begin{bmatrix} \sqrt{3} & -\sqrt{6} \\ 4\sqrt{3} & 0 \end{bmatrix}$ **62.** $\begin{bmatrix} \sqrt{7} & 0 \\ 2 & \sqrt{28} \end{bmatrix}\begin{bmatrix} 2\sqrt{3} & -\sqrt{7} \\ 0 & -6 \end{bmatrix}$

63. $\begin{bmatrix} -3 & 0 & 2 & 1 \\ 4 & 0 & 2 & 6 \end{bmatrix}\begin{bmatrix} -4 & 2 \\ 0 & 1 \end{bmatrix}$ **64.** $\begin{bmatrix} -1 & 2 & 4 & 1 \\ 0 & 2 & -3 & 5 \end{bmatrix}\begin{bmatrix} 1 & 2 & 4 \\ -2 & 5 & 1 \end{bmatrix}$

65. $\begin{bmatrix} -2 & 4 & 1 \end{bmatrix}\begin{bmatrix} 3 & -2 & 4 \\ 2 & 1 & 0 \\ 0 & -1 & 4 \end{bmatrix}$ **66.** $\begin{bmatrix} 0 & 3 & -4 \end{bmatrix}\begin{bmatrix} -2 & 6 & 3 \\ 0 & 4 & 2 \\ -1 & 1 & 4 \end{bmatrix}$

67. $\begin{bmatrix} -2 & -3 & -4 \\ 2 & -1 & 0 \\ 4 & -2 & 3 \end{bmatrix}\begin{bmatrix} 0 & 1 & 4 \\ 1 & 2 & -1 \\ 3 & 2 & -2 \end{bmatrix}$ **68.** $\begin{bmatrix} -1 & 2 & 0 \\ 0 & 3 & 2 \\ 0 & 1 & 4 \end{bmatrix}\begin{bmatrix} 2 & -1 & 2 \\ 0 & 2 & 1 \\ 3 & 0 & -1 \end{bmatrix}$

Given $A = \begin{bmatrix} 4 & -2 \\ 3 & 1 \end{bmatrix}$, $B = \begin{bmatrix} 5 & 1 \\ 0 & -2 \\ 3 & 7 \end{bmatrix}$, and $C = \begin{bmatrix} -5 & 4 & 1 \\ 0 & 3 & 6 \end{bmatrix}$, find each product, if possible. **See Examples 5–7.**

69. BA **70.** AC **71.** BC **72.** CB

73. AB **74.** CA **75.** A^2 **76.** A^3
(Hint: $A^3 = A^2 \cdot A$)

77. *Concept Check* Compare the answers to **Exercises 69 and 73, 71 and 72,** and **70 and 74.** How do they show that matrix multiplication is not commutative?

78. *Concept Check* Why is it not possible to find C^2 for matrix C defined as follows.

$$C = \begin{bmatrix} -5 & 4 & 1 \\ 0 & 3 & 6 \end{bmatrix}$$

*For each pair of matrices A and B, find (**a**) AB and (**b**) BA. **See Example 7.***

79. $A = \begin{bmatrix} 3 & 4 \\ -2 & 1 \end{bmatrix}, B = \begin{bmatrix} 6 & 0 \\ 5 & -2 \end{bmatrix}$

80. $A = \begin{bmatrix} 0 & -5 \\ -4 & 2 \end{bmatrix}, B = \begin{bmatrix} 3 & -1 \\ -5 & 4 \end{bmatrix}$

81. $A = \begin{bmatrix} 0 & 1 & -1 \\ 0 & 1 & 0 \\ 0 & 0 & 1 \end{bmatrix}, B = \begin{bmatrix} 1 & 0 & 0 \\ 0 & 1 & 0 \\ 0 & 0 & 1 \end{bmatrix}$

82. $A = \begin{bmatrix} -1 & 0 & 1 \\ 0 & 1 & 1 \\ -1 & -1 & 0 \end{bmatrix}, B = \begin{bmatrix} 0 & 0 & 1 \\ 0 & 1 & 0 \\ 1 & 0 & 0 \end{bmatrix}$

83. *Concept Check* In **Exercise 81**, $AB = A$ and $BA = A$. For this pair of matrices, B acts the same way for matrix multiplication as the number _____ acts for multiplication of real numbers.

84. *Concept Check* Find AB and BA for the following matrices.

$$A = \begin{bmatrix} a & b \\ c & d \end{bmatrix} \quad \text{and} \quad B = \begin{bmatrix} 1 & 0 \\ 0 & 1 \end{bmatrix}$$

Matrix B acts as the multiplicative _____ element for 2×2 square matrices.

*Solve each problem. **See Example 8.***

85. *Income from Yogurt* Yagel's Yogurt sells three types of yogurt: nonfat, regular, and super creamy, at three locations. Location I sells 50 gal of nonfat, 100 gal of regular, and 30 gal of super creamy each day. Location II sells 10 gal of nonfat, and Location III sells 60 gal of nonfat each day. Daily sales of regular yogurt are 90 gal at Location II and 120 gal at Location III. At Location II, 50 gal of super creamy are sold each day, and 40 gal of super creamy are sold each day at Location III.

(**a**) Write a 3×3 matrix that shows the sales figures for the three locations, with the rows representing the three locations.

(**b**) The incomes per gallon for nonfat, regular, and super creamy are \$12, \$10, and \$15, respectively. Write a 1×3 or 3×1 matrix displaying the incomes.

(**c**) Find a matrix product that gives the daily income at each of the three locations.

(**d**) What is Yagel's Yogurt's total daily income from the three locations?

86. *Purchasing Costs* The Bread Box, a small neighborhood bakery, sells four main items: sweet rolls, bread, cakes, and pies. The amount of each ingredient (in cups, except for eggs) required for these items is given by matrix A.

	Eggs	Flour	Sugar	Shortening	Milk	
Rolls (doz)	1	4	$\frac{1}{4}$	$\frac{1}{4}$	1	
Bread (loaf)	0	3	0	$\frac{1}{4}$	0	$= A$
Cake	4	3	2	1	1	
Pie (crust)	0	1	0	$\frac{1}{3}$	0	

The cost (in cents) for each ingredient when purchased in large lots or small lots is given by matrix B.

	Cost		
	Large Lot	Small Lot	
Eggs	5	5	
Flour	8	10	
Sugar	10	12	$= B$
Shortening	12	15	
Milk	5	6	

(**a**) Use matrix multiplication to find a matrix giving the comparative cost per bakery item for the two purchase options.

(b) Suppose a day's orders consist of 20 dozen sweet rolls, 200 loaves of bread, 50 cakes, and 60 pies. Write the orders as a 1×4 matrix, and, using matrix multiplication, write as a matrix the amount of each ingredient needed to fill the day's orders.

(c) Use matrix multiplication to find a matrix giving the costs under the two purchase options to fill the day's orders.

87. *(Modeling) Northern Spotted Owl Population*
Mathematical ecologists created a model to analyze population dynamics of the endangered northern spotted owl in the Pacific Northwest. The ecologists divided the female owl population into three categories: juvenile (up to 1 yr old), subadult (1 to 2 yr old), and adult (over 2 yr old). They concluded that the change in the makeup of the northern spotted owl population in successive years could be described by the following matrix equation.

$$\begin{bmatrix} j_{n+1} \\ s_{n+1} \\ a_{n+1} \end{bmatrix} = \begin{bmatrix} 0 & 0 & 0.33 \\ 0.18 & 0 & 0 \\ 0 & 0.71 & 0.94 \end{bmatrix} \begin{bmatrix} j_n \\ s_n \\ a_n \end{bmatrix}$$

The numbers in the column matrices give the numbers of females in the three age groups after n years and $n + 1$ years. Multiplying the matrices yields the following.

$j_{n+1} = 0.33a_n$ Each year 33 juvenile females are born for each 100 adult females.

$s_{n+1} = 0.18j_n$ Each year 18% of the juvenile females survive to become subadults.

$a_{n+1} = 0.71s_n + 0.94a_n$ Each year 71% of the subadults survive to become adults, and 94% of the adults survive.

(Source: Lamberson, R. H., R. McKelvey, B. R. Noon, and C. Voss, "A Dynamic Analysis of Northern Spotted Owl Viability in a Fragmented Forest Landscape," *Conservation Biology,* Vol. 6, No. 4.)

(a) Suppose there are currently 3000 female northern spotted owls made up of 690 juveniles, 210 subadults, and 2100 adults. Use the matrix equation to determine the total number of female owls for each of the next 5 yr.

(b) Using advanced techniques from linear algebra, we can show that in the long run,

$$\begin{bmatrix} j_{n+1} \\ s_{n+1} \\ a_{n+1} \end{bmatrix} \approx 0.98359 \begin{bmatrix} j_n \\ s_n \\ a_n \end{bmatrix}.$$

What can we conclude about the long-term fate of the northern spotted owl?

(c) In the model, the main impediment to the survival of the northern spotted owl is the number 0.18 in the second row of the 3×3 matrix. This number is low for two reasons.

• The first year of life is precarious for most animals living in the wild.

• Juvenile owls must eventually leave the nest and establish their own territory. If much of the forest near their original home has been cleared, then they are vulnerable to predators while searching for a new home.

Suppose that, thanks to better forest management, the number 0.18 can be increased to 0.3. Rework part (a) under this new assumption.

88. *(Modeling) Predator-Prey Relationship* In certain parts of the Rocky Mountains, deer provide the main food source for mountain lions. When the deer population is large, the mountain lions thrive. However, a large mountain lion population reduces the size of the deer population. Suppose the fluctuations of the two populations from year to year can be modeled with the matrix equation

$$\begin{bmatrix} m_{n+1} \\ d_{n+1} \end{bmatrix} = \begin{bmatrix} 0.51 & 0.4 \\ -0.05 & 1.05 \end{bmatrix} \begin{bmatrix} m_n \\ d_n \end{bmatrix}.$$

The numbers in the column matrices give the numbers of animals in the two populations after n years and $n + 1$ years, where the number of deer is measured in hundreds.

(a) Give the equation for d_{n+1} obtained from the second row of the square matrix. Use this equation to determine the rate at which the deer population will grow from year to year if there are no mountain lions.

(b) Suppose we start with a mountain lion population of 2000 and a deer population of 500,000 (that is, 5000 hundred deer). How large would each population be after 1 yr?   2 yr?

(c) Consider part (b) but change the initial mountain lion population to 4000. Show that the populations would both grow at a steady annual rate of 1.01.

89. *Northern Spotted Owl Population* Refer to **Exercise 87(b).** Show that the number 0.98359 is an approximate zero of the polynomial represented by

$$\begin{vmatrix} -x & 0 & 0.33 \\ 0.18 & -x & 0 \\ 0 & 0.71 & 0.94 - x \end{vmatrix}.$$

90. *Predator-Prey Relationship* Refer to **Exercise 88(c).** Show that the number 1.01 is a zero of the polynomial represented by

$$\begin{vmatrix} 0.51 - x & 0.4 \\ -0.05 & 1.05 - x \end{vmatrix}.$$

Use the following matrices, where all elements are real numbers, to show that each statement is true for 2×2 matrices.

$$A = \begin{bmatrix} a_{11} & a_{12} \\ a_{21} & a_{22} \end{bmatrix}, \quad B = \begin{bmatrix} b_{11} & b_{12} \\ b_{21} & b_{22} \end{bmatrix}, \quad and \quad C = \begin{bmatrix} c_{11} & c_{12} \\ c_{21} & c_{22} \end{bmatrix},$$

91. $A + B = B + A$
(commutative property)

92. $A + (B + C) = (A + B) + C$
(associative property)

93. $(AB)C = A(BC)$
(associative property)

94. $A(B + C) = AB + AC$
(distributive property)

95. $c(A + B) = cA + cB$,
for any real number c.

96. $(c + d)A = cA + dA$,
for any real numbers c and d.

97. $(cA)d = (cd)A$, for any real numbers c and d.

98. $(cd)A = c(dA)$, for any real numbers c and d.

5.8 Matrix Inverses

We have seen several parallels between the set of real numbers and the set of matrices. Another similarity is that both sets have identity and inverse elements for multiplication.

Identity Matrices By the identity property for real numbers,

$$a \cdot 1 = a \quad \text{and} \quad 1 \cdot a = a$$

for any real number a. If there is to be a multiplicative **identity matrix** I, such that

$$AI = A \quad \text{and} \quad IA = A,$$

for any matrix A, then A and I must be square matrices of the same dimension.

---

2 × 2 Identity Matrix

I_2 represents the 2×2 identity matrix.

$$I_2 = \begin{bmatrix} 1 & 0 \\ 0 & 1 \end{bmatrix}$$

To verify that I_2 is the 2×2 identity matrix, we must show that $AI = A$ and $IA = A$ for any 2×2 matrix A. Let

$$A = \begin{bmatrix} x & y \\ z & w \end{bmatrix}.$$

Then

$$AI = \begin{bmatrix} x & y \\ z & w \end{bmatrix}\begin{bmatrix} 1 & 0 \\ 0 & 1 \end{bmatrix} = \begin{bmatrix} x \cdot 1 + y \cdot 0 & x \cdot 0 + y \cdot 1 \\ z \cdot 1 + w \cdot 0 & z \cdot 0 + w \cdot 1 \end{bmatrix} = \begin{bmatrix} x & y \\ z & w \end{bmatrix} = A,$$

and

$$IA = \begin{bmatrix} 1 & 0 \\ 0 & 1 \end{bmatrix}\begin{bmatrix} x & y \\ z & w \end{bmatrix} = \begin{bmatrix} 1 \cdot x + 0 \cdot z & 1 \cdot y + 0 \cdot w \\ 0 \cdot x + 1 \cdot z & 0 \cdot y + 1 \cdot w \end{bmatrix} = \begin{bmatrix} x & y \\ z & w \end{bmatrix} = A.$$

Generalizing, there is an $n \times n$ identity matrix for every $n \times n$ square matrix. The $n \times n$ identity matrix has 1s on the main diagonal and 0s elsewhere.

---

n × n Identity Matrix

The $n \times n$ identity matrix is I_n.

$$I_n = \begin{bmatrix} 1 & 0 & \cdots & 0 \\ 0 & 1 & \cdots & 0 \\ \vdots & \vdots & a_{ij} & \vdots \\ 0 & 0 & \cdots & 1 \end{bmatrix}$$

The element $a_{ij} = 1$ when $i = j$ (the diagonal elements), and $a_{ij} = 0$ otherwise.

EXAMPLE 1 **Verifying the Identity Property of I_3**

Let $A = \begin{bmatrix} -2 & 4 & 0 \\ 3 & 5 & 9 \\ 0 & 8 & -6 \end{bmatrix}$. Give the 3×3 identity matrix I_3 and show that $AI_3 = A$.

ALGEBRAIC SOLUTION

The 3×3 identity matrix is

$$I_3 = \begin{bmatrix} 1 & 0 & 0 \\ 0 & 1 & 0 \\ 0 & 0 & 1 \end{bmatrix}.$$

Using matrix multiplication,

$$AI_3 = \begin{bmatrix} -2 & 4 & 0 \\ 3 & 5 & 9 \\ 0 & 8 & -6 \end{bmatrix} \begin{bmatrix} 1 & 0 & 0 \\ 0 & 1 & 0 \\ 0 & 0 & 1 \end{bmatrix}$$

$$= \begin{bmatrix} -2 & 4 & 0 \\ 3 & 5 & 9 \\ 0 & 8 & -6 \end{bmatrix}$$

$$= A.$$

GRAPHING CALCULATOR SOLUTION

The calculator screen in **Figure 27(a)** shows the identity matrix for $n = 3$. The screen in **Figure 27(b)** supports the algebraic result.

(a)   (b)

Figure 27

✔ **Now Try Exercise 7.**

Multiplicative Inverses   For every *nonzero* real number a, there is a multiplicative inverse $\frac{1}{a}$ that satisfies both of the following.

$$a \cdot \frac{1}{a} = 1 \quad \text{and} \quad \frac{1}{a} \cdot a = 1$$

(Recall: $\frac{1}{a}$ is also written a^{-1}.) In a similar way, if A is an $n \times n$ matrix, then its **multiplicative inverse,** written A^{-1}, must satisfy both of the following.

$$AA^{-1} = I_n \quad \text{and} \quad A^{-1}A = I_n$$

This means that only a square matrix can have a multiplicative inverse.

NOTE   Although $a^{-1} = \frac{1}{a}$ for any nonzero real number a, if A is a matrix, then $A^{-1} \neq \frac{1}{A}$. ***We do NOT use the symbol $\frac{1}{A}$ because 1 is a number and A is a matrix.***

To find the matrix A^{-1}, we use row transformations, introduced earlier in this chapter. As an example, we find the inverse of

$$A = \begin{bmatrix} 2 & 4 \\ 1 & -1 \end{bmatrix}.$$

Let the unknown inverse matrix be symbolized as follows.

$$A^{-1} = \begin{bmatrix} x & y \\ z & w \end{bmatrix}$$

By the definition of matrix inverse, $AA^{-1} = I_2$.

$$AA^{-1} = \begin{bmatrix} 2 & 4 \\ 1 & -1 \end{bmatrix} \begin{bmatrix} x & y \\ z & w \end{bmatrix} = \begin{bmatrix} 1 & 0 \\ 0 & 1 \end{bmatrix}$$

$$\begin{bmatrix} 2 & 4 \\ 1 & -1 \end{bmatrix} \begin{bmatrix} x & y \\ z & w \end{bmatrix} = \begin{bmatrix} 1 & 0 \\ 0 & 1 \end{bmatrix}$$

Use matrix multiplication for this product, which is repeated in the margin.

$$\begin{bmatrix} 2x + 4z & 2y + 4w \\ x - z & y - w \end{bmatrix} = \begin{bmatrix} 1 & 0 \\ 0 & 1 \end{bmatrix}$$

Set the corresponding elements equal to obtain a system of equations.

$$2x + 4z = 1 \quad (1)$$
$$2y + 4w = 0 \quad (2)$$
$$x - z = 0 \quad (3)$$
$$y - w = 1 \quad (4)$$

Because equations (1) and (3) involve only x and z, while equations (2) and (4) involve only y and w, these four equations lead to two systems of equations.

$$\begin{array}{cc} 2x + 4z = 1 & 2y + 4w = 0 \\ x - z = 0 & y - w = 1 \end{array} \quad \text{and}$$

Write the two systems as augmented matrices.

$$\begin{bmatrix} 2 & 4 & | & 1 \\ 1 & -1 & | & 0 \end{bmatrix} \quad \text{and} \quad \begin{bmatrix} 2 & 4 & | & 0 \\ 1 & -1 & | & 1 \end{bmatrix}$$

Each of these systems can be solved by the Gauss-Jordan method. However, since the elements to the left of the vertical bar are identical, the two systems can be combined into one matrix.

$$\begin{bmatrix} 2 & 4 & | & 1 \\ 1 & -1 & | & 0 \end{bmatrix} \quad \text{and} \quad \begin{bmatrix} 2 & 4 & | & 0 \\ 1 & -1 & | & 1 \end{bmatrix} \quad \text{yields} \quad \begin{bmatrix} 2 & 4 & | & 1 & 0 \\ 1 & -1 & | & 0 & 1 \end{bmatrix}$$

We can solve simultaneously using matrix row transformations. We need to change the numbers on the left of the vertical bar to the 2×2 identity matrix.

$$\begin{bmatrix} 1 & -1 & | & 0 & 1 \\ 2 & 4 & | & 1 & 0 \end{bmatrix} \qquad \text{Interchange R1 and R2 to introduce 1 in the upper left-hand corner.}$$

$$\begin{bmatrix} 1 & -1 & | & 0 & 1 \\ 0 & 6 & | & 1 & -2 \end{bmatrix} \qquad -2\text{R1} + \text{R2}$$

$$\begin{bmatrix} 1 & -1 & | & 0 & 1 \\ 0 & 1 & | & \frac{1}{6} & -\frac{1}{3} \end{bmatrix} \qquad \frac{1}{6}\text{R2}$$

$$\begin{bmatrix} 1 & 0 & | & \frac{1}{6} & \frac{2}{3} \\ 0 & 1 & | & \frac{1}{6} & -\frac{1}{3} \end{bmatrix} \qquad \text{R2} + \text{R1}$$

The numbers in the first column to the right of the vertical bar in the final matrix give the values of x and z. The second column gives the values of y and w. That is,

$$\begin{bmatrix} 1 & 0 & | & x & y \\ 0 & 1 & | & z & w \end{bmatrix} = \begin{bmatrix} 1 & 0 & | & \frac{1}{6} & \frac{2}{3} \\ 0 & 1 & | & \frac{1}{6} & -\frac{1}{3} \end{bmatrix}$$

so that

$$A^{-1} = \begin{bmatrix} x & y \\ z & w \end{bmatrix} = \begin{bmatrix} \frac{1}{6} & \frac{2}{3} \\ \frac{1}{6} & -\frac{1}{3} \end{bmatrix}.$$

CHECK Multiply A by A^{-1}. The result should be I_2.

$$AA^{-1} = \begin{bmatrix} 2 & 4 \\ 1 & -1 \end{bmatrix} \begin{bmatrix} \frac{1}{6} & \frac{2}{3} \\ \frac{1}{6} & -\frac{1}{3} \end{bmatrix}$$

$$= \begin{bmatrix} \frac{1}{3} + \frac{2}{3} & \frac{4}{3} - \frac{4}{3} \\ \frac{1}{6} - \frac{1}{6} & \frac{2}{3} + \frac{1}{3} \end{bmatrix}$$

$$= \begin{bmatrix} 1 & 0 \\ 0 & 1 \end{bmatrix}$$

$$= I_2 \checkmark$$

Thus, $\qquad\qquad A^{-1} = \begin{bmatrix} \frac{1}{6} & \frac{2}{3} \\ \frac{1}{6} & -\frac{1}{3} \end{bmatrix}$.

This process is summarized below.

Finding an Inverse Matrix

To obtain A^{-1} for any $n \times n$ matrix A for which A^{-1} exists, follow these steps.

Step 1 Form the augmented matrix $[A \,|\, I_n]$, where I_n is the $n \times n$ identity matrix.

Step 2 Perform row transformations on $[A \,|\, I_n]$ to obtain a matrix of the form $[I_n \,|\, B]$.

Step 3 Matrix B is A^{-1}.

NOTE To confirm that two $n \times n$ matrices A and B are inverses of each other, it is sufficient to show that $AB = I_n$. It is not necessary to show also that $BA = I_n$.

As illustrated by the examples, the most efficient order for the transformations in Step 2 is to make the changes column by column from left to right, so that for each column the required 1 is the result of the first change. Next, perform the steps that obtain 0s in that column. Then proceed to the next column.

EXAMPLE 2 Finding the Inverse of a 3 × 3 Matrix

Find A^{-1} if $A = \begin{bmatrix} 1 & 0 & 1 \\ 2 & -2 & -1 \\ 3 & 0 & 0 \end{bmatrix}$.

SOLUTION Use row transformations as follows.

Step 1 Write the augmented matrix $[A \,|\, I_3]$.

$$\left[\begin{array}{ccc|ccc} 1 & 0 & 1 & 1 & 0 & 0 \\ 2 & -2 & -1 & 0 & 1 & 0 \\ 3 & 0 & 0 & 0 & 0 & 1 \end{array} \right]$$

Step 2 There is already a 1 in the upper left-hand corner as desired. Begin by using the row transformation that will result in 0 for the first element in the second row. Multiply the elements of the first row by -2, and add the result to the second row.

$$\begin{bmatrix} 1 & 0 & 1 & | & 1 & 0 & 0 \\ 0 & -2 & -3 & | & -2 & 1 & 0 \\ 3 & 0 & 0 & | & 0 & 0 & 1 \end{bmatrix} \quad -2R1 + R2$$

We introduce 0 for the first element in the third row by multiplying the elements of the first row by -3 and adding to the third row.

$$\begin{bmatrix} 1 & 0 & 1 & | & 1 & 0 & 0 \\ 0 & -2 & -3 & | & -2 & 1 & 0 \\ 0 & 0 & -3 & | & -3 & 0 & 1 \end{bmatrix} \quad -3R1 + R3$$

To obtain 1 for the second element in the second row, multiply the elements of the second row by $-\frac{1}{2}$.

$$\begin{bmatrix} 1 & 0 & 1 & | & 1 & 0 & 0 \\ 0 & 1 & \frac{3}{2} & | & 1 & -\frac{1}{2} & 0 \\ 0 & 0 & -3 & | & -3 & 0 & 1 \end{bmatrix} \quad -\frac{1}{2}R2$$

We want 1 for the third element in the third row, so multiply the elements of the third row by $-\frac{1}{3}$.

$$\begin{bmatrix} 1 & 0 & 1 & | & 1 & 0 & 0 \\ 0 & 1 & \frac{3}{2} & | & 1 & -\frac{1}{2} & 0 \\ 0 & 0 & 1 & | & 1 & 0 & -\frac{1}{3} \end{bmatrix} \quad -\frac{1}{3}R3$$

The third element in the first row should be 0, so multiply the elements of the third row by -1 and add to the first row.

$$\begin{bmatrix} 1 & 0 & 0 & | & 0 & 0 & \frac{1}{3} \\ 0 & 1 & \frac{3}{2} & | & 1 & -\frac{1}{2} & 0 \\ 0 & 0 & 1 & | & 1 & 0 & -\frac{1}{3} \end{bmatrix} \quad -1R3 + R1$$

Finally, to introduce 0 as the third element in the second row, multiply the elements of the third row by $-\frac{3}{2}$ and add to the second row.

$$\begin{bmatrix} 1 & 0 & 0 & | & 0 & 0 & \frac{1}{3} \\ 0 & 1 & 0 & | & -\frac{1}{2} & -\frac{1}{2} & \frac{1}{2} \\ 0 & 0 & 1 & | & 1 & 0 & -\frac{1}{3} \end{bmatrix} \quad -\frac{3}{2}R3 + R2$$

A graphing calculator can be used to find the inverse of a matrix. This screen supports the result in **Example 2.** The elements of the inverse are expressed as fractions, so it is easier to compare with the inverse matrix found in the example.

Step 3 The last transformation shows the inverse.

$$A^{-1} = \begin{bmatrix} 0 & 0 & \frac{1}{3} \\ -\frac{1}{2} & -\frac{1}{2} & \frac{1}{2} \\ 1 & 0 & -\frac{1}{3} \end{bmatrix}$$

Confirm this by forming the product $A^{-1}A$ or AA^{-1}, each of which should equal the matrix I_3.

☑ **Now Try Exercises 17 and 25.**

If the inverse of a matrix exists, it is unique. That is, any given square matrix has no more than one inverse. The proof of this is left as **Exercise 69.**

EXAMPLE 3 Identifying a Matrix with No Inverse

Find A^{-1}, if possible, given that $A = \begin{bmatrix} 2 & -4 \\ 1 & -2 \end{bmatrix}$.

ALGEBRAIC SOLUTION

Using row transformations to change the first column of the augmented matrix

$$\begin{bmatrix} 2 & -4 & | & 1 & 0 \\ 1 & -2 & | & 0 & 1 \end{bmatrix}$$

results in the following matrices.

$$\begin{bmatrix} 1 & -2 & | & \frac{1}{2} & 0 \\ 1 & -2 & | & 0 & 1 \end{bmatrix} \quad \frac{1}{2}R1$$

$$\begin{bmatrix} 1 & -2 & | & \frac{1}{2} & 0 \\ 0 & 0 & | & -\frac{1}{2} & 1 \end{bmatrix} \quad -R1 + R2$$

At this point, the matrix should be changed so that the second row, second element will be 1. Because that element is now 0, there is no way to complete the desired transformation, so A^{-1} does not exist for this matrix A.

Just as there is no multiplicative inverse for the real number 0, not every matrix has a multiplicative inverse. Matrix A is an example of such a matrix.

GRAPHING CALCULATOR SOLUTION

If the inverse of a matrix does not exist, the matrix is called **singular,** as shown in **Figure 28** for matrix [A]. This occurs when the determinant of the matrix is 0.

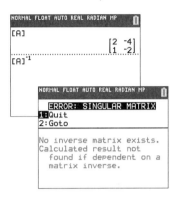

Figure 28

✔ **Now Try Exercise 21.**

Solution of Systems Using Inverse Matrices Matrix inverses can be used to solve *square linear systems of equations.* (A square system has the same number of equations as variables.) For example, consider the following linear system of three equations with three variables.

$$a_{11}x + a_{12}y + a_{13}z = b_1$$
$$a_{21}x + a_{22}y + a_{23}z = b_2$$
$$a_{31}x + a_{32}y + a_{33}z = b_3$$

The definition of matrix multiplication can be used to rewrite the system using matrices.

$$\begin{bmatrix} a_{11} & a_{12} & a_{13} \\ a_{21} & a_{22} & a_{23} \\ a_{31} & a_{32} & a_{33} \end{bmatrix} \cdot \begin{bmatrix} x \\ y \\ z \end{bmatrix} = \begin{bmatrix} b_1 \\ b_2 \\ b_3 \end{bmatrix} \quad (1)$$

(To see this, multiply the matrices on the left.)

If $A = \begin{bmatrix} a_{11} & a_{12} & a_{13} \\ a_{21} & a_{22} & a_{23} \\ a_{31} & a_{32} & a_{33} \end{bmatrix}$, $X = \begin{bmatrix} x \\ y \\ z \end{bmatrix}$, and $B = \begin{bmatrix} b_1 \\ b_2 \\ b_3 \end{bmatrix}$,

then the system given in (1) becomes $AX = B$. If A^{-1} exists, then each side of $AX = B$ can be multiplied on the left as shown on the next page.

$$A^{-1}(AX) = A^{-1}B \quad \text{Multiply each side on the left by } A^{-1}.$$
$$(A^{-1}A)X = A^{-1}B \quad \text{Associative property}$$
$$I_3X = A^{-1}B \quad \text{Inverse property}$$
$$X = A^{-1}B \quad \text{Identity property}$$

Matrix $A^{-1}B$ gives the solution of the system.

Solution of the Matrix Equation $AX = B$

Suppose A is an $n \times n$ matrix with inverse A^{-1}, X is an $n \times 1$ matrix of variables, and B is an $n \times 1$ matrix. The matrix equation

$$AX = B$$

has the solution $\qquad X = A^{-1}B.$

This method of using matrix inverses to solve systems of equations is useful when the inverse is already known or when many systems of the form $AX = B$ must be solved and only B changes.

EXAMPLE 4 **Solving Systems of Equations Using Matrix Inverses**

Solve each system using the inverse of the coefficient matrix.

(a) $2x - 3y = 4$
$ x + 5y = 2$

(b) $ x + z = -1$
$ 2x - 2y - z = 5$
$ 3x = 6$

ALGEBRAIC SOLUTION

(a) The system can be written in matrix form as

$$\begin{bmatrix} 2 & -3 \\ 1 & 5 \end{bmatrix} \begin{bmatrix} x \\ y \end{bmatrix} = \begin{bmatrix} 4 \\ 2 \end{bmatrix},$$

where $\quad A = \begin{bmatrix} 2 & -3 \\ 1 & 5 \end{bmatrix}, \quad X = \begin{bmatrix} x \\ y \end{bmatrix}, \quad$ and $\quad B = \begin{bmatrix} 4 \\ 2 \end{bmatrix}.$

An equivalent matrix equation is $AX = B$ with solution $X = A^{-1}B$. Use the methods described in this section to determine that

$$A^{-1} = \begin{bmatrix} \frac{5}{13} & \frac{3}{13} \\ -\frac{1}{13} & \frac{2}{13} \end{bmatrix},$$

and thus $A^{-1}B = \begin{bmatrix} \frac{5}{13} & \frac{3}{13} \\ -\frac{1}{13} & \frac{2}{13} \end{bmatrix} \begin{bmatrix} 4 \\ 2 \end{bmatrix} = \begin{bmatrix} 2 \\ 0 \end{bmatrix}.$

$X = A^{-1}B$, so

$$X = \begin{bmatrix} x \\ y \end{bmatrix} = \begin{bmatrix} 2 \\ 0 \end{bmatrix}.$$

The final matrix shows that the solution set of the system is $\{(2, 0)\}$.

GRAPHING CALCULATOR SOLUTION

(a) Enter $[A]$ and $[B]$ as defined in the algebraic solution, and then find the product $[A]^{-1}[B]$ as shown in **Figure 29.** The display indicates that the solution set is $\{(2, 0)\}$.

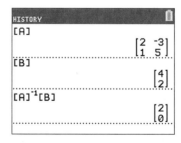

Figure 29

Notice that it is not necessary to actually compute $[A]^{-1}$ here. The calculator stores this inverse and then multiplies it by $[B]$ to obtain the column matrix that represents the solution.

(b) The coefficient matrix A for this system is

$$A = \begin{bmatrix} 1 & 0 & 1 \\ 2 & -2 & -1 \\ 3 & 0 & 0 \end{bmatrix},$$

and its inverse A^{-1} was found in **Example 2.** Let

$$X = \begin{bmatrix} x \\ y \\ z \end{bmatrix} \quad \text{and} \quad B = \begin{bmatrix} -1 \\ 5 \\ 6 \end{bmatrix}.$$

Because $X = A^{-1}B$, we have

$$\begin{bmatrix} x \\ y \\ z \end{bmatrix} = \underbrace{\begin{bmatrix} 0 & 0 & \frac{1}{3} \\ -\frac{1}{2} & -\frac{1}{2} & \frac{1}{2} \\ 1 & 0 & -\frac{1}{3} \end{bmatrix}}_{A^{-1} \text{ from } \textbf{Example 2}} \begin{bmatrix} -1 \\ 5 \\ 6 \end{bmatrix}$$

$$= \begin{bmatrix} 2 \\ 1 \\ -3 \end{bmatrix}.$$

The solution set is $\{(2, 1, -3)\}$.

(b) Figure 30 shows the coefficient matrix $[A]$ and the column matrix of constants $[B]$. Be sure to enter the product of $[A]^{-1}$ and $[B]$ in the correct order. Remember that matrix multiplication is not commutative.

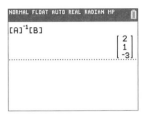

Figure 30

✔ **Now Try Exercises 35 and 49.**

5.8 Exercises

CONCEPT PREVIEW *Answer each question.*

1. What is the product of $\begin{bmatrix} 6 & 4 \\ -1 & 8 \end{bmatrix}$ and I_2 (in either order)?

2. What is the product $\begin{bmatrix} -5 & 7 & 4 \\ 2 & 3 & 0 \\ -1 & 6 & 6 \end{bmatrix} \begin{bmatrix} 1 & 0 & 0 \\ 0 & 1 & 0 \\ 0 & 0 & 1 \end{bmatrix}$?

3. It can be shown that the following matrices are inverses. What is their product (in either order)?

$$\begin{bmatrix} 5 & 1 \\ 4 & 1 \end{bmatrix} \quad \text{and} \quad \begin{bmatrix} 1 & -1 \\ -4 & 5 \end{bmatrix}$$

4. It can be shown that the following matrices are inverses. What is their product (in either order)?

$$\begin{bmatrix} 1 & 0 & 0 \\ 0 & -1 & 0 \\ -1 & 0 & 1 \end{bmatrix} \quad \text{and} \quad \begin{bmatrix} 1 & 0 & 0 \\ 0 & -1 & 0 \\ 1 & 0 & 1 \end{bmatrix}$$

5. What is the coefficient matrix of the following system?

$$3x - 6y = 8$$
$$-x + 3y = 4$$

6. What is the matrix equation form of the following system?

$$6x + 3y = 9$$
$$5x - y = 4$$

Provide a proof for each of the following.

7. Show that $I_3 A = A$ for

$$A = \begin{bmatrix} -2 & 4 & 0 \\ 3 & 5 & 9 \\ 0 & 8 & -6 \end{bmatrix} \quad \text{and} \quad I_3 = \begin{bmatrix} 1 & 0 & 0 \\ 0 & 1 & 0 \\ 0 & 0 & 1 \end{bmatrix}.$$

(This result, along with that of **Example 1,** illustrates that the commutative property holds when one of the matrices is an identity matrix.)

8. Let $A = \begin{bmatrix} a & b \\ c & d \end{bmatrix}$ and $I_2 = \begin{bmatrix} 1 & 0 \\ 0 & 1 \end{bmatrix}$. Show that $AI_2 = I_2 A = A$, thus proving that I_2 is the identity element for matrix multiplication for 2×2 square matrices.

Decide whether or not the given matrices are inverses of each other. (Hint: Check to see whether their products are the identity matrix I_n.)

9. $\begin{bmatrix} 5 & 7 \\ 2 & 3 \end{bmatrix}$ and $\begin{bmatrix} 3 & -7 \\ -2 & 5 \end{bmatrix}$

10. $\begin{bmatrix} 2 & 3 \\ 1 & 1 \end{bmatrix}$ and $\begin{bmatrix} -1 & 3 \\ 1 & -2 \end{bmatrix}$

11. $\begin{bmatrix} -1 & 2 \\ 3 & -5 \end{bmatrix}$ and $\begin{bmatrix} -5 & -2 \\ -3 & -1 \end{bmatrix}$

12. $\begin{bmatrix} 2 & 1 \\ 3 & 2 \end{bmatrix}$ and $\begin{bmatrix} 2 & 1 \\ -3 & 2 \end{bmatrix}$

13. $\begin{bmatrix} 0 & 1 & 0 \\ 0 & 0 & -2 \\ 1 & -1 & 0 \end{bmatrix}$ and $\begin{bmatrix} 1 & 0 & 1 \\ 1 & 0 & 0 \\ 0 & -1 & 0 \end{bmatrix}$

14. $\begin{bmatrix} 1 & 2 & 0 \\ 0 & 1 & 0 \\ 0 & 1 & 0 \end{bmatrix}$ and $\begin{bmatrix} 1 & -2 & 0 \\ 0 & 1 & 0 \\ 0 & -1 & 1 \end{bmatrix}$

15. $\begin{bmatrix} 1 & 0 & 0 \\ 0 & -1 & 0 \\ 1 & 0 & 1 \end{bmatrix}$ and $\begin{bmatrix} 1 & 0 & 0 \\ 0 & -1 & 0 \\ -1 & 0 & 1 \end{bmatrix}$

16. $\begin{bmatrix} 1 & 3 & 3 \\ 1 & 4 & 3 \\ 1 & 3 & 4 \end{bmatrix}$ and $\begin{bmatrix} 7 & -3 & -3 \\ -1 & 1 & 0 \\ -1 & 0 & 1 \end{bmatrix}$

Find the inverse, if it exists, for each matrix. **See Examples 2 and 3.**

17. $\begin{bmatrix} -1 & 2 \\ -2 & -1 \end{bmatrix}$

18. $\begin{bmatrix} 1 & -1 \\ 2 & 0 \end{bmatrix}$

19. $\begin{bmatrix} -1 & -2 \\ 3 & 4 \end{bmatrix}$

20. $\begin{bmatrix} 3 & -1 \\ -5 & 2 \end{bmatrix}$

21. $\begin{bmatrix} 5 & 10 \\ -3 & -6 \end{bmatrix}$

22. $\begin{bmatrix} -6 & 4 \\ -3 & 2 \end{bmatrix}$

23. $\begin{bmatrix} 1 & 0 & 1 \\ 0 & -1 & 0 \\ 2 & 1 & 1 \end{bmatrix}$

24. $\begin{bmatrix} 1 & 0 & 0 \\ 0 & -1 & 0 \\ 1 & 0 & 1 \end{bmatrix}$

25. $\begin{bmatrix} 2 & 3 & 3 \\ 1 & 4 & 3 \\ 1 & 3 & 4 \end{bmatrix}$

26. $\begin{bmatrix} -2 & 2 & 4 \\ -3 & 4 & 5 \\ 1 & 0 & 2 \end{bmatrix}$

27. $\begin{bmatrix} 2 & 2 & -4 \\ 2 & 6 & 0 \\ -3 & -3 & 5 \end{bmatrix}$

28. $\begin{bmatrix} 2 & 4 & 6 \\ -1 & -4 & -3 \\ 0 & 1 & -1 \end{bmatrix}$

29. $\begin{bmatrix} 1 & 1 & 0 & 2 \\ 2 & -1 & 1 & -1 \\ 3 & 3 & 2 & -2 \\ 1 & 2 & 1 & 0 \end{bmatrix}$

30. $\begin{bmatrix} 1 & -2 & 3 & 0 \\ 0 & 1 & -1 & 1 \\ -2 & 2 & -2 & 4 \\ 0 & 2 & -3 & 1 \end{bmatrix}$

31. $\begin{bmatrix} 3 & 2 & 0 & -1 \\ 2 & 0 & 1 & 2 \\ 1 & 2 & -1 & 0 \\ 2 & -1 & 1 & 1 \end{bmatrix}$

32. *Concept Check* Each graphing calculator screen shows A^{-1} for some matrix A. Find each matrix A. (*Hint:* $(A^{-1})^{-1} = A$)

(a)

(b)

Solve each system using the inverse of the coefficient matrix. **See Example 4.**

33. $-x + y = 1$
$2x - y = 1$

34. $x + y = 5$
$x - y = -1$

35. $2x - y = -8$
$3x + y = -2$

36. $x + 3y = -12$
$2x - y = 11$

37. $3x + 4y = -3$
$-5x + 8y = 16$

38. $2x - 3y = 10$
$2x + 2y = 5$

39. $6x + 9y = 3$
$-8x + 3y = 6$

40. $5x - 3y = 0$
$10x + 6y = -4$

41. $0.2x + 0.3y = -1.9$
$0.7x - 0.2y = 4.6$

42. $0.5x + 0.2y = 0.8$
$0.3x - 0.1y = 0.7$

43. $\dfrac{1}{2}x + \dfrac{1}{3}y = \dfrac{49}{18}$
$\dfrac{1}{2}x + 2y = \dfrac{4}{3}$

44. $\dfrac{1}{5}x + \dfrac{1}{7}y = \dfrac{12}{5}$
$\dfrac{1}{10}x + \dfrac{1}{3}y = \dfrac{5}{6}$

Concept Check *Show that the matrix inverse method cannot be used to solve each system.*

45. $7x - 2y = 3$
$14x - 4y = 1$

46. $x - 2y + 3z = 4$
$2x - 4y + 6z = 8$
$3x - 6y + 9z = 14$

Solve each system by using the inverse of the coefficient matrix. For Exercises 49–54, the inverses were found in **Exercises 25–30. See Example 4.**

47. $x + y + z = 6$
$2x + 3y - z = 7$
$3x - y - z = 6$

48. $2x + 5y + 2z = 9$
$4x - 7y - 3z = 7$
$3x - 8y - 2z = 9$

49. $2x + 3y + 3z = 1$
$x + 4y + 3z = 0$
$x + 3y + 4z = -1$

50. $-2x + 2y + 4z = 3$
$-3x + 4y + 5z = 1$
$x + 2z = 2$

51. $2x + 2y - 4z = 12$
$2x + 6y = 16$
$-3x - 3y + 5z = -20$

52. $2x + 4y + 6z = 4$
$-x - 4y - 3z = 8$
$y - z = -4$

53. $x + y + 2w = 3$
$2x - y + z - w = 3$
$3x + 3y + 2z - 2w = 5$
$x + 2y + z = 3$

54. $x - 2y + 3z = 1$
$y - z + w = -1$
$-2x + 2y - 2z + 4w = 2$
$2y - 3z + w = -3$

(Modeling) Solve each problem.

55. *Plate-Glass Sales* The amount of plate-glass sales S (in millions of dollars) can be affected by the number of new building contracts B issued (in millions) and automobiles A produced (in millions). A plate-glass company in California wants to forecast future sales using the past three years of sales. The totals for the three years are given in the table.

S	A	B
602.7	5.543	37.14
656.7	6.933	41.30
778.5	7.638	45.62

To describe the relationship among these variables, we can use the equation

$$S = a + bA + cB,$$

where the coefficients a, b, and c are constants that must be determined before the equation can be used. (*Source:* Makridakis, S., and S. Wheelwright, *Forecasting Methods for Management*, John Wiley and Sons.)

(a) Substitute the values for S, A, and B for each year from the table into the equation $S = a + bA + cB$, and obtain three linear equations involving a, b, and c.

(b) Use a graphing calculator to solve this linear system for a, b, and c. Use matrix inverse methods.

(c) Write the equation for S using these values for the coefficients.

(d) For the next year it is estimated that $A = 7.752$ and $B = 47.38$. Predict S. (The actual value for S was 877.6.)

(e) It is predicted that in 6 yr, $A = 8.9$ and $B = 66.25$. Find the value of S in this situation and discuss its validity.

56. *Tire Sales* The number of automobile tire sales is dependent on several variables. In one study the relationship among annual tire sales S (in thousands of dollars), automobile registrations R (in millions), and personal disposable income I (in millions of dollars) was investigated. The results for three years are given in the table.

S	R	I
10,170	112.9	307.5
15,305	132.9	621.63
21,289	155.2	1937.13

To describe the relationship among these variables, we can use the equation

$$S = a + bR + cI,$$

where the coefficients a, b, and c are constants that must be determined before the equation can be used. (*Source:* Jarrett, J., *Business Forecasting Methods*, Basil Blackwell, Ltd.)

(a) Substitute the values for S, R, and I for each year from the table into the equation $S = a + bR + cI$, and obtain three linear equations involving a, b, and c.

(b) Use a graphing calculator to solve this linear system for a, b, and c. Use matrix inverse methods.

(c) Write the equation for S using these values for the coefficients.

(d) If $R = 117.6$ and $I = 310.73$, predict S. (The actual value for S was 11,314.)

(e) If $R = 143.8$ and $I = 829.06$, predict S. (The actual value for S was 18,481.)

57. **Social Security Numbers** It is possible to find a polynomial that goes through a given set of points in the plane by using a process called **polynomial interpolation.** Recall that three points define a second-degree polynomial, four points define a third-degree polynomial, and so on. The only restriction on the points, because polynomials define functions, is that no two distinct points can have the same x-coordinate.

Using the SSN 539-58-0954, we can find an eighth-degree polynomial that lies on the nine points with x-coordinates 1 through 9 and y-coordinates that are digits of the SSN: $(1, 5)$, $(2, 3)$, $(3, 9)$, $\ldots$, $(9, 4)$. This is done by writing a system of nine equations with nine variables, which is then solved by the inverse matrix method. The graph of this polynomial is shown. Find such a polynomial using your own SSN.

58. Repeat **Exercise 57** but use $-1, -2, \ldots, -9$ for the x-coordinates.

Use a graphing calculator to find the inverse of each matrix. Give as many decimal places as the calculator shows. See Example 2.

59. $\begin{bmatrix} \frac{2}{3} & 0.7 \\ 22 & \sqrt{3} \end{bmatrix}$
60. $\begin{bmatrix} \sqrt{2} & 0.5 \\ -17 & \frac{1}{2} \end{bmatrix}$
61. $\begin{bmatrix} \frac{1}{2} & \frac{1}{4} & \frac{1}{3} \\ 0 & \frac{1}{4} & \frac{1}{3} \\ \frac{1}{2} & \frac{1}{2} & \frac{1}{3} \end{bmatrix}$
62. $\begin{bmatrix} 1.4 & 0.5 & 0.59 \\ 0.84 & 1.36 & 0.62 \\ 0.56 & 0.47 & 1.3 \end{bmatrix}$

Use a graphing calculator and the method of matrix inverses to solve each system. Give as many decimal places as the calculator shows. See Example 4.

63. $2.1x + y = \sqrt{5}$
    $\sqrt{2}x - 2y = 5$

64. $x - \sqrt{2}y = 2.6$
    $0.75x + y = -7$

65. $(\log 2)x + (\ln 3)y + (\ln 4)z = 1$
    $(\ln 3)x + (\log 2)y + (\ln 8)z = 5$
    $(\log 12)x + (\ln 4)y + (\ln 8)z = 9$

66. $\pi x + ey + \sqrt{2}z = 1$
    $ex + \pi y + \sqrt{2}z = 2$
    $\sqrt{2}x + ey + \pi z = 3$

Let $A = \begin{bmatrix} a & b \\ c & d \end{bmatrix}$, and let O be the 2×2 zero matrix. Show that each statement is true.

67. $A \cdot O = O \cdot A = O$

68. For square matrices A and B of the same dimension, if $AB = O$ and if A^{-1} exists, then $B = O$.

Work each problem.

69. Prove that any square matrix has no more than one inverse.

70. Give an example of two matrices A and B, where $(AB)^{-1} \neq A^{-1}B^{-1}$.

71. Suppose A and B are matrices, where A^{-1}, B^{-1}, and AB all exist. Show that $(AB)^{-1} = B^{-1}A^{-1}$.

72. Let $A = \begin{bmatrix} a & 0 & 0 \\ 0 & b & 0 \\ 0 & 0 & c \end{bmatrix}$, where a, b, and c are nonzero real numbers. Find A^{-1}.

73. Let $A = \begin{bmatrix} 1 & 0 & 0 \\ 0 & 0 & -1 \\ 0 & 1 & -1 \end{bmatrix}$. Show that $A^3 = I_3$, and use this result to find the inverse of A.

74. What are the inverses of I_n, $-A$ (in terms of A), and kA (k a scalar)?

Chapter 5 Test Prep

Key Terms

5.1 linear equation (first-degree equation) in *n* unknowns
system of equations
solutions of a system of equations
system of linear equations (linear system)
consistent system
independent equations
inconsistent system
dependent equations
equivalent system
ordered triple

5.2 matrix (matrices)
element (of a matrix)
augmented matrix
dimension (of a matrix)

5.3 determinant
minor
cofactor
expansion by a row or column
Cramer's rule

5.4 partial fraction decomposition
partial fraction

5.5 nonlinear system
5.6 half-plane
boundary
linear inequality in two variables
system of inequalities
linear programming
constraints
objective function
region of feasible solutions
vertex (corner point)

5.7 square matrix of order *n*
row matrix
column matrix
zero matrix
additive inverse (negative) of a matrix
scalar
5.8 identity matrix
multiplicative inverse (of a matrix)

New Symbols

(a, b, c)	ordered triple
$[A]$	matrix A (graphing calculator symbolism)
$\lvert A \rvert$	determinant of matrix A
a_{ij}	element in row i, column j of a matrix

D, D_x, D_y	determinants used in Cramer's rule
I_2, I_3	identity matrices
A^{-1}	multiplicative inverse of matrix A

Quick Review

Concepts	Examples

5.1 Systems of Linear Equations

Transformations of a Linear System

1. Interchange any two equations of the system.

2. Multiply or divide any equation of the system by a non-zero real number.

3. Replace any equation of the system by the sum of that equation and a multiple of another equation in the system.

A system may be solved by the substitution method, the elimination method, or a combination of the two methods.

Substitution Method
Use one equation to find an expression for one variable in terms of the other, and then substitute this expression into the other equation of the system.

Solve the system.

$$4x - y = 7 \quad (1)$$
$$3x + 2y = 30 \quad (2)$$

Solve for y in equation (1).

$$y = 4x - 7 \quad (3)$$

Substitute $4x - 7$ for y in equation (2), and solve for x.

$$3x + 2(4x - 7) = 30 \quad \text{(2) with } y = 4x - 7$$
$$3x + 8x - 14 = 30 \quad \text{Distributive property}$$
$$11x - 14 = 30 \quad \text{Combine like terms.}$$
$$11x = 44 \quad \text{Add 14.}$$
$$x = 4 \quad \text{Divide by 11.}$$

Substitute 4 for x in the equation $y = 4x - 7$ to find that $y = 9$. The solution set is $\{(4, 9)\}$.

Concepts	Examples

Elimination Method

Use multiplication and addition to eliminate a variable from one equation. To eliminate a variable, the coefficients of that variable in the equations must be additive inverses.

Solve the system.

$$x + 2y - z = 6 \quad (1)$$
$$x + y + z = 6 \quad (2)$$
$$2x + y - z = 7 \quad (3)$$

Add equations (1) and (2). The variable z is eliminated, and the result is $2x + 3y = 12$.

Eliminate z again by adding equations (2) and (3) to obtain $3x + 2y = 13$. Solve the resulting system.

Solving a Linear System with Three Unknowns

Step 1 Eliminate a variable from any two of the equations.

Step 2 Eliminate the ***same variable*** from a different pair of equations.

Step 3 Eliminate a second variable using the resulting two equations in two variables to obtain an equation with just one variable whose value we can now determine.

Step 4 Find the values of the remaining variables by substitution. Write the solution of the system as an ordered triple.

$$2x + 3y = 12 \quad (4)$$
$$3x + 2y = 13 \quad (5)$$

$$-6x - 9y = -36 \quad \text{Multiply (4) by } -3.$$
$$\underline{6x + 4y = 26} \quad \text{Multiply (5) by 2.}$$
$$-5y = -10 \quad \text{Add.}$$
$$y = 2 \quad \text{Divide by } -5.$$

Substitute 2 for y in equation (4).

$$2x + 3(2) = 12 \quad \text{(4) with } y = 2$$
$$2x + 6 = 12 \quad \text{Multiply.}$$
$$2x = 6 \quad \text{Subtract 6.}$$
$$x = 3 \quad \text{Divide by 2.}$$

Let $y = 2$ and $x = 3$ in any of the original equations to find $z = 1$. The solution set is $\{(3, 2, 1)\}$.

5.2 Matrix Solution of Linear Systems

Matrix Row Transformations

For any augmented matrix of a system of linear equations, the following row transformations will result in the matrix of an equivalent system.

1. Interchange any two rows.

2. Multiply or divide the elements of any row by a nonzero real number.

3. Replace any row of the matrix by the sum of the elements of that row and a multiple of the elements of another row.

Solve the system.

$$x + 3y = 7$$
$$2x + y = 4$$

$$\begin{bmatrix} 1 & 3 & | & 7 \\ 2 & 1 & | & 4 \end{bmatrix} \quad \text{Augmented matrix}$$

$$\begin{bmatrix} 1 & 3 & | & 7 \\ 0 & -5 & | & -10 \end{bmatrix} \quad -2R1 + R2$$

$$\begin{bmatrix} 1 & 3 & | & 7 \\ 0 & 1 & | & 2 \end{bmatrix} \quad -\frac{1}{5}R2$$

$$\begin{bmatrix} 1 & 0 & | & 1 \\ 0 & 1 & | & 2 \end{bmatrix} \quad -3R2 + R1$$

Gauss-Jordan Method

The Gauss-Jordan method is a systematic technique for applying matrix row transformations in an attempt to reduce a matrix to diagonal form, with 1s along the diagonal.

This leads to the system

$$x = 1$$
$$y = 2.$$

The solution set is $\{(1, 2)\}$.

Concepts	Examples

5.3 Determinant Solution of Linear Systems

Determinant of a 2×2 Matrix

If $A = \begin{bmatrix} a_{11} & a_{12} \\ a_{21} & a_{22} \end{bmatrix}$, then

$$|A| = \begin{vmatrix} a_{11} & a_{12} \\ a_{21} & a_{22} \end{vmatrix} = a_{11}a_{22} - a_{21}a_{12}.$$

Evaluate.

$$\begin{vmatrix} 3 & 5 \\ -2 & 6 \end{vmatrix} = 3(6) - (-2)5$$
$$= 28$$

Determinant of a 3×3 Matrix

If $A = \begin{bmatrix} a_{11} & a_{12} & a_{13} \\ a_{21} & a_{22} & a_{23} \\ a_{31} & a_{32} & a_{33} \end{bmatrix}$, then

$$|A| = \begin{vmatrix} a_{11} & a_{12} & a_{13} \\ a_{21} & a_{22} & a_{23} \\ a_{31} & a_{32} & a_{33} \end{vmatrix} = (a_{11}a_{22}a_{33} + a_{12}a_{23}a_{31} + a_{13}a_{21}a_{32})$$
$$- (a_{31}a_{22}a_{13} + a_{32}a_{23}a_{11} + a_{33}a_{21}a_{12})$$

In practice, we usually evaluate determinants by expansion by minors.

Evaluate by expanding about the second column.

$$\begin{vmatrix} 2 & -3 & -2 \\ -1 & -4 & -3 \\ -1 & 0 & 2 \end{vmatrix} = -(-3)\begin{vmatrix} -1 & -3 \\ -1 & 2 \end{vmatrix} + (-4)\begin{vmatrix} 2 & -2 \\ -1 & 2 \end{vmatrix}$$
$$- 0\begin{vmatrix} 2 & -2 \\ -1 & -3 \end{vmatrix}$$
$$= 3(-5) - 4(2) - 0(-8)$$
$$= -15 - 8 + 0$$
$$= -23$$

Cramer's Rule for Two Equations in Two Variables
Given the system

$$a_1x + b_1y = c_1$$
$$a_2x + b_2y = c_2,$$

if $D \neq 0$, then the system has the unique solution

$$x = \frac{D_x}{D} \quad \text{and} \quad y = \frac{D_y}{D},$$

where $D = \begin{vmatrix} a_1 & b_1 \\ a_2 & b_2 \end{vmatrix}$, $D_x = \begin{vmatrix} c_1 & b_1 \\ c_2 & b_2 \end{vmatrix}$, and $D_y = \begin{vmatrix} a_1 & c_1 \\ a_2 & c_2 \end{vmatrix}$.

Solve using Cramer's rule.

$$x - 2y = -1$$
$$2x + 5y = 16$$

$$x = \frac{D_x}{D} = \frac{\begin{vmatrix} -1 & -2 \\ 16 & 5 \end{vmatrix}}{\begin{vmatrix} 1 & -2 \\ 2 & 5 \end{vmatrix}} = \frac{-5 + 32}{5 + 4} = \frac{27}{9} = 3$$

$$y = \frac{D_y}{D} = \frac{\begin{vmatrix} 1 & -1 \\ 2 & 16 \end{vmatrix}}{\begin{vmatrix} 1 & -2 \\ 2 & 5 \end{vmatrix}} = \frac{16 + 2}{5 + 4} = \frac{18}{9} = 2$$

The solution set is $\{(3, 2)\}$.

General Form of Cramer's Rule
Let an $n \times n$ system have linear equations of the form

$$a_1x_1 + a_2x_2 + a_3x_3 + \cdots + a_nx_n = b.$$

Define D as the determinant of the $n \times n$ matrix of coefficients of the variables. Define D_{x_1} as the determinant obtained from D by replacing the entries in column 1 of D with the constants of the system. Define D_{x_i} as the determinant obtained from D by replacing the entries in column i with the constants of the system. If $D \neq 0$, then the unique solution of the system is

$$x_1 = \frac{D_{x_1}}{D}, \quad x_2 = \frac{D_{x_2}}{D}, \quad x_3 = \frac{D_{x_3}}{D}, \quad \ldots, \quad x_n = \frac{D_{x_n}}{D}.$$

Solve using Cramer's rule.

$$3x + 2y + z = -5$$
$$x - y + 3z = -5$$
$$2x + 3y + z = 0$$

Using the method of expansion by minors, it can be shown that $D_x = 45$, $D_y = -30$, $D_z = 0$, and $D = -15$.

$$x = \frac{D_x}{D} = \frac{45}{-15} = -3, \quad y = \frac{D_y}{D} = \frac{-30}{-15} = 2,$$

$$z = \frac{D_z}{D} = \frac{0}{-15} = 0$$

The solution set is $\{(-3, 2, 0)\}$.

Concepts	Examples

5.4 Partial Fractions

To solve for the constants in the numerators of a partial fraction decomposition, use either of the following methods or a combination of the two.

Method 1 **For Linear Factors**

Step 1 Multiply each side by the common denominator.

Step 2 Substitute the zero of each factor in the resulting equation. For repeated linear factors, substitute as many other numbers as necessary to find all the constants in the numerators. The number of substitutions required will equal the number of constants $A, B, \ldots$.

Method 2 **For Quadratic Factors**

Step 1 Multiply each side by the common denominator.

Step 2 Collect like terms on the right side of the resulting equation.

Step 3 Equate the coefficients of like terms to form a system of equations.

Step 4 Solve the system to find the constants in the numerators.

Find the partial fraction decomposition of $\dfrac{9}{2x^2 + 9x + 9}$.

$$\frac{9}{(2x + 3)(x + 3)} = \frac{A}{2x + 3} + \frac{B}{x + 3} \quad (1)$$

Multiply by $(2x + 3)(x + 3)$.

$$9 = A(x + 3) + B(2x + 3)$$
$$9 = Ax + 3A + 2Bx + 3B$$
$$9 = (A + 2B)x + (3A + 3B)$$

Now solve the system

$$A + 2B = 0$$
$$3A + 3B = 9$$

to obtain $A = 6$ and $B = -3$.

$$\frac{9}{2x^2 + 9x + 9} = \frac{6}{2x + 3} + \frac{-3}{x + 3} \quad \text{Substitute into (1).}$$

Check this result by combining the terms on the right.

5.5 Nonlinear Systems of Equations

Solving a Nonlinear System of Equations
A nonlinear system can be solved by the substitution method, the elimination method, or a combination of the two methods.

Solve the system.

$$x^2 + 2xy - y^2 = 14 \quad (1)$$
$$x^2 - y^2 = -16 \quad (2)$$

$$
\begin{array}{ll}
x^2 + 2xy - y^2 = 14 & (1) \\
\underline{-x^2 \qquad\quad + y^2 = 16} & \text{Multiply (2) by } -1. \\
2xy \qquad\quad = 30 & \text{Add to eliminate } x^2 \text{ and } y^2. \\
xy = 15 & \text{Divide by 2.} \\
y = \dfrac{15}{x} & \text{Solve for } y.
\end{array}
$$

Substitute $\frac{15}{x}$ for y in equation (2).

$$x^2 - \left(\frac{15}{x}\right)^2 = -16 \qquad \text{(2) with } y = \tfrac{15}{x}$$

$$x^2 - \frac{225}{x^2} = -16 \qquad \text{Square.}$$

$$x^4 + 16x^2 - 225 = 0 \qquad \begin{array}{l}\text{Multiply by } x^2.\\ \text{Add } 16x^2.\end{array}$$

$$(x^2 - 9)(x^2 + 25) = 0 \qquad \text{Factor.}$$

$$x^2 - 9 = 0 \quad \text{or} \quad x^2 + 25 = 0 \qquad \text{Zero-factor property}$$

$$x = \pm 3 \quad \text{or} \qquad x = \pm 5i \qquad \text{Solve each equation.}$$

Find corresponding y-values. The solution set is

$$\{(3, 5), (-3, -5), (5i, -3i), (-5i, 3i)\}.$$

Concepts	Examples

5.6 Systems of Inequalities and Linear Programming

Graphing an Inequality in Two Variables

Method 1

If the inequality is or can be solved for y, then the following hold.

- The graph of $y < f(x)$ consists of all the points that are *below* the graph of $y = f(x)$.

- The graph of $y > f(x)$ consists of all the points that are *above* the graph of $y = f(x)$.

Method 2

If the inequality is not or cannot be solved for y, then choose a test point not on the boundary.

- If the test point satisfies the inequality, then the graph includes all points on the *same* side of the boundary as the test point.

- If the test point does not satisfy the inequality, then the graph includes all points on the *other* side of the boundary.

Graph $y \geq x^2 - 2x + 3$.

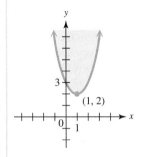

Graph the solution set of the system.

$$3x - 5y > -15$$
$$x^2 + y^2 \leq 25$$

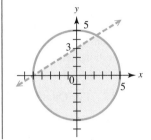

Solving Systems of Inequalities

To solve a system of inequalities, graph all inequalities on the same axes, and find the intersection of their solution sets.

Solving a Linear Programming Problem

Step 1 Write all necessary constraints and the objective function.

Step 2 Graph the region of feasible solutions.

Step 3 Identify all vertices (corner points).

Step 4 Find the value of the objective function at each vertex.

Step 5 The solution is given by the vertex producing the optimum value of the objective function.

Fundamental Theorem of Linear Programming

If an optimal value for a linear programming problem exists, then it occurs at a vertex of the region of feasible solutions.

The region of feasible solutions for the system below is given in the figure.

$$x + 2y \leq 14$$
$$3x + 4y \leq 36$$
$$x \geq 0$$
$$y \geq 0$$

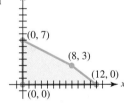

Maximize the objective function $8x + 12y$.

Vertex Point	Value of $8x + 12y$
$(0, 0)$	$8(0) + 12(0) = 0$
$(0, 7)$	$8(0) + 12(7) = 84$
$(8, 3)$	$8(8) + 12(3) = 100$ ← Maximum
$(12, 0)$	$8(12) + 12(0) = 96$

The maximum of 100 occurs at $(8, 3)$.

5.7 Properties of Matrices

Addition and Subtraction of Matrices

To add (subtract) matrices of the same dimension, add (subtract) corresponding elements.

Find the sum or difference.

$$\begin{bmatrix} 2 & 3 & -1 \\ 0 & 4 & 9 \end{bmatrix} + \begin{bmatrix} -8 & 12 & 1 \\ 5 & 3 & -3 \end{bmatrix} = \begin{bmatrix} -6 & 15 & 0 \\ 5 & 7 & 6 \end{bmatrix}$$

$$\begin{bmatrix} 5 & -1 \\ -8 & 8 \end{bmatrix} - \begin{bmatrix} -2 & 4 \\ 3 & -6 \end{bmatrix} = \begin{bmatrix} 7 & -5 \\ -11 & 14 \end{bmatrix}$$

Concepts	Examples

Scalar Multiplication

To multiply a matrix by a scalar, multiply each element of the matrix by the scalar.

Find the product.

$$3\begin{bmatrix} 6 & 2 \\ 1 & -2 \\ 0 & 8 \end{bmatrix} = \begin{bmatrix} 18 & 6 \\ 3 & -6 \\ 0 & 24 \end{bmatrix}$$
Multiply each element by the scalar 3.

Matrix Multiplication

The product AB of an $m \times n$ matrix A and an $n \times p$ matrix B is found as follows.

To find the ith row, jth column element of matrix AB, multiply each element in the ith row of A by the corresponding element in the jth column of B. The sum of these products will give the row i, column j element of AB.

Find the matrix product.

$$\begin{bmatrix} 1 & -2 & 3 \\ 5 & 0 & 4 \\ -8 & 7 & -7 \end{bmatrix}\begin{bmatrix} 1 \\ -2 \\ 3 \end{bmatrix} = \begin{bmatrix} 1(1) + (-2)(-2) + 3(3) \\ 5(1) + 0(-2) + 4(3) \\ -8(1) + 7(-2) + (-7)(3) \end{bmatrix}$$

$$\underbrace{}_{3 \times 3} \quad \underbrace{}_{3 \times 1}$$

$$= \begin{bmatrix} 14 \\ 17 \\ -43 \end{bmatrix}$$
$$3 \times 1$$

5.8 Matrix Inverses

Finding an Inverse Matrix

To obtain A^{-1} for any $n \times n$ matrix A for which A^{-1} exists, follow these steps.

Step 1 Form the augmented matrix $[A \mid I_n]$, where I_n is the $n \times n$ identity matrix.

Step 2 Perform row transformations on $[A \mid I_n]$ to obtain a matrix of the form $[I_n \mid B]$.

Step 3 Matrix B is A^{-1}.

Find A^{-1} if $A = \begin{bmatrix} 5 & 2 \\ 2 & 1 \end{bmatrix}$.

$$\begin{bmatrix} 5 & 2 & | & 1 & 0 \\ 2 & 1 & | & 0 & 1 \end{bmatrix}$$

$$\begin{bmatrix} 1 & 0 & | & 1 & -2 \\ 2 & 1 & | & 0 & 1 \end{bmatrix} \quad -2R2 + R1$$

$$\begin{bmatrix} 1 & 0 & | & 1 & -2 \\ 0 & 1 & | & -2 & 5 \end{bmatrix} \quad -2R1 + R2$$
$$\underbrace{}_{I_2} \quad \underbrace{}_{A^{-1}}$$

Therefore, $A^{-1} = \begin{bmatrix} 1 & -2 \\ -2 & 5 \end{bmatrix}$.

Chapter 5 Review Exercises

Use the substitution or elimination method to solve each system of equations. Identify any inconsistent systems or systems with infinitely many solutions. If a system has infinitely many solutions, write the solution set with y arbitrary.

1. $2x + 6y = 6$
$5x + 9y = 9$

2. $3x - 5y = 7$
$2x + 3y = 30$

3. $x + 5y = 9$
$2x + 10y = 18$

4. $\dfrac{1}{6}x + \dfrac{1}{3}y = 8$

$\dfrac{1}{4}x + \dfrac{1}{2}y = 12$

5. $y = -x + 3$
$2x + 2y = 1$

6. $0.2x + 0.5y = 6$
$0.4x + y = 9$

7. $3x - 2y = 0$
$9x + 8y = 7$

8. $6x + 10y = -11$
$9x + 6y = -3$

9. $2x - 5y + 3z = -1$
$x + 4y - 2z = 9$
$-x + 2y + 4z = 5$

10. $4x + 3y + z = -8$
$3x + y - z = -6$
$x + y + 2z = -2$

11. $5x - y = 26$
$4y + 3z = -4$
$3x + 3z = 15$

12. $x + z = 2$
$2y - z = 2$
$-x + 2y = -4$

Solve each problem.

13. *Concept Check* Create an inconsistent system of two equations.

14. *Connecting Graphs with Equations* Determine the system of equations illustrated in the graph. Write equations in standard form.

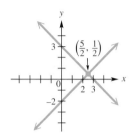

Solve each problem using a system of equations.

15. *Meal Planning* A cup of uncooked rice contains 15 g of protein and 810 calories. A cup of uncooked soybeans contains 22.5 g of protein and 270 calories. How many cups of each should be used for a meal containing 9.5 g of protein and 324 calories?

16. *Order Quantities* A company sells recordable CDs for $0.80 each and play-only CDs for $0.60 each. The company receives $76.00 for an order of 100 CDs. However, the customer neglected to specify how many of each type to send. Determine the number of each type of CD that should be sent.

17. *Indian Weavers* The Waputi Indians make woven blankets, rugs, and skirts. Each blanket requires 24 hr for spinning the yarn, 4 hr for dyeing the yarn, and 15 hr for weaving. Rugs require 30, 5, and 18 hr and skirts 12, 3, and 9 hr, respectively. If there are 306, 59, and 201 hr available for spinning, dyeing, and weaving, respectively, how many of each item can be made?

18. *(Modeling) Populations of Age Groups* The estimated resident populations (in percent) of young people (age 14 and under) and seniors (age 65 and over) in the United States for the years 2015–2050 are modeled by the following linear functions.

$$y_1 = -0.04x + 19.3 \quad \text{Young people}$$
$$y_2 = 0.17x + 16.4 \quad \text{Seniors}$$

In each case, x represents the number of years since 2015. (*Source:* U.S. Census Bureau.)

(a) Solve the system to find the year when these population percents will be equal.

(b) What percent, to the nearest tenth, of the U.S. resident population will be young people or seniors in the year found in part (a)? Answer may vary due to rounding.

(c) Use a calculator graph of the system to support the algebraic solution.

(d) Which population is increasing? (*Hint:* Consider the slopes of the lines.)

19. *(Modeling) Equilibrium Supply and Demand* Let the supply and demand equations for units of backpacks be

$$\text{supply: } p = \frac{3}{2}q \quad \text{and} \quad \text{demand: } p = 81 - \frac{3}{4}q.$$

(a) Graph these equations on the same axes.

(b) Find the equilibrium demand.

(c) Find the equilibrium price.

20. *(Modeling) Heart Rate* In a study, a group of athletes was exercised to exhaustion. Let x represent an athlete's heart rate 5 sec after stopping exercise and y this rate 10 sec after stopping. It was found that the maximum heart rate H for these athletes satisfied the two equations

$$H = 0.491x + 0.468y + 11.2$$
$$H = -0.981x + 1.872y + 26.4.$$

If an athlete had maximum heart rate $H = 180$, determine x and y graphically. Round to the nearest tenth. Interpret the answer. (*Source:* Thomas, V., *Science and Sport*, Faber and Faber.)

21. *(Modeling)* The table was generated using a function $y_1 = ax^2 + bx + c$. Use any three points from the table to find the equation for y_1.

NORMAL FLOAT AUTO REAL RADIAN MP
PRESS + FOR ΔTbl

X	Y1
0	1.5
1	-2.3
2	-1.3
3	4.5
4	15.1
5	30.5
6	50.7
7	75.7
8	105.5
9	140.1
10	179.5

X=0

22. *(Modeling)* The equation of a circle may be written in the form

$$x^2 + y^2 + ax + by + c = 0.$$

Find the equation of the circle passing through the points $(-3, -7)$, $(4, -8)$, and $(1, 1)$.

Solve each system in terms of the specified arbitrary variable.

23. $3x - 4y + z = 2$
$2x + y = 1$
(x arbitrary)

24. $2x - 6y + 4z = 5$
$5x + y - 3z = 1$
(z arbitrary)

Use the Gauss-Jordan method to solve each system.

25. $5x + 2y = -10$
$3x - 5y = -6$

26. $2x + 3y = 10$
$-3x + y = 18$

27. $3x + y = -7$
$x - y = -5$

28. $2x - y + 4z = -1$
$-3x + 5y - z = 5$
$2x + 3y + 2z = 3$

29. $x - z = -3$
$y + z = 6$
$2x - 3z = -9$

30. $2x - y + z = 4$
$x + 2y - z = 0$
$3x + y - 2z = 1$

Solve each problem using the Gauss-Jordan method to solve a system of equations.

31. *Mixing Teas* Three kinds of tea worth \$4.60, \$5.75, and \$6.50 per lb are to be mixed to get 20 lb of tea worth \$5.25 per lb. The amount of \$4.60 tea used is to be equal to the total amount of the other two kinds together. How many pounds of each tea should be used?

32. *Mixing Solutions* A 5% solution of a drug is to be mixed with some 15% solution and some 10% solution to make 20 ml of 8% solution. The amount of 5% solution used must be 2 ml more than the sum of the other two solutions. How many milliliters of each solution should be used?

33. *(Modeling) Master's Degrees* During the period 1975–2012, the numbers of master's degrees awarded to both males and females grew, but degrees earned by females grew at a greater rate. If $x = 0$ represents 1975 and $x = 37$ represents 2012, the number of master's degrees earned (in thousands) are closely modeled by the following system.

$$y = 3.79x + 128 \quad \text{Males}$$
$$y = 8.89x + 80.2 \quad \text{Females}$$

Solve the system to find the year in which males and females earned the same number of master's degrees. What was the total number, to the nearest thousand, of master's degrees earned in that year? (*Source:* U.S. Census Bureau.)

34. *(Modeling) Comparing Prices* One refrigerator sells for $700 and uses $85 worth of electricity per year. A second refrigerator is $100 more expensive but costs only $25 per year to run. Assuming that there are no repair costs, the costs to run the refrigerators over a 10-yr period are given by the following system of equations. Here, y represents the total cost in dollars, and x is time in years.

$$y = 700 + 85x$$
$$y = 800 + 25x$$

In how many years will the costs for the two refrigerators be equal? What are the equivalent costs at that time?

Evaluate each determinant.

35. $\begin{vmatrix} -1 & 8 \\ 2 & 9 \end{vmatrix}$

36. $\begin{vmatrix} -2 & 4 \\ 0 & 3 \end{vmatrix}$

37. $\begin{vmatrix} x & 4x \\ 2x & 8x \end{vmatrix}$

38. $\begin{vmatrix} -2 & 4 & 1 \\ 3 & 0 & 2 \\ -1 & 0 & 3 \end{vmatrix}$

39. $\begin{vmatrix} -1 & 2 & 3 \\ 4 & 0 & 3 \\ 5 & -1 & 2 \end{vmatrix}$

40. $\begin{vmatrix} -3 & 2 & 7 \\ 6 & -4 & -14 \\ 7 & 1 & 4 \end{vmatrix}$

Use Cramer's rule to solve each system of equations. If $D = 0$, use another method to determine the solution set.

41. $3x + 7y = 2$
$5x - y = -22$

42. $3x + y = -1$
$5x + 4y = 10$

43. $6x + y = -3$
$12x + 2y = 1$

44. $3x + 2y + z = 2$
$4x - y + 3z = -16$
$x + 3y - z = 12$

45. $x + y = -1$
$2y + z = 5$
$3x - 2z = -28$

46. $5x - 2y - z = 8$
$-5x + 2y + z = -8$
$x - 4y - 2z = 0$

Solve each equation.

47. $\begin{vmatrix} 3x & 7 \\ -x & 4 \end{vmatrix} = 8$

48. $\begin{vmatrix} 6x & 2 & 0 \\ 1 & 5 & 3 \\ x & 2 & -1 \end{vmatrix} = 2x$

Find the partial fraction decomposition for each rational expression.

49. $\dfrac{2}{3x^2 - 5x + 2}$

50. $\dfrac{11 - 2x}{x^2 - 8x + 16}$

51. $\dfrac{5 - 2x}{(x^2 + 2)(x - 1)}$

52. $\dfrac{x^3 + 2x^2 - 3}{x^4 - 4x^2 + 4}$

Solve each nonlinear system of equations.

53. $y = 2x + 10$
$x^2 + y = 13$

54. $x^2 = 2y - 3$
$x + y = 3$

55. $x^2 + y^2 = 17$
$2x^2 - y^2 = 31$

56. $2x^2 + 3y^2 = 30$
$x^2 + y^2 = 13$

57. $xy = -10$
$x + 2y = 1$

58. $xy + 2 = 0$
$y - x = 3$

59. $x^2 + 2xy + y^2 = 4$
$x - 3y = -2$

60. $x^2 + 2xy = 15 + 2x$
$xy - 3x + 3 = 0$

61. $2x^2 - 3y^2 = 18$
$2x^2 - 2y^2 = 14$

Solve each problem.

62. Find all values of b such that the straight line $3x - y = b$ touches the circle $x^2 + y^2 = 25$ at only one point.

63. Do the circle $x^2 + y^2 = 144$ and the line $x + 2y = 8$ have any points in common? If so, what are they?

64. Find the equation of the line passing through the points of intersection of the graphs of $x^2 + y^2 = 20$ and $x^2 - y = 0$.

Graph the solution set of each system of inequalities.

65. $x + y \le 6$
$2x - y \ge 3$

66. $y \le \dfrac{1}{3}x - 2$
$y^2 \le 16 - x^2$

67. Maximize the objective function $2x + 4y$ for the following constraints.

$$x \ge 0$$
$$y \ge 0$$
$$3x + 2y \le 12$$
$$5x + y \ge 5$$

68. *Connecting Graphs with Equations* Determine the system of inequalities illustrated in the graph. Write each inequality in standard form.

Solve each problem.

69. *Cost of Nutrients* Certain laboratory animals must have at least 30 g of protein and at least 20 g of fat per feeding period. These nutrients come from food A, which costs $0.18 per unit and supplies 2 g of protein and 4 g of fat; and from food B, which costs $0.12 per unit and has 6 g of protein and 2 g of fat. Food B is purchased under a long-term contract requiring that at least 2 units of B be used per serving. How much of each food must be purchased to produce the minimum cost per serving? What is the minimum cost?

70. *Profit from Farm Animals* A 4-H member raises only geese and pigs. She wants to raise no more than 16 animals, including no more than 10 geese. She spends $5 to raise a goose and $15 to raise a pig, and she has $180 available for this project. Each goose produces $6 in profit, and each pig produces $20 in profit. How many of each animal should she raise to maximize her profit? What is her maximum profit?

Find the values of the variables for which each statement is true, if possible.

71. $\begin{bmatrix} 5 & x+2 \\ -6y & z \end{bmatrix} = \begin{bmatrix} a & 3x-1 \\ 5y & 9 \end{bmatrix}$

72. $\begin{bmatrix} -6+k & 2 & a+3 \\ -2+m & 3p & 2r \end{bmatrix} + \begin{bmatrix} 3-2k & 5 & 7 \\ 5 & 8p & 5r \end{bmatrix} = \begin{bmatrix} 5 & y & 6a \\ 2m & 11 & -35 \end{bmatrix}$

Perform each operation, if possible.

73. $\begin{bmatrix} 3 \\ 2 \\ 5 \end{bmatrix} - \begin{bmatrix} 8 \\ -4 \\ 6 \end{bmatrix} + \begin{bmatrix} 1 \\ 0 \\ 2 \end{bmatrix}$

74. $4\begin{bmatrix} 3 & -4 & 2 \\ 5 & -1 & 6 \end{bmatrix} + \begin{bmatrix} -3 & 2 & 5 \\ 1 & 0 & 4 \end{bmatrix}$

75. $\begin{bmatrix} 2 & 5 & 8 \\ 1 & 9 & 2 \end{bmatrix} - \begin{bmatrix} 3 & 4 \\ 7 & 1 \end{bmatrix}$

76. $\begin{bmatrix} -3 & 4 \\ 2 & 8 \end{bmatrix}\begin{bmatrix} -1 & 0 \\ 2 & 5 \end{bmatrix}$

77. $\begin{bmatrix} 1 & 0 \\ 2 & 5 \end{bmatrix}\begin{bmatrix} -3 & 4 \\ 2 & 8 \end{bmatrix}$

78. $\begin{bmatrix} 1 & 2 \\ 3 & 0 \\ -6 & 5 \end{bmatrix}\begin{bmatrix} 4 & 8 \\ -1 & 2 \end{bmatrix}$

79. $\begin{bmatrix} 3 & 2 & -1 \\ 4 & 0 & 6 \end{bmatrix}\begin{bmatrix} -2 & 0 \\ 0 & 2 \\ 3 & 1 \end{bmatrix}$

80. $\begin{bmatrix} 1 & -2 & 4 & 2 \\ 0 & 1 & -1 & 8 \end{bmatrix}\begin{bmatrix} -1 \\ 2 \\ 0 \\ 1 \end{bmatrix}$

81. $\begin{bmatrix} -2 & 5 & 5 \\ 0 & 1 & 4 \\ 3 & -4 & -1 \end{bmatrix}\begin{bmatrix} 1 & 0 & -1 \\ -1 & 0 & 0 \\ 1 & 1 & -1 \end{bmatrix}$

82. $\begin{bmatrix} 0 & 1 & 4 \\ 7 & -2 & 9 \\ 10 & 0 & 1 \end{bmatrix}\begin{bmatrix} 3 & 2 & 1 \\ -4 & 7 & 6 \end{bmatrix}$

Find the inverse, if it exists, for each matrix.

83. $\begin{bmatrix} 2 & 1 \\ 5 & 3 \end{bmatrix}$

84. $\begin{bmatrix} -4 & 2 \\ 0 & 3 \end{bmatrix}$

85. $\begin{bmatrix} 2 & -1 & 0 \\ 1 & 0 & 1 \\ 1 & -2 & 0 \end{bmatrix}$

86. $\begin{bmatrix} 2 & 3 & 5 \\ -2 & -3 & -5 \\ 1 & 4 & 2 \end{bmatrix}$

Solve each system using the inverse of the coefficient matrix.

87. $5x - 4y = 1$
$\quad\; x + 4y = 3$

88. $2x + \;\, y = 5$
$\quad 3x - 2y = 4$

89. $3x + 2y + \;\; z = -5$
$\quad\; x - \;\, y + 3z = -5$
$\quad 2x + 3y + \;\; z = 0$

90. $x + y + z = 1$
$\quad 2x - y = -2$
$\quad\; 3y + z = 2$

Test

Use the substitution or elimination method to solve each system of equations. State whether it is an inconsistent system or has infinitely many solutions. If a system has infinitely many solutions, write the solution set with y arbitrary.

1. $3x - y = 9$
$x + 2y = 10$

2. $6x + 9y = -21$
$4x + 6y = -14$

3. $x - 2y = 4$
$-2x + 4y = 6$

4. $\dfrac{1}{4}x - \dfrac{1}{3}y = -\dfrac{5}{12}$
$\dfrac{1}{10}x + \dfrac{1}{5}y = \dfrac{1}{2}$

5. $2x + y + z = 3$
$x + 2y - z = 3$
$3x - y + z = 5$

Use the Gauss-Jordan method to solve each system.

6. $3x - 2y = 13$
$4x - y = 19$

7. $3x - 4y + 2z = 15$
$2x - y + z = 13$
$x + 2y - z = 5$

8. *Connecting Graphs with Equations* Find the equation $y = ax^2 + bx + c$ of the parabola through the given points.

9. *Ordering Supplies* A knitting shop orders yarn from three suppliers in Toronto, Montreal, and Ottawa. One month the shop ordered a total of 100 units of yarn from these suppliers. The delivery costs were $80, $50, and $65 per unit for the orders from Toronto, Montreal, and Ottawa, respectively, with total delivery costs of $5990. The shop ordered the same amount from Toronto and Ottawa. How many units were ordered from each supplier?

Evaluate each determinant.

10. $\begin{vmatrix} 6 & 8 \\ 2 & -7 \end{vmatrix}$

11. $\begin{vmatrix} 2 & 0 & 8 \\ -1 & 7 & 9 \\ 12 & 5 & -3 \end{vmatrix}$

Use Cramer's rule to solve each system of equations.

12. $2x - 3y = -33$
$4x + 5y = 11$

13. $x + y - z = -4$
$2x - 3y - z = 5$
$x + 2y + 2z = 3$

Find the partial fraction decomposition for each rational expression.

14. $\dfrac{9x + 19}{x^2 + 2x - 3}$

15. $\dfrac{x + 2}{x^3 + 2x^2 + x}$

Solve each nonlinear system of equations.

16. $2x^2 + y^2 = 6$
$x^2 - 4y^2 = -15$

17. $x^2 + y^2 = 25$
$x + y = 7$

Work each problem.

18. *Unknown Numbers* Find two numbers such that their sum is -1 and the sum of their squares is 61.

19. Graph the solution set.

$x - 3y \geq 6$
$y^2 \leq 16 - x^2$

20. Maximize the objective function $2x + 3y$ for the following constraints.

$x \geq 0$
$y \geq 0$
$x + 2y \leq 24$
$3x + 4y \leq 60$

21. *Jewelry Profits* The Schwab Company designs and sells two types of rings: the VIP and the SST. The company can produce up to 24 rings each day using up to 60 total hours of labor. It takes 3 hr to make one VIP ring and 2 hr to make one SST ring. How many of each type of ring should be made daily in order to maximize the company's profit, if the profit on one VIP ring is $30 and the profit on one SST ring is $40? What is the maximum profit?

22. Find the value of each variable for which the statement is true.

$$\begin{bmatrix} 5 & x + 6 \\ 0 & 4 \end{bmatrix} = \begin{bmatrix} y - 2 & 4 - x \\ 0 & w + 7 \end{bmatrix}$$

Perform each operation, if possible.

23. $3\begin{bmatrix} 2 & 3 \\ 1 & -4 \\ 5 & 9 \end{bmatrix} - \begin{bmatrix} -2 & 6 \\ 3 & -1 \\ 0 & 8 \end{bmatrix}$

24. $\begin{bmatrix} 1 \\ 2 \end{bmatrix} + \begin{bmatrix} 4 \\ -6 \end{bmatrix} + \begin{bmatrix} 2 & 8 \\ -7 & 5 \end{bmatrix}$

25. $\begin{bmatrix} 2 & 1 & -3 \\ 4 & 0 & 5 \end{bmatrix} \begin{bmatrix} 1 & 3 \\ 2 & 4 \\ 3 & -2 \end{bmatrix}$

26. $\begin{bmatrix} 2 & -4 \\ 3 & 5 \end{bmatrix} \begin{bmatrix} 4 \\ 2 \\ 7 \end{bmatrix}$

27. *Concept Check* Which of the following properties does not apply to multiplication of matrices?

A. commutative   **B.** associative   **C.** distributive   **D.** identity

Find the inverse, if it exists, for each matrix.

28. $\begin{bmatrix} -8 & 5 \\ 3 & -2 \end{bmatrix}$

29. $\begin{bmatrix} 4 & 12 \\ 2 & 6 \end{bmatrix}$

30. $\begin{bmatrix} 1 & 3 & 4 \\ 2 & 7 & 8 \\ -2 & -5 & -7 \end{bmatrix}$

Solve each system using the inverse of the coefficient matrix.

31. $2x + y = -6$
$3x - y = -29$

32. $x + y = 5$
$y - 2z = 23$
$x + 3z = -27$

Answers to Selected Exercises

To The Student

In this section we provide the answers that we think most students will obtain when they work the exercises using the methods explained in the text. If your answer does not look exactly like the one given here, it is not necessarily wrong. In many cases there are equivalent forms of the answer. For example, if the answer section shows $\frac{3}{4}$ and your answer is 0.75, you have obtained the correct answer but written it in a different (yet equivalent) form. Unless the directions specify otherwise, 0.75 is just as valid an answer as $\frac{3}{4}$. In general, if your answer does not agree with the one given in the text, see whether it can be transformed into the other form. If it can, then it is the correct answer. If you still have doubts, talk with your instructor.

Chapter R Review of Basic Concepts

R.1 Exercises

1. $\{1, 2, 3, 4, \ldots\}$ **3.** complement **5.** union
7. $\{1, 2, 3, 4, 5\}$ **9.** $\{16, 18\}$ **11.** finite; yes
13. infinite; no **15.** infinite; no **17.** infinite; no
19. $\{12, 13, 14, 15, 16, 17, 18, 19, 20\}$
21. $\left\{1, \frac{1}{2}, \frac{1}{4}, \frac{1}{8}, \frac{1}{16}, \frac{1}{32}\right\}$ **23.** $\{17, 22, 27, 32, 37, 42, 47\}$
25. $\{9, 10, 11, 12, 13, 14\}$ **27.** $\in$ **29.** $\notin$ **31.** $\in$
33. $\notin$ **35.** $\notin$ **37.** $\notin$ **39.** false **41.** true **43.** true
45. true **47.** false **49.** true **51.** true **53.** true
55. false **57.** true **59.** true **61.** false **63.** $\subseteq$
65. $\nsubseteq$ **67.** $\subseteq$ **69.** true **71.** false **73.** true
75. false **77.** true **79.** true **81.** $\{0, 2, 4\}$
83. $\{0, 1, 2, 3, 4, 5, 6, 7, 8, 9, 11, 13\}$ **85.** $\varnothing$; M and N are disjoint sets. **87.** $\{0, 1, 2, 3, 4, 5, 7, 9, 11, 13\}$
89. Q, or $\{0, 2, 4, 6, 8, 10, 12\}$ **91.** $\{10, 12\}$ **93.** $\varnothing$; $\varnothing$ and R are disjoint. **95.** N, or $\{1, 3, 5, 7, 9, 11, 13\}$; N and $\varnothing$ are disjoint. **97.** R, or $\{0, 1, 2, 3, 4\}$; M and N are disjoint. $(M \cap N)$ and R are disjoint.
99. $\{0, 1, 2, 3, 4, 6, 8\}$ **101.** R, or $\{0, 1, 2, 3, 4\}$
103. $\varnothing$; Q' and $(N' \cap U)$ are disjoint.
105. $\{1, 3, 5, 7, 9, 10, 11, 12, 13\}$
107. M, or $\{0, 2, 4, 6, 8\}$
109. Q, or $\{0, 2, 4, 6, 8, 10, 12\}$

R.2 Exercises

1. whole numbers **3.** base; exponent **5.** absolute value
7. 1000 **9.** 4 **11.** 1, 3 **13.** $-6, -\frac{12}{4}$ (or -3), 0, 1, 3
15. $-\sqrt{3}, 2\pi, \sqrt{12}$ **17.** -16 **19.** 16 **21.** -243
23. -162 **25.** -6 **27.** -60 **29.** -12 **31.** $-\frac{25}{36}$
33. $-\frac{6}{7}$ **35.** 28 **37.** $-\frac{1}{2}$ **39.** $-\frac{23}{20}$ **41.** $-\frac{25}{11}$ **43.** $-\frac{11}{3}$

45. $-\frac{13}{3}$ **47.** 6 **49.** distributive **51.** inverse
53. identity **55.** commutative **57.** associative
59. closure **61.** no; For example, $3 - 5 \neq 5 - 3$.
63. $20z$ **65.** $m + 11$ **67.** $\frac{2}{3}y + \frac{4}{9}z - \frac{5}{3}$
69. $(8 - 14)p = -6p$ **71.** $-4z + 4y$ **73.** 1700
75. 150 **77.** false; $|6 - 8| = |8| - |6|$ **79.** true
81. false; $|a - b| = |b| - |a|$ **83.** 10 **85.** $-\frac{4}{7}$
87. -8 **89.** 4 **91.** 16 **93.** 20 **95.** -1 **97.** -5
99. true **101.** false **103.** true **105.** 3 **107.** 9
109. x and y have the same sign. **111.** x and y have different signs. **113.** x and y have the same sign.
115. 19; This represents the number of strokes between their scores. **117.** 0.031 **119.** 0.026; Increased weight results in lower BAC. **121.** 113.2 **123.** 103.3 **125.** 97.4
127. 96.5 **129.** 95.3 **131.** 93.9 **133.** 122.5
135. 111.0

R.3 Exercises

1. 5 **3.** binomial **5.** FOIL **7.** true
9. false; $(x + y)^2 = x^2 + 2xy + y^2$ **11.** $-16x^7$ **13.** n^{11}
15. 9^8 **17.** $72m^{11}$ **19.** $-15x^5y^5$ **21.** $4m^3n^3$
23. 2^{10} **25.** $-216x^6$ **27.** $-16m^6$ **29.** $\frac{r^{24}}{s^6}$ **31.** $\frac{256m^8}{t^4p^8}$
33. -1 **35.** (a) B (b) C (c) B (d) C
37. polynomial; degree 11; monomial **39.** polynomial; degree 4; binomial **41.** polynomial; degree 5; trinomial
43. polynomial; degree 11; none of these
45. not a polynomial **47.** polynomial; degree 0; monomial
49. $x^2 - x + 2$ **51.** $12y^2 + 4$
53. $6m^4 - 2m^3 - 7m^2 - 4m$ **55.** $28r^2 + r - 2$
57. $15x^4 - \frac{7}{3}x^3 - \frac{2}{9}x^2$ **59.** $12x^5 + 8x^4 - 20x^3 + 4x^2$
61. $-2z^3 + 7z^2 - 11z + 4$
63. $m^2 + mn - 2n^2 - 2km + 5kn - 3k^2$
65. $16x^4 - 72x^2 + 81$ **67.** $x^4 - 2x^2 + 1$
69. $4m^2 - 9$ **71.** $16x^4 - 25y^2$
73. $16m^2 + 16mn + 4n^2$ **75.** $25r^2 - 30rt^2 + 9t^4$
77. $4p^2 - 12p + 9 + 4pq - 6q + q^2$
79. $9q^2 + 30q + 25 - p^2$
81. $9a^2 + 6ab + b^2 - 6a - 2b + 1$
83. $y^3 + 6y^2 + 12y + 8$
85. $q^4 - 8q^3 + 24q^2 - 32q + 16$ **87.** $p^3 - 7p^2 - p - 7$
89. $49m^2 - 4n^2$ **91.** $-14q^2 + 11q - 14$
93. $4p^2 - 16$ **95.** $11y^3 - 18y^2 + 4y$
97. $2x^5 + 7x^4 - 5x^2 + 7$ **99.** $4x^2 + 5x + 10 + \frac{21}{x - 2}$
101. $2m^2 + m - 2 + \frac{6}{3m + 2}$ **103.** $x^2 + 2 + \frac{5x + 21}{x^2 + 3}$

105. (a) $(x + y)^2$ **(b)** $x^2 + 2xy + y^2$ **(c)** The expressions are equivalent because they represent the same area. **(d)** special product for squaring a binomial

107. (a) 60,501,000 ft^3 **(b)** The shape becomes a rectangular box with a square base, with volume $V = b^2h$.

(c) If we let $a = b$, then $V = \frac{1}{3}h(a^2 + ab + b^2)$ becomes $V = \frac{1}{3}h(b^2 + bb + b^2)$, which simplifies to $V = hb^2$. Yes, the Egyptian formula gives the same result.

109. 5.3; 0.3 low **111.** 2.2; 0.2 low **113.** 1,000,000

115. 32 **117.** 9999 **118.** 3591 **119.** 10,404

120. 5041

R.4 Exercises

1. factoring **3.** multiplying **5.** sum of squares

7. (a) B **(b)** C **(c)** A **9.** B **11.** $12(m + 5)$

13. $8k(k^2 + 3)$ **15.** $xy(1 - 5y)$ **17.** $-2p^2q^4(2p + q)$

19. $4k^2m^3(1 + 2k^2 - 3m)$ **21.** $2(a + b)(1 + 2m)$

23. $(r + 3)(3r - 5)$ **25.** $(m - 1)(2m^2 - 7m + 7)$

27. The *completely* factored form is $4xy^3(xy^2 - 2)$.

29. $(2s + 3)(3t - 5)$ **31.** $(m^4 + 3)(2 - a)$

33. $(p^2 - 2)(q^2 + 5)$ **35.** $(2a - 1)(3a - 4)$

37. $(3m + 2)(m + 4)$ **39.** prime

41. $2a(3a + 7)(2a - 3)$ **43.** $(3k - 2p)(2k + 3p)$

45. $(5a + 3b)(a - 2b)$ **47.** $(4x + y)(3x - y)$

49. $2a^2(4a - b)(3a + 2b)$ **51.** $(3m - 2)^2$

53. $2(4a + 3b)^2$ **55.** $(2xy + 7)^2$ **57.** $(a - 3b - 3)^2$

59. $(3a + 4)(3a - 4)$ **61.** $(x^2 + 4)(x + 2)(x - 2)$

63. $(5s^2 + 3t)(5s^2 - 3t)$ **65.** $(a + b + 4)(a + b - 4)$

67. $(p^2 + 25)(p + 5)(p - 5)$ **69.** $(x - 4 + y)(x - 4 - y)$

71. $(y + x - 6)(y - x + 6)$ **73.** $(2 - a)(4 + 2a + a^2)$

75. $(5x - 3)(25x^2 + 15x + 9)$

77. $(3y^3 + 5z^2)(9y^6 - 15y^3z^2 + 25z^4)$

79. $r(r^2 + 18r + 108)$

81. $(3 - m - 2n)(9 + 3m + 6n + m^2 + 4mn + 4n^2)$

83. $9(7k - 3)(k + 1)$ **85.** $(3a - 7)^2$

87. $(a + 4)(a^2 - a + 7)$ **89.** $9(x + 1)(3x^2 + 9x + 7)$

91. $(m^2 - 5)(m^2 + 2)$ **93.** $(3t^2 + 5)(4t^2 - 7)$

95. $(2b + c + 4)(2b + c - 4)$ **97.** $(x + y)(x - 5)$

99. $(m - 2n)(p^4 + q)$ **101.** $(2z + 7)^2$

103. $(10x + 7y)(100x^2 - 70xy + 49y^2)$

105. $(5m^2 - 6)(25m^4 + 30m^2 + 36)$

107. $9(x + 2)(3x^2 + 4)$ **109.** $2y(3x^2 + y^2)$ **111.** prime

113. $4xy$ **115.** In general, a sum of squares is not factorable over the real number system. If there is a greatest common factor, as in $4x^2 + 16$, it may be factored out, as, here, to obtain $4(x^2 + 4)$. **117.** $\left(7x + \frac{1}{5}\right)\left(7x - \frac{1}{5}\right)$

119. $\left(\frac{5}{3}x^2 + 3y\right)\left(\frac{5}{3}x^2 - 3y\right)$ **121.** ± 36 **123.** 9

125. $(x - 1)(x^2 + x + 1)(x + 1)(x^2 - x + 1)$

126. $(x - 1)(x + 1)(x^4 + x^2 + 1)$

127. $(x^2 - x + 1)(x^2 + x + 1)$ **128.** additive inverse property (0 in the form $x^2 - x^2$ was added on the right.);

associative property of addition; factoring a perfect square trinomial; factoring a difference of squares; commutative property of addition **129.** They are the same.

130. $(x^4 - x^2 + 1)(x^2 + x + 1)(x^2 - x + 1)$

R.5 Exercises

1. rational expression **3.** 5 **5.** $\frac{4}{x}$ **7.** $\frac{10}{x}$

9. $\frac{13x}{20}$ **11.** $\{x \mid x \neq 6\}$ **13.** $\left\{x \mid x \neq -\frac{1}{2}, 1\right\}$

15. $\{x \mid x \neq -2, -3\}$ **17.** $\{x \mid x \neq -1\}$ **19.** $\{x \mid x \neq 1\}$

21. $\frac{2x + 4}{x}$ **23.** $\frac{-3}{t + 5}$ **25.** $\frac{8}{9}$ **27.** $\frac{m - 2}{m + 3}$ **29.** $\frac{2m + 3}{4m + 3}$

31. $x^2 - 4x + 16$ **33.** $\frac{25p^2}{9}$ **35.** $\frac{2}{9}$ **37.** $\frac{5x}{y}$

39. $\frac{2a + 8}{a - 3}$, or $\frac{2(a + 4)}{a - 3}$ **41.** 1 **43.** $\frac{m + 6}{m + 3}$ **45.** $\frac{x^2 - xy + y^2}{x^2 + xy + y^2}$

47. $\frac{x + 2y}{4 - x}$ **49.** B, C **51.** $\frac{19}{6k}$ **53.** $\frac{137}{30m}$ **55.** $\frac{a - b}{a^2}$

57. $\frac{5 - 22x}{12x^2y}$ **59.** 3 **61.** $\frac{2x}{(x + z)(x - z)}$ **63.** $\frac{4}{a - 2}$, or $\frac{-4}{2 - a}$

65. $\frac{3x + y}{2x - y}$, or $\frac{-3x - y}{y - 2x}$ **67.** $\frac{4x - 7}{x^2 - x + 1}$

69. $\frac{2x^2 - 9x}{(x - 3)(x + 4)(x - 4)}$ **71.** $\frac{x + 1}{x - 1}$ **73.** $\frac{-1}{x + 1}$

75. $\frac{(2 - b)(1 + b)}{b(1 - b)}$ **77.** $\frac{a + b}{a^2 - ab + b^2}$ **79.** $\frac{m^3 - 4m - 1}{m - 2}$

81. $\frac{p^3 - 16p + 3}{p + 4}$ **83.** $\frac{y^2 - 2y - 3}{y^2 + y - 1}$ **85.** $\frac{-1}{x(x + h)}$

87. $\frac{-2x - h}{(x^2 + 9)[(x + h)^2 + 9]}$ **89.** 0 mi

91. 20.1 (thousand dollars)

R.6 Exercises

1. true **3.** false; $\frac{a^6}{a^4} = a^2$ **5.** false; $(x + y)^{-1} = \frac{1}{x + y}$

7. (a) B **(b)** D **(c)** B **(d)** D **9. (a)** E **(b)** G

(c) F **(d)** F **11.** $-\frac{1}{64}$ **13.** $-\frac{1}{625}$ **15.** 9 **17.** $\frac{1}{16x^2}$

19. $\frac{4}{x^2}$ **21.** $-\frac{1}{a^3}$ **23.** 16 **25.** x^4 **27.** $\frac{1}{r^3}$ **29.** 6^6

31. $\frac{2r^3}{3}$ **33.** $\frac{4n^7}{3m^7}$ **35.** $-4r^6$ **37.** $\frac{625}{a^{10}}$ **39.** $\frac{p^4}{5}$

41. $\frac{1}{2pq}$ **43.** $\frac{4}{a^2}$ **45.** $\frac{5}{x^2}$ **47.** 13 **49.** 2 **51.** $-\frac{4}{3}$

53. This expression is not a real number. **55.** 4

57. 1000 **59.** -27 **61.** $\frac{256}{81}$ **63.** 9 **65.** 4 **67.** y

69. $k^{2/3}$ **71.** x^3y^8 **73.** $\frac{1}{x^{10/3}}$ **75.** $\frac{6}{m^{1/4}n^{3/4}}$ **77.** p^2

79. (a) 250 sec **(b)** $2^{-1.5} \approx 0.3536$

81. $y - 10y^2$ **83.** $-4k^{10/3} + 24k^{4/3}$ **85.** $x^2 - x$

87. $r - 2 + \frac{1}{r}$ **89.** $k^{-2}(4k + 1)$ **91.** $4t^{-4}(t^2 + 2)$

93. $z^{-1/2}(9 + 2z)$ **95.** $p^{-7/4}(p - 2)$ **97.** $4a^{-7/5}(-a + 4)$

99. $(p + 4)^{-3/2}(p^2 + 9p + 21)$

101. $6(3x + 1)^{-3/2}(9x^2 + 8x + 2)$

103. $2x(2x + 3)^{-5/9}(-16x^4 - 48x^3 - 30x^2 + 9x + 2)$

105. $b + a$ **107.** -1 **109.** $\frac{y(xy - 9)}{x^2y^2 - 9}$ **111.** $\frac{2x(1 - 3x^2)}{(x^2 + 1)^5}$

113. $\frac{1 + 2x^3 - 2x}{4}$ **115.** $\frac{3x - 5}{(2x - 3)^{4/3}}$ **117.** 27 **119.** 4

121. $\frac{1}{100}$

R.7 Exercises

1. $64^{1/3}$; 4   **3.** (a) F   (b) H   (c) G   (d) C   **5.** t

7. $5\sqrt{2}$ **9.** $-5\sqrt{xy}$ **11.** 5   **13.** 3   **15.** -5

17. This expression is not a real number.   **19.** 2   **21.** 2

23. $\sqrt[3]{m^2}$, or $\left(\sqrt[3]{m}\right)^2$ **25.** $\sqrt[3]{(2m+p)^2}$, or $\left(\sqrt[3]{2m+p}\right)^2$

27. $k^{2/5}$ **29.** $-3 \cdot 5^{1/2}p^{3/2}$ **31.** A   **33.** $x \ge 0$ **35.** $|x|$

37. $5k^2|m|$ **39.** $|4x - y|$ **41.** $3\sqrt[3]{3}$ **43.** $-2\sqrt[4]{2}$

45. $\sqrt{42pqr}$ **47.** $\sqrt[3]{14xy}$ **49.** $-\frac{3}{5}$ **51.** $-\frac{\sqrt[3]{5}}{2}$ **53.** $\frac{\sqrt[4]{m}}{n}$

55. -15 **57.** $32\sqrt[3]{2}$ **59.** $2x^2z^4\sqrt{2x}$ **61.** This expres-

sion cannot be simplified further.   **63.** $\frac{\sqrt{6x}}{3x}$ **65.** $\frac{x^2y\sqrt{xy}}{z}$

67. $\frac{2\sqrt[3]{x^2}}{x^2}$ **69.** $\frac{h\sqrt[4]{9g^3hr^2}}{3r^2}$ **71.** $\sqrt{3}$ **73.** $\sqrt[3]{2}$ **75.** $\sqrt[12]{2}$

77. $12\sqrt{2x}$ **79.** $7\sqrt[3]{3}$ **81.** $3x\sqrt[4]{x^2y^3} - 2x^2\sqrt[4]{x^2y^3}$

83. This expression cannot be simplified further.

85. -7 **87.** 10   **89.** $11 + 4\sqrt{6}$ **91.** $5\sqrt{6}$ **93.** $\frac{m\sqrt[3]{n^2}}{n}$

95. $\frac{x\sqrt[3]{2} - \sqrt[3]{5}}{x^3}$ **97.** $\frac{11\sqrt{2}}{8}$ **99.** $-\frac{25\sqrt[3]{9}}{18}$ **101.** $\frac{\sqrt{15} - 3}{2}$

103. $\frac{-7 + 2\sqrt{14} + \sqrt{7} - 2\sqrt{2}}{2}$ **105.** $\sqrt{p} - 2$

107. $\frac{3m\left(2 - \sqrt{m+n}\right)}{4 - m - n}$ **109.** $\frac{5\sqrt{x}\left(2\sqrt{x} - \sqrt{y}\right)}{4x - y}$

111. 17.7 ft per sec   **113.** $-12°$F   **115.** 2   **117.** 2

119. 3   **121.** It gives six decimal places of accuracy.

123. It first differs in the fourth decimal place.

Chapter R Review Exercises

1. $\{6, 8, 10, 12, 14, 16, 18, 20\}$ **3.** true   **5.** true

7. false   **9.** true   **11.** true   **13.** $\{2, 6, 9, 10\}$ **15.** $\varnothing$

17. $\varnothing$ **19.** $\{1, 2, 3, 4, 6, 8\}$ **21.** $\{1, 2, 3, 4, 5, 6, 7, 8,$

$9, 10\}$, or U **23.** $-12, -6, -\sqrt{4}$ (or -2), 0, 6

25. irrational number, real number   **27.** whole number,

integer, rational number, real number   **29.** The reciprocal of

a product is the product of the reciprocals.   **31.** A product

raised to a power is equal to the product of the factors to

that power.   **33.** A quotient raised to a power is equal to

the quotient of the numerator and the denominator to that

power.   **35.** commutative   **37.** associative   **39.** identity

41. 7.296 million   **43.** 32   **45.** $-\frac{37}{18}$ **47.** $-\frac{12}{5}$

49. -32 **51.** -13 **53.** $7q^3 - 9q^2 - 8q + 9$

55. $16y^3 + 42y^2 - 73y + 21$ **57.** $9k^2 - 30km + 25m^2$

59. $6m^2 - 3m + 5$ **61.** $3b - 8 + \frac{2}{b^2 + 4}$

63. $3(z - 4)^2(3z - 11)$ **65.** $(z - 8k)(z + 2k)$

67. $6a^6(4a + 5b)(2a - 3b)$ **69.** $(7m^4 + 3n)(7m^4 - 3n)$

71. $3(9r - 10)(2r + 1)$ **73.** $(x - 1)(y + 2)$

75. $(3x - 4)(9x - 34)$ **77.** $\frac{1}{2k^2(k - 1)}$ **79.** $\frac{x + 1}{x + 4}$

81. $\frac{p + 6q}{p + q}$ **83.** $\frac{2m}{m - 4}$, or $\frac{-2m}{4 - m}$ **85.** $\frac{q + p}{pq - 1}$ **87.** $\frac{16}{25}$

89. $-10z^8$ **91.** 1   **93.** $-8y^{11}p$ **95.** $\frac{1}{(p + q)^5}$

97. $-14r^{17/12}$ **99.** $y^{1/2}$ **101.** $\frac{1}{p^2q^4}$ **103.** $10\sqrt{2}$

105. $5\sqrt[4]{2}$ **107.** $-\frac{\sqrt[3]{50p}}{5p}$ **109.** $\sqrt[12]{m}$ **111.** 66

113. $-9m\sqrt{2m} + 5m\sqrt{m}$, or $m\left(-9\sqrt{2m} + 5\sqrt{m}\right)$

115. $\frac{6\left(3 + \sqrt{2}\right)}{7}$

In Exercises 117–125, we give only the corrected right
sides of the equations.

117. $x^3 + 5x$ **119.** m^6 **121.** $\frac{a}{2b}$ **123.** $(-2)^{-3}$

125. $\frac{7b}{8b + 7a}$

Chapter R Test

[R.1]   **1.** false   **2.** true   **3.** false   **4.** true

[R.2]   **5.** (a) $-13, -\frac{12}{4}$ (or -3), 0, $\sqrt{49}$ (or 7)

(b) $-13, -\frac{12}{4}$ (or -3), 0, $\frac{3}{5}$, 5.9, $\sqrt{49}$ (or 7)

(c) All are real numbers.   **6.** 4   **7.** (a) associative

(b) commutative   (c) distributive   (d) inverse   **8.** 104.7

[R.3]   **9.** $11x^2 - x + 2$ **10.** $36r^2 - 60r + 25$

11. $3t^3 + 5t^2 + 2t + 8$ **12.** $2x^2 - x - 5 + \frac{3}{x - 5}$

13. \$9992   **14.** \$11,727   [R.4]   **15.** $(3x - 7)(2x - 1)$

16. $(x^2 + 4)(x + 2)(x - 2)$ **17.** $2m(4m + 3)(3m - 4)$

18. $(x - 2)(x^2 + 2x + 4)(y + 3)(y - 3)$

19. $(a - b)(a - b + 2)$ **20.** $(1 - 3x^2)(1 + 3x^2 + 9x^4)$

[R.5]   **21.** $\frac{x^4(x + 1)}{3(x^2 + 1)}$ **22.** $\frac{x(4x + 1)}{(x + 2)(x + 1)(2x - 3)}$

23. $\frac{2a}{2a - 3}$, or $\frac{-2a}{3 - 2a}$ **24.** $\frac{y}{y + 2}$

[R.6, R.7]   **25.** $3x^2y^4\sqrt{2x}$ **26.** $2\sqrt{2x}$ **27.** $x - y$

28. $\frac{7\left(\sqrt{11} + \sqrt{7}\right)}{2}$ **29.** $\frac{y}{x}$ **30.** $\frac{9}{16}$ **31.** false

32. 2.1 sec

Chapter 1 Equations and Inequalities

1.1 Exercises

1. equation   **3.** first-degree equation   **5.** contradiction

7. true   **9.** false   **11.** $\{-4\}$ **13.** $\{1\}$ **15.** $\left\{-\frac{2}{7}\right\}$

17. $\left\{-\frac{7}{8}\right\}$ **19.** $\{-1\}$ **21.** $\{10\}$ **23.** $\{75\}$ **25.** $\{0\}$

27. $\{12\}$ **29.** $\{50\}$ **31.** identity; $\{$all real numbers$\}$

33. conditional equation; $\{0\}$ **35.** contradiction; $\varnothing$

37. identity; $\{$all real numbers$\}$ **39.** $l = \frac{V}{wh}$

41. $c = P - a - b$ **43.** $B = \frac{2\mathcal{A} - hb}{h}$, or $B = \frac{2\mathcal{A}}{h} - b$

45. $h = \frac{S - 2\pi r^2}{2\pi r}$, or $h = \frac{S}{2\pi r} - r$ **47.** $h = \frac{S - 2lw}{2w + 2l}$

Answers in Exercises 49–57 exist in equivalent forms.

49. $x = -3a + b$ **51.** $x = \frac{3a + b}{3 - a}$ **53.** $x = \frac{3 - 3a}{a^2 - a - 1}$

55. $x = \frac{2a^2}{a^2 + 3}$ **57.** $x = \frac{m + 4}{2m + 5}$ **59.** (a) \$63   (b) \$3213

61. 68°F   **63.** 10°C   **65.** 37.8°C   **67.** 463.9°C

69. 45°C

1.2 Exercises

1. 8 hr   **3.** $40   **5.** 90 L   **7.** A   **9.** D   **11.** 90 cm
13. 6 cm   **15.** 600 ft, 800 ft, 1000 ft   **17.** 4 ft
19. 50 mi   **21.** 2.7 mi   **23.** 45 min   **25.** 1 hr, 7 min,
34 sec; It is about $\frac{1}{2}$ the world record time.   **27.** 35 km per hr
29. $7\frac{1}{2}$ gal   **31.** 2 L   **33.** 4 mL   **35.** short-term note:
$100,000; long-term note: $140,000   **37.** $10,000 at 2.5%;
$20,000 at 3%   **39.** $50,000 at 1.5%; $90,000 at 4%
41. (a) $52   **(b)** $2500 **(c)** $5000   **43. (a)** $F = 14,000x$
(b) 1.9 hr   **45. (a)** 23.2 million   **(b)** 2020   **(c)** They
are quite close.   **(d)** 17.5 million   **(e)** When using the
model for predictions, it is best to stay within the scope of
the sample data.

1.3 Exercises

1. $\sqrt{-1}$; -1 **3.** complex conjugates   **5.** denominator
7. true   **9.** false; $-12 + 13i$ **11.** real, complex
13. pure imaginary, nonreal complex, complex
15. nonreal complex, complex   **17.** real, complex
19. pure imaginary, nonreal complex, complex
21. $5i$ **23.** $i\sqrt{10}$ **25.** $12i\sqrt{2}$ **27.** $-3i\sqrt{2}$ **29.** -13
31. $-2\sqrt{6}$ **33.** $\sqrt{3}$ **35.** $i\sqrt{3}$ **37.** $\frac{1}{2}$ **39.** -2
41. $-3 - i\sqrt{6}$ **43.** $2 + 2i\sqrt{2}$ **45.** $-\frac{1}{8} + \frac{\sqrt{2}}{8}i$
47. $12 - i$ **49.** 2   **51.** $1 - 10i$ **53.** $-13 + 4i\sqrt{2}$
55. $8 - i$ **57.** $-14 + 2i$ **59.** $5 - 12i$ **61.** 10   **63.** 13
65. 7   **67.** $25i$ **69.** $12 + 9i$ **71.** $20 + 15i$ **73.** $2 - 2i$
75. $\frac{3}{5} - \frac{4}{5}i$ **77.** $-1 - 2i$ **79.** $5i$ **81.** $8i$ **83.** $-\frac{2}{3}i$
85. $E = 2 + 62i$ **87.** $Z = 12 + 8i$ **89.** i **91.** -1
93. $-i$ **95.** 1   **97.** $-i$ **99.** $-i$

1.4 Exercises

1. G   **3.** C   **5.** H   **7.** D   **9.** D; $\left\{\frac{1}{3}, 7\right\}$ **11.** C; $\{-4, 3\}$
13. $\{2, 3\}$ **15.** $\left\{-\frac{2}{5}, 1\right\}$ **17.** $\left\{-\frac{3}{4}, 1\right\}$ **19.** $\{\pm 10\}$
21. $\left\{\frac{1}{2}\right\}$ **23.** $\left\{-\frac{3}{5}\right\}$ **25.** $\{\pm 4\}$ **27.** $\left\{\pm 3\sqrt{3}\right\}$
29. $\{\pm 9i\}$ **31.** $\left\{\frac{1 \pm 2\sqrt{3}}{3}\right\}$ **33.** $\left\{-5 \pm i\sqrt{3}\right\}$
35. $\left\{\frac{3}{5} \pm \frac{\sqrt{3}}{5}i\right\}$ **37.** $\{1, 3\}$ **39.** $\left\{-\frac{7}{2}, 4\right\}$ **41.** $\left\{1 \pm \sqrt{3}\right\}$
43. $\left\{-\frac{5}{2}, 2\right\}$ **45.** $\left\{\frac{2 \pm \sqrt{10}}{2}\right\}$ **47.** $\left\{1 \pm \frac{\sqrt{3}}{2}i\right\}$
49. He is incorrect because $c = 0$.   **51.** $\left\{\frac{1 \pm \sqrt{5}}{2}\right\}$
53. $\left\{3 \pm \sqrt{2}\right\}$ **55.** $\{1 \pm 2i\}$ **57.** $\left\{\frac{3}{2} \pm \frac{\sqrt{2}}{2}i\right\}$
59. $\left\{\frac{-1 \pm \sqrt{97}}{4}\right\}$ **61.** $\left\{\frac{-2 \pm \sqrt{10}}{2}\right\}$ **63.** $\left\{\frac{-3 \pm \sqrt{41}}{8}\right\}$
65. $\{5\}$ **67.** $\left\{2, -1 \pm i\sqrt{3}\right\}$ **69.** $\left\{-3, \frac{3}{2} \pm \frac{3\sqrt{3}}{2}i\right\}$
71. $t = \frac{\pm\sqrt{2sg}}{g}$ **73.** $v = \frac{\pm\sqrt{FrkM}}{kM}$ **75.** $t = \frac{\pm\sqrt{2a(r - r_0)}}{a}$
77. $t = \frac{v_0 \pm \sqrt{v_0{}^2 - 64h + 64s_0}}{32}$
79. (a) $x = \frac{y \pm \sqrt{8 - 11y^2}}{4}$ **(b)** $y = \frac{x \pm \sqrt{6 - 11x^2}}{3}$

81. (a) $x = \frac{-2y \pm \sqrt{10y^2 + 4}}{2}$ **(b)** $y = \frac{2x \pm \sqrt{10x^2 - 6}}{3}$
83. 0; one rational solution (a double solution)
85. 1; two distinct rational solutions   **87.** 84; two distinct
irrational solutions   **89.** -23; two distinct nonreal complex
solutions   **91.** 2304; two distinct rational solutions   **93.** no

In Exercises 95 and 97, there are other possible answers.
95. $a = 1, b = -9, c = 20$ **97.** $a = 1, b = -2, c = -1$

Chapter 1 Quiz

[1.1] 1. $\{2\}$ **2. (a)** contradiction; $\varnothing$ **(b)** identity;
{all real numbers}   **(c)** conditional equation; $\left\{\frac{11}{4}\right\}$
3. $y = \frac{3x}{a - 1}$ **[1.2] 4.** $10,000 at 2.5%; $20,000 at 3%
5. $6.59; The model predicts a wage that is $0.04 greater
than the actual wage.   **[1.3] 6.** $-\frac{1}{2} + \frac{\sqrt{6}}{4}i$ **7.** $\frac{3}{10} - \frac{8}{5}i$
[1.4] 8. $\left\{\frac{1}{6} \pm \frac{\sqrt{11}}{6}i\right\}$ **9.** $\left\{\pm\sqrt{29}\right\}$ **10.** $r = \frac{\pm\sqrt{2s4\theta}}{\theta}$

1.5 Exercises

1. A   **3.** D   **5.** A   **7.** B   **9.** 7, 8 or $-8, -7$ **11.** 12, 14
or $-14, -12$ **13.** 7, 9 or $-9, -7$ **15.** 9, 11 or $-11, -9$
17. 20, 22   **19.** 6, 8, 10   **21.** 7 in., 10 in.   **23.** 100 yd by
400 yd   **25.** 9 ft by 12 ft   **27.** 20 in. by 30 in.   **29.** 1 ft
31. 4   **33.** 3.75 cm   **35.** 5 ft   **37.** $10\sqrt{2}$ ft   **39.** 16.4 ft
41. 3000 yd   **43. (a)** 1 sec, 5 sec   **(b)** 6 sec
45. (a) It will not reach 80 ft.   **(b)** 2 sec   **47. (a)** 0.19 sec,
10.92 sec   **(b)** 11.32 sec   **49. (a)** $108.1 million   **(b)** 2004
51. (a) 19.2 hr   **(b)** 84.3 ppm (109.8 is not in the interval
$[50, 100]$.)   **53. (a)** 549.2 million metric tons   **(b)** 2019
55. $42,795 million   **57.** $80 - x$ **58.** $300 + 20x$
59. $R = (80 - x)(300 + 20x) = 24,000 + 1300x - 20x^2$
60. 10, 55; Because of the restriction, only $x = 10$ is valid.
The number of apartments rented is 70.   **61.** 80   **63.** 4

1.6 Exercises

1. rational equation   **3.** $\frac{1}{4}$ **5.** rational exponent   **7.** D
9. E   **11.** $-\frac{3}{2}, 6$ **13.** $2, -1$ **15.** 0   **17.** $\{-10\}$
19. $\varnothing$ **21.** $\varnothing$ **23.** $\{-9\}$ **25.** $\{-2\}$ **27.** $\varnothing$
29. $\left\{-\frac{5}{2}, \frac{1}{9}\right\}$ **31.** $\left\{\frac{3}{4}, 1\right\}$ **33.** $\{3, 5\}$ **35.** $\left\{-2, \frac{5}{4}\right\}$
37. $1\frac{7}{8}$ hr   **39.** 78 hr   **41.** $13\frac{1}{3}$ hr   **43.** 10 min   **45.** $\{3\}$
47. $\{-1\}$ **49.** $\{5\}$ **51.** $\{9\}$ **53.** $\{9\}$ **55.** $\varnothing$
57. $\{\pm 2\}$ **59.** $\{0, 3\}$ **61.** $\{-2\}$ **63.** $\left\{-\frac{2}{9}, 2\right\}$
65. $\{4\}$ **67.** $\{-2\}$ **69.** $\left\{\frac{2}{5}, 1\right\}$ **71.** $\{31\}$
73. $\{-3, 1\}$ **75.** $\{25\}$ **77.** $\{-27, 3\}$ **79.** $\{-29, 35\}$
81. $\left\{\frac{3}{2}\right\}$ **83.** $\left\{\frac{1}{4}, 1\right\}$ **85.** $\{0, 8\}$ **87.** $\left\{\pm 1, \pm\frac{\sqrt{10}}{2}\right\}$
89. $\left\{\pm\sqrt{3}, \pm i\sqrt{5}\right\}$ **91.** $\{-63, 28\}$ **93.** $\{0, 31\}$
95. $\left\{-\frac{5}{2}, -2, 0, \frac{1}{2}\right\}$ **97.** $\left\{\frac{-6 \pm 2\sqrt{3}}{3}, \frac{-4 \pm \sqrt{2}}{2}\right\}$

99. $\left\{-\frac{2}{7}, 5\right\}$ **101.** $\left\{-\frac{1}{27}, \frac{1}{8}\right\}$ **103.** $\left\{\pm\frac{1}{2}, \pm 4\right\}$

105. $h = \frac{d^2}{k^2}$ **107.** $m = (1 - n^{3/4})^{4/3}$ **109.** $e = \frac{Er}{R + r}$

111. $\{16\}$ **112.** $\{16\}$ **113.** Answers may vary.

114. $\{4\}$

Summary Exercises on Solving Equations

1. $\{3\}$ **2.** $\{-1\}$ **3.** $\left\{-3 \pm 3\sqrt{2}\right\}$ **4.** $\{2, 6\}$

5. $\varnothing$ **6.** $\{-31\}$ **7.** $\{-6\}$ **8.** $\{6\}$ **9.** $\left\{\frac{1}{5} \pm \frac{2}{5}i\right\}$

10. $\{-2, 1\}$ **11.** $\left\{-\frac{1}{243}, \frac{1}{3125}\right\}$ **12.** $\{-1\}$

13. $\{\pm i, \pm 2\}$ **14.** $\{-2.4\}$ **15.** $\{4\}$

16. $\left\{\frac{1}{3} \pm \frac{\sqrt{2}}{3}i\right\}$ **17.** $\left\{\frac{15}{7}\right\}$ **18.** $\{4\}$ **19.** $\{3, 11\}$

20. $\{1\}$ **21.** $\{x \mid x \neq 3\}$ **22.** $a = \pm\sqrt{c^2 - b^2}$

1.7 Exercises

1. F   **3.** A   **5.** I   **7.** B   **9.** E   **11.** A square bracket is used to show that a number is part of the solution set, and a parenthesis is used to indicate that a number is not part of the solution set.   **13.** $[-4, \infty)$ **15.** $[-1, \infty)$ **17.** $(-\infty, \infty)$

19. $(-\infty, 4)$ **21.** $\left[-\frac{11}{5}, \infty\right)$ **23.** $\left(-\infty, \frac{48}{7}\right)$ **25.** $[500, \infty)$

27. The product will never break even.   **29.** $(-5, 3)$

31. $[3, 6]$ **33.** $(4, 6)$ **35.** $[-9, 9]$ **37.** $\left(-\frac{16}{3}, \frac{19}{3}\right)$

39. $(-\infty, -2) \cup (3, \infty)$ **41.** $\left[-\frac{3}{2}, 6\right]$

43. $(-\infty, -3] \cup [-1, \infty)$ **45.** $[-2, 3]$ **47.** $[-3, 3]$

49. $\varnothing$ **51.** $\left[1 - \sqrt{2}, 1 + \sqrt{2}\right]$ **53.** A   **55.** $(-5, 3]$

57. $(-\infty, -2)$ **59.** $(-\infty, 6) \cup \left[\frac{15}{2}, \infty\right)$

61. $(-\infty, 1) \cup \left(\frac{9}{5}, \infty\right)$ **63.** $\left(-\infty, -\frac{3}{2}\right) \cup \left[-\frac{1}{2}, \infty\right)$

65. $(-2, \infty)$ **67.** $\left(0, \frac{4}{11}\right) \cup \left(\frac{1}{2}, \infty\right)$

69. $(-\infty, -2] \cup (1, 2)$ **71.** $(-\infty, 5)$ **73.** $\left[\frac{3}{2}, \infty\right)$

75. $\left(\frac{5}{2}, \infty\right)$ **77.** $\left[-\frac{8}{3}, \frac{3}{2}\right] \cup (6, \infty)$ **79. (a)** 2000

(b) 2008   **81.** between (and inclusive of) 4 sec and 9.75 sec

83. between (and inclusive of) 1 sec and 1.75 sec

85. between -1.5 sec and 4 sec   **87.** $\left\{\frac{4}{3}, -2, -6\right\}$

88. ↔ (number line)

choose $x = -10$, for example. It satisfies the original inequality. In the interval $(-6, -2)$, choose $x = -4$, for example. It does not satisfy the inequality. In the interval $\left(-2, \frac{4}{3}\right)$, choose $x = 0$, for example. It satisfies the original inequality. In the interval $\left(\frac{4}{3}, \infty\right)$, choose $x = 4$, for example. It does not satisfy the original inequality.

89. In the interval $(-\infty, -6)$,

90. (a) ↔ (number line)   **(b)** $(-\infty, -6] \cup \left[-2, \frac{4}{3}\right]$

91. $\left[-2, \frac{3}{2}\right] \cup [3, \infty)$ **93.** $(-\infty, -2] \cup [0, 2]$

95. $(-\infty, -1) \cup (-1, 3)$ **97.** $[-4, -3] \cup [3, \infty)$

99. $(-\infty, \infty)$

1.8 Exercises

1. F   **3.** D   **5.** G   **7.** C   **9.** $\left\{-\frac{1}{3}, 1\right\}$ **11.** $\left\{\frac{2}{3}, \frac{8}{3}\right\}$

13. $\{-6, 14\}$ **15.** $\left\{\frac{5}{2}, \frac{7}{2}\right\}$ **17.** $\left\{-\frac{4}{3}, \frac{2}{9}\right\}$ **19.** $\left\{-\frac{7}{3}, -\frac{1}{7}\right\}$

21. $\{1\}$ **23.** $(-\infty, \infty)$ **25.** A positive solution will return a negative quantity for $-5x$, and the absolute value of an expression cannot be negative.   **27.** $(-4, -1)$

29. $(-\infty, -4] \cup [-1, \infty)$ **31.** $\left(-\frac{3}{2}, \frac{5}{2}\right)$

33. $(-\infty, 0) \cup (6, \infty)$ **35.** $\left(-\infty, -\frac{2}{3}\right) \cup (4, \infty)$

37. $\left[-\frac{2}{3}, 4\right]$ **39.** $\left[-1, -\frac{1}{2}\right]$ **41.** $(-101, -99)$

43. $\left\{-1, -\frac{1}{2}\right\}$ **45.** $\{2, 4\}$ **47.** $\left(-\frac{4}{3}, \frac{2}{3}\right)$ **49.** $\left(-\frac{3}{2}, \frac{13}{10}\right)$

51. $\left(-\infty, \frac{3}{2}\right] \cup \left[\frac{7}{2}, \infty\right)$ **53.** $\varnothing$ **55.** $(-\infty, \infty)$ **57.** $\varnothing$

59. $\left\{-\frac{5}{8}\right\}$ **61.** $\varnothing$ **63.** $\left\{-\frac{1}{2}\right\}$ **65.** $\left(-\infty, -\frac{2}{3}\right) \cup \left(-\frac{2}{3}, \infty\right)$

In Exercises 67–73, the expression in absolute value bars may be replaced by its additive inverse. For example, in Exercise 67, $p - q$ may be written $q - p$.

67. $|p - q| = 2$ **69.** $|m - 7| \leq 2$ **71.** $|p - 9| < 0.0001$

73. $|r - 29| \geq 1$ **75.** $(0.9996, 1.0004)$ **77.** $[6.7, 9.7]$

79. $|F - 730| \leq 50$ **81.** $25.33 \leq R_L \leq 28.17$;

$36.58 \leq R_E \leq 40.92$ **83.** -6 or 6   **84.** $x^2 - x = 6$;

$\{-2, 3\}$ **85.** $x^2 - x = -6$; $\left\{\frac{1}{2} \pm \frac{\sqrt{23}}{2}i\right\}$

86. $\left\{-2, 3, \frac{1}{2} \pm \frac{\sqrt{23}}{2}i\right\}$ **87.** $\left\{-\frac{7}{3}, 2, -\frac{1}{6} \pm \frac{\sqrt{167}}{6}i\right\}$

89. $\left\{-\frac{1}{4}, 6\right\}$ **91.** $\{-1, 1\}$ **93.** $\varnothing$

95. $\left(-\infty, -\frac{1}{3}\right) \cup \left(-\frac{1}{3}, \infty\right)$

Chapter 1 Review Exercises

1. $\{6\}$ **3.** $\left\{-\frac{11}{3}\right\}$ **5.** $f = \frac{AB(p + 1)}{24}$ **7.** A, B

9. 13 in. on each side   **11.** $3\frac{3}{7}$ L   **13.** 560 km per hr

15. (a) $A = 36.525x$ **(b)** 2629.8 mg   **17. (a)** \$4.31; The model gives a figure that is \$0.51 more than the actual figure of \$3.80.   **(b)** 47.6 yr after 1956, which is mid-2003. This is close to the minimum wage changing to \$5.85 in 2007.

19. $13 - 3i$ **21.** $-14 + 13i$ **23.** $19 + 17i$ **25.** 146

27. $-30 - 40i$ **29.** $1 - 2i$ **31.** $-i$ **33.** i **35.** i

37. $\left\{-7 \pm \sqrt{5}\right\}$ **39.** $\left\{-3, \frac{5}{2}\right\}$ **41.** $\left\{-\frac{3}{2}, 7\right\}$

43. $\left\{2 \pm \sqrt{6}\right\}$ **45.** $\left\{\frac{\sqrt{5} \pm 3}{2}\right\}$ **47.** D

49. 76; two distinct irrational solutions

51. -124; two distinct nonreal complex solutions

53. 0; one rational solution (a double solution)

55. 6.25 sec and 7.5 sec   **57.** $\frac{1}{2}$ ft   **59.** \$801.9 billion

61. $\left\{\pm i, \pm\frac{1}{2}\right\}$ **63.** $\left\{-\frac{7}{24}\right\}$ **65.** $\varnothing$ **67.** $\{-239, 247\}$

69. $\{1, 4\}$ **71.** $\{-2, -1\}$ **73.** $\{3\}$ **75.** $\varnothing$
77. $\{-1\}$ **79.** $\{-\frac{7}{4}\}$ **81.** $\{-15, \frac{5}{2}\}$ **83.** $\left(-\frac{7}{13}, \infty\right)$
85. $(-\infty, 1]$ **87.** $[4, 5]$ **89.** $[-4, 1]$ **91.** $\left(-\frac{2}{3}, \frac{5}{2}\right)$
93. $(-\infty, -4] \cup [0, 4]$ **95.** $(-\infty, -2) \cup (5, \infty)$
97. $(-2, 0)$ **99.** $(-3, 1) \cup [7, \infty)$ **101. (a)** 79.8 ppb
(b) 87.7 ppb **103. (a)** 20 sec **(b)** between 2 sec and
18 sec **105.** The value 3 makes the denominator 0. Therefore, it must be excluded from the solution set.
107. $\{-11, 3\}$ **109.** $\left\{\frac{11}{27}, \frac{25}{27}\right\}$ **111.** $\left\{-\frac{2}{7}, \frac{4}{3}\right\}$
113. $[-6, -3]$ **115.** $\left(-\infty, -\frac{1}{7}\right) \cup (1, \infty)$
117. $\left(-\frac{17}{3}, 1\right)$ **119.** $(-\infty, \infty)$ **121.** $\{0, -4\}$
123. $|k - 6| = 12$ (or $|6 - k| = 12$)
125. $|t - 5| \geq 0.01$ (or $|5 - t| \geq 0.01$)

Chapter 1 Test
[1.1] **1.** $\{0\}$ **2.** $\{-12\}$ [1.4] **3.** $\left\{\frac{1}{2}, \frac{7}{3}\right\}$
4. $\left\{\frac{-1 \pm 2\sqrt{2}}{3}\right\}$ **5.** $\left\{-\frac{1}{3} \pm \frac{\sqrt{5}}{3}i\right\}$ [1.6] **6.** $\varnothing$
7. $\left\{-\frac{3}{4}\right\}$ **8.** $\{4\}$ **9.** $\{-3, 1\}$ **10.** $\{-2\}$
11. $\{\pm 1, \pm 4\}$ **12.** $\{-30, 5\}$ [1.8] **13.** $\left\{-\frac{5}{2}, 1\right\}$
14. $\left\{-6, \frac{4}{3}\right\}$ [1.1] **15.** $W = \frac{S - 2LH}{2H + 2L}$
[1.3] **16. (a)** $5 - 8i$ **(b)** $-29 - 3i$ **(c)** $55 + 48i$
(d) $6 + i$ **17. (a)** -1 **(b)** i **(c)** i
[1.2] **18. (a)** $A = 806,400x$ **(b)** 24,192,000 gal
(c) $P = 40.32x$; 40 pools **(d)** 25 days
19. length: 200 m; width: 110 m **20.** cashews: $23\frac{1}{3}$ lb;
walnuts: $11\frac{2}{3}$ lb [1.2] **21.** 15 mph [1.2] **22. (a)** 1.7% **(b)** 2007
[1.5] **23. (a)** 1 sec and 5 sec **(b)** 6 sec
[1.7] **24.** $(-3, \infty)$ **25.** $[-10, 2]$
26. $(-\infty, -1] \cup \left[\frac{3}{2}, \infty\right)$ **27.** $(-\infty, 3) \cup (4, \infty)$
[1.8] **28.** $(-2, 7)$ **29.** $(-\infty, -6] \cup [5, \infty)$ **30.** $\left\{-\frac{7}{3}\right\}$

Chapter 2 Graphs and Functions

2.1 Exercises
1. II **3.** 0 **5.** $(5, 0)$ **7.** true **9.** false; The midpoint
is a point with coordinates $(2, 2)$. **11.** any three of the
following: $(2, -5), (-1, 7), (3, -9), (5, -17), (6, -21)$
13. any three of the following: $(1999, 35), (2001, 29),$
$(2003, 22), (2005, 23), (2007, 20), (2009, 20)$
15. (a) 13 **(b)** $\left(1, -\frac{7}{2}\right)$ **17. (a)** $\sqrt{34}$ **(b)** $\left(\frac{11}{2}, \frac{7}{2}\right)$
19. (a) $3\sqrt{41}$ **(b)** $\left(0, \frac{5}{2}\right)$ **21. (a)** $\sqrt{133}$
(b) $\left(2\sqrt{2}, \frac{3\sqrt{5}}{2}\right)$ **23.** yes **25.** no **27.** yes **29.** yes
31. no **33.** no **35.** $(-3, 6)$ **37.** $(5, -4)$
39. $(2a - p, 2b - q)$ **41.** 26.1%; This estimate is very
close to the actual figure. **43.** $23,428

Other ordered pairs are possible in Exercises 47–57.

47. (a)

x	y
0	-2
4	0
2	-1

(b)

49. (a)

x	y
0	$\frac{5}{3}$
$\frac{5}{2}$	0
4	-1

(b)

51. (a)

x	y
0	0
1	1
2	4

(b)

53. (a)

x	y
3	0
4	1
7	2

(b)

55. (a)

x	y
4	2
-2	4
0	2

(b)

57. (a)

x	y
0	0
-1	-1
2	8

(b)

59. $(4, 0)$ **61.** III; I; IV; IV **63.** yes; no

2.2 Exercises
1. $(0, 0)$; 7 **3.** $(4, -7)$ **5.** B **7.** D
9. one (The point is $(0, 0)$.)
11. (a) $x^2 + y^2 = 36$ **(b)**

13. (a) $(x - 2)^2 + y^2 = 36$ **(b)**

$(x - 2)^2 + y^2 = 36$

15. (a) $x^2 + (y - 4)^2 = 16$ **(b)**

$x^2 + (y - 4)^2 = 16$

17. (a) $(x + 2)^2 + (y - 5)^2 = 16$ **(b)**

$(x + 2)^2 + (y - 5)^2 = 16$

19. (a) $(x - 5)^2 + (y + 4)^2 = 49$ **(b)**

$(x - 5)^2 + (y + 4)^2 = 49$

21. (a) $\left(x - \sqrt{2}\right)^2 + \left(y - \sqrt{2}\right)^2 = 2$
(b)

$(x - \sqrt{2})^2 + (y - \sqrt{2})^2 = 2$

23. (a) $(x - 3)^2 + (y - 1)^2 = 4$
(b) $x^2 + y^2 - 6x - 2y + 6 = 0$
25. (a) $(x + 2)^2 + (y - 2)^2 = 4$
(b) $x^2 + y^2 + 4x - 4y + 4 = 0$
27. yes; center: $(-3, -4)$; radius: 4 **29.** yes;
center: $(2, -6)$; radius: 6 **31.** yes; center: $\left(-\frac{1}{2}, 2\right)$;
radius: 3 **33.** no; The graph is nonexistent.
35. no; The graph is the point $(3, 3)$. **37.** yes;
center: $\left(\frac{1}{3}, -\frac{1}{3}\right)$; radius: $\frac{5}{3}$ **39.** $(3, 1)$ **41.** $(-2, -2)$
43. $(x - 3)^2 + (y - 2)^2 = 4$
45. $\left(2 + \sqrt{7}, 2 + \sqrt{7}\right), \left(2 - \sqrt{7}, 2 - \sqrt{7}\right)$
47. $(2, 3)$ and $(4, 1)$ **49.** $9 + \sqrt{119}, 9 - \sqrt{119}$
51. $\sqrt{113} - 5$ **53.** $(2, -3)$ **54.** $3\sqrt{5}$ **55.** $3\sqrt{5}$
56. $3\sqrt{5}$ **57.** $(x - 2)^2 + (y + 3)^2 = 45$

58. $(x + 2)^2 + (y + 1)^2 = 41$
59. $(x - 5)^2 + \left(y - \frac{9}{2}\right)^2 = \frac{169}{4}$
61. $(x - 3)^2 + \left(y - \frac{5}{2}\right)^2 = \frac{25}{4}$

2.3 Exercises

1. $\{3, 4, 10\}$ **3.** $x; y$ **5.** 10 **7.** $[0, \infty)$ **9.** $(-\infty, 3)$
11. function **13.** not a function **15.** function
17. function **19.** not a function; domain: $\{0, 1, 2\}$;
range: $\{-4, -1, 0, 1, 4\}$ **21.** function; domain:
$\{2, 3, 5, 11, 17\}$; range: $\{1, 7, 20\}$ **23.** function; domain:
$\{0, -1, -2\}$; range: $\{0, 1, 2\}$ **25.** function; domain:
$\{2010, 2011, 2012, 2013\}$; range: $\{64.9, 63.0, 65.1, 63.5\}$
27. function; domain: $(-\infty, \infty)$; range: $(-\infty, \infty)$
29. not a function; domain: $[3, \infty)$; range: $(-\infty, \infty)$
31. function; domain: $(-\infty, \infty)$; range: $(-\infty, \infty)$
33. function; domain: $(-\infty, \infty)$; range: $[0, \infty)$
35. not a function; domain: $[0, \infty)$; range: $(-\infty, \infty)$
37. function; domain: $(-\infty, \infty)$; range: $(-\infty, \infty)$
39. not a function; domain: $(-\infty, \infty)$; range: $(-\infty, \infty)$
41. function; domain: $[0, \infty)$; range: $[0, \infty)$
43. function; domain: $(-\infty, 0) \cup (0, \infty)$; range:
$(-\infty, 0) \cup (0, \infty)$ **45.** function; domain: $\left[-\frac{1}{4}, \infty\right)$;
range: $[0, \infty)$ **47.** function; domain: $(-\infty, 3) \cup (3, \infty)$;
range: $(-\infty, 0) \cup (0, \infty)$ **49.** B **51.** 4 **53.** -11
55. 3 **57.** $\frac{11}{4}$ **59.** $-3p + 4$ **61.** $3x + 4$ **63.** $-3x - 2$
65. $-6m + 13$ **67. (a)** 2 **(b)** 3 **69. (a)** 15 **(b)** 10
71. (a) 3 **(b)** -3 **73. (a)** 0 **(b)** 4 **(c)** 2 **(d)** 4
75. (a) -3 **(b)** -2 **(c)** 0 **(d)** 2
77. (a) $f(x) = -\frac{1}{3}x + 4$ **(b)** 3
79. (a) $f(x) = -2x^2 - x + 3$ **(b)** -18
81. (a) $f(x) = \frac{4}{3}x - \frac{8}{3}$ **(b)** $\frac{4}{3}$ **83.** $f(3) = 4$ **85.** -4
87. (a) $(-2, 0)$ **(b)** $(-\infty, -2)$ **(c)** $(0, \infty)$
89. (a) $(-\infty, -2); (2, \infty)$ **(b)** $(-2, 2)$ **(c)** none
91. (a) $(-1, 0); (1, \infty)$ **(b)** $(-\infty, -1); (0, 1)$ **(c)** none
93. (a) yes **(b)** $[0, 24]$ **(c)** 1200 megawatts
(d) at 17 hr or 5 p.m.; at 4 a.m. **(e)** $f(12) = 1900$;
At 12 noon, electricity use is 1900 megawatts.
(f) increasing from 4 a.m. to 5 p.m.; decreasing from
midnight to 4 a.m. and from 5 p.m. to midnight
95. (a) 12 noon to 8 p.m. **(b)** from midnight until about
6 a.m. and after 10 p.m. **(c)** about 10 a.m. and 8:30 p.m.
(d) The temperature is $40°$ from midnight to 6 a.m.,
when it begins to rise until it reaches a maximum of just
below $65°$ at 4 p.m. It then begins to fall until it reaches just
under $40°$ at midnight.

2.4 Exercises

1. B **3.** C **5.** A **7.** C **9.** D

In Exercises 11–29, we give the domain first and then the range.

11.

$(-\infty, \infty); (-\infty, \infty)$

13.

$(-\infty, \infty); (-\infty, \infty)$

15.

$(-\infty, \infty); (-\infty, \infty)$

17.

$(-\infty, \infty); \{-4\};$ constant function

19.

$(-\infty, \infty); \{0\};$ constant function

21.

$(-\infty, \infty); (-\infty, \infty)$

23.

$(-\infty, \infty); (-\infty, \infty)$

25.

$\{3\}; (-\infty, \infty)$

27.

$\{-2\}; (-\infty, \infty)$

29.

$\{5\}; (-\infty, \infty)$

31. A **33.** D

35.

37.

39. A, C, D, E **41.** $\frac{2}{5}$ **43.** -2 **45.** 0 **47.** 0
49. undefined

51. (a) $m = 3$
(b)

53. (a) $m = -\frac{3}{2}$
(b)

55. (a) $m = \frac{5}{2}$
(b)

57.

59.

61.

63.

65. $-\$4000$ per year; The value of the machine is decreasing $\$4000$ each year during these years. **67.** 0% per year (or no change); The percent of pay raise is not changing—it is 3% each year during these years. **69.** -78.8 thousand per year; The number of high school dropouts decreased by an average of 78.8 thousand per year from 1980 to 2012.
71. (a) The slope -0.0167 indicates that the average rate of change per year of the winning time for the 5000-m run is 0.0167 min less. It is negative because the times are generally decreasing as time progresses. **(b)** The Olympics were not held during World Wars I and II. **(c)** 13.05 min; The times differ by 0.30 min. **73.** 13,064 (thousands)
75. (a) -8.17 thousand mobile homes per year **(b)** The negative slope means that the number of mobile homes *decreased* by an average of 8.17 thousand each year from 2003 to 2013. **77. (a)** $C(x) = 10x + 500$ **(b)** $R(x) = 35x$
(c) $P(x) = 25x - 500$ **(d)** 20 units; do not produce
79. (a) $C(x) = 400x + 1650$ **(b)** $R(x) = 305x$
(c) $P(x) = -95x - 1650$ **(d)** $R(x) < C(x)$ for all positive x; don't produce, impossible to make a profit
81. 25 units; $\$6000$ **83.** 3 **84.** 3 **85.** the same
86. $\sqrt{10}$ **87.** $2\sqrt{10}$ **88.** $3\sqrt{10}$ **89.** The sum is $3\sqrt{10}$, which is equal to the answer in **Exercise 88.**
90. $B; C; A; C$ **91.** The midpoint is $(3, 3)$, which is the same as the middle entry in the table. **92.** 7.5

Chapter 2 Quiz

[2.1] **1.** $\sqrt{41}$ **2.** 2006: 6.76 million; 2010: 7.24 million
3.   [2.2] **4.**

$y = -x^2 + 4$

$x^2 + y^2 = 16$

5. radius: $\sqrt{17}$; center: $(2, -4)$ [2.3] **6.** 2
7. domain: $(-\infty, \infty)$; range: $[0, \infty)$ **8. (a)** $(-\infty, -3)$
(b) $(-3, \infty)$ **(c)** none   [2.4] **9. (a)** $\frac{3}{2}$ **(b)** 0
(c) undefined   **10.** 1320.5 thousand per year; The number
of new motor vehicles sold in the United States increased
an average of 1320.5 thousand per year from 2009 to 2013.

2.5 Exercises

1. 4; 3   **3.** x **5.** $-\frac{6}{7}$ **7.** D   **9.** C   **11.** $2x + y = 5$
13. $3x + 2y = -7$ **15.** $x = -8$ **17.** $y = -8$
19. $x - 4y = -13$ **21.** $y = \frac{2}{3}x - 2$ **23.** $x = -6$
(cannot be written in slope-intercept form)   **25.** $y = 4$
27. $y = 5x + 15$ **29.** $y = -4x - 3$ **31.** $y = \frac{3}{2}$
33. $(-2, 0)$; does not; undefined; $\left(0, \frac{1}{2}\right)$; does not; 0

35. slope: 3;   **37.** slope: 4;
y-intercept: $(0, -1)$ y-intercept: $(0, -7)$

39. slope: $-\frac{3}{4}$;   **41.** slope: $-\frac{1}{2}$;
y-intercept: $(0, 0)$ y-intercept: $(0, -2)$

43. slope: $\frac{3}{2}$;   **45. (a)** -2; $(0, 1)$; $\left(\frac{1}{2}, 0\right)$
y-intercept: $(0, 1)$ **(b)** $f(x) = -2x + 1$
47. (a) $-\frac{1}{3}$; $(0, 2)$; $(6, 0)$
(b) $f(x) = -\frac{1}{3}x + 2$
49. (a) -200; $(0, 300)$; $\left(\frac{3}{2}, 0\right)$
(b) $f(x) = -200x + 300$
51. (a) $x + 3y = 11$
(b) $y = -\frac{1}{3}x + \frac{11}{3}$
53. (a) $5x - 3y = -13$ **(b)** $y = \frac{5}{3}x + \frac{13}{3}$
55. (a) $y = 1$ **(b)** $y = 1$ **57. (a)** $y = 6$ **(b)** $y = 6$
59. (a) $-\frac{1}{2}$ **(b)** $-\frac{7}{2}$ **61. (a)** $y = 463.67x + 6312$
(b) \$8166.68; The result is \$96.68 more than the actual figure.

63. (a) $f(x) = 622.25x + 22{,}036$

$f(x) = 622.25x + 22{,}036$ To the nearest dollar, the average
tuition increase is \$622 per year
for the period because this is the
slope of the line.

(b) $f(3) = \$23{,}903$; This is a fairly good approximation.
(c) $f(x) = 653x + 21{,}634$
65. (a) $F = \frac{9}{5}C + 32$ **(b)** $C = \frac{5}{9}(F - 32)$ **(c)** $-40°$
67. (a) $C = 0.6516I + 2253$ **(b)** 0.6516   **69.** $\{3\}$
71. $\{-0.5\}$ **73. (a)** $\{12\}$ **(b)** The solution does not
appear in the x-values interval $[-10, 10]$. The minimum
and maximum values must include 12.   **75.** yes   **77.** no
79. $\sqrt{x_1{}^2 + m_1{}^2x_1{}^2}$ **80.** $\sqrt{x_2{}^2 + m_2{}^2x_2{}^2}$
81. $\sqrt{(x_2 - x_1)^2 + (m_2x_2 - m_1x_1)^2}$
83. $-2x_1x_2(m_1m_2 + 1) = 0$ **84.** Because $x_1 \neq 0$, $x_2 \neq 0$,
we have $m_1m_2 + 1 = 0$, implying that $m_1m_2 = -1$.
85. If two nonvertical lines are perpendicular, then the
product of the slopes of these lines is -1.

Summary Exercises on Graphs, Circles, Functions, and Equations

1. (a) $\sqrt{65}$ **(b)** $\left(\frac{5}{2}, 1\right)$ **(c)** $y = 8x - 19$ **2. (a)** $\sqrt{29}$
(b) $\left(\frac{3}{2}, -1\right)$ **(c)** $y = -\frac{2}{5}x - \frac{2}{5}$ **3. (a)** 5   **(b)** $\left(\frac{1}{2}, 2\right)$
(c) $y = 2$ **4. (a)** $\sqrt{10}$ **(b)** $\left(\frac{3\sqrt{2}}{2}, 2\sqrt{2}\right)$
(c) $y = -2x + 5\sqrt{2}$ **5. (a)** 2   **(b)** $(5, 0)$ **(c)** $x = 5$
6. (a) $4\sqrt{2}$ **(b)** $(-1, -1)$ **(c)** $y = x$ **7. (a)** $4\sqrt{3}$
(b) $\left(4\sqrt{3}, 3\sqrt{5}\right)$ **(c)** $y = 3\sqrt{5}$ **8. (a)** $\sqrt{34}$
(b) $\left(\frac{3}{2}, -\frac{3}{2}\right)$ **(c)** $y = \frac{5}{3}x - 4$
9. $y = -\frac{1}{3}x + \frac{1}{3}$ **10.** $y = 3$

11. $(x - 2)^2 + (y + 1)^2 = 9$ **12.** $x^2 + (y - 2)^2 = 4$

13. $y = -\frac{5}{6}x - \frac{5}{2}$ **14.** $x = -4$

15. $y = -\frac{2}{3}x$

16. $y = -\frac{4}{3}x$

33.

$$f(x) = \begin{cases} x^3 + 3 & \text{if } -2 \le x \le 0 \\ x + 3 & \text{if } 0 < x < 1 \\ 4 + x - x^2 & \text{if } 1 \le x \le 3 \end{cases}$$

17. yes; center: $(2, -1)$; radius: 3 **18.** no **19.** yes; center: $(6, 0)$; radius: 4 **20.** yes; center: $(-1, -8)$; radius: 2 **21.** no **22.** yes; center: $(0, 4)$; radius: 5

23. $\left(4 - \sqrt{7}, 2\right), \left(4 + \sqrt{7}, 2\right)$ **24.** 8

25. (a) domain: $(-\infty, \infty)$; range: $(-\infty, \infty)$

(b) $f(x) = \frac{1}{4}x + \frac{3}{2}$; 1 **26. (a)** domain: $[-5, \infty)$; range: $(-\infty, \infty)$ **(b)** y is not a function of x.

27. (a) domain: $[-7, 3]$; range: $[-5, 5]$

(b) y is not a function of x. **28. (a)** domain: $(-\infty, \infty)$;

range: $\left[-\frac{3}{2}, \infty\right)$ **(b)** $f(x) = \frac{1}{2}x^2 - \frac{3}{2}$; $\frac{1}{2}$

2.6 Exercises

1. E; $(-\infty, \infty)$ **3.** A; $(-\infty, \infty)$ **5.** F; $f(x) = x$

7. H; no **9.** B; $\{\ldots, -3, -2, -1, 0, 1, 2, 3, \ldots\}$

11. $(-\infty, \infty)$ **13.** $[0, \infty)$ **15.** $(-\infty, 3); (3, \infty)$

17. (a) -10 **(b)** -2 **(c)** -1 **(d)** 2

19. (a) -3 **(b)** 1 **(c)** 0 **(d)** 9

21.

$f(x) = \begin{cases} x - 1 & \text{if } x \le 3 \\ 2 & \text{if } x > 3 \end{cases}$

23.

$f(x) = \begin{cases} 4 - x & \text{if } x < 2 \\ 1 + 2x & \text{if } x \ge 2 \end{cases}$

25.

$f(x) = \begin{cases} -3 & \text{if } x \le 1 \\ -1 & \text{if } x > 1 \end{cases}$

27.

$f(x) = \begin{cases} 2 + x & \text{if } x < -4 \\ -x & \text{if } -4 \le x \le 5 \\ 3x & \text{if } x > 5 \end{cases}$

29.

$f(x) = \begin{cases} -\frac{1}{2}x^2 + 2 & \text{if } x \le 2 \\ \frac{1}{2}x & \text{if } x > 2 \end{cases}$

31.

$f(x) = \begin{cases} 2x & \text{if } -5 \le x < -1 \\ -2 & \text{if } -1 \le x < 0 \\ x^2 - 2 & \text{if } 0 \le x \le 2 \end{cases}$

In Exercises 35–41, we give one of the possible rules, then the domain, and then the range.

35. $f(x) = \begin{cases} -1 & \text{if } x \le 0 \\ 1 & \text{if } x > 0 \end{cases}$; $(-\infty, \infty)$; $\{-1, 1\}$

37. $f(x) = \begin{cases} 2 & \text{if } x \le 0 \\ -1 & \text{if } x > 1 \end{cases}$; $(-\infty, 0] \cup (1, \infty)$; $\{-1, 2\}$

39. $f(x) = \begin{cases} x & \text{if } x \le 0 \\ 2 & \text{if } x > 0 \end{cases}$; $(-\infty, \infty)$; $(-\infty, 0] \cup \{2\}$

41. $f(x) = \begin{cases} \sqrt[3]{x} & \text{if } x < 1 \\ x + 1 & \text{if } x \ge 1 \end{cases}$; $(-\infty, \infty)$; $(-\infty, 1) \cup [2, \infty)$

43. $(-\infty, \infty)$; **45.** $(-\infty, \infty)$;

$\{\ldots, -2, -1, 0, 1, 2, \ldots\}$ $\{\ldots, -2, -1, 0, 1, 2, \ldots\}$

$f(x) = [\![-x]\!]$ $f(x) = [\![2x]\!]$

47.

49.

51. (a) for $[0, 8]$: $y = 1.95x + 34.2$; for $(8, 13]$: $y = 0.48x + 45.96$

(b) $f(x) = \begin{cases} 1.95x + 34.2 & \text{if } 0 \le x \le 8 \\ 0.48x + 45.96 & \text{if } 8 < x \le 13 \end{cases}$

53. (a) 50,000 gal; 30,000 gal **(b)** during the first and fourth days **(c)** 45,000; 40,000 **(d)** 5000 gal per day

55. (a) $f(x) = 0.80 \left[\!\left[\frac{x}{2}\right]\!\right]$ if $6 \le x \le 18$ **(b)** \$3.20; \$5.60

2.7 Exercises

1. 3 **3.** left **5.** x **7.** 2; 3 **9.** y **11. (a)** B

(b) D **(c)** E **(d)** A **(e)** C **13. (a)** B **(b)** A

(c) G **(d)** C **(e)** F **(f)** D **(g)** H **(h)** E **(i)** I

15. (a) F **(b)** C **(c)** H **(d)** D **(e)** G **(f)** A

(g) E **(h)** I **(i)** B

17.

19.

21.

23.

25.

27.

29.

31.

33.

35. (a) $(4, 12)$ (b) $(8, 16)$
37. (a) $(2, 12)$ (b) $(32, 12)$

39.

41.

43. $x = 2$ **45.** y-axis **47.** x-axis, y-axis, origin **49.** origin
51. none of these **53.** odd **55.** even **57.** neither

59.

61.

63.

65.

67.

69.

71.

73.

75.

77.

79.

81.

83.

85.

87. (a)   The graph of $g(x)$ is reflected across the y-axis.

(b)   The graph of $g(x)$ is translated 2 units to the right.

(c)   The graph of $g(x)$ is reflected across the x-axis.

(d)   The graph of $g(x)$ is reflected across the x-axis and translated 2 units up.

89. It is the graph of $f(x) = |x|$ translated 1 unit to the left, reflected across the x-axis, and translated 3 units up. The equation is $y = -|x + 1| + 3$.   **91.** It is the graph of $g(x) = \sqrt{x}$ translated 1 unit to the right and translated 3 units down. The equation is $y = \sqrt{x - 1} - 3$.   **93.** It is the graph of $g(x) = \sqrt{x}$ translated 4 units to the left, stretched vertically by a factor of 2, and translated 4 units down. The equation is $y = 2\sqrt{x + 4} - 4$.   **95.** $f(-3) = -6$

97. $f(9) = 6$ **99.** $f(-3) = -6$ **101.** $g(x) = 2x + 13$

103. (a)   **(b)**

Chapter 2 Quiz

[2.5] **1. (a)** $y = 2x + 11$ **(b)** $\left(-\frac{11}{2}, 0\right)$ **2.** $y = -\frac{2}{3}x$

3. (a) $x = -8$ **(b)** $y = 5$ [2.6] **4. (a)** cubing function;
domain: $(-\infty, \infty)$; range: $(-\infty, \infty)$; increasing over
$(-\infty, \infty)$ **(b)** absolute value function; domain: $(-\infty, \infty)$;
range: $[0, \infty)$; decreasing over $(-\infty, 0)$; increasing over
$(0, \infty)$ **(c)** cube root function; domain: $(-\infty, \infty)$; range:
$(-\infty, \infty)$; increasing over $(-\infty, \infty)$ **5.** \$2.75

6.   [2.7] **7.**

$f(x) = \begin{cases} \sqrt{x} & \text{if } x \geq 0 \\ 2x + 3 & \text{if } x < 0 \end{cases}$

8.   **9.** $g(x) = -\sqrt{x + 4} - 2$

10. (a) even   **(b)** neither   **(c)** odd

$f(x) = 2|x - 1| + 3$

2.8 Exercises

1. 7   **3.** 12   **5.** 5   **7.** $(-\infty, \infty)$ **9.** $(-\infty, \infty)$
11. 12   **13.** -4 **15.** -38 **17.** $\frac{1}{2}$ **19.** $5x - 1; x + 9$;
$6x^2 - 7x - 20; \frac{3x + 4}{2x - 5}$; All domains are $(-\infty, \infty)$
except for that of $\frac{f}{g}$, which is $\left(-\infty, \frac{5}{2}\right) \cup \left(\frac{5}{2}, \infty\right)$.
21. $3x^2 - 4x + 3; x^2 - 2x - 3; 2x^4 - 5x^3 + 9x^2 - 9x$;
$\frac{2x^2 - 3x}{x^2 - x + 3}$; All domains are $(-\infty, \infty)$.   **23.** $\sqrt{4x - 1} + \frac{1}{x}$;
$\sqrt{4x - 1} - \frac{1}{x}; \frac{\sqrt{4x - 1}}{x}; x\sqrt{4x - 1}$; All domains are
$\left[\frac{1}{4}, \infty\right)$.   **25.** 280; 470; 750 (all in thousands)
27. 2008–2012   **29.** 6; It represents the dollars (in billions)
spent for general science in 2000.   **31.** space and
other technologies; 1995–2000 and 2010–2015
33. (a) 2   **(b)** 4   **(c)** 0   **(d)** $-\frac{1}{3}$ **35. (a)** 3
(b) -5 **(c)** 2   **(d)** undefined   **37. (a)** 5
(b) 5   **(c)** 0   **(d)** undefined

39.

x	$(f + g)(x)$	$(f - g)(x)$	$(fg)(x)$	$\left(\frac{f}{g}\right)(x)$
-2	6	-6	0	0
0	5	5	0	undefined
2	5	9	-14	-3.5
4	15	5	50	2

41. Both the slope formula and the difference quotient
represent the ratio of the vertical change to the horizontal
change. The slope formula is stated for a line, while the
difference quotient is stated for a function f.
43. (a) $2 - x - h$ **(b)** $-h$ **(c)** -1
45. (a) $6x + 6h + 2$ **(b)** $6h$ **(c)** 6
47. (a) $-2x - 2h + 5$ **(b)** $-2h$ **(c)** -2
49. (a) $\frac{1}{x + h}$ **(b)** $\frac{-h}{x(x + h)}$ **(c)** $\frac{-1}{x(x + h)}$
51. (a) $x^2 + 2xh + h^2$ **(b)** $2xh + h^2$ **(c)** $2x + h$
53. (a) $1 - x^2 - 2xh - h^2$ **(b)** $-2xh - h^2$ **(c)** $-2x - h$
55. (a) $x^2 + 2xh + h^2 + 3x + 3h + 1$ **(b)** $2xh + h^2 + 3h$
(c) $2x + h + 3$ **57.** -5 **59.** 7   **61.** 6   **63.** -1
65. 1   **67.** 9   **69.** 1   **71.** $g(1) = 9$, and $f(9)$ cannot
be determined from the table given.
73. (a) $-30x - 33; (-\infty, \infty)$ **(b)** $-30x + 52; (-\infty, \infty)$
75. (a) $\sqrt{x + 3}; [-3, \infty)$ **(b)** $\sqrt{x} + 3; [0, \infty)$
77. (a) $(x^2 + 3x - 1)^3; (-\infty, \infty)$
(b) $x^6 + 3x^3 - 1; (-\infty, \infty)$ **79. (a)** $\sqrt{3x - 1}; \left[\frac{1}{3}, \infty\right)$
(b) $3\sqrt{x} - 1; [1, \infty)$
81. (a) $\frac{2}{x + 1}; (-\infty, -1) \cup (-1, \infty)$
(b) $\frac{2}{x} + 1; (-\infty, 0) \cup (0, \infty)$
83. (a) $\sqrt{-\frac{1}{x} + 2}; (-\infty, 0) \cup \left[\frac{1}{2}, \infty\right)$
(b) $-\frac{1}{\sqrt{x + 2}}; (-2, \infty)$
85. (a) $\sqrt{\frac{1}{x + 5}}; (-5, \infty)$ **(b)** $\frac{1}{\sqrt{x} + 5}; [0, \infty)$
87. (a) $\frac{x}{1 - 2x}; (-\infty, 0) \cup \left(0, \frac{1}{2}\right) \cup \left(\frac{1}{2}, \infty\right)$
(b) $x - 2; (-\infty, 2) \cup (2, \infty)$
89.

x	$f(x)$	$g(x)$	$g(f(x))$
1	3	2	7
2	1	5	2
3	2	7	5

**In Exercises 97–101, we give only one of the many possible
ways.**

97. $g(x) = 6x - 2, f(x) = x^2$ **99.** $g(x) = x^2 - 1$,
$f(x) = \sqrt{x}$ **101.** $g(x) = 6x, f(x) = \sqrt{x} + 12$
103. $(f \circ g)(x) = 63{,}360x$; It computes the number of
inches in x miles.   **105. (a)** $\mathcal{A}(2x) = \sqrt{3}x^2$
(b) $64\sqrt{3}$ square units   **107. (a)** $(\mathcal{A} \circ r)(t) = 16\pi t^2$
(b) It defines the area of the leak in terms of the time t,
in minutes.   **(c)** 144π ft^2 **109. (a)** $N(x) = 100 - x$

(b) $G(x) = 20 + 5x$ **(c)** $C(x) = (100 - x)(20 + 5x)$
(d) $9600   **111. (a)** $g(x) = \frac{1}{2}x$ **(b)** $f(x) = x + 1$
(c) $(f \circ g)(x) = f(g(x)) = g(x) + 1 = \frac{1}{2}x + 1$
(d) $(f \circ g)(60) = \frac{1}{2}(60) + 1 = 31$ (dollars)

Chapter 2 Review Exercises

1. $\sqrt{85}; \left(-\frac{1}{2}, 2\right)$ **3.** $5; \left(-6, \frac{11}{2}\right)$ **5.** $(22, -6)$
7. $(x + 2)^2 + (y - 3)^2 = 225$
9. $(x + 8)^2 + (y - 1)^2 = 289$ **11.** $x^2 + y^2 = 34$
13. $x^2 + (y - 3)^2 = 13$ **15.** $(2, -3); 1$
17. $\left(-\frac{7}{2}, -\frac{3}{2}\right); \frac{3\sqrt{6}}{2}$ **19.** no; $[-6, 6]; [-6, 6]$
21. no; $(-\infty, \infty); (-\infty, -1] \cup [1, \infty)$
23. no; $[0, \infty); (-\infty, \infty)$ **25.** function of x
27. not a function of x **29.** $(-\infty, \infty)$
31. $(-\infty, 2]$ **33.** -15 **35.** -6

37.

39.

41.

43.

45.

47.

49. -2 **51.** 0 **53.** $-\frac{11}{2}$ **55.** undefined **57.** Initially, the car is at home. After traveling 30 mph for 1 hr, the car is 30 mi away from home. During the second hour, the car travels 20 mph until it is 50 mi away. During the third hour, the car travels toward home at 30 mph until it is 20 mi away. During the fourth hour, the car travels away from home at 40 mph until it is 60 mi away from home. During the last hour, the car travels 60 mi at 60 mph until it arrives home.
59. (a) $y = 4.35x + 30.7$; The slope 4.35 indicates that the percent of returns filed electronically rose an average of 4.35% per year during this period. **(b)** 65.5%
61. (a) $y = -2x + 1$ **(b)** $2x + y = 1$
63. (a) $y = 3x - 7$ **(b)** $3x - y = 7$ **65. (a)** $y = -10$
(b) $y = -10$ **67. (a)** not possible **(b)** $x = -7$

69.

71.

$f(x) = -(x + 1)^2 + 3$

73.

$f(x) = [\![x - 3]\!]$

75.

$f(x) = \begin{cases} -4x + 2 \text{ if } x \le 1 \\ 3x - 5 \text{ if } x > 1 \end{cases}$

77.

$f(x) = \begin{cases} |x| & \text{if } x < 3 \\ 6 - x & \text{if } x \ge 3 \end{cases}$

79. true **81.** false; For example, $f(x) = x^2$ is even, and $(2, 4)$ is on the graph but $(2, -4)$ is not. **83.** true
85. x-axis **87.** y-axis

89. none of these **91.** y-axis **93.** x-axis, y-axis, origin
95. Reflect the graph of $f(x) = |x|$ across the x-axis.
97. Translate the graph of $f(x) = |x|$ to the right 4 units and stretch it vertically by a factor of 2. **99.** $y = -3x - 4$
101. (a)

(b)

(c)

(d)

103. $3x^4 - 9x^3 - 16x^2 + 12x + 16$ **105.** 68 **107.** $-\frac{23}{4}$
109. $(-\infty, \infty)$ **111.** 2 **113.** $x - 2$ **115.** 1
117. undefined **119.** 8 **121.** -6 **123.** 2 **125.** 1
127. $f(x) = 36x; g(x) = 1760x; (f \circ g)(x) = f(g(x)) = f(1760x) = 36(1760x) = 63{,}360x$
129. $V(r) = \frac{4}{3}\pi(r + 3)^3 - \frac{4}{3}\pi r^3$

Chapter 2 Test

[2.3] **1. (a)** D **(b)** D **(c)** C **(d)** B **(e)** C **(f)** C
(g) C **(h)** D **(i)** D **(j)** C [2.4] **2.** $\frac{3}{5}$ [2.1] **3.** $\sqrt{34}$
4. $\left(\frac{1}{2}, \frac{5}{2}\right)$ [2.5] **5.** $3x - 5y = -11$ **6.** $f(x) = \frac{3}{5}x + \frac{11}{5}$
[2.2] **7. (a)** $x^2 + y^2 = 4$ **(b)** $(x - 1)^2 + (y - 4)^2 = 1$
8.

$x^2 + y^2 + 4x - 10y + 13 = 0$

[2.3] **9. (a)** not a function; domain: $[0, 4]$; range: $[-4, 4]$
(b) function; domain: $(-\infty, -1) \cup (-1, \infty)$; range: $(-\infty, 0) \cup (0, \infty)$; decreasing on $(-\infty, -1)$ and $(-1, \infty)$
[2.5] **10. (a)** $x = 5$ **(b)** $y = -3$

11. (a) $y = -3x + 9$ **(b)** $y = \frac{1}{3}x + \frac{7}{3}$

[2.3] 12. (a) $(2, \infty)$ **(b)** $(0, 2)$ **(c)** $(-\infty, 0)$
(d) $(-\infty, \infty)$ **(e)** $(-\infty, \infty)$ **(f)** $(-1, \infty)$

[2.6, 2.7]

13.

14.

15.

$f(x) = \begin{cases} 3 & \text{if } x < -2 \\ 2 - \frac{1}{2}x & \text{if } x \geq -2 \end{cases}$

16. (a)

(b)   **(c)**

(d)   **(e)**

17. It is translated 2 units to the left, stretched vertically by a factor of 2, reflected across the x-axis, and translated 3 units down. **18. (a)** yes **(b)** yes **(c)** yes

[2.8] 19. (a) $2x^2 - x + 1$ **(b)** $\frac{2x^2 - 3x + 2}{-2x + 1}$
(c) $\left(-\infty, \frac{1}{2}\right) \cup \left(\frac{1}{2}, \infty\right)$ **(d)** $4x + 2h - 3$
(e) 0 **(f)** -12 **(g)** 1 **20.** $\sqrt{2x - 6}$; $[3, \infty)$

21. $2\sqrt{x + 1} - 7$; $[-1, \infty)$

[2.4] 22. (a) $C(x) = 3300 + 4.50x$ **(b)** $R(x) = 10.50x$
(c) $R(x) - C(x) = 6.00x - 3300$ **(d)** 551

Chapter 3 Polynomial and Rational Functions

3.1 Exercises

1. 5 **3.** vertex **5.** -1 **7.** C **9.** D
11. (a) domain: $(-\infty, \infty)$; range: $[-4, \infty)$ **(b)** $(-3, -4)$
(c) $x = -3$ **(d)** $(0, 5)$ **(e)** $(-5, 0), (-1, 0)$
13. (a) domain: $(-\infty, \infty)$; range: $(-\infty, 2]$
(b) $(-3, 2)$ **(c)** $x = -3$ **(d)** $(0, -16)$
(e) $(-4, 0), (-2, 0)$ **15.** B **17.** D
19.

(d) The greater $|a|$ is, the narrower the parabola will be. The smaller $|a|$ is, the wider the parabola will be.

21.

(d) The graph of $y = (x - h)^2$ is translated h units to the right if h is positive and $|h|$ units to the left if h is negative.

23.

$f(x) = (x - 2)^2$

(a) $(2, 0)$ **(b)** $x = 2$
(c) $(-\infty, \infty)$ **(d)** $[0, \infty)$
(e) $(2, \infty)$ **(f)** $(-\infty, 2)$

25.

$f(x) = (x + 3)^2 - 4$

(a) $(-3, -4)$ **(b)** $x = -3$
(c) $(-\infty, \infty)$ **(d)** $[-4, \infty)$
(e) $(-3, \infty)$ **(f)** $(-\infty, -3)$

27.

$f(x) = -\frac{1}{2}(x + 1)^2 - 3$

(a) $(-1, -3)$ **(b)** $x = -1$
(c) $(-\infty, \infty)$ **(d)** $(-\infty, -3]$
(e) $(-\infty, -1)$ **(f)** $(-1, \infty)$

29.

$f(x) = x^2 - 2x + 3$

(a) $(1, 2)$ **(b)** $x = 1$
(c) $(-\infty, \infty)$ **(d)** $[2, \infty)$
(e) $(1, \infty)$ **(f)** $(-\infty, 1)$

31.

$f(x) = x^2 - 10x + 21$

(a) $(5, -4)$ **(b)** $x = 5$
(c) $(-\infty, \infty)$ **(d)** $[-4, \infty)$
(e) $(5, \infty)$ **(f)** $(-\infty, 5)$

33.

$f(x) = -2x^2 - 12x - 16$

(a) $(-3, 2)$ **(b)** $x = -3$
(c) $(-\infty, \infty)$ **(d)** $(-\infty, 2]$
(e) $(-\infty, -3)$ **(f)** $(-3, \infty)$

35.

$f(x) = -\frac{1}{2}x^2 - 3x - \frac{1}{2}$

(a) $(-3, 4)$ **(b)** $x = -3$
(c) $(-\infty, \infty)$ **(d)** $(-\infty, 4]$
(e) $(-\infty, -3)$ **(f)** $(-3, \infty)$

37. 3 **39.** none **41.** E **43.** D **45.** C
47. $f(x) = \frac{1}{4}(x - 2)^2 - 1$, or $f(x) = \frac{1}{4}x^2 - x$
49. $f(x) = -2(x - 1)^2 + 4$, or $f(x) = -2x^2 + 4x + 2$
51. linear; positive **53.** quadratic; positive
55. quadratic; negative **57. (a)** $f(t) = -16t^2 + 200t + 50$
(b) 6.25 sec; 675 ft **(c)** between 1.41 and 11.09 sec
(d) 12.75 sec **59. (a)** $640 - 2x$ **(b)** $0 < x < 320$
(c) $\mathcal{A}(x) = -2x^2 + 640x$ **(d)** between 57.04 ft and
85.17 ft or 234.83 ft and 262.96 ft **(e)** 160 ft by 320 ft;
The maximum area is 51,200 ft². **61. (a)** $2x$
(b) length: $2x - 4$; width: $x - 4$; $x > 4$
(c) $V(x) = 4x^2 - 24x + 32$ **(d)** 8 in. by 20 in.
(e) 13.05 in. to 14.22 in. **63. (a)** 23.32 ft per sec

(b) 12.88 ft   **65.** 10 and 10   **67.** 1 sec; 16 ft
69. $4.97   **71.** 2002
73. (a)

(b) $f(x) = -0.2857x^2 + 1.503x + 19.13$
(c)

The graph of f models the data closely.

(d) 19.5 million
75. (a)

(b) $f(x) = 0.0055(x - 40)^2 + 4.7$
(c)

The graph of f models the data closely.

(d) $g(x) = 0.0041x^2 - 0.3130x + 11.43$
(e) $f(89) = 17.9\%; g(89) = 16.0\%$
77. $c = 25$ **79.** $f(x) = \frac{1}{2}x^2 - \frac{7}{2}x + 5$ **81.** $(3, 6)$
83.

84. -4 and 2; $\{-4, 2\}$
85. $(-4, 2)$
86. $(-\infty, -4) \cup (2, \infty)$
87. $(-2, 3)$
89. $\left(-\infty, -\frac{3}{2}\right] \cup [6, \infty)$
91. $\left[2 - \sqrt{5}, 2 + \sqrt{5}\right]$

3.2 Exercises

1. $3; 4$ **3.** $1; 1$ **5.** $3; 6; 0$ **7.** $x^2 + 2x + 9$
9. $5x^3 + 2x - 3$ **11.** $x^3 + 2x + 1$
13. $x^4 + x^3 + 2x - 1 + \frac{3}{x + 2}$
15. $-9x^2 - 10x - 27 + \frac{-52}{x - 2}$ **17.** $\frac{1}{3}x^2 - \frac{1}{9}x + \frac{1}{27}$
19. $x^3 - x^2 - 6x$ **21.** $x^2 + x + 1$
23. $x^4 - x^3 + x^2 - x + 1$
25. $f(x) = (x + 1)(2x^2 - x + 2) - 10$
27. $f(x) = (x + 2)(x^2 + 2x + 1) + 0$
29. $f(x) = (x - 3)(4x^3 + 9x^2 + 7x + 20) + 60$
31. $f(x) = (x + 1)(3x^3 + x^2 - 11x + 11) + 4$

33. 0   **35.** -1 **37.** -6 **39.** $-6 - i$ **41.** 0   **43.** -5
45. 7   **47.** yes   **49.** yes   **51.** no; -9 **53.** yes
55. no; $\frac{357}{125}$ **57.** yes   **59.** no; $13 + 7i$ **61.** yes
63. no; $-2 + 7i$ **65.** $-12; (-2, -12)$ **66.** $0; (-1, 0)$
67. $\frac{15}{8}; \left(-\frac{1}{2}, \frac{15}{8}\right)$ **68.** $2; (0, 2)$ **69.** $0; (1, 0)$
70. $-\frac{5}{8}; \left(\frac{3}{2}, -\frac{5}{8}\right)$ **71.** $0; (2, 0)$ **72.** $8; (3, 8)$
73.   **74.**

3.3 Exercises

1. true   **3.** false; -2 is a zero of multiplicity 4. The number 2 is *not* a zero.   **5.** true   **7.** false; $\bar{z} = 7 + 6i$
9. yes   **11.** no   **13.** yes   **15.** no   **17.** no   **19.** yes
21. $f(x) = (x - 2)(2x - 5)(x + 3)$
23. $f(x) = (x + 3)(3x - 1)(2x - 1)$
25. $f(x) = (x + 4)(3x - 1)(2x + 1)$
27. $f(x) = (x - 3i)(x + 4)(x + 3)$
29. $f(x) = [x - (1 + i)](2x - 1)(x + 3)$
31. $f(x) = (x + 2)^2(x + 1)(x - 3)$
33. $-1 \pm i$ **35.** $3, 2 + i$ **37.** $i, \pm 2i$
39. (a) $\pm 1, \pm 2, \pm 5, \pm 10$ **(b)** $-1, -2, 5$
(c) $f(x) = (x + 1)(x + 2)(x - 5)$ **41. (a)** $\pm 1, \pm 2,$
$\pm 3, \pm 5, \pm 6, \pm 10, \pm 15, \pm 30$ **(b)** $-5, -3, 2$
(c) $f(x) = (x + 5)(x + 3)(x - 2)$ **43. (a)** $\pm 1, \pm 2,$
$\pm 3, \pm 4, \pm 6, \pm 12, \pm \frac{1}{2}, \pm \frac{3}{2}, \pm \frac{1}{3}, \pm \frac{2}{3}, \pm \frac{4}{3}, \pm \frac{1}{6}$
(b) $-4, -\frac{1}{3}, \frac{3}{2}$ **(c)** $f(x) = (x + 4)(3x + 1)(2x - 3)$
45. (a) $\pm 1, \pm 2, \pm 3, \pm 4, \pm 6, \pm 12, \pm \frac{1}{2}, \pm \frac{3}{2},$
$\pm \frac{1}{3}, \pm \frac{2}{3}, \pm \frac{4}{3}, \pm \frac{1}{4}, \pm \frac{3}{4}, \pm \frac{1}{6}, \pm \frac{1}{8}, \pm \frac{3}{8}, \pm \frac{1}{12}, \pm \frac{1}{24}$
(b) $-\frac{3}{2}, -\frac{2}{3}, \frac{1}{2}$ **(c)** $f(x) = 2(2x + 3)(3x + 2)(2x - 1)$
47. 2 (multiplicity 3), $\pm \sqrt{7}$ **49.** $0, 2, -3, 1, -1$
51. -2 (multiplicity 5), 1 (multiplicity 5), $1 - \sqrt{3}$
(multiplicity 2)   **53.** $f(x) = -3x^3 + 6x^2 + 33x - 36$
55. $f(x) = -\frac{1}{2}x^3 - \frac{1}{2}x^2 + x$
57. $f(x) = \frac{1}{6}x^3 + \frac{3}{2}x^2 + \frac{9}{2}x + \frac{9}{2}$
59. $f(x) = 5x^3 - 10x^2 + 5x$

In Exercises 61–77, we give only one possible answer.

61. $f(x) = x^2 - 10x + 26$
63. $f(x) = x^5 - 2x^4 + 3x^3 - 2x^2 + 2x$
65. $f(x) = x^3 - 3x^2 + x + 1$
67. $f(x) = x^4 - 6x^3 + 10x^2 + 2x - 15$
69. $f(x) = x^3 - 8x^2 + 22x - 20$

71. $f(x) = x^4 - 4x^3 + 5x^2 - 2x - 2$

73. $f(x) = x^4 - 16x^3 + 98x^2 - 240x + 225$

75. $f(x) = x^5 - 12x^4 + 74x^3 - 248x^2 + 445x - 500$

77. $f(x) = x^4 - 6x^3 + 17x^2 - 28x + 20$

79.

Positive	Negative	Nonreal Complex
2	1	0
0	1	2

81.

Positive	Negative	Nonreal Complex
3	0	0
1	0	2

83.

Positive	Negative	Nonreal Complex
1	1	2

85.

Positive	Negative	Nonreal Complex
4	0	0
2	0	2
0	0	4

87.

Positive	Negative	Nonreal Complex
2	3	0
2	1	2
0	3	2
0	1	4

89.

Positive	Negative	Nonreal Complex
2	3	0
2	1	2
0	3	2
0	1	4

91.

Positive	Negative	Nonreal Complex
4	2	0
4	0	2
2	2	2
2	0	4
0	2	4
0	0	6

93.

Positive	Negative	Nonreal Complex
0	5	0
0	3	2
0	1	4

95. $-5, 3, \pm 2i\sqrt{3}$ **97.** $1, 1, 1, -4$ **99.** $-3, -3, 0,$ $\dfrac{1 \pm i\sqrt{31}}{4}$ **101.** $2, 2, 2, \pm i\sqrt{2}$ **103.** $-\dfrac{1}{2}, 1, \pm 2i$

105. $-\dfrac{1}{5}, 1 \pm i\sqrt{5}$ **107.** $\pm 2i, \pm 5i$ **109.** $\pm i, \pm i,$

111. $0, 0, 3 \pm \sqrt{2}$ **113.** $3, 3, 1 \pm i\sqrt{7}$

115. $\pm 2, \pm 3, \pm 2i$ **121.** 2

3.4 Exercises

1. A **3.** one **5.** B and D **7.** $f(x) = x(x + 5)^2(x - 3)$

9.

11.

(a) $(0, \infty)$ **(b)** $(-\infty, 0)$ **(a)** none **(b)** $(-\infty, \infty)$

13.

15.

(a) $(-\infty, \infty)$ **(b)** none   **(a)** none **(b)** $(-\infty, \infty)$

17.

19.

(a) $(1, \infty)$ **(b)** $(-\infty, 1)$ **(a)** $(2, \infty)$ **(b)** $(-\infty, 2)$

21. **23.** **25.** **27.**

29.

$f(x) = x^3 + 5x^2 + 2x - 8$

31.

$f(x) = 2x(x - 3)(x + 2)$

33.

$f(x) = x^2(x - 2)(x + 3)^2$

35.

$f(x) = (3x - 1)(x + 2)^2$

37.

$f(x) = x^3 + 5x^2 - x - 5$

39.

$f(x) = x^3 - x^2 - 2x$

41.

$f(x) = 2x^3(x^2 - 4)(x - 1)$

43.

$f(x) = 2x^3 - 5x^2 - x + 6$

45.

$f(x) = 3x^4 - 7x^3 - 6x^2 + 12x + 8$

47. $f(2) = -2 < 0;$ $f(3) = 1 > 0$
49. $f(0) = 7 > 0;$ $f(1) = -1 < 0$
51. $f(1) = -6 < 0;$ $f(2) = 16 > 0$
53. $f(3.2) = -3.8144 < 0;$ $f(3.3) = 7.1891 > 0$
55. $f(-1) = -35 < 0;$ $f(0) = 12 > 0$
65. $f(x) = \frac{1}{2}(x + 6)(x - 2)(x - 5),$ or
$f(x) = \frac{1}{2}x^3 - \frac{1}{2}x^2 - 16x + 30$
67. $f(x) = (x - 1)^3(x + 1)^3,$ or $f(x) = x^6 - 3x^4 + 3x^2 - 1$
69. $f(x) = (x - 3)^2(x + 3)^2,$ or $f(x) = x^4 - 18x^2 + 81$
71. $f(x) = 2x(x - 3)(x + 2)$ **73.** $f(x) = (3x - 1)(x + 2)^2$

$f(1.25) = -14.21875$ $f(1.25) = 29.046875$

75. 2.7807764   **77.** 1.543689   **79.** $-3.0, -1.4, 1.4$
81. $-1.1, 1.2$ **83.** $(-0.09, 1.05)$ **85.** $(1.76, -5.34)$
87. $(-3.44, 26.15)$ **89.** Answers will vary.
91. (a) $0 < x < 6$ **(b)** $V(x) = x(18 - 2x)(12 - 2x),$ or
$V(x) = 4x^3 - 60x^2 + 216x$ **(c)** $x \approx 2.35; 228.16$ in.^{3}
(d) $0.42 < x < 5$ **93. (a)** $x - 1; (1, \infty)$
(b) $\sqrt{x^2 - (x - 1)^2}$ **(c)** $2x^3 - 5x^2 + 4x - 28{,}225 = 0$
(d) hypotenuse: 25 in.; legs: 24 in. and 7 in.   **95.** 3 ft

97. (a) 7.13 cm; The ball floats partly above the surface.
(b) The sphere is more dense than water and sinks below
the surface.   **(c)** 10 cm; The balloon is submerged with its
top even with the surface.
99. (a)   **(b)** $y = 33.93x + 113.4$

(c) $y = -0.0032x^3 + 0.4245x^2$ **(d)** linear: 1572 ft; cubic:
  $+ 16.64x + 323.1$ 1569 ft   **(e)** The cubic

                                              function is a slightly bet-
                                              ter fit because only one
                                              data point is not on the
                                              curve.   **101. B**

103.

$f(x) = x^3 - 3x^2 - 6x + 8$
$f(x) = (x - 4)(x - 1)(x + 2)$

(a) $\{-2, 1, 4\}$
(b) $(-\infty, -2) \cup (1, 4)$
(c) $(-2, 1) \cup (4, \infty)$

104.

$f(x) = x^3 + 4x^2 - 11x - 30$
$f(x) = (x - 3)(x + 2)(x + 5)$

(a) $\{-5, -2, 3\}$
(b) $(-\infty, -5) \cup (-2, 3)$
(c) $(-5, -2) \cup (3, \infty)$

105.

$f(x) = 2x^4 - 9x^3 - 5x^2 + 57x - 45$
$f(x) = (x - 3)^2(2x + 5)(x - 1)$

(a) $\{-2.5, 1, 3 \text{ (multiplicity 2)}\}$ **(b)** $(-2.5, 1)$
(c) $(-\infty, -2.5) \cup (1, 3) \cup (3, \infty)$

106.

$f(x) = 4x^4 + 27x^3 - 42x^2 - 445x - 300$
$f(x) = (x + 5)^2(4x + 3)(x - 4)$

(a) $\{-5 \text{ (multiplicity 2)}, -0.75, 4\}$ **(b)** $(-0.75, 4)$
(c) $(-\infty, -5) \cup (-5, -0.75) \cup (4, \infty)$

107.

$f(x) = -x^4 - 4x^3 + 3x^2 + 18x$
$f(x) = x(2 - x)(x + 3)^2$

(a) $\{-3 \text{ (multiplicity 2)}, 0, 2\}$
(b) $\{-3\} \cup [0, 2]$
(c) $(-\infty, 0] \cup [2, \infty)$

108.

$f(x) = -x^4 + 2x^3 + 8x^2$
$f(x) = x^2(4 - x)(x + 2)$

(a) $\{-2, 0 \text{ (multiplicity 2)}, 4\}$
(b) $[-2, 4]$
(c) $(-\infty, -2] \cup \{0\} \cup [4, \infty)$

Summary Exercises on Polynomial Functions, Zeros, and Graphs

1. (a)

Positive	Negative	Nonreal Complex
2	1	0
0	1	2

(b) $\pm 1, \pm 2, \pm 3, \pm 4, \pm 6, \pm 8, \pm 12, \pm 24,$
$\pm\frac{1}{2}, \pm\frac{3}{2}, \pm\frac{1}{3}, \pm\frac{2}{3}, \pm\frac{4}{3}, \pm\frac{8}{3}, \pm\frac{1}{6}$
(c) $-\frac{1}{2}, \frac{4}{3}, 6$ **(d)** no other complex zeros

2. (a)

Positive	Negative	Nonreal Complex
2	1	0
0	1	2

(b) $\pm 1,\ \pm 3,\ \pm\frac{1}{2},\ \pm\frac{3}{2}$ **(c)** $-1, \frac{1}{2}, 3$
(d) no other complex zeros

3. (a)

Positive	Negative	Nonreal Complex
4	0	0
2	0	2
0	0	4

(b) $\pm 1,\ \pm 2,\ \pm 4,\ \pm 8,\ \pm\frac{1}{3},\ \pm\frac{2}{3},\ \pm\frac{4}{3},\ \pm\frac{8}{3}$
(c) $\frac{2}{3}, 1$ **(d)** $-2i, 2i$

4. (a)

Positive	Negative	Nonreal Complex
3	1	0
1	1	2

(b) $\pm 1,\ \pm 2,\ \pm 3,\ \pm 6,\ \pm 9,\ \pm 18,\ \pm\frac{1}{2},\ \pm\frac{3}{2},\ \pm\frac{9}{2}$
(c) $-\frac{1}{2}, 2$ **(d)** $-3i, 3i$

5. (a)

Positive	Negative	Nonreal Complex
3	1	0
1	1	2

(b) $\pm 1,\ \pm 2,\ \pm\frac{1}{2},\ \pm\frac{1}{3},\ \pm\frac{2}{3},\ \pm\frac{1}{6}$ **(c)** $\frac{1}{3}, \frac{1}{2}$
(d) $-\sqrt{2}, \sqrt{2}$

6. (a)

Positive	Negative	Nonreal Complex
2	2	0
2	0	2
0	2	2
0	0	4

(b) $\pm 1,\ \pm 2,\ \pm 3,\ \pm 4,\ \pm 6,\ \pm 12,\ \pm\frac{1}{5},\ \pm\frac{2}{5},\ \pm\frac{3}{5},\ \pm\frac{4}{5},$
$\pm\frac{6}{5},\ \pm\frac{12}{5}$ **(c)** $-2, \frac{2}{5}$ **(d)** $-\sqrt{3}, \sqrt{3}$

7. (a)

Positive	Negative	Nonreal Complex
4	0	0
2	0	2
0	0	4

(b) $0,\ \pm 1,\ \pm 2,\ \pm 4,\ \pm 8,\ \pm 16$
(c) $0, 2$ (multiplicity 2) **(d)** $1 - i\sqrt{3}, 1 + i\sqrt{3}$

8. (a)

Positive	Negative	Nonreal Complex
1	3	0
1	1	2

(b) $\pm 1,\ \pm 3,\ \pm 9,\ \pm\frac{1}{2},\ \pm\frac{3}{2},\ \pm\frac{9}{2}$ **(c)** -3 (multiplicity 2)
(d) $\frac{2 - \sqrt{6}}{2}, \frac{2 + \sqrt{6}}{2}$

9. (a)

Positive	Negative	Nonreal Complex
1	3	0
1	1	2

(b) $\pm 1,\ \pm\frac{1}{2},\ \pm\frac{1}{4},\ \pm\frac{1}{8}$ **(c)** $-1, \frac{1}{2}$
(d) $-\frac{1}{4} - \frac{\sqrt{3}}{4}i,\ -\frac{1}{4} + \frac{\sqrt{3}}{4}i$

10. (a)

Positive	Negative	Nonreal Complex
3	2	0
3	0	2
1	2	2
1	0	4

(b) $\pm 1,\ \pm 2,\ \pm 3,\ \pm 6,\ \pm\frac{1}{2},\ \pm\frac{3}{2}$ **(c)** $-3, -2, \frac{1}{2},$
1 (multiplicity 2) **(d)** no other complex zeros

11. (a)

Positive	Negative	Nonreal Complex
1	3	0
1	1	2

(b) $\pm 1,\ \pm 2,\ \pm 3,\ \pm 6,$ **(c)** $-3, -1$ (multiplicity 2), 2
(d) no other real zeros **(e)** no other complex zeros
(f) $(-3, 0), (-1, 0), (2, 0)$ **(g)** $(0, -6)$
(h) $f(4) = 350; (4, 350)$ **(i)** ↰↱
(j)

$f(x) = x^4 + 3x^3 - 3x^2 - 11x - 6$

12. (a)

Positive	Negative	Nonreal Complex
3	2	0
3	0	2
1	2	2
1	0	4

(b) $\pm 1,\ \pm 3,\ \pm 5,\ \pm 9,\ \pm 15,\ \pm 45,\ \pm\frac{1}{2},\ \pm\frac{3}{2},\ \pm\frac{5}{2},$
$\pm\frac{9}{2},\ \pm\frac{15}{2},\ \pm\frac{45}{2}$ **(c)** $-3, \frac{1}{2}, 5$ **(d)** $-\sqrt{3}, \sqrt{3}$
(e) no other complex zeros **(f)** $(-3, 0), \left(\frac{1}{2}, 0\right), (5, 0),$
$\left(-\sqrt{3}, 0\right), \left(\sqrt{3}, 0\right)$ **(g)** $(0, 45)$ **(h)** $f(4) = 637;$
$(4, 637)$ **(i)** ↖↘ **(j)**

$f(x) = -2x^5 + 5x^4 + 34x^3 - 30x^2 - 84x + 45$

13. (a)

Positive	Negative	Nonreal Complex
4	1	0
2	1	2
0	1	4

(b) $\pm 1,\ \pm 5,\ \pm\frac{1}{2},\ \pm\frac{5}{2}$ **(c)** 5 **(d)** $-\frac{\sqrt{2}}{2}, \frac{\sqrt{2}}{2}$
(e) $-i, i$ **(f)** $\left(-\frac{\sqrt{2}}{2}, 0\right), \left(\frac{\sqrt{2}}{2}, 0\right), (5, 0)$ **(g)** $(0, 5)$
(h) $f(4) = -527; (4, -527)$ **(i)** ↙↗
(j)

$f(x) = 2x^5 - 10x^4 + x^3 - 5x^2 - x + 5$

14. (a)

Positive	Negative	Nonreal Complex
2	2	0
2	0	2
0	2	2
0	0	4

(b) $\pm 1, \pm 2, \pm 3, \pm 6, \pm 9, \pm 18, \pm\frac{1}{3}, \pm\frac{2}{3}$

(c) $-\frac{2}{3}, 3$ **(d)** $\frac{-1 + \sqrt{13}}{2}, \frac{-1 - \sqrt{13}}{2}$ **(e)** no other complex

zeros **(f)** $\left(-\frac{2}{3}, 0\right), (3, 0), \left(\frac{-1 \pm \sqrt{13}}{2}, 0\right)$ **(g)** $(0, 18)$

(h) $f(4) = 238; (4, 238)$ **(i)** ↿⇂ **(j)**

$f(x) = 3x^4 - 4x^3 - 22x^2 + 15x + 18$

15. (a)

Positive	Negative	Nonreal Complex
1	3	0
1	1	2

(b) $\pm 1, \pm 2, \pm\frac{1}{2}$ **(c)** $-1, 1$ **(d)** no other real zeros

(e) $-\frac{1}{4} + \frac{\sqrt{15}}{4}i, -\frac{1}{4} - \frac{\sqrt{15}}{4}i$ **(f)** $(-1, 0), (1, 0)$

(g) $(0, 2)$ **(h)** $f(4) = -570; (4, -570)$ **(i)** ⌢⌢

(j)

$f(x) = -2x^4 - x^3 + x + 2$

16. (a)

Positive	Negative	Nonreal Complex
0	4	0
0	2	2
0	0	4

(b) $0, \pm 1, \pm 3, \pm 9, \pm 27, \pm\frac{1}{2}, \pm\frac{3}{2}, \pm\frac{9}{2}, \pm\frac{27}{2},$

$\pm\frac{1}{4}, \pm\frac{3}{4}, \pm\frac{9}{4}, \pm\frac{27}{4}$ **(c)** $0, -\frac{3}{2}$ (multiplicity 2)

(d) no other real zeros **(e)** $\frac{1}{2} + \frac{\sqrt{11}}{2}i, \frac{1}{2} - \frac{\sqrt{11}}{2}i$

(f) $\left(-\frac{3}{2}, 0\right), (0, 0)$ **(g)** $(0, 0)$ **(h)** $f(4) = 7260;$

$(4, 7260)$ **(i)** ↗↗ **(j)**

$f(x) = 4x^5 + 8x^4 + 9x^3 + 27x^2 + 27x$

17. (a)

Positive	Negative	Nonreal Complex
1	1	2

(b) $\pm 1, \pm 5, \pm\frac{1}{3}, \pm\frac{5}{3}$ **(c)** no rational zeros **(d)** $-\sqrt{5},$

$\sqrt{5}$ **(e)** $-\frac{\sqrt{3}}{3}i, \frac{\sqrt{3}}{3}i$ **(f)** $\left(-\sqrt{5}, 0\right), \left(\sqrt{5}, 0\right)$

(g) $(0, -5)$ **(h)** $f(4) = 539; (4, 539)$ **(i)** ↿⇂

(j)

$f(x) = 3x^4 - 14x^2 - 5$

18. (a)

Positive	Negative	Nonreal Complex
2	3	0
2	1	2
0	3	2
0	1	4

(b) $\pm 1, \pm 3, \pm 9$ **(c)** $-3, -1$ (multiplicity 2), 1, 3

(d) no other real zeros **(e)** no other complex zeros

(f) $(-3, 0), (-1, 0), (1, 0), (3, 0)$ **(g)** $(0, -9)$

(h) $f(4) = -525; (4, -525)$ **(i)** ↖↘

(j)

$f(x) = -x^5 - x^4 + 10x^3 + 10x^2 - 9x - 9$

19. (a)

Positive	Negative	Nonreal Complex
4	0	0
2	0	2
0	0	4

(b) $\pm 1, \pm 2, \pm 3, \pm 4, \pm 6, \pm 12, \pm\frac{1}{3}, \pm\frac{2}{3}, \pm\frac{4}{3}$

(c) $\frac{1}{3}, 2$ (multiplicity 2), 3 **(d)** no other real zeros

(e) no other complex zeros **(f)** $\left(\frac{1}{3}, 0\right), (2, 0), (3, 0)$

(g) $(0, -12)$ **(h)** $f(4) = -44; (4, -44)$

(i) ⌢⌢ **(j)**

$f(x) = -3x^4 + 22x^3 - 55x^2 + 52x - 12$

20. For the function in **Exercise 12:** ± 1.732; for the function

in **Exercise 13:** ± 0.707; for the function in **Exercise 14:**

$-2.303, 1.303$; for the function in **Exercise 17:** ± 2.236

3.5 Exercises

1. $(-\infty, 0) \cup (0, \infty); (-\infty, 0) \cup (0, \infty)$ **3.** none;

$(-\infty, 0)$ and $(0, \infty)$; none **5.** $x = 3; y = 2$ **7.** even;

symmetry with respect to the y-axis **9.** A, B, C **11.** A

13. A **15.** A, C, D

17. To obtain the graph of f, stretch the graph of $y = \frac{1}{x}$ vertically by a factor of 2.
(a) $(-\infty, 0) \cup (0, \infty)$
(b) $(-\infty, 0) \cup (0, \infty)$ **(c)** none
(d) $(-\infty, 0)$ and $(0, \infty)$

19. To obtain the graph of f, shift the graph of $y = \frac{1}{x}$ to the left 2 units.
(a) $(-\infty, -2) \cup (-2, \infty)$
(b) $(-\infty, 0) \cup (0, \infty)$ **(c)** none
(d) $(-\infty, -2)$ and $(-2, \infty)$

21. To obtain the graph of f, shift the graph of $y = \frac{1}{x}$ up 1 unit.
(a) $(-\infty, 0) \cup (0, \infty)$
(b) $(-\infty, 1) \cup (1, \infty)$ **(c)** none
(d) $(-\infty, 0)$ and $(0, \infty)$

23. To obtain the graph of f, stretch the graph of $y = \frac{1}{x^2}$ vertically by a factor of 2 and reflect across the x-axis.
(a) $(-\infty, 0) \cup (0, \infty)$ **(b)** $(-\infty, 0)$
(c) $(0, \infty)$ **(d)** $(-\infty, 0)$

25. To obtain the graph of f, shift the graph of $y = \frac{1}{x^2}$ to the right 3 units.
(a) $(-\infty, 3) \cup (3, \infty)$ **(b)** $(0, \infty)$
(c) $(-\infty, 3)$ **(d)** $(3, \infty)$

27. To obtain the graph of f, shift the graph of $y = \frac{1}{x^2}$ to the left 2 units, reflect across the x-axis, and shift 3 units down.
(a) $(-\infty, -2) \cup (-2, \infty)$
(b) $(-\infty, -3)$ **(c)** $(-2, \infty)$
(d) $(-\infty, -2)$

29. D **31.** G **33.** E **35.** F

In selected Exercises 37–59, V.A. represents vertical asymptote, H.A. represents horizontal asymptote, and O.A. represents oblique asymptote.

37. V.A.: $x = 5$; H.A.: $y = 0$
39. V.A.: $x = -\frac{1}{2}$; H.A.: $y = -\frac{3}{2}$
41. V.A.: $x = -3$; O.A.: $y = x - 3$
43. V.A.: $x = -2$, $x = \frac{5}{2}$; H.A.: $y = \frac{1}{2}$
45. V.A.: none; H.A.: $y = 1$
47. (a) $f(x) = \frac{2x - 5}{x - 3}$ **(b)** $\frac{5}{2}$ **(c)** H.A.: $y = 2$; V.A.: $x = 3$
49. (a) $y = x + 1$ **(b)** at $x = 0$ and $x = 1$ **(c)** above
51. A **53.** V.A.: $x = 2$; H.A.: $y = 4$; $(-\infty, 2) \cup (2, \infty)$
55. V.A.: $x = \pm 2$; H.A.: $y = -4$; $(-\infty, -2) \cup (-2, 2) \cup (2, \infty)$ **57.** V.A.: none; H.A.: $y = 0$; $(-\infty, \infty)$
59. V.A.: $x = -1$; O.A.: $y = x - 1$; $(-\infty, -1) \cup (-1, \infty)$

61.

$f(x) = \frac{x + 1}{x - 4}$

63.

$f(x) = \frac{x + 2}{x - 3}$

65.

$f(x) = \frac{4 - 2x}{8 - x}$

67.

$f(x) = \frac{3x}{x^2 - x - 2}$

69.

$f(x) = \frac{5x}{x^2 - 1}$

71.

$f(x) = \frac{(x + 6)(x - 2)}{(x + 3)(x - 4)}$

73.

$f(x) = \frac{3x^2 + 3x - 6}{x^2 - x - 12}$

75.

$f(x) = \frac{9x^2 - 1}{x^2 - 4}$

77.

$f(x) = \frac{(x - 3)(x + 1)}{(x - 1)^2}$

79.

$f(x) = \frac{x}{x^2 - 9}$

81.

$f(x) = \frac{1}{x^2 + 1}$

83.

$f(x) = \frac{(x + 4)^2}{(x - 1)(x + 5)}$

85.

$f(x) = \frac{20 + 6x - 2x^2}{8 + 6x - 2x^2}$

87.

$f(x) = \frac{x^2 + 1}{x + 3}$

89.
$$f(x) = \frac{x^2 + 2x}{2x - 1}$$

91.
$$f(x) = \frac{x^2 - 9}{x + 3}$$

93.
$$f(x) = \frac{2x^2 - 5x - 2}{x - 2}$$

95.
$$f(x) = \frac{x^2 - 1}{x^2 - 4x + 3}$$

97.
$$f(x) = \frac{(x^2 - 9)(2 + x)}{(x^2 - 4)(3 + x)}$$

99.
$$f(x) = \frac{x^4 - 20x^2 + 64}{x^4 - 10x^2 + 9}$$

101. $f(x) = \dfrac{(x - 3)(x + 2)}{(x - 2)(x + 2)}$, or $f(x) = \dfrac{x^2 - x - 6}{x^2 - 4}$

103. $f(x) = \dfrac{x - 2}{x(x - 4)}$, or $f(x) = \dfrac{x - 2}{x^2 - 4x}$

105. $f(x) = \dfrac{-x(x - 2)}{(x - 1)^2}$, or $f(x) = \dfrac{-x^2 + 2x}{x^2 - 2x + 1}$

107. Several answers are possible. One answer is $f(x) = \dfrac{(x - 3)(x + 1)}{(x - 1)^2}$.

109.
$$f(x) = \frac{x + 1}{x - 4}$$
$$f(1.25) = -0.8\overline{1}$$

111.
$$f(x) = \frac{x^2 + 2x}{2x - 1}$$
$$f(1.25) = 2.708\overline{3}$$

113. (a) 26 per min **(b)** 5 park attendants

For $r = x$, $y = T(x) = \dfrac{2x - 25}{2x^2 - 50x}$ $y = 0.5$

115. (a) 52 mph

(b)

x	$d(x)$	x	$d(x)$
20	34	50	273
25	56	55	340
30	85	60	415
35	121	65	499
40	164	70	591
45	215		

(c) It more than doubles, compare the values of $d(20)$ and $d(40)$, for example.   **(d)** It would be linear.

117. All answers are given in tens of millions.
(a) \$65.5 **(b)** \$64 **(c)** \$60 **(d)** \$40 **(e)** \$0

119. $y = 1$ **120.** $(x + 4)(x + 1)(x - 3)(x - 5)$

121. (a) $(x - 1)(x - 2)(x + 2)(x - 5)$

(b) $f(x) = \dfrac{(x + 4)(x + 1)(x - 3)(x - 5)}{(x - 1)(x - 2)(x + 2)(x - 5)}$

122. (a) $x - 5$ **(b)** 5   **123.** $(-4, 0), (-1, 0), (3, 0)$

124. $(0, -3)$ **125.** $x = 1, x = 2, x = -2$

126. $\left(\dfrac{7 \pm \sqrt{241}}{6}, 1\right)$

127.

$$f(x) = \frac{x^4 - 3x^3 - 21x^2 + 43x + 60}{x^4 - 6x^3 + x^2 + 24x - 20}$$

128. (a) $(-4, -2) \cup (-1, 1) \cup (2, 3)$
(b) $(-\infty, -4) \cup (-2, -1) \cup (1, 2) \cup (3, 5) \cup (5, \infty)$

Chapter 3 Quiz

[3.1] **1. (a)**

$$f(x) = -2(x + 3)^2 - 1$$

vertex: $(-3, -1)$;
axis: $x = -3$;
domain: $(-\infty, \infty)$;
range: $(-\infty, -1]$;
increasing on $(-\infty, -3)$;
decreasing on $(-3, \infty)$

(b)

$$f(x) = 2x^2 - 8x + 3$$

vertex: $(2, -5)$; axis: $x = 2$;
domain: $(-\infty, \infty)$; range: $[-5, \infty)$;
increasing on $(2, \infty)$;
decreasing on $(-\infty, 2)$

2. (a) $s(t) = -16t^2 + 64t + 200$
(b) between 0.78 sec and 3.22 sec

[3.2] **3.** no; 38   **4.** yes

[3.3] **5.** $f(x) = x^4 - 7x^3 + 10x^2 + 26x - 60$

[3.4] **6.**

$$f(x) = x(x - 2)^3(x + 2)^2$$

7.

$$f(x) = 2x^4 - 9x^3 - 5x^2 + 57x - 45$$
$$f(x) = (x - 3)^2(2x + 5)(x - 1)$$

8. [3.5] **9.**

$f(x) = -4x^5 + 16x^4 + 13x^3 - 76x^2 - 3x + 18$
$f(x) = -(x + 2)(2x + 1)(2x - 1)(x - 3)^2$

$f(x) = \dfrac{3x + 1}{x^2 + 7x + 10}$

10.

$y = x + 3$

$f(x) = \dfrac{x^2 + 2x + 1}{x - 1}$

Summary Exercises on Solving Equations and Inequalities

1. (a) $(-1, 1) \cup (1, \infty)$ **(b)** $(-\infty, -1] \cup \{1\}$
2. (a) $(-\infty, -1) \cup (-1, 1) \cup (1, \infty)$ **(b)** $\varnothing$
3. (a) $\{0\}$ **(b)** $(-3, 0) \cup (3, \infty)$
4. (a) $\{1, 3\}$ **(b)** $[1, 2) \cup (2, 3]$
5. $\left\{\frac{4}{3}\right\}$ **6.** $\{20\}$ **7.** $\left\{\frac{9}{2}, \frac{14}{3}, \frac{16}{3}, \frac{11}{2}\right\}$ **8.** $\{25, 64\}$
9. $\{3, 7\}$ **10.** $\{2\}$ **11.** $\left\{\pm\frac{1}{8}\right\}$ **12.** $\{13\}$
13. inequality; $\left(-\infty, \frac{2}{5}\right) \cup \left(\frac{2}{5}, \infty\right)$ **14.** inequality;
$(-\infty, -1] \cup [4, \infty)$ **15.** equation; $\{-1, 2\}$
16. inequality; $\{-3\} \cup [1, \infty)$ **17.** equation; $\{-4, 0, 3\}$
18. inequality; $\left(-\sqrt{3}, -\frac{1}{2}\right) \cup \left(\frac{1}{2}, \sqrt{3}\right)$
19. inequality; $(-\infty, 0) \cup \left[\frac{3}{2}, 3\right) \cup [5, \infty)$
20. inequality; $(-2, -1) \cup (2, \infty)$ **21.** equation; $\{1\}$
22. inequality; $(-2, 0) \cup (2, \infty)$

3.6 Exercises

1. increases; decreases **3.** 60 **5.** 6 **7.** -30 **9.** 60
11. $\frac{3}{2}$ **13.** $\frac{12}{5}$ **15.** $\frac{18}{125}$ **17.** C **19.** A **21.** The circumference of a circle varies directly as (or is proportional to) its radius. **23.** The speed varies directly as (or is proportional to) the distance traveled and inversely as the time.
25. The strength of a muscle varies directly as (or is proportional to) the cube of its length. **27.** 69.08 in. **29.** 850 ohms
31. 12 km **33.** 16 in. **35.** 90 revolutions per minute
37. 0.0444 ohm **39.** \$875 **41.** 800 lb **43.** $\frac{8}{9}$ metric ton
45. $\frac{66\pi}{17}$ sec **47.** 21 **49.** 365.24 **51.** 92; undernourished
53. y is half as large as before. **55.** y is one-third as large as before. **57.** p is $\frac{1}{32}$ as large as before.

Chapter 3 Review Exercises

1. $f(x) = 3(x + 4)^2 - 5$

vertex: $(-4, -5)$; axis: $x = -4$;
x-intercepts: $\left(\dfrac{-12 \pm \sqrt{15}}{3}, 0\right)$;
y-intercept: $(0, 43)$;
domain: $(-\infty, \infty)$; range: $[-5, \infty)$;
increasing on $(-4, \infty)$;
decreasing on $(-\infty, -4)$

3.

vertex: $(-2, 11)$; axis: $x = -2$;
x-intercepts: $\left(\dfrac{-6 \pm \sqrt{33}}{3}, 0\right)$;
y-intercept: $(0, -1)$;
domain: $(-\infty, \infty)$; range: $(-\infty, 11]$;
increasing on $(-\infty, -2)$;
decreasing on $(-2, \infty)$

$f(x) = -3x^2 - 12x - 1$

5. (h, k) **7.** $k \le 0$; $\left(h \pm \sqrt{\frac{-k}{a}}, 0\right)$ **9.** 90 m by 45 m
11. (a) 120 **(b)** 40 **(c)** 22 **(d)** 84 **(e)** 146
(f)

The minimum occurs in August when $x = 8$.

13. Because the discriminant is 67.3033, a positive number, there are two x-intercepts. **15. (a)** the open interval $(-0.52, 2.59)$ **(b)** $(-\infty, -0.52) \cup (2.59, \infty)$
17. $x^2 + 4x + 1 + \dfrac{-7}{x - 3}$ **19.** $2x^2 - 8x + 31 + \dfrac{-118}{x + 4}$
21. $(x - 2)(5x^2 + 7x + 16) + 26$ **23.** -1 **25.** 28
27. yes **29.** $7 - 2i$

In Exercises 31–35, other answers are possible.
31. $f(x) = x^3 - 10x^2 + 17x + 28$
33. $f(x) = x^4 - 5x^3 + 3x^2 + 15x - 18$
35. $f(x) = x^4 - 6x^3 + 9x^2 - 6x + 8$ **37.** $\frac{1}{2}, -1, 5$
39. (a) $f(-1) = -10 < 0$; $f(0) = 2 > 0$
(b) $f(2) = -4 < 0$; $f(3) = 14 > 0$ **(c)** 2.414
41. (a) The numbers in the bottom row of synthetic division are all positive. **(b)** The numbers in the bottom row of synthetic division alternate positive and negative.
43. yes **45.** $f(x) = -2x^3 + 6x^2 + 12x - 16$
47. $1, -\frac{1}{2}, \pm 2i$ **49.** $\frac{13}{2}$
51. Any polynomial that can be factored into $a(x - b)^3$ satisfies the conditions. One example is $f(x) = 2(x - 1)^3$.

$f(x) = 2(x - 1)^3$

53. (a) $(-\infty, \infty)$ **(b)** $(-\infty, \infty)$ **(c)** $f(x) \to \infty$ as
$x \to \infty$, $f(x) \to -\infty$ as $x \to -\infty$:
(d) at most seven **(e)** at most six

55.

$f(x) = (x - 2)^2(x + 3)$

57.

$f(x) = 2x^3 + x^2 - x$

59.

$f(x) = x^4 + x^3 - 3x^2 - 4x - 4$

61. C **63.** E **65.** B
67. 7.6533119, 1, −0.6533119

69. (a)

(b) $f(x) = -0.0109x^2 + 0.8693x + 11.85$
(c) $f(x) = -0.00087x^3 + 0.0456x^2 - 0.2191x + 17.83$
(d) $f(x) = -0.0109x^2 + 0.8693x + 11.85$

(e) Both functions approximate the data well. The quadratic function is probably better for prediction, because it is unlikely that the percent of out-of-pocket spending would decrease after 2025 (as the cubic function shows) unless changes were made in Medicare law.

71. 12 in. × 4 in. × 15 in.

73.

75.

77.

79.

81. (a)

(b) One possibility is
$f(x) = \dfrac{(x - 2)(x - 4)}{(x - 3)^2}$.

83. $f(x) = \dfrac{-3x + 6}{x - 1}$

85. (a)

(b) $127.3 thousand

87. 35 **89.** 0.75 **91.** 27 **93.** 33,750 units

Chapter 3 Test

[3.1] **1.**

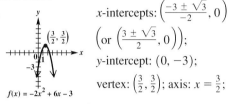

x-intercepts: $\left(\dfrac{-3 \pm \sqrt{3}}{-2}, 0\right)$
$\left(\text{or } \left(\dfrac{3 \pm \sqrt{3}}{2}, 0\right)\right)$;
y-intercept: $(0, -3)$;
vertex: $\left(\dfrac{3}{2}, \dfrac{3}{2}\right)$; axis: $x = \dfrac{3}{2}$;

domain: $(-\infty, \infty)$; range: $\left(-\infty, \dfrac{3}{2}\right]$; increasing on $\left(-\infty, \dfrac{3}{2}\right)$;

decreasing on $\left(\dfrac{3}{2}, \infty\right)$

2. (a) 2.75 sec **(b)** 169 ft **(c)** 0.7 sec and 4.8 sec
(d) 6 sec [3.2] **3.** $3x^2 - 2x - 5 + \dfrac{16}{x + 2}$
4. $2x^2 - x - 5$ **5.** 53 [3.3] **6.** It is a factor. The other
factor is $6x^3 + 7x^2 - 14x - 8$. **7.** $-2, -3 - 2i, -3 + 2i$
8. $f(x) = 2x^4 - 2x^3 - 2x^2 - 2x - 4$ **9.** Because
$f(x) > 0$ for all x, the graph never intersects or touches the
x-axis. Therefore, $f(x)$ has no real zeros.

[3.3, 3.4] **10. (a)** $f(1) = 5 > 0$; $f(2) = -1 < 0$

(b)

Positive	Negative	Nonreal Complex
2	1	0
0	1	2

(c) 4.0937635, 1.8370381, −0.9308016

[3.4] **11.**

$g(x) = -2(x + 5)^4 + 3$

To obtain the graph of g, translate the graph of f 5 units to the left, stretch by a factor of 2, reflect across the x-axis, and translate 3 units up.

12. C **13.**

$f(x) = x^3 - 5x^2 + 3x + 9$

14.

$f(x) = 2x^2(x - 2)^2$

15.

$f(x) = -x^3 - 4x^2 + 11x + 30$

16. $f(x) = 2(x - 2)^2(x + 3)$, or
$f(x) = 2x^3 - 2x^2 - 16x + 24$
17. (a) 270.08 **(b)** increasing
from $t = 0$ to $t = 5.9$ and $t = 9.5$
to $t = 15$; decreasing from $t = 5.9$
to $t = 9.5$

[3.5] **18.** **19.**

20. (a) $y = 2x + 3$ **(e)**

(b) $(-2, 0), \left(\frac{3}{2}, 0\right)$

(c) $(0, 6)$

(d) $x = 1$

[3.6] **21.** 60

22. $\frac{640}{9}$ kg

Chapter 4 Inverse, Exponential, and Logarithmic Functions

4.1 Exercises

1. one-to-one **3.** one-to-one **5.** range; domain

7. $\sqrt[3]{x}$ **9.** -3 **11.** one-to-one **13.** not one-to-one

15. one-to-one **17.** one-to-one **19.** not one-to-one

21. one-to-one **23.** one-to-one **25.** not one-to-one

27. one-to-one **29.** no **31.** untying your shoelaces

33. leaving a room **35.** unscrewing a light bulb

37. inverses **39.** not inverses **41.** inverses

43. not inverses **45.** inverses **47.** not inverses

49. inverses **51.** $\{(6, -3), (1, 2), (8, 5)\}$

53. not one-to-one **55.** inverses **57.** not inverses

59. (a) $f^{-1}(x) = \frac{1}{3}x + \frac{4}{3}$ **(b)**

(c) Domains and ranges of both f and f^{-1} are $(-\infty, \infty)$.

61. (a) $f^{-1}(x) = -\frac{1}{4}x + \frac{3}{4}$ **(b)**

(c) Domains and ranges of both f and f^{-1} are $(-\infty, \infty)$.

63. (a) $f^{-1}(x) = \sqrt[3]{x - 1}$ **(b)**

(c) Domains and ranges of both f and f^{-1} are $(-\infty, \infty)$.

65. not one-to-one

67. (a) $f^{-1}(x) = \frac{1}{x}, x \neq 0$ **(b)**

(c) Domains and ranges of both f and f^{-1} are $(-\infty, 0) \cup (0, \infty)$.

69. (a) $f^{-1}(x) = \frac{1 + 3x}{x}, x \neq 0$ **(b)**

(c) Domain of f = range of $f^{-1} = (-\infty, 3) \cup (3, \infty)$. Domain of f^{-1} = range of $f = (-\infty, 0) \cup (0, \infty)$.

71. (a) $f^{-1}(x) = \frac{3x + 1}{x - 1}, x \neq 1$ **(b)**

(c) Domain of f = range of $f^{-1} = (-\infty, 3) \cup (3, \infty)$. Domain of f^{-1} = range of $f = (-\infty, 1) \cup (1, \infty)$.

73. (a) $f^{-1}(x) = \frac{3x + 6}{x - 2}, x \neq 2$ **(b)**

(c) Domain of f = range of $f^{-1} = (-\infty, 3) \cup (3, \infty)$. Domain of f^{-1} = range of $f = (-\infty, 2) \cup (2, \infty)$.

75. (a) $f^{-1}(x) = x^2 - 6, x \geq 0$ **(b)**

(c) Domain of f = range of $f^{-1} = [-6, \infty)$. Domain of f^{-1} = range of $f = [0, \infty)$.

77.   **79.**

81.

83. 4 **85.** 2 **87.** -2

89. It represents the cost, in dollars, of building 1000 cars.

91. $\frac{1}{a}$ **93.** not one-to-one

95. one-to-one; $f^{-1}(x) = \frac{-5 - 3x}{x - 1}, x \neq 1$

97. $f^{-1}(x) = \frac{1}{3}x + \frac{2}{3}$; MIGUEL HAS ARRIVED

99. 6858   124   2743   63   511   124   1727   4095;
$f^{-1}(x) = \sqrt[3]{x+1}$

4.2 Exercises

1. $16; \frac{1}{16}$ **3.** falls   **5.** $\frac{1}{8}$; 1; 8 **7.** $\{-3\}$ **9.** $\{2540.22\}$

11. 9 **13.** $\frac{1}{9}$ **15.** $\frac{1}{16}$ **17.** 16 **19.** $3\sqrt{3}$ **21.** $\frac{1}{8}$

23. 13.076 **25.** 10.267

27.   **29.**

31.   **33.**

35.   **37.**

39.   **41.**
$(-\infty, \infty); (1, \infty)$ $(-\infty, \infty); (0, \infty)$

43.   **45.**
$(-\infty, \infty); (-\infty, 0)$ $(-\infty, \infty); (0, \infty)$

47.   **49.**
$(-\infty, \infty); (2, \infty)$ $(-\infty, \infty); (-4, \infty)$

51.   **53.**
$(-\infty, \infty); (-2, \infty)$ $(-\infty, \infty); (0, \infty)$

55.   **57.**
$(-\infty, \infty); (0, \infty)$ $(-\infty, \infty); (0, \infty)$

59.   **61.**
$(-\infty, \infty); (2, \infty)$ $(-\infty, \infty); (-1, \infty)$

63. $f(x) = 3^x - 2$ **65.** $f(x) = 2^{x+3} - 1$
67. $f(x) = -2^{x+2} + 3$ **69.** $f(x) = 3^{-x} + 1$

71. $\left\{\frac{1}{2}\right\}$ **73.** $\{-2\}$ **75.** $\{0\}$ **77.** $\left\{\frac{1}{2}\right\}$

79. $\left\{\frac{1}{5}\right\}$ **81.** $\{-7\}$ **83.** $\{\pm 8\}$ **85.** $\{4\}$

87. $\{\pm 2\}$ **89.** $\{-27\}$ **91.** $\left\{-\frac{2}{3}\right\}$ **93.** $\left\{\frac{4}{3}\right\}$

95. $\{3\}$ **97. (a)** $\$11,643.88$; $\$2737.34$
(b) $\$11,667.25$; $\$2760.71$ **99.** $\$22,902.04$
101. $\$3528.81$ **103.** 2.5% **105.** Bank A (even though it
has the greatest stated rate)
107. (a)   **(b)** exponential

(c)

$P(x) = 1013e^{-0.0001341x}$

(d) $P(1500) \approx 828$ mb; $P(11,000) \approx 232$ mb
109. (a) $63,000$ **(b)** $42,000$ **(c)** $21,000$
111. $\{0.9\}$ **113.** $\{-0.5, 1.3\}$ **115.** The variable is
located in the base of a power function and in the exponent
of an exponential function.   **117.** $f(x) = 2^x$
119. $f(x) = \left(\frac{1}{4}\right)^x$ **121.** $f(t) = 27 \cdot 9^t$ **123.** $f(t) = \left(\frac{1}{3}\right)9^t$
125. 2.717 (A calculator gives 2.718.)
127. yes; an inverse function
128.   **129.** $x = a^y$ **130.** $x = 10^y$
131. $x = e^y$ **132.** (q, p)

4.3 Exercises

1. (a) C **(b)** A **(c)** E **(d)** B **(e)** F **(f)** D

3. $2^3 = 8$ **5.** $\left\{\frac{4}{9}\right\}$ **7.**

$(0, \infty);\ (-\infty, \infty)$

9. $\log_{10} 2 + \log_{10} x - \log_{10} 7$ **11.** $\log_3 81 = 4$

13. $\log_{2/3} \frac{27}{8} = -3$ **15.** $6^2 = 36$ **17.** $\left(\sqrt{3}\right)^8 = 81$

19. $\{-4\}$ **21.** $\left\{\frac{1}{2}\right\}$ **23.** $\left\{\frac{1}{4}\right\}$ **25.** $\{8\}$

27. $\{9\}$ **29.** $\left\{\frac{1}{5}\right\}$ **31.** $\{64\}$ **33.** $\left\{\frac{2}{3}\right\}$ **35.** $\{243\}$

37. $\{13\}$ **39.** $\{3\}$ **41.** $\{5\}$

43.

$f(x) = (\log_2 x) + 3$

$(0, \infty);\ (-\infty, \infty)$

45.

$x = -3$

$f(x) = |\log_2(x+3)|$

$(-3, \infty);\ [0, \infty)$

47.

$f(x) = \log_{1/2}(x - 2)$

$x = 2$

$(2, \infty);\ (-\infty, \infty)$

49. E **51.** B **53.** F

55.

$f(x) = \log_5 x$

57.

$x = -1$

$f(x) = \log_5(x+1)$

59.

$x = 1$

$f(x) = \log_{1/2}(1 - x)$

61.

$x = 1$

$f(x) = \log_3(x - 1) + 2$

63.

$x = -3$

$f(x) = \log_{1/2}(x+3) - 2$

65. $f(x) = \log_2(x + 1) - 3$
67. $f(x) = \log_2(-x + 3) - 2$
69. $f(x) = -\log_3(x - 1)$
71. $\log_2 6 + \log_2 x - \log_2 y$
73. $1 + \frac{1}{2}\log_5 7 - \log_5 3$

75. This cannot be simplified.

77. $\frac{1}{2}(\log_2 5 + 3 \log_2 r - 5 \log_2 z)$

79. $\log_2 a + \log_2 b - \log_2 c - \log_2 d$

81. $\frac{1}{2}\log_3 x + \frac{1}{3}\log_3 y - 2 \log_3 w - \frac{1}{2}\log_3 z$

83. $\log_a \frac{xy}{m}$ **85.** $\log_a \frac{m}{nt}$ **87.** $\log_b\left(x^{-1/6} y^{11/12}\right)$

89. $\log_a\left[(z + 1)^2(3z + 2)\right]$ **91.** $\log_5 \frac{5^{1/3}}{m^{1/3}}$, or $\log_5 \sqrt[3]{\frac{5}{m}}$

93. 0.7781 **95.** 0.1761 **97.** 0.3522 **99.** 0.7386

101. (a) **(b)** logarithmic

103. (a) -4 **(b)** 6 **(c)** 4 **(d)** -1
107. $\{0.01, 2.38\}$

Summary Exercises on Inverse, Exponential, and Logarithmic Functions

1. They are inverses. **2.** They are not inverses.

3. They are inverses. **4.** They are inverses.

5.

$y = x$

6.

$y = x$

7. It is not one-to-one.

8.

$y = x$

9. B **10.** D **11.** C **12.** A

13. The functions in **Exercises 9 and 12** are inverses of one another. The functions in **Exercises 10 and 11** are inverses of one another. **14.** $f^{-1}(x) = 5^x$

15. $f^{-1}(x) = \frac{1}{3}x + 2$; Domains and ranges of both f and f^{-1} are $(-\infty, \infty)$. **16.** $f^{-1}(x) = \sqrt[3]{\frac{x}{2}} - 1$; Domains and ranges of both f and f^{-1} are $(-\infty, \infty)$.

17. f is not one-to-one. **18.** $f^{-1}(x) = \frac{5x + 1}{2 + 3x}$; Domain of f = range of $f^{-1} = \left(-\infty, \frac{5}{3}\right) \cup \left(\frac{5}{3}, \infty\right)$. Domain of f^{-1} = range of $f = \left(-\infty, -\frac{2}{3}\right) \cup \left(-\frac{2}{3}, \infty\right)$.

19. f is not one-to-one. **20.** $f^{-1}(x) = \sqrt{x^2 + 9}$, $x \ge 0$; Domain of f = range of $f^{-1} = [3, \infty)$. Domain of f^{-1} = range of $f = [0, \infty)$. **21.** $\log_{1/10} 1000 = -3$

22. $\log_a c = b$ **23.** $\log_{\sqrt{3}} 9 = 4$ **24.** $\log_4 \frac{1}{8} = -\frac{3}{2}$

25. $\log_2 32 = x$ **26.** $\log_{27} 81 = \frac{4}{3}$ **27.** $\{2\}$ **28.** $\{-3\}$

29. $\{-3\}$ **30.** $\{25\}$ **31.** $\{-2\}$ **32.** $\left\{\frac{1}{3}\right\}$

33. $(0, 1) \cup (1, \infty)$ **34.** $\left\{\frac{3}{2}\right\}$ **35.** $\{5\}$ **36.** $\{243\}$

37. $\{1\}$ **38.** $\{-2\}$ **39.** $\{1\}$ **40.** $\{2\}$ **41.** $\{2\}$

42. $\left\{\frac{1}{9}\right\}$ **43.** $\left\{-\frac{1}{3}\right\}$ **44.** $(-\infty, \infty)$

4.4 Exercises

1. increasing **3.** $f^{-1}(x) = \log_5 x$ **5.** natural; common

7. There is no power of 2 that yields a result of 0.

9. $\log 8 = 0.90308999$ **11.** 12   **13.** -1 **15.** 1.7993
17. -2.6576 **19.** 3.9494   **21.** 0.1803   **23.** 3.9494
25. 0.1803   **27.** The logarithm of the product of two numbers is equal to the sum of the logarithms of the numbers.   **29.** 3.2   **31.** 8.4   **33.** 2.0×10^{-3}
35. 1.6×10^{-5} **37.** poor fen   **39.** bog   **41.** rich fen
43. (a) 2.60031933   **(b)** 1.60031933   **(c)** 0.6003193298
(d) The whole number parts will vary, but the decimal parts will be the same.   **45.** 1.6   **47.** -2 **49.** $\frac{1}{2}$ **51.** 3.3322
53. -8.9480 **55.** 10.1449   **57.** 2.0200   **59.** 10.1449
61. 2.0200   **63. (a)** 20   **(b)** 30   **(c)** 50   **(d)** 60
(e) 3 decibels   **65. (a)** 3   **(b)** 6   **(c)** 8
67. $631{,}000{,}000 I_0$ **69.** 106.6 thousand; We must assume that the model continues to be logarithmic.
71. (a) 2   **(b)** 2   **(c)** 2   **(d)** 1   **73.** 1
75. between 7°F and 11°F   **77.** 1.13 billion yr
79. 2.3219   **81.** -0.2537 **83.** -1.5850 **85.** 0.8736
87. 1.9376   **89.** -1.4125 **91.** $4v + \frac{1}{2}u$ **93.** $\frac{3}{2}u - \frac{5}{2}v$
95. (a) 4   **(b)** 25   **(c)** $\frac{1}{e}$ **97. (a)** 6   **(b)** $\ln 3$ **(c)** $\ln 9$
99. D   **101.** domain: $(-\infty, 0) \cup (0, \infty)$; range: $(-\infty, \infty)$; symmetric with respect to the y-axis   **103.** $f(x) = 2 + \ln x$, so it is the graph of $g(x) = \ln x$ translated 2 units up.
105. $f(x) = \ln x - 2$, so it is the graph of $g(x) = \ln x$ translated 2 units down.

Chapter 4 Quiz

[4.1] **1.** $f^{-1}(x) = \frac{x^3 + 6}{3}$ [4.2] **2.** $\{4\}$

3.

[4.3] **4.**

domain: $(-\infty, \infty)$;   domain: $(-2, \infty)$;
range: $(-\infty, 0)$ range: $(-\infty, \infty)$
[4.2] **5. (a)** \$18,563.28   **(b)** \$18,603.03   **(c)** \$18,612.02
(d) \$18,616.39   [4.4] **6. (a)** 1.5386   **(b)** 3.5427
[4.3] **7.** The expression $\log_6 25$ represents the exponent to which 6 must be raised to obtain 25.   **8. (a)** $\{4\}$ **(b)** $\{5\}$
(c) $\left\{\frac{1}{16}\right\}$ [4.4] **9.** $\frac{1}{2}\log_3 x + \log_3 y - \log_3 p - 4\log_3 q$
10. 7.8137   **11.** 3.3578   **12.** 12

4.5 Exercises

1. B   **3.** E   **5.** D   **7.** $\log_7 19$; $\frac{\log 19}{\log 7}$; $\frac{\ln 19}{\ln 7}$

9. $\log_{1/2} 12$; $\frac{\log 12}{\log \frac{1}{2}}$; $\frac{\ln 12}{\ln \frac{1}{2}}$ **11.** $\{1.771\}$ **13.** $\{-2.322\}$

15. $\{-6.213\}$ **17.** $\{-1.710\}$ **19.** $\{3.240\}$
21. $\{\pm 2.146\}$ **23.** $\{9.386\}$ **25.** $\varnothing$ **27.** $\{32.950\}$
29. $\{7.044\}$ **31.** $\{25.677\}$ **33.** $\{2011.568\}$
35. $\{\ln 2, \ln 4\}$ **37.** $\left\{\ln \frac{3}{2}\right\}$ **39.** $\{\log_5 4\}$ **41.** $\{e^2\}$

43. $\left\{\frac{e^{1.5}}{4}\right\}$ **45.** $\{2 - \sqrt{10}\}$ **47.** $\{16\}$ **49.** $\{3\}$
51. $\{e\}$ **53.** $\{-6, 4\}$ **55.** $\{-8, 0\}$ **57.** $\{5\}$
59. $\{-5\}$ **61.** $\varnothing$ **63.** $\{-2\}$ **65.** $\{0\}$ **67.** $\{12\}$
69. $\{25\}$ **71.** $\varnothing$ **73.** $\left\{\frac{5}{2}\right\}$ **75.** $\{3\}$ **77.** $\left\{\frac{1 + \sqrt{41}}{4}\right\}$
79. $\{6\}$ **81.** $\{4\}$ **83.** $\{1, 100\}$ **85.** Proposed solutions that cause *any argument of a logarithm* to be negative or zero must be rejected. The statement is not correct. For example, the solution set of $\log(-x + 99) = 2$ is $\{-1\}$.
87. $x = e^{k/(p-a)}$ **89.** $t = -\frac{1}{k}\log\left(\frac{T - T_0}{T_1 - T_0}\right)$
91. $t = -\frac{2}{R}\ln\left(1 - \frac{RI}{E}\right)$ **93.** $x = \dfrac{\ln\left(\frac{A + B - y}{B}\right)}{-C}$
95. $A = \frac{B}{x^C}$ **97.** $t = \dfrac{\log \frac{A}{P}}{n \log\left(1 + \frac{r}{n}\right)}$ **99.** \$11,611.84
101. 2.6 yr   **103.** 2.64%   **105. (a)** 10.9%   **(b)** 35.8%
(c) 84.1%   **107.** 2019   **109. (a)** 62%   **(b)** 1989
111. (a) $P(x) = 1 - e^{-0.0034 - 0.0053x}$
(b) $P(x) = 1 - e^{-0.0034 - 0.0053x}$

(c) $P(60) \approx 0.275$, or 27.5%; The reduction in carbon emissions from a tax of \$60 per ton of carbon is 27.5%.   **(d)** $x = \$130.14$
113. $f^{-1}(x) = \ln x + 5$; domain: $(0, \infty)$; range: $(-\infty, \infty)$
115. $f^{-1}(x) = \ln(x + 4) - 1$; domain: $(-4, \infty)$;
range: $(-\infty, \infty)$ **117.** $f^{-1}(x) = \frac{1}{3}e^{x/2}$; domain: $(-\infty, \infty)$;
range: $(0, \infty)$ **119.** $\{1.52\}$ **121.** $\{0\}$
123. $\{2.45, 5.66\}$ **125.** When dividing each side by $\log \frac{1}{3}$, the direction of the inequality symbol should be reversed because $\log \frac{1}{3}$ is negative.

4.6 Exercises

1. B   **3.** C   **5.** B   **7.** C   **9.** $\frac{1}{3}\ln \frac{1}{3}$ **11.** $\frac{1}{100}\ln \frac{1}{2}$
13. $\frac{1}{2}\ln \frac{1}{4}$ **15. (a)** 440 g   **(b)** 387 g   **(c)** 264 g
(d) 22 yr   **17.** 1600 yr   **19.** 3.6 g   **21.** 16 days
23. 9000 yr   **25.** 15,600 yr   **27.** 6.25°C
29. (a) 3% compounded quarterly   **(b)** \$826.95
31. 27.73 yr   **33.** 36.62 yr   **35. (a)** 315   **(b)** 229
(c) 142   **37. (a)** $P_0 = 1$; $a \approx 1.01355$ **(b)** 1.3 billion
(c) 2030   **39. (a)** \$8412   **(b)** 2010   **41. (a)** \$14,542
(b) \$16,162   **(c)** \$17,494   **43. (a)** $y = 14{,}225e^{0.034x}$
(b) 2025–26   **45. (a)** 15,000   **(b)** 9098   **(c)** 5249
47. (a) 611 million   **(b)** 746 million   **(c)** 1007 million
49. 13.2 hr   **51.** 2020   **53.** 6.9 yr   **55.** 11.6 yr
57. (a) 0.065; 0.82; Among people age 25, 6.5% have some CHD, while among people age 65, 82% have some CHD.   **(b)** 48 yr

Summary Exercises on Functions: Domains and Defining Equations

1. $(-\infty, \infty)$ 2. $\left[\frac{7}{2}, \infty\right)$ 3. $(-\infty, \infty)$

4. $(-\infty, 6) \cup (6, \infty)$ 5. $(-\infty, \infty)$

6. $(-\infty, -3] \cup [3, \infty)$ 7. $(-\infty, -3) \cup (-3, 3) \cup (3, \infty)$

8. $(-\infty, \infty)$ 9. $(-4, 4)$ 10. $(-\infty, -7) \cup (3, \infty)$

11. $(-\infty, -1] \cup [8, \infty)$ 12. $(-\infty, 0) \cup (0, \infty)$

13. $(-\infty, \infty)$ 14. $(-\infty, -5) \cup (-5, \infty)$ 15. $[1, \infty)$

16. $\left(-\infty, -\sqrt{5}\right) \cup \left(-\sqrt{5}, \sqrt{5}\right) \cup \left(\sqrt{5}, \infty\right)$

17. $(-\infty, \infty)$ 18. $(-\infty, -1) \cup (-1, 1) \cup (1, \infty)$

19. $(-\infty, 1)$ 20. $(-\infty, 2) \cup (2, \infty)$

21. $(-\infty, \infty)$ 22. $[-2, 3] \cup [4, \infty)$

23. $(-\infty, -2) \cup (-2, 3) \cup (3, \infty)$

24. $[-3, \infty)$ 25. $(-\infty, 0) \cup (0, \infty)$

26. $\left(-\infty, -\sqrt{7}\right) \cup \left(-\sqrt{7}, \sqrt{7}\right) \cup \left(\sqrt{7}, \infty\right)$

27. $(-\infty, \infty)$ 28. $\varnothing$ 29. $[-2, 2]$ 30. $(-\infty, \infty)$

31. $(-\infty, -7] \cup (-4, 3) \cup [9, \infty)$ 32. $(-\infty, \infty)$

33. $(-\infty, 5]$ 34. $(-\infty, 3)$ 35. $(-\infty, 4) \cup (4, \infty)$

36. $(-\infty, \infty)$ 37. $(-\infty, -5] \cup [5, \infty)$ 38. $(-\infty, \infty)$

39. $(-2, 6)$ 40. $(0, 1) \cup (1, \infty)$ 41. A   42. B

43. C   44. D   45. A   46. B   47. D   48. C

49. C   50. B

Chapter 4 Review Exercises

1. not one-to-one   3. one-to-one   5. not one-to-one

7. $f^{-1}(x) = \sqrt[3]{x + 3}$ 9. It represents the number of years after 2004 for the investment to reach \$50,000.

11. one-to-one   13. B   15. C   17. $\log_2 32 = 5$

19. $\log_{3/4} \frac{4}{3} = -1$ 21. $10^3 = 1000$ 23. $e^{1/2} = \sqrt{e}$

25. 2   27. $\log_3 m + \log_3 n - \log_3 5 - \log_3 r$

29. -1.3862 31. 11.8776 33. 1.1592 35. $\left\{\frac{22}{5}\right\}$

37. $\{3.667\}$ 39. $\{-13.257\}$ 41. $\{-0.485\}$

43. $\{2.102\}$ 45. $\{-2.487\}$ 47. $\{3\}$ 49. $\{\ln 3\}$

51. $\{6.959\}$ 53. $\{e^{13/3}\}$ 55. $\left\{\frac{\sqrt[4]{10} - 7}{2}\right\}$ 57. $\{3\}$

59. $\left\{-\frac{4}{3}, 5\right\}$ 61. $\left\{1, \frac{10}{3}\right\}$ 63. $\{2\}$

65. $\{-3\}$ 67. $I_0 = \frac{I}{10^{d/10}}$ 69. $\{1.315\}$

71. (a) $1{,}000{,}000 I_0$ (b) $158{,}500{,}000 I_0$

(c) 158.5 times greater   73. 3.0%   75. \$20,363.38

77. 17.3 yr   79. 2016   81. (a) \$15,207   (b) \$10,716

(c) \$4491   (d) They are the same.

Chapter 4 Test

[4.1] 1. (a) $(-\infty, \infty); (-\infty, \infty)$ (b) The graph is a stretched translation of $y = \sqrt[3]{x}$, which passes the horizontal line test and is thus a one-to-one function.

(c) $f^{-1}(x) = \frac{x^3 + 7}{2}$ (d) $(-\infty, \infty); (-\infty, \infty)$

(e)   The graphs are reflections of each other across the line $y = x$.

[4.2, 4.3] 2. (a) B   (b) A   (c) C   (d) D

3. $\left\{\frac{1}{2}\right\}$ 4. (a) $\log_4 8 = \frac{3}{2}$ (b) $8^{2/3} = 4$

[4.1–4.3] 5.

$$f(x) = \left(\frac{1}{2}\right)^x$$
$$g(x) = \log_{1/2} x$$

They are inverses.

[4.3] 6. $2 \log_7 x + \frac{1}{4} \log_7 y - 3 \log_7 z$ [4.4] 7. 3.3780

8. 7.7782   9. 1.1674   [4.2] 10. $\{\perp 125\}$ 11. $\{0\}$

[4.5] 12. $\{0.631\}$ [4.2] 13. $\{4\}$ [4.5] 14. $\{12.548\}$

15. $\{2.811\}$ 16. $\left\{0, \ln \frac{3}{2}\right\}; \{0, 0.405\}$ [4.3] 17. $\left\{\frac{3}{4}\right\}$

[4.5] 18. $\{0, 6\}$ 19. $\{2\}$ 20. $\varnothing$ 21. $\left\{\frac{7}{2}\right\}$

[4.4] 22. The expression $\log_5 27$ represents the exponent to which 5 must be raised in order to obtain 27. To approximate it with a calculator, use the change-of-base theorem.

$\log_5 27 = \frac{\log 27}{\log 5} \approx 2.0478$ [4.6] 23. 10 sec

24. (a) 42.8 yr   (b) 42.7 yr   25. 39.2 yr

26. (a) 329.3 g   (b) 13.9 days

Chapter 5 Systems and Matrices

5.1 Exercises

1. 4   3. -11 5. infinitely many   7. $\{(-4, 1)\}$

9. $\{(48, 8)\}$ 11. $\{(1, 3)\}$ 13. $\{(-1, 3)\}$

15. $\{(3, -4)\}$ 17. $\{(0, 2)\}$ 19. $\{(0, 4)\}$

21. $\{(1, 3)\}$ 23. $\{(4, -2)\}$ 25. $\{(2, -2)\}$

27. $\{(4, 6)\}$ 29. $\{(5, 2)\}$ 31. $\varnothing$; inconsistent system

33. $\left\{\left(\frac{y + 9}{4}, y\right)\right\}$; infinitely many solutions   35. $\varnothing$; inconsistent system   37. $\left\{\left(\frac{-2y + 6}{7}, y\right)\right\}$; infinitely many solutions   39. $\varnothing$; inconsistent system   41. $x - 3y = -3$
$3x + 2y = 6$

43. $\{(0.820, -2.508)\}$ 45. $\{(0.892, 0.453)\}$

47. $\{(1, 2, -1)\}$ 49. $\{(2, 0, 3)\}$

51. $\{(1, 2, 3)\}$ 53. $\{(4, 1, 2)\}$ 55. $\left\{\left(\frac{1}{2}, \frac{2}{3}, -1\right)\right\}$

57. $\{(-3, 1, 6)\}$ 59. $\left\{\left(\frac{-z + 4}{3}, \frac{4z - 7}{3}, z\right)\right\}$

61. $\left\{\left(\frac{-5z + 69}{5}, -z + 15, z\right)\right\}$

63. $\{(-9z + 47, 7z - 32, z)\}$ 65. $\varnothing$; inconsistent system

67. $\left\{\left(-\frac{z}{9}, \frac{z}{9}, z\right)\right\}$; infinitely many solutions

69. $\{(2, 2)\}$ **71.** $\left\{\left(\frac{1}{5}, 1\right)\right\}$ **73.** $\{(4, 6, 1)\}$

75. $k \neq -6; k = -6$ **77.** $y = -3x - 5$

79. $y = \frac{3}{4}x^2 + \frac{1}{4}x - \frac{1}{2}$ **81.** $y = 3x - 1$

83. $y = -\frac{1}{2}x^2 + x + \frac{1}{4}$ **85.** $x^2 + y^2 - 4x + 2y - 20 = 0$

87. $x^2 + y^2 + x - 7y = 0$

89. $x^2 + y^2 - \frac{7}{5}x + \frac{27}{5}y - \frac{126}{5} = 0$

91. **(a)** $a = 0.0118, b = 0.8633, c = 317$;
$C = 0.0118x^2 + 0.8633x + 317$ **(b)** 2091

93. 2006–2013   **95.** (2005.2, 1.26)   **97.** year; population
(in millions)   **99.** 23, 24   **101.** baseball: $212.47; football:
$478.60   **103.** 120 gal of $9.00; 60 gal of $3.00; 120 gal of
$4.50   **105.** 28 in.; 17 in.; 14 in.   **107.** $100,000 at 3%;
$40,000 at 2.5%; $60,000 at 1.5%   **109.** $\{(40, 15, 30)\}$

111. 11.92 lb of Arabian Mocha Sanani; 14.23 lb of
Organic Shade Grown Mexico; 23.85 lb of Guatemala
Antigua   **113.** **(a)** $16   **(b)** $11   **(c)** $6   **114.** **(a)** 8
(b) 4   **(c)** 0   **115.** See the answer to **Exercise 117.**

116. **(a)** 0   **(b)** $\frac{40}{3}$ **(c)** $\frac{80}{3}$

117.

118. price: $6; demand: 8

5.2 Exercises

1. 2 rows, 4 columns; 2×4 **3.** $\begin{bmatrix} -3 & 5 & | & 2 \\ 6 & 2 & | & 7 \end{bmatrix}$

5. $\begin{bmatrix} 3 & 2 & 0 & | & 5 \\ -9 & 0 & 6 & | & 1 \\ 0 & -8 & 1 & | & 4 \end{bmatrix}$ **7.** $\begin{bmatrix} 1 & 4 \\ 0 & -9 \end{bmatrix}$ **9.** $\begin{bmatrix} 1 & 5 & 6 \\ 0 & 13 & 11 \\ 4 & 7 & 0 \end{bmatrix}$

11. $\begin{bmatrix} 2 & 3 & | & 11 \\ 1 & 2 & | & 8 \end{bmatrix}$; 2×3 **13.** $\begin{bmatrix} 2 & 1 & 1 & | & 3 \\ 3 & -4 & 2 & | & -7 \\ 1 & 1 & 1 & | & 2 \end{bmatrix}$; 3×4

15. $3x + 2y + z = 1$ **17.** $x = 2$ **19.** $x + y = 3$
$\quad\quad 2y + 4z = 22$ $\quad\quad y = 3$ $\quad\quad 2y + z = -4$
$\quad\quad -x - 2y + 3z = 15$ $\quad\quad z = -2$ $\quad\quad x - z = 5$

21. $\{(2, 3)\}$ **23.** $\{(-3, 0)\}$ **25.** $\left\{\left(1, \frac{2}{3}\right)\right\}$

27. $\varnothing$ **29.** $\left\{\left(\frac{4y + 7}{3}, y\right)\right\}$ **31.** $\{(-1, -2, 3)\}$

33. $\{(-1, 23, 16)\}$ **35.** $\{(2, 4, 5)\}$ **37.** $\left\{\left(\frac{1}{2}, 1, -\frac{1}{2}\right)\right\}$

39. $\{(2, 1, -1)\}$ **41.** $\varnothing$ **43.** $\left\{\left(\frac{-15z - 12}{23}, \frac{z - 13}{23}, z\right)\right\}$

45. $\{(1, 1, 2, 0)\}$ **47.** $\{(0, 2, -2, 1)\}$

49. $\{(0.571, 7.041, 11.442)\}$ **51.** none   **53.** $A = \frac{1}{2}$,
$B = -\frac{1}{2}$ **55.** $A = \frac{1}{2}, B = \frac{1}{2}$ **57.** day laborer: $152;
concrete finisher: $160   **59.** 12, 6, 2   **61.** 9.6 cm³ of 7%;
30.4 cm³ of 2%   **63.** $5000 at 1.5%; $10,000 at 2.2%;
$10,000 at 2.4%   **65.** 44.4 g of A; 133.3 g of B; 222.2 g of C

67. **(a)** 65 or older: $y = 0.0017x + 0.148$; ages 25–39:
$y = -0.0028x + 0.2$ **(b)** $\{(11.5556, 0.1676)\}$;
2026; 16.8%   **(c)** The *percent* of people in the U.S.
population aged 25–39 is decreasing, but not necessarily
the *number* of people in this category.   **69.** **(a)** using the
first equation, 245 lb; using the second equation, 253 lb
(b) for the first, 7.46 lb; for the second, 7.93 lb
(c) 118 lb and 66 in.

71.

n	T
3	28
6	191
10	805
29	17,487
100	681,550
200	5,393,100
400	42,906,200
1000	668,165,500
5000	8.3×10^{10}
10,000	6.7×10^{11}
100,000	6.7×10^{14}

73. no; It increases by almost a factor of 8.

75. $a + 871b + 11.5c + 3d = 239$
$\quad a + 847b + 12.2c + 2d = 234$
$\quad a + 685b + 10.6c + 5d = 192$
$\quad a + 969b + 14.2c + 1d = 343$

76. $\begin{bmatrix} 1 & 871 & 11.5 & 3 & | & 239 \\ 1 & 847 & 12.2 & 2 & | & 234 \\ 1 & 685 & 10.6 & 5 & | & 192 \\ 1 & 969 & 14.2 & 1 & | & 343 \end{bmatrix}$; $\begin{matrix} a \approx -715.457, \\ b \approx 0.348, \\ c \approx 48.659, \\ d \approx 30.720 \end{matrix}$

77. $F = -715.457 + 0.348A + 48.659P + 30.720W$
78. 323; This estimate is very close to the actual value
of 320.

5.3 Exercises

1. 0   **3.** $x^2 - 12$ **5.** -3 or 3   **7.** -31 **9.** 7   **11.** 0
13. -26 **15.** 0   **17.** 2, -6, 4   **19.** -6, 0, -6 **21.** 186
23. 17   **25.** 166   **27.** 0   **29.** 0   **31.** 1   **33.** 2
35. $-144 - 8\sqrt{10}$ **37.** -5.5 **39.** -3 **41.** 15   **43.** 3
45. 0   **47.** 0   **49.** 16   **51.** 17   **53.** 54   **55.** 0   **57.** 0
59. -88 **61.** 298   **63.** $\{(2, 2)\}$ **65.** $\{(2, -5)\}$
67. $\{(2, 0)\}$ **69.** Cramer's rule does not apply because
$D = 0; \varnothing$ **71.** Cramer's rule does not apply because
$D = 0; \left\{\left(\frac{-2y + 4}{3}, y\right)\right\}$ **73.** $\{(-4, 12)\}$ **75.** $\{(-3, 4, 2)\}$
77. Cramer's rule does not apply because $D = 0; \varnothing$
79. Cramer's rule does not apply because $D = 0$;
$\left\{\left(\frac{19z - 32}{4}, \frac{-13z + 24}{4}, z\right)\right\}$ **81.** $\{(0, 0, -1)\}$ **83.** $\{(0, 4, 2)\}$
85. $\left\{\left(\frac{31}{5}, \frac{19}{10}, -\frac{29}{10}\right)\right\}$ **87.** $W_1 = W_2 = \frac{100\sqrt{3}}{3} \approx 58$ lb
89. 1 unit²   **91.** 9.5 units²   **93.** 19,328.3 ft²   **95.** $\left\{-\frac{4}{3}\right\}$
97. $\{-1, 4\}$ **99.** $\{-4\}$ **101.** $\{13\}$ **103.** $\left\{-\frac{1}{2}\right\}$

105. $\{(-a - b, a^2 + ab + b^2)\}$ **107.** $\{(1, 0)\}$
109. $\{(-1, 2)\}$ **112.** 102   **113.** 102; yes   **114.** no

5.4 Exercises

1. $3x(2x + 1)$ **3.** $(x + 4)(3x^2 + 1)$ **5.** $x(2x^2 + 1)^2$
7. $\frac{5}{3x} + \frac{-10}{3(2x + 1)}$ **9.** $\frac{6}{5(x + 2)} + \frac{8}{5(2x - 1)}$
11. $\frac{5}{6(x + 5)} + \frac{1}{6(x - 1)}$ **13.** $\frac{4}{x} + \frac{4}{1 - x}$ **15.** $\frac{15}{x} + \frac{-5}{x + 1} + \frac{-6}{x - 1}$
17. $\frac{2}{(x + 2)^2} + \frac{-3}{(x + 2)^3}$ **19.** $1 + \frac{-2}{x + 1} + \frac{1}{(x + 1)^2}$
21. $\frac{1}{9} + \frac{-1}{x} + \frac{25}{18(3x + 2)} + \frac{29}{18(3x - 2)}$ **23.** $\frac{-3}{5x^2} + \frac{3}{5(x^2 + 5)}$
25. $\frac{-2}{7(x + 4)} + \frac{6x - 3}{7(3x^2 + 1)}$ **27.** $\frac{1}{4x} + \frac{-8}{19(2x + 1)} + \frac{-9x - 24}{76(3x^2 + 4)}$
29. $x^3 - x^2 + \frac{-1}{3(2x + 1)} + \frac{2}{3(x + 2)}$
31. $\frac{-1}{x} + \frac{2x}{2x^2 + 1} + \frac{2x + 3}{(2x^2 + 1)^2}$
33. $\frac{-1}{x + 2} + \frac{3}{(x^2 + 4)^2}$ **35.** $5x^2 + \frac{3}{x} + \frac{-1}{x + 3} + \frac{2}{x - 1}$
37. $\frac{-1}{4(x + 1)} + \frac{1}{4(x - 1)} + \frac{1}{2(x^2 + 1)}$ **39.** $\frac{2}{x} + \frac{-1}{x - 1} + \frac{3}{x + 2}$

Chapter 5 Quiz

[5.1] **1.** $\{(-2, 0)\}$ **2.** $\left\{\left(x, \frac{-x + 2}{2}\right)\right\}$, or $\{(-2y + 2, y)\}$
3. $\varnothing$ **4.** $\{(3, -4)\}$ [5.2] **5.** $\{(-5, 2)\}$
[5.3] **6.** $\{(-3, 6)\}$ [5.1] **7.** $\{(-2, 1, 2)\}$
[5.2] **8.** $\{(2, 1, -1)\}$ [5.3] **9.** Cramer's rule does not
apply because $D = 0$; $\left\{\left(\frac{2y + 6}{9}, y, \frac{23y + 6}{9}\right)\right\}$
[5.1] **10.** at home: \$3961.20; away from home: \$2640.80
11. \$1000 at 2%; \$1500 at 3%; \$2500 at 4%
[5.3] **12.** -3 **13.** 59   [5.4] **14.** $\frac{7}{x - 5} + \frac{3}{x + 4}$
15. $\frac{-1}{x + 2} + \frac{6}{x + 4} + \frac{-3}{x - 1}$

5.5 Exercises

1. 4   **3.** -1 **5.** $x^2 + x = 2$ **13.** Consider the graphs. A
line and a parabola cannot intersect in more than two points.
15. $\{(1, 1), (-2, 4)\}$ **17.** $\left\{(2, 1), \left(\frac{1}{3}, \frac{4}{9}\right)\right\}$
19. $\{(2, 12), (-4, 0)\}$ **21.** $\left\{\left(-\frac{3}{5}, \frac{7}{5}\right), (-1, 1)\right\}$
23. $\{(2, 2), (2, -2), (-2, 2), (-2, -2)\}$ **25.** $\{(0, 0)\}$
27. $\left\{\left(i, \sqrt{6}\right), \left(-i, \sqrt{6}\right), \left(i, -\sqrt{6}\right), \left(-i, -\sqrt{6}\right)\right\}$
29. $\{(1, -1), (-1, 1), (1, 1), (-1, -1)\}$ **31.** $\varnothing$
33. $\left\{\left(\sqrt{6}, 0\right), \left(-\sqrt{6}, 0\right)\right\}$ **35.** $\left\{(-3, 5), \left(\frac{15}{4}, -4\right)\right\}$
37. $\left\{\left(4, -\frac{1}{8}\right), \left(-2, \frac{1}{4}\right)\right\}$ **39.** $\{(3, 4), (-3, -4),$
$\left(\frac{4\sqrt{3}}{3}i, -3i\sqrt{3}\right), \left(-\frac{4\sqrt{3}}{3}i, 3i\sqrt{3}\right)\}$ **41.** $\{(\sqrt{5}, 0),$
$(-\sqrt{5}, 0), (\sqrt{5}, \sqrt{5}), (-\sqrt{5}, -\sqrt{5})\}$
43. $\{(3, 5), (-3, -5), (5i, -3i), (-5i, 3i)\}$
45. $\{(5, 0), (-5, 0), (0, -5)\}$ **47.** $\{(3, -3), (3, 3)\}$
49. $\{(2, 2), (-2, -2), (2, -2), (-2, 2)\}$

51. $\{(-0.79, 0.62), (0.88, 0.77)\}$ **53.** $\{(0.06, 2.88)\}$
55. -14 and -3 **57.** 8 and 6, 8 and -6, -8 and 6,
-8 and -6 **59.** 27 and 6, -27 and -6
61. 5 m and 12 m   **63.** yes   **65.** $y = 9$
67. *First answer:* length = width: 6 ft; height: 10 ft
Second answer: length = width: 12.780 ft; height: 2.204 ft
71. (a) 160 units   **(b)** \$3   **73. (a)** Revenue for public
colleges from state sources is decreasing, and revenue from
tuition is increasing.   **(b)** 2007 to 2008   **(c)** The sources
of revenue were equal in 2011, when they each contributed
about 24%.   **75.** Translate the graph of $y = |x|$ one unit to
the right.   **76.** Translate the graph of $y = x^2$ four units down.
77. $y = \begin{cases} x - 1 & \text{if } x \geq 1 \\ 1 - x & \text{if } x < 1 \end{cases}$ **78.** $x^2 - 4 = x - 1$ ($x \geq 1$);
$x^2 - 4 = 1 - x$ ($x < 1$)   **79.** $\frac{1 + \sqrt{13}}{2}; \frac{-1 - \sqrt{21}}{2}$
80. $\left\{\left(\frac{1 + \sqrt{13}}{2}, \frac{-1 + \sqrt{13}}{2}\right), \left(\frac{-1 - \sqrt{21}}{2}, \frac{3 + \sqrt{21}}{2}\right)\right\}$

Summary Exercises on Systems of Equations

1. $\{(-3, 2)\}$ **2.** $\{(-5, -4)\}$
3. $\left\{\left(0, \sqrt{5}\right), \left(0, -\sqrt{5}\right)\right\}$ **4.** $\{(-10, -6, 25)\}$
5. $\left\{\left(\frac{4}{3}, 3\right), \left(-\frac{1}{2}, -8\right)\right\}$ **6.** $\{(-1, 2, -4)\}$
7. $\{(5z - 13, -3z + 9, z)\}$ **8.** $\left\{(-2, -2), \left(\frac{14}{5}, -\frac{2}{5}\right)\right\}$
9. $\varnothing$ **10.** $\{(-3, 3, -3)\}$ **11.** $\left\{\left(-3 + i\sqrt{7}, 3 + i\sqrt{7}\right),\right.$
$\left.\left(-3 - i\sqrt{7}, 3 - i\sqrt{7}\right)\right\}$ **12.** $\{(2, -5, 3)\}$
13. $\{(1, 2), (-5, 14)\}$ **14.** $\varnothing$ **15.** $\{(0, 1, 0)\}$
16. $\left\{\left(\frac{3}{8}, -1\right)\right\}$ **17.** $\{(-1, 0, 0)\}$
18. $\{(-8z - 56, z + 13, z)\}$ **19.** $\left\{(0, 3), \left(-\frac{36}{17}, -\frac{3}{17}\right)\right\}$
20. $\{(2y + 3, y)\}$ **21.** $\{(1, 2, 3)\}$
22. $\left\{\left(2 + 2i\sqrt{2}, 5 - i\sqrt{2}\right), \left(2 - 2i\sqrt{2}, 5 + i\sqrt{2}\right)\right\}$
23. $\{(2, 2, 5)\}$ **24.** $\varnothing$ **25.** $\{(4, 2), (-1, -3)\}$
26. $\{(-2, -4, 0)\}$ **27.** $\left\{\left(\sqrt{13}, i\sqrt{2}\right), \left(-\sqrt{13}, i\sqrt{2}\right),\right.$
$\left.\left(\sqrt{13}, -i\sqrt{2}\right), \left(-\sqrt{13}, -i\sqrt{2}\right)\right\}$ **28.** $\{(-z - 6, 5, z)\}$
29. $\{(1, -3, 4)\}$ **30.** $\left\{\left(-\frac{16}{3}, -\frac{11}{3}\right), (4, 1)\right\}$
31. $\{(-1, -2, 5)\}$ **32.** $\varnothing$ **33.** $\{(1, -3), (-3, 1)\}$
34. $\{(1, -6, 2)\}$ **35.** $\{(-6, 9), (-1, 4)\}$
36. $\{(2, -3, 1)\}$

5.6 Exercises

1. C   **3.** B   **5.** does not   **7.** $(2, -5)$ **9.** dashed
11.

13.

15.

17.

19.

21.

23.

25.

27.

29. above   **31.** B   **33.** C   **35.** A

37.

39.

41.

43.

45.

47.

49.

51. The solution set is $\varnothing$.

53.

55.

57.

59.

61.

63.

65.

67.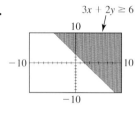

69. $x + y \ge 2, x + y \le 6$ **71.** $x + 2y \le 4$

$3x - 4y \le 12$
73. $x^2 + y^2 \le 16$
$y \ge 2$

75. $x > 0$ **77.** maximum of 65 at $(5, 10)$;
$y > 0$ minimum of 8 at $(1, 1)$
$x^2 + y^2 \le 4$ **79.** maximum of 66 at $(7, 9)$;
$y > \frac{3}{2}x - 1$ minimum of 3 at $(1, 0)$

81. maximum of 100 at $(1, 10)$; minimum of 0 at $(1, 0)$

83. Let $x =$ the number of Brand X pills and $y =$ the number of Brand Y pills. Then $3000x + 1000y \ge 6000$, $45x + 50y \ge 195$, $75x + 200y \ge 600$, $x \ge 0$, $y \ge 0$.

85. 300 cartons of food and 400 cartons of clothes; 6200 people

87. 8 of A and 3 of B, for a maximum storage capacity of 100 ft³

89. $3\frac{3}{4}$ servings of A and $1\frac{7}{8}$ servings of B, for a minimum cost of $1.69

5.7 Exercises

1. 2; 0   **3.** 6; -1; -1; 12   **5.** 3×6 **7.** 2×2; square
9. 3×4 **11.** 2×1; column   **13.** $a = 0, b = 4, c = -3,$
$d = 5$ **15.** $x = -4, y = 14, z = 3, w = -2$
17. This cannot be true.   **19.** $w = 5, x = 6, y = -2, z = 8$
21. $z = 18, r = 3, s = 3, p = 3, a = \frac{3}{4}$ **23.** Be sure that the two matrices have the same dimension. The sum will have this dimension as well. To find the elements of the sum, add the corresponding elements of the two matrices.

25. $\begin{bmatrix} -2 & -5 \\ 17 & 4 \end{bmatrix}$ **27.** $\begin{bmatrix} -2 & -7 & 7 \\ 10 & -2 & 7 \end{bmatrix}$

29. They cannot be added. **31.** $\begin{bmatrix} -6 & 8 \\ 4 & 2 \end{bmatrix}$

33. $\begin{bmatrix} 4 \\ -5 \\ 4 \end{bmatrix}$ **35.** They cannot be subtracted.

37. $\begin{bmatrix} -\sqrt{3} & -13 \\ 4 & -2\sqrt{5} \\ -1 & -\sqrt{2} \end{bmatrix}$ **39.** $\begin{bmatrix} 5x+y & y \\ 6x+2y & x+3y \end{bmatrix}$

41. $\begin{bmatrix} -4 & 8 \\ 0 & 6 \end{bmatrix}$ **43.** $\begin{bmatrix} -9 & 3 \\ 6 & 0 \end{bmatrix}$ **45.** $\begin{bmatrix} 14 & 2 \\ -12 & 6 \end{bmatrix}$

47. $\begin{bmatrix} -1 & -3 \\ 2 & -3 \end{bmatrix}$ **49.** yes; 2×5 **51.** no **53.** yes; 3×2

55. $\begin{bmatrix} 13 \\ 25 \end{bmatrix}$ **57.** $\begin{bmatrix} -17 \\ -1 \end{bmatrix}$ **59.** $\begin{bmatrix} 17\sqrt{2} & -4\sqrt{2} \\ 35\sqrt{3} & 26\sqrt{3} \end{bmatrix}$

61. $\begin{bmatrix} 3+4\sqrt{4} & -3\sqrt{2} \\ 2\sqrt{15}+12\sqrt{6} & -2\sqrt{30} \end{bmatrix}$

63. They cannot be multiplied. **65.** $\begin{bmatrix} 2 & 7 & -4 \end{bmatrix}$

67. $\begin{bmatrix} -15 & -16 & 3 \\ -1 & 0 & 9 \\ 7 & 6 & 12 \end{bmatrix}$ **69.** $\begin{bmatrix} 23 & -9 \\ -6 & -2 \\ 33 & 1 \end{bmatrix}$

71. $\begin{bmatrix} -25 & 23 & 11 \\ 0 & -6 & -12 \\ -15 & 33 & 45 \end{bmatrix}$ **73.** They cannot be multiplied.

75. $\begin{bmatrix} 10 & -10 \\ 15 & -5 \end{bmatrix}$ **77.** $BA \neq AB$, $BC \neq CB$, $AC \neq CA$

79. (a) $\begin{bmatrix} 38 & -8 \\ -7 & -2 \end{bmatrix}$ **(b)** $\begin{bmatrix} 18 & 24 \\ 19 & 18 \end{bmatrix}$

81. (a) $\begin{bmatrix} 0 & 1 & -1 \\ 0 & 1 & 0 \\ 0 & 0 & 1 \end{bmatrix}$ **(b)** $\begin{bmatrix} 0 & 1 & -1 \\ 0 & 1 & 0 \\ 0 & 0 & 1 \end{bmatrix}$ **83.** 1

85. (a) $\begin{bmatrix} 50 & 100 & 30 \\ 10 & 90 & 50 \\ 60 & 120 & 40 \end{bmatrix}$ **(b)** $\begin{bmatrix} 12 \\ 10 \\ 15 \end{bmatrix}$ **(c)** $\begin{bmatrix} 2050 \\ 1770 \\ 2520 \end{bmatrix}$

(d) \$6340 **87.** Answers will vary a little if intermediate steps are rounded. **(a)** 2940, 2909, 2861, 2814, 2767 **(b)** The northern spotted owl will become extinct. **(c)** 3023, 3051, 3079, 3107, 3135

5.8 Exercises

1. $\begin{bmatrix} 6 & 4 \\ -1 & 8 \end{bmatrix}$ **3.** $\begin{bmatrix} 1 & 0 \\ 0 & 1 \end{bmatrix}$, or I_2 **5.** $\begin{bmatrix} 3 & -6 \\ -1 & 3 \end{bmatrix}$

9. yes **11.** no **13.** no **15.** yes

17. $\begin{bmatrix} -\frac{1}{5} & -\frac{2}{5} \\ \frac{2}{5} & -\frac{1}{5} \end{bmatrix}$ **19.** $\begin{bmatrix} 2 & 1 \\ -\frac{3}{2} & -\frac{1}{2} \end{bmatrix}$ **21.** The inverse does not exist.

23. $\begin{bmatrix} -1 & 1 & 1 \\ 0 & -1 & 0 \\ 2 & -1 & -1 \end{bmatrix}$ **25.** $\begin{bmatrix} \frac{7}{8} & -\frac{3}{8} & -\frac{3}{8} \\ -\frac{1}{8} & \frac{5}{8} & -\frac{3}{8} \\ -\frac{1}{8} & -\frac{3}{8} & \frac{5}{8} \end{bmatrix}$

27. $\begin{bmatrix} -\frac{15}{4} & -\frac{1}{4} & -3 \\ \frac{5}{4} & \frac{1}{4} & 1 \\ -\frac{3}{2} & 0 & -1 \end{bmatrix}$ **29.** $\begin{bmatrix} \frac{1}{2} & 0 & \frac{1}{2} & -1 \\ \frac{1}{10} & -\frac{2}{5} & \frac{3}{10} & -\frac{1}{5} \\ -\frac{7}{10} & \frac{4}{5} & -\frac{11}{10} & \frac{12}{5} \\ \frac{1}{5} & \frac{1}{5} & -\frac{2}{5} & \frac{3}{5} \end{bmatrix}$

31. $\begin{bmatrix} 0 & -\frac{1}{3} & \frac{1}{3} & \frac{2}{3} \\ \frac{1}{3} & \frac{2}{3} & -\frac{1}{3} & -1 \\ \frac{2}{3} & 1 & -\frac{4}{3} & -\frac{4}{3} \\ -\frac{1}{3} & \frac{1}{3} & \frac{1}{3} & 0 \end{bmatrix}$ **33.** $\{(2, 3)\}$

35. $\{(-2, 4)\}$ **37.** $\left\{\left(-2, \frac{3}{4}\right)\right\}$ **39.** $\left\{\left(-\frac{1}{2}, \frac{2}{3}\right)\right\}$

41. $\{(4, -9)\}$ **43.** $\left\{\left(6, -\frac{5}{6}\right)\right\}$ **45.** The inverse of

$\begin{bmatrix} 7 & -2 \\ 14 & -4 \end{bmatrix}$ does not exist. **47.** $\{(3, 1, 2)\}$

49. $\left\{\left(\frac{5}{4}, \frac{1}{4}, -\frac{3}{4}\right)\right\}$ **51.** $\{(11, -1, 2)\}$ **53.** $\{(1, 0, 2, 1)\}$

55. (a) $602.7 = a + 5.543b + 37.14c$
$656.7 = a + 6.933b + 41.30c$
$778.5 = a + 7.638b + 45.62c$
(b) $a = -490.5$, $b = -89$, $c = 42.72$
(c) $S = -490.5 - 89A + 42.72B$ **(d)** $S = 843.6$
(e) $S = 1547.6$; Predictions made beyond the scope of the data are valid only if current trends continue.

57. Answers will vary.

59. $\begin{bmatrix} -0.1215875322 & 0.0491390161 \\ 1.544369078 & -0.046799063 \end{bmatrix}$

61. $\begin{bmatrix} 2 & -2 & 0 \\ -4 & 0 & 4 \\ 3 & 3 & -3 \end{bmatrix}$

63. $\{(1.68717058, -1.306990242)\}$
65. $\{(13.58736702, 3.929011993, -5.342780076)\}$

73. $A^{-1} = A^2 = \begin{bmatrix} 1 & 0 & 0 \\ 0 & -1 & 1 \\ 0 & -1 & 0 \end{bmatrix}$

Chapter 5 Review Exercises

1. $\{(0, 1)\}$ **3.** $\{(-5y + 9, y)\}$; infinitely many solutions
5. $\varnothing$; inconsistent system **7.** $\left\{\left(\frac{1}{3}, \frac{1}{2}\right)\right\}$ **9.** $\{(3, 2, 1)\}$
11. $\{(5, -1, 0)\}$ **13.** One possible answer is $\begin{array}{l} x + y = 2 \\ x + y = 3 \end{array}$.
15. rice: $\frac{1}{3}$ cup; soybeans: $\frac{1}{5}$ cup **17.** 5 blankets; 3 rugs; 8 skirts **19. (a)** **(b)** 36 **(c)** \$54

21. $y_1 = \frac{12}{5}x^2 - \frac{31}{5}x + \frac{3}{2}$, or $y_1 = 2.4x^2 - 6.2x + 1.5$

23. $\{(x, -2x + 1, -11x + 6)\}$ **25.** $\{(-2, 0)\}$

27. $\{(-3, 2)\}$ **29.** $\{(0, 3, 3)\}$ **31.** 10 lb of $4.60 tea; 8 lb of $5.75 tea; 2 lb of $6.50 tea **33.** 1984; 327 thousand

35. -25 **37.** 0 **39.** -1 **41.** $\{(-4, 2)\}$

43. Cramer's rule does not apply because $D = 0$; $\varnothing$

45. $\{(14, -15, 35)\}$ **47.** $\left\{\frac{8}{19}\right\}$ **49.** $\frac{2}{x-1} - \frac{6}{3x-2}$

51. $\frac{1}{x-1} - \frac{x+3}{x^2+2}$ **53.** $\{(-3, 4), (1, 12)\}$

55. $\{(-4, 1), (-4, -1), (4, -1), (4, 1)\}$

57. $\left\{(5, -2), \left(-4, \frac{5}{2}\right)\right\}$ **59.** $\{(-2, 0), (1, 1)\}$

61. $\left\{\left(\sqrt{3}, 2i\right), \left(\sqrt{3}, -2i\right), \left(-\sqrt{3}, 2i\right), \left(-\sqrt{3}, -2i\right)\right\}$

63. yes; $\left\{\left(\frac{8-8\sqrt{41}}{5}, \frac{16+4\sqrt{41}}{5}\right), \left(\frac{8+8\sqrt{41}}{5}, \frac{16-4\sqrt{41}}{5}\right)\right\}$

65.

$x + y \le 6$
$2x - y \ge 3$

67. maximum of 24 at $(0, 6)$

69. 3 units of food A and 4 units of food B; Minimum cost is $1.02 per serving. **71.** $a = 5$, $x = \frac{3}{2}$, $y = 0$, $z = 9$

73. $\begin{bmatrix} -4 \\ 6 \\ 1 \end{bmatrix}$ **75.** They cannot be subtracted.

77. $\begin{bmatrix} 3 & -4 \\ 4 & 48 \end{bmatrix}$ **79.** $\begin{bmatrix} -9 & 3 \\ 10 & 6 \end{bmatrix}$ **81.** $\begin{bmatrix} -2 & 5 & -3 \\ 3 & 4 & -4 \\ 6 & -1 & -2 \end{bmatrix}$

83. $\begin{bmatrix} 3 & -1 \\ -5 & 2 \end{bmatrix}$ **85.** $\begin{bmatrix} \frac{2}{3} & 0 & -\frac{1}{3} \\ \frac{1}{3} & 0 & -\frac{2}{3} \\ -\frac{2}{3} & 1 & \frac{1}{3} \end{bmatrix}$

87. $\left\{\left(\frac{2}{3}, \frac{7}{12}\right)\right\}$ **89.** $\{(-3, 2, 0)\}$

Chapter 5 Test

[5.1] **1.** $\{(4, 3)\}$ **2.** $\left\{\left(\frac{-3y-7}{2}, y\right)\right\}$; infinitely many solutions **3.** $\varnothing$; inconsistent system **4.** $\{(1, 2)\}$

5. $\{(2, 0, -1)\}$ [5.2] **6.** $\{(5, 1)\}$ **7.** $\{(5, 3, 6)\}$

[5.1] **8.** $y = 2x^2 - 8x + 11$ **9.** 22 units from Toronto; 56 units from Montreal; 22 units from Ottawa

[5.3] **10.** -58 **11.** -844 **12.** $\{(-6, 7)\}$

13. $\{(1, -2, 3)\}$ [5.4] **14.** $\frac{2}{x+3} + \frac{7}{x-1}$

15. $\frac{2}{x} + \frac{-2}{x+1} + \frac{-1}{(x+1)^2}$ [5.5] **16.** $\{(1, 2), (-1, 2),$

$(1, -2), (-1, -2)\}$ **17.** $\{(3, 4), (4, 3)\}$ **18.** 5 and -6

[5.6] **19.**

$x - 3y \ge 6$
$y^2 \le 16 - x^2$

20. maximum of 42 at $(12, 6)$ **21.** 0 VIP rings and 24 SST rings; Maximum profit is $960.

[5.7] **22.** $x = -1$, $y = 7$, $w = -3$ **23.** $\begin{bmatrix} 8 & 3 \\ 0 & -11 \\ 15 & 19 \end{bmatrix}$

24. The matrices cannot be added. **25.** $\begin{bmatrix} -5 & 16 \\ 19 & 2 \end{bmatrix}$

26. The matrices cannot be multiplied. **27.** A

[5.8] **28.** $\begin{bmatrix} -2 & -5 \\ -3 & -8 \end{bmatrix}$ **29.** The inverse does not exist.

30. $\begin{bmatrix} -9 & 1 & -4 \\ -2 & 1 & 0 \\ 4 & -1 & 1 \end{bmatrix}$ **31.** $\{(-7, 8)\}$ **32.** $\{(0, 5, -9)\}$

Photo Credits

Index